T0201865

Strongly Interacting Quantum Systems out of Equilibrium

Lecture Notes of the Les Houches Summer School:

Volume 99, 30th July – 24th August 2012

Strongly Interacting Quantum Systems out of Equilibrium

Edited by

Thierry Giamarchi, Andrew J. Millis, Olivier Parcollet,
Hubert Saleur, Leticia F. Cugliandolo

OXFORD
UNIVERSITY PRESS

OXFORD
UNIVERSITY PRESS

Great Clarendon Street, Oxford, OX2 6DP,
United Kingdom

Oxford University Press is a department of the University of Oxford.
It furthers the University's objective of excellence in research, scholarship,
and education by publishing worldwide. Oxford is a registered trade mark of
Oxford University Press in the UK and in certain other countries

First Edition published in 2016
Reprinted 2016
Impression: 2

Published in the United States of America by Oxford University Press
198 Madison Avenue, New York, NY 10016, United States of America

British Library Cataloguing in Publication Data
Data available

Library of Congress Control Number: 2016934102

ISBN 978–0–19–876816–6

Printed and bound by
CPI Group (UK) Ltd, Croydon, CR0 4YY

École de Physique des Houches

Service inter-universitaire commun
à l'Université Joseph Fourier de Grenoble
et à l'Institut National Polytechnique de Grenoble

Subventionné par l'Université Joseph Fourier de Grenoble,
le Centre National de la Recherche Scientifique,
le Commissariat à l'Énergie Atomique

Directeur:
Leticia F. Cugliandolo, Sorbonne Universités, Université Pierre et Marie Curie,
Laboratoire de Physique Théorique et Hautes Energies, Paris, France

Directeurs scientifiques de la session:
Thierry Giamarchi, Département de Physique, Université de Genève, Suisse
Andrew J. Millis, Physics Department, Columbia University, USA
Olivier Parcollet, IPhT CEA Saclay, France
Hubert Saleur, IPhT CEA Saclay, France and Physics Department, USC,
Los Angeles, USA
Leticia F. Cugliandolo, Sorbonne Universités, Université Pierre et Marie Curie,
Laboratoire de Physique Théorique et Hautes Energies, Paris, France

Previous sessions

Publishers

- Session VIII: Dunod, Wiley, Methuen
- Sessions IX and X: Herman, Wiley
- Session XI: Gordon and Breach, Presses Universitaires
- Sessions XII–XXV: Gordon and Breach
- Sessions XXVI–LXVIII: North Holland
- Session LXIX–LXXVIII: EDP Sciences, Springer
- Session LXXIX–LXXXVIII: Elsevier
- Session LXXXIX– : Oxford University Press

Preface

Much of our understanding of physics is based on the remarkable idea that in many cases one can replace an average over the time configurations of a system by an ensemble average with respect to a special measure, now called the Boltzmann distribution after the Austrian physicist Ludwig Boltzmann. The interchangeability of time and ensemble averages may be taken to define the notion of equilibrium, and is the cornerstone of the wonderful field of statistical physics that successfully describes a wealth of physical phenomena.

However, the world around us is in many respects nonequilibrium: from the flow of current through a wire to the evolution of the universe from the moment of the Big Bang, we observe behavior that goes beyond Boltzmannian equilibrium. In the domain of classical mechanics, nonequilibrium phenomena have been studied for many years, but in the quantum domain, the study of strongly nonequilibrium phenomena has been much less developed, especially for the crucial case of strongly correlated quantum systems that cannot be understood on the single-particle level.

Over the last decade or so, the situation has changed: new experimental tools and theoretical concepts are providing new insights into the collective nonequilibrium behavior of quantum systems. The exquisite control provided by laser trapping and cooling techniques allows us to observe the behavior of condensed Bose and degenerate Fermi gases under nonequilibrium drive or after "quenches" in which a Hamiltonian parameter is suddenly or slowly changed. On the solid state front, high-intensity short-time pulses and fast (femtosecond) probes allow solids to be put into highly excited states and probed before relaxation and dissipation occur. Experimental developments are matched by progress in theoretical techniques ranging from exact solutions of strongly interacting nonequilibrium models to new approaches to nonequilibrium numerics.

The summer school "Strongly Interacting Quantum Systems out of Equilibrium" was designed to summarize this progress, lay out the open questions, and define directions for future work. The school, Session XCIX of the Les Houches School of Physics, was held in the beautiful and stimulating environment of Les Houches, France, from 30 July to 24 August 2012. A total of 51 students coming from 12 countries in North and South America, Europe, and Asia attended. Nineteen professors working in France, Germany, Israel, Switzerland, the UK, and the USA lectured or gave specialized seminars, covering practically all aspects of nonequilibrium physics. The high quality and intense focus of the students, helped by the splendid weather and spectacular mountain scenery, gave the school a splendid scientific atmosphere, with lively discussions continuing outside the lecture halls, over meals, and often late into the night.

The chapters in this volume are based on the lecture notes for most of the long and short courses given at the school.

- Ehud Altman presented various aspects of out-of-equilibrium physics in the context of cold atomic systems.
- Jürgen Berges gave a general introduction to the theory of nonequilibrium physics and the concept of thermalization.
- Giulio Biroli covered the classical aspects of non-equilibrium dynamics, such as coarsening and thermal quenches, and related them to their quantum counterparts.
- Uli Schollwöck discussed numerical methods such as the density matrix renormalization group that allow one to deal with out-of-equilibrium physics in one-dimensional systems.
- Natan Andrei and Hubert Saleur discussed the exact solutions that allow one to study out-of-equilibrium and steady-state transport in low-dimensional structures.
- Bernard Doyon discussed systems close to critical points and out of equilibrium, as well as ways to deal with such systems.
- Thomas Gasenzer discussed turbulence in Bose gases and how it leads to nonthermal fixed points.
- Patrycja Paruch discussed the propagation and out-of-equilibrium physics of disordered elastic systems such as domain walls in magnetic and ferroelectric materials.
- Luca Perfetti described ultrafast probes and related phenomena in solid state systems.

We close this preface by thanking the institutions that provided financial support for the school: Manep, the Swiss National Science Foundation (FNS), ICTP, ERC, CNRS, the Les Houches Theoretical Physics School, and the Joseph Fourier University. Finally, special thanks go to the administrative staff of the Les Houches Center, who provided such excellent support to the participants and organizers of the school. We warmly thank Brigitte Rousset, Murielle Gardette, and Isabel Lelièvre for their professional expertise and limitless help.

Thierry Giamarchi
Andrew J. Millis
Olivier Parcollet
Hubert Saleur
Leticia F. Cugliandolo

Contents

List of participants

ORGANIZERS

GIAMARCHI THIERRY
DPMC–MaNEP, University of Geneva, Switzerland
MILLIS ANDREW
Department of Physics, Columbia University, USA
PARCOLLET OLIVIER
Institut de Physique Théorique, CEA Saclay, France
SALEUR HUBERT
Institut de Physique Théorique, CEA Saclay, France

LECTURERS

ALEINER IGOR
Department of Physics, Columbia University, USA
ALTMAN EHUD
Department of Condensed Matter Physics, The Weizmann Institute of Science, Israel
ANDREI NATAN
Department of Physics and Astronomy, Rutgers University, USA
BERGES JUERGEN
Institute for Theoretical Physics, University of Heidelberg, Germany
BIROLI GIULIO
Institut de Physique Théorique, CEA Saclay, France
BLOCH IMMANUEL
Max Planck Institute, Garching, Germany
GLATTLI CHRISTIAN
Institut de Physique Théorique, CEA Saclay, France
PARUCH PATRYCJA
DPMC, University of Geneva, Switzerland
PERFETTI LUCA
Laboratoire des Solides Irradiés, Ecole Polytechnique, France
SCHOLLWÖCK ULRICH
University of Munich, Germany
SON DAM THANH
Institute for Nuclear Theory, University of Washington, USA

SEMINAR SPEAKERS

DOYON BENJAMIN
Department of Mathematics, King's College London, UK

GASENZER THOMAS
Institute for Theoretical Physics, University of Heidelberg, Germany

GRITSEV VLADIMIR
University of Fribourg, Switzerland

KOLLATH CORINNA
University of Geneva, Switzerland

LeHUR KARINE
CPHT, Ecole Polytechnique, France

LOTH SEBASTIAN
Max Planck Institute, Hamburg, Germany

SAVONA VINCENZO
EPFL Lausanne, Switzerland

WERNER PHILIPP
University of Fribourg, Switzerland

STUDENTS

ANTIPOV ANDREY
Max Planck Institute, Dresden, Germany

ARON CAMILLE
Department of Physics and Astronomy, Rutgers University, USA

AYRAL THOMAS
Institut de Physique Théorique, CEA Saclay, France

BAGROV ANDREY
Lorentz Institute for Theoretical Physics, Leiden University, Netherlands

BENOIST TRISTAN
LPT, Ecole Normale Supérieur, France

BERA SOUMYA
Institut Néel and Universtité Joseph Fourier, France

BINDER MORITZ
Theoretical and Mathematical Physics, Ludwig Maximilians University and University of Munich, Germany

BONNES LARS
Institute for Theoretical Physics, University of Innsbruck, Austria

CHOU YANG-ZHI
Department of Physics and Astronomy, Rice University, USA

COHEN Guy
Tel Aviv University, Israel

COLLET Tim
Karlsruhe Institute for Technology, TKM, Germany

DALLA TORRE Emanuele
Department of Physics, Jefferson Laboratory, Harvard University, USA

FILIPPONE Michele
Laboratoire Pierre Aigrain, Ecole Normale Supérieur, France

FOINI Laura
Laboratoire de Physique Théorique et Hautes Energies, Université Pierre et Marie Curie, France

FOKKEMA Thessa
Institute of Physics, University of Amsterdam, Netherlands

FOTSO Herbert F.
Department of Physics, Georgetown University, USA

FRETON Loïc
Laboratoire Matériaux et Phénomènes Quantiques, Université Denis Diderot, France

GANAHL Martin
Graz University, Austria

HOFMANN Johannes
DAMTP, Centre for Mathematical Sciences, University of Cambridge, UK

HOOGEVEEN Marianne
Department of Mathematics, King's College London, UK

IVASHKO Artem
Lorentz Institute for Theoretical Physics, Leiden University, Netherlands

IYER Deepak
Department of Physics and Astronomy, Rutgers University, USA

JANOT Alexander
Institute for Theoretical Physics, University of Leipzig, Germany

KAMINISHI Eriko
Department of Physics, Ochanomizu University, Japan

KHEMANI Vedika
Princeton University, USA

LEVY Tal
School of Chemistry, The Sackler Faculty of Exact Sciences, Tel Aviv University, Israel

LIM Lih-King
Laboratoire de Physique des Solides, Université Paris Sud, France

MATHEY Steven
Institute for Theoretical Physics, University of Heidelberg, Germany

MESSIO Laura
Institut de Physique Théorique, CEA Saclay, France

MORI Takashi
Department of Physics, University of Tokyo, Japan

NESSI Nicolas
La Plata Physics Institute, Universidad Nacional de La Plata, Argentina

OLSCHLAGER Christoph
University of Hamburg, Germany

PETKOVIC Aleksandra
Laboratoire de Physique Théorique, Ecole Normale Supérieure, France

PIELAWA Susanne
Department of Condensed Matter Physics, Weizmann Institute of Science, Israel

PRICE Hannah
Theory of Condensed Matter Group, Cavendish Laboratory, University of Cambridge, UK

RESHETNYAK Igor
CPHT, Ecole Polytechnique, France

RIEDER Maria-Theresa
Dahlem Center for Complex Quantum Systems, Fachbereich Physik, Free University of Berlin, Germany

RISTIVOJEVIC Zoran
Laboratoire de Physique Théorique, Ecole Normale Supérieure, France

ROY Bitan
National High Magnetic Field Laboratory, Theoretical Condensed Matter Group, Florida University, USA

SANDRI Matteo
SISSA, International School for Advanced Studies, Trieste, Italy

SCHECTER Michael
School of Physics and Astronomy, University of Minnesota, Minneapolis, USA

SCHIRO Marco
Princeton Center for Theoretical Sciences, USA

SCIOLLA Bruno
Institut de Physique Théorique, CEA Saclay, France

SHASHI Aditya
Physics and Astronomy Department, Rice University, USA

SHIRAI Tatsuhiko
Department of Physics, University of Tokyo, Japan

SMACCHIA Pietro
SISSA, International School for Advanced Studies, Trieste, Italy

SNIZHKO Kyrylo
Department of Physics, Lancaster University, UK

SZECSENYI Istvan
Institute of Theoretical Physics, Eötvös Lorand University, Hungary

TSUJI Naoto
Institute for Theoretical Physics, ETH Zürich, Switzerland
van LANGE Arjon
Debye Institute for Nanomaterials Science, University of Utrecht, Netherlands
YANG Bo
Department of Physics, Princeton University, USA

Part I

Lectures

1

Nonequilibrium quantum dynamics in ultracold quantum gases

Ehud ALTMAN

Department of Condensed Matter Physics
Weizmann Institute of Science
Rehovot 76100
Israel

Strongly Interacting Quantum Systems out of Equilibrium. First Edition. Thierry Giamarchi et al.
© Oxford University Press 2016. Published in 2016 by Oxford University Press.

Chapter Contents

1.1 Introduction to ultracold quantum gases

Much of the recent work on ultracold atomic systems has focused on using them as simulators to investigate fundamental problems in condensed matter systems. However, it is important to note that many-body systems of ultracold atoms are not one-to-one analogs of known solid state systems. These are independent physical systems that have their own special features and present new types of challenges and opportunities. One of the main differences between ensembles of ultracold atoms and solid state systems lies in the experimental tools available for characterizing many-body states. While many traditional techniques of condensed matter physics are not easily available (e.g., transport measurements), there are several tools that are unique to ultracold atoms. These includes the time-of-flight expansion technique, interference experiments, molecular association spectroscopy, and many more. In some cases, these tools have been used to obtain unique information about quantum systems. One of the most striking examples is single-site resolution in optical lattices [14, 109], which has allowed exploration of the superfluid-to-Mott insulator transition at an unprecedented level. Another example is measurement of the full distribution functions of the contrast of interference fringes in low-dimensional condensates, which has provided information about high-order correlation functions [48, 49, 51, 89].

Ultracold atoms can make unique contributions to our understanding of the nonequilibrium dynamics of quantum many-body systems. Several factors make ultracold atoms ideally suited for the study of dynamical phenomena:

- *Convenient timescales*
  The characteristic frequencies of many-body systems of ultracold atoms are in the kilohertz range. System parameters can be modified, and the resulting dynamical properties can be measured at these timescales (or faster). It is useful to contrast this with the dynamics of solid state systems, which usually take place at giga- and terahertz frequencies, which are extremely difficult for experimental analysis.
- *Isolation from the environment*
  Ultracold atomic gases are unique in being essentially closed, almost completely decoupled from an external environment. More traditional condensed matter systems, by contrast, are always strongly coupled to some kind of thermal bath, usually made of the phonons in the crystal. This distinction does not matter much if a system is at thermal equilibrium, because of the equivalence of ensembles. However, it leads to crucial differences in the nonequilibrium dynamics of the two systems. While the dynamics in standard materials is generically overdamped and classical in nature, atomic systems can follow coherent quantum dynamics over long timescales. A corollary of this is that it is also much harder for systems of ultracold atoms to attain thermal equilibrium. Hence, understanding dynamics is important for interpreting experiments even when ultracold atoms are used to explore equilibrium phases.
- *Rich toolbox*
  One of the problems encountered in studying the nonequilibrium dynamics of many-body systems is the difficulty of characterizing complicated transient states. Experimental techniques that have recently been developed for ultracold atoms

are well suited to this challenging task. Useful tools include local resolution [14, 109], measurements of quantum noise [3, 39], and interference fringe statistics [48, 49, 51, 89].

The following are some fundamental questions that can be addressed with ultracold atomic systems and that will be touched upon in this chapter:

1. *Emergent phenomena in quantum dynamics*
   The main theme in the study of complex systems during the last century has been the identification of emergent phenomena and the development of an understanding of the universal behavior that they exhibit. Examples of successful theories of universal phenomena in equilibrium phases include the Ginzburg–Landau–Wilson theory of broken-symmetry phases and critical phenomena, the theory of Fermi liquids, and, more recently, the study of quantum Hall states, spin liquids, and other topological phases. In addition, there is a growing understanding of emergent universal phenomena in classical systems out of equilibrium: turbulence, coarsening dynamics near phase transitions, the formation of large-scale structures such as sand dunes and Raleigh–Bénard cells, the dynamics of active matter such as flocking of animals, and many more. Yet there is very little understanding of nonequilibrium many-body phenomena that are inherently quantum in nature. It is an interesting open question whether the dynamics of complex systems can exhibit emergent universal phenomena in which quantum interference and many-body entanglement play an important role.

2. *Thermalization and prethermalization in closed systems*
   Intimately related to the establishment of emergent phenomena is the question of thermalization and the approach to equilibrium in closed quantum systems. Cold atomic systems are ideal for testing the common belief that generic many-body systems ultimately approach thermal equilibrium. There are, however, exceptions to this rule. Integrable models, for example, fail to come to thermal equilibrium, because the dynamics is constrained by an infinite set of integrals of motion. In general, integrable models require extremely fine tuning. Nevertheless, systems of ultracold atoms in one-dimensional confining potentials often realize very nearly integrable models. The reason for this is that the interactions naturally realize almost pure two-body contact interactions. This fact has been used to demonstrate breaking of integrability in the Lieb–Liniger model of one-dimensional bosons with contact interactions [62]. This experiment has motivated many theoretical and numerical investigations to characterize the nonthermal steady state that is reached in such cases. The latter is usually characterized using a generalized Gibbs ensemble (GGE) that seeks the equilibrium (maximum-entropy) state subject to the infinite number of constraints that are set by the values of the integrals of motion in the initial state [95].
   The systems that realize nearly integrable models usually include small perturbations that break integrability. It is believed (though not proved) that these perturbations will eventually lead to true thermalization. Hence, the GGE is considered to be a long-lived prethermalized state. An interesting question that we will touch upon concerns the timescale for relaxational dynamics to take over and lead to true thermal equilibrium.

3. *Absence of thermalization and many-body localization*

    A body of recent work starting with a theoretical suggestion [19, 44] (and a much earlier conjecture by Anderson [7]) points to quantum many-body systems with quenched disorder as providing a generic alternative to thermalization. That is, in contrast to integrable models, lack of thermalization is supposed to be robust to a large class of local perturbations in such closed systems. This phenomenon is known as many-body localization (MBL). Important questions that are beginning to be addressed with ultracold atomic systems [64, 101] concern the dynamical behavior in the MBL state and the nature of the transition from MBL to conventional ergodic dynamics. If there is a critical point controlling the transition from a thermalizing to a nonthermalizing state, does it involve singularities in correlation functions similar to those that dominate quantum critical phenomena at equilibrium?

This chapter focuses on universal phenomena in dynamics that can and have been investigated using ultracold atomic systems. Of course, it cannot cover everything, so some illustrative examples have been picked that are interesting from a fundamental point of view and that have also been directly investigated experimentally using ultracold atomic systems. In Section 1.2, we discuss the dynamics of ultracold atom interferometers, primarily in one dimension, as case studies of prethermalization and thermalization dynamics. In Section 1.3, we turn to the dynamics of bosons in optical lattices. we review the theoretical understanding of the quantum phase transition from a superfluid to a Mott insulator and then discuss emergent dynamical phenomena that have been studied in the vicinity of this transition. In particular, we discuss the nature of the Higgs amplitude mode that emerges near the critical point, as well as the modes for decay of supercurrents due to the enhanced quantum fluctuations near the Mott transition. In Section 1.4, we discuss far-from-equilibrium dynamics involving a quench of system parameters across the superfluid-to-Mott insulator phase transition. In Section 1.5, we review some of the recent work on quenches in one-dimensional dimensional optical lattices. Finally, in Section 1.6, we turn to review recent progress in the theoretical understanding of MBL as well as recent experiments that have investigated this phenomenon. We have included here the very significant recent progress on MBL that was made after the lectures on which this chapter is based were delivered.

We have tried to make each section self-contained so that it can be read independently of the others. However, if the reader is not familiar with the physics of bosons on optical lattices and the Mott transition it would help to read Section 1.3 before Section 1.4.

1.2   Dynamics of ultracold atom interferometers: quantum phase diffusion

Atomic condensates were used early on as interferometers, utilizing matter waves to make precision measurements of acceleration. Such investigations provide a natural and simple example of a dynamical quench experiment, which, besides its practical applications, raises fundamental questions in nonequilibrium quantum physics. Hence, this example will serve here to illustrate several key concepts that appear frequently in nonequilibrium studies of cold atoms.

An idealized interferometer experiment consists of a macroscopic condensate separated into two wells and prepared with a well-defined relative phase between the wells. When the wells are disconnected, the relative phase is expected to evolve in time under the influence of the potential difference between the two wells. Following this evolution, the phase is measured directly by releasing the atoms from the trap and observing the interference pattern established after a time of flight.

One method to realize such a preparation protocol is to start with one condensate in an elongated potential well and raise a radiofrequency-induced potential barrier to split the condensate into two one-dimensional condensates [102]. A schematic setup of this nature is illustrated in Fig. 1.1. Alternatively, instead of using two separate wells as the two arms of the interferometer, one can use two internal states of the atoms, as was done in [117]. The system is prepared with a condensate in a single internal state, say $|\uparrow\rangle$. A two-photon transition is used to induce a $\pi/2$ rotation to the state $(|\uparrow\rangle + |\downarrow\rangle)/\sqrt{2}$; this is the analog of the coherent splitting in the double-well realization. Thereafter, the system is allowed to evolve freely under the influence of the Hamiltonian, with no coupling between the two internal states. Finally, the coherence between the two states remaining after a time t is determined through a Ramsey-type measurement, that is, by measuring how much of the $|\uparrow\rangle$ state population is restored following a $-\pi/2$ rotation of the internal state.

In reality, the relative phase is not determined solely by the external potential difference between the two wells (or two internal spin states). The phase field of an interacting condensate should be viewed as a quantum operator with its own dynamics. The intrinsic quantum evolution leads to uncertainty in the relative phase, which grows in time and eventually limits the accuracy of the interferometric measurement. Thus, even if the condensates are subject to exactly the same potential, the relative phase measured by the interference pattern between them will have a growing random component. The rate and functional form with which this uncertainty grows is a

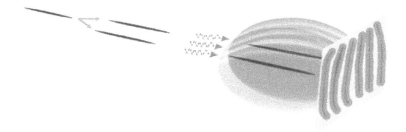

Fig. 1.1 Typical setup for an interferometer measurement. A condensate is split into two decoupled wells such that the relative condensate phase between the wells is initially well defined, $\theta_- = 0$. The two condensates are allowed to evolve under the influence of the local potentials acting on them for a time t_m. At time t_m, they are released from the trap and their expansion leads to an interference pattern that is observed on a light absorption image.

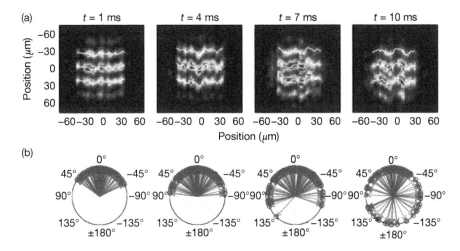

Fig. 1.2 (a) Example of fringes from interfering one-dimensional condensates taken at varying times after the condensate had been coherently split [53]. (b) The distribution of the relative phase extracted from the interference fringes grows with time. (Reprinted by permission from Macmillan Publishers Ltd: *Nature* [53], copyright (2007).)

hallmark of the many-body dynamics. Figure 1.2 is an example of an experiment performed by the Vienna group [53] that shows the distribution of relative phase determined in repeated measurements after different times from the initial preparation.

It is interesting to note the essential difference between the problem we consider here and the dephasing of single-particle interference effects as seen, for example, in mesoscopic electron systems. In the latter case, the dephasing of a single-electron wavefunction is a result of the interaction of the single electron with a thermal bath that consists of the other electrons in the Fermi liquid or of phonons. Because the bath is thermal and the energy of the injected electron is low, the dephasing problem is ultimately recast in terms of linear response theory [111].

By contrast, dynamic splitting of the condensate in the ultracold atom interferometer takes the system far from equilibrium, and the question of phase coherence is then essentially one of quantum dynamics. The system is prepared in an initial state determined by the splitting scheme, which then evolves under the influence of a completely different Hamiltonian, that of the split system. This is the first example we encounter of a quantum quench. Dephasing, from this point of view, is the process that takes the system to a new steady (or quasisteady) state. In this respect, the ultracold atom interferometer is a useful tool for the study of nonequilibrium quantum dynamics. One of our goals is to classify this dynamics into different universality classes.

1.2.1 Phase diffusion in the single-mode approximation

Let us start the theoretical discussion with the simplest case of perfect single-mode condensates initially coupled by a large Josephson coupling that locks their phases.

The Hamiltonian that describes the dynamics of the relative phase φ is given in the number–phase representation by

$$H = \tfrac{1}{2}Un^2 - J\cos\varphi. \tag{1.1}$$

Here $U = \mu/N$ is the charging energy of the condensate, where μ is the chemical potential and $N = \langle N_1 \rangle = \langle N_2 \rangle$ is the average particle number in the condensate. Alternatively, we can express the charging energy using the contact interaction u and the condensate volume L^d as $U = u/L^d$. The relative particle number $n = (N_1 - N_2)/2$ and relative phase φ are conjugate operators obeying the commutation relation $[\varphi, n] = i$.

At time $t = 0$, the Josephson coupling J is shut off and the time evolution begins subject to the interaction Hamiltonian alone. The system is probed by releasing the atoms from the trap after a time evolution over a time t subject to the Hamiltonian of decoupled wells, $J = 0$. The aim is to infer the coherence between the two condensates from the emerging oscillating pattern in the density profile.

Here it is important to note that such information can only be inferred from averaging the density profile (interference fringes) over many repetitions of the experiment. A single time-of-flight image originating from a pair of macroscopic Bose condensates exhibits density modulations $\rho(x) \sim A_0 \cos(q_0 x + \varphi)$ of undiminishing intensity $A_0 \propto N$ but with possibly a completely random phase [10, 29, 89]. If the condensates were completely in phase with each other, then φ would be fixed and the average interference pattern taken over many experimental runs would be the same that seen in a single shot. On the other hand, if the relative phase φ were completely undetermined, the average interference pattern would vanish. Therefore, the correct measure of coherence is the average interference amplitude remaining after averaging over many shots, which is directly related to the quantum expectation value $\mathcal{A}(t) \sim \langle \cos\varphi \rangle$.

The time evolution of the relative phase subject to the Hamiltonian (1.1) with $J = 0$ follows the obvious analogy to the spreading of a wavepacket of a particle with mass $m = 1/U = L^d/u$, with the relative phase fluctuation following a ballistic evolution

$$\langle \varphi(t)^2 \rangle = \langle \varphi(0)^2 \rangle + \frac{\langle \delta n(0)^2 \rangle u^2}{L^{2d}} t^2. \tag{1.2}$$

This implies a Gaussian decay of the fringe amplitude with time:

$$\mathcal{A}(t) \approx A_0 e^{-t^2/t_D^2}, \tag{1.3}$$

with the dephasing timescale $\tau_D = (L^d/u)/\sqrt{\langle \delta n(0)^2 \rangle} = (N/\mu)/\sqrt{\langle \delta n(0)^2 \rangle}$. Several important insights may already be gleaned from this simple model:

1. Quantum phase diffusion in a true condensate is a finite-size effect. Indeed, in the proper thermodynamic limit, the broken-symmetry state with well-defined phase is infinitely long-lived.

2. Quantum phase diffusion is driven by interaction. Setting $u = 0$ eliminates it.
3. The rate of phase diffusion depends on the initial state. A narrower initial particle number distribution entails a longer dephasing time.

The last point implies that the dephasing time t_D depends crucially on the preparation protocol of the interferometer. In one extreme, the initial state is prepared with maximal coherence, that is, as a state with all particles in a symmetric superposition between the two wells. In the limit of large particle number N, we can take this state to have a Gaussian relative number distribution between the wells with a variance $\langle \delta n(0)^2 \rangle = N/2$. Accordingly, the phase diffusion time in this case is $\tau_D = \sqrt{2N}/\mu$.

As we shall see, such a state can be prepared by a rapid ramp-down of J on a scale faster than the characteristic scale set by the chemical potential μ. We can increase the coherence time by preparing an initial state with a narrower number distribution. Owing to the number–phase uncertainty relation, we must pay by having a larger initial phase distribution and hence achieve lower accuracy in phase determination at short times. In quantum optics terminology, this is called a squeezed state.

The simplest way to achieve squeezing is to split the condensates more slowly. In the slow extreme of adiabatic splitting, the system ends up in the ground state with completely undetermined relative phase (and vanishing number fluctuations). This state is infinitely long-lived, but of course useless for phase determination. More generally, we can split the condensates on a timescale τ_s. The frequency scale with which we should compare in order to assess whether this drive is slow or fast is the instantaneous gap in the system, given roughly by the Josephson frequency $\omega_J(t) = \sqrt{UJ(t)}$. As long as $\omega(t) > 1/\tau_s$, the system to a good approximation remains in the instantaneous ground state and the dynamics is essentially adiabatic. However, from some point during the split when $\omega(t) < 1/\tau_s$, the dynamics should rather be viewed as a sudden split. The effective initial state is the ground state of the system at the break-point $\omega(t^*) = 1/\tau_s$.

Using a harmonic approximation of the junction, $H_{\mathrm{osc}} = (\mu/2N)n^2 + \frac{1}{2}J\varphi^2$, the number fluctuation in the initial state is easily found to be $\sqrt{\langle \delta n^2 \rangle} \sim \sqrt{\omega(t^*)N/\mu} = \sqrt{N/\mu\tau_s}$. We then obtain the phase-diffusion time of the squeezed state, $\tau_D = \sqrt{N\tau_s/\mu}$. Because of the squeezing, the minimal phase uncertainty is now $\sqrt{\varphi(0)^2} \sim \sqrt{\mu\tau_s/N}$, which is larger than the uncertainty $1/\sqrt{N}$ in the case of the (fast) maximally coherent split.

The single-mode phase diffusion considered in this section was first discussed in the context of Bose condensates by Leggett and Sols [70]. It is, however, a special case of the more general problem of spontaneous symmetry breaking as discussed originally by Anderson [6]. A single-mode Hamiltonian emerges as a description of the dynamics of the uniform broken symmetry order parameter. In this case, it is the dynamics of the $U(1)$ order parameter, or condensate phase. The dynamics is governed by an effective mass that increases with the system volume. Hence, the phase becomes static because the infinite mass in the thermodynamic limit allows spontaneous symmetry breaking. In Section 1.2.2, we will determine the conditions under which the condensate zero mode can indeed be safely separated from the linearly dispersing Goldstone modes.

In particular, we will show how the multimode dynamics can dominate the dephasing in interferometers composed of low-dimensional condensates.

1.2.2 Many modes: hydrodynamic theory of phase diffusion in low-dimensional systems

The single-mode approximation described above is expected to be a reasonable description of bulk three-dimensional condensates. In this case, internal phase fluctuations, quantum or thermal, are innocuous since they do not destroy the broken symmetry. The mode responsible for restoring the symmetry in a finite system is the uniform mode and therefore only its dynamics has to be considered. However, many interference experiments have been performed with highly elongated, essentially one-dimensional, condensates [53, 60, 102, 117]. In this case, zero-point fluctuations due to long-wavelength phonons destroy the broken symmetry even in the thermodynamic limit. It is expected that these phonons will have a crucial impact on the loss of phase coherence.

The model Hamiltonian to consider is a direct generalization of the single-mode Hamiltonian considered in Section 1.2.1:

$$H = H_1 + H_2 - J(t) \int d^d x \cos(\varphi_1 - \varphi_2). \tag{1.4}$$

Here $H_{1,2}$ are the Hamiltonians of the individual one- (or higher-) dimensional condensates and the Josephson coupling $J(t)$ now operates along the entire length (or area) of the condensates. As in the single-mode case, the system is prepared with strong Josephson coupling, which is thereafter rapidly shut off. In the way in which this scheme is commonly implemented [60, 102], the system actually starts as a one-dimensional condensate in a single tube, which is then split into a double tube over a timescale τ_s slow compared with transverse energy levels in the tube. At the same time, τ_s can be fast compared with the chemical potential of the condensate, leading to preparation of a coherent state of the relative phase [102]. Alternatively, τ_s can be larger than the inverse chemical potential, leading to preparation of a squeezed phase state [60].

We want to track the ensuing dynamics of the relative phase $\varphi_1 - \varphi_2$ following this preparation stage. Again, this is done by releasing the atoms from the trap at fixed times t after the split and inspecting the resulting "matter wave" interference fringes in the density profile. As in the single-mode case, interference fringes can be seen even when the two condensates are independent. The correct measure of the coherence between the two condensates is the average of the fringe amplitude over many repeats of the experiment, directly related to the expectation value $A(x, t) \sim \langle \cos[\varphi_1(x,t) - \varphi_2(x,t)] \rangle$. What can be conveniently measured in an experiment is in fact the integral of this quantity over a certain averaging length:

$$\mathcal{A}_L(t) = \langle A_L(t) \rangle \equiv \int_{-L/2}^{L/2} dx \, \langle \cos[\varphi_1(x,t) - \varphi_2(x,t)] \rangle. \tag{1.5}$$

This quantity of course vanishes for independent condensates. It is important to distinguish $\mathcal{A}_L(t)$ from the quantities $\langle |A_L(t)| \rangle$ and $\langle |A_L(t)|^2 \rangle$ that would give a nonvanishing value (dependent on L) even for independent condensates [89]. Figure 1.2 shows an example of interference fringes and the distribution of phases of the quantity $A_L(t)$ measured in repeated experiments.

Here, in computing the time evolution of the average fringe amplitude $\mathcal{A}_L(t)$, we will closely follow the analysis in [23]. There, a detailed theoretical description of the dynamics is obtained by approximating the exact Hamiltonians $H_{1,2}$ by the low-energy harmonic-fluid description of the phase fluctuations (phonons) in the superfluid [92]:

$$H_\alpha = \int dx \left[\rho_s (\nabla \theta_\alpha)^2 + \kappa^{-1} \Pi_\alpha^2 \right]. \tag{1.6}$$

Here θ_α is the phase field of the condensate $\alpha = 1$ or 2 and Π_α is the canonical conjugate smooth density field. ρ_s and κ are the macroscopic phase stiffness and compressibility of the condensate, respectively. These should in general be considered as phenomenological parameters extracted from experiment. However, for weakly interacting Bose gases, the compressibility is only weakly renormalized from the bare value given by the inverse contact interaction $\kappa = g^{-1} = \rho/\mu$, while the stiffness is directly related to the average density as $\rho_s = \rho_0/m$. The short-range cutoff of this hydrodynamic theory is the healing length of the condensate, $\xi_h = 1/\sqrt{4mg\rho_0}$. It will sometimes be convenient to express quantities in terms of the Luttinger parameter $K = \pi\sqrt{\rho_s\kappa}$ that determines the power-law decay of correlations in one dimension.

We note that the anharmonic terms neglected here are irrelevant for the asymptotic low-energy response in the ground state, yet they may influence the decoherence dynamics at long times. This will be considered in Section 1.2.3.

It is convenient to transform to the relative $\varphi = \theta_1 - \theta_2$ and "center-of-mass" $\theta = (\theta_1 + \theta_2)/2$ phase variables and their respective conjugate momenta $\Pi = \Pi_1 + \Pi_2$ and $n = (\Pi_1 - \Pi_2)/2$. Within the harmonic-fluid description, the relative and center-of-mass coordinates are decoupled. This is not changed by the dynamic Josephson term, which does not involve the center-of-mass modes at all. Since the observable in which we are interested, (1.5), involves only the relative phase, we can forget about the center-of-mass modes at this level of approximation.

After the Josephson coupling is turned off, the relative phase evolves under the influence of a purely harmonic Hamiltonian. In Fourier space, the different spatial modes of the relative degrees of freedom are conveniently decoupled and we have

$$H_- = \tfrac{1}{2} \sum_q \left(2\kappa^{-1}|\Pi_q|^2 + \tfrac{1}{2}\rho_s q^2 |\varphi_q|^2 \right). \tag{1.7}$$

Assuming that the initial state is a Gaussian wavefunction, the time evolution of the mean fringe amplitude, our proxy for phase coherence, is easily computed from the harmonic theory:

$$\mathcal{A}(t) = \rho\langle e^{i\varphi(x)}\rangle = \rho \exp\left[-\frac{1}{2L^d}\sum_q \langle|\varphi_q(t)|^2\rangle \right] = \rho e^{-g(t)}, \tag{1.8}$$

$$\langle \varphi_q(t)^2 \rangle = \langle \varphi_q(0)^2 \rangle \cos^2(cqt) + \frac{1}{\langle \varphi_q(0)^2\rangle}\left(\frac{\pi}{qK}\right)^2 \sin^2(cqt). \tag{1.9}$$

To make further progress, we need the form of the initial phase distribution, which depends on the preparation scheme.

Fast-split scheme

As in the single-mode case, we begin with a discussion of a fast split. In this case, all particles remain in the symmetric superposition between the two wells. Such an initial state is particularly simple to write in the number–phase representation if there are many particles per length unit ξ_h that serves as a cutoff to the hydrodynamic theory. This is equivalent to the weak-coupling requirement $K \gg 1$. In this case, the Poisson statistics of the particle number is well approximated by Gaussian statistics with $\langle \Pi_q(0)^2 \rangle \approx \rho$. Accordingly, the phase wavefunction is Gaussian in this limit and the variance $\langle \varphi_q(0)^2 \rangle \approx 1/2\rho$ should be plugged into (1.11).

We convert the sum over all $\mathbf{q} \neq 0$ in (1.8) to an integral, while separating out the uniform $q = 0$ component, which leads to a unique contribution. Following a simple change of variables, we then have

$$g(t) \approx \frac{t^2}{\tau_L^2} + \frac{S_d}{(2\pi)^d} \int_0^{\mu t} dz\, z^{d-1} \left[\frac{1}{4\rho(ct)^d} \cos^2 z + \rho \left(\frac{\pi}{K} \right)^2 (ct)^{2-d} \frac{\sin^2 z}{z^2} \right], \quad (1.10)$$

where $S_d = 2\pi^{d/2}/\Gamma(n/2)$ is the surface area of a unit sphere in d dimensions and $\tau_L = \mu^{-1}\sqrt{\rho L} = \mu^{-1}\sqrt{N}$. The first term in the integrand saturates to a constant at long times. The second term depends crucially on the spatial dimension d. The contribution of the zero mode together with the integral over the internal phonons then gives

$$\mathcal{A}(t) \propto \mathcal{A}_0 e^{-t^2/\tau_L^2} \times \begin{cases} e^{-t/\tau_Q}, & d = 1, \\ (\mu t)^{-\alpha}, & d = 2, \\ \text{const}, & d = 3, \end{cases} \quad (1.11)$$

where

$$\tau_Q = \frac{2}{\rho c} \left(\frac{K}{\pi} \right)^2 = \frac{2}{\mu} \frac{K}{\pi^2}. \quad (1.12)$$

Let us pause to discuss this result. The contribution of the zero mode is the same in all dimensions. It gives rise to a Gaussian decay on a timescale that diverges in the thermodynamic limit. This is precisely our result obtained in Section 1.2.1 from the single-mode approximation. Now, in addition, we have included the internal modes at the Gaussian level. In three dimensions, we see from (1.11) that the phase fluctuations do not contribute to the time dependence at long times. This is tantamount to the statement that in three dimensions broken symmetry is dynamically stable in the thermodynamic limit. On the other hand, in $d = 1, 2$, the coherence decays even in the thermodynamic limit owing to the internal phonons. This is the dynamical counterpart of the Mermin–Wagner theorem, showing how the internal (Gaussian) phase fluctuations dynamically destroy long-range order. The decay can be traced back to an infrared divergence of the q integral.

Prethermalization in the fast-split scheme

The harmonic model we are considering now is the simplest example of an integrable model. The different q modes are decoupled from each other, and their occupations are conserved quantities. Hence, the model does not in general relax to thermal equilibrium. It is interesting to ask what is the nature of the steady state the system reaches after it has dephased. Approach to such a nonthermal steady state is commonly referred to as prethermalization. The name stems from the expectation that there is a much longer timescale over which the dephased state would finally relax to a true thermal state. Thus, the prethermalized state is a long-lived quasisteady state established in the system before the onset of true thermalization. In our case, the final relaxation to the thermal state is governed by the enharmonic terms. This analysis is postponed to Section 1.2.3.

Within the Gaussian approximation, the full information on the steady state is held in the two-point correlation function (in the relative phase sector):

$$C(x,t) = \rho^2 \langle e^{i[\varphi(x,t)-\varphi(0,t)]} \rangle = \rho^2 e^{-\langle \varphi(x,t)^2 \rangle + \langle \varphi(x,t)\varphi(0,t) \rangle}$$

$$= \rho^2 e^{-(1/L^d)\sum_q \langle |\varphi_q(t)|^2 \rangle [1-\cos(qx)]} = e^{-f(x,t)} \tag{1.13}$$

This expression is directly analogous to the formula for the fringe amplitude (1.8). Similarly, the important contribution to the integral $f(x,t)$ comes from the second term in (1.11), leading, for $d = 1$, to

$$f(x,t) \approx \frac{\pi S_d}{K^2} \rho c t \int_0^{\mu t} dz \, \frac{\sin^2 z}{z^2} \left[1 - \cos\left(z\frac{x}{ct} \right) \right]. \tag{1.14}$$

This integral exhibits completely different behavior depending on whether the point at the position x and time t is inside or outside the "light cone" emanating from the other point, which we have set to $x = 0$ at the time of the split $t = 0$. For $ct < x$, the last cosine term is rapidly oscillating and averages to zero. In this case, we have $f(t,x) \approx 2g(t)$ and therefore $C(x,t) \approx \mathcal{A}(t)^2$. This agrees with the expectation that, for $t < x/c$, there was no time for information to propagate between the two points and therefore the correlation function can be decoupled into the independent expectation values $\mathcal{A}(t)$ at the two points.

On the other hand, at long times $t \gg x/c$, we make a different change of variables using $\alpha = qx = zx/ct$ as an integration variable to obtain

$$f(x,t) \approx \frac{\pi S_d}{K^2} \rho x \int_0^{x/\xi_h} d\alpha \, \frac{\sin^2(\alpha/2)}{\alpha^2} \left[1 - \cos\left(\alpha\frac{2ct}{x} \right) \right]. \tag{1.15}$$

Now the last cosine term is rapidly oscillating $(ct/x \gg 1)$ and we have a time-independent result $f(x,t) \approx (\pi/2K)^2 \rho x$. This implies an exponential decay of the phase correlations in the steady state with a correlation length $\xi_\varphi = 4K^2/\pi^2 \rho$.

Such steady-state behavior mimics the exponential decay of phase correlations in equilibrium at a finite temperature, which is characterized by the correlation length $\xi_\varphi^{eq} = cK/\pi T = \mu K^2/\pi \rho T$. Thus, we can assign an effective temperature

$T_{\text{eff}} = \pi\mu/4$ to the prethermalized state by comparing the correlation length in it with the correlation length in thermal equilibrium.

It is important to note that the effective temperature T_{eff} is completely independent and, in fact, has nothing to do with the real temperature of the condensate prior to the split. Following the rapid split, the original, thermal degrees of freedom of that condensate are projected onto the symmetric density and phase modes of the split condensate. The state of the relative degrees of freedom, accessible to interference experiments, is a pure state fully determined by the splitting process. Within the harmonic-fluid theory, the symmetric and antisymmetric sectors are decoupled, and so within this approximation the temperature of the original condensate does not affect the dynamics.

The prethermalized state reached following a fast split of a one-dimensional condensate has been seen in beautiful experiments by the Vienna group using atom chips [48, 110]. These results are based on an earlier theoretical proposal by Kitagawa et al. [63] showing that the prethermalized state can be fully characterized by the distribution of the amplitude of fringes seen in an interference experiment. These experiments confirm the applicability of the harmonic theory on the accessible timescales.

Finite split rate

A dynamic split done on a timescale $\tau_s < \mu^{-1}$ can be treated in complete analogy to our treatment of a single mode. For simplicity, we separate the two stages of the dynamics. First, we treat the splitting process in which the Josephson coupling is gradually turned off. The outcome of this process provides the initial state for calculating the dynamics at later times under the influence of the Hamiltonian of the decoupled tubes. Here we give a brief summary of a detailed calculation along these lines that can be found in [22].

To address the splitting dynamics, we expand the Josephson coupling to quadratic order to obtain a time-dependent mass term for the relative phase mode, $H_J \approx \sum_q \Delta(t)|\varphi_q|^2$. We then assess for each mode separately whether the splitting on a timescale τ_s is slow or fast compared with the characteristic frequency $\omega_q = cq$ of this mode. For modes at wavevectors $q > 1/c\tau_s$, the split is effectively adiabatic. Therefore, these modes end up in the quantum ground state of the final, fully split, Hamiltonian. Modes at lower wavevectors develop adiabatically only to the point that their frequency is $\omega_q \sim 1/\tau$ and they remain frozen for the rest of the splitting process.

1.2.3 Beyond the harmonic-fluid description of one-dimensional phase diffusion

The harmonic-fluid Hamiltonian (1.6) is a fixed point of the renormalization group in a one-dimensional system, which describes a stable quantum phase of matter known as a Luttinger liquid [52]. Hence, as long as the macroscopic (fully renormalized) values of κ and ρ_s are used, the harmonic Hamiltonian provides an asymptotically exact description of the ground state and low-energy excitations of the system. The neglected anharmonic terms affect the long-distance correlation and low-frequency response only through the renormalization of the quadratic coefficients κ and ρ_s from their bare values.

But, in spite of being irrelevant for the linear response of the system *in the ground state*, the nonlinear terms can influence the long-time dynamics of the interferometer, which starts in a state of finite energy density after the split. As discussed above, the splitting scheme leads to excitation of modes at all wavevectors, which then evolve independently within the quadratic theory (1.6). Only the nonlinear terms can break this integrability through the coupling they induce between modes at different wavevectors. Moreover, the nonlinear terms lead to coupling between the relative and total density fluctuations of the two split condensates. Recall that the relative fluctuations are produced in a pure state by the splitting scheme, while the total density fluctuations already existed in the original single condensate and are presumed to be thermal. Thus, the nonlinear terms give rise, effectively, to coupling with a thermal bath.

Self-consistent phonon damping

The long-time limit of the evolution of the relative phase due to the equilibration with the thermal bath was considered by Burkov et al. [26]. The central assumption in this approach is that the relative mode has nearly reached thermal equilibrium with the center-of-mass modes at a final temperature T_f. We note that the timescale taken by the system to reach this near-equilibrium regime is set by different processes [11]; this will be commented on later.

The damped relative phase mode near thermal equilibrium can be described by phenomenological Langevin dynamics

$$\ddot{\varphi}_k + c^2 k^2 \varphi_k + 2\gamma_k \dot{\varphi}_k = 2\zeta_k. \tag{1.16}$$

Because the system is near thermal equilibrium, the noise satisfies the fluctuation–dissipation theorem with $\langle \zeta_k(t)\zeta_{-k}(t')\rangle = 2(\mu/\rho)T_f\gamma_k\delta(t-t')$. Equation (1.16) is simply a model of damped phonons, with γ_k being the width of the phonon peaks in the structure factor. We expect that the final temperature T_f will be somewhat higher than the initial temperature T_i of the unsplit condensate because of the energy added to the system in the split process. For a sudden split, the added energy density is $\sim \frac{1}{2}\mu/\xi_h$, i.e., the chemical potential per healing length.

The kinematic conditions for linearly dispersing phonons in one dimension lead to divergence of the damping rate γ_k^0 calculated to one-loop order. A finite result is obtained, however, if a damping rate is inserted into the phonon Green's function at the outset and then calculated self-consistently. Such a self-consistent calculation, first done by Andreev [9] (see also [26]), gives a nonanalytic dependence of γ_k on k:

$$\gamma_k \approx \frac{1}{2\pi}\sqrt{\mu T \mathcal{K}}\left(\frac{c|k|}{\mu}\right)^{3/2}. \tag{1.17}$$

This expression is valid for wavevectors k in the linearly dispersing regime, i.e., $ck \ll \mu$, and for weak interactions such that the Luttinger parameter $\mathcal{K} \gg 1$.

The solution for $\varphi_k(t)$ can be formally written in terms of the Green's function of the damped harmonic oscillator:

$$\varphi_k(t) = \int_0^t dt_1 \, \mathcal{G}(k, t - t_1) \zeta_k(t_1), \tag{1.18}$$

where

$$\mathcal{G}(t) = \frac{1}{2\pi} \int d\omega \, e^{i\omega t} \frac{1}{\omega^2 - \epsilon_k + i\omega\gamma_k} = e^{-\gamma_k t/2} \frac{\sin\left[\sqrt{\epsilon_k^2 - (\gamma_k/2)^2}\, t\right]}{\sqrt{\epsilon_k^2 - (\gamma_k/2)^2}}. \tag{1.19}$$

Now by plugging in the δ-correlated noise near equilibrium we can compute the relative phase fluctuations:

$$\langle \varphi_k(t)^2 \rangle = \frac{2T\mu}{\rho} \gamma_k \int_0^t dt_1 \, e^{-\gamma_k(t-t_1)} \frac{\sin^2\left[\sqrt{\epsilon_k^2 - (\gamma_k/2)^2}\,(t-t_1)\right]}{\epsilon_k^2 - (\gamma_k/2)^2}$$

$$= \frac{T\mu}{\rho\,\epsilon_k^2}\left[1 - e^{-\gamma_k t}\left(1 + \frac{\gamma_k}{2\epsilon_k}\sin(2\epsilon_k t)\right)\right] + O\big((\gamma_k/\epsilon_k)^2\big) \tag{1.20}$$

Using the result $\gamma_k = \eta\epsilon_k^{3/2}$ (with $\eta = \sqrt{T}/(2\pi\mu)$) from (1.17), we find

$$\langle \varphi(x,t)^2 \rangle = \int_0^\mu d\epsilon \, \nu_{1d}(\epsilon) \langle \varphi_\epsilon(t)^2 \rangle \approx \frac{T\mu}{\pi\rho c} \int_0^\mu d\epsilon \, \frac{1}{\epsilon^2}\left(1 - e^{-\eta\epsilon^{3/2}t}\right)$$

$$\approx \frac{T}{\mathcal{K}}\left(\int_0^{(\eta t)^{-2/3}} \frac{\eta t}{\sqrt{\epsilon}} d\epsilon + \int_{(\eta t)^{-2/3}}^\infty \frac{d\epsilon}{\epsilon^2}\right) = \frac{3T}{\mathcal{K}}(\eta t)^{2/3}, \tag{1.21}$$

where we have used the relation valid for weak interaction, $\pi\rho c/\mu = \pi\sqrt{\mathcal{K}\rho_s} = \mathcal{K}$. This result implies decay of phase coherence as

$$\mathcal{A}(t) = \mathcal{A}_0 e^{-(t/t_T)^{2/3}}, \tag{1.22}$$

with $t_T = (2\pi\mu/T^2)\mathcal{K}$. Here $\mathcal{K} = (\pi/2)\sqrt{\rho/gm}$ is the Luttinger parameter in the weak-coupling limit. The nonanalytic time dependence of the phase coherence stems directly from the nonanalyticity of the momentum-dependent damping rate.

It is interesting to note that the decoherence driven by thermal fluctuations, (1.22), being a stretched exponential, is slower than the dephasing driven by quantum dynamics, which, as we have seen, depends exponentially on time. The physical reason behind this somewhat counterintuitive result is that the thermally excited phonons provide a damping mechanism that slows down the unitary phase evolution.

Kardar–Parisi–Zhang scaling

The result (1.22) relies on the scaling of the phonon damping rate $\gamma_k \propto |k|^{3/2}$ derived using a self-consistent diagrammatic approach [9]. Because the approach is not rigorously controlled, it would be good to understand the scaling in a different way.

An alternative viewpoint was provided by Kulkarni and Lamacraft [66], who suggested a possible connection between the one-dimensional condensate dynamics at finite temperature and the Kardar–Parisi–Zhang (KPZ) equation commonly used to describe randomly growing interfaces. If indeed the dynamics belonged to the same universality class, it would immediately imply as a consequence the anomalous dynamical scaling obtained in the self-consistent calculation.

Following [66], we illustrate the relation to KPZ dynamics, starting from the hydrodynamic equations of the condensate:

$$\partial_t \rho + \partial_x(v\rho) = 0,$$
$$\partial_t v + v\partial_x v + (g/m)\partial_x \rho = 0. \tag{1.23}$$

These are the Euler and continuity equations, with $v(x,t)$ the velocity field, $\rho(x,t)$ the density, and g the contact interaction. Linearizing these equations with respect to the velocity field and the deviation $n(x,t) = \rho(x,t) - \rho_0$ from the average density leads to a wave equation describing phonons of the system moving to the right or left at the speed of sound $\pm c_0 = \sqrt{\rho_0 g/m}$. Note that beyond the linear approximation, the local speed of sound of the fluid depends on the local density $c_\rho = \sqrt{\rho g/m} = c_0\sqrt{\rho/\rho_0}$. Now define the "chiral velocities" as the fluid velocities measured with respect to the local sound velocity of left- and right-moving waves: $v_\pm = v \pm c_\rho$. In terms of the chiral velocities, the equations of motion can be written as

$$v_\pm + v_\pm \partial_x v_\pm = \tfrac{1}{3}(\partial_t + v_\pm \partial_x)v_\mp. \tag{1.24}$$

Kulkarni and Lamacraft pointed to the similarity between these equations and two copies of the so-called noisy Burgers equation. This would be precisely the situation if we could replace the right-hand sides of the equations by fields $\zeta_\pm(x,t)$ providing Gaussian white noise to the left-hand sides. As it stands, however, we can think of the left-moving modes v_-, randomly occupied with the Bose distribution at temperature T, as a random force field acting on the right-movers and similarly of the right-movers as providing the noise to the dynamics of the left-moving modes. With this approximation, the coupled equations reduce to two effectively independent noisy Burgers equations.

The connection between the noisy Burgers equation and the KPZ equation is well known. It is easily demonstrated by representing the chiral velocities as spatial derivatives of chiral phases $\varphi_\pm = (1/m)\partial_x\varphi_\pm$, just as the actual velocity is related to the condensate phase $v = (1/m)\partial_x\varphi$. Plugging these relations into the Burgers equations and integrating over x gives $\partial_t\varphi_\pm + (1/m)(\partial_x\varphi_\pm)^2 = \zeta_\pm$. There is also a dissipative term $D\partial_x^2\varphi_\pm$ generally present in the KPZ equation. It is absent here because we started from nondissipative Hamiltonian dynamics. Dissipation, however, would be generated upon coarse graining because of the coupling between high- and low-wavevector modes induced by the nonlinearity. Hence, in the long-wavelength limit, we expect the dynamics to be governed by the KPZ equation:

$$\partial_t \varphi_\pm = D\partial_x^2 \varphi_\pm + \tfrac{1}{2}\lambda(\partial_x\varphi_\pm)^2 + \zeta_\pm. \tag{1.25}$$

The height field (here the chiral phases $\varphi_\pm$) is known to obey the dynamic scaling $\langle[\varphi_\pm(x,t) - \varphi_\pm(x,0)]^2\rangle \sim t^{2/3}$. Hence, the phase field $\varphi = (\varphi_+ + \varphi_-)/2$ obeys the same scaling, which leads to the stretched exponential decay of the coherence, (1.22).

1.2.4   Discussion and experimental situation

Above, we have arrived at two seemingly conflicting results. The harmonic-fluid theory of quantum phase diffusion predicts exponential decay (1.8) of phase coherence in a long one-dimensional system, while the hydrodynamic theory of thermal relaxation predicts a stretched exponential decay (1.22). Which will be seen in experiments? Direct comparison between the thermal and quantum timescales t_T and t_Q suggests that as long as the temperature of the relevant bath is smaller than the chemical potential, the quantum phase diffusion should dominate. By the time the thermal relaxation sets in, the phase system is essentially dephased.

Moreover, the calculation of the thermal phase relaxation time t_T summarized in Section 1.2.3 assumes that the process can be described in terms of an effective hydrodynamic theory of a nearly thermalized system. However, the system in question is initially very far from equilibrium and, being a one-dimensional system of bosons with short-range interactions, it is very nearly described by an integrable model (the Lieb–Liniger model). Clearly, the equilibration rate should be proportional to some measure of integrability breaking, which is absent in the expression for t_T. Breaking of integrability in an elongated Bose gas with tight transverse confinement stems from effective three-particle interactions mediated by virtual occupation of transverse excited states [75]. Therefore, the integrability breaking should be controlled by the ratio $\mu/\omega_\perp$ of the interaction energy (chemical potential) to the transverse confinement frequency.

Arzamasovs et al. [11] have computed the thermalization rate of a nearly integrable, weakly interacting one-dimensional Bose gas by expanding around the integrable Lieb–Liniger model. Their result can be expressed as

$$\frac{1}{\tau_R} \sim \mu\left(\frac{\mu}{\omega_\perp}\right)^2 \left(\frac{T}{\mu}\right)^7 \frac{1}{K^3}. \tag{1.26}$$

From this, it is clear that in the relevant experimental regime of $K \sim 10$ and temperature range $T \lesssim \mu$, the system does not equilibrate. The dynamics should be well described by quantum phase diffusion leading to a prethermalized state.

The first experiments that measured the dephasing dynamics seemed to contradict this conclusion, reporting initially a stretched exponential decay of the phase coherence [53] with the exponent 2/3 expected from the thermal dephasing process. However, improved experiments by the same group done with a very similar apparatus found excellent agreement with prethermalization driven by the harmonic-fluid dynamics [48] (for a detailed reanalysis explaining the problems in the original experiments that led to the misinterpretation, see [110]).

1.3   Dynamics of ultracold bosons in optical lattices

The natural regime of ultracold Bose gases is that of weak interactions. In this regime, the gas is well described by the dynamics of a nonfluctuating classical field through the Gross–Pitaevskii equation. Even in low dimensions, where fluctuation effects are inevitably important at long wavelengths, they are hard to observe in small traps. One approach to enhance fluctuation effects and potentially reach novel quantum phases and dynamics is to load the atoms into optical lattice potentials generated by standing waves of laser light.

The lattice has two important effects on the quantum gas. One effect is to increase the effective mass of the atoms and thereby quench the kinetic energy with respect to the interactions. The second important effect is to break Galilean invariance. This liberates the quantum gas from strong constraints, thus opening the way to the realization of alternative quantum phases and new modes of dynamics. In this regard, the observation of a quantum phase transition of bosons in an optical lattice from a superfluid to a Mott insulator [45] has been an important milestone in the study of ultracold atomic systems. This experiment has lead to intense theoretical and experimental work to understand the dynamics of strongly correlated lattice bosons, and later also of fermions.

The superfluid-to-insulator transition of lattice bosons was first discussed by Giamarchi and Schulz [42] in the context of one-dimensional systems and soon after by Fisher et al. [38] for higher-dimensional systems. Investigation of this physics with ultracold atoms in optical lattices was first proposed by Jaksch et al. [58]. In most of this section, we review the basic theory of the superfluid-to-Mott insulator transition and describe recent theoretical and experimental work that has advanced our understanding of the universal dynamical response near the critical point. Discussion of far-from-equilibrium dynamics in this regime is postponed to Sections 1.3.3 and 1.4.

1.3.1   Mott transition in two- and three-dimensional lattices

Superfluidity in two and three dimensions is perturbatively stable to the introduction of a weak lattice potential regardless of the interaction strength and independent of the commensurability of the potential with the particle density. To allow the establishment of an insulating phase, it is therefore important to work with a strong lattice potential with wide gaps between the lowest and the second Bloch bands. In this case, if both the temperature and the interaction strength are smaller than the band gap, we can use the well-known Bose–Hubbard model as a microscopic description of bosons in the lowest Bloch band:

$$H = -J\sum_{\langle ij\rangle}(b_i^\dagger b_j + \text{h.c.}) + \tfrac{1}{2}U\sum_i n_i(n_i-1) - \mu\sum_i n_i. \qquad (1.27)$$

The chemical potential μ is used to control the average lattice filling n. For simplicity, we will not consider here the effects due to the global trap potential. The values of the model parameters expressed in terms of the microscopic couplings are given by

$$J \approx \sqrt{\frac{16}{\pi}} E_r \left(\frac{V_0}{E_r}\right)^{3/4} e^{-2\sqrt{V_0/E_r}}, \tag{1.28}$$

$$U \approx \sqrt{\frac{8}{\pi}} ka E_r \left(\frac{V_0}{E_r}\right)^{3/4}, \tag{1.29}$$

where k is the wavevector of the lattice, a is the s-wave scattering length corresponding to the two-body interaction between atoms, and $E_r = \hbar^2 k^2/2m$ is the recoil energy. For details on the effect of the trap and derivation of the model parameters, see [24, 58].

The fact that there must be a phase transition in the ground state of the model as a function of the relative strength of interaction at integer lattice filling can be established by considering the two extreme limits. First consider the limit of weak interactions. For vanishing interaction strength (and $T = 0$), all the bosons necessarily condense at the bottom of the tight-binding band. Interactions that are weak compared with the bandwidth (i.e., $Un \ll 4zJ$) cannot excite particles far from the band bottom. Therefore, in this regime, the system can be described by an effective continuum action (Gross–Pitaevskii)

$$S_{GP} = \int d^d x \, dt \left(i\psi^\star \partial_t \psi - \frac{1}{2m_*}|\nabla\psi|^2 + \mu|\psi|^2 - u|\psi|^4 \right), \tag{1.30}$$

where m_* is the effective mass in the lowest Bloch band, $u = Ua^d$, and a is the lattice constant. Hence, the physics in this regime is identical to that of a weakly interacting superfluid in the continuum. The effect of the lattice is only to renormalize parameters.

Consider now the opposite regime of strong interactions or small hopping J. In particular, let us start with the extreme limit of decoupled sites (i.e., vanishing hopping, $J = 0$). At integer lattice filling n, the system of decoupled sites has a unique ground state in which each site is occupied by exactly n bosons. The elementary excitations above this ground state are gapped (with a gap U), and consist of particles (sites with $n+1$ bosons) and holes (sites with $n-1$ bosons). The gap makes this state robust against the introduction of a small hopping matrix element J. The elementary excitations gain a dispersion with bandwidth $W \sim 2dnJ$. However, these excitations remain gapped and retain their $U(1)$ charge quantum number as long as $W < U$.

The gap to charge excitations implies that the zero-temperature Mott phase is incompressible. A chemical potential couples with an opposite sign to particle and to hole excitations, decreasing the gap of one species while increasing it for the other. But as long as both excitations remain gapped, the lattice filling cannot change with chemical potential. Thus, the compressibility $\kappa = \partial n/\partial \mu = 0$.

From the above consideration, one can infer that the Mott phases are established as lobes in the space of chemical potential μ and hopping J, characterized by a constant integer filling (see Fig. 1.3). On changing the chemical potential, the upper (or lower) boundary of the phase is reached when either the particle (or hole) excitations

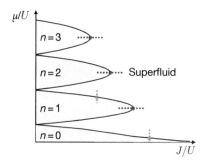

Fig. 1.3 Zero-temperature phase diagram of the Bose–Hubbard model. The gray areas are the incompressible Mott phases with integer filling n. The horizontal dashed lines correspond to phase transitions occurring at constant commensurate filling; they are described by the relativistic critical theory (1.33). In contrast, the transitions tuned along the vertical dashed lines are described by the Gross–Pitaevskii action (1.30), which also describes the weak-interaction regime. The latter can be thought of as being near a transition tuned by chemical potential from a vacuum state to the weakly interacting Bose–Einstein condensate.

condense. The density begins to vary continuously on further increase (or decrease) of the chemical potential beyond the boundary with the compressible phase. The critical theory that describes this transition, tuned by the chemical potential, is just the field theory (1.30), where the Bose field here describes the low-energy particle or hole excitations at the upper or lower phase boundary, respectively.

The transition can also be tuned by varying the tunneling strength (or the interaction) at a fixed integer density. In this case, the particle and hole excitations must condense simultaneously, which enforces particle–hole symmetry at the critical point. This in turn implies an emergent relativistic (i.e., Lorentz-invariant) theory in the vicinity of the quantum phase transition, which in the gapped phase must indeed describe particle and antiparticle excitations of equal mass.

More directly, the emergence of Lorentz invariance at the tip of the Mott lobe can be established as follows (see, e.g., [97]). We write down a more general theory than (1.30):

$$S = \int d^d x \, dt \left[iK'\psi^* \partial_t \psi + K|\partial_t \psi|^2 - c^2|\nabla\psi|^2 + r(\mu)|\psi|^2 - u|\psi|^4 \right]. \tag{1.31}$$

In general, if both K' and K are nonvanishing, we can neglect K at the critical point, since it is irrelevant by simple power counting compared with the K' term. However, we now show that K' must vanish at the tip of the Mott lobe, leaving us with a relativistic theory. By requiring invariance under the uniform gauge transformation (i.e., redefinition of the energy)

$$\psi \rightarrow \psi e^{-i\phi},$$
$$\mu \rightarrow \mu + \partial_t \phi, \tag{1.32}$$

we find that $K' = \partial r / \partial \mu$. Thus, at the tips of the Mott lobes where $r(\mu)$ reaches a minimum as a function of μ, the coefficient K' must vanish. Although the special transition at the tip may seem highly fine-tuned, it is actually realized naturally in a uniform system with a fixed (integer) average filling.

1.3.2   The Higgs resonance near the superfluid-to-Mott insulator transition

We now discuss how the existence of a quantum phase transition from a superfluid to a Mott insulator impacts the dynamics of bosons in an optical lattice. As discussed above, near the critical point, where the diverging correlation length far exceeds the lattice spacing, the dynamics is described by an effective relativistic field theory

$$S_{\text{eff}} = \int d^d x \, dt \left(|\partial_t \psi|^2 - c^2 |\nabla \psi|^2 - r|\psi|^2 - \tfrac{1}{2} u |\psi|^4 \right), \tag{1.33}$$

where c is the sound velocity. One of the most interesting consequences of this emergent critical theory is the appearance of a new gapped excitation analogous to the Higgs resonance in particle physics.

    The collective mode structure in the superfluid phase near the Mott transition can be understood by considering the classical oscillations of the order parameter field about its equilibrium broken-symmetry state. Note that both the critical action (1.33) valid near the superfluid-to-Mott transition at integer filling and the Galilean-invariant action (1.30) of the weakly interacting condensate describe the motion of the order parameter field on a Mexican hat potential as illustrated in Fig. 1.4. Naively, it looks as if in both cases there are two modes: a soft Goldstone mode corresponding to fluctuations of the order parameter along the degenerate minimum of the potential

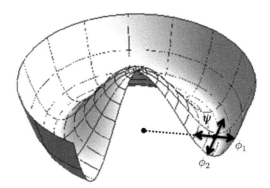

Fig. 1.4 Mexican hat potential that describes the dynamics of the order parameter in the superfluid phase. The longitudinal and transverse fluctuations ϕ_1 and ϕ_2 are two independent modes in the effective relativistic description of the critical point (1.33): the gapless Goldstone mode and the gapped Higgs excitation. In the Gross–Pitaevskii action (1.30), by contrast, the same two fluctuations form a canonically conjugate pair and therefore make up a single mode: the Goldstone mode of the superfluid.

and a massive fluctuation of the order parameter amplitude. This static picture is misleading. It hides a crucial difference in the dynamics that stems from the different kinetic terms.

In the Galilean-invariant (Gross–Pitaevskii) dynamics, the amplitude of the order parameter is proportional to the particle density, i.e., $\psi \sim \sqrt{\rho}e^{i\varphi}$. Plugging this into the kinetic term of (1.30), we get $i\psi^*\dot{\psi} = \rho\dot{\varphi}$. Hence, the amplitude and phase mode are in this case a canonically conjugate pair that together make up only one excitation mode: the gapless Goldstone mode (or phonon) of the superfluid. On the other hand, in the relativistic theory, the phase and amplitude fluctuations are not canonical conjugates. This is easy to see by expanding in small fluctuations around a classical symmetry-breaking solution. We take $\psi = \psi_0 + \phi_1 + i\phi_2$, where $\psi_0 = \sqrt{-r/u}$ is a real classical saddle-point solution and the real fields ϕ_1 and ϕ_2 represent fluctuations in the massive and soft directions of the potential. A quadratic expansion of the action (1.33) in these fluctuation fields gives

$$S_0 = \int d^dx \, dt \left\{ \left[\dot{\phi}_1^2 - c^2(\nabla\phi_1)^2 + 2r\phi_1^2 \right] + \left[\dot{\phi}_2^2 - c^2(\nabla\phi_2)^2 \right] \right\}. \qquad (1.34)$$

Hence, in the relativistic theory (1.34), ϕ_1 and ϕ_2 represent two independent harmonic modes: a massless Goldstone mode ϕ_2 and a massive amplitude mode ϕ_1 with a mass that vanishes at the critical point. The decoupling between phase and amplitude (at least in the quadratic part of the theory) is possible because, near the Mott transition, *the order parameter amplitude is unrelated to the particle density*. Indeed, the amplitude vanishes even as we tune the system to the Mott transition at constant particle density.

The order parameter amplitude mode is closely analogous to the famous Higgs particle in the Standard Model of particle physics. In the Standard Model, the order parameter is charged under a local gauge symmetry; therefore, the Goldstone mode is replaced, through the Anderson–Higgs mechanism, by a massive gauge boson. What is left of the order parameter dynamics is then only the amplitude fluctuations, which make up the Higgs boson. In our case, because the condensate is gauge-neutral, there is no "Higgs mechanism," and Goldstone modes remain and coexist with the Higgs amplitude mode.

The gapless Goldstone modes provide a decay channel, for example through the coupling $u\phi_1\phi_2^2$ obtained upon expanding the original action to cubic order, which can lead to broadening of the Higgs resonance. This observation naturally raises the question of whether the amplitude mode is observable as a sharp resonance near the critical point. The answer to this question, which turns out to be interesting and subtle, was formulated only recently [41, 85, 86]. The theory was confirmed when the Higgs mode was first observed in a system of ultracold atoms near the superfluid-to-Mott insulator transition [36]. Below, we review the theoretical understanding and the measurement scheme used to detect and characterize the Higgs mode.

As a first attempt to address the questions of decay of the Higgs mode and assess whether it is visible as a peak in some linear response measurement, we may be tempted to compute the self-energy of the longitudinal mode ϕ_1. To lowest order in the cubic coupling $u\phi_1\phi_2^2$, the Matsubara self-energy is given by

$$\Sigma_1(q) = u^2 \int \frac{d^{d+1}k}{(2\pi)^d} \frac{1}{k^2(k+q)^2} \sim \begin{cases} \dfrac{u^2}{\log q}, & d = 3, \\ \dfrac{1}{q}, & d = 2. \end{cases} \tag{1.35}$$

In real frequency, this implies $\text{Im}\,\Sigma_1(\omega, \mathbf{q} = 0) \sim 1/|\omega|$ in two dimensions. One may worry that the low-frequency divergence in the above self-energy would wash away or completely mask any peaked response associated with the Higgs mode. But this conclusion is incorrect. As will be explained below, this self-energy does not reflect the intrinsic decay of the amplitude mode.

To clarify whether a particular mode would show up as a resonance in a linear response measurement, it is important to first specify what is the perturbation to which the system is responding. A common probe in some systems with broken symmetry is one that couples directly to the order parameter. We can write a perturbation of this type as $H_{\text{vex}} = \vec{h}(t) \cdot \vec{\psi}$, where $\vec{\psi}$ is a vector (e.g., magnetic) order parameter. In magnetic systems, neutron scattering acts as a vector probe of this type. In the case of a superfluid, we can think of the complex order parameter $\psi = \psi_1 + i\psi_2$ as a two-component vector $\vec{\psi} = (\psi_1, \psi_2)$. The component of the external vector field acting parallel to the ordering direction $\vec{\psi}_0$ (longitudinal component) couples to the amplitude fluctuation σ. Thus, the self-energy (1.35), with infrared-divergent spectrum, is related to the linear response of the system to a longitudinal vector probe, i.e., $\text{Im}\,\Sigma_1(\omega) \sim \chi_\parallel''(\omega)$. But the vector perturbation is not a natural one to apply in ultracold atomic systems, because such a perturbation has to violate charge conservation.

A simple and direct scheme to measure the dynamical response of ultracold atoms involves periodic modulation of the optical lattice potential [100]. For atoms in the lowest Bloch band, described by the Hubbard model (1.27), the lattice modulation translates to a modulation of the hopping amplitude J. Since the lattice modulation does not break the $U(1)$ symmetry, this probe can only couple to scalar terms in the critical action (1.33). In particular, because the lattice strength is the microscopic parameter used to tune the transition, its modulation translates to a modulation of the tuning parameter r, i.e., the perturbation Hamiltonian is $H_s = \lambda(t)|\psi|^2$.

In order to describe the response to the scalar probe, it is convenient to use the polar representation of the order parameter, $\psi = (\psi_0 + \sigma)e^{i\theta}$. At quadratic order, the amplitude and phase fluctuations σ and θ play the same roles as the longitudinal and transverse fluctuations ϕ_1 and ϕ_2, respectively. The scalar probe λ couples linearly to σ in the same way as the longitudinal probe couples to ϕ_1. However, the cubic coupling of the amplitude mode to the phase fluctuations is of the form $\sigma\partial_\mu\theta\partial^\mu\theta$, i.e., the amplitude fluctuations couple to gradients of the phase and not directly to the phase. Indeed, this must be the case, by the $U(1)$ symmetry. The self-energy of the scalar amplitude fluctuation is then given by a similar loop diagram as that of the longitudinal fluctuation (1.35), but the additional gradients now cancel the denominators to give an infrared-convergent result $\Sigma_s(\omega) \sim \omega^3$.

It is instructive to reconsider the longitudinal susceptibility in the polar representation as well. As before, let us assume, without loss of generality, that the broken-symmetry order parameter is real. The longitudinal fluctuation of the order parameter can be written directly in the polar representation:

$$\phi_1 = \mathrm{Re}\,\psi - \psi_0 = (\psi_0 + \sigma)\cos\theta - \psi_0 = \sigma + \tfrac{1}{2}\psi_0\theta^2 + O(\sigma\theta^2). \qquad (1.36)$$

It is clear from this expression that the longitudinal fluctuation is not a pure amplitude fluctuation σ; it is contaminated by pairs of Goldstone modes through the θ^2 term. A longitudinal probe therefore directly excites this gapless continuum, leading to the infrared-divergent response we have obtained above. Compared with the longitudinal probe, the more physical scalar probe also gives a cleaner signature of the Higgs amplitude mode [85].

Having shown that the response to a scalar probe is infrared-convergent, we now briefly review what is known about the full line shape and how the Higgs resonance manifests in it. Of particular interest is the question whether a peak associated to the Higgs resonance appears in the scaling limit. We broaden our discussion a bit and consider the response to the scalar probe on both sides of the critical point. On the Mott side, there is only a gapped excitation at $q = 0$, which therefore cannot decay and corresponds to a real pole at $\Delta \equiv (g_c - g)^\nu$. We shall now use Δ rather than $|g - g_c|$ to parametrize the deviation from the critical point on both sides of the transition.

The pertinent question concerning the response on the (ordered) superfluid side of the transition is whether a peak appears also on this side in the scaling limit. If it does, then the frequency at the peak must also vanish as $|g - g_c|^\nu$ upon approaching the critical point. Equivalently, the scalar susceptibility of the superfluid is expected to follow a scaling form $\chi(\omega, \Delta) \sim \Delta^{D-2/\nu}\Phi(\omega/\Delta)$, with $\Phi(x)$ a universal scaling function with a peak at $x = x_p \sim 1$ [86]. Recent numerical results [41, 91, 94], as well as theoretical analyses [61, 86], indeed find such a a scaling form. It is important to note, however, that neither method is fully controlled. In particular, the quantum Monte Carlo simulations performed for this system can give a Matsubara response function, while the analytic continuation to real frequencies requires uncontrolled approximations or otherwise is an exponentially hard problem. Therefore, an experimental test of the theoretical and numerical predictions is needed.

An experiment aimed at testing some of the above predictions was reported in [36] using bosons in a two-dimensional optical lattice. Here, the heating rate $\dot{E} \sim \omega\chi_s''(\omega)$ due to weak lattice modulations was measured to high accuracy. Typical response spectra and the characteristic mode frequency obtained from them are reproduced here in Fig. 1.5. Note that these spectra exhibit a sharp edge followed by a continuum rather than a peak. This behavior is associated with the inhomogeneity of the trap. The leading edge stems from the response in the middle of the trap, which is closest to the critical point. The scaling of the leading edge upon approaching the transition from either side agrees quantitatively with the numerical predictions. In particular, the ratio of the excitation frequency in the superfluid to that in the Mott side is found to be $\omega_p/\Delta \approx 2$, consistent with state-of-the-art numerics [41, 94] and markedly different from the mean-field prediction $\omega_p/\Delta = \sqrt{2}$. The conclusion from the theoretical

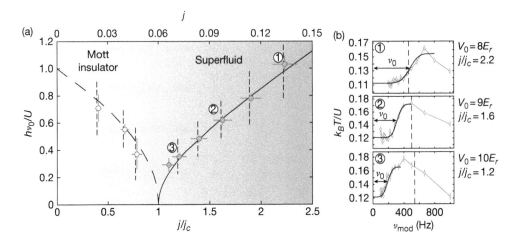

Fig. 1.5 (a) Energy of the Higgs excitation in the superfluid phase ($j/j_c > 1$) and of the gapped particle–hole excitation in the Mott insulator ($j/j_c < 1$) measured in [36] shows softening of the modes on approaching the critical point. (b) Examples of the measured spectra. The Higgs mode is seen as a sharp leading edge rather than a peak because of inhomogeneous broadening in the harmonic trap. (Reprinted by permission from Macmillan Publishers Ltd: *Nature* [36], copyright (2012).)

analyses and the experiment is that the Higgs mode in two spatial dimensions is visible and the spectral peak associated with it survives in the scaling limit.

1.3.3 Dynamical instability and decay of superflow in an optical lattice

Bosons tend to condense and become superfluids at low temperature. In Section 1.3.2, we have seen that bosons in an optical lattice can lose their superfluid properties at zero temperature through a phase quantum transition into the Mott insulating state. We have discussed how this transition is manifested in the dynamical linear response of the system at equilibrium. In this subsection, we will discuss the breakdown of superfluidity that occurs owing to flow of a supercurrent in the lattice. This is a highly nonequilibrium route for breakdown of superfluidity. In particular, we will focus on the interplay between the Mott transition and the critical current in the superfluid phase.

If a superfluid in a Galilean-invariant system is carrying a current, then this current can be removed by moving to the Galilean frame of the superfluid. Hence, without disorder or a lattice potential to break translational symmetry, a state with any magnitude of current is necessarily stable. Here we will focus on the effect of the lattice. In particular, we will seek the critical current above which superfluidity breaks down in the different regimes and the conditions under which this critical current is indeed sharply defined.

Mean-field critical current

Let us start with a mean-field description of supercurrents. Consider a condensate described by the order parameter $\psi_i = \sqrt{n_s}\, e^{i\theta_i}$, where i is a lattice site. A state with uniform current, say in the $\hat{x}$ direction, is described by a uniform phase twist, $\theta_i = p x_i$. The supercurrent density is then

$$ I = \frac{1}{2im_*}\left(\psi^*_{i+\hat{x}}\psi_i - \psi^*_i \psi_{i+\hat{x}}\right) = \frac{n_s}{m_*}\sin p, \qquad (1.37) $$

where m_* is the effective mass of a particle on the lattice and $p \equiv \theta_{i+\hat{x}} - \theta_i$ is the phase change across a link in the $\hat{x}$ direction.

By looking into this expression, we can anticipate many of the results that we will discuss in more detail below. In a weakly interacting condensate at $T = 0$, the superfluid density is just the total atom density $n_s = n$, which is essentially independent of the phase twist p. Then the maximal current that the system can carry is $I_c = n/m_*$, and this occurs when the phase twists by $\pi/2$ over a lattice constant ($p_c = \pi/2$). This result can be understood by considering the single-particle dispersion $\epsilon(p) = -m_*^{-1}\cos p$, where p is the lattice momentum. A condensate with phase twist p is obtained by condensing the bosons into the state with lattice momentum p. When p exceeds $\pi/2$, the local effective mass at that momentum, $m(p)^{-1} = \partial^2\epsilon/\partial p^2$, turns negative. The situation near the Mott transition is different. The system there is highly sensitive to an increase in the effective mass m_*, which leads to vanishing of n_s. It would, in the same way, be sensitive to increasing the local mass $m(p)$. Hence, in this regime, n_s is a decreasing function of p and we reach the critical (maximal) current at a much smaller value of the phase twist $p_c < \pi/2$.

To see how the instability occurs in the weakly interacting limit, we have to consider the relevant equation of motion in this regime, which is the lattice Gross–Pitaevskii equation

$$ i\frac{d\psi_i}{dt} = -J\sum_{\hat{\delta}}\psi_{i+\delta} + U|\psi_i|^2\psi_i. \qquad (1.38) $$

This mean-field description provides an accurate description of the dynamics for $U \ll Jn$. The idea now is to linearize the equation in the fluctuations δn and φ around the uniform-current solution $\psi_i = \sqrt{n + \delta n}\, e^{ipx_i + \varphi}$. For $p = 0$, solving these linear equations simply gives the linearly dispersing phonons (Bogoliubov modes) of the superfluid. However, beyond the critical value $p = \pi/2$, there are eigenmodes that take imaginary values, indicating a dynamical instability [118].

It is interesting to convert the critical current for dynamical instability to a critical flow velocity. In doing so, we get $v_c = p_c/m_* = (\pi/2)Ja$. This is much larger than Landau's criterion, which gives the much smaller sound velocity $v_s = \sqrt{JUn}\,a$ as the maximal stable flow velocity. In terms of the collective modes, Landau's criterion corresponds to the point where the excitations with negative momenta turn negative because of the Doppler shift. Negative frequencies, unlike imaginary ones, do not by themselves imply an instability. The idea of Landau's criterion is that if we add static impurities to the system, then they will induce scattering that creates pairs of negative-and positive-frequency modes that will lead to decay of the supercurrent. But in a pure lattice system, the current will decay only when the dynamical instability is reached. The dynamical instability in a weakly interacting lattice condensate has been seen experimentally [37].

Let us now turn to the strongly interacting regime, near the Mott transition, where the correlation length $\xi \gg a$. Here the dynamics is described by the effective continuum relativistic action (1.33). Treating this theory within the mean-field approximation, we obtain the saddle-point equation of motion

$$\ddot{\psi} = \nabla^2\psi + \tilde{r}\psi - |\psi|^2\psi, \tag{1.39}$$

where time and the field have been rescaled: $t \to ct$, $\psi \to c^{-1}\sqrt{u}\,\psi$, and $\tilde{r} = r/c^2$. In these units, we can write the correlation length as $\xi = 1/\sqrt{\tilde{r}}$. This equation admits the stationary solutions

$$\psi(x, \mathbf{z}) = \sqrt{r - p^2}\,e^{ipx}, \tag{1.40}$$

where $\mathbf{z}$ denotes the spatial coordinates transverse to $\hat{x}$. From this, it is obvious that the solutions disappear, and therefore the current-carrying states cannot be stable for a phase twist $p \geq \sqrt{r} = 1/\xi$. To find the precise critical twist for a dynamical instability, following [4], we expand (1.39) in small fluctuations around the stationary solutions. In the long-wavelength limit $\mathbf{q} \to 0$, this gives two modes with frequencies

$$\omega_1^2(\mathbf{q}) = 2(r - p^2) + \frac{r + p^2}{r - p^2}q_x^2 + \mathbf{q}_\perp^2, \tag{1.41}$$

$$\omega_2^2(\mathbf{q}) = q_x^2\frac{r - 3p^2}{r - p^2} + \mathbf{q}_\perp^2 \tag{1.42}$$

The first mode ω_1 is the generalization of the amplitude (Higgs) mode to the finite-current ($p \neq 0$) situation. It is gapped and has positive frequency unless $p^2 > r$. The second mode ω_2 is the gapless phase (phonon) mode. This mode becomes unstable for $p > p_c = \sqrt{r/3} = 1/(\sqrt{3}\,\xi)$. We see that the critical momentum vanishes as $1/\xi$ upon approaching the critical current.

The behavior of the critical current near the Mott transition in a three-dimensional optical lattice has been investigated experimentally [78]. A current was induced by moving the optical lattice, while the main observable was the existence of a sharp peak in the momentum distribution function as measured in a standard time-of-flight expansion scheme. The critical momentum was then determined as the flow momentum at which the condensate peak disappeared from the time-of-flight absorption image. Figure 1.6 shows the measured critical current versus the lattice depth, which is in excellent agreement with the mean-field result discussed above.

Current decay below the critical current

So far, we have discussed the breakdown of superfluidity at the critical current within a mean-field analysis. This seems to be a good enough approximation to describe

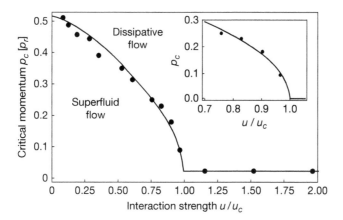

Fig. 1.6 Measured critical momentum (phase twist) as a function of the lattice strength near the superfluid-to-Mott insulator transition. The continuous line is the mean-field approximation for the critical momentum discussed in the text. (Adapted with permission from [78]. Copyrighted by the American Physical Society.)

experiments in three-dimensional lattices [78]. However, it is of fundamental interest to ask whether there is a possibility for current to decay, owing to fluctuations, even when the flow is slower than the critical flow.

The fact that the flow below the critical current is a stable solution of the classical equation of motion implies that in order to decay, the field configuration must *tunnel* through a classically forbidden region of phase space. In field theory, such tunneling of a macroscopic field configuration out of a metastable state was first termed by Coleman [33] "decay of the false vacuum." Here, the false vacuum is the state with a twist; it must go over an action barrier corresponding to the creation of a phase slip or a soliton in space–time to unwind the twist. A similar approach was taken earlier to describe the current decay due to thermal activation of phase slips in superconducting wires below the critical current [69, 76].

For now, calculation of the tunneling probability of a multidimensional field configuration appears to be a formidable problem. However, following [90], we will simplify it by restricting ourselves to the asymptotic behavior of the decay rate near the classical critical current, i.e., as $p \to p_c$ from below, where the action barrier is low. We will show that in this regime the pertinent information about the solution can be extracted from a simple scaling ansatz.

To understand how the scaling approach works, consider first the toy problem of escape from a metastable state in single-particle quantum mechanics (or field theory in $d = 0$ dimensions). As usual, to facilitate a semiclassical approximation of the tunneling path, we rotate the action to imaginary time:

$$S_{0d} = \int d\tau \left[\dot{\phi}^2 + V(\phi) \right]. \tag{1.43}$$

Here $V[\phi]$ is the potential, which for a small barrier can generically be written as a cubic function $V(\phi) = \epsilon\phi^2 - \phi^3$. The limit $\epsilon \to 0$, at which the barrier vanishes and the particle at $\phi = 0$ becomes classically unstable, is analogous to reaching the classical critical current in the field theory discussed above. Within the semiclassical theory, the decay rate is given by $\Gamma \sim \Gamma_0 \exp[-\hbar^{-1}S_{\rm sp}(\epsilon)]$, where $S_{\rm sp}(\epsilon)$ is the saddle-point action and Γ_0 is an "attempt rate" obtained from integrating over the Gaussian fluctuations around the saddle point (for a complete pedagogical treatment of this problem, see, e.g., [2]).

We now pull the ϵ dependence out of the action (1.43) by applying the rescaling $\phi \to \phi/\epsilon$ and $\tau \to \sqrt{\epsilon}\,\tau$. This leads to the action

$$S_{0d} = \epsilon^{5/2} \int d\tau \left(\dot{\phi}^2 + \phi^2 - \phi^3\right). \tag{1.44}$$

Consequently, the decay rate in the semiclassical approximation is given by $\Gamma_{\rm sc} \sim \Gamma_0 \exp(-\hbar^{-1}\epsilon^{5/2}\tilde{s}_{\rm sp})$, where $\tilde{s}_{\rm sp}$ is just a number independent of ϵ. In this way, we have managed to obtain the parametric dependence on the vanishing barrier height without solving the saddle-point equations. This will be needed only to obtain the number $\tilde{s}_{\rm sp}$. In the case of the toy model, this number can be computed exactly because there is an exact solution to the saddle-point equations, while in more interesting situations it can be approximated using a variational ansatz.

Let us now turn to the real situation at hand. The first task is to write down an action, analogous to (1.43), that can describe the tunneling of the field configuration out of the current-carrying state close to the critical current. We start with the effective action of the Mott transition in imaginary time, using units such that $c = 1$. Furthermore, we rescale all lengths and times with the correlation length, i.e., $x_\alpha \to x_\alpha/\xi$. In these units,

$$S = \int d^{d+1}x \left(|\nabla\psi|^2 + r|\psi|^2 + \tfrac{1}{2}|\psi|^4\right). \tag{1.45}$$

Next, following [90], we express ψ using the fluctuations around the stable current-carrying state with phase twist k (note that we are using $k = p\xi$, which describes the twist, or momentum, in units of $1/\xi$ appropriate for the above action):

$$\psi(x, \mathbf{z}) = \sqrt{1 - k^2}\left[1 + \eta(x, \mathbf{z})\right]e^{ikx + i\phi(x, \mathbf{z})}. \tag{1.46}$$

Here the coordinate x is the direction of the current, while $\mathbf{z}$ denotes all other coordinates, including imaginary time, transverse to the current. In order to allow for the possibility of tunneling across the small barrier, we must expand the action to cubic order in the fluctuations η and ϕ. The two fluctuations are decoupled at the quadratic level after the transformation

$$\eta \to \eta - \frac{k}{1 - k^2}\partial_x\eta. \tag{1.47}$$

Now the gapped amplitude fluctuation can be disregarded and we are left with a cubic action that describes the action barrier in terms of the phase mode:

$$S = \int d^d z\, dx \left[\tfrac{1}{2}(\partial_z \partial_x \phi)^2 + \tfrac{2}{3}(\partial_z \phi)^2 + \tfrac{1}{2}(\partial_x^2 \phi)^2 + 2\sqrt{3}\,\epsilon(\partial_x \phi)^2 - 2\tfrac{1}{\sqrt{3}}(\partial_x \phi)^3 \right], \quad (1.48)$$

where we have denoted $\epsilon \equiv k_c - k$. After applying the rescaling

$$x \to \frac{x}{2 \cdot 3^{1/4}\sqrt{\epsilon}}, \qquad z \to \frac{z}{6\epsilon}, \qquad \phi \to \frac{3^{3/4}}{2}\sqrt{\epsilon}\,\phi. \qquad (1.49)$$

we obtain, to leading order in ϵ,

$$S = \frac{3^{9/4-d}}{2^d}(k_c - k)^{2.5-d} \int d^d z\, dx \left[(\nabla \phi)^2 + (\partial_x^2 \phi)^2 - (\partial_x \phi)^3 \right]. \qquad (1.50)$$

We conclude from this that the current decay rate is

$$\Gamma_d = \Gamma_0 \exp\left[-C_d(k_c - k)^{2.5-d} \right], \qquad (1.51)$$

where C_d is a number, to be discussed below, that can be obtained from a variational calculation. The physical picture behind this scaling solution is as follows. The rescaling was done after identifying natural scales in the action (1.48) that become singular at the classical critical twist. The natural length in the direction of the current is $x_{||} \sim 1/\sqrt{k_c - k}$, and in the transverse and time directions it is $x_\perp \sim 1/(k_c - k)$. Rescaling with these lengths amounts to the postulate that the critical instanton has these spatial extents in the respective directions of space–time. In addition, the energy barrier scales as $E_{\text{inst}} \sim \epsilon^3$. Putting all this together, we see that the instanton action must behave as $S_{\text{inst}} \sim E_{\text{inst}} \times x_{||} \times x_\perp^d \sim (k_c - k)^{2.5-d}$.

The precise value of the action depends on the detailed shape (functional form) of the instanton, which is encapsulated in the constants C_d. These constants have been obtained using a variational calculation in [90]. The results are $C_1 = 73$ and $C_2 = 67$. In three dimensions, the tunneling action appears to diverge as $k \to k_c$; that is, creating critical instanton solutions incurs a diverging action cost, whereas in one and two dimensions the cost of the critical instantons vanishes. Therefore, in three dimensions, the system would actually take the alternative route of creating finite-size (noncritical instantons) that incur a finite action cost in the limit $k \to k_c$. A variational calculation in three dimensions gives $\Gamma_3 \sim e^{-4.3}$ [90]. So, while in lower dimensions the probability for current decay goes to 1 continuously on approaching the critical current, in three dimensions it jumps from a small decay probability below the critical current to 1 above it. This explains why the mean-field picture worked so well in describing the experiments done in three-dimensional optical lattices [78].

1.3.4 The Mott transition in one dimension

So far, our discussion has been aimed at lattices in two or higher spatial dimensions. The Mott transition in a one-dimensional system is different in several important aspects. Here we give a very brief review of the essential universal features of the one-dimensional problem. More detailed accounts of the theory and relation to experiments with ultracold atoms, can be found in recent reviews, such as [31].

The arguments given above for the existence of an incompressible Mott phase in higher dimensions relied on having a strong lattice potential. In particular, the extreme limit of decoupled sites makes no sense otherwise. More generally, it can be shown that a superfluid in two or higher dimensions is perturbatively stable to the presence of a *weak* lattice potential, regardless of the commensurability. In one dimension, by contrast, the Mott transition can occur in the presence of an arbitrarily weak lattice potential commensurate with the particle density if the repulsive interactions are strong enough.

This can be seen by starting from the universal long-wavelength description of a Galilean-invariant superfluid, the harmonic-fluid theory already encountered in Section 1.2:

$$H_0 = \frac{1}{2} \int dx \left[\kappa^{-1} \left(\frac{1}{\pi} \partial_x \phi \right)^2 + \rho_s (\partial_x \theta)^2 \right]. \tag{1.52}$$

Here $\theta(x)$ is the condensate phase field, $\pi^{-1} \partial \phi$ is the smooth density field conjugate to the phase $([\theta(x), \partial \phi(x')] = i\pi \delta(x - x'))$, $\rho_s = \rho_0/m$ is the superfluid stiffness of the Galilean-invariant superfluid, and κ is the macroscopic compressibility, which in a weakly interacting gas is simply the inverse interaction constant. The algebraic decay of long-distance correlations is completely controlled by the Luttinger parameter $K = \pi \sqrt{\kappa \rho_s}$.

The full density operator can be written in terms of the operator $\phi(x)$ as [52]

$$\rho(x) = (\rho_0 - \pi^{-1} \partial_x \phi) \sum_{m=0}^{\infty} \cos[2m\pi\rho_0 - 2m\phi(x)]. \tag{1.53}$$

Accordingly, a commensurate lattice potential $V(x) = V_0 \cos(2\pi\rho_0 x)$, which couples to the particle density, has a nonoscillating contribution to the Hamiltonian:

$$H_{\text{lat}} = \int dx \, V(x) \rho(x) = V_0 \int dx \cos(2\phi). \tag{1.54}$$

The low-energy Hamiltonian is thus given by the well-known sine–Gordon model. Scaling analysis shows that the lattice potential embodied in the cosine term becomes a relevant perturbation if the interaction is sufficiently strong that the Luttinger parameter $K < 2$. In this case, the field $\phi(x)$ is locked to a minimum of the cosine potential even for arbitrarily small values of V_0. $\phi(x)$ can be viewed as the displacement of atoms from a putative lattice arrangement, where a uniform shift of $\phi \to \phi + \pi$ corresponds to translating all atoms by one lattice constant. Therefore, locking of ϕ to 0 or π corresponds to the same Mott state.

It is interesting that the mechanism that drives localization in the regime of weak lattice potential is an interference effect, namely, coherent backscattering enhanced by interactions. This is rather different from the mechanism discussed in the case of a strong lattice, which stems from a local gap apparent in the atomic (single-site) limit. Nonetheless, it should be emphasized that the long-wavelength physics effective near the critical point in one dimension is universal and always described by the sine–Gordon mode (1.54), even when the microscopic physics is modeled by the one-band Hubbard model (1.27). Of course, the coefficients of the sine–Gordon model have in general a very complex connection to the parameters of the microscopic model, and usually should be regarded as phenomenological parameters. But the crucial point is that any other nonlinear terms allowed by the $U(1)$ and lattice translational symmetries are less relevant under rescaling than the cosine term in (1.54).

1.4 Fast-quench dynamics of bosons in optical lattices

Here we will concentrate on quantum quenches, in which the system parameters are varied in a way that dynamically drives the system across a quantum phase transition. There are important theoretical motivations to study such quenches. In many cases, the quantum states on the two sides of the transition are conventional in the sense that they have a classical mean-field description That is, they can be represented well by a nonentangled wavefunction that can be factorized as a direct product of single-site wavefunctions. For example, the Mott phase admits such a classical description in terms of the site occupation numbers, whereas the superfluid phase is classical when represented in terms of the condensate phase. But because of the nontrivial commutation relations between the phase and the occupation number, the two representations are not compatible with each other and neither can provide an effective classical description of the time evolution when the system is driven across the phase transition. This makes the dynamics nontrivial and inherently quantum mechanical. Nonetheless, thanks to the universality associated with quantum critical points, there is hope of gaining theoretical insight into some aspects of the many-body quantum dynamics despite its complexity. In the following subsections, we will discuss several issues that have both been addressed theoretically and motivated by experiments with ultracold atoms.

1.4.1 Collapse and revival of the condensate on quenching to decoupled sites

An extreme example of a quench from a superfluid to a Mott insulator was demonstrated in an early experiment by Greiner et al. [46] soon after the observation of the superfluid-to-Mott insulator quantum phase transition. The experimental protocol was as follows:

1. The system was prepared deep in the superfluid phase.
2. Starting from the superfluid state, the optical lattice was suddenly ramped up, essentially shutting off the tunneling between sites completely.
3. The system was probed by releasing the atoms from the trap (and simultaneously shutting down the optical lattice) at varying times after the quench and observing the resulting time-of-flight image.

The initial state, being a Bose condensate, is expected to exhibit sharp interference peaks at zero momentum and at all other reciprocal lattice vectors. When the system is probed at later times of the evolution, the peaks disappear and reappear periodically over a time period set by the on-site interaction.

The result is easily understood in the extreme case studied in [46]. Because hopping was turned off in the quench, the final Hamiltonian consists only of the on-site interaction term in decoupled sites:

$$H_f = \tfrac{1}{2}U \sum_i n_i(n_i - 1) - \mu n_i. \tag{1.55}$$

Furthermore, if the initial state is deep in the superfluid phase, then it is described well by a site-factorizable state with a coherent-state wavefunction $|z\rangle = \exp(-|z|^2 - zb^\dagger)\,|0\rangle$ at each lattice site. Solving for the time evolution therefore reduces to a single-site problem, with the wavefunction of the site being given by

$$e^{-iH_f t}\,|\psi(0)\rangle = e^{-|z|^2} \sum_{n=0}^{\infty} \frac{z^n}{\sqrt{n!}} e^{-i\frac{1}{2}Un(n-1)t + i\mu n t}\,|n\rangle. \tag{1.56}$$

Here $|n\rangle$ is the state of a site occupied by n bosons. Recall that $|z|^2 = \langle n \rangle$. The weight of the sharp peaks in the momentum distribution is proportional to the squared modulus of the condensate order parameter, which is easily computed from the above wavefunction:

$$|\langle a_i \rangle|^2 = |z|^2 e^{-4|z|^2 \sin^2(Ut/2)}. \tag{1.57}$$

According to this expression, the interference peaks decay (or "collapse") on a timescale $t_C \approx (U|z|)^{-1} = 1/(U\sqrt{\langle n \rangle})$. They then fully revive at integer multiples of the time $t_R = 2\pi/U$. The two timescales t_C and t_R are widely separated when the site occupation is large, $|z|^2 \gg 1$. Only in this case is there a near-total collapse of the interference peak.

Note that the collapse time, and in fact the whole process, is identical to the single-mode phase diffusion calculated in Section 1.2.1. During the collapse, the coherent state $|z\rangle$ with a well-defined phase $\varphi = \arg(z)$ gradually becomes a superposition of a growing number of coherent states with concomitantly growing phase uncertainty. However, analysis of the problem in the coherent-state basis reveals an interesting effect occurring exactly at half the time between revivals, when the interference peaks are at their lowest. To see the effect, consider the "paired" order parameter

$$|\langle a_i^2 \rangle| = |z|^4 e^{-4|z|^2 \sin^2(Ut)}. \tag{1.58}$$

We see that at times $t_q = (2q + 1)\pi/U$, with q an integer, there are revivals of the pair order parameter $\langle a \rangle$, whereas the condensate $\langle a \rangle$ is fully collapsed. Indeed, the wavefunction at these times is a superposition of precisely two coherent states: $|\psi\rangle = (|z\rangle + |-z\rangle)/2$. This state therefore breaks the $U(1)$ symmetry, as does the initial state, but has a residual $\mathbb{Z}_2$ symmetry associated with phase rotations by π.

The time-of-flight images at the times t_q, when paired condensate order is established, are not expected to exhibit any interference peaks. This special order can be detected, however, by looking at the noise correlations in the time-of-flight image [3]. Specifically, it would be manifested as sharp peaks in the correlation between fluctuations at k and $-k$.

1.4.2 Beyond decoupled sites: final Hamiltonian near the transition

So far, we have discussed the extreme quench of the superfluid to a lattice of decoupled sites. Now we turn to the more general case in which the lattice strength is suddenly increased, but to a point where there is still non-negligible tunneling between the sites. One way to approach this problem is to limit ourselves to the vicinity of the critical point, considering quenches in which both the initial and final Hamiltonians are close to the quantum phase transition. In this regime, one might expect the dynamics to be described by the effective field theory (1.33).

The classical equation of motion implied by this effective theory is the nonlinear wave equation

$$\ddot{\phi} = \rho_s \nabla^2 \phi - r\psi - u|\phi|^2\phi. \tag{1.59}$$

If we assume, for simplicity, that the initial state was a homogeneous mean-field superfluid, then we can drop the spatial derivatives. This is then the equation of motion of a classical particle of mass $m = 1$ moving on the Mexican hat potential. The quench amounts to preparing the particle with some value ϕ_i that is larger in modulus than the equilibrium value. In a small quench, the new equilibrium value $\phi_0 = \sqrt{-r/u}$ is only slightly smaller than the initial value. The field ϕ will undergo nearly harmonic oscillations around the new minimum position. This is precisely the Higgs, or amplitude, mode. In the other extreme of a large quench, the final Hamiltonian is in the Mott-insulator regime, so the new equilibrium field is $\phi_0 = 0$ ($r > 0$).

Within the simple single-mode picture, there is a "dynamical transition" between the large-quench and small-quench regimes [103]. For a fixed initial state, the transition occurs when the energy of the initial (static) field configuration in the new potential is just enough to climb to the top of the Mexican hat. At this point, the time of oscillation diverges. However, this singularity may well be a peculiarity of the classical mean-field dynamics. Suppose we allow for quantum fluctuations (uncertainty) in the initial field configuration and its conjugate momenta (we shall see below how this can be done), but evolve each instance of the distribution according to the classical equations of motion. This can be shown to be the leading correction in the semiclassical approximation [87]. In this case, only one special instance of zero measure in the entire distribution follows an evolution with a diverging timescales, while generic instances of the ensemble are always off of the critical trajectory.

1.4.3 Sudden quench from the Mott insulator to the superfluid side

This model with a single complex field (i.e., O(2)) describes the the superfluid-to-Mott insulator transition at commensurate filling. During the quench, the tuning parameter of the transition is rapidly changed. If we are in the disordered phase initially, then $\phi(x)$ starts with small quantum fluctuations around that average value $\phi = 0$. When the sign of the tuning parameter r is changed, these fluctuations start to grow and we expect them to eventually develop into a superfluid order parameter. Our discussion of such quenches across symmetry-breaking transitions will follow closely a nice recent review by Lamacraft and Moore [68].

Before we go into greater detail, a few general remarks are appropriate about what can be expected from such evolution. Quantum mechanics is certainly crucial in the initial stage of the order parameter growth dynamics when the quench is from a disordered state (e.g., a Mott insulator) to an ordered phase. However, there is quite general agreement that when the order parameter is already locally formed, its subsequent dynamics becomes essentially classical. The degree of freedom that grows and becomes an order parameter is intrinsically quantum in that it is composed of non-commuting components. However, as its expectation value grows, the effect of the commutator becomes negligible compared with the expectation values and is therefore irrelevant in the determination of the correlation functions. Second, the coherent degrees of freedom that we denote as the order parameter are supplemented by a continuum of "incoherent degrees of freedom" such as phonons that are also generated in the quench. Owing to the presence of interaction terms in the action, the order parameter rapidly becomes entangled with the phonons, and so we should no longer discuss superpositions of order parameter configurations. Thus, we expect the order parameter to become classical both in the single-particle sense that commutators are unimportant and in the many-body sense due to the absence of superposition states. In other words, we expect the system to revert to effective classical (Langevin) dynamics in which the "incoherent" degrees of freedom serve as a bath for the order parameter dynamics. The description of the quantum evolution is important, however, to set up the correct initial state for the later classical coarsening. An interesting question for future investigation regards the extent to which the classical correlations that emerge in the later dynamics can retain signatures of nontrivial quantum correlations that were present in the initial state.

1.4.4 Model system: inverted harmonic oscillators

We will now discuss an effective model, following [67], that illustrates how the order parameter dynamics becomes classical at rather short times after a quantum quench from a disordered phase into the broken-symmetry phase. In the initial stages of order parameter growth, while the fluctuations ϕ^2 are still small, it is reasonable to neglect the interaction term. In this case, we can decouple the modes in momentum space, where each mode at momentum q is described by an independent quadratic theory:

$$H_0 = \tfrac{1}{2} \sum_q \left[|\Pi_q|^2 + (r + q^2)|\phi_q|^2 \right]. \tag{1.60}$$

When we quench from a positive to a negative value of r, a subset of the modes with sufficiently small q are suddenly subject to an inverted harmonic potential and therefore become unstable. Although the fate of these modes is very simple, we can use it to illustrate many important concepts. First and foremost, we can show precisely how these modes become effectively classical.

Single-mode case

Let us first concentrate on a single unstable mode described by

$$H = \tfrac{1}{2}p^2 + \tfrac{1}{2}\omega^2 x^2, \tag{1.61}$$

which is a toy model for the evolution of the uniform $q = 0$ mode, which happens to be the most unstable. The Heisenberg evolution of the quantum operators is given by

$$x(t) = x(0)\coth(|\omega|t) + \frac{p(0)}{|\omega|}\sinh(|\omega|t)$$

$$\to \tfrac{1}{2}e^{|\omega|t}\left[x(0) + \frac{p(0)}{|\omega|}\right], \tag{1.62}$$

$$p(t) = p(0)\coth(|\omega|t) + x(0)|\omega|\sinh(|\omega|t)$$

$$\to \tfrac{1}{2}e^{|\omega|t}[p(0) + |\omega|x(0)] = |\omega|x(t).$$

As a first sign that the dynamics becomes classical, we see that in the long-time limit, the operators become effectively commuting (or, more precisely, the noncommuting part becomes negligible in the computation of correlations). Using the above expressions, we can easily calculate any correlation function at time t in terms of correlations in the initial state. For example, the fluctuation of the position is

$$\langle x(t)^2 \rangle = \langle x(0)^2 \rangle \coth^2(|\omega|t) + \langle p(0)^2 \rangle \sinh^2(|\omega|t)$$

$$\to \frac{1}{4}\left[\langle x(0)^2 \rangle + \frac{\langle p(0)^2 \rangle}{|\omega|^2}\right]e^{2|\omega|t}. \tag{1.63}$$

Assuming the initial state is the ground state of a stable harmonic oscillator with real frequency ω_0, then

$$\langle x(t)^2 \rangle = \frac{1}{2\omega_0}\coth^2(|\omega|t) + \frac{\omega_0}{2|\omega|^2}\sinh^2(|\omega|t)$$

$$\to \frac{1}{8}\left(\frac{1}{\omega_0} + \frac{\omega_0}{|\omega|^2}\right)e^{2|\omega|t}. \tag{1.64}$$

We can interpret this time dependence as the classical evolution $x_c(t) = x_0 e^{|\omega|t}$, with the initial position x_0 being a Gaussian random variable with variance

$$x_0^2 = \frac{1}{8}\left(\frac{1}{\omega_0} + \frac{\omega_0}{|\omega|^2}\right). \tag{1.65}$$

The momentum at the same late time will be fully determined by the position. This claim can be substantiated by inspecting the Wigner distribution associated with the wavefunction at late times.

The Schrödinger wavefunction in the position representation is given by

$$\psi(x,t) \rightarrow A(t) \exp\left[-i\frac{x^2|\omega|}{2}\left(1 - 2e^{i\theta}e^{-2|\omega|t}\right)\right], \tag{1.66}$$

where $A(t)$ is a normalization constant and $\theta = \tan^{-1}(\omega_0/|\omega|)$. This can be seen by showing that $\psi(x,t)$ is annihilated by the long-time limit of $a(t) = x(t) + ip(t)/\omega_i$. The Wigner distribution associated with the wavefunction at late times is

$$f(p,x,t) = \int dx'\, \psi^*\left(x + \tfrac{1}{2}x',t\right)\psi\left(x - \tfrac{1}{2}x',t\right)e^{-ipx'}$$

$$= B(t)\exp\left\{-\alpha(t)x^2 - \frac{[p - \beta(t)x]^2}{\alpha(t)}\right\}, \tag{1.67}$$

where $\alpha(t) = 2|\omega|\sin 2\theta e^{-2|\omega|t}$ and $\beta(t) = |\omega|(1 - 2\cos 2\theta e^{-2|\omega|t})$. Because of the decay of $\beta(t)$, the Wigner distribution becomes increasingly squeezed in one direction. Specifically, the p distribution approaches a delta function $\delta(p - |\omega|x)$, concentrated on the phase-space trajectory of a classical particle.

Many-mode harmonic model

We now proceed to discuss the many-mode harmonic model given by (1.60). In this case, each mode is described by a different harmonic oscillator with a frequency dispersing as $\omega_q = \sqrt{r + q^2}$. We see that only a subset of the modes having $q < \sqrt{|r|}$ become unstable. The scale $r^{-1/2}$ gives a natural lengthscale for the inhomogeneity of the order parameter immediately after the quench. Because the order parameters at two points separated by more than $r^{-1/2}$ grow in random uncorrelated directions, this also sets the distance between topological defects in the order parameter field immediately after the quench. From the simple arguments above, we would predict a vortex density of $n_v \sim r$.

Quantum coarsening

To substantiate this point and derive the subsequent evolution, we can compute the time dependence of the field correlation function in the harmonic approximation and infer the evolution of the vortex density. This analysis was carried out by Lamacraft [67] in an attempt to explain experiments with spinor condensates rapidly quenched across a quantum phase transition to an easy plane ferromagnetic state [98].

The off-diagonal correlation function

$$C(x,t) = \langle \phi^*(x,t)\phi(x,t)\rangle = \int \frac{d^d k}{(2\pi)^d} e^{ikx}\langle \phi_k \phi_{-k}\rangle \tag{1.68}$$

can be evaluated directly from (1.64), which holds for each k mode. It is convenient for simplicity to consider the case where we quench from deep in the insulating phase into the ordered phase not too far from the transition. In this case, $r_0 \gg r$ and we get a simple expression

$$\langle \phi^*(x,t)\phi(x',t)\rangle \approx \frac{\sqrt{r_0}}{2} \int \frac{d^d k}{(2\pi)^d} e^{ik(x-x')} \frac{\sinh^2\left(t\sqrt{|r|-k^2}\right)}{|r|-k^2}. \tag{1.69}$$

We see that the value of this correlation function is determined to a large extent by causality. First, at $t = 0$, two different points x and x' are uncorrelated owing to the deep disordered initial state we have taken. Furthermore, for any value of t such that $2kt > x - x'$, i.e., outside of the forward light cone, we can close the contour of integration in the upper half-plane and, since there are no poles in the upper half-plane, the integral remains precisely zero. The correlations at distances within the light cone $x - x' < 2kt$ grow exponentially, reflecting the exponential order parameter growth.

From the behavior of the correlation function, we can infer the density of topological defects. Note that since we have taken the initial state to be uncorrelated, there is an initial large density of topological defects regulated only by a high-energy cutoff that starts to fall with time. The density of defects (generic zeroes) in a Gaussian wavefunction is given by the Halperin–Liu–Mazenko formula [72]:

$$n_v = \frac{1}{2\pi} \frac{C''(0)}{C(0)} = \frac{1}{2\pi C(0)} \int \frac{d^d k}{(2\pi)^d} k^2 \langle \phi_k \phi_{-k} \rangle. \tag{1.70}$$

Note that this formula can be understood as the average square wavevector, weighted by the mode occupation, of the inhomogeneous field. Applying this formula to the correlation function found above, we have

$$n_v(t) = \frac{\sqrt{|r|}}{4\pi t}. \tag{1.71}$$

Of course this type of coarsening does not continue for ever. When $|\phi|^2 \sim |r|/u$, the nonlinear term in the action becomes important, leading to two effects. First, it stops the exponential growth of the order parameter and, second, it affects the coupling between the modes, leading to dissipation. Given the exponential growth of the order parameter (1.69), this occurs at time

$$t^* \approx \frac{1}{2\sqrt{|r|}} \log\left(\frac{8r_f^2}{u\sqrt{r_i}}\right). \tag{1.72}$$

This leads to a formula for the topological defect density at the characteristic time t^*:

$$n_v(t^*) \sim r \log^{-1}\left(\frac{8r_f^2}{u\sqrt{r_i}}\right), \tag{1.73}$$

which agrees with the anticipated result up to the logarithmic factor.

Dynamics at later times

The quadratic approximation has to break down at times longer than t^*. First, the exponential growth is halted by the nonlinear terms. Second, the coupling between the modes will give rise to effective noise and dissipation that is expected to eventually lead to thermalization. Because the order parameter field is by now essentially classical and moreover coupled to a continuum of phonon modes, the latter stage of the dynamics is expected be described by classical Langevin dynamics. Exactly which model to use from the alphabetical classification of [54] is determined by the conservation laws.

1.4.5   Defect density beyond the Gaussian theory: Kibble–Zurek scaling

The fact that the quench takes place in the vicinity of a quantum critical point suggests using the scaling properties of the critical point to make statements valid beyond the Gaussian theory. However, taking the initial state of the sudden quench to be far from the critical point as we did in Section 1.4.4 is not compatible with such an analysis. Here we discuss an alternative scenario that will be amenable to scaling analysis. The analysis presented below is due to Polkovnikov [88] and is closely related to scaling arguments by Zurek [120] for quench across a classical phase transition.

We consider a system with continuous time dependence described by the Hamiltonian

$$H = H_0 + \lambda(t)V. \tag{1.74}$$

Here H_0 denotes the Hamiltonian at the quantum critical point, V is a relevant operator that allows tuning across the critical point, and

$$\lambda(t) = v\frac{t^r}{r!} \tag{1.75}$$

is the time-dependent field that drives the transition. The special case $\alpha = 1$ describes a linear ramp across the transition. On the other hand, the limit $\alpha \to \infty$ describes a sudden quench with an amplitude set by the coefficient v.

The basic idea behind the Kibble–Zurek scaling analysis is to compare the instantaneous rate of change $R = \dot{\lambda}/\lambda$ with the instantaneous gap Δ. When we are still far from the critical point, the time evolution can be considered as adiabatic as long as $R < \Delta$. As we approach the critical point, the gap vanishes as $\Delta \sim \lambda^{\nu z}$, so eventually the rate of change becomes fast relative to the gap $R > \Delta$. From here, the time evolution can be considered as adiabatic until the gap reemerges on the other side of the transition. According to this reasoning, the crossover from adiabatic to sudden occurs at the value of λ where the equation $\dot{\lambda}/\lambda = \Delta$ holds, or

$$\alpha v^{1/\alpha}\lambda_*^{-1/\alpha} = \lambda_*^{\nu z}. \tag{1.76}$$

Therefore,

$$\lambda_* = \alpha^{\alpha/(\alpha\nu z+1)}v^{1/(\alpha\nu z+1)}. \tag{1.77}$$

The order parameter configuration (correlations) freezes to the structure it had at the point $\lambda(t_*) = \lambda_*$, so the vortex density is set by the correlation length at this point, $n_v = \xi_*^{-2}$. Using the scaling $\xi = 1/\lambda^\nu$, we obtain the vortex density:

$$n_v(t_*) \sim v^{2\nu/(\alpha\nu z+1)}. \tag{1.78}$$

The Kibble–Zurek arguments involving a crossover from sudden to adiabatic evolution help to illustrate the physics; however, we can obtain the same result from more general pure scaling arguments, which will allow further generalization. First note that the scaling of ξ_* with v can be found if we know the scaling dimension $[v]$ of v, since $v \sim 1/\xi^{[v]}$. Now we use $\lambda = vt^\alpha$, so $[v] = [\lambda] - \alpha[t] = \nu^{-1} + \alpha z$. Inverting the relation, we have the characteristic scale $\xi_* = v^{-1/(\nu^{-1}+\alpha z)}$, leading to the Kibble–Zurek result for the vortex density. We can generalize this to describe the behavior of any correlation function after the quench. For example, a two-point equal-time correlation function must be proportional to a scaling function:

$$\langle \hat{O}(x)\hat{O}(0)\rangle = \frac{1}{\xi_*^{2\eta}}\mathcal{F}\left(\frac{x}{\xi_*}\right), \tag{1.79}$$

where η is the scaling dimension of the operator O and ξ_* was found above. As an example, we can take the order parameter correlations in the one-dimensional Ising model, $\langle \sigma^z(x,t)\sigma^z(0,t)\rangle$. The scaling dimension of σ^z at the Ising critical point is $\eta = 1/8$, $z = 1$, and $\nu = 1$. Moreover, since correlations decay exponentially in the disordered phase, we expect $\mathcal{F}(y) \sim e^{-y}$ at long distances. Using the the scaling of ξ_* with v, we expect the following behavior of correlations at long distances as a function of the ramp rate v:

$$\langle \sigma^z(x,t)\sigma^z(0,t)\rangle = v^{\frac{1}{4}(1+\alpha)^{-1}}\exp\left(-xv^{(1+\alpha)^{-1}}\right). \tag{1.80}$$

1.5   Quench dynamics and equilibration in closed one-dimensional systems

In Section 1.4, we discussed a dynamic quench that takes the system across a quantum critical point. The discussion of the quantum dynamics was confined to the behavior of the correlations at short times following the quench. We pointed to reasons why the dynamics becomes classical at long times. The unitary quantum dynamics gives way to a simpler coarse-grained description in terms of classical Langevin equations as the system approaches thermal equilibrium.

One-dimensional systems could be an exception to this rule for at least the following reasons:

1. Quantum fluctuations in one dimension generally inhibit the growth of the order parameter, eliminating one reason for classical behavior.
2. There are integrable one-dimensional models that cannot thermalize. Even when integrability is broken, routes to relaxation are more restricted in one dimension than in higher dimensions; therefore, one may expect a parametrically long period of time separating the initial simple unitary evolution and eventual thermalization. Such a state is often referred to as a pre-thermalization.

Significant theoretical efforts have therefore been made in recent years to understand the evolution of correlations following a quench in one dimension. But even in one dimension, the problem is very hard and in general intractable. Strictly speaking, the Hilbert space needed for calculation grows exponentially with the system size. The problem is especially acute in one dimension, since no single dominant mode grows to macroscopic values and dominates the dynamics. On the other hand, we know of an efficient way to simulate quantum ground states in one dimension using matrix product states or the density matrix renormalization group (DMRG); see the review by Schollwöck [99] and his contribution to this volume (Chapter 4). This method relies on theorems that guarantee a low entanglement entropy in the ground state. However, during quantum dynamics remote parts of the system gradually get more entangled giving rise to growth of entanglement entropy that rapidly inhibits efficient simulation.

1.5.1   Growth of entanglement entropy

It is natural to take the initial state in a quantum quench as a ground state of some local Hamiltonian. Such an initial state is characterized by low entanglement entropy, i.e., at most an area law up to logarithmic corrections. When this state starts to evolve under the influence of a different Hamiltonian, the entanglement entropy will in general grow. The way it grows is important and interesting for several reasons. First, the growth of entanglement entropy limits the ability to simulate the time evolution using the matrix product state, time-dependent DMRG, or related methods. Second, if the entanglement entropy is considered from the information-theoretic perspective, its time evolution measures how fast information propagates between different regions of the system following the quench. Finally, the growth of entanglement entropy can tell us something about the approach to equilibrium and the nature of the long-time equilibrium state itself. For instance, does the system approach equilibrium on a finite timescale, independently of the system size? Does the entanglement entropy of a finite subsystem approach the same value expected if it had reached true thermal equilibrium?

The Lieb–Robinson bounds [71] on propagation of correlations also limit the growth of the entanglement entropy to be linear in time [25, 27]. In clean systems, as far as we know, this limit is always saturated. The intuitive picture is that excitation of the system by the quench leads to uniform parametric excitation of quasiparticle pairs (e.g., particle–hole pairs) that travel in opposite directions. Since each particle–hole pair is in general entangled, the crossing of the interface by one of the partners adds one unit to the entanglement entropy between the two halves of the system. Let us consider only the fastest of these, which travel at the Lieb–Robinson velocity v_m. Then the number of particle–hole pairs shared between the two halves of the system grows linearly with time, and hence the entropy grows as $S(t) \sim v_m t$.

1.5.2   Evolution of correlations following a quench in one dimension

As mentioned above, the rapid growth of entanglement entropy in the evolution following a quench prohibits direct numerical simulation of generic models in extended systems, at least for long times. Some progress can be made by resorting to exactly solvable models and benchmarking against numerical simulations on short to inter-mediate timescales. Because of the relevance to recent experiments with ultracold atoms in optical lattices (see, e.g., [32]), this approach has become somewhat of an in-dustry. So far, however, it appears that it is hard to draw general conclusions from the results. We will therefore present only a brief review of this topic, focusing on examples where exactly solvable models can be used to demonstrate a more general rule.

Decay of antiferromagnetic order in a spin-$\frac{1}{2}$ chain: gapless final state

A natural model in which to consider the decay of order parameter correlations in one dimension is a quantum spin chain. For example, we want to look at the spin-$\frac{1}{2}$ xxz chain:

$$H = J \sum_i \left(S_i^x S_{i+1}^x + S_i^y S_{i+1}^y + \Delta S_i^z S_{i+1}^z \right). \tag{1.81}$$

Note that the spin model can be thought of as a hard-core boson model with hopping $J/2$, where $b_i^\dagger = S_i^+$. Alternatively, this model is mapped exactly through a Jordan–Wigner transformation to a model of fermions with hopping $J/2$ and nearest-neighbor interaction Δ. The ground state of this model is a z-antiferromagnet with gapped excitations for $\Delta > 1$ and a gapless phase with power-law correlations (Luttinger liquid) for $0 \leq \Delta < 1$.

Let us consider a scenario in which a system, prepared with antiferromagnetic order, is quenched to the state with $0 \leq \Delta < 1$. The question we ask concerns the decay of the antiferromagnetic order parameter

$$m(t) = \frac{1}{N} \sum_j (-1)^j \langle S_j^z \rangle. \tag{1.82}$$

A closely related problem has been addressed directly by a recent experiment using bosons in optical lattices [112], with the boson density in the lattice playing the role of the S^z component of the spin.

Of course, the experiment is not described by exactly the model presented above; however, both systems have the same long-wavelength limit. It is therefore tempt-ing to take a universal approach to this class of problems using the long-wavelength continuum limit, in this case a Luttinger liquid, to describe the evolution following the quench. The general framework for doing this using conformal field theory (CFT) has been set up by Calabrese and Cardy [28] (see also [30] for the specific case of a Luttinger liquid).

In the case under study, here we can also make a direct calculation within the harmonic Luttinger liquid framework. In terms of the bosonized fields, the order par-ameter is given by $\cos 2\phi$. Initially, it is concentrated near one of the two degenerate minima. The time evolution of this operator is now in direct correspondence with the decay of phase coherence $\cos \theta$ in an interferometer. The result, by duality to the latter problem, is

$$m_s(t) \sim e^{-t/\tau}, \tag{1.83}$$

where $\tau = 2/(\pi K \Delta_0)$.

In order to assess the validity of the low-energy approach, we shall benchmark it against an exactly solvable model [17, 18]. Consider for this purpose a quench to a final Hamiltonian with $\Delta = 0$, i.e., the "xx model." The xx model can be mapped through a Jordan–Wigner transformation to a model of free fermions in which we can calculate the order parameter evolution exactly.

We shall take the initial state to be a charge density wave (of the fermion density):

$$|\Psi_0\rangle = \prod_{k=-\pi/2}^{\pi/2} \left(u_k + v_k c_{k+\pi}^\dagger c_k\right)|FS\rangle = \prod_k \left(u_k c_k^\dagger + v_k c_{k+\pi}^\dagger\right)|0\rangle. \tag{1.84}$$

Note that the case $u_k = v_k = 1/\sqrt{2}$ corresponds to a perfect Néel state. A more general state arising from a self-consistent mean-field theory of the gapped phase has, in complete analogy with a BCS state,

$$\langle c_{k+\pi}^\dagger c_k \rangle_0 = u_k v_k = \frac{\Delta_0}{2\sqrt{\epsilon_k^2 + \Delta_0^2}}, \tag{1.85}$$

where $\epsilon_k = -J \cos k$ is the fermion dispersion. Using the exact Heisenberg evolution of the fermions, $c_k(t) = c_k e^{-it\epsilon_k}$, we can compute the time dependence of the order parameter:

$$m_s(t) = \frac{1}{N} \sum_k e^{i\, 2\epsilon_k t} \langle \psi_0 | c_k^\dagger c_{k+\pi} |\psi_0\rangle$$

$$= \frac{1}{\pi} \int_{-J}^0 d\epsilon \, \frac{\cos(2t\epsilon)}{\sqrt{J^2 - \epsilon^2}} \frac{\Delta_0}{\sqrt{\epsilon^2 + \Delta_0^2}}. \tag{1.86}$$

Let us first consider the case of a perfect Néel initial state. In this case, the factor $u_k v_k$ is a constant, $\frac{1}{2}$, and we can do the integral exactly, with the result

$$m_s(t) = \tfrac{1}{2} J_0(2Jt) \xrightarrow{Jt \gg 1} \frac{1}{\sqrt{4\pi Jt}} \cos\left(2Jt - \tfrac{1}{4}\pi\right). \tag{1.87}$$

This result has no trace of the exponential decay implied by the analysis of the CFT in the long-wavelength limit. In fact, since the Néel state is obtained as an initial state with $\Delta_0 \to \infty$, the CFT would predict a decay on a vanishing timescale. The source of the discrepancy is that the initial state occupies many high-energy modes at the band edges and is therefore far from the range where the low-energy theory is expected to hold. The above integral and hence the decay of the staggered moment are dominated by the van Hove singularity at the band edge. Note that an analysis using a linearized spectrum yields $(\Lambda t)^{-1} \sin(\Lambda t)$, where Λ is the high-frequency cutoff.

For a better comparison with the CFT result, we now consider an initial state that is a weak charge density wave characterized by $\Delta_0 \ll J$. This state is much closer to the gapless Luttinger liquid. In this case, there are two main contributions to the integral, the first coming from $\epsilon \approx 0$ near the Fermi points and the second from the square-root singularities at $\epsilon = \pm J$. Summing these two contributions, we have

$$m_s(t) \approx \frac{\Delta_s}{\pi J} K_0(2\Delta_0 t) + \frac{\Delta_0}{2J} J_0(2Jt)$$

$$\xrightarrow{\Delta_0 t \gg 1} \frac{1}{\sqrt{4\pi Jt}} \left[\sqrt{\frac{\Delta_0}{J}} e^{-2\Delta_0 t} + \frac{\Delta_0}{J} \cos\left(2Jt - \tfrac{1}{4}\pi\right) \right]. \qquad (1.88)$$

In addition to the oscillatory decay identical to that found in the case of a Néel state, we find here a nonoscillatory exponential decay, with a timescale $\tau = (2\Delta_0)^{-1}$, in agreement with the CFT result. Comparing the two terms, we see, however, that at times $t > \Delta_0^{-1} \ln(J/\Delta_0)$, the oscillatory decay dominates the order parameter dynamics. Nevertheless, in contrast to the case of a Néel initial state, the low-energy modes contribute to the dynamics over a parametrically long period of time.

There are a few more comments that should be made regarding the applicability of low-energy theories. First, it is interesting to note a symmetry of the dynamics, which allows us to map the dynamics subject to the Hamiltonian with $\Delta > 0$ exactly onto the dynamics subject to the Hamiltonian with $\Delta < 0$. In particular, the time evolution of an initial Néel state under the antiferromagnetic Hamiltonian ($\Delta = 1$) is identical to its evolution under the ferromagnetic Hamiltonian ($\Delta = -1$). This is in spite of the fact that the low-energy description associated with the two Hamiltonians are completely different. In the antiferromagnetic interaction, the low-energy modes are linear in k, whereas in the ferromagnet interaction, they are quadratic. This result is a corollary of a more general theorem to be proved below concerning the dynamics under the influence of a Hamiltonian H versus the dynamics under $-H$.

1.6 Quantum dynamics in random systems: many-body localization

1.6.1 Introduction

The quantum dynamics of isolated random systems, such as systems of ultracold atoms in a random potential, has attracted a great deal of attention recently as a possible alternative paradigm to thermalization. The conventional wisdom is that closed many-body systems, following unitary time evolution, ultimately reach thermal equilibrium. In a quantum system, the process of thermalization involves the entanglement of local degrees of freedom with many distant ones and thereby the loss of any quantum information that may have been stored locally in the initial state. In the absence of accessible quantum correlations, the long-time dynamics of such thermalizing systems is described well by classical hydrodynamics (i.e., slow fluctuations of conserved quantities and order parameter fields) [54, 73] It is of fundamental interest to find systems that defy the conventional paradigm and in which quantum correlations do persist and affect the long-time behavior.

Integrable quantum systems, mentioned in Section 1.5, evade thermalization because their time evolution is constrained by an infinite set of integrals of motion.

These are, however, highly fine tuned models. Any small perturbation from the exact integrable Hamiltonian will lead to eventual thermalization. Indeed, the main theme of both Sections 1.2 and 1.5 was prethermalization. Namely, on short to intermediate timescales, the system reaches a quasisteady state that can be understood based on the integrable models, while the slower processes that lead to thermalization dominate only at parametrically longer timescales. In contrast, it is expected that systems with quenched randomness provide a generic alternative to thermalization. That is, the nonthermal behavior would be robust to moderate changes to system coupling constants (disorder, interactions, etc.).

The idea that isolated many-body systems subject to sufficiently strong quenched randomness would fail to thermalize was first conjectured by Anderson in his original paper on localization [7], although in his calculations he could only address a one-particle problem. This idea was revitalized in the last decade when convincing theoretical arguments were given for the stability of Anderson localization to moderate interaction between the particles even in highly excited many-body states (i.e., states with nonvanishing energy density) [19, 44, 57].

Theoretical studies in the last few years have shown that the MBL state can be viewed as a distinct dynamical phase of matter, which exhibits clear universal signatures in dynamics [1, 8, 15, 16, 20, 79, 82, 105–107, 113, 114, 119]. In particular, these studies show that, similarly to integrable systems, such localized Hamiltonians are characterized by an infinite set of integrals of motion. In this case, they are local only up to possible exponentially decaying tails. However, in contrast to integrable systems, the conservation laws can be slightly modified, but not destroyed, by moderate changes in the interaction parameters. The integrals of motion are destroyed, or become fully delocalized, only in a phase transition that marks a singularity in the system dynamics. This localization transition defines a sharp boundary between a macroscopic system dominated by quantum physics even at long times to one that is governed by emergent classical dynamics. Some understanding of the transition is beginning to emerge based on general arguments [50] and novel renormalization group approaches [93, 115].

The physics of MBL is beginning to be addressed experimentally [64, 81]. In an interesting recent development, the breaking of ergodicity due to MBL was directly observed for the first time and studied in detail using ultracold atoms in an optical lattice [101]. The main focus of this section will be on explaining the current experimental effort. We will briefly review the essential theoretical notions needed to understand the experimental developments in MBL. More detailed introductions to the theory can be found in recent reviews [5, 79].

1.6.2 Stability of Anderson localization to interactions and the many-body mobility edge

Let us for a start consider the system of weakly interacting fermions discussed by Basko et al. [19]. As was done in that work, we will assume that the particles are subject to sufficiently strong disorder that all of the single-particle eigenstates are localized. For example, the localizing single-particle part of the Hamiltonian could be a tight-binding model supplemented by strong on-site disorder:

$$H = -t \sum_{\langle ij \rangle} c_i^\dagger c_j + \text{h.c.} + \sum_i V_i n_i + \sum_{\langle ij \rangle} U n_i n_j, \qquad (1.89)$$

where h.c. indicates the Hermitian conjugate of the preceding term. Without the interaction U, such a system is definitely nonergodic if all the single-particle states are localized. In particular, a density fluctuation present in the initial state will not be able to relax to the equilibrium distribution. This is true regardless of the energy of the initial state. The question, as formulated for example in [19, 44], is whether this nonergodic state is robust to the addition of weak interactions between the particles.

To elucidate the effect of interactions, it is convenient to work with the Fock basis $|\{n_\alpha\}\rangle$ describing occupations of the single-particle localized eigenstates $|\alpha\rangle$ localized around real-space positions $\vec{x}_\alpha$:

$$H = \sum_\alpha \epsilon_\alpha n_\alpha + \sum_{\alpha\beta\gamma\delta} \tilde{U}_{\alpha\beta\gamma\delta} c_\alpha^\dagger c_\beta^\dagger c_\gamma c_\delta. \qquad (1.90)$$

The noninteracting part of the Hamiltonian is represented here by the diagonal first term, where ϵ_α are the random energies of the single-particle localized states. The short-range density–density interaction $U n_i n_j$, which is diagonal in the original lattice basis, has off-diagonal terms in the single-particle eigenstate basis which can hop particles between localized states centered at different locations. Note that $\alpha, \beta, \gamma, \delta$ should be centered within a localization distance ξ away from each other to have a significant matrix element.

Although it can mediate hopping, the interaction does not necessarily delocalize the particles. At strong disorder, the interaction-induced hopping often takes a particle to a state that is highly off-resonant with the initial state. If the system is at a finite temperature (or energy density), the missing energy for a hop may be obtained by de-exciting, at the same time, a particle–hole excitation with the right energy. Hence, the particle–hole excitations can potentially provide an effective intrinsic bath that can enable incoherent hopping processes just as phonons do when the system is not isolated. The difference from a phonon bath, however, is that the particle–hole excitations are themselves localized and therefore have a discrete local spectrum. A crude way to assess whether the particle–hole excitations can indeed give rise to incoherent hopping is to compute the putative hopping rate using Fermi's Golden Rule and then verify that the rate thus obtained is larger than the level spacing of the putative bath, in which case the particle–hole excitations may be considered as an effective continuum bath. This leads to the following criterion for delocalization valid at low temperatures: $T\tilde{U}/\Delta_\xi^2 > 1$, where T is the temperature and Δ_ξ is the level spacing within a localization volume. In other words, there is a critical temperature $T_c \approx \Delta_\xi^2/\tilde{U}$ below which the system is strictly localized according to this argument.

The above argument is perturbative, and one might worry that a higher-order process involving a large number of particle–hole excitations may ultimately delocalize the particles. To address this issue, Basko et al. [19] invoked an approximate mapping to a single-particle problem, taking the localized Fock states $|\{n_\alpha\}\rangle$ to be sites in a multidimensional random graph containing 2^N sites, where N is the number of real-space sites. The random local energy associated with each site is the energy of the Fock state computed with the noninteracting Hamiltonian (i.e., $E[\{n_\alpha\}] = \sum_\alpha \epsilon_\alpha n_\alpha$). The interaction matrix elements $U_{\alpha\beta\gamma\delta}$ can be viewed as an effective hopping that connects the sites of the multidimensional lattice. A system in a given (extensive) energy explores only the states of the Fock space lattice that are compatible with that energy. Hence, the effective connectivity of the Fock space lattice grows sharply with the energy density. Finally, the approximate localization criterion $T_c \approx \Delta_\xi^2/\tilde{U}$ is obtained (for low temperatures) by applying Anderson's criterion for delocalization [7] to the Fock space lattice with a connectivity set by the temperature (or energy density).

It is important to note the crucial difference between the many-body mobility edge discussed above and the conventional notion of a mobility edge in the single-particle spectrum. In a (weakly interacting) system with a single-particle mobility edge, there will always be a finite density of particles thermally excited above the mobility edge, which will thus contribute to thermally activated transport. This is not the case with a many-body mobility edge. The critical energy corresponding to the many-body mobility edge is generically extensive, like almost all the many-body spectrum. Now suppose that the system is prepared in a thermal state (say by coupling to an external bath and then decoupling it) with average energy below the many-body mobility edge. Because the energy fluctuations in the thermal ensemble are sub-extensive, the probability of finding the system with energy above the mobility edge vanishes in the thermodynamic limit. Hence, below the critical energy density or critical temperature, the response of the system must be strictly insulating.

Before closing this section, we note that a Hamiltonian similar to (1.90) can also be written for spins:

$$H_{\text{spin}} = \sum_i h_i S_i^z + \sum_{ij} J_{ij}^{\alpha\beta} S_i^\alpha S_j^\beta. \tag{1.91}$$

Here the first term describes noninteracting spins subject to a local random magnetic field. The interaction term can potentially give rise to energy transport. But the same arguments as given for the fermion system imply that the localized state should be stable toward moderate interactions between the spins. Note that the spin Hamiltonian does not have to conserve spin; dipolar interactions, for example, are not spin-symmetric. But localization is still well defined via the absence of energy transport.

1.6.3 Numerical studies of the localized phase

Models with bounded spectrum, such as spin chains and lattice fermions, are convenient for doing numerical simulations. It was first pointed out in [80] that in such models the entire many-body spectrum can be localized for sufficiently strong disorder when the mobility edges coming from the bottom and top of the many-body spectrum join in the middle of it. The situation in which the entire spectrum is localized can be viewed as localization at infinite temperature; that is, the ensemble in which ever energy eigenstate is equally likely. By tuning the disorder or interaction strength, we may drive an MBL transition at infinite temperature.

The case of a fully localized spectrum is also highly relevant to studies of time evolution following a quench from a simple product state. Most simple initial states are composed mostly of eigenstates in the middle of the spectrum. Hence, the long-time dynamics will change sharply when the many-body mobility edge closes in the middle of the spectrum.

We start by describing DMRG calculations of the time evolution following a quench in an XXZ model with random field [16, 119]:

$$H = -J \sum_i \left[(S_i^+ S_{i+1}^- + \text{h.c.}) + J_z S_i^z S_{i+1}^z \right] + \sum_i h_i S_i^z. \qquad (1.92)$$

Note that this model is equivalent, through a Jordan–Wigner transformation, to a model of interacting spineless fermions in a random lattice potential with J_z the interaction strength. The time evolution dictated by this Hamiltonian was calculated starting from a product state of the spins randomly oriented up or down along S^z.

In the clean systems encountered in previous sections, such dynamics could only be computed for very short times. The problem was the rapid growth of entanglement entropy, which soon prohibited an efficient description in terms of matrix product states. In a localized system, however, one might expect that no correlations would be formed across the middle partition over distances larger than some (localization) lengthscale ξ (here ξ is given in units of the lattice constant). If this were true, then the entanglement entropy would be bounded from above by a value of order ξ, thus enabling DMRG calculations to long times.

The numerical calculations, however, found a surprising result [16, 119]. The expected saturation of the entanglement entropy occurred only for $J_z = 0$, which corresponds to the special case of Anderson localization of noninteracting fermions. For any other value of J_z, the entanglement entropy grew as $S(t) \sim \log t$ at long times. While the entanglement growth is unbounded, it is nonetheless rather slow and therefore allows DMRG calculations to be taken to long times. For small values of the interaction J_z, the onset of the logarithmic growth of $S(t)$ was delayed to a time of order $\hbar/J_z$. On the other hand, the prefactor of the logarithmic growth was roughly independent of interactions. Now, the logarithmic growth of the entanglement entropy is understood to be a ubiquitous feature of MBL. Below, we discuss how it can be understood within a simple theory of this state.

The fact that the entanglement entropy grows in an unbounded way, even if very slowly, immediately raises the question of whether the system ultimately thermalizes. This question was addressed using numerical simulations in [16], where it was shown that the entanglement entropy of a finite subsystem with length L saturates to a constant $S(L) = s_\infty L$. The saturation time is exponentially large in L. Crucially, however, the saturation value, although it follows a volume law, was found to be significantly lower than the value expected at thermal equilibrium, given the system energy.

1.6.4   Effective theory of the MBL phase

The simple universal behavior, such as the ubiquitous slow logarithmic growth of the entanglement entropy, seen in numerical simulations of the MBL state calls for a simple effective description of the long-time dynamics in this state. In the more familiar context of low-energy equilibrium physics, the renormalization group provides a conceptual framework to describe the universal features of quantum phases and phase transitions. The idea is that if we are interested only in low energies and in measurements done with low resolution, then systems that may be microscopically very different can be described by the same effective theory, provided they are in the same phase. In principle, these effective theories are obtained as fixed points of the renormalization group: a successive elimination of short-wavelength (or high-energy) fluctuations and rescaling. The fixed-point theory is universal and in most cases simpler than the original microscopic theory. A case in point is Fermi liquid theory, which is essentially an integrable theory. As found to be the correct renormalization group fixed point [108], or simply as postulated by Landau, it can be used as a powerful phenomenological description of the phase.

It is natural to ask if there is an analogous framework to describe the long-time dynamics in the localized phase. The problem is that the quench dynamics involves all energy scales, so one cannot integrate out high-energy modes and focus on the low energies. Indeed, as remarked above, the dynamics is dominated by very high-energy eigenstates in the middle of the many-body spectrum. Nonetheless, a renormalization group framework can be applied as shown in [84, 114, 115]. The basic idea is to integrate out high-frequency modes in order to gain a simpler description of the evolution at long times, which is similar to renormalization group schemes applied to classical dynamical systems (see. e.g., [74]). In the quantum context, targeting low frequencies corresponds to focusing on small energy differences between eigenstates rather than on low absolute energies as in the equilibrium scheme. Application of the renormalization group leads to a simple fixed-point Hamiltonian that can be used as an effective theory of the MBL state. In generic cases, the form of the effective theory can also be postulated based on general arguments and used as a phenomenological description of the state [55, 106].

Before focusing on the effective theory of the localized phase, we give a brief qualitative description of the renormalization group scheme. For more details of how exactly it works, see [5, 84, 114, 115]. The idea is similar to the real-space strong-disorder renormalization group (SDRG) scheme developed by Dasgupta and Ma [34]. Given a Hamiltonian such as (1.92), we pick out at every step the fastest spins in the chain, namely, those subject to the largest couplings. These couplings give rise to a local gap or frequency scale Ω, which is our running high-frequency cutoff. If the disorder is strong, then neighboring spins will typically be affected by much weaker couplings and therefore are essentially frozen on the timescale $1/\Omega$. The effective Hamiltonian for the dynamics, coarse-grained to longer times, is obtained through a unitary transformation of the Hamiltonian (moving to the rotating frame of the fast spins), which freezes out the fast spins and generates effective weaker couplings between the remaining slow ones nearby. The transformation that decouples the fast spins is found perturbatively in J_{typ}/Ω, where J_{typ} is the value of typical couplings on the chain. Because the renormalization group is implemented as a series of unitary transformations, degrees of freedom are not eliminated from the Hilbert space. Instead, at every step, another local operator is transformed (frozen) into an integral of motion. The fixed-point Hamiltonian is therefore finally obtained as a Hamiltonian written entirely in terms of conserved quantities that are perturbatively related to the bare local operators, such as S_i^z in (1.92).

Thus, lack of thermalization in the MBL state is seen to be the result of an infinite number of (quasi) local conservation rules that are revealed in the course of the renormalization group flow. As in the case of integrable models, expectation values of integrals of motion that are constrained by the initial conditions prevent relaxation to a true equilibrium. The important difference of the MBL state from conventional integrable models, however, is that the effective theory is a stable renormalization group fixed point and it is therefore robust to moderate changes in the microscopic coupling constants (disorder strength, interaction strength, etc.). Hence, in contrast to integrable models, the MBL state proves to be a generic alternative to thermalization dynamics.

An effective theory written entirely using integrals of motion has also been postulated, based on general arguments, as a phenomenological description of the MBL phase [55, 106]. It is simple to understand how such a theory emerges by considering the weakly interacting fermion model (1.90). In this basis, it is clear that the occupations of the single-particle localized states, $n_\alpha = \sum_{ij} \phi_\alpha^\star(x_i)\phi_\alpha(x_j)c_i^\dagger c_j$, are conserved quantities of the noninteracting model. Note that these are not strictly local, because the single-particle wavefunctions $\phi_\alpha(x)$ have exponential tails away from their center at x_α.

Starting from the noninteracting system, one can derive an effective Hamiltonian perturbatively in U. The off-diagonal terms of the interaction U in the Fock basis are off-resonant and hence do not contribute, at lowest order, to the effective Hamiltonian. To first order in U, the diagonal matrix elements of the interaction give rise to an effective interaction $\sum_{\alpha\beta} V_{\alpha\beta} n_\alpha n_\beta$, where $V_{\alpha\beta} \sim U e^{-|\mathbf{x}_\alpha - \mathbf{x}_\beta|/\xi}$. Physically, an interaction between electrons occupying distant localized states is generated owing to the exponentially small spatial overlap of the two localized wavefunctions. Higher orders in perturbation theory renormalize the integrals of motion $n_\alpha \to \tilde{n}_\alpha$, where $\tilde{n}_\alpha$ may now contain terms higher than quadratic in the fermion operators. High orders in perturbation theory also produce interaction terms involving multiple integrals of motion, such as $V_{\alpha\beta\gamma}\tilde{n}_\alpha\tilde{n}_\beta\tilde{n}_\gamma$.

Similar considerations apply to the spin model (1.92) and suggest an effective model of the form [55, 106]

$$H_{\text{eff}} = \sum_i \tilde{h}_i \tau_i^z + \sum_{i,j}\sum_n \tilde{J}_{ij} \tau_i^z \hat{B}_{ij} \tau_j^z. \tag{1.93}$$

Here the integrals of motion τ_i^z are Pauli operators that are related by a quasilocal unitary transformation to the original Pauli operators σ_i^z of the original spin. Following [55], we will term the quasilocal τ operators l-bits or "logical bits," while the strictly local σ operators will be termed p-bits or "physical bits." The $\hat{B}_{ij}$ are general operators including contributions from all possible strings of τ_k^z and τ_k^0 (unit matrices) extending between sites i and j. The nonlocal interactions $\tilde{J}_{ij}$ fall off exponentially with distance as $J^z e^{-|x_i - x_j|/\xi}$.

The effective model (1.93) provides a simple explanation of the logarithmic growth of entanglement entropy with time [55, 107] and, furthermore, allows the prediction of a host of other intriguing properties of the MBL state. The growth of entanglement entropy can be understood as follows. Consider two l-bits from the two sides of a given partition of the system that are separated by a long distance r. In the initial state, these l-bits would generally not be in their eigenstates even if the local p-bits were prepared in a well-defined state. Therefore, the interaction between the l-bits generates entanglement after a characteristic time of the order of $t = J_{ij}^{-1} \approx J_z^{-1} e^{|r|/\xi}$. By this time, all the l-bits within the interval of length r between the two chosen l-bits will have had a chance to entangle with each other as much as the constraints set by the initial state allow. Hence, the entanglement entropy at this time would grow to $S = s_* r$. That is, s_* is the average diagonal entropy of single l-bits in the initial state, $s_* = \rho_{\uparrow\uparrow} \log \rho_{\uparrow\uparrow} + \rho_{\downarrow\downarrow} \log \rho_{\downarrow\downarrow}$. Now, inverting the relation between t and r gives $S(t) = s_* \xi \log(J_z t)$. The von Neumann entanglement entropy of a finite subsystem of length L saturates to a volume law $S(L) = s_* L$ with a coefficient s_* that depends on the initial state and will generally be lower than the thermal value.

We see that the logarithmic growth of entanglement stems from dephasing of oscillations in the presence of a diagonal interaction term that is present only in the case of interacting fermions. Hence, this logarithmic growth is a feature distinguishing between a noninteracting Anderson-localized state and an interacting MBL state. The entanglement entropy is, however, not readily accessible to experimental probes. It is desirable to identify indications of this physics in quantities that are directly measurable. Specifically, I consider below the decay of fluctuations in the relaxation dynamics of observables after a quench in the MBL phase. These results were obtained in Ref. [105], while related theoretical arguments were also given in Ref. [113].

For simplicity, we start by considering the relaxation dynamics of the l-bits τ_i although they are not directly accessible in experiments. we will then generalize to the accessible physical bits σ_i. The relaxation dynamics of the integrals of motion τ_i^z is trivial; being integrals of motion, they simply do not decay, i.e., $\langle \tau_i^z(t) \rangle = \langle \tau_i^z(0) \rangle$. What about the nonconserved components $\langle \tau_i^{x,y}(t) \rangle$? If there are no interactions, then $\langle \tau_i^{x,y} \rangle$ simply precess around the τ^z axis with random independent frequencies set by the local fields h_i. With interactions, the time evolution is more complicated because of the generated entanglement. Assume that the system starts in a product state

$$|\Psi_0\rangle = \prod_{j=1}^{L} \left(A_\uparrow^j |\uparrow\rangle_j + A_\downarrow^j |\downarrow\rangle_j \right) = \sum_{\tau_1,\ldots,\tau_L} \prod_{j=1}^{L} A_{\tau_i}^j |\{\tau\}\rangle. \tag{1.94}$$

The time-dependent state, evolving under the influence of the effective Hamiltonian (1.93), is

$$\sum_{\{\tau\}} \prod_{j=1}^{L} A_{\tau_i}^j e^{-iE[\{\tau\}]t} |\{\tau\}\rangle, \tag{1.95}$$

where $E[\{\tau\}] = \sum_i h_i \tau_i + \sum_{i,j} V_{ij} \tau_i \tau_j$ and $\{\tau\}$ is short for $\tau_1, \ldots, \tau_L$. The expectation value of the l-bit operator in this state is

$$\langle \tau_i^x(t) \rangle = \sum_{\{\tau\}_{j \neq i}} e^{i\omega_i[\{\tau\}]\, t} (A_\uparrow^i)^* A_\downarrow^i \prod_{j \neq i} |A_{\tau_j}^j|^2 + \text{c.c.}, \qquad (1.96)$$

where $\omega_i[\{\tau\}] \equiv E[\uparrow_i, \{\tau\}] - E[\downarrow_i, \{\tau\}]$ is the energy change due to flipping spin i given the configuration $\{\tau\}$ of all other spins. The reason this is a sum over all 2^L basis states of the chain is that the spin i could in principle be entangled with the entire chain. But we know that after evolution time t, the spin can only be significantly entangled with the $2l(t) \sim 2\xi \log(J_z t)$ spins nearest to it. Mathematically, this is seen from (1.96) by noting that the phase $\omega_i[\{\tau\}]t$ is essentially unchanged by flipping τ_j if $|x_j - x_i| \gg l(t)$. Hence, the phase factor can be taken out of the sums over such remote spins τ_j and these sums then simply give 1. We remain with a sum over $N_\omega(t) = 2^{2l(t)}$ terms oscillating at independent random frequencies. Hence, $\langle \tau_i^x(t) \rangle$ will appear as random noise with a fluctuation amplitude $\sim 1/\sqrt{N_\omega}$. Because N_ω grows with time, the noise strength will correspondingly decrease in time as

$$\left[\overline{\langle \tau_i^x(t) \rangle^2} \right]^{1/2} = N_\omega(t)^{-1/2} \sim e^{-\xi \ln 2 \ln(J_z t)} = \left(\frac{1}{J_z t} \right)^{\xi \ln 2}. \qquad (1.97)$$

In practice, the system can be prepared with well-defined states of the physical bits σ_j rather than the l-bits τ_j. However, these two sets of operators are perturbatively related, so we may write $\sigma_i^z = Z \tau_i^z + \ldots$. If we prepare the system with nonvanishing expectation value $\langle \sigma_i^z \rangle_0$, then a component of this expectation value cannot decay, and we expect $\langle \sigma^z \rangle(t)$ to relax to a nonvanishing stationary value at long times $\sim Z \langle \tau^z \rangle_0 \sim |Z|^2$. The nonconserved component of σ_i^z, on the other hand, is expected to undergo oscillations at $2^{2l(t)}$ independent frequencies just as discussed above for the nonconserved operator τ_i^x.

To summarize, we expect local physical operators, such as the local density in a fermion system, that are perturbatively related to conserved quantities, to attain a nonequilibrium stationary value at long times. Moreover, the same dephasing process that gives rise to the logarithmic growth of entanglement entropy leads to a power-law decay of the temporal noise in these observables. The exponent associated with this power-law decay is not universal (it depends on the initial state and he disorder strength); it is directly related to the prefactor of the logarithmic growth of entanglement entropy.

Before proceeding, we briefly mention other features of the MBL phase that can be understood within the effective theories. The first intriguing property is the persistence of quantum coherence in the MBL phase. Suppose we want to use the accessible p-bits as q-bits. We have seen that if we prepare such a q-bit in a state with definite σ_i^z, then some memory of the initial state remains forever. On the other hand if we prepare the q-bit to point in the xy plane, then it starts to dephase and entangle with many other spins. We might worry that the quantum coherence that existed in the initial state (the phase, or the direction, on the xy plane) is lost irreversibly. However, using the effective Hamiltonian (1.93), it has been shown that the coherence, i.e., the initial state of the q-bit, can be partially retrieved even after infinite time using protocols that address only the accessible p-bit [13, 104].

Another intriguing feature of MBL is that there can be more than one MBL phase [56]. These can exhibit, for example, broken symmetry and even topological order and protected edge states [13]. Note that these orders occur in high-energy states, even where they are not allowed to exist in an equilibrium situation. As an example, consider the one-dimensional random transverse field Ising model:

$$H = \sum_i h_i \sigma_i^x + \sum_i J_i \sigma_i^z \sigma_{i+1}^z + V \sum_i \sigma_i^x \sigma_{i+1}^x. \tag{1.98}$$

Note that with the added coupling V, the model cannot be mapped to free fermions. There are two simple limits to this model that lead to two distinct prototype dynamical phases. The first is the free-spin limit where the couplings J_i and V_i are set to zero. In this case, σ_i^x are clearly integrals of motion and the eigenstates of this system do not break the $\mathbb{Z}_2$ symmetry. This "paramagnetic" state persists upon adding moderate interactions J_i and V_i, as the integrals of motion σ_i^x transform to closely related l-bits τ_i^x. The other simple limit is where $h_i = V_i = 0$. In this case, the eigenstates are frozen random spin configurations in the σ^z basis, such as $|\uparrow\uparrow, \downarrow\uparrow\uparrow\uparrow\downarrow \ldots \downarrow\rangle$, which clearly break the $\mathbb{Z}_2$ symmetry. For each eigenstate, we have $\langle \psi_n | \sigma_i^z | \psi_n \rangle \neq 0$. Again, the integrals of motion do not change qualitatively on the addition of moderate couplings h_i and V_i. This state can be viewed as an eigenstate glass because of the frozen spin order in eigenstates. The two distinct localized phases—eigenstate glass and paramagnet—are separated by a dynamical quantum phase transition that can be described using the dynamical renormalization group approach mentioned above [84, 115].

Note that while there is spin order in eigenstates, i.e., $\langle \psi_n | \sigma_i^z | \psi_n \rangle \neq 0$, the order vanishes in a thermal average. Hence, from a thermodynamic perspective, the glass is just a paramagnet. However, it is in principle easy to detect the glass order dynamically. Because σ_i^z have an overlap with true conserved quantities, the autocorrelation function $\langle \sigma_i^z(t) \sigma^z(0) \rangle$ must approach a nonvanishing constant at long times. This is true for any initial state, including when the system is prepared in a thermal state.

1.6.5   The MBL transition

We have argued above that the MBL regime should be viewed as a distinct quantum dynamical phase of matter that provides a generic alternative to thermalization dynamics. These two phases must be separated by a dynamical phase transition, which represents a sharp boundary between ergodic dynamics that ultimately becomes classical and dynamics that is inherently quantum even at long times. Theories of the MBL transition are only starting to appear [93, 116] and the picture is far from complete. Yet there is agreement on some interesting features of the transition, which could be highly relevant to experiments with ultracold atoms. In this subsection, we briefly review recent progress in understanding some features of the MBL transition in one-dimensional systems.

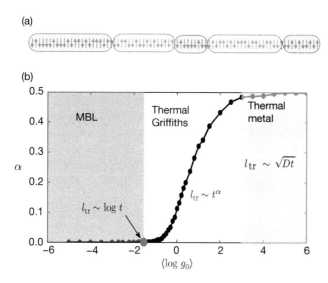

Fig. 1.7 (a) Schematic picture of the system near the MBL transition. According to this picture, the system is fragmented into incipient insulating and metallic puddles. The ultimate fate of the sample is decided on large scales depending on whether the metallic puddles are sufficiently large and dense to thermalize the intervening insulators. This competition and the resulting scaling near the critical point are described by a renormalization group analysis. (b) One outcome of the renormalization group is the scaling relation between length and time, $l_{tr} \sim t^\alpha$, which describes the length over which an energy fluctuation can be transported over a time t. Here α is plotted as a function of a microscopic tuning parameter (roughly disorder strength) for a one-dimensional system. The continuous vanishing of α at the critical point implies the existence of a subdiffusive delocalized phase between the insulator and normal thermal fluid.

A plausible coarse-grained picture of the system close to the MBL transition is shown in Fig. 1.7(a). Because of fluctuations in the disorder strength, some regions (subchains containing many local degrees of freedom) may behave locally as insulators, whereas others are incipient conductors. Because these regions are finite, they behave as insulators or conductors according to their local properties only if they are observed at some finite timescale. At longer scales, locally insulating regions may be thermalized by larger nearby conducting blocks, or, conversely, isolated thermal regions may be absorbed into larger insulators around them. A semi-phenomenological renormalization group scheme that captures the thermalization or absence of thermalization between the different regions at increasing scales has been presented by Vosk et al. [116]. Whether a block is insulating or thermal is assumed to be determined by a single parameter $g = \Gamma/\Delta$, where Γ is the is the inverse time for entanglement propagation across the block and Δ is the many-body level spacing of the block. This scheme leads to a flow of the distribution of g with two possible stable fixed points—insulating and thermalizing—in which g respectively decreases or grows exponentially with the lengthscale.

One result of the renormalization group study is the relation between the length l of a cluster (subchain) and the time τ it takes to transport energy across it $\tau \sim l^z$. In a conventional thermalizing system, we expect diffusive transport, which implies $z = 2$ (i.e., $l = \sqrt{D\tau}$). So, one might expect that the transition to the localized phase could be characterized by how the diffusion constant D vanishes upon approaching the critical point. But this turns out to be a misguided expectation. Instead, we find that the dynamical exponent z diverges at the critical point, or, as shown in Fig. 1.7(b), $\alpha = z^{-1}$ vanishes continuously as the transition is approached. This result implies that the delocalized phase near the critical point exhibits subdiffusive transport, as also supported by numerical studies of small systems using DMRG and exact diagonalization [1, 15]. At the critical point and in the insulating phase, we have $\tau \sim \tau_0 e^l / l_0$, where l_0 and τ_0 are microscopic time- and lengthscales, respectively (in a weakly interacting fermion system, for example, l_0 could be the single-particle localization length).

It is easy to understand the origin of the subdiffusive behavior even without delving into details of the renormalization group scheme. Let us assume that the MBL transition is controlled by a scale-invariant critical point, as pointed to by the renormalization group. As we approach this critical point from the delocalized side, there must be a diverging correlation length ξ. At scales below ξ, the system has not yet "decided" if it is going to be insulating or localized on larger scales, and therefore we are likely to find local insulating behavior on this scale. Insulating regions much larger than ξ have a low probability per unit length, $p(l) \sim \xi^{-1} e^{-l/\xi}$, but they can occur with high probability in a sufficiently long sample. In a system of length L, close to the critical point the transport will be dominated by the slowest (i.e., the longest) insulating cluster on the chain. We can estimate the length l_m of this cluster by requiring the probability $Lp(l_m)$ to be approximately 1. Thus, the longest bottleneck for transport is typically of size $l_m \approx \xi \ln(L/\xi)$, leading to a transport time

$$\tau_{\text{tr}} \sim \tau_0 e^{l_m/l_0} \approx \tau_0 (L/\xi)^{\xi/l_0}. \tag{1.99}$$

From this expression, we can read off the dynamical exponent $z = \xi/l_0$, which indeed diverges at the critical point. Such a phase, dominated by rare slow clusters, is called a Griffiths phase [47].

The Griffiths phase terminates at the critical point marking the transition to the MBL phase. Interestingly, a good scaling variable to characterize this critical point is the entanglement entropy of half the system [116]. In a finite system of size L, the results indicate a universal crossover from area-law entanglement entropy in the localized phase to fully thermal volume-law entanglement in the delocalized phase. At the critical point itself, the fluctuations of the half-system entanglement entropy diverge and scale with the volume L (whereas they approach constants at large L on either side of the transition). A corollary of this result is that the Griffiths phase described above is ergodic; it is characterized by a fully thermal volume-law entropy with vanishing fluctuations.

1.6.6 The experimental situation

Because systems of ultracold atoms are almost perfectly isolated from the environment, they provide a promising platform to investigate the physics of MBL. Until recently, experiments have focused on Anderson localization of noninteracting particles [21, 59, 65, 96]. A typical experiment used a quench scheme to study the expansion of particles in a random potential background [21]. The particles were prepared in a highly confined state induced by a dipole trap; then the cloud was released and expanded in a potential created by a random speckle pattern. The interaction between the expanding particles was negligible compared with the large kinetic energy acquired from the initial confined state. A direct signature of localization was obtained by observing the density profile of the expanding cloud at long times, which is a proxy of the single-particle wavefunction $|\psi(x,t)|^2$. Another set of experiments studied the zero-temperature quantum phase transition of interacting bosons from a superfluid to a disordered Bose insulator with linear response transport-like measurements [35, 40, 83], using a quantum quench across the transition [77].

The above experiments did not address the generic MBL problem. The goal in this case would be to demonstrate a robust nonergodic state, which exists at finite energy and over a range of interaction strengths. An early experiment to show evidence for MBL measured the global mass transport of interacting fermions on a disordered three-dimensional lattice [64]. The relevant signature was apparent immobility of the atomic cloud, which persisted for a range of temperatures. One problem in measuring global transport, however, is that conserved quantities are slowest to equilibrate. Even in a thermalizing system, one may therefore need exceedingly long times to distinguish between slow transport and complete absence thereof. In addition, global measurements are potentially sensitive to the inhomogeneity of the trap. For example, a localized weakly interacting region at the wings of the cloud can block or slow down movement of a more mobile core.

Some of these issues were overcome in a recent experiment designed to study MBL by observing the relaxation of local observables [101]. The experiment was done with a system of interacting fermions in a quasiperiodic one-dimensional lattice, well described by the following model Hamiltonian:

$$\hat{H} = - J \sum_{i,\sigma} \left(\hat{c}_{i,\sigma}^{\dagger} \hat{c}_{i+1,\sigma} + \text{h.c.} \right)$$
$$+ \Delta \sum_{i,\sigma} \cos(2\pi\beta i + \phi) \, \hat{c}_{i,\sigma}^{\dagger} \hat{c}_{i,\sigma} + U \sum_{i} \hat{n}_{i,\uparrow} \hat{n}_{i,\downarrow}. \tag{1.100}$$

Here the magnitude Δ of the incommensurate lattice plays the role of effective disorder strength. Although this is not a true random potential, the model without interactions exhibits Anderson localization above a critical effective disorder $\Delta_c = 2J$ [12]. The difference between this model and a one-dimensional model with a true random potential is that in the latter all states are localized for arbitrarily weak disorder. Numerical results suggest that the interacting system with sufficiently large effective randomness also exhibits MBL with the same universal properties as the generic MBL phase.

The MBL state is identified in the experiment and its properties are studied by implementing a quench scheme and observing the consequent relaxation dynamics of observables. Specifically, the system is prepared in a far-from-equilibrium density-wave state as depicted in Fig. 1.8(a). This initial state then begins to evolve subject to the full Hamiltonian modeled by (1.100) and we monitor the time dependence of the density-wave order. The latter is captured by the normalized imbalance between the number of particles in even sites (N_e) and that in odd sites (N_o):

$$\mathcal{I} = \frac{N_e - N_o}{N_e + N_o}. \tag{1.101}$$

In an ergodic phase, the system would relax to thermal equilibrium corresponding to a high temperature, in which the symmetry between even and odd sites must be restored so that $\mathcal{I} \to 0$. Instead, the experimental time traces show, for a wide regime of interaction and disorder strengths, an initial rapid drop but then saturation of the imbalance to a stationary value different from zero at times that are long compared with the hopping rate. Examples of such time traces are shown in Fig. 1.8(b). This result is a direct signature of robust ergodicity breaking; it shows that there is a phase characterized by an infinite set of local conserved quantities, closely related to the local densities.

The phase diagram shown in Fig. 1.8(c) is constructed after measuring time traces of the imbalance for a wide range of interaction and disorder strengths. The light gray flow lines depict the postulated renormalization group flow, which controls the localization transitions on this phase diagram. The axis corresponding to $U = 0$ exhibits the nongeneric critical point of the noninteracting Aubry–Andrey model [12]. On the other hand, for any finite interaction strength, the transition is expected to be controlled by the generic fixed point addressed in [116]. The interactions also have a noticeable effect on the phase boundary. Moderate interactions have a delocalizing effect that leads to an increase in the critical disorder strength. However, the critical disorder strength drops again when the interaction exceeds the single-particle bandwidth.

The behavior at large U has a strong dependence on the nature of the initial state, which is controlled in the experiment by changing the strength of interaction U_i in the preparation stage, when the atoms are loaded into the optical lattice. When U_i is large and positive, there are almost no doubly occupied sites in the initial state, while if U_i is large and negative, there are many double occupancies. Now, if the interaction effective during time evolution is strong, $U \to \pm\infty$, then, owing to energy conservation, doublons present in the initial state cannot decay and new ones cannot form. The bold black phase boundary in Fig. 1.8(c) corresponds to the doubloon-free situation corresponding to large positive U_i. In this case, the dynamics of a state with no doublons can be mapped exactly to the dynamics of noninteracting (spinless) fermions [101], which explains the symmetry between the $U = 0$ and $U \to \infty$ limits. On the other hand, if there are doublons in the initial state, then they become extremely heavy particles (with effective hopping $J_D \sim J^2/U$ at large U) and are therefore easy to localize. The corresponding critical disorder is the lower boundary of the striped area in the phase diagram.

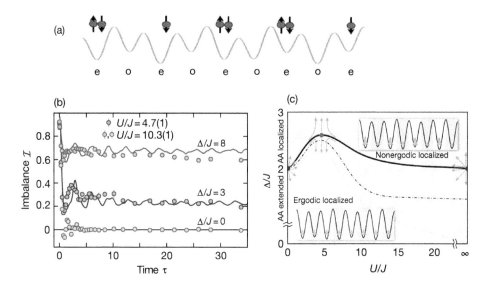

Fig. 1.8 (a) Schematic illustration of the initial state for the quench dynamics studied experimentally in [101]; the system is prepared in a density-wave configuration where every second site is empty. (b) Time evolution of the relative density imbalance between even and odd sites, $\mathcal{I} = (\langle N_e - N_o \rangle / (N_e + N_o))$. Relaxation of the imbalance to a nonvanishing stationary value is a direct demonstration of ergodicity breaking from [101]. (c) Experimentally determined phase boundary between the ergodic fluid and the nonergodic MBL phase. The bold line corresponds to an initial state with no (or very few) doublons, while the lower line corresponds to an initial state with many doublons. (Parts (b) and (c) are from [101]. Reprinted with permission from AAAS.)

The above results have provided evidence for MBL, but more experiments are needed to substantiate them and to characterize the dynamics in the MBL phase and phase transition. One issue that came up in the recent experiment [101] is that on timescales much larger than shown in Fig. 1.8(b), the imbalance slowly decays owing to the slow destruction of MBL by scattering and loss of atoms interacting with the laser light. Recall that MBL relies crucially on having a closed system and is therefore very sensitive to the above loss processes that, in effect, open it. It may be interesting to learn about the MBL phase by investigating how susceptible it is to the addition of such dissipative couplings. This can be particularly useful in studying the MBL phase transition. Here an analogy with conventional quantum phase transitions (QPTs) can be useful. A QPT is sharply defined only at zero temperature, just as the MBL transition exists only for vanishing coupling to a bath. Nonetheless, a QPT has a dramatic influence on the physics at finite temperatures, where, sufficiently close to the critical point, the temperature provides the only infrared cutoff scale. Similarly, we expect the MBL transition to broaden by the dissipative coupling into a universal crossover controlled by the putative localization critical point. The dissipative coupling provides the only relevant infrared cutoff in this case. Hence, by controlling it, experiments should be able to access the critical properties, just as the temperature dependence of observables reveals the critical properties of quantum phase transitions.

Further experiments are also needed in order to characterize the dynamics in the insulating phase as well as the delocalized Griffiths phase. For example, as discussed in Section 1.6.4, the temporal fluctuations of the imbalance about its long-time stationary value is expected to be suppressed as a power of the time in the MBL phase owing to the correlated oscillations of a growing number of entangled particles [104]. By contrast, in a noninteracting localized system, the fluctuations, due to independent oscillations of nonentangled particles, do not decay in time. Anomalous power laws are also expected to occur in the frequency dependence of dynamic-response functions due to Griffiths-like effects in the insulating regime [43, 84].

References

[1] Agarwal, K., Gopalakrishnan, S., Knap, M., Mueller, M., and Demler, E. (2015). Anomalous diffusion and Griffiths effects near the many-body localization transition. *Phys. Rev. Lett.*, **114**, 160401.

[2] Altland, A. and Simons, B. D. (2010). *Condensed Matter Field Theory* (2nd edn). Cambridge: Cambridge University Press.

[3] Altman, E., Demler, E., and Lukin, M. (2004). Probing many-body states of ultracold atoms via noise correlations. *Phys. Rev. A*, **70**, 013603.

[4] Altman, E., Polkovnikov, A., Demler, E., Halperin, B. I., and Lukin, M. D. (2005). Superfluid–insulator transition in a moving system of interacting bosons. *Phys. Rev. Lett.*, **95**, 020402.

[5] Altman, E. and Vosk, R. (2015). Universal dynamics and renormalization in many-body-localized systems. *Annu. Rev. Condens. Matter Phys.*, **6**, 383–409.

[6] Anderson, P. W. (1952). An approximate quantum theory of the antiferromagnetic ground state. *Phys. Rev.*, **86**, 694–701.

[7] Anderson, P. W. (1958). Absence of diffusion in certain random lattices. *Phys. Rev.*, **109**, 1492–1505.

[8] Andraschko, F., Enss, T., and Sirker, J. (2014). Purification and many-body localization in cold atomic gases. *Phys. Rev. Lett.*, **113**, 217201.

[9] Andreev, D. (1980). Quantum dynamics of impurities in a 1D Bose gas. *Zh. Eksp. Teor. Fiz.*, **78**, 2064.

[10] Andrews, M. R., Townsend, C. G., Miesner, H.-J., Durfee, D. S., Kurn, D. M., and Ketterle, W. (1997). Observation of interference between two bose condensates. *Science*, **275**, 637–641.

[11] Arzamasovs, M., Bovo, F., and Gangardt, D. M. (2013). Kinetics of mobile impurities and correlation functions in one-dimensional superfluids at finite temperature. *Phys. Rev. Lett*, **112**, 170602.

[12] Aubry, S. and André, G. (1980). Analyticity breaking and Anderson localization in incommensurate lattices. *Ann. Isr. Phys. Soc*, **3**, 133.

[13] Bahri, Y., Vosk, R., Altman, E., and Vishwanath, A. (2015). Localization and topology protected quantum coherence at the edge of hot matter. *Nature Commun.* **6**, 7341.

[14] Bakr, W. S., Gillen, J. I., Peng, A., Fölling, S., and Greiner, M. (2009). A quantum gas microscope for detecting single atoms in a Hubbard-regime optical lattice. *Nature*, **462**, 74–77.

[15] Bar Lev, Y. and Reichman, D. R. (2014). Dynamics of many-body localization. *Phys. Rev. B*, **89**, 220201.

[16] Bardarson, J. H., Pollmann, F., and Moore, J. E. (2012). Unbounded growth of entanglement in models of many-body localization. *Phys. Rev. Lett.*, **109**, 017202.

[17] Barmettler, P., Punk, M., Gritsev, V., Demler, E., and Altman, E. (2009). Relaxation of antiferromagnetic order in spin-1/2 chains following a quantum quench. *Phys. Rev. Lett.*, **102**, 130603.

[18] Barmettler, P., Punk, M., Gritsev, V., Demler, E., and Altman, E. (2010). Quantum quenches in the anisotropic spin-1/2 Heisenberg chain: different approaches to many-body dynamics far from equilibrium. *New J. Phys.*, **12**, 055017.

[19] Basko, D. M, Aleiner, I. L, and Altshuler, B. L. (2006). Metal–insulator transition in a weakly interacting many-electron system with localized single-particle states. *Ann. Phys. (NY)*, **321**, 1126–1205.

[20] Bauer, B. and Nayak, C. (2013). Area laws in a many-body localized state and its implications for topological order. *J. Stat. Mech.*, **2013**, P09005.

[21] Billy, J., Josse, V., Zuo, Z., Bernard, A., Hambrecht, B., Lugan, P., Clément, D., Sanchez-Palencia, L., Bouyer, P., and Aspect, A. (2008). Direct observation of Anderson localization of matter waves in a controlled disorder. *Nature*, **453**, 891–894.

[22] Bistritzer, R. (2007). Low dimensional ultracold interferometers & Coulomb drag in high Landau levels. PhD Thesis, Weizmann Institute of Science.

[23] Bistritzer, R. and Altman, E. (2007). Intrinsic dephasing in one-dimensional ultracold atom interferometers. *Proc. Natl Acad. Sci. USA*, **104**, 9955–9959.

[24] Bloch, I., Dalibard, J., and Zwerger, W. (2008). Many-body physics with ultracold gases. *Rev. Mod. Phys.*, **80**, 885–964.

[25] Bravyi, S., Hastings, M. B, and Verstraete, F. (2006). Lieb–Robinson bounds and the generation of correlations and topological quantum order. *Phys. Rev. Lett.*, **97**, 050401.

[26] Burkov, A., Lukin, M., and Demler, E. (2007). Decoherence dynamics in low-dimensional cold atom interferometers. *Phys. Rev. Lett.*, **98**, 200404.

[27] Calabrese, P. and Cardy, J. (2005). Evolution of entanglement entropy in one-dimensional systems. *J. Stat. Mech.*, **2005**, P04010.

[28] Calabrese, P. and Cardy, J. (2006). Time dependence of correlation functions following a quantum quench. *Phys. Rev. Lett.*, **96**, 136801.

[29] Castin, Y. and Dalibard, J. (1997). Relative phase of two Bose–Einstein condensates. *Phys. Rev. A*, **55**, 4330–4337.

[30] Cazalilla, M. A. (2006). Effect of suddenly turning on interactions in the Luttinger model. *Phys. Rev. Lett.*, **97**, 156403.

[31] Cazalilla, M. A., Citro, R., Giamarchi, T., Orignac, E., and Rigol, M. (2011, Dec). One dimensional bosons: from condensed matter systems to ultracold gases. *Rev. Mod. Phys.*, **83**, 1405–1466.

[32] Cheneau, M., Barmettler, P., Poletti, D., Endres, M., Schauß, P., Fukuhara, T., Gross, C., Bloch, I., Kollath, C., and Kuhr, S. (2012). Light-cone-like spreading of correlations in a quantum many-body system. *Nature*, **481**, 484–487.

[33] Coleman, S. (1977). Fate of the false vacuum: semiclassical theory. *Phys. Rev. D*, **15**, 2929–2936.

[34] Dasgupta, C. and Ma, S.-K (1980). Low-temperature properties of the random Heisenberg antiferromagnetic chain. *Phys. Rev. B*, **22**, 1305–1319.

[35] D'Errico, C., Lucioni, E., Tanzi, L., Gori, L., Roux, G., McCulloch, I. P., Giamarchi, T., Inguscio, M., and Modugno, G. (2014). Observation of a disordered bosonic insulator from weak to strong interactions. *Phys. Rev. Lett.*, **113**, 095301.

[36] Endres, M., Fukuhara, T., Pekker, D., Cheneau, M., Schauss, P., Gross, C., Demler, E., Kuhr, S., and Bloch, I. (2012). The "Higgs" amplitude mode at the two-dimensional superfluid/Mott insulator transition. *Nature*, **487**, 454–458.

[37] Fallani, L., De Sarlo, L., Lye, J. E., Modugno, M., Saers, R., Fort, C., and Inguscio, M. (2004). Observation of dynamical instability for a Bose–Einstein condensate in a moving 1D optical lattice. *Phys. Rev. Lett.*, **93**, 140406.

[38] Fisher, D. S. (1992). Random transverse field Ising spin chains. *Phys. Rev. Lett.*, **69**, 534–537.

[39] Fölling, S., Gerbier, F., Widera, A., Mandel, O., Gericke, T., and Bloch, I. (2005). Spatial quantum noise interferometry in expanding ultracold atom clouds. *Nature*, **434**, 481–484.

[40] Gadway, B., Pertot, D., Reeves, J., Vogt, M., and Schneble, D. (2011). Glassy behavior in a binary atomic mixture. *Phys. Rev. Lett.*, **107**, 145306.

[41] Gazit, S., Podolsky, D., and Auerbach, A. (2013). Fate of the Higgs mode near quantum criticality. *Phys. Rev. Lett.*, **110**, 140401.

[42] Giamarchi, T. and Schulz, H. J. (1988). Anderson localization and interactions in one-dimensional metals. *Phys. Rev. B*, **37**, 325–340.

[43] Gopalakrishnan, S., Mueller, M., Khemani, V., Knap, M., Demler, E., and Huse, D. A. (2015). Low-frequency conductivity in many-body localized systems. *Phys. Rev. B*, **92**, 104202.

[44] Gornyi, I. V., Mirlin, A. D., and Polyakov, D. G. (2005). Interacting electrons in disordered wires: Anderson localization and low-t transport. *Phys. Rev. Lett.*, **95**, 206603.

[45] Greiner, M., Mandel, O., Esslinger, T., Hänsch, T. W., and Bloch, I. (2002). Quantum phase transition from a superfluid to a Mott insulator in a gas of ultracold atoms. *Nature*, **415**, 39–44.

[46] Greiner, M., Mandel, O., Hänsch, T. W., and Bloch, I. (2002). Collapse and revival of the matter wave field of a Bose–Einstein condensate. *Nature*, **419**, 51–54.

[47] Griffiths, R. B. (1969). Nonanalytic behavior above the critical point in a random Ising ferromagnet. *Phys. Rev. Lett.*, **23**, 17–19.

[48] Gring, M., Kuhnert, M., Langen, T., Kitagawa, T., Rauer, B., Schreitl, M., Mazets, I., Smith, D. A., Demler, E., and Schmiedmayer, J. (2012). Relaxation and prethermalization in an isolated quantum system. *Science*, **337**, 1318–1322.

[49] Gritsev, V., Altman, E., Demler, E., and Polkovnikov, A. (2006). Full quantum distribution of contrast in interference experiments between interacting one-dimensional Bose liquids. *Nature Phys.*, **2**, 705–709.

[50] Grover, T. (2014). Certain general constraints on the many-body localization transition. arXiv:1405.1471 [cond-mat.dis-nn].

[51] Hadzibabic, Z., Krüger, P., Cheneau, M., Battelier, B., and Dalibard, J. (2006). Berezinskii–Kosterlitz–Thouless crossover in a trapped atomic gas. *Nature*, **441**, 1118–1121.

[52] Haldane, F. D. M. (1981). Effective harmonic-fluid approach to low-energy properties of one-dimensional quantum fluids. *Phys. Rev. Lett.*, **47**, 1840–1843.

[53] Hofferberth, S., Lesanovsky, I., Fischer, B., Schumm, T., and Schmiedmayer, J. (2007). nonequilibrium coherence dynamics in one-dimensional Bose gases. *Nature*, **449**, 324–327.

[54] Hohenberg, P. and Halperin, B. (1977). Theory of dynamic critical phenomena. *Rev. Mod. Phys.*, **49**, 435–479.

[55] Huse, D. A., Nandkishore, R., and Oganesyan, V. (2014). Phenomenology of fully many-body-localized systems. *Phys. Rev. B*, **90**, 174202.

[56] Huse, D. A., Nandkishore, R., Oganesyan, V., Pal, A., and Sondhi, S. L. (2013). Localization-protected quantum order. *Phys. Rev. B*, **88**, 014206.

[57] Imbrie, J. Z. (2014). On many-body localization for quantum spin chains. arXiv:1403.7837 [math-ph].

[58] Jaksch, D., Bruder, C., Cirac, J. I., Gardiner, C. W., and Zoller, P. (1998). Cold bosonic atoms in optical lattices. *Phys. Rev. Lett.*, **81**, 3108–3111.

[59] Jendrzejewski, F., Bernard, A., Müller, K., Cheinet, P., Josse, V., Piraud, M., Pezzé, L., Sanchez-Palencia, L., Aspect, A., and Bouyer, P. (2012). Three-dimensional localization of ultracold atoms in an optical disordered potential. *Nature Phys.*, **8**, 398–403.

[60] Jo, G.-B., Shin, Y., Will, S., Pasquini, T. A., Saba, M., Ketterle, W., Pritchard, D. E., Vengalattore, M., and Prentiss, M. (2007). Long phase coherence time and number squeezing of two Bose–Einstein condensates on an atom chip. *Phys. Rev. Lett.*, **98**, 030407.

[61] Katan, Y. T. and Podolsky, D. (2015). Spectral function of the Higgs mode in $4 - \epsilon$ dimensions. *Phys. Rev. B*, **91**, 075132.

[62] Kinoshita, T., Wenger, T., and Weiss, D. S. (2006). A quantum Newton's cradle. *Nature*, **440**, 900–903.

[63] Kitagawa, T., Imambekov, A., Schmiedmayer, J., and Demler, E. (2011). The dynamics and prethermalization of one-dimensional quantum systems probed through the full distributions of quantum noise. *New J. Phys.*, **13**, 073018.

[64] Kondov, S. S., McGehee, W. R., Xu, W., and DeMarco, B. (2015). Disorder-induced localization in a strongly correlated atomic Hubbard gas. *Phys. Rev. Lett.*, **114**, 083002.

[65] Kondov, S. S., McGehee, W. R., Zirbel, J. J., and DeMarco, B. (2011). Three-dimensional aAnderson localization of ultracold matter. *Science*, **334**, 66–68.

[66] Kulkarni, M. and Lamacraft, A. (2013). Finite-temperature dynamical structure factor of the one-dimensional Bose gas: from the Gross–Pitaevskii equation to the

Kardar–Parisi–Zhang universality class of dynamical critical phenomena. *Phys. Rev. A*, **88**, 021603.

[67] Lamacraft, A. (2007). Quantum quenches in a spinor condensate. *Phys. Rev. Lett.*, **98**, 160404.

[68] Lamacraft, A. and Moore, J. (2012). Potential insights into nonequilibrium behavior from atomic physics. In *Ultracold Bosonic and Fermionic Gases* (ed. K. Levin, A. L. Fetter, and D. M. Stamper-Kurn), pp. 177–202. Oxford: Elsevier.

[69] Langer, J. S. and Ambegaokar, V. (1967). Intrinsic resistive transition in narrow superconducting channels. *Phys. Rev.*, **164**, 498–510.

[70] Leggett, A. J. and Sols, F. (1991). On the concept of spontaneously broken gauge symmetry in condensed matter physics. *Found. Phys.*, **21**, 353–364.

[71] Lieb, E. H. and Robinson, D. W. (1972). The finite group velocity of quantum spin systems. *Commun. Math. Phys*, **28**, 251–257.

[72] Liu, F. and Mazenko, G. F. (1992). Defect–defect correlation in the dynamics of first-order phase transitions. *Phys. Rev. B*, **46**, 5963–5971.

[73] Lux, J., Müller, J., Mitra, A., and Rosch, A. (2014). Hydrodynamic long-time tails after a quantum quench. *Phys. Rev. A*, **89**, 053608.

[74] Mathey, L. and Polkovnikov, A. (2010). Light cone dynamics and reverse Kibble–Zurek mechanism in two-dimensional superfluids following a quantum quench. *Phys. Rev. A*, **81**, 033605.

[75] Mazets, I., Schumm, T., and Schmiedmayer, J. (2008). Breakdown of integrability in a quasi-1D ultracold bosonic gas. *Phys. Rev. Lett.*, **100**, 2–5.

[76] McCumber, D. E. and Halperin, B. I. (1970). Time scale of intrinsic resistive fluctuations in thin superconducting wires. *Phys. Rev. B*, **1**, 1054–1070.

[77] Meldgin, C., Ray, U., Russ, P., Ceperley, D., and DeMarco, B. (2015). Probing the Bose-glass–superfluid transition using quantum quenches of disorder. arXiv:1502.02333 [cond-mat.quant-gas].

[78] Mun, J., Medley, P., Campbell, G. K., Marcassa, L. G., Pritchard, D. E., and Ketterle, W. (2007). Phase diagram for a Bose–Einstein condensate moving in an optical lattice. *Phys. Rev. Lett.*, **99**, 150604.

[79] Nandkishore, R. and Huse, D. A. (2015). Many-body localization and thermalization in quantum statistical mechanics. *Annu. Rev. Condens. Matter*, **6**, 15–38.

[80] Oganesyan, V. and Huse, D. A. (2007). Localization of interacting fermions at high temperature. *Phys. Rev. B*, **75**, 155111.

[81] Ovadia, M., Kalok, D., Tamir, I., Mitra, S., Sacépé, B., and Shahar, D. (2015). Evidence for a finite-temperature insulator. *Sci. Rep.* **5**, 13503.

[82] Pal, A.and Huse, D. A. (2010). Many-body localization phase transition. *Phys. Rev. B*, **82**, 174411.

[83] Pasienski, M., McKay, D., White, M., and Demarco, B. (2010). A disordered insulator in an optical lattice. *Nature Phys.*, **6**, 677–680.

[84] Pekker, D., Refael, G., Altman, E., Demler, E., and Oganesyan, V. (2014). Hilbert-glass transition: new universality of temperature-tuned many-body dynamical quantum criticality. *Phys. Rev. X*, **4**, 011052.

[85] Podolsky, D., Auerbach, A., and Arovas, D. P. (2011). Visibility of the amplitude (Higgs) mode in condensed matter. *Phys. Rev. B*, **84**, 174522.

[86] Podolsky, D. and Sachdev, S. (2012). Spectral functions of the Higgs mode near two-dimensional quantum critical points. *Phys. Rev. B*, **86**, 054508.

[87] Polkovnikov, A. (2003). Evolution of the macroscopically entangled states in optical lattices. *Phys. Rev. A*, **68**, 033609.

[88] Polkovnikov, A. (2005). Universal adiabatic dynamics in the vicinity of a quantum critical point. *Phys. Rev. B*, **72**, 161201.

[89] Polkovnikov, A., Altman, E., and Demler, E. (2006). Interference between independent fluctuating condensates. *Proc. Natl Acad. Sci. USA*, **103**, 6125–6129.

[90] Polkovnikov, A., Altman, E., Demler, E., Halperin, B., and Lukin, M. D. (2005, Jun). Decay of superfluid currents in a moving system of strongly interacting bosons. *Phys. Rev. A*, **71**, 063613.

[91] Pollet, L. and Prokof'ev, N. (2012, Jul). Higgs mode in a two-dimensional superfluid. *Phys. Rev. Lett.*, **109**, 010401.

[92] Popov, V. N. (1972). On the theory of the superfluidity of two- and one-dimensional Bose systems. *Theor. Math. Phys.*, **11**, 565–573.

[93] Potter, A. C., Vasseur, R., and Parameswaran, S. A. (2015). Theory of phase transitions from quantum glasses to thermal fluids. arXiv:1501.03501 [cond-mat.dis-nn].

[94] Rançon, A. and Dupuis, N. (2014). Higgs amplitude mode in the vicinity of a $(2+1)$-dimensional quantum critical point. *Phys. Rev. B*, **89**, 180501.

[95] Rigol, M., Dunjko, V., and Olshanii, M. (2008). Thermalization and its mechanism for generic isolated quantum systems. *Nature*, **452**, 854–858.

[96] Roati, G., D'Errico, C., Fallani, L., Fattori, M., Fort, C., Zaccanti, M., Modugno, G., Modugno, M., and Inguscio, M. (2008). Anderson localization of a noninteracting Bose–Einstein condensate. *Nature*, **453**, 895–898.

[97] Sachdev, S. (2001). *Quantum Phase Transitions*. Cambridge: Cambridge University Press.

[98] Sadler, L. E., Higbie, J. M., Leslie, S. R., Vengalattore, M., and Stamper-Kurn, D. M. (2006). Spontaneous symmetry breaking in a quenched ferromagnetic spinor Bose–Einstein condensate. *Nature*, **443**, 312–315.

[99] Schollwöck, U. (2011). The density-matrix renormalization group in the age of matrix product states. *Ann. Phys. (NY)*, **326**, 96–192.

[100] Schori, C., Stöferle, T., Moritz, H., Köhl, M., and Esslinger, T. (2004). Excitations of a superfluid in a three-dimensional optical lattice. *Phys. Rev. Lett.*, **93**, 240402.

[101] Schreiber, M., Hodgman, S. S., Bordia, P., Lüschen, H. P., Fischer, M. H., Vosk, R., Altman, E., Schneider, U., and Bloch, I. (2015). Observation of many-body localization of interacting fermions in a quasi-random optical lattice. *Science* **349**, 849–845.

[102] Schumm, T., Hofferberth, S., Andersson, L. M., Wildermuth, S., Groth, S., Bar-Joseph, I., Schmiedmayer, J., and Krüger, P. (2005). Matter–wave interferometry in a double well on an atom chip. *Nature Phys.*, **1**, 57–62.

[103] Sciolla, B. and Biroli, G. (2010). Quantum quenches and off-equilibrium dynamical transition in the infinite-dimensional Bose–Hubbard model. *Phys. Rev. Lett.*, **105**, 220401.

[104] Serbyn, M., Knap, M., Gopalakrishnan, S., Papić, Z., Yao, N. Y., Laumann, C. R., Abanin, D. A., Lukin, M. D., and Demler, E. A. (2014). Interferometric probes of many-body localization. *Phys. Rev. Lett.*, **113**, 147204.

[105] Serbyn, M., Papić, Z., and Abanin, D. A. (2014). Quantum quenches in the many-body localized phase. *Phys. Rev. B* **90**, 174302.

[106] Serbyn, M, Papić, Z., and Abanin, D. A. (2013). Local conservation laws and the structure of the many-body localized states. *Phys. Rev. Lett.*, **111**, 127201.

[107] Serbyn, Maksym, Papić, Z., and Abanin, D. A. (2013). Universal slow growth of entanglement in interacting strongly disordered systems. *Phys. Rev. Lett.*, **110**, 260601.

[108] Shankar, R. (1994). Renormalization-group approach to interacting fermions. *Rev. Mod. Phys.*, **66**, 129–192.

[109] Sherson, J. F., Weitenberg, C., Endres, M., Cheneau, M., Bloch, I., and Kuhr, S. (2010). Single-atom-resolved fluorescence imaging of an atomic Mott insulator. *Nature*, **467**, 68–72.

[110] Smith, D. A., Gring, M., Langen, T., Kuhnert, M., Rauer, B., Geiger, R., Kitagawa, T., Mazets, I., Demler, E., and Schmiedmayer, J. (2013). Prethermalization revealed by the relaxation dynamics of full distribution functions. *New J. Phys.*, **15**, 075011.

[111] Stern, A., Aharonov, Y., and Imry, Y. (1990). Phase uncertainty and loss of interference: a general picture. *Phys. Rev. A*, **41**, 3436.

[112] Trotzky, S., Chen, Y.-A., Flesch, A., McCulloch, I. P., Schollwöck, U., Eisert, J., and Bloch, I. (2012). Probing the relaxation towards equilibrium in an isolated strongly correlated one-dimensional Bose gas. *Nature Phys.*, **8**, 325–330.

[113] Vasseur, R., Parameswaran, S. A., and Moore, J. E. (2015). Quantum revivals and many-body localization. *Phys. Rev. B*, **91**, 140202(R).

[114] Vosk, R. and Altman, E. (2013). Many-body localization in one dimension as a dynamical renormalization group fixed point. *Phys. Rev. Lett.*, **110**, 067204.

[115] Vosk, R. and Altman, E. (2014). Dynamical quantum phase transitions in random spin chains. *Phys. Rev. Lett.*, **112**, 217204.

[116] Vosk, R., Huse, D. A., and Altman, E. (2015). Theory of the many-body localization transition in one dimensional systems. *Phys. Rev. X*, **5**, 031032.

[117] Widera, A., Trotzky, S., Cheinet, P., Fölling, S., Gerbier, F., Bloch, I., Gritsev, V., Lukin, M. D., and Demler, E. (2008). Quantum spin dynamics of mode-squeezed Luttinger liquids in two-component atomic gases. *Phys. Rev. Lett.*, **100**, 140401.

[118] Wu, B. and Niu, Q. (2001). Landau and dynamical instabilities of the superflow of Bose–Einstein condensates in optical lattices. *Phys. Rev. A*, **64**, 061603.

[119] Žnidarič, M., Prosen, T., and Prelovšek, P. (2008). Many-body localization in the Heisenberg XXZ magnet in a random field. *Phys. Rev. B*, **77**, 064426.

[120] Zurek, W. H. (1985). Cosmological experiments in superfluid helium? *Nature*, **317**, 505–508.

2

Nonequilibrium quantum fields: from cold atoms to cosmology

Jürgen BERGES

Institute for Theoretical Physics
Heidelberg University
Germany

Strongly Interacting Quantum Systems out of Equilibrium. First Edition. Thierry Giamarchi et al.
© Oxford University Press 2016. Published in 2016 by Oxford University Press.

Chapter Contents

Preface

This chapter is based on lecture notes and supplementary material for the school on Strongly Interacting Quantum Systems out of Equilibrium, held from 30 July to 24 August 2012 at École de Physique des Houches, Les Houches, France.

2.1   Introduction

2.1.1   Units

We will mostly work with natural units where Planck's constant divided by 2π, the speed of light, and Boltzmann's constant are set to one:

$$\hbar = c = k_B = 1. \tag{2.1}$$

As a consequence, for instance, the mass of a particle (m) is equal to its rest energy (mc^2) and to its inverse Compton wavelength ($mc/\hbar$). More generally, the units of

$$[\text{energy}] = [\text{mass}] = [\text{temperature}] = [\text{length}]^{-1} = [\text{time}]^{-1}$$

may all be expressed in terms of the basic unit of energy, where we take the electron-volt (eV), i.e., the amount of energy gained by the charge of a single electron moved across an electric potential difference of one volt. Recall that

$$\text{GeV} = 10^3 \, \text{MeV} = 10^6 \, \text{keV} = 10^9 \, \text{eV}.$$

The conversion to other units, such as length, can be obtained from $\hbar c/\text{MeV} \simeq 197.33 \times 10^{-15}$ m $\equiv 197.33$ fm, with $c = 2.9979 \times 10^8 \, \text{m s}^{-1}$ and $\hbar = 6.5822 \times 10^{-22}$ MeV s. For later use, for the conversion to length, time, temperature, and mass, we have

$$\text{MeV}^{-1} = 197.33 \, \text{fm},$$
$$\text{MeV}^{-1} = 6.5822 \times 10^{-22} \, \text{s},$$
$$\text{MeV} = 1.1605 \times 10^{10} \, \text{K},$$
$$\text{MeV} = 1.7827 \times 10^{-27} \, \text{g}.$$

2.1.2   Isolated quantum systems in extreme conditions

In recent years, we have witnessed a dramatic convergence of research on nonequilibrium quantum systems in extreme conditions across traditional lines of specialization. Prominent examples include the evolution of the early universe shortly after a period of strongly accelerated expansion called inflation, the initial stages in collisions of ultrarelativistic nuclei at giant laboratory facilities, and table-top experiments with degenerate quantum gases far from equilibrium. Even though the typical energy scales of these systems differ vastly, they can show very similar dynamical properties. Certain characteristic numbers can even be quantitatively the same, and one may use this *universality* to learn from table-top experiments with ultracold atoms something about dynamics in the very early universe.

Here the focus lies on isolated quantum systems, whose dynamics is governed by *unitary time evolution*. This sets them apart from other condensed matter systems where the influence of the environment is largely unavoidable. Isolated systems offer the possibility to study fundamental aspects of quantum statistical mechanics, such as *nonequilibrium instabilities* at early times and late-time *thermalization*, from first principles. A quantum many-body system in thermal equilibrium is independent of its history in time and characterized by a few conserved quantities only. Therefore, any thermalization process starting from a nonequilibrium initial state requires an effective loss of details of the initial conditions at sufficiently long times.

An effective *partial memory loss* of the initial state can already be observed at earlier characteristic stages of the nonequilibrium unitary time evolution. The corresponding *prethermalization* is characterized in terms of approximately conserved quantities, which evolve in a quasistationary manner even though the system is still far from equilibrium. An extreme case occurs if the nonequilibrium dynamics becomes *self-similar*. This amounts to an enormous reduction of the sensitivity to details of the underlying theory and initial conditions. The time evolution in this self-similar regime is described in terms of universal scaling exponents and scaling functions associated with *nonthermal fixed points*, which is similar in spirit to the description of critical phenomena in thermal equilibrium. While the notion of thermal fixed points describing different universality classes is well established for systems in thermal equilibrium, the classification of far-from-equilibrium universality classes is a rapidly progressing research topic.

In the following, we will outline these concepts for the examples of relativistic inflaton dynamics in the early universe in Section 2.1.3 and for the nonrelativistic dynamics of an ultracold Bose gas in Section 2.1.4. These sections are meant to provide an overview of the different stages, starting far from equilibrium such that the system approaches a nonthermal fixed point before eventual thermalization. A schematic view is given in Fig. 2.1.

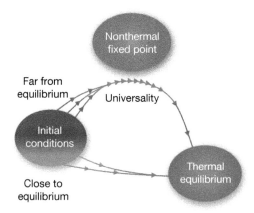

Fig. 2.1 Schematic evolution toward thermal equilibrium as exemplified in Sections 2.1.3 and 2.1.4. Starting far from equilibrium, the system is attracted toward a nonthermal fixed point, thus acquiring its universal properties before relaxing toward thermal equilibrium. (K. Boguslavski.)

After introducing the basic ingredients for nonequilibrium quantum field theory in more detail in Section 2.2, we analyze the notion of thermalization in isolated quantum systems using functional integral techniques in Section 2.3. Section 2.4 explains the extent to which nonequilibrium quantum field theory can be mapped onto a classical statistical field theory problem, which provides powerful simulation techniques. Section 2.5 is devoted to nonequilibrium instabilities, and the subsequent evolution toward nonthermal fixed points is explained in Section 2.6.

2.1.3   Heating the universe after inflation

Inflation

One of the most striking developments in our understanding of the universe is that its properties at the smallest length scales described by quantum field theory are tightly connected to its properties at the largest scales up to the size of the observable universe today. In order to discuss some aspects of this, which will be relevant for this chapter, we recall that the universe is expanding with cosmological time t. The change is described by the Hubble parameter

$$H = \frac{\dot{a}(t)}{a(t)}, \tag{2.2}$$

which is defined in terms of the scale factor $a(t)$ for a homogeneous and isotropic universe and the dot denotes the time derivative. The scale factor relates the proper distance between a pair of objects such that space itself is expanding with time. At the present moment t_0, we take $a(t_0) = 1$, where $t_0 \simeq 1/H_0 \simeq 14\,\mathrm{Gyr} \simeq 4.4 \times 10^{17}$ s.

The expansion has not been the same at all times. Inflation denotes a period at very early times where the expansion was strongly accelerated with $\ddot{a} > 0$. During this period, a tiny region of space expanded to a huge size. As a consequence, spatial curvature was decreased and the universe acquired a practically *flat geometry*. Furthermore, the accelerated expansion became so fast that points on the sky that are not in causal contact today were indeed correlated before inflation. This explains why we can see cosmic microwave radiation with practically the same temperature from all directions of the sky, thus providing a solution of the so-called *horizon problem*. Most strikingly, quantum fluctuations stretched to macroscopic sizes by the expansion can explain the *origin of initial perturbations* in the matter density. The gravitational instability of these initial perturbations then leads to large-scale structure formation into clusters and superclusters of galaxies along with fluctuations of the cosmic microwave background radiation, all in remarkable agreement with observations.

Knowledge about the amplitude of the perturbations can give important information about the energy scale of inflation. A typical characteristic energy density may be around $\epsilon_{\mathrm{inflation}} \sim (10^{15}\text{--}10^{16}\,\mathrm{GeV})^4$, which is reminiscent of the grand unification scale above which the electroweak and strong forces may become similar in strength. However, many different scenarios still exist, and in the following we will concentrate on some generic aspects. The energy density at a given time, ϵ, determines the Hubble parameter by the Friedmann equation

$$H^2 = \frac{8\pi G}{3}\epsilon. \tag{2.3}$$

Here G denotes Newton's gravitational constant and we define the reduced Planck mass $M_{\rm P} \equiv (8\pi G)^{-1/2} \simeq 2.435 \times 10^{18}$ GeV.

In the simplest model of inflation, the dominant contribution to the energy density at time t is described in terms of a spatially homogeneous scalar field $\phi(t)$, which can be decomposed as

$$\epsilon_\phi = \tfrac{1}{2}\dot{\phi}^2 + V(\phi) \tag{2.4}$$

into a kinetic energy part $\tfrac{1}{2}\dot{\phi}^2$ and a potential energy part $V(\phi)$. Likewise, the pressure is given by the difference

$$P_\phi = \tfrac{1}{2}\dot{\phi}^2 - V(\phi). \tag{2.5}$$

As a consequence, a time-independent field has *negative pressure*: $P_\phi = -\epsilon_\phi$ with *constant energy density*. According to (2.3), in this case $H = \text{const}$ such that

$$\frac{\dot{a}}{a} = \text{const}, \qquad \text{i.e.,} \quad a(t) \sim e^{Ht}. \tag{2.6}$$

To be consistent with the observed properties of the cosmic microwave radiation, one may need about 60 e-folds of growth before the end of inflation. This requires the specification of a suitable potential $V(\phi)$ along with initial conditions that allow for such dynamics. It should be stressed, however, that there are many models of inflation with different degrees of freedom and levels of sophistication. Since this will not be relevant for our purposes, we continue for the moment with our simple single-field model.

Coherent field evolution

Generically, the previous matter (with pressure $P_m = 0$) and radiation ($P_r = \tfrac{1}{3}\epsilon_r$ describing relativistic particles) content of the universe is diluted away during inflation according to

$$\epsilon_m \sim \frac{1}{a^3}, \qquad \epsilon_r \sim \frac{1}{a^4}. \tag{2.7}$$

The suppression factor for the radiation can be understood from the fact that its energy density $\epsilon_r \sim n_r \omega_r$ is given by the number density, which scales with inverse volume as $n_r \sim 1/a^3$, times the mean energy per particle, which is redshifted as $\omega_r \sim 1/a$ since the wavelength is stretched by the expansion. As a consequence of the dramatic dilution, after inflation all the energy is stored in the large coherent field amplitude of the inflaton. Inflation ends once the time dependence of the initially slowly moving field $\phi(t)$ can no longer be neglected. In this case, there is no constant energy density dominating, and the solution (2.6) is no longer valid.

To discuss the subsequent time evolution, we first note that the *classical equation of motion* for the scalar field is

$$\ddot{\phi} + 3H\dot{\phi} + \frac{\mathrm{d}V}{\mathrm{d}\phi} = 0. \tag{2.8}$$

This may be seen as the equation of motion for a relativistic scalar field with a "friction" term $3H\dot{\phi}$ due to the expansion of the universe. For the following, we also specialize to a quartic potential of the form

$$V(\phi) = \frac{1}{2}m^2\phi^2 + \frac{\lambda}{4!}\phi^4. \tag{2.9}$$

Furthermore, we note that for relativistic particles dominating the energy density, the scale factor evolves as

$$a(t) \sim t^{1/2}, \qquad H \sim \frac{1}{t}, \tag{2.10}$$

in agreement with $H^2 \sim \epsilon_r \sim 1/a^4$ according to (2.3). In this case, the equation of motion for the field can be rewritten in a form that is very useful for later analysis. We introduce *conformal time* t_c and a *field rescaling* as

$$t_\mathrm{c} = \int \frac{\mathrm{d}t}{a}, \qquad \phi_\mathrm{c} = a\phi, \tag{2.11}$$

where the integration boundaries run from some initial time to the final time (today). Then the field equation (2.8) for the potential (2.9) becomes

$$\phi_\mathrm{c}'' + \left(m^2 a^2 - \frac{a''}{a}\right)\phi_\mathrm{c} + \frac{\lambda}{6}\phi_\mathrm{c}^3 = 0, \tag{2.12}$$

where primes denote derivatives with respect to t_c. For radiation domination, we have $a(t) \sim t^{1/2}$ and with (2.11) that $t_\mathrm{c} \sim 2t^{1/2}$. Consequently, $a \sim t_\mathrm{c}$ leads to $a'' = 0$. From (2.12), we observe that for massless fields, where $m^2 = 0$, the dynamics in expanding space is conformally equivalent to dynamics in Minkowski space–time with no expansion in this case, i.e.,

$$\boxed{\phi_\mathrm{c}'' + \frac{\lambda}{6}\phi_\mathrm{c}^3 = 0.} \tag{2.13}$$

For a given nonzero initial field amplitude, this equation describes an oscillating field in a quartic potential. For notational simplicity, in the following, we will drop index labels and employ $\phi(t)$ to denote the rescaled field in conformal time.

Nonequilibrium instabilities and preheating

As a classical homogeneous field, the solution of (2.13) would continue oscillating. However, in a *quantum field theory*, the macroscopic inflaton field amplitude $\phi(t)$

corresponds to the expectation value of a (real) scalar Heisenberg field operator $\Phi(t, \mathbf{x})$:

$$\phi(t) = \langle \Phi(t, \mathbf{x}) \rangle \equiv \mathrm{Tr}\,\{\varrho_0\,\Phi(t, \mathbf{x})\}, \qquad (2.14)$$

which involves the density operator ϱ_0, as will be explained in more detail in Section 2.2. After inflation ends at some time, which we may set to $t = 0$, the initial state for the subsequent evolution is approximately described by a vacuum-like, pure-state density operator $\varrho_0 \equiv \varrho(t\!=\!0)$ with trace $\mathrm{Tr}\,\varrho_0 = \mathrm{Tr}\,\varrho_0^2 = 1$. Typical physical normalizations of the inflaton model employ an initially large field amplitude $\phi(t = 0) \sim 0.3 M_{\mathrm{P}}$ with a very weak coupling of $\lambda \sim 10^{-13}$. We note that the fluctuating quantum field $\Phi(t, \mathbf{x})$ depends explicitly on the spatial variable $\mathbf{x}$, while its expectation value $\phi(t)$ is homogeneous in our case.

It is a central topic of this chapter to show how the transition from a pure state described by a coherent macroscopic field amplitude proceeds toward radiation, thus connecting to the thermal history of the universe at later times. This can involve dramatic far-from-equilibrium phenomena at intermediate times. Indeed, the situation of the classically oscillating macroscopic field $\phi(t)$ turns out to be unstable in a quantum world, which will be shown in detail in Section 2.5. Here we will outline the main aspects of the dynamics in order to introduce some important notions of nonequilibrium quantum field theory. Seeded by quantum fluctuations, a *parametric resonance instability* occurs. This instability is signaled by an exponential growth of correlation functions, which corresponds to very fast particle production called *preheating*.

While the macroscopic field is described by a one-point function $\langle \Phi(t, \mathbf{x}) \rangle$, particle production can be extracted from *two-point correlation functions*. More precisely, one considers the symmetrized correlator $\langle \Phi(t, \mathbf{x})\Phi(t', \mathbf{x}') + \Phi(t', \mathbf{x}')\Phi(t, \mathbf{x}) \rangle$, which is the anticommutator expectation value of two fields evaluated at equal times $t = t'$. The spatially homogeneous correlation function may be used to define a time-dependent occupation number distribution $f(t, \mathbf{p})$ of particle modes with momentum $\mathbf{p}$ after Fourier transformation:

$$\frac{f(t, \mathbf{p}) + \frac{1}{2}}{\omega(t, \mathbf{p})} + (2\pi)^3 \delta(\mathbf{p})\phi^2(t) \equiv \frac{1}{2}\int \mathrm{d}^3 x\, e^{-i\mathbf{p}\cdot\mathbf{x}}\langle \Phi(t, \mathbf{x})\Phi(t, 0) + \Phi(t, 0)\Phi(t, \mathbf{x})\rangle \quad (2.15)$$

for a given (in general time-dependent) dispersion relation $\omega(t, \mathbf{p})$. Because of spatial isotropy, the functions $f(t, \mathbf{p})$ and $\omega(t, \mathbf{p})$ depend on the modulus of momentum, and we will frequently write $f(t, |\mathbf{p}|)$, etc. Here the term $\sim \phi^2(t)$ coming together with the Dirac δ-function denotes the coherent field part of the correlator at zero momentum. We note that in a finite volume V, we have $(2\pi)^3\delta(\mathbf{0}) \to V$ and the scaling of the field term with volume just reflects the presence of a coherent zero mode spreading over the entire volume. For the following, all integrals will be considered to be suitably regularized, say, by some high-momentum cutoff if necessary.

After inflation, all previous particle content is diluted away and the absence of particles corresponds to an initial state with $f(t\!=\!0, \mathbf{p})\!=\!0$ for the subsequent evolution. Apart from the coherent field zero-mode, in this case only the "quantum half" coming from the vacuum in (2.15) contributes. These quantum fluctuations seed the parametric resonance instability, which leads to a decay of the macroscopic field ϕ together with an exponentially growing mode occupancy after inflation:

$$\boxed{f(t,Q) \sim \exp(\gamma_Q t).}$$ (2.16)

Here, γ_Q denotes a real and positive growth rate for some range of momenta around a fast-growing characteristic mode Q. The far-from-equilibrium dynamics of an instability with exponentially growing modes may be seen as the opposite of what happens during the relaxation of some close-to-equilibrium modes, which decay exponentially. It is important to note that for transient phenomena the presence of instabilities is as common as relaxation is for the late-time approach to thermal equilibrium.

The exponential growth (2.16) leads to an unusually large occupancy of modes at the characteristic momentum scale Q. Most importantly, this over-occupied system becomes strongly correlated despite the presence of a very weak coupling $\lambda \ll 1$. Qualitatively, this may be understood from a "mean-field" approximation for quantum corrections. In this approximation, the quartic interaction term in the potential (2.9) is replaced in the quantum theory by

$$\frac{\lambda}{4!}\Phi^4(t,\mathbf{x}) \rightarrow \frac{\lambda}{4!}6\langle\Phi^2(t,\mathbf{x})\rangle\Phi^2(t,\mathbf{x}).$$ (2.17)

Here, $\langle\Phi^2(t,\mathbf{x})\rangle$ means the anticommutator expectation value appearing also in (2.15), now evaluated at equal space–time points. Applying proper renormalization conditions as in standard vacuum quantum field theory, we concentrate on the finite part of the term $\lambda\langle\Phi^2(t,\mathbf{x})\rangle$ given by

$$\lambda\left[\int \frac{\mathrm{d}^3p}{(2\pi)^3}\frac{f(t,\mathbf{p})}{\omega(t,\mathbf{p})} + \phi^2(t)\right]$$ (2.18)

according to (2.15).

Strong correlations may be expected when interaction effects become comparable in size to the contributions from the typical kinetic energies of particles. For our case of particles with characteristic momentum Q, this happens at the time t_* when their occupation number $f(t_*,Q)$ grows so high that their contribution to the mean-field shift (2.18) is of the same order as their kinetic term $\sim Q^2$. The latter is characteristic for the relativistic theory, which is second order in space–time derivatives. Thus, altogether, we have

$$\lambda\int \frac{\mathrm{d}^3p}{(2\pi)^3}\frac{f(t_*,\mathbf{p})}{\omega(t_*,\mathbf{p})} \overset{!}{\sim} Q^2.$$ (2.19)

Parametrically, we can estimate the integral for a relativistic dispersion $\omega \simeq |\mathbf{p}|$ as

$$\lambda\int^Q \mathrm{d}p\,|\mathbf{p}|^2\frac{f(t_*,\mathbf{p})}{|\mathbf{p}|} \sim \lambda f(t_*,Q)Q^2.$$ (2.20)

This is of order $\sim Q^2$ if the occupancy is as large as

$$\boxed{f(t_*, Q) \sim \frac{1}{\lambda}.}$$

(2.21)

Taking into account (2.16), such an extreme nonequilibrium condition is expected to occur at the time

$$t_* \simeq \frac{1}{\gamma_Q} \ln\left(\frac{1}{\lambda}\right)$$

(2.22)

after inflation.

The maximally amplified mode turns out to be given by $Q \sim \sqrt{\lambda}\,\phi(t = 0)$ and approximately $Q/\gamma_Q \sim \mathcal{O}(10)$. It is striking that the quantum correction in (2.17) cannot be neglected, no matter how small the coupling λ is. Although this is a robust feature of the preheating dynamics, we will see in Section 2.5 that for a quantitative analysis one has to go beyond a mean-field analysis. In particular, strongly nonlinear quantum corrections have to be taken into account even before t_*, which are not included in the mean-field approach. Apart from the primary growth (2.16) starting early, nonlinear effects lead to an important *secondary amplification* period of fluctuations with enhanced growth rates for higher-momentum modes. As a consequence, a wide range of growing modes lead to a *prethermalization* of the equation of state, given by an almost constant ratio of pressure over energy density, at the end of this early stage while the distribution function itself is still far from equilibrium.

Nonthermal fixed points and turbulence

The exponentially fast transfer of the zero-mode energy into fluctuations during preheating stops around t_*, when the energy densities stored in the coherent field and in fluctuations become of the same order. Quantum statistical corrections far beyond the mean-field approximation govern the subsequent dynamics of the strongly correlated system. If everything is strongly correlated, one might naively expect a fast evolution. However, the opposite turns out to be true and the evolution slows down considerably after t_*. In contrast to the earlier instability regime, which exhibits a characteristic scale for the growth of fluctuations, the subsequent evolution is governed by a *nonthermal fixed point*. This is an attractor solution with *self-similar scaling behavior* of correlation functions.

A somewhat similar phenomenon is well known close to thermal equilibrium, where strong correlations near continuous phase transitions lead to scaling behavior and a corresponding critical slowing down of the dynamics. However, the far-from-equilibrium situation after preheating is rather different. The typical momentum of a relativistic system in thermal equilibrium is given by the temperature T, and the Bose–Einstein distribution is of order one for momenta $|\mathbf{p}| \sim T$. Moreover, thermal equilibrium respects *detailed balance*, where each process is equilibrated by its reverse process as will be discussed in Section 2.3. In contrast, the strong overoccupation (2.21) of typical momenta $\sim Q$ represents an extreme nonequilibrium distribution of modes, which leads to a net flux of energy and particles across momentum scales, thus violating detailed balance.

To get a first understanding of this point, we may compare the characteristic energy density $\epsilon \sim \int^Q [d^3 p/(2\pi)^3]\, |\mathbf{p}| f(\mathbf{p}) \sim Q^4/\lambda$ of the far-from-equilibrium state with a Stefan–Boltzmann law $\sim T^4$ in thermal equilibrium. We see that the nonequilibrium distribution exhibits a smaller characteristic momentum $Q \sim \lambda^{1/4} T$ than the corresponding thermal system for the same energy density. Therefore, we expect a transport of energy from lower- to higher-momentum scales on the way toward thermalization. Indeed, the underlying mechanism for the slow dynamics is related to the transport of conserved quantities, which leads to power-law behavior as will be discussed in detail in Section 2.6.

The situation is further complicated in the presence of additional conserved quantities, apart from energy conservation. Though total particle number is a priori not conserved for the relativistic system, an effectively conserved particle number emerges during the transient nonequilibrium evolution. To get a simple illustration of the consequences of both energy and number conservation, we may think of an idealized three-scales problem: We consider particles that are "injected" at the characteristic momentum scale Q as a consequence of the decay of the coherent field as described above. We then ask for the particles and the energy transported to a characteristic lower momentum scale K and to a higher scale Λ with $K \ll Q \ll \Lambda$. We let $\dot{n}_Q$ denote the injection rate for the number density at the scale Q, while $\dot{n}_K$ and $\dot{n}_\Lambda$ characterize the corresponding "ejection" rates at K and Λ. Then number and energy conservation in this simple setup correspond to

$$\dot{n}_Q = \dot{n}_K + \dot{n}_\lambda, \qquad Q\dot{n}_Q = K\dot{n}_K + \Lambda\dot{n}_\Lambda. \tag{2.23}$$

These can be solved for $\dot{n}_K$ and $\dot{n}_\lambda$ as

$$\dot{n}_K = \frac{\Lambda - Q}{\Lambda - K}\dot{n}_Q \simeq \dot{n}_Q, \qquad \dot{n}_\Lambda = \frac{Q - K}{\Lambda - K}\dot{n}_Q \simeq \frac{Q}{\Lambda}\dot{n}_Q, \tag{2.24}$$

where the latter approximate equalities exploit the separation of scales. According to the first equation, the particles are transported to lower scales, while the second says $\Lambda\dot{n}_\Lambda \simeq Q\dot{n}_Q$ such that the energy transport occurs toward higher scales.

To illustrate this further, Fig. 2.2 shows the schematic behavior of the distribution function $f(t, \mathbf{p})$ in the vicinity of the nonthermal fixed point. The distribution is given as a function of momentum for two subsequent times $t = t_1$ and $t_2 > t_1$. We observe that the particle transport toward low momenta is part of a *dual cascade*, in which energy is also transferred toward higher momenta. It turns out that there is no single scaling solution conserving both energy and particle number in this case. Instead, a dual cascade emerges such that in a given inertial range of momenta only one conservation law governs the scaling behavior.

The inverse particle cascade occurs in a highly occupied infrared range of momenta with $|\mathbf{p}| \ll Q$. In this nonperturbative regime with $f(t, \mathbf{p}) \gtrsim 1/\lambda$, the effectively conserved particle number receives a dominant contribution at the time-dependent scale $K(t)$, where the integrand for particle number $\sim |\mathbf{p}|^2 f(t, \mathbf{p})$ has its maximum.

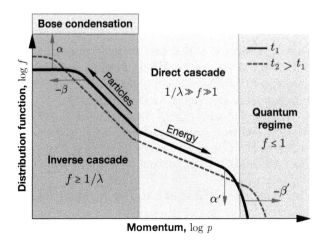

Fig. 2.2 Schematic behavior of the occupation number distribution near the nonthermal fixed point as a function of momentum $|\mathbf{p}|$ for two times $t = t_1$ with $t_2 > t_1$. The scaling exponents α and β characterize the rate and direction of the inverse particle cascade, while α' and β' describe the direct energy cascade during the self-similar evolution. (A. Piñeiro Orioli.)

The cascade exhibits an approximate power-law behavior in an inertial range of momenta. This power law is similar to the one known from stationary-wave turbulence in the presence of external sources or sinks. However, we are dealing with an isolated system. Instead of simple power laws, the transport in isolated systems is described in terms of the more general notion of a *self-similar evolution*, where the distribution function acquires a scaling form:

$$f(t, \mathbf{p}) = t^\alpha f_S(t^\beta \mathbf{p}).$$
(2.25)

Here, all quantities are considered to be dimensionless by use of some suitable momentum scale. The values of the scaling exponents α and β, as well as the form of the time-independent *fixed-point distribution* $f_S(t^\beta \mathbf{p})$ are universal. More precisely, all models in the same universality class can be related by a multiplicative rescaling of t and $|\mathbf{p}|$. Quantities that are invariant under this rescaling are universal.

The exponents for the infrared particle cascade are described well by

$$\alpha = \beta d, \qquad \beta = \tfrac{1}{2},$$
(2.26)

where d denotes the spatial dimension (here $d = 3$). The positive values for α and β determine the rate and direction of the particle number transport toward low momenta, since a given characteristic momentum scale $K(t_1) = K_1$ evolves as $K(t) = K_1(t/t_1)^{-\beta}$ with amplitude $f(t, K(t)) \sim t^\alpha$ according to (2.25). The fixed relation between α and β reflects the conservation of particle number density $n = \int d^d p \, f(t, \mathbf{p}) \sim t^{\alpha - d\beta}$ in this inertial range by using the self-similarity (2.25).

The emergence of an effectively conserved particle number for infrared modes can be explained by the dynamical generation of a mass gap for the (massless) relativistic theory, which arises from an intriguing interplay of condensation and medium effects. Indeed, the inverse particle cascade toward the infrared continuously (re-)populates the zero mode, which leads to a power-law behavior also for the condensate field $\phi(t)$. In particular, if we would start the evolution from overoccupation (2.21) without an initial coherent field part ($\phi(t = 0) = 0$), the inverse cascade would lead to the formation of a Bose condensate in this far-from-equilibrium state, which is shown in Section 2.6.

The particle transport toward low momenta is accompanied by energy transport toward higher momenta. In Fig. 2.2, we indicate that in the direct energy cascade regime other scaling exponents α' and β' with a different scaling function f'_S are found than for the inverse particle cascade. In the perturbative higher-momentum range, we have $1/\lambda \gg f(\mathbf{p}) \gg 1$ and the energy is dominated by the scale $\Lambda(t)$. For the hard modes with $|\mathbf{p}| \sim \Lambda$, the negative exponent $\beta = -1/5$ determines the evolution of characteristic momenta and $\alpha = -4/5$ that of their occupancy. The latter exponent determines the parametric time $t \sim t_* \lambda^{-5/4}$ at which the occupancies of the hard modes $f(t, \Lambda(t))$, which are of order $1/\lambda$ at t_*, drop to become comparable to the "quantum half." At this stage, a subsequent approach to a thermal equilibrium distribution with the help of elastic and inelastic scattering processes sets in.

2.1.4 Ultracold quantum gases far from equilibrium

Experimentally, the investigation of quantum systems in extreme conditions is boosted by the physics of ultracold quantum gases. Because these gases are very cold and dilute, they can be used to obtain clean realizations of microscopic Hamiltonians built from low-energy elementary scattering and interactions with applied fields. By using optical or atom chip traps, they provide a flexible testbed with tunable interactions, symmetries, and dimensionality, with connections to a wide variety of systems. Setups employing ultracold quantum gases can be largely isolated, in contrast to many other condensed matter systems, where significant couplings to the environment are mostly inevitable. This offers the possibility to study fundamental aspects of far-from-equilibrium dynamics and the thermalization process in a highly isolated environment.

We have seen in Section 2.1.3 that the dynamics of the early universe after inflation can have different characteristic stages. After a very rapid evolution caused by nonequilibrium instabilities, the dynamics becomes self-similar. The time evolution in this self-similar regime is described in terms of universal scaling exponents and scaling functions. The universality observed opens the exciting possibility to find other systems in the same universality class, which may be more easily accessible experimentally. Ultracold quantum gases at nanokelvins can differ in their characteristic energy scale by more than 38 orders of magnitude from inflationary physics. Nevertheless, they can show very similar dynamical behavior, and their universal properties can even be quantitatively the same. Therefore, one may learn from experiments with cold atoms aspects about the dynamics during the early stages of our universe.

We will consider here interacting bosons with s-wave scattering length a, and leave the inclusion of fermions to later sections. For a gas of density n, the average interatomic distance is $n^{-1/3}$. Together with the scattering length a, this can be used to define a dimensionless "diluteness parameter"

$$\zeta = \sqrt{na^3}. \tag{2.27}$$

For a typical scattering length of, say, $a \simeq 5\,\text{nm}$ and bulk density $n \simeq 10^{14}\,\text{cm}^{-3}$, the diluteness parameter $\zeta \simeq 10^{-3}$ is very small, and in the following we will always assume $\zeta \ll 1$. The density and scattering length can also be used to define a characteristic "coherence length," whose inverse is described by the momentum scale

$$Q = \sqrt{16\pi an}. \tag{2.28}$$

To observe the dynamics near nonthermal fixed points for the interacting Bose gas, an unusually large occupancy of modes at the inverse coherence length scale Q has to be prepared. More precisely, for a weakly coupled gas of average density $n = \int [\mathrm{d}^3 p/(2\pi)^3]\, f_{\mathrm{nr}}(\mathbf{p})$, this requires a characteristic mode occupancy as large as

$$\boxed{f_{\mathrm{nr}}(Q) \sim \frac{1}{\zeta}.} \tag{2.29}$$

This represents a far-from-equilibrium distribution of modes, and we note that the diluteness parameter ζ plays the corresponding role of the coupling in (2.21). Such an extreme condition may be obtained, for instance, from a quench or nonequilibrium instabilities similar to the previously discussed relativistic case.

Most importantly, the system in this overoccupied regime is strongly correlated. These properties may be understood from a Gross–Pitaevskii equation for a nonrelativistic complex Bose field χ:

$$i\partial_t \chi(t, \mathbf{x}) = \left[-\frac{\nabla^2}{2m} + g|\chi(t, \mathbf{x})|^2 \right] \chi(t, \mathbf{x}). \tag{2.30}$$

Here the coupling g is not dimensionless and is determined from the mass m and scattering length as $g = 4\pi a/m$. The total number of particles is given by $\int \mathrm{d}^3 x\, |\chi(t, \mathbf{x})|^2$ and is conserved.

In the mean-field approximation, the effect of the interaction term in the Gross–Pitaevskii equation is a constant energy shift for each particle,

$$\Delta E = 2g\langle|\chi|^2\rangle = 2gn = 2g \int \frac{\mathrm{d}^3 p}{(2\pi)^3} f_{\mathrm{nr}}(\mathbf{p}), \tag{2.31}$$

which can be absorbed in a redefinition of the chemical potential. However, we note that for the very high occupancy (2.29) of the typical momentum Q, the shift in energy is not small compared with the relevant kinetic energy $Q^2/2m$, i.e., $2gn \sim Q^2/2m$. Parametrically, this can be directly verified using (2.29):

$$g \int \mathrm{d}^3 p\, f_{\mathrm{nr}}(\mathbf{p}) \sim gQ^3 f_{\mathrm{nr}}(Q) \sim g\frac{Q^3}{\zeta} \sim g\frac{Q^3}{mgQ} \sim \frac{Q^2}{m}. \tag{2.32}$$

Here we have used the fact that with $a = mg/(4\pi)$, (2.28) implies $Q = 2\sqrt{mgn}$ and (2.27) gives $\zeta = mgQ/(16\pi^{3/2})$. Most importantly, the energy shift $2gn$ is of the order of the kinetic energy $Q^2/2m$, irrespective of the coupling strength g. This already hints at a strongly correlated system, where the dependence on the details of the underlying model parameters is lost.

To study the nonequilibrium evolution of such a system, we may define a time-dependent occupation number distribution $f_{\mathrm{nr}}(t, \mathbf{p})$ from the equal-time two-point correlation function:

$$f_{\mathrm{nr}}(t, \mathbf{p}) + \tfrac{1}{2} + (2\pi)^3 \delta(\mathbf{p})|\chi_0|^2(t)$$
$$\equiv \frac{1}{2} \int \mathrm{d}^3 x\, \mathrm{e}^{-\mathrm{i}\mathbf{p}\cdot\mathbf{x}} \langle \chi(t, \mathbf{x})\chi^*(t, 0) + \chi(t, 0)\chi^*(t, \mathbf{x}) \rangle. \quad (2.33)$$

The term $|\chi_0|^2(t)$ coming together with the Dirac δ-function determines the condensate fraction at zero momentum. These definitions are in complete analogy with (2.15) for the relativistic theory. The only major difference is the appearance of the dispersion in the definition for the relativistic case, which is a consequence of the second-order differential equation (2.13), while the nonrelativistic equation (2.30) is first-order in time.

Since the far-from-equilibrium dynamics of the strongly correlated system is expected to become insensitive to the details of the initial conditions, we may choose the initial condensate fraction to be zero. Moreover, it turns out that in the overoccupied regime, where $f_{\mathrm{nr}}(t, Q)$ is much larger than the "quantum half," the quantum statistical dynamics is essentially classical statistical such that the quantum and the corresponding classical field theories belong to the same universality class. The classical aspects of nonequilibrium quantum theories will be addressed in Section 2.4.

Figure 2.3 presents a striking example of universality far from equilibrium: it shows the normalized fixed-point distribution $f_{S,\mathrm{nr}}(t^\beta \mathbf{p}) = t^{-\alpha} f_{\mathrm{nr}}(t, \mathbf{p})$ in the infrared, after evolving from overoccupied initial conditions for the nonrelativistic theory (circles) using the simulation techniques of Section 2.4. The corresponding fixed-point distribution can be extracted for relativistic theories, such as discussed in Section 2.1.3, which we can generalize to include N-component inflaton models as done in Section 2.5. If the relativistic and nonrelativistic theories belong to the same universality class, then the normalized fixed-point distribution has to agree. That this is indeed the case in the infrared scaling regime is exemplified in Fig. 2.3 for the example of the relativistic $N = 2$ component field theory (squares).

A crucial ingredient for the universality observed is the presence of strong correlations, which can occur even for weakly coupled systems if they are overoccupied. These extreme occupancies of modes can be found in a variety of isolated systems in extreme conditions, and the identification of nonequilibrium universality classes represents a crucial step for their understanding. This is of particular relevance also for nonequilibrium gauge theories, such as those employed in the description of the initial stages of ultrarelativistic heavy-ion collision experiments. Although we cannot cover this in depth in this chapter, we suggest some further reading below.

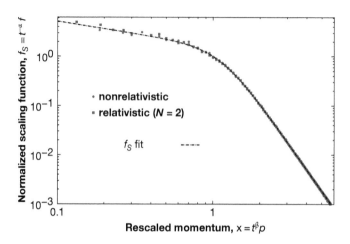

Fig. 2.3 Normalized fixed-point distribution $f_S(t^\beta \mathbf{p}) = t^{-\alpha} f(t, \mathbf{p})$ in the infrared regime of the inverse particle cascade. The simulation results for the relativistic (squares) inflaton model with quartic interaction and the nonrelativistic (circles) Gross–Pitaevskii field theory agree to very good accuracy.

2.1.5 Bibliography

- Part of the presentation here about nonthermal universality classes follows J. Berges, K. Boguslavski, S. Schlichting, and R. Venugopalan. Universality far from equilibrium: from superfluid Bose gases to heavy-ion collisions. *Phys. Rev. Lett.* **114** (2015) 061601, where a version of Fig. 2.1 appeared at https://journals.aps.org/prl/issues/114/6 to highlight Editors' Suggestions. Figures 2.2 and 2.3 are taken with permission from A. Piñeiro Orioli, K. Boguslavski, and J. Berges, Universal self-similar dynamics of relativistic and nonrelativistic field theories near nonthermal fixed points. *Phys. Rev. D* **92** (2015) 025041 (Copyright (2015) by the American Physical Society). Part of the discussion about self-similar evolution is taken from the same reference.
- Part of this section is based on J. Berges, Introduction to nonequilibrium quantum field theory. *AIP Conf. Proc.* **739** (2005) 3 (arXiv:hep-ph/0409233). The latter covers the phenomenon of (pre)thermalization and aspects of renormalization in detail, while here we discuss in particular the important notion of self-similar evolution near nonthermal fixed points.
- For overviews in cosmology about aspects of nonequilibrium inflaton dynamics, see L. Kofman. Preheating after inflation. *Lect. Notes Phys.* **738** (2008) 55.
- The preheating phenomenon was first developed in L. Kofman, A. D. Linde, and A. A. Starobinsky. Reheating after inflation. *Phys. Rev. Lett.* **73** (1994) 3195. The subsequent direct energy cascade in the perturbative regime was shown in this context in R. Micha and I. Tkachev. Relativistic turbulence: a long way from preheating to equilibrium. *Phys. Rev. Lett.* **90** (2003) 121301. The inverse particle cascade in the nonperturbative regime and the underlying nonthermal

fixed point have been found in J. Berges, A. Rothkopf, and J. Schmidt. Nonthermal fixed points: effective weak-coupling for strongly correlated systems far from equilibrium. *Phys. Rev. Lett.* **101** (2008) 041603.

- For an overview of nonthermal fixed points in Bose gases, see Chapter 7 of this volume.
- For an overview of nonthermal fixed points in heavy-ion collisions, see J. Berges, K. Boguslavski, S. Schlichting, and R. Venugopalan. Universal attractor in a highly occupied non-Abelian plasma. *Phys. Rev. D* **89** (2014) 114007.
- A perspective on cold atoms and cosmology with related references may be found in J. Schmiedmayer and J. Berges. Cold atom cosmology. *Science* **341** (2013) 1188.

2.2   Nonequilibrium quantum field theory

2.2.1   How to describe nonequilibrium quantum fields

There are very few ingredients for the description of nonequilibrium quantum fields. *Nonequilibrium dynamics typically requires the specification of an initial state* at some given time t_0. This may include a density operator $\varrho_0 \equiv \varrho(t_0)$ in a mixed $(\mathrm{Tr}\, \varrho_0^2 < 1)$ or pure $(\mathrm{Tr}\, \varrho_0^2 = 1)$ state. "Nonequilibrium" means that ϱ_0 does not correspond to a thermal equilibrium density operator: $\varrho_0 \neq \varrho^{\mathrm{(eq)}}$ with, for instance, $\varrho^{\mathrm{(eq)}} \sim \mathrm{e}^{-\beta H}$ for the case of a canonical thermal ensemble with inverse temperature β for a given Hamilton operator H.

Completely equivalent to the specification of the initial density operator ϱ_0 is the knowledge of all initial correlation functions: the initial one-point function $\mathrm{Tr}\,\{\varrho_0 \Phi(t_0, \mathbf{x})\}$, the two-point function $\mathrm{Tr}\,\{\varrho_0 \Phi(t_0, \mathbf{x})\Phi(t_0, \mathbf{y})\}$, the three-point function, etc. To be specific, here $\Phi(x)$ may denote a scalar Heisenberg field operator depending on time and space, $x = (x^0, \mathbf{x})$. Typically, the "experimental setup" requires only knowledge about the few lowest correlation functions at the time t_0, whereas complicated higher correlation functions may build up at later times. From the behavior of the correlation functions for times $x^0 > t_0$, one can extract the time evolution of all other quantities.

Once the nonequilibrium initial state has been specified, for *closed systems* the time evolution is completely determined by the Hamiltonian. Equivalently, for a given classical action S, the dynamics can be described in terms of a functional integral. From the latter, one obtains the *effective action* Γ, which is the generating functional for all correlation functions of the quantum theory. *There are no further ingredients involved concerning the dynamics than what is known from standard vacuum quantum field theory.* It should be stressed that during the nonequilibrium time evolution, there is no loss of information in any strict sense. The important process of thermalization is a nontrivial question in a calculation from first principles. Thermal equilibrium retains no memory about the time history except for the values of a few conserved charges. Equilibrium is time-translation-invariant and cannot be reached from a nonequilibrium evolution on a fundamental level. It is striking to observes that the evolution can go very closely toward thermal equilibrium without ever starting again to deviate from it for accessible times. The observed *effective loss of details about the initial conditions* can mimic very accurately the irreversible dynamics obtained from effective descriptions in their range of applicability.

For out-of-equilibrium calculations, there are additional complications that do not appear in vacuum or thermal equilibrium. A major new aspect for approximate descriptions concerns *secularity*: the standard perturbative time evolution suffers from the presence of spurious, so-called secular, terms, which grow with time and invalidate the expansion even in the presence of a weak coupling. Here it is important to note that the very same problem can also appear for nonperturbative approximation schemes such as standard $1/N$ expansions beyond leading order, where N denotes the number of field components. To obtain a uniform approximation in time, where an expansion parameter controlling the error at early times is valid also at late times, requires selective summation of an infinite series of contributions.

If a nonequilibrium initial state is evolved from early to very late times, it is in general crucial that energy be conserved exactly for the given approximation. If thermal equilibrium is approached, then the late-time result is insensitive to the details of the initial conditions and becomes uniquely determined by the energy density and further conserved charges. This property can be conveniently implemented if the dynamics is obtained from an effective action by a variational principle. The analogue in classical mechanics is well known: if the equations of motion can be derived from the principle of least action, then they will not admit any friction term without further approximations.

Many of these requirements can indeed be achieved using efficient functional integral techniques based on so-called n-particle irreducible (nPI) effective actions. They often provide a practical means to describe far-from-equilibrium dynamics as well as thermalization from first principles. Closely related are approximations based on identities relating different n-point functions such as Dyson–Schwinger equations, or identities between derivatives of correlators described by the functional renormalization group.

Local symmetries as encountered in gauge theories pose even stronger restrictions on possible approximation schemes. Here, the derivation of a sequence of effective theories that disentangle dynamics on different time and distance scales can be very efficient. In particular, for weakly coupled theories close to equilibrium at high temperature, a wealth of information can be obtained from effective descriptions such as kinetic theory.

Some aspects of nonequilibrium systems are often successfully described using classical statistical field theory methods. Important examples involve the dynamics of nonequilibrium instabilities (e.g., parametric resonance) and universal properties of dynamic critical phenomena near second-order phase transitions. Classical statistical field theory can be implemented on a space–time lattice in Minkowski space and simulated on computers. Classical Rayleigh–Jeans divergences and the lack of genuine quantum effects limit their use. It is therefore important to understand for which time and distance scales classical simulations can adequately reproduce the behavior of the underlying quantum theory of interest.

Of course, this list of methods is not complete, and the discussion here is restricted to just some of the major field-theoretic techniques and applications. In the following, we will introduce the functional integral formulation of nonequilibrium quantum field theory, which provides a very efficient starting point to derive the different approaches that will be discussed in later sections.

2.2.2 Nonequilibrium generating functional

We consider first a scalar quantum field theory with Heisenberg field operator $\Phi(x)$. Although most of the developments are completely equivalent for relativistic as well as nonrelativistic theories, in the following we will always assume a relativistic setup for definiteness and point out the relevant differences where necessary.

Knowledge of all correlation functions, which are represented as traces over time-dependent Heisenberg field operators for given initial density operator ϱ_0, completely specifies the quantum system at any time. Therefore, all information about nonequilibrium quantum field theory is contained in the *generating functional* for correlation functions,

$$ Z[J, R; \varrho_0] = \mathrm{Tr} \left\{ \varrho_0 T_{\mathcal{C}} \exp\left[i \left(\int_{x,\mathcal{C}} J(x)\Phi(x) + \tfrac{1}{2} \int_{xy,\mathcal{C}} \Phi(x)R(x,y)\Phi(y) \right) \right] \right\}, \quad (2.34) $$

for given ϱ_0. It is the generalization of the *partition function* for nonequilibrium systems in the presence of sources $J(x)$ and $R(x,y)$. Since this is a generating functional for the trace over Heisenberg field operators, the time argument x^0 of the field operator $\Phi(x^0, \mathbf{x})$ is evaluated along a *closed real-time contour* $\mathcal{C}$ appearing in the integrals, where we write $\int_{x,\mathcal{C}} \equiv \int_{\mathcal{C}} \mathrm{d}x^0 \int \mathrm{d}^d x$ in d spatial dimensions. This will be explained in more detail in Section 2.2.3 and should be distinguished from computations restricted to scattering matrix elements between asymptotic incoming and outgoing states, which can be formulated without a closed time contour $\mathcal{C}$. Employing the latter is more general, in the sense that vacuum quantum field theory as well as thermal and nonequilibrium systems can be described. The time path $\mathcal{C}$ starts at time t_0 running forward along the real-time axis and then runs back to t_0. Graphically, $\mathcal{C}$ is depicted as

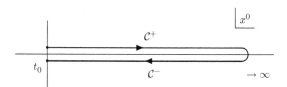

where we have shifted parts of the curve slightly away from the real axis purely for visualization purposes. Contour time ordering along this real-time path is denoted by $T_{\mathcal{C}}$. It corresponds to usual time ordering along the forward piece $\mathcal{C}^+$ and reversed ordering on the backward piece $\mathcal{C}^-$. In particular, any time on $\mathcal{C}^-$ is considered later than any time on $\mathcal{C}^+$.

We have introduced the generating functional with two source terms on the contour, $J(x)$ and $R(x, y)$, which can be extended straightforwardly to take into account further source terms if necessary. Standard functional differentiation with respect to sources is extended to include time arguments along the closed time path. In particular,

$$\frac{\delta J(x)}{\delta J(y)} = \delta_{\mathcal{C}}(x - y) \equiv \delta_{\mathcal{C}}(x^0 - y^0)\delta(\mathbf{x} - \mathbf{y}), \tag{2.35}$$

where the Dirac $\delta_{\mathcal{C}}(x^0 - y^0)$ is defined on the closed time contour to be zero everywhere except at $x^0 = y^0$ if either x^0 and y^0 are both on $\mathcal{C}^+$ or both are on $\mathcal{C}^-$, where it is infinite with

$$\int_{\mathcal{C}} dx^0\, \delta_{\mathcal{C}}(x^0) = 1. \tag{2.36}$$

Setting all source terms to zero in (2.34), we obtain the normalized partition sum

$$Z[J, R; \varrho_0]\big|_{J,R=0} = \mathrm{Tr}\,\{\varrho_0\} = 1. \tag{2.37}$$

Nonequilibrium correlation functions correspond to expectation values of products of Heisenberg field operators. These can be obtained by functional differentiation of (2.34). For instance, the one-point function or *macroscopic field* ϕ reads

$$\frac{\delta Z[J, R; \varrho_0]}{i\delta J(x)}\bigg|_{J,R=0} = \mathrm{Tr}\,\{\varrho_0 \Phi(x)\} \equiv \langle \Phi(x) \rangle \equiv \phi(x). \tag{2.38}$$

Two-point functions are obtained from second functional derivatives, and n-point functions involve n derivatives. Together with the time-ordered two-point function

$$\frac{\delta^2 Z[J, R; \varrho_0]}{i\delta J(x)\, i\delta J(y)}\bigg|_{J,R=0} = \mathrm{Tr}\,\{\varrho_0 T_{\mathcal{C}} \Phi(x)\Phi(y)\} \equiv \langle T_{\mathcal{C}} \Phi(x)\Phi(y) \rangle, \tag{2.39}$$

we can introduce the *connected two-point function* or *propagator* as

$$G(x, y) \equiv \langle T_{\mathcal{C}} \Phi(x)\Phi(y) \rangle - \phi(x)\phi(y). \tag{2.40}$$

Here $G(x, y)$ is defined on the contour $\mathcal{C}$ and the time ordering of $T_{\mathcal{C}}\Phi(x)\Phi(y)$ has to be evaluated according to the positions of the time arguments of the fields. In terms of the Heaviside step function $\theta(x^0 - y^0)$, this reads explicitly

$$T_{\mathcal{C}}\Phi(x)\Phi(y)$$

$$= \begin{cases} \Phi(x)\Phi(y)\theta(x^0 - y^0) + \Phi(y)\Phi(x)\theta(y^0 - x^0) & \text{for } x^0,\, y^0 \text{ on } \mathcal{C}^+ \\ \Phi(x)\Phi(y)\theta(y^0 - x^0) + \Phi(y)\Phi(x)\theta(x^0 - y^0) & \text{for } x^0,\, y^0 \text{ on } \mathcal{C}^- \\ \Phi(y)\Phi(x) & \text{for } x^0 \text{ on } \mathcal{C}^+,\, y^0 \text{ on } \mathcal{C}^- \\ \Phi(x)\Phi(y) & \text{for } x^0 \text{ on } \mathcal{C}^-,\, y^0 \text{ on } \mathcal{C}^+ \end{cases}$$

$$\equiv \Phi(x)\Phi(y)\theta_{\mathcal{C}}(x^0 - y^0) + \Phi(y)\Phi(x)\theta_{\mathcal{C}}(y^0 - x^0), \tag{2.41}$$

where the last equation defines the contour step function $\theta_{\mathcal{C}}(x^0 - y^0)$.

For an efficient notation, the field $\Phi(x)$ may be written as $\Phi^{\pm}(x^0, \mathbf{x})$, where the $\pm$ superscript indicates the part of the contour $\mathcal{C}^{\pm}$ on which the time argument x^0 is located. Then the contour integration for the linear source term in (2.34) takes the form

$$\int_{x,\mathcal{C}} \Phi(x)J(x) \equiv \int_{t_0}^{\infty} dx^0 \int d^dx \left[\Phi^+(x)J^+(x) - \Phi^-(x)J^-(x) \right], \qquad (2.42)$$

where the minus sign comes from the reversed time integration along $\mathcal{C}^-$. Similarly, the bilinear source term in (2.34) can be decomposed into four terms by introducing $R^{++}(x,y)$, $R^{--}(x,y)$, $R^{+-}(x,y)$, and $R^{-+}(x,y)$ according to the different possibilities to locate x^0 and y^0 on $\mathcal{C}^{\pm}$. For instance, the one-point function for vanishing sources is obtained as

$$\left. \frac{\delta Z[J, R; \varrho_0]}{i\delta J^+(x)} \right|_{J,R=0} = \phi(x). \qquad (2.43)$$

We emphasize that the field expectation value is the same if obtained from a derivative with respect to either J^+ or J^- in the absence of sources, since time ordering plays no role for a one-point function. Taking the second functional derivative of the generating functional (2.34) with respect to the source J^+ and setting J and R to zero afterward, we obtain

$$\left. \frac{\delta^2 Z[J, R; \varrho_0]}{i\delta J^+(x)\, i\delta J^+(y)} \right|_{J,R=0} = \langle \Phi(x)\Phi(y)\theta(x^0 - y^0) + \Phi(y)\Phi(x)\theta(y^0 - x^0) \rangle$$

$$\equiv G^{++}(x,y) + \phi(x)\phi(y). \qquad (2.44)$$

Here we have used the fact that contour ordering corresponds to standard time ordering if all time arguments are on $\mathcal{C}^+$. We have also introduced the notation $G^{++}(x,y)$ in order to distinguish this correlator from the other possible second functional derivatives with respect to the sources J^+ and J^-, setting $J, R = 0$ afterward. These can be written as

$$\left. \frac{\delta^2 Z[J, R; \varrho_0]}{i\delta J^-(x)\, i\delta J^-(y)} \right|_{J,R=0} = \langle \Phi(x)\Phi(y)\theta(y^0 - x^0) + \Phi(y)\Phi(x)\theta(x^0 - y^0) \rangle$$

$$\equiv G^{--}(x,y) + \phi(x)\phi(y),$$

$$\left. \frac{\delta^2 Z[J, R; \varrho_0]}{i\delta J^+(x)\, i\delta J^-(y)} \right|_{J,R=0} = \langle \Phi(y)\Phi(x) \rangle \equiv G^{+-}(x,y) + \phi(x)\phi(y), \qquad (2.45)$$

$$\left. \frac{\delta^2 Z[J, R; \varrho_0]}{i\delta J^-(x)\, i\delta J^+(y)} \right|_{J,R=0} = \langle \Phi(x)\Phi(y) \rangle \equiv G^{-+}(x,y) + \phi(x)\phi(y).$$

While any nonequilibrium two-point correlation function can be written in terms of the components (2.44) and (2.45), not all of them are independent. In particular, using the property $\theta(x^0 - y^0) + \theta(y^0 - x^0) = 1$ of the Heaviside step function, one obtains the following algebraic identity for $J, R = 0$:

$$G^{++}(x, y) + G^{--}(x, y) = G^{+-}(x, y) + G^{-+}(x, y). \tag{2.46}$$

Later, we will observe that a further so-called fluctuation–dissipation relation exists for the special case of vacuum or thermal equilibrium density matrices. However, for general out-of-equilibrium situations, this is not the case.

2.2.3   Functional integral representation

In the above discussion, the time ordering along the closed contour $\mathcal{C}$ appears as a bookkeeping device that allows one to conveniently describe different components of nonequilibrium correlation functions. To simplify the evaluation of correlation functions, in the following, we write the generating functional (2.34) in terms of a functional integral representation. In this construction, the closed time path appears because we want to compute correlation functions that are given as the *trace* over the density operator with time-ordered products of Heisenberg field operators. Representing the trace as a path integral will require a time path where the initial and final times are identified.

We evaluate the trace using eigenstates of the field operator $\Phi^{\pm}$ at time t_0:

$$\Phi^{\pm}(t_0, \mathbf{x}) |\varphi^{\pm}\rangle = \varphi_0^{\pm}(\mathbf{x}) |\varphi^{\pm}\rangle, \tag{2.47}$$

such that (2.34) may be written as

$$
\begin{aligned}
&Z[J, R; \varrho_0] \\
&= \int [d\varphi_0^+] \, \langle \varphi^+ | \, \varrho_0 \, T_{\mathcal{C}} \exp\left\{ i \left[\int_{x, \mathcal{C}} \Phi(x) J(x) + \frac{1}{2} \int_{x, y, \mathcal{C}} \Phi(x) R(x, y) \Phi(y) \right] \right\} | \varphi^+ \rangle.
\end{aligned}
\tag{2.48}
$$

The integration measure is

$$\int [d\varphi_0^{\pm}] \equiv \int \prod_{\mathbf{x}} d\varphi_0^{\pm}(\mathbf{x}). \tag{2.49}$$

With the insertion

$$\int [d\varphi_0^-] \, |\varphi^-\rangle\langle\varphi^-| = \mathbb{1}, \tag{2.50}$$

we may bring (2.48) to the form

$$Z[J, R; \varrho_0] = \int [d\varphi_0^+][d\varphi_0^-] \, \langle \varphi^+ | \, \varrho_0 \, |\varphi^-\rangle \, (\varphi^-, t_0 \, | \, \varphi^+, t_0)_{J, R}. \tag{2.51}$$

Here the transition amplitude in the presence of the sources is given by

$$
\left(\varphi^-, t_0 \,|\, \varphi^+, t_0\right)_{J,R}
$$
$$
\equiv \langle \varphi^- | \, T_{\mathcal{C}} \exp\left\{ i \left[\int_{x,\mathcal{C}} \Phi(x) J(x) + \frac{1}{2} \int_{x,y,\mathcal{C}} \Phi(x) R(x,y) \Phi(y) \right] \right\} | \varphi^+ \rangle. \quad (2.52)
$$

This matrix element can be written as a functional integral over "classical" fields $\varphi(x)$ on the closed time path $\mathcal{C}$, whose derivation is given below. For a quantum theory with classical action $S[\varphi]$, it reads

$$
\left(\varphi^-, t_0 \,|\, \varphi^+, t_0\right)_{J,R}
$$
$$
= \int_{\varphi_0^+}^{\varphi_0^-} \mathscr{D}'\varphi \exp\left\{ i \left[S[\varphi] + \int_{x,\mathcal{C}} \varphi(x) J(x) + \frac{1}{2} \int_{x,y,\mathcal{C}} \varphi(x) R(x,y) \varphi(y) \right] \right\}. \quad (2.53)
$$

Here the classical action includes the time contour integral over the respective Lagrangian density $\mathcal{L}(x)$, i.e., $S = \int_{\mathcal{C}} \mathrm{d}x^0 \int \mathrm{d}^d x\, \mathcal{L}(x)$. The functional integration goes over the field configurations $\varphi(x)$ depending on space and times $x^0 > t_0$ that satisfy the boundary conditions $\varphi^\pm(x^0 = t_0, \mathbf{x}) = \varphi_0^\pm(\mathbf{x})$ for x^0 on $\mathcal{C}^\pm$. The prime on the functional measure indicates that the integration over the fields at $x^0 = t_0$ is excluded.

Putting everything together, we may write the generating functional (2.34) as

$$
Z[J, R; \varrho_0] = \underbrace{\int [\mathrm{d}\varphi_0^+][\mathrm{d}\varphi_0^-] \, \langle \varphi^+ | \, \varrho_0 \, | \varphi^- \rangle}_{\text{initial conditions}}
$$

$$
\times \underbrace{\int_{\varphi_0^+}^{\varphi_0^-} \mathscr{D}'\varphi \exp\left\{ i \left[S[\varphi] + \int_{x,\mathcal{C}} J(x)\varphi(x) + \frac{1}{2} \int_{xy,\mathcal{C}} \varphi(x) R(x,y) \varphi(y) \right] \right\}}_{\text{quantum dynamics}}.
$$
$$(2.54)$$

This expression displays two important ingredients entering nonequilibrium quantum field theory: the quantum fluctuations described by the functional integral with action S and the statistical fluctuations encoded in the averaging procedure with the matrix elements of the initial density operator ϱ_0.

Formulations of closed time path generating functionals typically employ a time interval starting at some t_0 and extending to the far future as mentioned above. Causality implies that for any n-point function with finite time arguments, the contributions of an infinite time path cancel for times exceeding the largest time argument of the n-point function. We will see this explicitly when we derive time evolution equations for correlation functions in Section 2.2.6. To avoid unnecessary cancellations of infinite time path contributions, we will often consider finite time paths. The largest time of the path can be kept as a parameter and is evolved in the time evolution equations.

Details of construction of the path integral

To simplify the evaluation of correlation functions, we wrote the above generating functional in terms of a functional integral representation using (2.53). Here we describe the construction of the corresponding transition amplitude following standard presentations and refer to the literature for more information.

We start from a Schrödinger picture where the field operators are independent of time. For the scalar field theory, the Hamiltonian $H[\Pi, \Phi]$ is expressed in terms of the field $\Phi(\mathbf{x})$ and the conjugate momentum field operator $\Pi(\mathbf{x})$. We work with basis sets of eigenstates for which $\Phi(\mathbf{x})$ or $\Pi(\mathbf{x})$ are multiplicative operators

$$\Phi(\mathbf{x})|\varphi\rangle = \varphi(\mathbf{x})|\varphi\rangle, \qquad \Pi(\mathbf{x})|\pi\rangle = \pi(\mathbf{x})|\pi\rangle,$$

respectively. The completeness and orthogonality conditions read

$$\int [\mathrm{d}\varphi]\, |\varphi\rangle\langle\varphi| = \mathbb{1}, \qquad \langle\varphi_j|\varphi_i\rangle = \delta\,[\varphi_j - \varphi_i] \tag{2.55}$$

and

$$\int \left[\frac{\mathrm{d}\pi}{2\pi}\right] |\pi\rangle\langle\pi| = \mathbb{1}, \qquad \langle\pi_j|\pi_i\rangle = \delta[\pi_j - \pi_i]. \tag{2.56}$$

To become familiar with the notation, we recall that the δ-functional may be represented as a product of individual δ-functions, one for each point $\mathbf{x}$ in space:

$$\delta[\varphi_j - \varphi_i] = \prod_{\mathbf{x}} \delta\left(\varphi_j(\mathbf{x}) - \varphi_i(\mathbf{x})\right). \tag{2.57}$$

If we write each δ-function in the momentum representation

$$\delta\left(\varphi_j(\mathbf{x}) - \varphi_i(\mathbf{x})\right) = \int \frac{\mathrm{d}\pi_i(\mathbf{x})}{2\pi} \exp\left\{\mathrm{i}\pi_i(\mathbf{x})\left[\varphi_j(\mathbf{x}) - \varphi_i(\mathbf{x})\right]\right\}, \tag{2.58}$$

we obtain

$$\begin{aligned}\delta[\varphi_j - \varphi_i] &= \prod_{\mathbf{x}} \int \frac{\mathrm{d}\pi_i(\mathbf{x})}{2\pi} \exp\left\{\mathrm{i}\pi_i(\mathbf{x})\left[\varphi_j(\mathbf{x}) - \varphi_i(\mathbf{x})\right]\right\} \\ &= \int \left[\frac{\mathrm{d}\pi_i}{2\pi}\right] \exp\left\{\mathrm{i}\int \mathrm{d}^d x\,\pi_i(\mathbf{x})\left[\varphi_j(\mathbf{x}) - \varphi_i(\mathbf{x})\right]\right\}.\end{aligned} \tag{2.59}$$

Writing the same δ-functional using orthogonality and completeness conditions as $\delta[\varphi_j - \varphi_i] = \int [\mathrm{d}\pi_i/(2\pi)]\,\langle\varphi_j|\pi_i\rangle\langle\pi_i|\varphi_i\rangle$, we recover by comparison with (2.59) the familiar overlap

$$\langle\varphi_j|\pi_i\rangle = \exp\left[\mathrm{i}\int \mathrm{d}^d x\,\pi_i(\mathbf{x})\varphi_j(\mathbf{x})\right]. \tag{2.60}$$

If the dynamics is described by a Hamiltonian H, then the state $|\varphi_0\rangle$ at time t_0 evolves into $e^{-iH(t_f-t_0)}|\varphi_0\rangle$ at time t_f. The transition amplitude for the field in configuration $\varphi_0(\mathbf{x})$ at time t_0 to evolve to $\varphi_f(\mathbf{x})$ at time t_f is thus

$$(\varphi_f, t_f \mid \varphi_0, t_0) \equiv \langle\varphi_f| e^{-iH(t_f-t_0)} |\varphi_0\rangle. \tag{2.61}$$

To conveniently evaluate this amplitude, we write it as a path integral. For this, we divide the interval $t_f - t_0$ into $N + 1$ equal periods. Next, we insert a complete set of states at each division and consider the limit $N \to \infty$:

$$(\varphi_f, t_f \mid \varphi_0, t_0) = \lim_{N\to\infty} \int \prod_{i=1}^{N} [d\varphi_i] \langle\varphi_f| e^{-iH(t_f-t_N)} |\varphi_N\rangle\langle\varphi_N| e^{-iH(t_N-t_{N-1})} |\varphi_{N-1}\rangle$$
$$\cdots \langle\varphi_1| e^{-iH(t_1-t_0)} |\varphi_0\rangle. \tag{2.62}$$

As N gets large, $\Delta t = (t_{i+1} - t_i) \to 0$ and we may expand the exponentials:

$$\langle\varphi_{i+1}| e^{-iH\Delta t} |\varphi_i\rangle \simeq \langle\varphi_{i+1}| 1 - iH\Delta t |\varphi_i\rangle$$
$$= \delta\left[\varphi_{i+1} - \varphi_i\right] - i\Delta t \langle\varphi_{i+1}| H |\varphi_i\rangle. \tag{2.63}$$

The first term in the above sum is a δ-functional, which can be written in terms of its momentum representation (2.59). It remains to evaluate the operators in the matrix element $\langle\varphi_{i+1}| H[\Pi, \Phi] |\varphi_i\rangle$. We consider first the case of a matrix element for a contribution $h_\pi[\Pi]$ that only depends on the conjugate field momentum. We write, inserting a complete set of states,

$$\langle\varphi_{i+1}| h_\pi[\Pi] |\varphi_i\rangle = \int \left[\frac{d\pi_i}{2\pi}\right] \langle\varphi_{i+1}| h_\pi[\Pi] |\pi_i\rangle\langle\pi_i|\varphi_i\rangle$$
$$= \int \left[\frac{d\pi_i}{2\pi}\right] h_\pi[\pi_i] \exp\left\{i \int d^d x\, \pi_i(\mathbf{x}) \left[\varphi_{i+1}(\mathbf{x}) - \varphi_i(\mathbf{x})\right]\right\}, \tag{2.64}$$

where (2.60) is employed for the second equality. Typical Hamiltonians can be written as a sum of a field and a conjugate momentum term, i.e., $H[\Pi, \Phi] = h_\varphi[\Phi] + h_\pi[\Pi]$. For the field contribution $h_\varphi[\Phi]$, we employ a symmetric operator ordering. In general, matrix elements including also products of Φ and Π operators can be evaluated by a symmetric operator ordering. For instance, the symmetrization of a term like $\Phi\Pi \to \{\Phi, \Pi\}/2 \equiv (\Phi\Pi + \Pi\Phi)/2$ leads to the matrix element

$$\langle\varphi_{i+1}| \frac{\Phi\Pi + \Pi\Phi}{2} |\varphi_i\rangle = \frac{\varphi_{i+1} + \varphi_i}{2} \langle\varphi_{i+1}| \Pi |\varphi_i\rangle$$
$$= \bar\varphi_i \int \left[\frac{d\pi_i}{2\pi}\right] \pi_i \exp\left\{i \int d^d x\, \pi_i(\mathbf{x}) \left[\varphi_{i+1}(\mathbf{x}) - \varphi_i(\mathbf{x})\right]\right\}, \tag{2.65}$$

where

$$\bar\varphi_i \equiv \frac{\varphi_{i+1} + \varphi_i}{2}. \tag{2.66}$$

In general, we can write

$$\langle \varphi_{i+1} | H[\Pi, \Phi] | \varphi_i \rangle = \int \left[\frac{\mathrm{d}\pi_i}{2\pi} \right] H[\pi_i, \bar{\varphi}_i] \exp \left\{ \mathrm{i} \int \mathrm{d}^d x \, \pi_i(\mathbf{x}) \left[\varphi_{i+1}(\mathbf{x}) - \varphi_i(\mathbf{x}) \right] \right\}.$$

$$(2.67)$$

Therefore, (2.63) may be represented as

$$\langle \varphi_{i+1} | \mathrm{e}^{-\mathrm{i}H\Delta t} | \varphi_i \rangle = \int \left[\frac{\mathrm{d}\pi_i}{2\pi} \right] \left\{ 1 - \mathrm{i}\,\Delta t \, H[\pi_i, \bar{\varphi}_i] + \mathcal{O}(\Delta t^2) \right\}$$

$$\times \exp \left\{ \mathrm{i} \int \mathrm{d}^d x \, \pi_i(\mathbf{x}) \left([\varphi_{i+1}(\mathbf{x}) - \varphi_i(\mathbf{x})] \right) \right\}$$

$$\simeq \int \left[\frac{\mathrm{d}\pi_i}{2\pi} \right] \exp \left\{ \mathrm{i} \int \mathrm{d}^d x \, \pi_i(\mathbf{x}) \left[\varphi_{i+1}(\mathbf{x}) - \varphi_i(\mathbf{x}) \right] - \mathrm{i}\,\Delta t \, H[\pi_i, \bar{\varphi}_i] \right\}. \quad (2.68)$$

Putting everything together gives

$$\langle \varphi_f, t_f | \varphi_0, t_0 \rangle = \lim_{N \to \infty} \int \prod_{i=1}^{N} [\mathrm{d}\varphi_i] \prod_{j=0}^{N} \left[\frac{\mathrm{d}\pi_i}{2\pi} \right]$$

$$\times \exp \left\{ \mathrm{i}\,\Delta t \sum_{j=0}^{N} \left(\int \mathrm{d}^d x \, \pi_j(\mathbf{x}) \frac{\varphi_{j+1}(\mathbf{x}) - \varphi_j(\mathbf{x})}{\Delta t} - H[\pi_j, \bar{\varphi}_j] \right) \right\}$$

$$\equiv \int_{\varphi(t_0, \mathbf{x}) = \varphi_0(\mathbf{x})}^{\varphi(t_f, \mathbf{x}) = \varphi_f(\mathbf{x})} \mathcal{D}'\varphi \, \mathcal{D}\pi \, \exp \left\{ \mathrm{i} \int_{t_0}^{t_f} \mathrm{d}x^0 \left(\int \mathrm{d}^d x \, \pi(x) \frac{\partial \varphi(x)}{\partial x^0} - H[\pi, \varphi] \right) \right\}.$$

$$(2.69)$$

Above, we have defined $\varphi_{N+1}(\mathbf{x}) = \varphi_f(\mathbf{x})$ and identify $(\varphi_i(\mathbf{x}), t_i) = \varphi(t_i, \mathbf{x})$ employed for the continuum notation. We emphasize that all references to operators are gone in (2.69).

If the Hamiltonian is quadratic in $\pi(x)$, as for a real scalar field theory with

$$H[\pi, \varphi] = \int \mathrm{d}^d x \left[\frac{1}{2}\pi^2 + \frac{1}{2}(\nabla\varphi)^2 + \frac{m^2}{2}\varphi^2 + V(\varphi) \right], \qquad (2.70)$$

with mass parameter m and interaction part $V(\varphi)$, then we can carry out the functional integration over π in (2.69). Completing the squares, we find from the Gaussian integral

$$\int \frac{\mathrm{d}\pi_j(\mathbf{x})}{2\pi} \exp \left[\mathrm{i}\,\pi_j(\varphi_{j+1} - \varphi_j) - \frac{1}{2}\pi_j^2 \Delta t \right] = (2\pi\mathrm{i}\,\Delta t)^{-1/2} \exp \left[\mathrm{i} \frac{(\varphi_{j+1} - \varphi_j)^2}{2\Delta t} \right]$$

$$(2.71)$$

that the net effect is to replace π by the time derivative of φ in the exponential of (2.69). Applying the continuum notation, the matrix element then becomes

$$
(\varphi_f, t_f \,|\, \varphi_0, t_0) = \int_{\varphi(t_0,\mathbf{x})=\varphi_0(\mathbf{x})}^{\varphi(t_f,\mathbf{x})=\varphi_f(\mathbf{x})} \mathscr{D}'\varphi \, e^{iS[\varphi]}, \tag{2.72}
$$

where the classical action reads

$$
S[\varphi] = \int_{t_0}^{t_f} dx^0 \int d^d x \left[\frac{1}{2}\left(\frac{\partial \varphi}{\partial x^0}\right)^2 - \frac{1}{2}(\nabla\varphi)^2 - \frac{m^2}{2}\varphi^2 - V(\varphi)\right]. \tag{2.73}
$$

Here the measure means

$$
\int \mathscr{D}'\varphi = \lim_{N\to\infty} \int \prod_{i=1}^{N}\prod_{\mathbf{x}} d\varphi(t_i, \mathbf{x})\,(2\pi i\,\Delta t)^{-1/2}, \qquad N\Delta t = t_f - t_0. \tag{2.74}
$$

The final time t_f in the above matrix elements is, of course, arbitrary. Applying the same construction to the closed time path of Section 2.2.2, first to the forward piece $\mathcal{C}^+$ and subsequently to the backward piece $\mathcal{C}^-$, is straightforward. Adding to the Hamiltonian appropriate source terms as employed in Section 2.2.2 leads to the generating functional for correlation functions.

2.2.4 Initial conditions

To understand in more detail how the initial density matrix enters calculations, we consider first the example of a *Gaussian density matrix* for a real scalar field theory. For simplicity, we neglect for a moment the spatial dependences, setting $d = 0$, i.e., we consider quantum mechanics. The generalization to higher d will typically be straightforward.[1] In this case, the most general form of a Gaussian density matrix can be parametrized in terms of *five real parameters* ϕ_0, $\dot{\phi}_0$, ξ, η, and σ:

$$
\langle \varphi^+ |\varrho_0| \varphi^- \rangle = \frac{1}{\sqrt{2\pi\xi^2}} \exp\left\{ i\dot{\phi}_0(\varphi_0^+ - \varphi_0^-) - \frac{\sigma^2+1}{8\xi^2}\left[(\varphi_0^+ - \phi_0)^2 + (\varphi_0^- - \phi_0)^2\right]\right.
$$
$$
\left. + i\frac{\eta}{2\xi}\left[(\varphi_0^+ - \phi_0)^2 - (\varphi_0^- - \phi_0)^2\right] + \frac{\sigma^2-1}{4\xi^2}(\varphi_0^+ - \phi_0)(\varphi_0^- - \phi_0)\right\}, \tag{2.75}
$$

whose values will be further discussed below. Gaussianity refers here to the fact that the highest power of $\varphi_0^{\pm}$ appearing in the exponential of (2.75) is two. Therefore, density matrix averages of field operators at initial time only involve Gaussian integrals.

---

[1] We note that for homogeneous field expectation values, taking into account spatial dependences essentially amounts to adding a momentum label in Fourier space (see also below).

In order to interpret the above parameters and to see that we indeed consider the most general Gaussian density matrix, we first note that (2.75) is equivalent to the following set of initial conditions for one- and two-point functions:

$$\phi_0 = \mathrm{Tr}\,\{\varrho_0 \Phi(t)\}|_{t=t_0}, \qquad \dot{\phi}_0 = \mathrm{Tr}\,\{\varrho_0 \partial_t \Phi(t)\}|_{t=t_0}, \tag{2.76}$$

$$\xi^2 = \mathrm{Tr}\,\{\varrho_0 \Phi(t)\Phi(t')\}|_{t=t'=t_0} - \phi_0\phi_0, \tag{2.77}$$

$$\xi\eta = \tfrac{1}{2}\mathrm{Tr}\,\{\varrho_0 [\partial_t \Phi(t)\Phi(t') + \Phi(t)\partial_{t'} \Phi(t')]\}|_{t=t'=t_0} - \dot{\phi}_0\phi_0, \tag{2.78}$$

$$\eta^2 + \frac{\sigma^2}{4\xi^2} = \mathrm{Tr}\,\{\varrho_0 \partial_t \Phi(t)\partial_{t'} \Phi(t')\}|_{t=t'=t_0} - \dot{\phi}_0\dot{\phi}_0. \tag{2.79}$$

In contrast to the symmetrized combination (2.78), we note that the antisymmetrized initial correlator involving the commutator of Φ and $\partial_t \Phi$ at t_0 is not independent, because of the operator commutation relation

$$[\Phi(t), \partial_t \Phi(t)] = \mathrm{i}. \tag{2.80}$$

The equivalence between the initial density matrix and the initial conditions for the correlators can be verified explicitly. For instance,

$$\mathrm{Tr}\,\{\varrho_0\} = \int_{-\infty}^{\infty} \mathrm{d}\varphi_0^+ \, \langle \varphi^+ | \varrho_0 | \varphi^+ \rangle$$

$$= \frac{1}{\sqrt{2\pi\xi^2}} \int_{-\infty}^{\infty} \mathrm{d}\varphi_0^+ \exp\left[-\frac{1}{2\xi^2}(\varphi_0^+ - \phi_0)^2\right] = 1, \tag{2.81}$$

where we note that the right-hand side of (2.75) does not depend on $\dot{\phi}_0$, σ, and η for $\varphi_0^- = \varphi_0^+$. Similarly,

$$\mathrm{Tr}\,\{\varrho_0 \Phi(t_0)\} = \frac{1}{\sqrt{2\pi\xi^2}} \int_{-\infty}^{\infty} \mathrm{d}\varphi_0^+ \, \varphi_0^+ \exp\left[-\frac{1}{2\xi^2}(\varphi_0^+ - \phi_0)^2\right] = \phi_0, \tag{2.82}$$

which is simply obtained by the shift $\varphi_0^+ \to \varphi_0^+ + \phi_0$, etc. In the same way, since only Gaussian integrations appear, we find that all initial n-point functions with $n > 2$ can be expressed in terms of products of the one- and two-point functions.

Of course, *higher initial time derivatives are not independent*, as can be observed from the field equation of motion. For instance, for the above scalar theory Hamiltonian (2.70) with mass m and the interaction part $V(\Phi) = \lambda\Phi^4/4!$, the corresponding field equation reads

$$\langle \partial_t^2 \Phi \rangle = -m^2 \langle \Phi \rangle - \frac{\lambda}{6} \langle \Phi^3 \rangle. \tag{2.83}$$

Since $\langle \Phi^3 \rangle$ is given at the initial time in terms of one- and two-point functions for Gaussian ϱ_0, also second and higher time derivatives are not independent. We conclude that for our case the most general Gaussian density matrix is indeed described by the five parameters appearing in (2.75). In particular, all observable information contained in the density matrix can be conveniently expressed in terms of the correlation functions (2.76)–(2.79).

For further interpretation of the initial conditions, we note that

$$
\begin{aligned}
\mathrm{Tr}\,\{\varrho_0^2\} &= \int_{-\infty}^{\infty} \mathrm{d}\varphi_0^+\, \mathrm{d}\varphi_0^-\, \langle \varphi^+|\,\varrho_0\,|\varphi^-\rangle \langle \varphi^-|\,\varrho_0\,|\varphi^+\rangle \\
&= \frac{1}{2\pi\xi^2} \int_{-\infty}^{\infty} \mathrm{d}\varphi_0^+\, \mathrm{d}\varphi_0^-\, \exp\left\{ -\frac{\sigma^2+1}{4\xi^2}\left[(\varphi_0^+)^2 + (\varphi_0^-)^2\right] + \frac{\sigma^2-1}{2\xi^2}\varphi_0^+\varphi_0^- \right\} \\
&= \frac{1}{\sigma},
\end{aligned}
\tag{2.84}
$$

where again we have shifted the integration variables by ϕ_0 to arrive at the second equality. The latter shows that for $\sigma > 1$, the density matrix describes a mixed state, which may be stated in terms of a nonzero occupation number[2] f with $\sigma = 1 + 2f$. For $\sigma = 1$, the "mixing term" in (2.75) is absent and we obtain a pure-state density matrix of the product form

$$
\varrho_0 = |\Psi\rangle\langle\Psi|,
\tag{2.85}
$$

with Schrödinger wavefunction

$$
\langle \varphi^+|\Psi\rangle = \frac{1}{(2\pi\xi^2)^{1/4}} \exp\left[i\dot{\phi}_0\varphi_0^+ - \left(\frac{1}{4\xi^2} + i\frac{\eta}{2\xi}\right)(\varphi_0^+ - \phi_0)^2 \right].
\tag{2.86}
$$

In order to go beyond Gaussian initial density matrices and to field theory in d spatial dimensions, we may generalize the above example and parametrize the most general density matrix as

$$
\langle \varphi^+|\varrho_0|\varphi^-\rangle = \mathcal{N}e^{ih_{\mathcal{C}}[\varphi]},
\tag{2.87}
$$

with normalization $\mathcal{N}$ and $h_{\mathcal{C}}[\varphi]$ expanded in powers of the fields:

$$
\begin{aligned}
h_{\mathcal{C}}[\varphi] &= \alpha_0 + \int_{x,\mathcal{C}} \alpha_1(x)\varphi(x) + \frac{1}{2}\int_{xy,\mathcal{C}} \alpha_2(x,y)\varphi(x)\varphi(y) \\
&\quad + \frac{1}{3!}\int_{xyz,\mathcal{C}} \alpha_3(x,y,z)\varphi(x)\varphi(y)\varphi(z) \\
&\quad + \frac{1}{4!}\int_{xyzw,\mathcal{C}} \alpha_4(x,y,z,w)\varphi(x)\varphi(y)\varphi(z)\varphi(w) + \ldots
\end{aligned}
\tag{2.88}
$$

Here the integrals have to be evaluated along the forward piece, $\mathcal{C}^+$, and the backward piece, $\mathcal{C}^-$, of the closed time path with $\varphi^+(t_0, \mathbf{x}) = \varphi_0^+(\mathbf{x})$ and $\varphi^-(t_0, \mathbf{x}) = \varphi_0^-(\mathbf{x})$ as described in Sections 2.2.2 and 2.2.3. Since the density matrix ϱ_0 is specified at time t_0

---

[2] For the special case of thermal equilibrium, the occupation number is given by the Bose–Einstein distribution.

only, all time integrals in (2.88) can contribute only at the endpoints of the closed time contour. As a consequence, the coefficients $\alpha_1(x)$, $\alpha_2(x,y)$, $\alpha_3(x,y,z)$, ... vanish identically for times different than t_0. More precisely, we have

$$\int_{x,\mathcal{C}} \alpha_1(x)\varphi(x) \equiv \int d^d x \left[\alpha_1^+(\mathbf{x})\,\varphi_0^+(\mathbf{x}) + \alpha_1^-(\mathbf{x})\,\varphi_0^-(\mathbf{x})\right],$$

$$\int_{xy,\mathcal{C}} \alpha_2(x,y)\varphi(x)\varphi(y) \equiv \int d^d x\, d^d y \left[\alpha_2^{++}(\mathbf{x},\mathbf{y})\varphi_0^+(\mathbf{x})\varphi_0^+(\mathbf{y})\right.$$

$$+ \alpha_2^{+-}(\mathbf{x},\mathbf{y})\,\varphi_0^+(\mathbf{x})\varphi_0^-(\mathbf{y})$$

$$+ \alpha_2^{-+}(\mathbf{x},\mathbf{y})\,\varphi_0^-(\mathbf{x})\varphi_0^+(\mathbf{y})$$

$$\left. + \alpha_2^{--}(\mathbf{x},\mathbf{y})\,\varphi_0^-(\mathbf{x})\varphi_0^-(\mathbf{y})\right] \tag{2.89}$$

and so on. We note that α_0 appearing in (2.88) is an irrelevant constant that can be taken into account by a rescaling of the overall normalization $\mathcal{N}$ in (2.87). For a physical density matrix, the other coefficients are, of course, not arbitrary. Hermiticity of the density matrix, $\varrho_0 = \varrho_0^\dagger$, implies $i\alpha_1^+ = (i\alpha_1^-)^*$, $i\alpha_2^{++} = (i\alpha_2^{--})^*$, $i\alpha_2^{+-} = (i\alpha_2^{-+})^*$, etc. If the initial state is invariant under symmetries, there are further constraints. For example, for an initial state that is invariant under $\Phi \to -\Phi$, all $\alpha_n(x_1,\ldots,x_n)$ with odd n vanish. If the initial state is homogeneous in space, the $\alpha_n(x_1,\ldots,x_n)$ are invariant under space translations $\mathbf{x}_i \to \mathbf{x}_i + \mathbf{a}$ for $i = 1,\ldots,n$ and arbitrary spatial vector $\mathbf{a}$. In this case, they can be conveniently described in spatial momentum space:

$$\alpha_n^{\pm\cdots}(\mathbf{x}_1,\ldots,\mathbf{x}_n)$$

$$= \int \frac{d^d p_1}{(2\pi)^d} \cdots \frac{d^d p_n}{(2\pi)^d} \exp\left(i\sum_{j=1}^n \mathbf{p}_j \mathbf{x}_j\right)(2\pi)^d \delta(\mathbf{p}_1 + \ldots + \mathbf{p}_n)\alpha_n^{\pm\cdots}(\mathbf{p}_1,\ldots,\mathbf{p}_n). \tag{2.90}$$

Using the parametrization (2.87) and (2.88) for the most general initial density matrix, we observe that the generating functional (2.54) introduced above may be written as

$$Z[J,R;\varrho_0] = \int \mathcal{D}\varphi \exp\left\{i\left[S[\varphi] + \int_{x,\mathcal{C}} J(x)\varphi(x) + \frac{1}{2}\int_{xy,\mathcal{C}} R(x,y)\varphi(x)\varphi(y)\right.\right.$$

$$\left.\left. + \frac{1}{3!}\int_{xyz,\mathcal{C}} \alpha_3(x,y,z)\varphi(x)\varphi(y)\varphi(z) + \ldots\right]\right\}. \tag{2.91}$$

Here we have neglected an irrelevant normalization constant and rescaled the sources in (2.54) as $J(x) + \alpha_1(x) \to J(x)$ and $R(x,y) + \alpha_2(x,y) \to R(x,y)$. The sources J and R may therefore be conveniently used to absorb the lower linear and quadratic contributions coming from the density matrix specifying the initial state. This exploits the fact that *the initial density matrix is completely encoded in terms of initial-time sources* for the functional integral.

The generating functional (2.91) can be used to describe situations involving initial density matrices for arbitrarily complex physical situations. However, often the initial conditions of an experiment may be described by only a few lowest n-point functions. For many practical purposes, the initial density matrix is well described by a Gaussian one. For instance, the initial conditions for the reheating dynamics in the early universe at the end of inflation are described by a Gaussian density matrix to rather good accuracy. Also many effective theories for the description of nonequilibrium dynamics, such as Boltzmann equations, exploit an assumption about Gaussianity of initial conditions.

From (2.91), we observe that for Gaussian initial density matrices, for which $\alpha_i \equiv 0$ for $i \geq 3$, we have

$$Z\big[J, R; \varrho_0^{(\text{Gauss})}\big] \to Z[J, R]. \tag{2.92}$$

As a consequence, in this case, the nonequilibrium generating functional corresponds to the generating functional with linear and bilinear source terms introduced in (2.34) for a closed time path.

In field theory, non-Gaussian initial density matrices pose no problems in principle, but they do require that additional initial-time sources be taken into account. An alternative to initial-time sources is to represent a non-Gaussian initial density matrix with the help of an additional imaginary piece of the time contour preceding the closed time path $\mathcal{C}$. We refer to the literature at the end of this section for further details concerning this alternative representation.

2.2.5 Effective actions

The functional $Z[J, R; \varrho_0]$ given by (2.91) is the nonequilibrium quantum field theoretical generalization of the thermodynamic partition function in the presence of source terms. In thermodynamics, Legendre transforms of the logarithm of the partition function lead to equivalent descriptions of the physics. Similarly, in nonequilibrium quantum field theory, Legendre transforms with respect to the various source terms lead to different representations of a free energy functional Γ. The Legendre transform with respect to the linear source term, $\sim J$, will lead to the so-called one-particle irreducible (1PI) effective action $\Gamma[\phi]$ parametrized by the one-point function ϕ. An additional Legendre transform with respect to the bilinear source term, $\sim R$, gives the two-particle irreducible (2PI) effective action $\Gamma[\phi, G]$, which is parametrized in addition by the two-point function or propagator G. In the presence of higher source terms, this procedure can, in principle, be continued to n-particle irreducible (nPI) effective actions up to arbitrarily high n.

In the absence of approximations, all these generating functionals give equivalent descriptions of the physics and any choice is only a matter of convenience or efficiency. For nonequilibrium initial-value problems, it is often particularly helpful to employ effective actions, since they are described in terms of correlation functions whose initial values are typically much more easily accessible than those of source terms. Furthermore, the ability to find suitable approximation schemes in practice can depend strongly on the appropriate choice of the functional representation. In Section 2.3, it will be demonstrated how 2PI effective actions can be applied to describe important processes such as thermalization in quantum field theory. Later sections of this chapter also include discussions of further applications, such as nonequilibrium instabilities and an efficient derivation of kinetic theory in its range of validity.

In this section, we discuss the construction of 1PI and 2PI effective actions. The latter provides a powerful starting point for approximations in nonequilibrium quantum field theory with Gaussian initial density matrix as specified in Section 2.2.4. It should be emphasized again that the use of a 2PI effective action for Gaussian initial density matrices represents a priori no approximation for the dynamics—in an interacting quantum field theory, higher irreducible correlations build up in general corresponding to non-Gaussian density matrices for times $t > t_0$. It only restricts the possible setup described by the initial conditions for correlation functions.

To be able to explain a variety of approximation schemes in the following, we consider here the example of a quantum field theory for a real N-component scalar field φ_a ($a = 1, \ldots, N$) with $O(N)$-symmetric classical action

$$S[\varphi] = \int_{x,\mathcal{C}} \left\{ \frac{1}{2} \partial^\mu \varphi_a(x) \partial_\mu \varphi_a(x) - \frac{m^2}{2} \varphi_a(x)\varphi_a(x) - \frac{\lambda}{4!N} [\varphi_a(x)\varphi_a(x)]^2 \right\}. \quad (2.93)$$

Here summation over repeated indices is implied, as well as the integration over the closed time path $\mathcal{C}$ as introduced in Section 2.2.2. Starting from $Z[J,R]$, given by (2.91) with $\alpha_i \equiv 0$ for $i \geq 3$, we write

$$Z[J,R] \equiv \exp\left(iW[J,R] \right)$$
$$= \int \mathcal{D}\varphi \, \exp\left\{ i \left[S[\varphi] + \int_{x,\mathcal{C}} J_a(x)\varphi_a(x) + \frac{1}{2} \int_{xy,\mathcal{C}} R_{ab}(x,y)\varphi_a(x)\varphi_b(y) \right] \right\},$$
$$(2.94)$$

with real $J_a(x)$, and $R_{ab}(x,y) = R_{ba}(y,x)$.

The *macroscopic field* ϕ_a and the *connected two-point function* G_{ab} in the presence of the source terms $\sim J_a(x)$ and $\sim R_{ab}(x,y)$ can be obtained by variation of $W[J,R]$:

$$\frac{\delta W[J,R]}{\delta J_a(x)} = \frac{-i \, \delta \ln Z[J,R]}{\delta J_a(x)} = \frac{1}{Z[J,R]} \frac{\delta Z[J,R]}{i \, \delta J_a(x)} \equiv \phi_a(x), \quad (2.95)$$

$$\frac{\delta W[J,R]}{\delta R_{ab}(x,y)} \equiv \frac{1}{2}[\phi_a(x)\phi_b(y) + G_{ab}(x,y)]. \quad (2.96)$$

These definitions are in accordance with (2.38) and (2.40) for vanishing sources, i.e., $J, R = 0$. However, we keep J and R nonvanishing in (2.96) and, to ease the notation, the dependence of $\phi = \phi(J,R)$ and $G = G(J,R)$ on the sources is suppressed.

Before constructing the 2PI effective action, we consider first the *1PI effective action*. It is constructed via a Legendre transform with respect to the source term linear in the field:

$$\Gamma^R[\phi] = W[J,R] - \int_{x,\mathcal{C}} \frac{\delta W[J,R]}{\delta J_a(x)} J_a(x) = W[J,R] - \int_{x,\mathcal{C}} J_a(x)\phi_a(x). \quad (2.97)$$

To complete this construction, we note that the relation between ϕ and J is R-dependent, i.e., inverting $\phi = \delta W[J, R]/\delta J$ yields $J = J^R(\phi)$ on the right-hand side of (2.97). Taking the functional derivative with respect to the field, we find

$$\frac{\delta \Gamma^R[\phi]}{\delta \phi_a(x)} = \int_{y,\mathcal{C}} \frac{\delta W[J, R]}{\delta J_b(y)} \frac{\delta J_b(y)}{\delta \phi_a(x)} - \int_{y,\mathcal{C}} \phi_b(y) \frac{\delta J_b(y)}{\delta \phi_a(x)} - J_a(x) = -J_a(x), \qquad (2.98)$$

and the source derivative yields

$$\frac{\delta \Gamma^R[\phi]}{\delta R_{ab}(x, y)} = \frac{\delta W[J, R]}{\delta R_{ab}(x, y)} + \int_{z,\mathcal{C}} \frac{\delta W[J, R]}{\delta J_c(z)} \frac{\delta J_c(z)}{\delta R_{ab}(x, y)} - \int_{z,\mathcal{C}} \phi_c(z) \frac{\delta J_c(z)}{\delta R_{ab}(x, y)}$$

$$= \frac{\delta W[J, R]}{\delta R_{ab}(x, y)}. \qquad (2.99)$$

The second equalities in both (2.98) and (2.99) arise from cancellations using (2.95).

The standard 1PI effective action $\Gamma[\phi]$ is obtained from (2.97) for $R = 0$, i.e., $\Gamma[\phi] = \Gamma^{R=0}[\phi]$. For nonzero R, the functional $\Gamma^R[\phi]$ may be viewed as the 1PI effective action for a theory governed by a modified classical action $S^R[\varphi]$, with

$$S^R[\varphi] \equiv S[\varphi] + \frac{1}{2} \int_{xy,\mathcal{C}} R_{ab}(x, y) \varphi_a(x) \varphi_b(y). \qquad (2.100)$$

As a consequence, it is straightforward to recover for $\Gamma^R[\phi]$ "textbook" relations for the 1PI effective action taking into account R. For instance, evaluating $\Gamma^R[\phi]$ in a saddle-point approximation ("one-loop") around the macroscopic field configuration $\phi_a(x)$, we find, up to irrelevant constants,

$$\Gamma^{R(\text{1loop})}[\phi] = S^R[\phi] + \frac{i}{2} \text{Tr}_\mathcal{C} \ln[G_0^{-1}(\phi) - iR], \qquad (2.101)$$

where the trace involves the integration over space–time coordinates on the closed time path $\mathcal{C}$ as well as summation over field indices. We derive (2.101) below, which is indeed the standard one-loop result for the 1PI effective action with $S[\phi] \to S^R[\phi]$ and $G_0^{-1}(\phi) \to G_0^{-1}(\phi) - iR$. Here the classical inverse propagator $iG_{0,ab}^{-1}(x, y; \phi) \equiv \delta^2 S[\phi]/\delta\phi_a(x)\delta\phi_b(y)$ for the action (2.93) reads

$$iG_{0,ab}^{-1}(x, y; \phi) = -\left[\Box_x + m^2 + \frac{\lambda}{6N}\phi_c(x)\phi_c(x)\right]\delta_{ab}\delta_\mathcal{C}(x - y)$$

$$- \frac{\lambda}{3N}\phi_a(x)\phi_b(x)\delta_\mathcal{C}(x - y). \qquad (2.102)$$

It is instructive to compare this with the result obtained from the approximation (2.101) for the propagator by taking the functional derivative with respect to R: $\delta\Gamma^{R(\text{1loop})}[\phi]/\delta R = \frac{1}{2}\{\phi\phi + [G_0^{-1}(\phi) - iR]^{-1}\}$. Since this has to agree with $\frac{1}{2}(\phi\phi + G)$ using (2.99) and the definition (2.97), we obtain $G^{-1} = G_0^{-1}(\phi) - iR$ in this approximation.

We now perform a further Legendre transform of $\Gamma^R[\phi]$ with respect to the source R to arrive at the *2PI effective action*, $\Gamma[\phi, G]$, which is a functional of the field $\phi_a(x)$ and the propagator $G_{ab}(x, y)$:

$$\Gamma[\phi, G] = \Gamma^R[\phi] - \int_{xy,\mathcal{C}} \underbrace{\frac{\delta \Gamma^R[\phi]}{\delta R_{ab}(x, y)}}_{\frac{\delta W[J,R]}{\delta R_{ab}(x,y)} = \frac{1}{2}[\phi_a(x)\phi_b(y) + G_{ab}(x,y)]} R_{ab}(x, y)$$

$$= \Gamma^R[\phi] - \frac{1}{2} \int_{xy,\mathcal{C}} [\phi_a(x)\phi_b(y) + G_{ab}(x, y)] R_{ab}(x, y). \qquad (2.103)$$

Here we have used first (2.99) and then the definition (2.96). Of course, the two subsequent Legendre transforms, which have been used to arrive at (2.103), agree with a simultaneous Legendre transform of $W[J, R]$ with respect to both source terms:

$$\Gamma[\phi, G] = W[J, R] - \int_{x,\mathcal{C}} \frac{\delta W[J, R]}{\delta J_a(x)} J_a(x) - \int_{xy,\mathcal{C}} \frac{\delta W[J, R]}{\delta R_{ab}(x, y)} R_{ab}(x, y)$$

$$= W[J, R] - \int_{x,\mathcal{C}} \phi_a(x) J_a(x) - \frac{1}{2} \int_{xy,\mathcal{C}} [\phi_a(x)\phi_b(y) + G_{ab}(x, y)] R_{ab}(x, y). \qquad (2.104)$$

Taking the field and propagator dependence of the sources into account, we obtain from (2.104), after cancellations, the stationarity conditions

$$\frac{\delta \Gamma[\phi, G]}{\delta \phi_a(x)} = -J_a(x) - \int_{y,\mathcal{C}} R_{ab}(x, y)\phi_b(y), \qquad (2.105)$$

$$\frac{\delta \Gamma[\phi, G]}{\delta G_{ab}(x, y)} = -\frac{1}{2} R_{ab}(x, y), \qquad (2.106)$$

which are the *quantum equations of motion for ϕ and G*.

To gain familiarity with the 2PI effective action, we note that the approximate expression

$$\Gamma^{(1\text{loop})}[\phi, G] = S[\phi] + \frac{i}{2}\text{Tr}_{\mathcal{C}} \ln G^{-1} + \frac{i}{2}\text{Tr}_{\mathcal{C}}\{[G_0^{-1}(\phi) - G^{-1}]G\}$$

$$= \Gamma^{R(1\text{loop})}[\phi] - \frac{1}{2} \int_{xy,\mathcal{C}} [\phi_a(x)\phi_b(y) + G_{ab}(x, y)] R_{ab}(x, y) \qquad (2.107)$$

indeed corresponds to the Legendre transform of the one-loop result (2.101) for the 1PI effective action $\Gamma^R[\phi]$ as written in the second line of (2.107). To see this, we first observe that the stationarity equation (2.106) yields in this case

$$\frac{\delta \Gamma^{(1\text{loop})}[\phi, G]}{\delta G_{ab}(x, y)} = -\frac{i}{2}G_{ab}^{-1}(x, y) + \frac{i}{2}G_{0,ab}^{-1}(x, y; \phi) = -\frac{1}{2}R_{ab}(x, y), \qquad (2.108)$$

or $G^{-1} = G_0^{-1}(\phi) - iR$. Using this to replace $G_0^{-1}(\phi) - G^{-1}$ in the first line of (2.107) yields, with the definitions (2.100) and (2.103), the result.

To go beyond this approximation, it is convenient to write the exact $\Gamma[\phi, G]$ as the one-loop-type expression in the first line of (2.107) and the *"rest,"* $\Gamma_2[\phi, G]$:

$$\boxed{\Gamma[\phi, G] = S[\phi] + \frac{i}{2} \mathrm{Tr}\, _c \ln G^{-1} + \frac{i}{2} \mathrm{Tr}\, _c G_0^{-1}(\phi) G + \Gamma_2[\phi, G] + \mathrm{const.}} \qquad (2.109)$$

Here we have written $\mathrm{Tr}\, _c G^{-1} G$ as an irrelevant constant that can be adjusted for normalization. To get an understanding of $\Gamma_2[\phi, G]$, we vary this expression with respect to G, which yields

$$G_{ab}^{-1}(x, y) = G_{0,ab}^{-1}(x, y; \phi) - iR_{ab}(x, y) - \Sigma_{ab}(x, y; \phi, G), \qquad (2.110)$$

where we have defined

$$\boxed{\Sigma_{ab}(x, y; \phi, G) \equiv 2i \frac{\delta \Gamma_2[\phi, G]}{\delta G_{ab}(x, y)}.} \qquad (2.111)$$

This gives the *proper self-energy*, $\Sigma_{ab}(x, y; \phi, G)$, in terms of a functional derivative of $\Gamma_2[\phi, G]$ with respect to $G_{ab}(x, y)$.

Inverting (2.110), we may express for further interpretation the full propagator G as an infinite series using a compact matrix notation:

$$G = (G_0^{-1} - iR)^{-1} + (G_0^{-1} - iR)^{-1} \Sigma (G_0^{-1} - iR)^{-1}$$
$$+ (G_0^{-1} - iR)^{-1} \Sigma (G_0^{-1} - iR)^{-1} \Sigma (G_0^{-1} - iR)^{-1} + \dots \qquad (2.112)$$

A term is called *one-particle irreducible* (1PI) if it remains connected when an arbitrary internal propagator $(G_0^{-1} - iR)^{-1}$ is removed. For instance, removing in the expression $(G_0^{-1} - iR)^{-1} \Sigma (G_0^{-1} - iR)^{-1} \Sigma (G_0^{-1} - iR)^{-1}$ the inner propagator, we end up with two disconnected pieces $(G_0^{-1} - iR)^{-1} \Sigma$ and $\Sigma (G_0^{-1} - iR)^{-1}$. Therefore, that contribution is *one-particle reducible*. If we were to start classifying all one-particle reducible and irreducible structures contributing to G accordingly, we would end up with the series (2.112) with the self-energy Σ containing all 1PI contributions. If this were not the case, then we could always write G as the series in (2.112) plus a correction ΔG. However, since the sum of the series is known to be given by (2.110), this correction must be zero, i.e., $\Delta G = 0$.

Most importantly, from the fact that $\Sigma(\phi, G)$ contains only 1PI contributions, we can conclude the important property of $\Gamma_2[\phi, G]$ that it only contains *two-particle irreducible* (2PI) contributions with respect to the full propagator G. A contribution is said to be two-particle irreducible if it does not become disconnected by removing two inner propagators. Suppose $\Gamma_2[\phi, G]$ had a *two-particle reducible* contribution. The latter could be written as $\tilde{\Gamma} G G \tilde{\Gamma}'$, where GG denotes in matrix notation two propagators connecting two parts $\tilde{\Gamma}$ and $\tilde{\Gamma}'$ of a contribution. Then $\Sigma(\phi, G)$ would contain a contribution of the form $\tilde{\Gamma} G \tilde{\Gamma}'$, since it is given by a derivative of Γ_2 with respect to G. Such a structure is *one-particle reducible* and cannot occur for the proper self-energy. Therefore, two-particle reducible contributions to $\Gamma_2[\phi, G]$ have to be absent.

This can be conveniently illustrated using a diagrammatic language, where propagators are associated with lines that can meet in the presence of interactions. For instance, for a quartic interaction, this is exemplified for a two- and a three-loop graph contributing to Γ_2 below. Diagrammatically, the graphs contributing to $\Sigma(\phi, G)$ are obtained by opening one propagator line in graphs contributing to $\Gamma_2[\phi, G]$, where the corresponding contributions are shown schematically on the right:

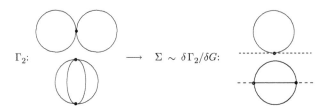

In these diagrams, all lines are associated with the propagator G. Using the expansion (2.112), we see that each diagram corresponds to an infinite series of diagrams with lines associated with the "classical" propagator $(G_0^{-1} - iR)^{-1}$, and the above two- and three-loop diagrams contain, for example, so-called "daisies" and "ladder" resummations:

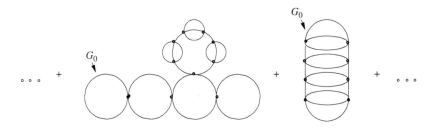

The diagrammatic loop expansion of the 2PI effective action will be discussed in detail in Section 2.3.

Details of the one-loop effective action

We derive the approximate result (2.101) for the effective action $\Gamma^R[\phi]$ using a saddle-point approximation. Starting from the Legendre transform (2.97), we have

$$\Gamma^R[\phi] = -i \ln Z[J, R] - \int_x J_a(x)\phi_a(x). \tag{2.113}$$

Using the defining functional integral (2.94), this can be written as

$$
e^{i\Gamma^R[\phi]} = \int \mathcal{D}\varphi \exp\left\{ i\left[S^R[\varphi] + \int_x J_a(x)[\varphi_a(x) - \phi_a(x)] \right] \right\}
$$

$$
= \int \mathcal{D}\varphi \exp\left\{ i\left[S^R[\phi + \varphi] + \int_x J_a(x)\varphi_a(x) \right] \right\}
$$

$$
= e^{iS^R[\phi]} \int \mathcal{D}\varphi \exp\left\{ i\left[S^R[\phi + \varphi] - S^R[\phi] + \int_x J_a(x)\varphi_a(x) \right] \right\}, \quad (2.114)
$$

where for the second equality we have employed a field shift

$$
\varphi_a(x) \to \varphi_a(x) + \phi_a(x). \quad (2.115)
$$

Separating the classical part, we may write

$$
\Gamma^R[\phi] = S^R[\varphi] + \Gamma_1^R[\phi], \quad (2.116)
$$

where the fluctuation part reads

$$
\Gamma_1^R[\phi] = -i \ln \int \mathcal{D}\varphi \exp\left\{ i\left[S^R[\phi + \varphi] - S^R[\phi] + \int_x J_a(x)\varphi_a(x) \right] \right\}. \quad (2.117)
$$

Furthermore, using the definitions (2.100) and (2.102), we can verify that

$$
S^R[\phi + \varphi] - S^R[\phi] = \frac{1}{2} \int_{xy} \varphi_a(x)\left[iG_{0,ab}^{-1}(x, y; \phi) + R_{ab}(x, y) \right]\varphi_b(y)
$$

$$
+ S_{\text{int}}[\varphi, \phi] + \int_x \varphi_a(x)\frac{\delta S^R[\phi]}{\delta \phi_a(x)}. \quad (2.118)
$$

Here we have introduced the interaction term

$$
S_{\text{int}}[\varphi, \phi] = -\frac{\lambda}{6N} \int_x \phi_a(x)\varphi_a(x)\varphi_b(x)\varphi_b(x) - \frac{\lambda}{4!N} \int_x [\varphi_a(x)\varphi_a(x)]^2. \quad (2.119)
$$

Starting from a classical action (2.93) with a quartic self-interaction, we observe from (2.119) that an effective cubic interaction term appears in the functional integral for the effective action in the presence of a nonzero field ϕ.

From plugging (2.118) into (2.117), we obtain with the help of the field equation (2.98), which reads

$$
\frac{\delta S^R[\phi]}{\delta \phi_a(x)} + \frac{\delta \Gamma_1^R[\phi]}{\delta \phi_a(x)} = -J_a(x), \quad (2.120)
$$

an exact functional integro-differential equation for $\Gamma_1^R[\phi]$:

$$\Gamma_1^R[\phi] = -\mathrm{i}\ln \int \mathscr{D}\varphi \exp\left\{\mathrm{i}\left[\frac{1}{2}\int_{xy}\varphi_a(x)\left[\mathrm{i}G_{0,ab}^{-1}(x,y;\phi)+R_{ab}(x,y)\right]\varphi_b(y)\right.\right.$$

$$\left.\left. + S_{\mathrm{int}}[\varphi,\phi] - \int_x\varphi_a(x)\frac{\delta\Gamma_1^R[\phi]}{\delta\phi_a(x)}\right]\right\}. \tag{2.121}$$

The exponential in the integrand contains a term that is quadratic in the fields, a contribution S_{int} with cubic and quartic field powers, and a linear source-like term $\sim \delta\Gamma_1^R[\phi]/\delta\phi$. The linear term represents a so-called tadpole, which guarantees a vanishing expectation value of the fluctuating field, $\langle\varphi\rangle_{J,R} = 0$, after the shift (2.115). Its role is taken into account by taking ϕ to describe the macroscopic field. If we neglect the cubic and quartic field powers contained in S_{int}, we arrive at the approximate saddle-point or one-loop expression

$$\Gamma_1^{R(1\mathrm{loop})}[\phi] = -\mathrm{i}\ln \int \mathscr{D}\varphi \exp\left\{-\frac{1}{2}\int_{xy}\varphi_a(x)\left[G_{0,ab}^{-1}(x,y;\phi)-\mathrm{i}R_{ab}(x,y)\right]\varphi_b(y)\right\}$$

$$= -\mathrm{i}\ln\left\{\det\left[G_0^{-1}(\phi)-\mathrm{i}R\right]\right\}^{-1/2} + \mathrm{const}$$

$$= \frac{\mathrm{i}}{2}\mathrm{Tr}\left\{\ln\left[G_0^{-1}(\phi)-\mathrm{i}R\right]\right\} + \mathrm{const}, \tag{2.122}$$

which leads to (2.101) up to an irrelevant constant.

2.2.6   Nonequilibrium evolution equations

Propagator evolution equations

The functional representation of the nonequilibrium 2PI effective action, $\Gamma[\phi,G]$, employs a one-point function ϕ and a two-point function G, whose physical values may be computed for all times of interest starting from the initial time t_0. The equations of motion are given by the stationarity conditions (2.105) and (2.106). We will consider first the scalar field theory in the symmetric regime, where $\Gamma[\phi=0,G] \equiv \Gamma[G]$ is sufficient. We come back to the case of a nonvanishing field expectation value and consider also additional field degrees of freedom, including fermionic and gauge fields, later.

The stationarity condition (2.106) for the propagator equation leads to (2.110), which reads $G_{ab}^{-1}(x,y) = G_{0,ab}^{-1}(x,y) - \Sigma_{ab}(x,y) - \mathrm{i}R_{ab}(x,y)$. Alternatively to (2.112), we may invert this equation by writing

$$G_{ab}(x,y) = G_{0,ab}(x,y) + \int_{zw,\mathcal{C}} G_{0,ac}(x,z)\left[\Sigma_{cd}(z,w)+\mathrm{i}R_{cd}(z,w)\right]G_{db}(w,y). \tag{2.123}$$

This can be directly verified with the help of (2.110) by acting with G^{-1} on the left of (2.123) to get $GG^{-1} = \mathbb{1}$ and with $G_0^{-1} - \Sigma - \mathrm{i}R$ on the right of (2.123) to get, in matrix notation,

$$[G_0 + G_0(\Sigma + \mathrm{i}R)G](G_0^{-1} - \Sigma - \mathrm{i}R)$$
$$= G_0 G_0^{-1} - G_0(\Sigma + \mathrm{i}R) + G_0(\Sigma + \mathrm{i}R)G \underbrace{(G_0^{-1} - \Sigma - \mathrm{i}R)}_{G^{-1}} = \mathbb{1}. \quad (2.124)$$

Following Section 2.2.4, the initial Gaussian density matrix leads to a nonzero source term $R(x, y)$ at initial times $x^0, y^0 = t_0$ only. Therefore, (2.123) may be used to relate source terms to initial values of the propagator and its first derivatives for the relativistic theory (2.93). However, typically, the initial values for correlation functions are known and it would be favorable to prescribe them directly without recourse to sources. In particular, we may rewrite the equation of motion as a *partial differential equation*. For instance, starting from (2.110), convolution with G using $\int_{z,\mathcal{C}} G_{ac}^{-1}(x, z) G_{cb}(z, y) = \delta_{ab} \delta_{\mathcal{C}}(x - y)$ leads to

$$\int_{z,\mathcal{C}} G_{0,ac}^{-1}(x, z) G_{cb}(z, y) - \int_{z,\mathcal{C}} \left[\Sigma_{ac}(x, z) + \mathrm{i}R_{ac}(x, z) \right] G_{cb}(z, y) = \delta_{ab} \delta_{\mathcal{C}}(x - y).$$
$$(2.125)$$

In the absence of a macroscopic field, we have

$$\mathrm{i}G_{0,ab}^{-1}(x, y) = -(\Box_x + m^2)\delta_{ab}\delta_{\mathcal{C}}(x - y), \quad (2.126)$$

and we obtain the evolution equation for the propagator for $\phi = 0$:

$$\boxed{\left(\Box_x + m^2 \right) G_{ab}(x, y) + \mathrm{i} \int_{z,\mathcal{C}} \left[\Sigma_{ac}(x, z) + \mathrm{i}R_{ac}(x, z) \right] G_{cb}(z, y) = -\mathrm{i}\delta_{ab}\delta_{\mathcal{C}}(x - y),}$$
$$(2.127)$$

which is an exact equation for known self-energy Σ. Since a source term representing the initial density matrix has support at initial time only, it gives a vanishing contribution to $\int_{z,\mathcal{C}} R_{ac}(x, z) G_{cb}(z, y)$ for $x^0 > t_0$. To solve this equation for some given approximation of $\Sigma_{ab}(x, y)$, the values for $G_{ab}(x, y)$ and first derivatives have to be prescribed at $x^0, y^0 = t_0$. Then (2.127) is used to evolve $G_{ab}(x, y)$ for times larger than t_0. We use this to describe in the following the evolution equation for physical propagators, setting $R = 0$ in (2.127).

It is convenient to introduce a decomposition of the two-point function G into spectral and statistical components. The corresponding evolution equations for the spectral function and statistical propagator are fully equivalent to the evolution equation for G, but have a simple physical interpretation. While the spectral function encodes the spectrum of the theory, the statistical propagator gives information about occupation numbers. Loosely speaking, the decomposition makes explicit what states are available and how often they are occupied. This interpretation will become clearer as we proceed.

For the neutral scalar field theory, there are two linearly independent real-valued two-point functions, which may be associated to the real and the imaginary part of G. More precisely, we consider the expectation value of the commutator and the anticommutator of two fields,

$$\text{commutator:} \qquad \rho_{ab}(x,y) = i\langle[\Phi_a(x), \Phi_b(y)]\rangle, \qquad (2.128)$$

$$\text{anticommutator } (\phi = 0): \qquad F_{ab}(x,y) = \tfrac{1}{2}\langle\{\Phi_a(x), \Phi_b(y)\}\rangle. \qquad (2.129)$$

Here $\rho_{ab}(x,y)$ denotes the *spectral function* and $F_{ab}(x,y)$ the *statistical two-point function*. For neutral scalar fields, the real functions obey $F_{ab}(x,y) = F_{ba}(y,x)$ and $\rho_{ab}(x,y) = -\rho_{ba}(y,x)$. We note from (2.128) that the spectral function ρ encodes the equal-time commutation relations

$$\rho_{ab}(x,y)|_{x^0 = y^0} = 0, \qquad \partial_{x^0}\rho_{ab}(x,y)|_{x^0 = y^0} = \delta_{ab}\delta(\mathbf{x} - \mathbf{y}). \qquad (2.130)$$

The spectral function is also directly related to the retarded or advanced propagator, $G_{ab}^R(x,y) = \rho_{ab}(x,y)\Theta(x^0 - y^0) = G_{ab}^A(y,x)$, respectively.

The decomposition identity for spectral and statistical components of the propagator in the absence of sources reads

$$\boxed{G_{ab}(x,y) = F_{ab}(x,y) - \frac{i}{2}\rho_{ab}(x,y)\text{sgn}_\mathcal{C}(x^0 - y^0).} \qquad (2.131)$$

Here $\text{sgn}_\mathcal{C}(x^0 - y^0) = \theta_\mathcal{C}(x^0 - y^0) - \theta_\mathcal{C}(y^0 - x^0)$, using the contour step function introduced in (2.41). With this, the above decomposition is directly understood as

$$
\begin{aligned}
G_{ab}(x,y) \overset{(\phi=0)}{=} & \langle\Phi_a(x)\Phi_b(y)\rangle\theta_\mathcal{C}(x^0 - y^0) + \langle\Phi_b(y)\Phi_a(x)\rangle\theta_\mathcal{C}(y^0 - x^0) \\
= & \frac{1}{2}\langle\{\Phi_a(x), \Phi_b(y)\}\rangle\left(\theta_\mathcal{C}(x^0 - y^0) + \theta_\mathcal{C}(y^0 - x^0)\right) \\
& - \frac{i}{2}i\langle[\Phi_a(x), \Phi_b(y)]\rangle\underbrace{\left(\theta_\mathcal{C}(x^0 - y^0) - \theta_\mathcal{C}(y^0 - x^0)\right)}_{\text{sgn}_\mathcal{C}(x^0-y^0)}.
\end{aligned} \qquad (2.132)
$$

Making the contour ordering explicit, we distinguish the four propagators

$$
\begin{aligned}
G_{ab}^{++}(x,y) &= F_{ab}(x,y) - \frac{i}{2}\rho_{ab}(x,y)\,\text{sgn}(x^0 - y^0), \\[4pt]
G_{ab}^{--}(x,y) &= F_{ab}(x,y) + \frac{i}{2}\rho_{ab}(x,y)\,\text{sgn}(x^0 - y^0), \\[4pt]
G_{ab}^{+-}(x,y) &= F_{ab}(x,y) + \frac{i}{2}\rho_{ab}(x,y), \\[4pt]
G_{ab}^{-+}(x,y) &= F_{ab}(x,y) - \frac{i}{2}\rho_{ab}(x,y),
\end{aligned} \qquad (2.133)
$$

which correspond to the definitions (2.44) and (2.45) with the standard sign function $\text{sgn}(x^0 - y^0) = \theta(x^0 - y^0) - \theta(y^0 - x^0)$. Since $F_{ab}(x,y)$ and $\rho_{ab}(x,y)$ are not time-ordered correlation functions, no distinction between $\mathcal{C}^+$ and $\mathcal{C}^-$ for the location of their time arguments has to be taken into account here.

To obtain a similar decomposition for the self-energy, we separate Σ into a "local" and a "nonlocal" part according to

$$\Sigma_{ab}(x,y) = -i\Sigma_{ab}^{(0)}(x)\delta(x-y) + \overline{\Sigma}_{ab}(x,y). \qquad (2.134)$$

Since $\Sigma^{(0)}$ just corresponds to a space–time-dependent mass shift, it is convenient for the following to introduce the notation

$$M_{ab}^2(x) \overset{(\phi=0)}{=} m^2\delta_{ab} + \Sigma_{ab}^{(0)}(x). \qquad (2.135)$$

To make the time ordering for the nonlocal part of the self-energy, $\overline{\Sigma}_{ab}(x,y)$, explicit we can use the same identity as for the propagator (2.131) to decompose:

$$\overline{\Sigma}_{ab}(x,y) = \Sigma_{ab}^F(x,y) - \frac{i}{2}\Sigma_{ab}^\rho(x,y)\operatorname{sgn}_{\mathcal{C}}(x^0-y^0). \qquad (2.136)$$

Out of equilibrium, we have to follow the time evolution both for the statistical propagator F and for the spectral function ρ. The evolution equations are obtained from (2.127) with the help of the identities (2.131) and (2.136). Most importantly, once expressed in terms of F and ρ, the time ordering is explicit and the respective sign functions appearing in the time-ordered propagator can be conveniently evaluated along the closed real-time contour $\mathcal{C}$.

With the notation (2.134), the time-evolution equation for the time-ordered propagator (2.127) reads

$$\left[\Box_x\delta_{ac} + M_{ab}^2(x)\right]G_{cb}(x,y) + i\int_{z,\mathcal{C}}\overline{\Sigma}_{ac}(x,z)G_{cb}(z,y) = -i\delta_{ab}\delta_{\mathcal{C}}(x-y), \qquad (2.137)$$

where we have set $R=0$. For the evaluation along the time contour $\mathcal{C}$ involved in the integration $\int_z \equiv \int_{\mathcal{C}} dz^0 \int d^d z$, we employ (2.131) and (2.136):

$$i\int_{z,\mathcal{C}}\overline{\Sigma}_{ac}(x,z)G_{cb}(z,y) = i\int_{z,\mathcal{C}}\left[\Sigma_{ac}^F(x,z)F_{cb}(z,y)\right.$$

$$-\frac{i}{2}\Sigma_{ac}^F(x,z)\rho_{cb}(z,y)\operatorname{sgn}_{\mathcal{C}}(z^0-y^0)$$

$$-\frac{i}{2}\Sigma_{ac}^\rho(x,z)F_{cb}(z,y)\operatorname{sgn}_{\mathcal{C}}(x^0-z^0)$$

$$\left.-\frac{1}{4}\Sigma_{ac}^\rho(x,z)\rho_{cb}(z,y)\operatorname{sgn}_{\mathcal{C}}(x^0-z^0)\operatorname{sgn}_{\mathcal{C}}(z^0-y^0)\right]. $$

$$(2.138)$$

The first term on the right-hand side vanishes because of integration along the *closed time contour* $\mathcal{C}$. To proceed for the second and third terms, we split the contour integral such that the sign functions have definite values, for instance,

$$\int_{\mathcal{C}} dz^0\, \operatorname{sgn}_{\mathcal{C}}(z^0-y^0) = \int_{t_0}^{y^0} dz^0\,(-1) + \int_{y^0}^{t_0} dz^0 = -2\int_{t_0}^{y^0} dz^0 \qquad (2.139)$$

for the closed contour with initial time t_0. We emphasize that the contributions from times later than y^0 simply cancel. To evaluate the last term on the right-hand side of (2.138), it is convenient to distinguish the following cases:

(a) $\theta_{\mathcal{C}}(x^0 - y^0) = 1$:

$$\int_{\mathcal{C}} dz^0 \, \mathrm{sgn}_{\mathcal{C}}(x^0 - z^0) \, \mathrm{sgn}_{\mathcal{C}}(z^0 - y^0) = \int_{t_0}^{y^0} dz^0(-1) + \int_{y^0}^{x^0} dz^0 + \int_{x^0}^{t_0} dz^0(-1);$$

(2.140)

(b) $\theta_{\mathcal{C}}(y^0 - x^0) = 1$:

$$\int_{\mathcal{C}} dz^0 \, \mathrm{sgn}_{\mathcal{C}}(x^0 - z^0) \, \mathrm{sgn}_{\mathcal{C}}(z^0 - y^0) = \int_{t_0}^{x^0} dz^0(-1) + \int_{x^0}^{y^0} dz^0 + \int_{y^0}^{t_0} dz^0(-1).$$

(2.141)

It can be seen that (a) and (b) differ only by an overall sign factor $\sim \mathrm{sgn}_{\mathcal{C}}(x^0 - y^0)$. Combining the integrals therefore gives

$$i \int_{z,\mathcal{C}} \overline{\Sigma}_{ac}(x, z) G_{cb}(z, y) = \int d^d z \left\{ \int_{t_0}^{x^0} dz^0 \, \Sigma_{ac}^{\rho}(x, z) F_{cb}(z, y) \right.$$
$$\left. - \int_{t_0}^{y^0} dz^0 \, \Sigma_{ac}^{F}(x, z) \rho_{cb}(z, y) - \frac{i}{2} \mathrm{sgn}_{\mathcal{C}}(x^0 - y^0) \int_{y^0}^{x^0} dz^0 \, \Sigma_{ac}^{\rho}(x, z) \rho_{cb}(z, y) \right\}.$$

(2.142)

We finally employ

$$\Box_x G_{ab}(x, y) = \Box_x F_{ab}(x, y) - \frac{i}{2} \mathrm{sgn}_{\mathcal{C}}(x^0 - y^0) \Box_x \rho_{ab}(x, y) - i\delta_{ab}\delta_{\mathcal{C}}(x - y) \quad (2.143)$$

such that the δ-term cancels with the respective term on the right-hand side of the evolution equation (2.137). Here we have used

$$-\frac{i}{2}\partial_{x^0}^2 \left[\rho_{ab}(x, y) \, \mathrm{sgn}_{\mathcal{C}}(x^0 - y^0)\right] = -\frac{i}{2} \mathrm{sgn}_{\mathcal{C}}(x^0 - y^0)\partial_{x^0}^2 \rho_{ab}(x, y)$$
$$\underbrace{-i\delta_{\mathcal{C}}(x^0 - y^0)\partial_{x^0}\rho_{ab}(x, y)}_{-i\delta_{ab}\delta_{\mathcal{C}}(x-y)}, \quad (2.144)$$

where (2.130) is employed in the second line and the term $\sim \rho_{ab}(x, y)\delta_{\mathcal{C}}(x^0 - y^0)$ is observed to vanish since $\sim \rho_{ab}(x, y)|_{x^0=y^0} = 0$. Comparing coefficients, which here corresponds to separating real and imaginary parts, we find from (2.142) and (2.143) the equations for $F_{ab}(x, y)$ and $\rho_{ab}(x, y)$.

Using the abbreviated notation $\int_{t_1}^{t_2} dz \equiv \int_{t_1}^{t_2} dz^0 \int_{-\infty}^{\infty} d^d z$, we arrive at the *coupled evolution equations for the statistical propagator and the spectral function:*

$$
\begin{aligned}
\left[\Box_x \delta_{ac} + M_{ac}^2(x) \right] F_{cb}(x,y) &= - \int_{t_0}^{x^0} dz \, \Sigma_{ac}^\rho(x,z) F_{cb}(z,y) \\
&\quad + \int_{t_0}^{y^0} dz \, \Sigma_{ac}^F(x,z) \rho_{cb}(z,y), \\[1em]
\left[\Box_x \delta_{ac} + M_{ac}^2(x) \right] \rho_{cb}(x,y) &= - \int_{y^0}^{x^0} dz \, \Sigma_{ac}^\rho(x,z) \rho_{cb}(z,y).
\end{aligned}
\tag{2.145}
$$

For the considered case with $\phi = 0$, the self-energies depend only on F and ρ, i.e., $M^2 = M^2(F)$, $\Sigma^F = \Sigma^F(\rho, F)$, and $\Sigma^\rho = \Sigma^\rho(\rho, F)$. We note that the local self-energy correction (2.135) encoded in M^2 does not depend on the spectral function, because the latter vanishes for equal-time arguments.

It can be seen that (2.145) are *causal equations* with characteristic "*memory*" *integrals*, which integrate over the time history of the evolution. We emphasize that the presence of memory integrals is a property of the exact theory and in accordance with all symmetries. In particular, the equations describe a unitary time evolution without further approximations. The equations themselves do not single out a direction of time and they should be clearly distinguished from phenomenological nonequilibrium equations, where irreversibility is typically put in by hand. Since the nonequilibrium evolution equations are exact for known self-energies, they are fully equivalent to any kind of identity for the two-point functions and are often called Kadanoff–Baym or Schwinger–Dyson equations.

Note that the initial-time properties of the spectral function have to comply with the equal-time commutation relations (2.130). In contrast, for $F_{ab}(x,y)$ as well as its first derivatives, the initial conditions at t_0 need to be supplied in order to solve these equations. To make contact with the discussion of initial conditions in Section 2.2.4, we consider for a moment the spatially homogeneous case, for which $F_{ab}(x,y) = F_{ab}(x^0, y^0; \mathbf{x} - \mathbf{y}) = \int [d^d p/(2\pi)^d] \exp[i\mathbf{p} \cdot (\mathbf{x} - \mathbf{y})] F_{ab}(x^0, y^0; \mathbf{p})$, and equivalently for $\rho_{ab}(x,y)$. In terms of the Fourier components $F_{ab}(t, t'; \mathbf{p})$, the integro-differential equations (2.145) can be solved with the following initial conditions respecting the $O(N)$ symmetry of the theory:

$$
F_{ab}(t, t'; \mathbf{p})|_{t = t' = t_0} = \xi_{\mathbf{p}}^2 \, \delta_{ab},
$$

$$
\tfrac{1}{2} [\partial_t F_{ab}(t, t'; \mathbf{p}) + \partial_{t'} F_{ab}(t, t'; \mathbf{p})]|_{t = t' = t_0} = \xi_{\mathbf{p}} \eta_{\mathbf{p}} \delta_{ab},
$$

$$
\partial_t \partial_{t'} F_{ab}(t, t'; \mathbf{p})|_{t = t' = t_0} = \left(\eta_{\mathbf{p}}^2 + \frac{\sigma_{\mathbf{p}}^2}{4\xi_{\mathbf{p}}^2} \right) \delta_{ab}.
\tag{2.146}
$$

Here we have used the fact that the required correlators at initial time are identical to those given in (2.78) for the considered case $\phi \equiv 0$ if the momentum labels are attached for $d > 0$. Accordingly, these are the very same parameters that have to be specified for the corresponding Gaussian initial density matrix (2.75). We emphasize that the initial conditions for the spectral function equation are completely fixed by the properties of the theory itself: the equal-time commutation relations (2.130) specify $\rho_{ab}(t, t'; \mathbf{p})|_{t=t'=t_0} = 0$, $\partial_t \rho_{ab}(t, t'; \mathbf{p})|_{t=t'=t_0} = \delta_{ab}$, and $\partial_t \partial_{t'} \rho_{ab}(t, t'; \mathbf{p})|_{t=t'=t_0} = 0$ for the antisymmetric spectral function.

Nonvanishing field expectation value

In the presence of a nonzero field expectation value, $\phi \neq 0$, the same decompositions (2.131) and (2.136) apply. Equivalently, one can always view (2.131) as defining F and ρ from the *connected* propagator G. This means that the symmetric part, which is described by a statistical two-point function, becomes, in the presence of a macroscopic field,

$$F_{ab}(x, y) = \tfrac{1}{2}\langle\{\Phi_a(x), \Phi_b(y)\}\rangle - \phi_a(x)\phi_b(y), \tag{2.147}$$

whereas the antisymmetric spectral function is given by the same expression as before, $\rho_{ab}(x, y) = i\langle[\Phi_a(x), \Phi_b(y)]\rangle$. The general form of the scalar evolution equations for the spectral and statistical function (2.145) remains the same for $\phi \neq 0$. The only change compared with the symmetric regime is that the functional dependence now includes field-dependent terms $M^2 = M^2(\phi, F)$, $\Sigma_F = \Sigma_F(\phi, \rho, F)$, and $\Sigma_\rho = \Sigma_\rho(\phi, \rho, F)$.

For the N-component scalar field theory (2.93) with the field-dependent inverse classical propagator (2.102) acting on the left-hand side of the evolution equations for $F_{ab}(x, y)$ and $\rho_{ab}(x, y)$, we have ($\phi^2 \equiv \phi_a \phi_a$)

$$M_{ab}^2(x) = \left[m^2 + \frac{\lambda}{6N}\phi^2(x)\right]\delta_{ab} + \frac{\lambda}{3N}\phi_a(x)\phi_b(x) + \Sigma_{ab}^{(0)}(x). \tag{2.148}$$

In this case, the evolution equations for the spectral function and statistical two-point function (2.145) are supplemented by a differential equation for ϕ given by the stationarity condition (2.105). For this, we have to compute the functional derivative of (2.109), i.e.,

$$\frac{\delta \Gamma}{\delta \phi_a(x)} = \frac{\delta S}{\delta \phi_a(x)} + \frac{i}{2}\frac{\delta \operatorname{Tr}\{G_0^{-1}(\phi)G\}}{\delta \phi_a(x)} + \frac{\delta \Gamma_2}{\delta \phi_a(x)} = -J_a(x), \tag{2.149}$$

according to (2.105) for $R = 0$. Again, since a source term representing the initial density matrix has support only at time t_0, we have $J_a(x) = 0$ for $x^0 > t_0$ in the absence of external sources. With

$$\frac{i}{2}\operatorname{Tr}\{G_0^{-1}(\phi)G\}$$
$$= -\frac{1}{2}\int_x \left\{\left[\Box_x + m^2 + \frac{\lambda}{6N}\phi^2(x)\right]\delta_{ab} + \frac{\lambda}{3N}\phi_a(x)\phi_b(x)\right\}F_{ba}(x, x) \tag{2.150}$$

for the N-component scalar field theory, this yields the *field evolution equation*

$$\left\{ \left(\Box_x + m^2 + \frac{\lambda}{6N}[\phi^2(x) + F_{cc}(x,x)] \right) \delta_{ab} + \frac{\lambda}{3N} F_{ab}(x,x) \right\} \phi_b(x) = \frac{\delta\Gamma_2}{\delta\phi_a(x)}.$$

(2.151)

The solution of this equation requires specification of the field and its first derivative at the initial time. In the context of the above discussion for spatially homogeneous fields, this just corresponds to specifying $\phi_a(t_0)$ and $\partial_t\phi_a(x^0)|_{x^0=t_0}$.

Evolution equations for fermions and gauge fields

The different dynamical roles of spectral and statistical components is a generic property of nonequilibrium field theory and is not specific to scalar field degrees of freedom. In terms of spectral and statistical components, the equations for *fermionic fields* or *gauge fields* have very similar structures as well. To be specific, we consider Dirac fermions $\Psi(x)$ and $\bar{\Psi} = \Psi^\dagger\gamma^0$, in matrix notation with Dirac matrices γ^μ for $\mu = 0,\ldots,3$.[3] The time-ordered fermion propagator is $\Delta(x,y) = \langle\Psi(x)\bar{\Psi}(y)\rangle\theta_C(x^0 - y^0) - \langle\bar{\Psi}(y)\Psi(x)\rangle\theta_C(y^0 - x^0)$, where the minus sign is a consequence of the anticommuting property of fermionic fields. Correspondingly, in contrast to bosons, for fermions, the field anticommutator is associated with the spectral function:[4]

anticommutator: $\qquad \rho^{(f)}(x,y) = i\langle\{\Psi(x), \bar{\Psi}(y)\}\rangle,$ (2.152)

commutator: $\qquad F^{(f)}(x,y) = \frac{1}{2}\langle[\Psi(x), \bar{\Psi}(y)]\rangle.$ (2.153)

In terms of spectral and statistical components, the propagator reads

$$\Delta(x,y) = F^{(f)}(x,y) - \frac{i}{2}\rho^{(f)}(x,y)\,\mathrm{sgn}_C(x^0 - y^0).$$

(2.154)

The equal-time anticommutation relations for the fields are again encoded in $\rho^{(f)}(x,y)$. For instance, for Dirac fermions,

$$\gamma^0\rho^{(f)}(x,y)|_{x^0=y^0} = i\delta(\mathbf{x} - \mathbf{y}),$$

(2.155)

which will uniquely specify the initial conditions for the evolution equation for $\rho^{(f)}$ similar to what is observed for bosons.

For Dirac fermions with mass $m^{(f)}$, the free inverse propagator reads

$$i\Delta_0^{-1}(x,y) = \left(i\slashed{\partial}_x - m^{(f)}\right)\delta(x - y),$$

(2.156)

where $\slashed{\partial} \equiv \gamma^\mu\partial_\mu$. Similarly to the bosonic case, we may write, in the absence of sources,

$$\Delta^{-1}(x,y) = \Delta_0^{-1}(x,y) - \Sigma^{(f)}(x,y),$$

(2.157)

---

[3] Dirac matrices obey $\{\gamma^\mu, \gamma^\nu\} = 2g^{\mu\nu}$, with $g^{\mu\nu} = \mathrm{diag}(1, -1, -1, -1)$.
[4] We use $\langle\Psi\rangle = \langle\bar{\Psi}\rangle = 0$ in the absence of sources.

which defines the proper fermion self-energy $\Sigma^{(f)}(x, y)$. The corresponding decomposition for the fermion self-energy reads

$$\Sigma^{(f)}(x, y) = \Sigma_F^{(f)}(x, y) - \frac{i}{2}\Sigma_\rho^{(f)}(x, y)\,\mathrm{sgn}_\mathcal{C}(x^0 - y^0). \tag{2.158}$$

If there is a local contribution to the proper self-energy, this is to be separated in complete analogy to the scalar equation (2.134), and the decomposition (2.158) is taken for the nonlocal part of the self-energy.

From the equation of motion for the time-ordered fermion propagator (2.157), we obtain a suitable time evolution equation by convoluting with Δ. For a free inverse propagator as in (2.156) for Dirac fermions, this yields

$$\left(i\partial\!\!\!/_x - m^{(f)}\right)\Delta(x, y) - i\int_{z,\mathcal{C}} \Sigma^{(f)}(x, z)\Delta(z, y) = i\delta_\mathcal{C}(x - y). \tag{2.159}$$

Following the lines of the earlier discussion for scalars, we find for the fermion case the coupled evolution equations

$$\left(i\partial\!\!\!/_x - m^{(f)}\right)F^{(f)}(x, y) = \int_{t_0}^{x^0} dz\,\Sigma_\rho^{(f)}(x, z)F^{(f)}(z, y) - \int_{t_0}^{y^0} dz\,\Sigma_F^{(f)}(x, z)\rho^{(f)}(z, y), \tag{2.160}$$

$$\left(i\partial\!\!\!/_x - m^{(f)}\right)\rho^{(f)}(x, y) = \int_{y^0}^{x^0} dz\,\Sigma_\rho^{(f)}(x, z)\rho^{(f)}(z, y). \tag{2.161}$$

Similarly, the nonequilibrium evolution equations for gauge fields can be obtained as well. For instance, for a gauge field $A^\mu(x)$ with Lorentz indices $\mu = 0, \ldots, 3$, the full time-ordered propagator $D^{\mu\nu}(x, y) = \langle T_\mathcal{C} A^\mu(x)A^\nu(y)\rangle$ may be written as

$$D^{\mu\nu}(x, y) = F_D^{\mu\nu}(x, y) - \frac{i}{2}\rho_D^{\mu\nu}(x, y)\,\mathrm{sgn}_\mathcal{C}(x^0 - y^0). \tag{2.162}$$

For a theory with free inverse gauge field propagator given by

$$iD_{0,\mu\nu}^{-1}(x, y) = \left[g_{\mu\nu}\Box - (1 - \xi^{-1})\partial_\mu\partial_\nu\right]_x \delta(x - y) \tag{2.163}$$

for covariant gauges with gauge-fixing parameter ξ and vanishing macroscopic field, $\langle A^\mu(x)\rangle = 0$, we find the respective equations from (2.145) by

$$(\Box_x + M^2)\rho(x, y) \longrightarrow - \left[g^\mu{}_\gamma\Box - (1 - \xi^{-1})\partial^\mu\partial_\gamma\right]_x \rho_D^{\gamma\nu}(x, y), \tag{2.164}$$

and equivalently for $F_D^{\gamma\nu}(x, y)$. Of course, the respective Lorentz and internal indices have to be attached to the corresponding self-energies on the right-hand sides of the equations.

Details of functional integral for fermions

In Section 2.2.3, we derived a bosonic functional integral in terms of eigenvalues of the bosonic field operators. Fermionic operators anticommute, which is intimately connected with the Pauli exclusion principle. In order to represent fermionic integrals, we will have to deal with anticommuting eigenvalues of fermionic field operators. Anticommuting numbers and functions are called Grassmann variables. Here we recall some basic aspects that we need for our purposes and refer to the literature for more extensive discussions.

For a single anticommuting variable η, we have

$$\{\eta, \eta\} = 0, \tag{2.165}$$

such that $\eta\eta = -\eta\eta$, and thus $\eta^2 = 0$. The derivative operator $\mathrm{d}/\mathrm{d}\eta$ can be defined in an analogous way as for ordinary numbers and we write

$$\left\{\frac{\mathrm{d}}{\mathrm{d}\eta}, \eta\right\} = 1. \tag{2.166}$$

As a consequence of (2.165), the power series expansion of any suitable function $f(\eta)$ may be written as

$$f(\eta) = a + b\eta, \tag{2.167}$$

with ordinary commuting numbers a and b. Therefore, $\mathrm{d}^2 f(\eta)/\mathrm{d}\eta^2 = 0$ and we have $\{\mathrm{d}/\mathrm{d}\eta, \mathrm{d}/\mathrm{d}\eta\} = 0$.

In order to define integration for Grassmann variables, we start by asking what could be the analogue of the bosonic Gaussian integral, $\int \mathrm{d}x \exp(-ax^2) = \sqrt{\pi/a}$? An important property is certainly the invariance under the shift of the integration variable $x \to x + b$. To carry over such a property to the fermionic integral, we define

$$\int \mathrm{d}\eta \, (a + b\eta) \overset{!}{=} b. \tag{2.168}$$

In fact, since $\eta^2 = 0$, the integral must be linear in η and the only linear function that respects the property of shift invariance is a constant. Consequently, integration for Grassmann variables acts just like differentiation. To write a Gaussian integral, we consider a complex anticommuting variable $\eta = (\eta_\mathrm{R} + i\eta_\mathrm{I})/\sqrt{2}$, where η_R and η_I are real-valued anticommuting variables. With $\eta^* = (\eta_\mathrm{R} - i\eta_\mathrm{I})/\sqrt{2}$, we find, for instance, $\eta\eta^* = (\eta_\mathrm{R}^2 + i\eta_\mathrm{I}\eta_\mathrm{R} - i\eta_\mathrm{R}\eta_\mathrm{I} + \eta_\mathrm{I}^2)/2 = -i\eta_\mathrm{R}\eta_\mathrm{I} = -\eta^*\eta$. The Gaussian integral is

$$\int \mathrm{d}\eta^* \mathrm{d}\eta \, e^{-b\eta^*\eta} = \int \mathrm{d}\eta^* \mathrm{d}\eta \, (1 - b\eta^*\eta)$$

$$= b \int \mathrm{d}\eta^* \mathrm{d}\eta \, \eta\eta^* = b, \tag{2.169}$$

where the sign change in the second equality comes from changing the order of the variables to be able to use first $\int d\eta\, \eta = 1$ and then $\int d\eta^*\, \eta^* = 1$.

Next, we want to consider the space of anticommuting functions $\eta(x)$. Generalizing (2.166), functional differentiation is defined as

$$\left\{\frac{\delta}{\delta\eta(x)}, \eta(y)\right\} = \delta(x - y). \tag{2.170}$$

The corresponding Gaussian functional is

$$\int \mathscr{D}\eta^*\, \mathscr{D}\eta \, \exp\left[-\int_{xy} \eta^*(x)B(x,y)\eta(y)\right] = \det B, \tag{2.171}$$

where $\int \mathscr{D}\eta^*\, \mathscr{D}\eta = \int \prod_x d\eta^*(x)\, d\eta(x)$.

The construction of the 2PI effective action for fermionic fields proceeds along very similar lines as for bosons. However, one has to take into account the anticommuting behavior of the fermion fields. A main differences compared with the bosonic case can be observed from the one-loop part ($\Gamma_2 \equiv 0$) of the 2PI effective action. Using again the Dirac fermions introduced in the previous section and, for comparison, a real scalar field φ with vanishing field expectation values, it involves the integrals

$$\text{fermions:} \quad -i \ln \int \mathscr{D}\bar{\psi}\, \mathscr{D}\psi\, e^{iS_0^{(f)}} = -i \ln(\det \Delta_0^{-1}) = -i \operatorname{Tr} \ln \Delta_0^{-1},$$

$$\text{bosons:} \quad -i \ln \int \mathscr{D}\varphi\, e^{iS_0} = -i \ln(\det G_0^{-1})^{-1/2} = \frac{i}{2} \operatorname{Tr} \ln G_0^{-1}. \tag{2.172}$$

Here $S_0^{(f)} = \int d^4x\, d^4y\, \bar{\psi}(x) i\Delta_0^{-1}(x,y)\psi(y)$ denotes a fermion action that is bilinear in the Grassmann fields. For Dirac fermions, the free inverse propagator $\Delta_0^{-1}(x,y)$ is given by (2.156) and $\bar{\psi} = \psi^\dagger\gamma^0$. For the bosons, S_0 is given by the quadratic part of (2.93). Comparing the two integrals, we see that the factor $\frac{1}{2}$ for the bosonic case is replaced by -1 for the fermion fields because of their anticommuting property. With this difference, following along similar lines as for the bosonic case, we find that the 2PI effective action for fermions can be written in complete analogy to (2.109). Accordingly, for the case of vanishing fermion field expectation values, $\langle\Psi\rangle = \langle\bar{\Psi}\rangle = 0$, we have

$$\boxed{\Gamma[\Delta] = -i \operatorname{Tr}\{\ln \Delta^{-1}\} - i \operatorname{Tr}\{\Delta_0^{-1}\Delta\} + \Gamma_2[\Delta] + \text{const.}} \tag{2.173}$$

Here $\Gamma_2[\Delta]$ *contains all 2PI diagrams* with lines associated with the contour time-ordered propagator $\Delta(x,y) = \langle T_{\mathcal{C}}\Psi(x)\bar{\Psi}(y)\rangle$. The trace "Tr" includes integration over time and spatial coordinates, as well as summation over field indices.

As for the bosonic case of (2.106), the equation of motion for Δ in the absence of external sources is obtained by extremizing the effective action:

$$\frac{\delta\Gamma[\Delta]}{\delta\Delta(x,y)} = 0. \tag{2.174}$$

Using (2.173), this stationarity condition can be written as

$$\Delta^{-1}(x,y) = \Delta_0^{-1}(x,y) - \Sigma^{(\mathrm{f})}(x,y;\Delta), \tag{2.175}$$

with the proper fermion self-energy

$$\boxed{\Sigma^{(\mathrm{f})}(x,y;\Delta) \equiv -\mathrm{i}\frac{\delta\Gamma_2[\Delta]}{\delta\Delta(y,x)}} \tag{2.176}$$

2.2.7   The special case of thermal equilibrium

For the neutral scalar field theory, the canonical thermal equilibrium density matrix is

$$\varrho_\beta = \frac{1}{Z_\beta}\mathrm{e}^{-\beta H}, \tag{2.177}$$

with the partition sum $Z_\beta = \mathrm{Tr}\{\mathrm{e}^{-\beta H}\}$. For some time t_0, we take $\varrho(t_0) = \varrho_\beta$ as the density matrix in the generating functional (2.34). The construction of a path-integral representation then follows closely Section 2.2.2. The main difference is that $\mathrm{e}^{-\beta H} = \mathrm{e}^{-\mathrm{i}(-\mathrm{i}\beta)H}$ may be interpreted as an evolution operator in imaginary time. This simplifies taking into account a thermal density matrix that is not Gaussian in general. Following the steps of Section 2.2.2, we evaluate the trace using eigenstates of the Heisenberg field operator $\Phi(x)$ at time $x^0 = t_0$, $\Phi(t_0,\mathbf{x})|\varphi\rangle = \varphi_0(\mathbf{x})|\varphi\rangle$. The partition function may then be written as

$$Z_\beta = \int[\mathrm{d}\varphi_0]\,\langle\varphi|\,\varrho_\beta\,|\varphi\rangle. \tag{2.178}$$

The interpretation of the canonical equilibrium density matrix as an evolution operator in imaginary time allows us to consider

$$\langle\varphi|\,\mathrm{e}^{-\beta H}\,|\varphi\rangle \equiv (\varphi,t_0-\mathrm{i}\beta\,|\,\varphi,t_0) \tag{2.179}$$

as a transition amplitude. As a consequence, we can use (2.72) applied to imaginary times in order to represent it as a path integral. Since we are calculating a trace, we have to identify

$$\varphi(t_0-\mathrm{i}\beta,\mathbf{x}) = \varphi(t_0,\mathbf{x}) = \varphi_0(\mathbf{x}). \tag{2.180}$$

This periodicity of the field in thermal equilibrium represents an important constraint that is, in general, not present out of equilibrium.

In order to write down a generating functional also for real-time correlation functions in thermal equilibrium, we extend with $T_{\mathcal{C}_\beta}$ the time ordering to include the imaginary-time branch from t_0 to $t_0 - i\beta$. Here any time on the imaginary branch is considered to be later than any time on the previously discussed real-time contour $\mathcal{C}$. Graphically, $\mathcal{C}_\beta$ is as follows:

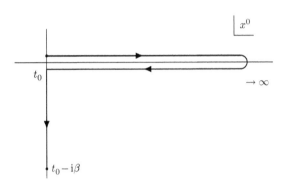

We also include sources $J(x)$ and $R(x,y)$ and use the notation $\int_{x,\mathcal{C}_\beta} \equiv \int_{\mathcal{C}_\beta} dx^0 \int d^d x$. The generating functional for contour time-ordered correlation functions is then given by

$$Z_\beta[J,R] =$$
$$\int [d\varphi_0] \, \langle \varphi | \, \varrho_\beta T_{\mathcal{C}_\beta} \exp \left\{ i \left[\int_{x,\mathcal{C}_\beta} \Phi(x) J(x) + \frac{1}{2} \int_{xy,\mathcal{C}_\beta} \Phi(x) R(x,y) \Phi(y) \right] \right\} | \varphi \rangle.$$

$$(2.181)$$

Inserting complete sets of states along the contour, following Section 2.2.2, we obtain

$$Z_\beta[J,R] =$$
$$\int_{\text{periodic}} \mathscr{D}\varphi \, \exp \left\{ i \left[S_\beta[\varphi] + \int_{x,\mathcal{C}_\beta} J(x)\varphi(x) + \frac{1}{2} \int_{xy,\mathcal{C}_\beta} R(x,y)\varphi(x)\varphi(y) \right] \right\}, \quad (2.182)$$

where "periodic" here refers to the periodicity condition for the fields (2.180). The thermal equilibrium contour action S_β for a theory with quartic self-interaction reads

$$S_\beta[\varphi] = \int_{x,\mathcal{C}_\beta} \left\{ \frac{1}{2} \partial^\mu \varphi(x) \partial_\mu \varphi(x) - \frac{m^2}{2} \varphi(x)\varphi(x) - \frac{\lambda}{4!} [\varphi(x)\varphi(x)]^2 \right\}. \quad (2.183)$$

For the specific case of calculating static quantities, the real-time extent of $\mathcal{C}_\beta$ can be taken to zero. Static quantities can then be obtained from the corresponding generating functional with the imaginary-time interval $[0, -i\beta]$ only, which is often formulated in "Euclidean" space–time with the introduction of a Euclidean time $\tau = it$

and Euclidean action $S_E = -\mathrm{i}S$. Because of the periodicity condition (2.180), the Euclidean time dimension is compact. In Fourier space, this leads to a discrete set of so-called Matsubara frequencies, which we will not consider here. Quantities like the spectral function are, however, difficult to obtain from such a formulation since they would require analytic continuation. The above time contour including the real-time axis gives direct access to these quantities. We emphasize that in either formulation— including real times or not—all correlation functions are computed here with the thermal density matrix (2.177) such that the system is always in equilibrium.

Contour time-ordered correlation functions in thermal equilibrium can be obtained from the generating functional (2.182) by functional differentiation. For instance, the two-point function is given by

$$\frac{1}{Z_\beta}\frac{\delta^2 Z_\beta[J,R]}{\delta J(x)\delta J(y)}\bigg|_{J,R=0} = \langle T_{\mathcal{C}_\beta}\Phi(x)\Phi(y)\rangle = G^{(\mathrm{eq})}(x-y) + \phi\phi. \qquad (2.184)$$

For the last equality, we have used the fact that thermal equilibrium is translation-invariant, such that the one-point function ϕ is homogeneous and, accordingly, the two-point function depends only on relative coordinates, $G^{(\mathrm{eq})}(x,y) = G^{(\mathrm{eq})}(x-y)$. Similar to (2.131), we may write

$$G^{(\mathrm{eq})}(x-y) = F^{(\mathrm{eq})}(x-y) - \frac{\mathrm{i}}{2}\rho^{(\mathrm{eq})}(x-y)\,\mathrm{sgn}_{\mathcal{C}_\beta}(x^0-y^0), \qquad (2.185)$$

with the contour sign function $\mathrm{sgn}_{\mathcal{C}_\beta}(x^0-y^0) = \theta_{\mathcal{C}_\beta}(x^0-y^0) - \theta_{\mathcal{C}_\beta}(y^0-x^0)$.

We now derive an important consequence of the periodicity condition (2.180) for correlation functions. At the initial time t_0 of the contour $\mathcal{C}_\beta$, we have

$$\langle T_{\mathcal{C}_\beta}\Phi(x)\Phi(y)\rangle|_{x^0=t_0} = \langle \Phi(y)\Phi(x)\rangle|_{x^0=t_0} \qquad (2.186)$$

and at the endpoint $t_0 - \mathrm{i}\beta$,

$$\langle T_{\mathcal{C}_\beta}\Phi(x)\Phi(y)\rangle|_{x^0=t_0-\mathrm{i}\beta} = \langle \Phi(x)\Phi(y)\rangle|_{x^0=t_0-\mathrm{i}\beta}. \qquad (2.187)$$

The periodicity condition for the fields in thermal equilibrium then implies[5]

$$\langle \Phi(y)\Phi(x)\rangle|_{x^0=t_0} = \langle \Phi(x)\Phi(y)\rangle|_{x^0=t_0-\mathrm{i}\beta}. \qquad (2.188)$$

Using the decomposition (2.185), this reads

$$\left[F^{(\mathrm{eq})}(x-y) + \frac{\mathrm{i}}{2}\rho^{(\mathrm{eq})}(x-y)\right]\bigg|_{x^0=t_0} = \left[F^{(\mathrm{eq})}(x-y) - \frac{\mathrm{i}}{2}\rho^{(\mathrm{eq})}(x-y)\right]\bigg|_{x^0=t_0-\mathrm{i}\beta}.$$
$$(2.189)$$

---

[5] The periodicity condition is also called the Kubo–Martin–Schwinger ("KMS") condition.

Translation invariance also makes it convenient to consider the Fourier transform with *real* four-momentum $(\omega, \mathbf{p})$:

$$F^{(\text{eq})}(x-y) = \int \frac{d\omega \, d^d p}{(2\pi)^{d+1}} e^{-i\omega(x^0-y^0)+i\mathbf{p}\cdot(\mathbf{x}-\mathbf{y})} F^{(\text{eq})}(\omega, \mathbf{p}), \tag{2.190}$$

and equivalently for the spectral part. Taking for a moment for granted that these integrals can be properly regularized and defined for the considered quantum field theory, (2.189) reads in Fourier space

$$F^{(\text{eq})}(\omega, \mathbf{p}) + \frac{i}{2}\rho^{(\text{eq})}(\omega, \mathbf{p}) = e^{-\beta\omega}\left[F^{(\text{eq})}(\omega, \mathbf{p}) - \frac{i}{2}\rho^{(\text{eq})}(\omega, \mathbf{p})\right]. \tag{2.191}$$

Solving for $F^{(\text{eq})}(\omega, \mathbf{p})$ and using $(1+e^{-\beta\omega})/(1-e^{\beta\omega}) = (e^{\beta\omega}+1)/(e^{\beta\omega}-1) = 1+2/(e^{\beta\omega}-1)$, the periodicity condition for bosons can be written as[6]

$$\boxed{F^{(\text{eq})}(\omega, \mathbf{p}) = -i\left[\tfrac{1}{2} + f_\beta(\omega)\right]\rho^{(\text{eq})}(\omega, \mathbf{p})} \tag{2.192}$$

with $f_\beta(\omega) = (e^{\beta\omega}-1)^{-1}$ denoting the Bose–Einstein distribution function. Equation (2.192) relates the spectral function to the statistical propagator. This is also called the *fluctuation–dissipation relation* and will be discussed further in later sections. While $\rho^{(\text{eq})}$ encodes the information about the spectrum of the theory, we see from (2.192) that the function $F^{(\text{eq})}$ encodes the statistical aspects in terms of the occupation number distribution $f_\beta(\omega)$. In the same way, we can obtain for the Fourier transforms of the spectral and statistical components of the self-energy the thermal equilibrium relation

$$\Sigma_F^{(\text{eq})}(\omega, \mathbf{p}) = -i\left[\tfrac{1}{2} + f_\beta(\omega)\right]\Sigma_\rho^{(\text{eq})}(\omega, \mathbf{p}). \tag{2.193}$$

From (2.192) and (2.193), we obtain the identity

$$\Sigma_\rho^{(\text{eq})}(\omega, \mathbf{p})F^{(\text{eq})}(\omega, \mathbf{p}) - \Sigma_F^{(\text{eq})}(\omega, \mathbf{p})\rho^{(\text{eq})}(\omega, \mathbf{p}) = 0. \tag{2.194}$$

Later, we will see that this condition can be rewritten as the difference of a "gain" and "loss" term in kinetic descriptions in their range of applicability.

For fermions, the contour-ordered propagator reads, in thermal equilibrium, $\Delta^{(\text{eq})}(x-y) = \langle\Psi(x)\bar{\Psi}(y)\rangle\theta_{\mathcal{C}_\beta}(x^0-y^0) - \langle\bar{\Psi}(y)\Psi(x)\rangle\theta_{\mathcal{C}_\beta}(y^0-x^0)$, with the minus sign coming from the anticommuting property of fermions. If we proceed along the same lines as for bosons above, i.e., by evaluating the contour-ordered correlation function once at the initial time and once at the final point of the contour, this minus sign leads to the *antiperiodicity condition*

$$\langle\bar{\Psi}(y)\Psi(x)\rangle|_{x^0=t_0} = -\langle\Psi(x)\bar{\Psi}(y)\rangle|_{x^0=t_0-i\beta}. \tag{2.195}$$

[6] In our conventions, the Fourier transform of the real-valued antisymmetric function $\rho(x,y)$ is purely imaginary, while that of the symmetric function $F(x,y)$ is real.

Using the decomposition (2.154) for fermions, we can translate this again into a relation between the spectral and statistical two-point function. The difference is that the Bose–Einstein distribution in (2.192) is replaced by the Fermi–Dirac distribution $f_\beta^{(f)}(\omega) = (e^{\beta\omega}+1)^{-1}$ according to $\frac{1}{2} + f_\beta(\omega) \to \frac{1}{2} - f_\beta^{(f)}(\omega)$ in the respective relation.

It is important to realize that for a general out-of-equilibrium situation, the spectral and statistical components of correlation functions are not related by a fluctuation–dissipation relation. Such a relation is a manifestation of the tremendous simplification that happens if the system can be characterized in terms of thermal equilibrium correlation functions. An even more stringent reduction occurs for the vacuum, where $f_\beta(\omega) \equiv 0$ such that the spectral and statistical function agree up to a normalization. In this respect, nonequilibrium quantum field theory is more complicated, since it admits the description of more general situations and encompasses the thermal equilibrium or vacuum theory as special cases.

2.2.8 Nonrelativistic quantum field theory limit

For the nonrelativistic limit of a relativistic quantum field theory, one considers processes with typical momenta of particles that are small compared with their mass m. Consequently, we can write $\sqrt{\mathbf{p}^2 + m^2} \simeq m + \mathbf{p}^2/(2m)$. A related important aspect of the nonrelativistic limit is the absence of antiparticles, such that the typical chemical potential associated with the difference between the particle and antiparticle numbers is about $\mu \simeq m$. Therefore, in a nonrelativistic context, it is often convenient to introduce the different chemical potential

$$\mu_{\mathrm{nr}} = \mu - m. \tag{2.196}$$

To be specific, we consider a relativistic quantum theory for a complex scalar field $\varphi(t, \mathbf{x})$. This theory can be directly related to the field theory for $a = 1, \ldots, N$ real field components $\varphi_a(t, \mathbf{x})$ of Section 2.2.5 by taking $N = 2$ and identifying

$$\varphi = \frac{1}{\sqrt{2}}\left(\varphi_1 + i\varphi_2\right). \tag{2.197}$$

In the nonrelativistic regime, this complex scalar field theory can be effectively described in terms of the Gross–Pitaevskii field theory employed already in Section 2.1.4 for the physics of ultracold Bose gases.

In order to relate the relativistic and nonrelativistic descriptions, it is convenient to start from the corresponding functional-integral expression (2.69), which represents the partition function before the conjugate momentum field is integrated out. Since the derivations in thermal equilibrium and out of equilibrium follow mostly the same lines using the results of the previous sections—with the main difference being that the time integration runs along the closed time path for the nonequilibrium system and includes the interval $[0, -i\beta]$ in thermal equilibrium—in the following, we do not distinguish them in the notation. This will be sufficient to derive, in particular, the nonrelativistic version of the relativistic evolution equations (2.145).

In terms of the complex field $\varphi(t, \mathbf{x})$ and its conjugate momentum field $\pi(t, \mathbf{x})$, the functional integral for the partition function reads

$$Z[\mu] = \int \mathcal{D}\varphi^* \, \mathcal{D}\varphi \, \mathcal{D}\pi^* \, \mathcal{D}\pi \, \exp\left\{ i \int dt \left[\int d^d x \, (\pi \partial_t \varphi + \pi^* \partial_t \varphi^*) - (H - \mu Q) \right] \right\},$$

(2.198)

with the Hamiltonian

$$H = \int d^3 x \left[\pi^* \pi + (\nabla \varphi^*) \cdot (\nabla \varphi) + m^2 \varphi^* \varphi + \frac{\lambda}{12} (\varphi^* \varphi)^2 \right]$$

(2.199)

and charge

$$Q = i \int d^3 x \, (\varphi^* \pi^* - \pi \varphi).$$

(2.200)

Here, for brevity, we write the partition function without sources, which can always be added along the lines of Section 2.2.2.

The term in the exponential of (2.198) is essentially the classical action for the interacting theory. We first consider the quadratic part of that action by setting $\lambda = 0$ in (2.199), which can be written in a compact matrix notation in spatial Fourier space as

$$i \int dt \int \frac{d^d p}{(2\pi)^d} \left(\sqrt{\omega} \varphi^*(t, \mathbf{p}), \; \frac{-i}{\sqrt{\omega}} \pi(t, \mathbf{p}) \right) \begin{pmatrix} -\omega & i\partial_t + \mu \\ i\partial_t + \mu & -\omega \end{pmatrix} \begin{pmatrix} \sqrt{\omega}\varphi(t, \mathbf{p}) \\ \frac{i}{\sqrt{\omega}} \pi^*(t, \mathbf{p}) \end{pmatrix},$$

(2.201)

with $\omega = \sqrt{\mathbf{p}^2 + m^2}$. To proceed, we look for a canonical transformation such that the above quadratic part is diagonal in the new fields. This is achieved by introducing

$$\chi = \sqrt{\frac{\omega}{2}} \varphi + \frac{i}{\sqrt{2\omega}} \pi^*, \qquad \bar{\chi} = \sqrt{\frac{\omega}{2}} \varphi^* + \frac{i}{\sqrt{2\omega}} \pi$$

(2.202)

together with their complex conjugates. Inverted, these read $\varphi = (\chi + \bar{\chi}^*)/\sqrt{2\omega}$ and $\pi = i\sqrt{\omega/2}(\chi^* - \bar{\chi})$. In terms of the new fields, the quadratic part (2.201) becomes the sum of a particle $(+\mu)$ and an antiparticle $(-\mu)$ contribution:

$$i \int dt \int \frac{d^d p}{(2\pi)^d} [\chi^*(t, \mathbf{p})(i\partial_t - \omega + \mu)\chi(t, \mathbf{p}) + \bar{\chi}^*(t, \mathbf{p})(i\partial_t - \omega - \mu)\bar{\chi}(t, \mathbf{p})].$$

(2.203)

Using the fact that $\omega \simeq m + \mathbf{p}^2/(2m)$ and (2.196), we can write for the particle contribution in configuration space

$$i \int dt \int d^d x \left[\chi^*(t, \mathbf{x}) \left(i\partial_t + \frac{\nabla^2}{2m} + \mu_{nr} \right) \chi(t, \mathbf{x}) \right].$$

(2.204)

Of course, taking into account the interaction in (2.199) for $\lambda \neq 0$, with

$$\frac{\lambda}{12}(\varphi^*\varphi)^2 = \frac{\lambda}{48\omega^2}(\chi^*\chi + \chi\bar{\chi} + \chi^*\bar{\chi}^* + \bar{\chi}^*\bar{\chi})^2, \qquad (2.205)$$

we see that particles and antiparticles do not decouple in general. However, since in the nonrelativistic limit all characteristic scales are taken to be much smaller than the mass m, with $\mu - m \ll m \simeq \mu$, the term $\sim \bar{\chi}^*(m + \mu)\bar{\chi} \simeq 2m\bar{\chi}^*\bar{\chi}$ dominates the quadratic antiparticle part of the action. Loop corrections involving the "heavy" antiparticle modes are suppressed and the coupling terms involving antiparticle fields may be approximately neglected. In Section 2.6, we show that this is an excellent approximation for the infrared scaling regime near nonthermal fixed points.

With these approximations, the final result for the nonrelativistic limit of $Z[\mu] \simeq Z_{\mathrm{nr}}[\mu_{\mathrm{nr}}]$ reads

$$Z_{\mathrm{nr}}[\mu_{\mathrm{nr}}] = \int \mathscr{D}\chi^* \, \mathscr{D}\chi \exp\left\{ \mathrm{i} \int \mathrm{d}t \int \mathrm{d}^d x \left[\chi^* \left(\mathrm{i}\partial_t + \frac{\nabla^2}{2m} + \mu_{\mathrm{nr}} \right) \chi - \frac{g}{2}(\chi^*\chi)^2 \right] \right\}.$$

$$(2.206)$$

Here the nonrelativistic coupling g is not dimensionless and is given by

$$g = \frac{\lambda}{24m^2}. \qquad (2.207)$$

To study the nonequilibrium evolution, we can follow the developments of previous sections, now starting from the nonrelativistic action for the complex field $\chi(x)$ with $t \equiv x^0$ on the closed time path $\mathcal{C}$, which, for $\mu_{\mathrm{nr}} = 0$, reads

$$S[\chi, \chi^*] = \int_{\mathcal{C}} \mathrm{d}x^0 \int \mathrm{d}^d x \left[\chi^* \left(\mathrm{i}\partial_{x^0} + \frac{\nabla^2}{2m} \right) \chi - \frac{g}{2}(\chi^*\chi)^2 \right]. \qquad (2.208)$$

Varying this classical action with respect to the field gives the Gross–Pitaevskii equation employed in (2.30). The second derivative gives the inverse classical propagator. For a complex field, there are four combinations of the two derivatives with respect to χ or χ^*. These can be efficiently described using the index notation

$$\chi_1(x) \equiv \chi(x), \qquad \chi_2(x) \equiv \chi^*(x), \qquad (2.209)$$

such that the inverse classical propagator matrix reads, in terms of χ_a for $a = 1, 2$,

$$\frac{\delta^2 S[\chi, \chi^*]}{\delta\chi_a(x)\delta\chi_b^*(y)} = \left\{ \mathrm{i}\sigma_{ab}^3 \partial_{x^0} + \delta_{ab} \left[-\frac{\nabla^2}{2m} + \frac{g}{2}\chi_c(x)\chi_c^*(x) \right] + g\chi_a(x)\chi_b^*(x) \right\} \delta_{\mathcal{C}}(x - y).$$

$$(2.210)$$

Here $\sigma^3 = \mathrm{diag}(1, -1)$ denotes the third Pauli matrix.

Following the corresponding steps from Section 2.2.6, we can write down the quantum evolution equations. For instance, the equations for the nonrelativistic statistical two-point function $F_{ab}^{(\mathrm{nr})}(x,y)$ and the spectral function $\rho_{ab}^{(\mathrm{nr})}(x,y)$ in the absence of a condensate read[7]

$$
\left[i\sigma_{ac}^3 \partial_{x^0} - \Omega_{ac}(x)\right] F_{cb}^{(\mathrm{nr})}(x,y) = \int_{t_0}^{x^0} dz\, \Sigma_{ac}^{\rho(\mathrm{nr})}(x,z) F_{cb}^{(\mathrm{nr})}(z,y)
$$

$$
- \int_{t_0}^{y^0} dz\, \Sigma_{ac}^{F(\mathrm{nr})}(x,z) \rho_{cb}^{(\mathrm{nr})}(z,y), \qquad (2.211)
$$

$$
\left[i\sigma_{ac}^3 \partial_{x^0} - \Omega_{ac}(x)\right] \rho_{cb}^{(\mathrm{nr})}(x,y) = \int_{y^0}^{x^0} dz\, \Sigma_{ac}^{\rho(\mathrm{nr})}(x,z) \rho_{cb}^{(\mathrm{nr})}(z,y), \qquad (2.212)
$$

where

$$
\Omega_{ab}(x) = \delta_{ab}\left[-\frac{\nabla^2}{2m} + \frac{g}{2} F_{cc}^{(\mathrm{nr})}(x,x)\right] + g\, F_{ab}^{(\mathrm{nr})}(x,x) \qquad (2.213)
$$

and $\Sigma_{ab}^{F(\mathrm{nr})}(x,y)$ and $\Sigma_{ab}^{\rho(\mathrm{nr})}(x,y)$ denote the corresponding nonrelativistic statistical and spectral parts of the self-energies, respectively.

2.2.9 Bibliography

- 2PI or so-called Φ-derivable approximation schemes have been put forward in J. M. Luttinger and J. C. Ward. Ground state energy of a many fermion system. 2. *Phys. Rev.* **118** (1960) 1417; in G. Baym. Selfconsistent approximation in many body systems. *Phys. Rev.* **127** (1962) 1391; and in J. M. Cornwall, R. Jackiw, and E. Tomboulis. Effective action for composite operators. *Phys. Rev. D* **10** (1974) 2428.
- General discussions on 2PI or higher effective actions include H. Kleinert. Higher effective actions for Bose systems. *Fortschr. Phys.* **30** (1982) 187; A. N. Vasiliev. *Functional Methods in Quantum Field Theory and Statistical Physics.* Gordon & Breach (1998); and J. Berges. nPI effective action techniques for gauge theories. *Phys. Rev. D* **70** (2004) 105010.
- In the context of nonequilibrium dynamics, apart from L. P. Kadanoff and G. Baym, *Quantum Statistical Mechanics.* Benjamin (1962), some reviews are K. Chou, Z. Su, B. Hao, and L. Yu. Equilibrium and nonequilibrium formalisms made unified. *Phys. Rept.* **118** (1985) 1; E. Calzetta and B. L. Hu. Nonequilibrium quantum fields: closed-time-path effective action, Wigner function, and Boltzmann equation. *Phys. Rev. D* **37** (1988) 2878; P. Danielewicz. Quantum theory of nonequilibrium processes, I. *Ann. Phys. (NY)* **152** (1984) 239; and J. Berges, Introduction to nonequilibrium quantum field theory. *AIP Conf. Proc.* **739** (2005) 3 (arXiv:hep-ph/0409233).

---

[7] In full analogy to (2.128) and (2.129), we can write in terms of the Heisenberg field operators $\hat{\chi}_a$ the two-point functions $F_{ab}^{(\mathrm{nr})}(x,y) = \frac{1}{2}\langle\{\hat{\chi}_a(x), \hat{\chi}_b^\dagger(y)\}\rangle$ and $\rho_{ab}^{(\mathrm{nr})}(x,y) = i\langle[\hat{\chi}_a(x), \hat{\chi}_b^\dagger(y)]\rangle$.

- An introduction to the close relation between these techniques and the nonequilibrium functional renormalization group is given in J. Berges and D. Mesterhazy. Introduction to the nonequilibrium functional renormalization group. *Nucl. Phys. B Proc. Suppl.* **228** (2012) 37–60 (arXiv:1204.1489 [hep-ph]).
- Aspects of renormalization are discussed in H. van Hees and J. Knoll. Renormalization in self-consistent approximations schemes at finite temperature. I: Theory. *Phys. Rev. D* **65** (2012) 025010; J. P. Blaizot, E. Iancu, and U. Reinosa. Renormalization of phi-derivable approximations in scalar field theories. *Nucl. Phys. A* **736** (2004) 149; J. Berges, S. Borsanyi, U. Reinosa, and J. Serreau. Nonperturbative renormalization for 2PI effective action techniques. *Ann. Phys. (NY)* **320** (2005) 344; and U. Reinosa and J. Serreau. 2PI effective action for gauge theories: renormalization. *JHEP* **0607** (2006) 028. The renormalization of initial-value problems is specifically addressed in M. Garny and M. M. Muller. Kadanoff–Baym equations with non-Gaussian initial conditions: the equilibrium limit. *Phys. Rev. D* **80** (2009) 085011.
- For a discussion of the relation between relativistic and nonrelativistic theories, see, e.g., T. S. Evans. The condensed matter limit of relativistic QFT. In *Proc. 4th Workshop on Thermal Field Theories and their Applications, Dalian, China, 1995* (ed. Y. X. Gui et al.). World Scientific (1996) (arXiv:hep-ph/9510298).

2.3 Thermalization

2.3.1 Two-particle irreducible loop or coupling expansion

Expansion of the effective action

Loop or coupling expansions of the 2PI effective action proceed along the same lines as the corresponding expansions for the standard 1PI effective action, with the only differences being that

- contributions are parametrized in terms of "dressed" propagators, which are obtained from the stationarity condition as in (2.106), instead of classical propagators;
- only 2PI contributions are kept.

Using the decomposition (2.109), the 2PI effective action contains the contributions from the classical action supplemented by a saddle-point ("one-loop") correction and a 2PI contribution Γ_2. For the example of the $O(N)$-symmetric classical action (2.93) for the scalar field $\varphi_a(x)$, the contributions to $\Gamma_2[\phi, G]$ are constructed from the effective interaction

$$iS_{\text{int}}[\phi, \varphi] = -\int_{x,\mathcal{C}} i\frac{\lambda}{6N}\phi_a(x)\varphi_a(x)\varphi_b(x)\varphi_b(x) - \int_{x,\mathcal{C}} i\frac{\lambda}{4!N}[\varphi_a(x)\varphi_a(x)]^2. \quad (2.214)$$

As discussed in Section 2.2.5, this is obtained from the classical action by shifting $\varphi_a(x) \to \phi_a(x) + \varphi_a(x)$ and collecting all terms cubic and quartic in the fluctuating field $\varphi_a(x)$. As for the case of the 1PI effective action, in addition to the quartic interaction, there is thus an effective cubic interaction for nonvanishing field expectation value. To obtain loop or coupling expansions, we Taylor expand (2.214) using $e^x = \sum_n x^n/n!$ so that only Gaussian functional integrals have to be performed to compute the corrections.

To be specific, we consider first the scalar field theory for a one-component field, i.e., $N = 1$ in (2.214). To lowest order, we have $\Gamma_2[\phi, G] = 0$ and we recover the one-loop result given in (2.107). Further corrections can be very efficiently classified with a standard graphical notation. The propagator G is associated with a line and the interactions are represented as points where four (three) lines meet for a four-(three-)vertex:

$$G(x,y) \quad = \quad x \; \text{———} \; y,$$

$$\lambda \int_{x,\mathcal{C}} \quad = \quad \times_x, \qquad (2.215)$$

$$\lambda \int_{x,\mathcal{C}} \phi(x) \quad = \quad \downarrow_x \,.$$

Since $\Gamma_2[\phi, G]$ is a functional, which associates a number to the fields ϕ and G, only diagrams with closed loops of propagator lines and vertices can contribute. We can classify the contributions to $\Gamma_2[\phi, G]$ according to their number of closed loops, i.e.,

$$\Gamma_2[\phi, G] = \Gamma_2^{(2\text{loop})}[\phi, G] + \Gamma_2^{(3\text{loop})}[\phi, G] + \dots \qquad (2.216)$$

At two-loop order, there are two contributions: $\Gamma_2^{(2\text{loop})}[\phi, G] = \Gamma_2^{(2\text{a})}[G] + \Gamma_2^{(2\text{b})}[\phi, G]$. The field-independent contribution is

$$\Gamma_2^{(2\text{a})}[G] = -i\, 3 \left(-i\frac{\lambda}{4!}\right) \int_{x,\mathcal{C}} G^2(x,x) = -\frac{\lambda}{8} \; \text{OO}, \qquad (2.217)$$

where we have made explicit the different factors coming from the overall $-i$ in the defining functional integral for $\Gamma[\phi, G]$ (see (2.94)), the factor 3 from the three different ways to form closed loops with a single four-vertex, and the factor from the vertex itself. The field-dependent two-loop contribution is given by

$$\Gamma_2^{(2\text{b})}[\phi, G] = -i\, 6\frac{1}{2} \int_{xy,\mathcal{C}} \left(-i\frac{\lambda}{6}\phi(x)\right) \left(-i\frac{\lambda}{6}\phi(y)\right) G^3(x,y) = i\frac{\lambda^2}{12} \; \ominus, \qquad (2.218)$$

where the number of different ways to form closed loops from two three-vertices is 6, and the factor $\frac{1}{2}$ comes from the Taylor expansion of the exponential to second order. At three-loop order, there are three contributions, $\Gamma_2^{(3\text{loop})}[\phi, G] = \Gamma_2^{(3a)}[G] + \Gamma_2^{(3b)}[\phi, G] + \Gamma_2^{(3c)}[\phi, G]$, with

$$\Gamma_2^{(3a)}[G] = \mathrm{i}\,\frac{\lambda^2}{48}\int_{xy,\mathcal{C}} G^4(x,y) = \mathrm{i}\,\frac{\lambda^2}{48}\;\bigcirc\!\!\!\bigcirc\;,$$

$$\Gamma_2^{(3b)}[\phi, G] = \frac{\lambda^3}{8}\int_{xyz,\mathcal{C}} G^2(x,y)G^2(y,z)G(z,x)\phi(x)\phi(z) = \frac{\lambda^3}{8}\;\ominus\;,$$

(2.219)

$$\Gamma_2^{(3c)}[\phi, G] = -\,\mathrm{i}\,\frac{\lambda^4}{24}\int_{xyzw,\mathcal{C}} G(x,y)G(y,z)G(z,x)G(x,w)G(w,y)G(w,z)$$

$$\times\,\phi(x)\phi(y)\phi(z)\phi(w) = -\mathrm{i}\,\frac{\lambda^4}{24}\;\triangle\;.$$

Of course, care has to be taken when estimating the order of the coupling for a given contribution. For instance, at the classical level, for the simple case of a space–time-independent field ϕ, we can already see from $\delta S/\delta\phi(x) = 0$ that $m^2\phi+\lambda\phi^3/6 = 0$. As a consequence, for the case with spontaneous symmetry breaking with $\phi \neq 0$ and $m^2 < 0$, we find $\phi = \pm\sqrt{-6m^2/\lambda}$. Since, parametrically, $\phi \sim 1/\sqrt{\lambda}$, both of the above two-loop contributions should be taken into account to first order in the coupling λ. Similarly, all three-loop contributions should be included at the next order. For general out-of-equilibrium situations, the power counting can be much more involved, even in the presence of a small coupling. We will consider explicit examples of time-dependent or "dynamical" power counting schemes for far-from-equilibrium dynamics in later sections.

The 2PI loop expansion of course exhibits many fewer topologically distinct diagrams than the respective 1PI expansion. For instance, in the symmetric phase ($\phi = 0$), up to four loops, only a single diagram contributes at each order. At fifth order, there are two distinct diagrams, which are shown along with the lower-loop graphs in Fig. 2.4, with prefactors suppressed. For later discussion, we give here the results up to five loops for the $O(N)$-symmetric field theory with classical action (2.93) for general N. For $\phi = 0$, by virtue of $O(N)$ rotations, the propagator can be taken to be diagonal:

$$G_{ab}(x,y) = G(x,y)\delta_{ab}.\qquad(2.220)$$

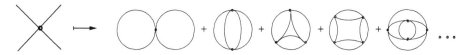

Fig. 2.4 Topologically distinct diagrams in the 2PI loop expansion shown to five-loop order for $\phi = 0$. The suppressed prefactors are given in (2.221)–(2.224).

This represents the most general form of the propagator for the symmetric regime if the initial conditions respect the $O(N)$ symmetry. To five-loop order with $\Gamma_2[G] \equiv \Gamma_2[\phi = 0, G]$, we find

$$\Gamma_2^{(2\text{loop})}[G]\Big|_{G_{ab} = G\delta_{ab}} = -\frac{\lambda}{8}\frac{(N+2)}{3}\int_{x,\mathcal{C}} G^2(x,x), \tag{2.221}$$

$$\Gamma_2^{(3\text{loop})}[G]\Big|_{G_{ab} = G\delta_{ab}} = \frac{i\lambda^2}{48}\frac{(N+2)}{3N}\int_{xy,\mathcal{C}} G^4(x,y), \tag{2.222}$$

$$\Gamma_2^{(4\text{loop})}[G]\Big|_{G_{ab} = G\delta_{ab}} = \frac{\lambda^3}{48}\frac{(N+2)(N+8)}{27N^2}\int_{xyz,\mathcal{C}} G^2(x,y)G^2(x,z)G^2(z,y), \tag{2.223}$$

$$\Gamma_2^{(5\text{loop})}[G]\Big|_{G_{ab} = G\delta_{ab}} =$$

$$-\frac{i\lambda^4}{128}\frac{(N+2)(N^2+6N+20)}{81N^3}\int_{xyzw,\mathcal{C}} G^2(x,y)G^2(y,z)G^2(z,w)G^2(w,x)$$

$$-\frac{i\lambda^4}{32}\frac{(N+2)(5N+22)}{81N^3}\int_{xyzw,\mathcal{C}} G^2(x,y)G(x,z)G(x,w)G^2(z,w)G(y,z)G(y,w). \tag{2.224}$$

Self-energies

In Section 2.2.6, we have derived the coupled evolution equations (2.145) for the statistical propagator F and the spectral function ρ, as well as the field equation (2.151). A systematic approximation to the exact equations can be obtained from the above loop or coupling expansion of the 2PI effective action. This determines all the required self-energies to a given order in the expansion from the variation of the 2PI effective action using (2.106). We emphasize that *all classifications of contributions are done for the effective action*. Once an approximation order is specified on the level of the effective action, one obtains a *closed set of equations* and no further approximations on the level of the evolution equations are required. This ensures the "conserving" properties of 2PI expansions such as energy conservation, since all approximate equations of motion are obtained from a variational principle.

For the computation of self-energies with

$$\Sigma(x,y) \equiv -i\Sigma^{(0)}(x)\delta_{\mathcal{C}}(x-y) + \overline{\Sigma}(x,y) = 2i\frac{\delta\Gamma_2[\phi,G]}{\delta G(x,y)}, \tag{2.225}$$

we again start with the one-component scalar field theory using the self-energy decomposition (2.134). The field-independent two-loop contribution (2.217) gives

$$2i\frac{\delta\Gamma_2^{(2a)}[G]}{\delta G(x,y)} = -i\frac{\lambda}{2}G(x,x)\delta_{\mathcal{C}}(x-y). \tag{2.226}$$

With the above self-energy decomposition, we obtain

$$\Sigma^{(0)} = \frac{\lambda}{2}F(x, x),\tag{2.227}$$

where we have used the fact that $\rho(x, x) = 0$. This contribution corresponds to a space–time-dependent mass shift in the evolution equations. Similarly, for the field-dependent contribution, we find

$$2\mathrm{i}\,\frac{\delta\Gamma_2^{(2\mathrm{b})}[\phi, G]}{\delta G(x, y)} = -\frac{\lambda^2}{2}G^2(x, y)\phi(x)\phi(y).\tag{2.228}$$

To write this in terms of statistical and spectral two-point functions, we use (2.131) and consider

$$G^2(x, y) = \left[F(x, y) - \frac{\mathrm{i}}{2}\mathrm{sgn}_{\mathcal{C}}(x^0 - y^0)\rho(x, y)\right]^2$$

$$= F^2(x, y) - \frac{1}{4}\rho^2(x, y) - \mathrm{i}\,\mathrm{sgn}_{\mathcal{C}}(x^0 - y^0)F(x, y)\rho(x, y),\tag{2.229}$$

where $\mathrm{sgn}_{\mathcal{C}}^2(x^0 - y^0) = 1$ has been employed. With the self-energy decomposition (2.136), this leads to

$$\Sigma^F(x, y) = -\frac{\lambda^2}{2}\left[F^2(x, y) - \frac{1}{4}\rho^2(x, y)\right]\phi(x)\phi(y),$$
$$\Sigma^\rho(x, y) = -\lambda^2 F(x, y)\rho(x, y)\phi(x)\phi(y).\tag{2.230}$$

It remains to compute the loop contributions to the field evolution equation (2.151), i.e.,

$$\frac{\delta\Gamma_2^{(2\mathrm{b})}[\phi, G]}{\delta\phi(x)} = \mathrm{i}\,\frac{\lambda^2}{6}\int_{y,\mathcal{C}}G^3(x, y)\phi(y).\tag{2.231}$$

Similar to the above steps, we consider

$$G^3(x, y) = \left[F(x, y) - \frac{\mathrm{i}}{2}\mathrm{sgn}_{\mathcal{C}}(x^0 - y^0)\rho(x, y)\right]^3$$

$$= F^3(x, y) - \frac{3\mathrm{i}}{2}\mathrm{sgn}_{\mathcal{C}}(x^0 - y^0)F^2(x, y)\rho(x, y)$$

$$- \frac{3}{4}F(x, y)\rho^2(x, y) + \frac{\mathrm{i}}{8}\mathrm{sgn}_{\mathcal{C}}(x^0 - y^0)\rho^3(x, y),\tag{2.232}$$

using, in particular, $\mathrm{sgn}_{\mathcal{C}}^3(x^0 - y^0) = \mathrm{sgn}_{\mathcal{C}}(x^0 - y^0)$. Following the discussion in Section 2.2.6, we split the contour integration as

$$\int_{\mathcal{C}} dy^0\, G^3(x,y) = \int_{t_0}^{x^0} dy^0 \left[-\frac{3i}{2} F^2(x,y)\rho(x,y) + \frac{i}{8}\rho^3(x,y) \right]$$

$$+ \int_{x_0}^{t^0} dy^0 \left[\frac{3i}{2} F^2(x,y)\rho(x,y) - \frac{i}{8}\rho^3(x,y) \right]$$

$$= -3i \int_{t_0}^{x^0} dy^0\, \rho(x,y) \left[F^2(x,y) - \frac{1}{12}\rho^2(x,y) \right]. \tag{2.233}$$

Putting everything together gives

$$\frac{\delta\Gamma_2^{(2b)}[\phi,G]}{\delta\phi(x)} = \frac{\lambda^2}{2} \int_{t_0}^{x^0} dy\, \rho(x,y) \left[F^2(x,y) - \frac{1}{12}\rho^2(x,y) \right] \phi(y), \tag{2.234}$$

where we have again used the abbreviated notation $\int_{t_1}^{t_2} dz \equiv \int_{t_1}^{t_2} dz^0 \int d^d z$. The evolution equations (2.145) and (2.151), together with the loop contributions (2.227), (2.230), and (2.234), represent a closed set of equations for $F(x,y)$, $\rho(x,y)$, and $\phi(x)$. A corresponding procedure can be carried out starting from the above three-loop expressions for $\Gamma_2[\phi,G]$, and similarly at higher loop orders.

For later discussions, we present here also the case of the scalar $O(N)$-symmetric field theory with a vanishing field expectation value starting from the three-loop Γ_2. The propagator in the symmetric regime can be written as (2.220). From $\Gamma_2[G]|_{G_{ab}=G\delta_{ab}}$, we obtain the self-energy as

$$\Sigma_{ab}(x,y) = 2i \frac{\delta\Gamma_2[G]|_{G_{ab}=G\delta_{ab}}}{\delta G_{ab}(x,y)} = 2i \frac{\delta_{ab}}{N} \frac{\delta\Gamma_2[G]|_{G_{ab}=G\delta_{ab}}}{\delta G(x,y)}, \tag{2.235}$$

where we have employed $G(x,y) = G_{ab}(x,y)\delta_{ab}/N$ for the last equality. Writing $\Sigma_{ab}(x,y) = \Sigma(x,y)\delta_{ab}$ and using the decomposition (2.134), we thus have

$$\Sigma(x,y) = \frac{2i}{N} \frac{\delta\Gamma_2[G]|_{G_{ab}=G\delta_{ab}}}{\delta G(x,y)}. \tag{2.236}$$

From the two-loop contribution to $\Gamma_2[G]$ given in (2.221), we find

$$\frac{2i}{N} \frac{\delta\Gamma_2^{(2)}[G]|_{G_{ab}=G\delta_{ab}}}{\delta G(x,y)} = -i\lambda \frac{(N+2)}{6N} G(x,x)\delta_{\mathcal{C}}(x-y). \tag{2.237}$$

Using the propagator decomposition (2.131) in spectral and statistical components, this leads to the one-loop self-energy

$$\Sigma^{(0)}(x) = \lambda \frac{N+2}{6N} F(x,x), \tag{2.238}$$

which corresponds to (2.227) for $N = 1$. Similarly, from $\Gamma_2[G]$ at three loops, we find

$$\overline{\Sigma}^{(2\text{loop})}(x,y) = -\lambda^2 \frac{N+2}{18N^2} G^3(x,y). \tag{2.239}$$

Using the decomposition (2.136) for the self-energy, we have

$$\Sigma_F(x,y) = -\lambda^2 \frac{N+2}{18N^2} F(x,y) \left[F^2(x,y) - \frac{3}{4}\rho^2(x,y) \right], \qquad (2.240)$$

$$\Sigma_\rho(x,y) = -\lambda^2 \frac{N+2}{6N^2} \rho(x,y) \left[F^2(x,y) - \frac{1}{12}\rho^2(x,y) \right], \qquad (2.241)$$

which enter (2.135) and (2.145).

Solving nonequilibrium evolution equations

The nonequilibrium evolution equations obtained from the loop expansion of the 2PI effective action can be used as a starting point for a wealth of further approximations that give analytical insight into the dynamics. For instance, we will later show that already two-loop self-energies as in (2.240) and (2.241) include and go beyond standard kinetic or Boltzmann equations describing two-to-two scattering of particles. However, it is very instructive to consider their numerical solution without further approximations. In particular, this will allow us to observe the striking process of thermalization in quantum field theory from first principles.

Despite the apparent complexity of the evolution equations (2.145) and (2.151), they can be very efficiently implemented and solved on a computer. Here it is important to note that all equations are explicit in time, i.e., all quantities at some time t can be obtained by integration over the explicitly known functions for earlier times $t_{\text{past}} < t$ for given initial conditions. The time-evolution equations (2.145) for $\rho(x,y)|_{x^0 = t_1, y^0 = t_2}$ and $F(x,y)|_{x^0 = t_1, y^0 = t_2}$ do not depend on the right-hand sides on $\rho(x,y)$ and $F(x,y)$ for times $x^0 \geq t_1$ and $y^0 \geq t_2$, and similarly for the field evolution equation (2.151). To see this, we note that the integrands vanish identically for the upper time limits of the memory integrals because of the antisymmetry of the spectral components, with $\rho(x,y)|_{x^0 = y^0} \equiv 0$ and $\Sigma_\rho(x,y)|_{x^0 = y^0} \equiv 0$. As a consequence, only explicitly known quantities at earlier times determine the time evolution of the unknowns at later times. *The numerical implementation therefore only involves sums over known functions.*

Solving the evolution equations requires the specification of (Gaussian) initial conditions as discussed in Section 2.2.4. We consider first the one-component scalar field theory for spatially homogeneous systems and Fourier transform with respect to the spatial coordinates. The equal-time commutation relations fix the Fourier modes of the spectral function $\rho(x^0, x^0, \mathbf{p}) = 0$, $\partial_{x^0}\rho(x^0, y^0; \mathbf{p}) = 1$, and $\partial_{x^0}\partial_{y^0}\rho(x^0, x^0; \mathbf{p}) = 0$ also at the initial time. Here we take

$$F(x^0, y^0; \mathbf{p})|_{x^0 = y^0 = 0} = \frac{1}{\omega_\mathbf{p}} \left[f_\mathbf{p}(0) + \tfrac{1}{2} \right],$$

$$\partial_{x^0}\partial_{y^0} F(x^0, y^0; \mathbf{p})|_{x^0 = y^0 = 0} = \omega_\mathbf{p} \left[f_\mathbf{p}(0) + \tfrac{1}{2} \right], \qquad (2.242)$$

$$\partial_{x^0} F(x^0, y^0; \mathbf{p})|_{x^0 = y^0 = 0} = 0, \qquad \phi(x^0)|_{x^0 = 0} = 0, \qquad \partial_{x^0}\phi(x^0)|_{x^0 = 0},$$

with $\omega_{\mathbf{p}} = \sqrt{\mathbf{p}^2 + m^2}$ as initial conditions. Since the initial macroscopic field and its derivative are taken to be zero, they remain so at all times for the considered $\mathbb{Z}_2$-symmetric theory with the corresponding invariance of the action under $\phi \to -\phi$. Here $f_{\mathbf{p}}(0)$ plays the role of an initial occupation number distribution, which, for instance, may be taken as

$$f_{\mathbf{p}}(0) = \mathcal{N} \exp\left[-\frac{1}{2\sigma^2}(|\mathbf{p}| - |\mathbf{p}_{\mathrm{ts}}|)^2 \right]. \tag{2.243}$$

The real parameter σ controls the width of the initial distribution and $\mathcal{N}$ is a normalization constant. The initial condition is clearly far from thermal equilibrium and reminiscent of two colliding high-energy wavepackets. More precisely, it describes a spatially homogeneous collection of particles that move with approximately the same momentum peaked around $\mathbf{p}_{\mathrm{ts}}$ and $-\mathbf{p}_{\mathrm{ts}}$. Of course, other initial distributions can be considered, which we will also do in the following.

In addition, we have to state renormalization conditions in order to define the quantum field theory, as for any renormalizable theory in vacuum or thermal equilibrium. From this perspective, it is rather simple to consider first a field theory in one spatial dimension. In this case, the one-loop correction (2.227) evaluated with a momentum regulator Λ shows a logarithmic regulator dependence. This can be analyzed at the initial time, where, for the spatially homogeneous initial conditions (2.242), we find for $f_{\mathbf{p}}(0) = 0$ that the large-momentum behavior for $d = 1$ is given by

$$\Sigma^{(0)}(t_0) = \frac{\lambda}{2} \int^{\Lambda} \frac{dp}{2\pi} \frac{1}{2\omega_{\mathbf{p}}} \sim \ln \Lambda. \tag{2.244}$$

We absorb this regulator dependence in a bare mass parameter μ with the replacement $m^2 \mapsto \mu^2 = m^2 - \delta m^2$ in (2.135). The counterterm δm^2 cancels the divergent vacuum contribution coming from the one-loop graph. The finite part of δm^2 is fixed by requiring that the renormalized one-loop mass parameter in vacuum ($f_{\mathbf{p}}(0) \equiv 0$) be equal to m and we express all dimensionful scales in units of m. Since higher loop corrections are finite for one spatial dimension as Λ is sent to infinity, the renormalization procedure is straightforward in this case.[8] A crucial question of nonequilibrium dynamics is how quickly the system *effectively* loses the details about the initial conditions. Thermal equilibrium keeps no memory about the time history except for the values of a few conserved charges. As a consequence, if the real scalar field theory approaches thermal equilibrium, then the late-time result will be uniquely determined by energy density.

As a first example, we show in Fig. 2.5 the time dependence of the equal-time propagator $F(t, t; \mathbf{p})$ for three Fourier modes $|\mathbf{p}|/m = 0, 3, 5$ and three very *different initial conditions with the same energy density*. The coupling is chosen for all runs

---

[8] The situation is different for higher spatial dimensions, where a consistent renormalization will require that non-Gaussian corrections be taken into account. For further reading on the renormalization of 2PI effective actions, see the literature in Section 2.2.9.

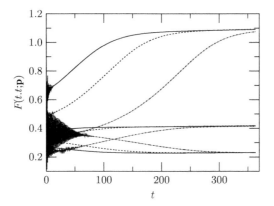

Fig. 2.5 Evolution of the equal-time two-point function $F(t, t; \mathbf{p})$ with Fourier modes $|\mathbf{p}| = 0$, 3, and 5 from the 2PI three-loop effective action. The evolution is shown for three very different nonequilibrium initial conditions with the same energy density. (All in units of m.)

to be $\lambda/m^2 = 10$. The large value suggests important corrections beyond the two-loop self-energy approximation; however, for the moment, it allows us to discuss the relevant qualitative properties in a pronounced way and we will consider the question of convergence in more detail later. For the solid line in Fig. 2.5, the initial conditions are close to a mean field thermal solution with inverse temperature $\beta = 0.1/m$, whereas the initial mode distributions for the dashed and the dashed–dotted lines deviate more and more substantially from a thermal equilibrium distribution. It can be seen that propagator modes with very different initial values but the same momentum $|\mathbf{p}|$ approach the same large-time value. The asymptotic behavior of the two-point function modes is uniquely determined by the initial energy density. Concerning the underlying nonequilibrium processes, we will see that this is also a very interesting observation in view of strong kinematic restrictions for scattering processes in one spatial dimension.

The results shown are obtained from a rather simple discretization of the equations for the Fourier components $F(t, t'; \mathbf{p})$ and $\rho(t, t'; \mathbf{p})$. They are based on a time discretization $t = na_t$, $t' = ma_t$ with step size a_t such that $F(t, t') \mapsto F(n, m)$, and

$$\partial_t^2 F(t, t') \mapsto \frac{1}{a_t^2}[F(n+1, m) + F(n-1, m) - 2F(n, m)], \qquad (2.245)$$

$$\int_0^t dt\, F(t, t') \mapsto a_t \left[F(0, m)/2 + \sum_{l=1}^{n-1} F(l, m) + F(n, m)/2 \right], \qquad (2.246)$$

where we have suppressed the momentum labels in the notation. Here the second derivative is replaced by a finite-difference expression that is symmetric in $a_t \leftrightarrow -a_t$. It is obtained by subsequently employing the lattice "forward derivative" $[F(n+1, m) - F(n, m)]/a_t$ and "backward derivative" $[F(n, m) - F(n-1, m)]/a_t$. The integral is approximated using the trapezoidal rule with the average function value

$[F(n, m) + F(n + 1, m)]/2$ in an interval of length a_t. The time-discretized version of (2.145) then reads, for the one-component theory,

$$
\begin{aligned}
F(n + 1, m; \mathbf{p}) =&\, 2F(n, m; \mathbf{p}) - F(n - 1, m; \mathbf{p}) \\
&- a_t^2 \left[\mathbf{p}^2 + m^2 + \frac{\lambda}{2} \int_{\mathbf{k}} F(n, n; \mathbf{k}) \right] F(n, m; \mathbf{p}) \\
&- a_t^3 \Bigg\{ \Sigma_\rho(n, 0; \mathbf{p}) F(0, m; \mathbf{p})/2 - \Sigma_F(n, 0; \mathbf{p}) \rho(0, m; \mathbf{p})/2 \\
&\quad + \sum_{l=1}^{m-1} \left[\Sigma_\rho(n, l; \mathbf{p}) F(l, m; \mathbf{p}) - \Sigma_F(n, l; \mathbf{p}) \rho(l, m; \mathbf{p}) \right] \\
&\quad + \sum_{l=m}^{n-1} \Sigma_\rho(n, l; \mathbf{p}) F(l, m; \mathbf{p}) \Bigg\},
\end{aligned}
\tag{2.247}
$$

and similarly for the spectral function. These equations are explicit in time: starting with $n = 1$, for the time step $n + 1$ one computes successively all entries with $m = 0, \ldots, n + 1$ from known functions at earlier times. This discretization leads already to stable numerics for small enough step size a_t, but the convergence properties may be easily improved with more sophisticated standard estimators if required.

As for the continuum case, the propagators satisfy the symmetry properties $F(n, m) = F(m, n)$ and $\rho(n, m) = -\rho(m, n)$. Consequently, only "half" of the (n, m)-matrices have to be computed and $\rho(n, n) \equiv 0$. Similarly, since the self-energy Σ_ρ is antisymmetric in time, one can exploit the fact that $\Sigma_\rho(n, n)$ vanishes identically. As initial conditions, one has to specify $F(0, 0; \mathbf{p})$, $F(1, 0; \mathbf{p})$, and $F(1, 1; \mathbf{p})$, while $\rho(0, 0; \mathbf{p})$, $\rho(1, 0; \mathbf{p})$, and $\rho(1, 1; \mathbf{p})$ are fixed by the equal-time commutation relations (2.130).

It is crucial for an efficient numerical implementation that each step forward in time does not involve the solution of a self-consistent or gap equation. This is manifest in the above discretization. The main numerical limitation of the approach is set by the time integrals ("memory integrals"), which grow with time and therefore slow down the numerical evaluation. Typically, the influence of early times on the late-time behavior is suppressed and can be neglected numerically in a controlled way. In this case, it is often sufficient to only take into account the contributions from the memory integrals for times much larger then a characteristic inverse damping rate. An error estimate then involves a series of runs with increasing memory time.

For scalars, one may use for the spatial dependence a standard lattice discretization with periodic boundary conditions. For a spatial volume $V = (N_s a_s)^d$ in d dimensions with lattice spacing a_s, one finds for the momenta

$$
\mathbf{p}^2 \mapsto \sum_{i=1}^{d} \frac{4}{a_s^2} \sin^2 \left(\frac{a_s p_i}{2} \right), \qquad p_i = \frac{2\pi n_i}{N_s a}, \tag{2.248}
$$

where $n_i = 0, \ldots, N_s - 1$. This can be easily understood from acting with the corresponding finite-difference expression (2.245) for space components: $\partial_x^2 e^{-ipx} \mapsto$

$e^{-ipx}(e^{ipa_s} + e^{-ipa_s} - 2)/a_s^2 = -e^{-ipx} 4\sin^2(pa_s/2)/a_s^2$. Exploiting the lattice symmetries reduces the number of independent lattice sites. Taken this explicitly into account becomes more important the larger the space dimension. For instance, on the lattice, there is only a subgroup of the rotational symmetry generated by the permutations of p_x, p_y, p_z and the reflections $p_x \leftrightarrow -p_x$, etc. for $d = 3$. Exploiting these lattice symmetries reduces the number of independent lattice sites to $(N_s+1)(N_s+3)(N_s+5)/48$. The self-energies may be calculated in coordinate space, where they are given by products of coordinate-space correlation functions, and then transformed back to momentum space. The coordinate-space correlation functions are available through fast Fourier transformation routines.

The lattice introduces a momentum cutoff π/a_s; however, the renormalized quantities should be insensitive to cutoff variations for sufficiently large π/a_s. In order to study the infinite-volume limit one has to remove finite-size effects. This may be done by increasing the volume until convergence of the results is observed. For time-evolution problems, the volume that is necessary to reach the infinite-volume limit to a given accuracy can depend on the timescale. This is, in particular, due to the fact that finite systems can show characteristic recurrence times after which an initial effective damping of oscillations can be reversed.

Nonequilibrium evolution of the spectral function

In order to understand the nonequilibrium dynamics of the spectral function ρ versus the statistical two-point function F, it is helpful to consider for a moment the free field theory. With $\Sigma = 0$ in the nonequilibrium evolution equations (2.145), the above initial conditions (2.242) lead to the plane-wave solutions

$$F^{(\text{free})}(x^0, y^0; \mathbf{p}) = \frac{1}{\omega_\mathbf{p}}\left[f_\mathbf{p}(0) + \tfrac{1}{2}\right]\cos\left[\omega_\mathbf{p}(x^0 - y^0)\right],$$
$$\rho^{(\text{free})}(x^0, y^0; \mathbf{p}) = \frac{1}{\omega_\mathbf{p}}\sin\left[\omega_\mathbf{p}(x^0 - y^0)\right]$$

(2.249)

with frequency $\omega_\mathbf{p}$. We may bring this into a more suggestive form by introducing the center coordinate $X^0 = (x^0 + y^0)/2$ and the relative coordinate $s^0 = x^0 - y^0$, with respect to which a Fourier transformation to momentum space is performed (Wigner transformation). To analyze the spectral function, we may perform a Wigner transformation and write

$$i\tilde\rho(X^0; \omega, \mathbf{p}) = \int_{-2X^0}^{2X^0} ds^0\, e^{i\omega s^0}\rho(X^0 + s^0/2, X^0 - s^0/2; \mathbf{p}).$$

(2.250)

The factor of i on the left-hand side is introduced so that $\tilde\rho(X^0; \omega, \mathbf{p})$ is real. Since we consider an initial-value problem with $x^0, y^0 \geq 0$, the time integral over s^0 is bounded by its maximum value for $y^0 = 0$ where $s^0 = 2X^0$ and its minimum value for $x^0 = 0$

where $s^0 = -2X^0$. After performing a Wigner transformation of the free spectral function (2.249), we find

$$\tilde{\rho}^{(\text{free})}(X^0; \omega, \mathbf{p}) = \frac{\sin[(\omega - \omega_{\mathbf{p}})2X^0]}{\omega_{\mathbf{p}}(\omega - \omega_{\mathbf{p}})} - \frac{\sin[(\omega + \omega_{\mathbf{p}})2X^0]}{\omega_{\mathbf{p}}(\omega + \omega_{\mathbf{p}})}. \tag{2.251}$$

For finite X^0, this spectral function shows a rapidly oscillating behavior, while its envelope is peaked at $\omega = \pm\omega_{\mathbf{p}}$. In the limit $X^0 \to \infty$, it reduces to

$$\tilde{\rho}^{(\text{free})}(\omega, \mathbf{p}) = 2\pi \, \text{sgn}(\omega) \, \delta \left(\omega^2 - \omega_{\mathbf{p}}^2 \right), \tag{2.252}$$

which describes the familiar form of the free spectral function in Fourier space. In general, the positivity condition $\text{sgn}(\omega)\tilde{\rho}(X^0; \omega, \mathbf{p}) \geq 0$ can only be shown to hold in the special case that the initial density matrix commutes with the full Hamiltonian, such as in thermal equilibrium. In this case, the system is of course stationary and independent of X^0. As a consequence, the interpretation of the Wigner-transformed nonequilibrium spectral function as the density of states should be taken with care. Nevertheless, as a consequence of the equal-time commutation relation, the Wigner transform obeys the sum rule $\int [d\omega/(2\pi)] \, \omega \tilde{\rho}(X^0; \omega, \mathbf{p}) = 1$ for the free as well as the interacting theory.

Going beyond the free theory, we know that the equilibrium spectral function would acquire a "width" in the presence of a nonzero imaginary part of the self-energy, i.e., $\Sigma^\rho \neq 0$.[9] Such a width in Fourier space is related to a damping of correlation functions in time. To obtain a qualitative understanding of this statement, we may consider the case of some spectral function that is assumed to be strictly exponentially damped, i.e., $\tilde{\rho}(x^0, y^0; \mathbf{p}) = e^{-\gamma_{\mathbf{p}}|x^0 - y^0|} E_{\mathbf{p}}^{-1} \sin[E_{\mathbf{p}}(x^0 - y^0)]$. The effective frequency $E_{\mathbf{p}}$ and rate $\gamma_{\mathbf{p}}$ are allowed to depend on the time X^0. The corresponding Wigner transform reads $\tilde{\rho}(X^0; \omega, \mathbf{p}) = \tilde{\rho}_{\text{BW}}(X^0; \omega, \mathbf{p}) + \delta\tilde{\rho}(X^0; \omega, \mathbf{p})$, where $\tilde{\rho}_{\text{BW}}$ denotes the Breit–Wigner function

$$\tilde{\rho}_{\text{BW}}(X^0; \omega, \mathbf{p}) = \frac{2\omega\Gamma_{\mathbf{p}}(X^0)}{[\omega^2 - E_{\mathbf{p}}^2(X^0)]^2 + \omega^2\Gamma_{\mathbf{p}}^2(X^0)} \tag{2.253}$$

with a width $\Gamma_{\mathbf{p}}(X^0) = 2\gamma_{\mathbf{p}}(X^0)$. The additional contribution $\delta\tilde{\rho}(X^0; \omega, \mathbf{p})$ vanishes exponentially as $\exp(-\Gamma_{\mathbf{p}}X^0)$, such that the finite-time effect is indeed suppressed for $X^0 \gg 1/\Gamma_{\mathbf{p}}$.

For the numerical solution, we solve the evolution equations (2.145) using the above two-loop self-energy contributions. The initial conditions are given by (2.242) with the initial distribution (2.243). For the plots, we have used a space lattice with spacing $ma = 0.3$ and a time lattice $a_0/a = 0.25$. The system size is $mL = 24$. In Fig. 2.6, we display the Wigner transform $\tilde{\rho}(X^0; \omega, p)$ for the zero-momentum mode for $mX^0 = 35.1$. It is clear that the interacting theory has a continuous spectrum

---

[9] For a vanishing ω dependence, the function $\Gamma^{(\text{eq})}(\omega, \mathbf{p}) \equiv \Sigma_\rho^{(\text{eq})}(\omega, \mathbf{p})/2\omega$ plays the role of a decay rate for one-particle excited states with momentum $\mathbf{p}$.

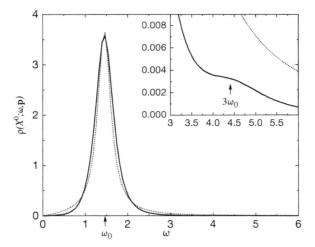

Fig. 2.6 Wigner transform of the spectral function as a function of ω at $X^0 = 35.1$ for $\mathbf{p} = 0$ in units of m. Also shown is a fit to a Breit–Wigner function (dotted) with $(\omega_0, \Gamma_0) = (1.46, 0.37)$. The inset shows a blow-up around the three-particle threshold $3\omega_0$. The expected bump is small but visible. Here the coupling is $\lambda/m^2 = 4$.

described by a peaked spectral function with a nonzero width. The inset shows a blow-up of the zero mode around the three-particle threshold $3\omega_0/m = 4.38$. The expected enhancement in the spectral function is small but visible. In the figure, we also present a fit to a Breit–Wigner spectral function. While the position of the peak can be fitted easily, the overall shape and width are only qualitatively captured. In particular, the slope of $\rho(X^0; \omega, p)$ for small ω is quantitatively different. We also see that the Breit–Wigner fit gives a narrower spectral function (smaller width) and therefore would predict a somewhat slower exponential relaxation in real time.

2.3.2 Two-particle irreducible $1/N$ expansion

Power counting

The loop or coupling expansion described in Section 2.3.1 is restricted, a priori, to weakly coupled systems. However, even in the presence of a weak coupling, the dynamics can be strongly correlated. Important examples concern systems with high occupation numbers, where the statistical correlation function F grows large. In this subsection, we discuss a nonperturbative approximation scheme for the 2PI effective action. It classifies the contributions to the 2PI effective action according to their scaling with powers of $1/N$, where N denotes the number of field components:

$$\Gamma_2[\phi, G] = \underbrace{\Gamma_2^{\text{LO}}[\phi, G]}_{\sim N^1} + \underbrace{\Gamma_2^{\text{NLO}}[\phi, G]}_{\sim N^0} + \underbrace{\Gamma_2^{\text{NNLO}}[\phi, G]}_{\sim N^{-1}}. \tag{2.254}$$

Each subsequent contribution Γ_2^{LO}, Γ_2^{NLO}, Γ_2^{NNLO}, etc. is down by an additional factor of $1/N$. The importance of an expansion in powers of $1/N$ stems from the fact that it provides a controlled expansion parameter that is not based on weak couplings. It can be applied to describe physics characterized by nonperturbatively large fluctuations, such as are encountered near second-order phase transitions in thermal equilibrium, or for extreme nonequilibrium phenomena such as parametric resonance and the subsequent emergence of strong turbulence, to mention just two examples that are discussed in later sections. For the latter cases, a 2PI coupling or loop expansion is not applicable. The method can be applied to bosonic or fermionic theories alike if a suitable field number parameter is available, and we exemplify it here for the case of the scalar $O(N)$-symmetric theory with classical action (2.93).

We present a classification scheme based on $O(N)$ invariants that parametrize the 2PI diagrams contributing to $\Gamma[\phi, G]$. The interaction term of the classical action in (2.93) is written such that $S[\phi]$ scales proportional to N. From the fields ϕ_a alone, one can construct only one independent invariant under $O(N)$ rotations, which can be taken as $\text{Tr}\{\phi\phi\} \equiv \phi^2 = \phi_a\phi_a \sim N$. The minimum ϕ_0 of the classical effective potential for this theory is given by $\phi_0^2 = N(-6m^2/\lambda)$ for negative mass-squared m^2 and scales proportional to N. Similarly, the trace with respect to the field indices of the classical propagator G_0 is of order N.

The 2PI effective action is a singlet under $O(N)$ rotations and parametrized by the two fields ϕ_a and G_{ab}. To write down the possible $O(N)$ invariants that can be constructed from these fields, we note that the number of ϕ fields has to be even in order to construct an $O(N)$ singlet. For a compact notation, we use $(\phi\phi)_{ab} \equiv \phi_a\phi_b$. All functions of ϕ and G, which are singlets under $O(N)$, can be built from the irreducible (i.e., nonfactorizable in field-index space) invariants

$$\phi^2, \qquad \text{Tr}\{G^n\}, \qquad \text{Tr}\{\phi\phi G^n\}. \tag{2.255}$$

We note that for given N, only the invariants with $n \leq N$ are irreducible—there cannot be more independent invariants than fields. We will see below that for lower orders in the $1/N$ expansion and for sufficiently large N, one has $n < N$. In particular, for the next-to-leading order approximation, one finds that only invariants with $n \leq 2$ appear, which makes the expansion scheme appealing from a practical point of view.

Since each single graph contributing to $\Gamma[\phi, G]$ is an $O(N)$ singlet, we can express them with the help of the set of invariants in (2.255). The factors of N in a given graph have two origins:

- Each irreducible invariant is taken to scale proportionally to N, since it contains exactly one trace over the field indices.
- Each vertex provides a factor of $1/N$.

The expression (2.109) for the 2PI effective action contains, beside the classical action, the one-loop contribution proportional to $\text{Tr}\{\ln G^{-1}\} + \text{Tr}\{G_0^{-1}(\phi)G\}$ and a nonvanishing $\Gamma_2[\phi, G]$ if higher loops are taken into account. The one-loop term contains both leading-order (LO) and next-to-leading-order (NLO) contributions. The logarithmic term corresponds, in the absence of other terms, simply to the free field

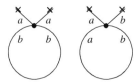

Fig. 2.7 Graphical representation of the ϕ-dependent contributions for $\Gamma_2 \equiv 0$. The crosses denote field insertions $\sim \phi_a \phi_a$ for the left diagram, which contributes at leading order, and $\sim \phi_a \phi_b$ for the right diagram, contributing at next-to-leading order.

effective action and scales proportionally to the number of field components N. To separate the LO and NLO contributions at the one-loop level, consider the second term $\text{Tr}\{G_0^{-1}(\phi)G\}$. From the form of the classical propagator (2.102), we see that it can be decomposed into a term proportional to $\text{Tr}\{G\} \sim N$ and terms $\sim (\lambda/6N)[\text{Tr}\{\phi\phi\}\text{Tr}\{G\} + 2\text{Tr}\{\phi\phi G\}]$. This can be seen as the sum of two "2PI one-loop graphs" with field insertions $\sim \phi_a \phi_a$ and $\sim \phi_a \phi_b$, respectively, as shown in Fig. 2.7. Counting the factors of N coming from the traces and the prefactor, we see that only the first contributes at LO, while the second is NLO.

According to the above rules, one draws all topologically distinct 2PI diagrams and counts the number of closed lines as well as the number of lines connecting two field insertions in a diagram following the indices. For instance, the left diagram of Fig. 2.7 admits one line connecting two field insertions and one closed line. In contrast, the right diagram admits only one line connecting two field insertions by following the indices. Therefore, the right diagram exhibits one factor of N less and becomes subleading. The same applies to the following two-loop diagram:

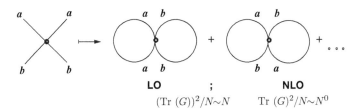

The same approach can be applied to higher orders. We consider first the contributions to $\Gamma_2[\phi = 0, G] \equiv \Gamma_2[G]$, i.e. for a vanishing field expectation value (we shall discuss $\phi \neq 0$ below). The LO contribution to $\Gamma_2[G]$ consists of only one two-loop graph, whereas to NLO, there is an infinite series of contributions, which can be analytically summed:

$$\Gamma_2^{\text{LO}}[G] = -\frac{\lambda}{4!N}\int_x G_{aa}(x,x)G_{bb}(x,x), \tag{2.256}$$

$$\Gamma_2^{\text{NLO}}[G] = \frac{i}{2}\text{Tr}\{\ln B(G)\}, \tag{2.257}$$

$$B(x,y;G) = \delta(x-y) + i\frac{\lambda}{6N}G_{ab}(x,y)G_{ab}(x,y). \tag{2.258}$$

To see that (2.257) with (2.258) sums the infinite series of diagrams

we can expand as follows:

$$
\mathrm{Tr}\left\{\ln B(G)\right\} = \int_x \left[\mathrm{i}\frac{\lambda}{6N}G_{ab}(x,x)G_{ab}(x,x) \right]
$$
$$
- \frac{1}{2}\int_{xy}\left[\mathrm{i}\frac{\lambda}{6N}G_{ab}(x,y)G_{ab}(x,y) \right]\left[\mathrm{i}\frac{\lambda}{6N}G_{a'b'}(y,x)G_{a'b'}(y,x) \right]
$$
$$
+ \dots \tag{2.259}
$$

The first term on the right-hand side of (2.259) corresponds to the two-loop graph with the index structure exhibiting one trace such that the contribution scales as $\mathrm{Tr}\left\{G^2\right\}/N \sim N^0$. We see that each additional contribution also scales proportionally to $(\mathrm{Tr}\left\{G^2\right\}/N)^n \sim N^0$ for all $n \geq 2$. Thus, all terms contribute at the same order.

The terms appearing in the presence of a nonvanishing field expectation value are obtained from the effectively cubic interaction term in (2.214) for $\phi \neq 0$. We first note that there is no ϕ-dependent graph contributing at LO. To NLO, there is again an infinite series of diagrams $\sim N^0$, which can be summed:

$$
\Gamma_2^{\mathrm{LO}}[\phi, G] \equiv \Gamma_2^{\mathrm{LO}}[G], \tag{2.260}
$$

$$
\Gamma_2^{\mathrm{NLO}}[\phi, G] = \Gamma_2^{\mathrm{NLO}}[\phi \equiv 0, G] + \mathrm{i}\frac{\lambda}{6N}\int_{xy} I(x,y;G)\phi_a(x)G_{ab}(x,y)\phi_b(y), \tag{2.261}
$$

$$
I(x,y;G) = \frac{\lambda}{6N}G_{ab}(x,y)G_{ab}(x,y) - \mathrm{i}\frac{\lambda}{6N}\int_z I(x,z;G)G_{ab}(z,y)G_{ab}(z,y). \tag{2.262}
$$

The series of terms contained in (2.261) with (2.262) corresponds to the diagrams

The functions $I(x,y;G)$ and the inverse of $B(x,y;G)$ are closely related by

$$
B^{-1}(x,y;G) = \delta(x-y) - \mathrm{i}I(x,y;G), \tag{2.263}
$$

which follows from convoluting (2.258) with B^{-1} and using (2.262). We note that B and I do not depend on ϕ, and $\Gamma_2[\phi, G]$ is only quadratic in ϕ at NLO.

It is straightforward to apply the above description of an expansion of the 2PI effective action in the number of field components to boson or fermion field theories with "vector-like" N-component fields. It is very helpful that for the N-component theory discussed above, one can analytically sum the infinite series of NLO contributions analytically. The situation is different for theories with "matrix" fields, such as the $SU(N)$ gauge theories relevant for QCD, where no corresponding closed expression is known.

Self-energies in the symmetric regime

We consider first the case of the scalar $O(N)$-symmetric field theory with a vanishing field expectation value such that $F_{ab}(x,y) = F(x,y)\delta_{ab}$ and $\rho_{ab}(x,y) = \rho(x,y)\delta_{ab}$. The case $\phi \neq 0$ is treated below. In the $1/N$ expansion of the 2PI effective action to NLO, the effective mass term $M^2(x; G)$ appearing in the evolution equations (2.145) is given by

$$M^2(x; F) = m^2 + \lambda \frac{N+2}{6N} F(x,x). \tag{2.264}$$

We see that this local self-energy part receives LO and NLO contributions. In contrast, the non-ocal part of the self-energy (2.134) is nonvanishing only at NLO, $\overline{\Sigma}(x,y;G) = -(\lambda/3N)G(x,y)I(x,y)$, and, using the decomposition identities (2.131) and (2.136), we find

$$\Sigma_F(x,y) = -\frac{\lambda}{3N}\left[F(x,y)I_F(x,y) - \tfrac{1}{4}\rho(x,y)I_\rho(x,y)\right], \tag{2.265}$$

$$\Sigma_\rho(x,y) = -\frac{\lambda}{3N}\left[F(x,y)I_\rho(x,y) + \rho(x,y)I_F(x,y)\right]. \tag{2.266}$$

Here the summation function (2.262) reads, in terms of its statistical and spectral components,[10]

$$I_F(x,y) = \Pi_F(x,y) - \int_{t_0}^{x^0} dz\, I_\rho(x,z)\Pi_F(z,y) + \int_{t_0}^{y^0} dz\, I_F(x,z)\Pi_\rho(z,y),$$

$$I_\rho(x,y) = \Pi_\rho(x,y) - \int_{y^0}^{x^0} dz\, I_\rho(x,z)\Pi_\rho(z,y), \tag{2.267}$$

where

$$\Pi_F(x,y) = \frac{\lambda}{6N}\left[F_{ab}(x,y)F_{ab}(x,y) - \tfrac{1}{4}\rho_{ab}(x,y)\rho_{ab}(x,y)\right],$$

$$\Pi_\rho(x,y) = \frac{\lambda}{3N}F_{ab}(x,y)\rho_{ab}(x,y), \tag{2.268}$$

---

[10] This follows from using the decomposition identity for the propagator (2.131) and $I(x,y) = I_F(x,y) - \tfrac{1}{2}iI_\rho(x,y)\,\mathrm{sgn}_{\mathcal{C}}(x^0 - y^0)$.

using the abbreviated notation $\int_{t_1}^{t_2} \mathrm{d}z \equiv \int_{t_1}^{t_2} \mathrm{d}z^0 \int_{-\infty}^{\infty} \mathrm{d}^d z$. Here we have kept in (2.268) the general notation without summing over field indices, since the same expressions will be used below for the case with a nonvanishing field expectation value. We note that $F(x,y)$ and $\rho(x,y)$, along with the other statistical and spectral components of the self-energies, are real functions.

Nonvanishing field expectation value

In the presence of a nonzero field expectation value ϕ_a the most general propagator can no longer be evaluated for the diagonal configuration (2.220). For the N-component scalar field theory (2.93), we have, with $\phi^2 \equiv \phi_a \phi_a$,

$$M_{ab}^2(x; \phi, F) = \left\{ m^2 + \frac{\lambda}{6N}[F_{cc}(x,x) + \phi^2(x)] \right\} \delta_{ab} + \frac{\lambda}{3N}[F_{ab}(x,x) + \phi_a(x)\phi_b(x)].$$

$$(2.269)$$

The self-energies $\Sigma_{ab}^F(x,y) \equiv \Sigma_{ab}^F(x,y; \phi, \rho, F)$ and $\Sigma_{ab}^\rho(x,y) \equiv \Sigma_{ab}^\rho(x,y; \phi, \rho, F)$ are obtained from (2.261) as

$$\Sigma_{ab}^F(x,y) = -\frac{\lambda}{3N}\left\{ I_F(x,y)[\phi_a(x)\phi_b(y) + F_{ab}(x,y)] - \tfrac{1}{4}I_\rho(x,y)\rho_{ab}(x,y) \right.$$
$$\left. + P_F(x,y)F_{ab}(x,y) - \tfrac{1}{4}P_\rho(x,y)\rho_{ab}(x,y) \right\},$$
$$(2.270)$$

$$\Sigma_{ab}^\rho(x,y) = -\frac{\lambda}{3N}\left\{ I_\rho(x,y)[\phi_a(x)\phi_b(y) + F_{ab}(x,y)] + I_F(x,y)\rho_{ab}(x,y) \right.$$
$$\left. + P_\rho(x,y)F_{ab}(x,y) + P_F(x,y)\rho_{ab}(x,y) \right\}.$$
$$(2.271)$$

The functions $I_F(x,y) \equiv I_F(x,y; \rho, F)$ and $I_\rho(x,y) \equiv I_\rho(x,y; \rho, F)$ satisfy the corresponding equations as for the case of a vanishing macroscopic field given above. The respective ϕ-dependent summation functions $P_F(x,y) \equiv P_F(x,y; \phi, \rho, F)$ and $P_\rho(x,y) \equiv P_\rho(x,y; \phi, \rho, F)$ are given by

$$P_F(x,y) = -\frac{\lambda}{3N}\left\{ H_F(x,y) - \int_{t_0}^{x^0} \mathrm{d}z \, [H_\rho(x,z)I_F(z,y) + I_\rho(x,z)H_F(z,y)] \right.$$

$$+ \int_{t_0}^{y^0} \mathrm{d}z \, [H_F(x,z)I_\rho(z,y) + I_F(x,z)H_\rho(z,y)]$$

$$- \int_{t_0}^{x^0} \mathrm{d}z \int_{t_0}^{y^0} \mathrm{d}v \, I_\rho(x,z)H_F(z,v)I_\rho(v,y)$$

$$+ \int_{t_0}^{x^0} \mathrm{d}z \int_{t_0}^{z^0} \mathrm{d}v \, I_\rho(x,z)H_\rho(z,v)I_F(v,y)$$

$$\left. + \int_{t_0}^{y^0} \mathrm{d}z \int_{z^0}^{y^0} \mathrm{d}v \, I_F(x,z)H_\rho(z,v)I_\rho(v,y) \right\},$$
$$(2.272)$$

$$P_\rho(x,y) = -\frac{\lambda}{3N}\left\{H_\rho(x,y) - \int_{y^0}^{x^0} dz\,[H_\rho(x,z)I_\rho(z,y) + I_\rho(x,z)H_\rho(z,y)]\right.$$

$$\left. + \int_{y^0}^{x^0} dz \int_{y^0}^{z^0} dv\, I_\rho(x,z)H_\rho(z,v)I_\rho(v,y)\right\}, \tag{2.273}$$

with

$$H_F(x,y) \equiv -\phi_a(x)F_{ab}(x,y)\phi_b(y), \qquad H_\rho(x,y) \equiv -\phi_a(x)\rho_{ab}(x,y)\phi_b(y). \tag{2.274}$$

The time evolution equation for the field (2.151) for the 2PI effective action to NLO (2.261) is given by

$$\left\{\left[\Box_x + \frac{\lambda}{6N}\phi^2(x)\right]\delta_{ab} + M_{ab}^2(x;\phi=0,F)\right\}\phi_b(x)$$

$$= \frac{\lambda}{3N}\int_{t_0}^{x^0} dy\,[I_\rho(x,y)F_{ab}(x,y) + I_F(x,y)\rho_{ab}(x,y)]\,\phi_b(y)$$

$$= -\int_{t_0}^{x^0} dy\,\Sigma_{ab}^\rho(x,y;\phi=0,F,\rho)\phi_b(y). \tag{2.275}$$

Dephasing at leading order

For simplicity, we consider spatially homogeneous field expectation values $\phi_a(t) = \langle\Phi_a(t,\mathbf{x})\rangle$, such that we can use the modes $F_{ab}(t,t';\mathbf{p})$ and $\rho_{ab}(t,t';\mathbf{p})$ after spatial Fourier transformation to describe the dynamics.

The LO contribution to the 2PI effective action (2.256) adds a time-dependent mass shift to the free field evolution equation. The resulting effective mass term, given by (2.269) for $N \to \infty$, is the same for all Fourier modes and consequently each mode propagates "collisionlessly." There are no further corrections, since, according to (2.257) and (2.261), the self-energies Σ_F and Σ_ρ are $\mathcal{O}(1/N)$ and vanish in this limit. Therefore, the evolution equations (2.145) for this approximation read

$$[\partial_t^2 + \mathbf{p}^2 + M^2(t;\phi,F)]\,F_{ab}(t,t';\mathbf{p}) = 0,$$

$$[\partial_t^2 + \mathbf{p}^2 + M^2(t;\phi,F)]\,\rho_{ab}(t,t';\mathbf{p}) = 0, \tag{2.276}$$

$$\left[\partial_t^2 + \frac{\lambda}{6N}\phi^2(t) + M^2(t;0,F)\right]\phi_b(t) = 0,$$

with

$$M^2(t;\phi,F) \equiv m^2 + \frac{\lambda}{6N}\left[\int_{\mathbf{p}} F_{cc}(t,t;\mathbf{p}) + \phi^2(t)\right], \tag{2.277}$$

where $\int_{\mathbf{p}} \equiv \int d^d p/(2\pi)^d$. In this case, we can see that the evolution of F and ϕ is decoupled from ρ. Similar to the free field theory limit, at LO the spectral function does not influence the time evolution of the statistical propagator. The reason is that in this approximation the spectrum consists only of "quasiparticle" modes of energy $\omega_\mathbf{p}(t) = \sqrt{\mathbf{p}^2 + M^2(t)}$ with an infinite lifetime. The associated mode particle numbers are conserved for each momentum separately. In contrast, in the interacting quantum field theory beyond LO, direct scattering processes occur such that mode occupancies can change. We note that as a consequence of this approximation, there are also no memory integrals appearing on the right-hand sides of (2.276).

As an example, we consider here the dynamics for $d = 1$ with the following initial conditions at $t_0 = 0$:

$$F(0,0;\mathbf{p}) = \frac{f_\mathbf{p}(0) + \frac{1}{2}}{\sqrt{\mathbf{p}^2 + M^2(0)}}, \qquad \partial_t F(t,0;\mathbf{p})|_{t=0} = 0,$$

$$\qquad\qquad (2.278)$$

$$F(0,0;\mathbf{p})\partial_t\partial_{t'} F(t,t';\mathbf{p})|_{t=t'=0} = \left[f_\mathbf{p}(0) + \tfrac{1}{2}\right]^2,$$

$$\phi(0) = \partial_t\phi(t)|_{t=0} = 0. \qquad\qquad (2.279)$$

We refer to the suggested literature in Section 2.3.4 for further studies in dimensions $d > 1$ and for different initial conditions. Here $F_{ab}(t,t';\mathbf{p}) = F(t,t';\mathbf{p})\delta_{ab}$, which is valid for all times with these initial conditions. The mass term $M^2(0)$ is given by the gap equation (2.277) in the presence of the initial nonthermal particle number distribution (2.243).

As the renormalization condition we choose the initial renormalized mass in vacuum, $m_R \equiv M(0)|_{f(0)=0} = 1$, as our dimensionful scale. In these units, the particle number is peaked around $|\mathbf{p}| = p_\mathrm{ts} = 5m_R$ with a width determined by $\sigma = 0.5m_R$ and amplitude $\mathcal{N} = 10$. We consider the effective coupling $\lambda/(6m_R^2) = 1$.

Figure 2.8(a) shows the time evolution of the equal-time correlation modes $F(t,t;\mathbf{p})$ for different momenta: zero momentum, a momentum close to the maximally populated momentum p_ts, and about twice p_ts. We can see that the equal-time correlations at LO are strictly constant in time. This behavior can be understood from the fact that for the initial condition employed the evolution starts at a time-translation-invariant nonthermal solution of the LO equations. There is an infinite number of LO solutions that are constant in time, depending on the chosen initial condition details. This is in sharp contrast to the well-founded expectation that the late-time behavior may be well described by thermal equilibrium physics. Scattering effects included in the NLO approximation indeed drive the evolution toward thermal equilibrium, which is discussed in detail in Section 2.3.2.

A remaining question is what happens at LO if the time evolution does not start from a time-translation-invariant nonthermal solution of the LO equations, as was the case for the initial condition employed above. Figure 2.9(a) shows the evolution of $M^2(t)$ in the LO approximation as a function of time t, following a "quench" described by an instant drop in the effective mass term from $2M^2$ to M^2 at initial time. As a consequence, the mass term appearing in the initial conditions (2.279) is not the same

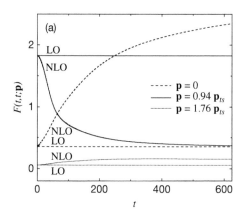
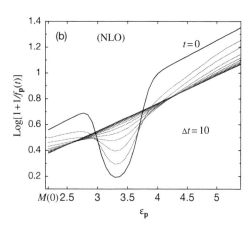

Fig. 2.8 (a) Comparison of the LO and NLO time dependences of the equal-time correlation modes $F(t, t; \mathbf{p})$ for the initial condition (2.278). The importance of scattering included in the NLO approximation is apparent: the evolution deviates from the nonthermal LO solution and the initially peaked distribution decays, approaching thermal equilibrium at late times. (b) Effective particle number distribution for a peaked initial distribution in the presence of a thermal background. The solid line shows the initial distribution, which for low and for high momenta follows a Bose–Einstein distribution, i.e., $\ln[1 + 1/f_{\mathbf{p}}(0)] \simeq \epsilon_{\mathbf{p}}(0)/T_0$. At late times, the nonthermal distribution equilibrates and approaches a straight line with inverse slope $T_{\mathrm{eq}} > T_0$.

as the one appearing in the evolution equations (2.276). The initial particle number distribution is

$$f_{\mathbf{p}}(0) = \frac{1}{\exp\left[\sqrt{\mathbf{p}^2 + 2M^2(0)}/T_0\right] - 1}, \qquad (2.280)$$

with $T_0 = 2M(0)$ and $\phi(0) = \dot{\phi}(0) = 0$. The sudden change in the effective mass term drives the system out of equilibrium, and one can study its relaxation.

In Fig. 2.9(a), we plot $M^2(t)$ for three different couplings $\lambda = \lambda_0 \equiv 0.5M^2(0)$ (bottom curve), $\lambda = 10\lambda_0$ (middle curve), and $\lambda = 40\lambda_0$ (top curve). All quantities are given in units of appropriate powers of $M(0)$. Therefore, all curves in Fig. 2.9(a) start at 1. The time-dependent mass squared $M^2(t)$ shoots up in response to the "quench" and stays below the value $2M(0)^2$ of the initial distribution. The amplitude of initial oscillations is quickly reduced and the evolution rapidly approaches constant asymptotic values. Since there are no direct scatterings included in this approximation, the damping is due to a simple dephasing phenomenon of harmonic oscillations with a time-dependent frequency shift. Correspondingly, the correlator $F(t, t'; \mathbf{p})$, which is a function of the relative time $t - t'$ and of $t + t'$ at early times, loses its dependence on $t + t'$ asymptotically such that $F(t, t; \mathbf{p})$ becomes a constant.

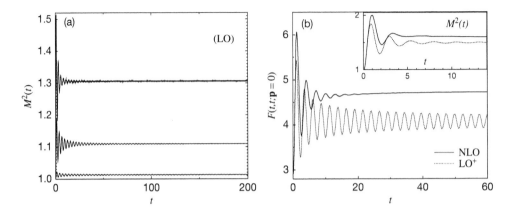

Fig. 2.9 (a) Time-dependent mass term $M^2(t)$ in the LO approximation for three different couplings following a "quench." All quantities are given in units of appropriate powers of $M(0)$. (b) Time dependence of the equal-time zero-mode $F(t, t; \mathbf{p} = 0)$ after a "quench." The inset shows the mass term $M^2(t)$, which includes a sum over all modes. The dotted lines represent the Hartree approximation (LO$^+$), while the solid lines give the NLO results. The coupling is $\lambda/6N = 0.17M^2(0)$ for $N = 4$.

To understand this in more detail, we compare the asymptotic values with the self-consistent solution of the LO mass equation (2.277) for constant mass squared M^2_{gap} and given particle number distribution $f_{\mathbf{p}}(0)$:[11]

$$M^2_{\text{gap}} = m^2 + \frac{\lambda}{6} \int \frac{\mathrm{d}p}{2\pi} \left[f_{\mathbf{p}}(0) + \tfrac{1}{2} \right] \frac{1}{\sqrt{\mathbf{p}^2 + M^2_{\text{gap}}}}. \tag{2.281}$$

The result from this gap equation is $M^2_{\text{gap}} = \{1.01, 1.10, 1.29\}M^2(0)$ for the three values of λ, respectively. For this wide range of couplings, the values are in good numerical agreement with the corresponding dynamical large-time results, which can be inferred from Fig. 2.9(a) as $\{1.01, 1.11, 1.31\}M^2(0)$. We can conclude that the asymptotic behavior at LO is well described in terms of the initial particle number distribution $f_{\mathbf{p}}(0)$. We emphasize that the latter is not a thermal distribution for the late-time mass terms with values smaller than $2M^2(0)$.

The question of how strongly the LO late-time results deviate from thermal equilibrium depends of course on the details of the initial conditions. Typically, time- and/or momentum-averaged quantities are better determined by the LO approximation than quantities characterizing a specific momentum mode. This is related to the prethermalization of characteristic bulk quantities, and we refer for further discussions of this topic to the literature listed in Section 2.3.4.

---

[11] Here the logarithmic divergence of the one-dimensional integral is absorbed into the bare mass parameter m^2 using the same renormalization condition as for the dynamical evolution in the LO approximation, i.e., $m^2 + \frac{1}{6}\lambda \int (\mathrm{d}p/2\pi) \left[f_{\mathbf{p}}(0) + \tfrac{1}{2} \right] \left[\mathbf{p}^2 + M^2(0) \right]^{-1/2} = M^2(0)$.

Figure 2.9(b) shows the equal-time zero-mode $F(t, t; \mathbf{p} = 0)$ along with $M^2(t)$, which includes the sum over all modes for comparison. Here we employ a "quench" with a larger drop in the effective mass term from $2.9M^2(0)$ to $M^2(0)$. The initial particle number distribution is

$$f_{\mathbf{p}}(0) = \frac{1}{\exp\left[\sqrt{\mathbf{p}^2 + M^2(0)}/T_0\right] - 1}, \tag{2.282}$$

with $T_0 = 8.5M(0)$. The dotted curves show the dynamics obtained from an "improved" LO (Hartree) approximation LO^+ that takes into account the local part of the NLO self-energy contribution and is often employed in the literature. The resulting equations have the very same structure as the LO ones, however, with the LO and NLO contribution to the mass term $M^2(t)$ included as given by (2.264). The large-time limit of the mass term in the LO^+ approximation is determined by the LO^+ stationary solution in complete analogy to the above discussion. Figure 2.9(b) also shows the NLO results, which are discussed below.

The effective loss of details about the initial conditions is a prerequisite for thermalization. At LO, this is obstructed by an infinite number of spurious conserved quantities (mode particle numbers), which retain initial-time information. An important quantity in this context is the unequal-time two-point function $F(t, 0; \mathbf{p})$, which characterizes the correlations with the initial time. Clearly, if thermal equilibrium is approached, then these correlations should be damped. In Fig. 2.10(a), the dotted curve shows the unequal-time zero-mode $F(t, 0; \mathbf{p} = 0)$ following the same "quench"

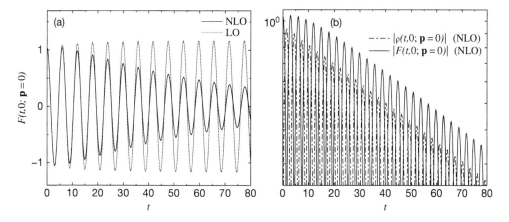

Fig. 2.10 (a) Evolution of the unequal-time correlation $F(t, 0; \mathbf{p} = 0)$ after a "quench." Unequal-time correlation functions approach zero in the NLO approximation and correlations with early times are effectively suppressed ($\lambda/6N = (5/6N \simeq 0.083)M^2(0)$ for $N = 10$). In contrast, there is no decay of correlations with earlier times for the LO approximation. (b) The logarithmic plot of $|\rho(t, 0; \mathbf{p} = 0)|$ and $|F(t, 0; \mathbf{p} = 0)|$ as a function of time t shows an oscillation envelope that rapidly approaches a straight line. At NLO, the correlation modes therefore approach an exponentially damped behavior. (All in units of $M(0)$.)

at LO as for Fig. 2.9(a). We can see no decay of correlations with earlier times for the LO approximation. Scattering effects entering at NLO are crucial for a sufficient effective loss of memory about the initial conditions, which is discussed next.

Thermalization at NLO

In contrast to the LO approximation, at NLO, the self-energies Σ_F and $\Sigma_\rho \sim \mathcal{O}(1/N)$ do not vanish. For the initial conditions (2.278), all correlators are diagonal in field index space and $\phi \equiv 0$ for all times. In this case, the evolution equations derived from the NLO 2PI effective action (2.257) are given by (2.145) with the self-energies (2.264)–(2.267). As discussed above, the NLO evolution equations are causal equations with characteristic "memory" integrals, which integrate over the time history of the evolution, taken to start at time $t_0 = 0$.

We consider first the same initial condition (2.278) as for the LO case discussed above. The result is shown in Fig. 2.8(a) for $N = 10$. We can see that the NLO corrections quickly lead to a decay of the initially high population of modes around p_{ts}. On the other hand, low-momentum modes get populated such that thermal equilibrium is approached at late times.

In order to make this apparent, we can plot the results in a different way. For this, we note that according to (2.278), the statistical propagator corresponds to the ratio of the *particle number distribution*

$$f_{\mathbf{p}}(t) + \tfrac{1}{2} = \left[F(t, t; \mathbf{p}) K(t, t; \mathbf{p}) - Q^2(t, t; \mathbf{p}) \right]^{1/2} \tag{2.283}$$

at initial time $t = 0$ and the corresponding mode energy. Here we have defined

$$
\begin{aligned}
K(t, t'; \mathbf{p}) &\equiv \partial_t \partial_{t'} F(t, t'; \mathbf{p}), \\
Q(t, t'; \mathbf{p}) &\equiv \tfrac{1}{2} \left[\partial_t F(t, t'; \mathbf{p}) + \partial_{t'} F(t, t'; \mathbf{p}) \right].
\end{aligned}
\tag{2.284}
$$

Since we employed $Q(0, 0; \mathbf{p}) = 0$, for the initial conditions (2.278), the *mode energy* is given by

$$\epsilon_{\mathbf{p}}(t) = \left[\frac{K(t, t; \mathbf{p})}{F(t, t; \mathbf{p})} \right]^{1/2}, \tag{2.285}$$

at initial time, such that $F(t, t; \mathbf{p}) = \left[f_{\mathbf{p}}(t) + \tfrac{1}{2} \right] / \epsilon_{\mathbf{p}}(t)$. For illustration of the results, we may use (2.283) and (2.285) for times $t > 0$ in order to *define an effective mode particle number and energy*, where we set $Q(t, t'; \mathbf{p}) \equiv 0$ in (2.283) for the moment.

The behavior of the effective particle number is illustrated in Fig. 2.8(b), where we plot $\ln[1 + 1/f_{\mathbf{p}}(t)]$ as a function of $\epsilon_{\mathbf{p}}(t)$. For this plot, we have employed an initial peaked distribution at $p_{\text{ts}}/m_{\text{R}} = 2.5$, where we have added an initial "thermal background" distribution with temperature $T_0/m_{\text{R}} = 4$.[12] Correspondingly, from the solid line ($t = 0$) in Fig. 2.8(b), we see the initial "thermal background" as a straight line

---

[12] The initial mass term is $M(0)/m_{\text{R}} = 2.24$ and $\lambda/6N = 0.5m_{\text{R}}^2$ for $N = 4$.

distorted by the nonthermal peak. The curves represent snapshots at equidistant time steps $\Delta t\, m_{\mathrm{R}} = 10$. After rapid changes in $f_{\mathbf{p}}(t)$ at early times, the subsequent curves converge to a straight line to high accuracy, with inverse slope $T_{\mathrm{eq}}/T_0 = 1.175$. The initial high occupation number in a small momentum range decays quickly with time. More and more low-momentum modes become populated, and the particle distribution approaches a thermal shape.

The crucial importance of the NLO corrections for the nonequilibrium dynamics can also be observed for other initial conditions. Figure 2.9(b) shows the equal-time zero-mode $F(t,t;\mathbf{p}=0)$, along with $M^2(t)$, including the sum over all modes, following a "quench" as described in Section 2.3.2. While the dynamics for vanishing self-energies Σ_F and Σ_ρ is quickly dominated by the spurious LO stationary solutions, this is no longer the case once the NLO self-energy corrections are included.

In particular, we observe a very efficient damping of oscillations at NLO. This becomes even more pronounced for unequal-time correlators, as shown for $F(t,0;\mathbf{p}=0)$ in Fig. 2.10(a). In Fig. 2.10(b), the approach to an approximately exponential behavior is demonstrated for $|\rho(t,0;\mathbf{p}=0)|$ and $|F(t,0;\mathbf{p}=0)|$ with the same parameters as for Fig. 2.10(a). The logarithmic plot shows that after a nonexponential period at early times, the envelope of oscillations can be well approximated by a straight line.

The strong qualitative difference between LO and NLO appears because an infinite number of spurious conserved quantities are removed once scattering is taken into account. It should be emphasized that the step going from LO to NLO is qualitatively very different to the one going from NLO to NNLO or further. In order to understand better what happens on going from LO to NLO, we consider again the effective particle number (2.283). It is straightforward by taking the time derivative on both sides of (2.283) to obtain an evolution equation for $f_{\mathbf{p}}(t)$ with the help of the relations (2.145):

$$
\begin{aligned}
&\left[f_{\mathbf{p}}(t) + \tfrac{1}{2}\right]\partial_t f_{\mathbf{p}}(t) \\
&= \int_{t_0}^t dt'' \left\{ \left[\Sigma_\rho(t,t'';\mathbf{p})F(t'',t;\mathbf{p}) - \Sigma_F(t,t'';\mathbf{p})\rho(t'',t;\mathbf{p})\right]\partial_t F(t,t';\mathbf{p})|_{t=t'} \right. \\
&\quad \left. - \left[\Sigma_\rho(t,t'';\mathbf{p})\partial_t F(t'',t;\mathbf{p}) - \Sigma_F(t,t'';\mathbf{p})\partial_t \rho(t'',t;\mathbf{p})\right]F(t,t;\mathbf{p}) \right\}.
\end{aligned}
$$

$$(2.286)$$

Here t_0 denotes the initial time, which was set to zero in (2.145) without loss of generality. Since $\Sigma_F \sim \mathcal{O}(1/N)$ as well as Σ_ρ, we see directly that at LO, i.e., for $N \to \infty$, the particle number for each momentum mode is strictly conserved: $\partial_t f_{\mathbf{p}}(t) \equiv 0$ at LO. Stated differently, (2.283) just specifies the infinite number of additional constants of motion that appear at LO. In contrast, once corrections beyond LO are taken into account, (2.283) no longer represent conserved quantities.

2.3.3 Transport equations

We have seen in the previous sections that after a transient rapid evolution, which is characterized by damped oscillations of correlation functions with frequency determined by the renormalized mass for zero spatial momentum, the subsequent evolution of equal-time correlation functions such as $F(t, t; \mathbf{p})$ becomes comparably smooth. Therefore, after the oscillations become effectively damped out, an approximate description taking into account only low orders in an expansion in derivatives with respect to the center coordinates $\sim (t + t')$ is expected to be suitable. This will be used in the following to obtain a set of equations, which form the basis of kinetic theory and which will be used to describe the transport of conserved quantities in Section 2.6.

Gradient expansion

We first note that the spectral function (2.128) is directly related to the retarded propagator G_R, or the advanced one G_A, by

$$G_R(x, y) = \Theta(x^0 - y^0)\rho(x, y), \qquad G_A(x, y) = -\Theta(y^0 - x^0)\rho(x, y). \qquad (2.287)$$

Similarly, the retarded and advanced self-energies are

$$\Sigma_R(x, y) = \Theta(x^0 - y^0)\Sigma_\rho(x, y), \qquad \Sigma_A(x, y) = -\Theta(y^0 - x^0)\Sigma_\rho(x, y). \qquad (2.288)$$

Here we consider the case of a vanishing field expectation value, $\phi(x) = 0$. With the help of the above notation, interchanging x and y in the evolution equation (2.145) for $F(x, y)$, and subtraction, we obtain

$$\left[\Box_x - \Box_y + M^2(x) - M^2(y)\right] F(x, y)$$
$$= \int \mathrm{d}^{d+1} z \, \theta(z^0) \left[F(x, z)\Sigma_A(z, y) + G_R(x, z)\Sigma_F(z, y) \right.$$
$$\left. - \Sigma_R(x, z)F(z, y) - \Sigma_F(x, z)G_A(z, y) \right]. \qquad (2.289)$$

The same procedure yields for the spectral function

$$\left[\Box_x - \Box_y + M^2(x) - M^2(y)\right] \rho(x, y)$$
$$= \int \mathrm{d}^{d+1} z \left[G_R(x, z)\Sigma_\rho(z, y) + \rho(x, z)\Sigma_A(z, y) \right.$$
$$\left. - \Sigma_\rho(x, z)G_A(z, y) - \Sigma_R(x, z)\rho(z, y) \right]. \qquad (2.290)$$

So far, the equations (2.289) and (2.290) are fully equivalent to the exact equations (2.145).

Transport equations are obtained by prescribing F, ρ, and derivatives at a *finite* time using the equations with $t_0 \to -\infty$ as an approximate description. Furthermore, one employs a gradient expansion to (2.289) and (2.290). In practice, this expansion is carried out to low order in the number of derivatives with respect to the center coordinates

$$X^\mu \equiv \frac{x^\mu + y^\mu}{2} \qquad (2.291)$$

and powers of the relative coordinates

$$s^\mu \equiv x^\mu - y^\mu. \tag{2.292}$$

Even for finite X^0, one assumes that the relative-time coordinate s^0 ranges from $-\infty$ to ∞ in order to achieve a convenient description in Wigner space, i.e., in Fourier space with respect to the relative coordinates (2.292). This requires a loss of information about the details of the initial state, which enters in the derivation as an assumption, and limits the use of the approximate equations to not too early times.

For the description in Wigner space, we introduce the Fourier transforms with respect to the relative coordinates, such as

$$
\begin{aligned}
F(X,\omega,\mathbf{p}) &= \int_{-2X^0}^{2X^0} ds^0\, e^{i\omega s^0} \int_{-\infty}^{\infty} d^d s\, e^{-i\mathbf{p}\cdot\mathbf{s}} F\left(X + \tfrac{1}{2}s, X - \tfrac{1}{2}s\right) \\
&= 2\int_{0}^{2X^0} ds^0\, \cos(\omega s^0) \int_{-\infty}^{\infty} d^d s\, e^{-i\mathbf{p}\cdot\mathbf{s}} F\left(X + \tfrac{1}{2}s, X - \tfrac{1}{2}s\right), \tag{2.293}
\end{aligned}
$$

using the symmetry property $F(x,y) = F(y,x)$ for the second line. We emphasize that the time integral over s^0 is bounded by $\pm 2X^0$. The time-evolution equations are initialized at time $x^0 = y^0 = 0$ such that $x^0 \geq 0$ and $y^0 \geq 0$. According to (2.292), the minimum value of the relative coordinate s^0 is then given by $-y^0 = -2X^0$ for $x^0 = 0$, while its maximum value is $x^0 = 2X^0$ for $y^0 = 0$, as illustrated in Fig. 2.11. Similarly, we define

$$
\begin{aligned}
\tilde{\rho}(X,\omega,\mathbf{p}) &= -i\int_{-2X^0}^{2X^0} ds^0\, e^{i\omega s^0} \int_{-\infty}^{\infty} d^d s\, e^{-i\mathbf{p}\cdot\mathbf{s}} \rho\left(X + \tfrac{1}{2}s, X - \tfrac{1}{2}s\right) \\
&= 2\int_{0}^{2X^0} ds^0\, \sin(\omega s^0) \int_{-\infty}^{\infty} d^d s\, e^{-i\mathbf{p}\cdot\mathbf{s}} \rho\left(X + \tfrac{1}{2}s, X - \tfrac{1}{2}s\right), \tag{2.294}
\end{aligned}
$$

where a factor of i has been included in the definition so that $\tilde{\rho}(X,\omega,\mathbf{p})$ is real and we have used $\rho(x,y) = -\rho(y,x)$. The equivalent transformations are done to obtain the self-energies $\Sigma_F(X,\omega,\mathbf{p})$ and $\tilde{\Sigma}_\rho(X,\omega,\mathbf{p})$.

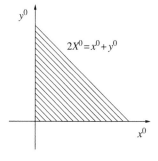

Fig. 2.11 For given finite $2X^0 = x^0 + y^0$, the relative coordinate $s^0 = x^0 - y^0$ has a finite range from $-2X^0$ to $2X^0$.

In order to exploit the convenient properties of a Fourier transform, it is a standard procedure to extend the limits of the integrals over the relative-time coordinate in (2.293) and (2.294) to $\pm\infty$. For instance, using the chain rule

$$\Box_x - \Box_y = 2\frac{\partial}{\partial s_\mu}\frac{\partial}{\partial X^\mu} \tag{2.295}$$

and the gradient expansion of the mass terms in (2.289) to NLO,[13]

$$M^2\left(X + \tfrac{1}{2}s\right) - M^2\left(X - \tfrac{1}{2}s\right) \simeq s^\mu \frac{\partial M^2(X)}{\partial X^\mu}, \tag{2.296}$$

the left-hand side of (2.289) becomes, in Wigner space,

$$\int_{-\infty}^{\infty} d^{d+1}s\, e^{ip_\mu s^\mu} \left[2\frac{\partial}{\partial s_\mu}\frac{\partial}{\partial X^\mu} + s^\mu \frac{\partial M^2(X)}{\partial X^\mu}\right] F\left(X + \tfrac{1}{2}s, X - \tfrac{1}{2}s\right)$$

$$= -i\left[2p^\mu \frac{\partial}{\partial X^\mu} + \frac{\partial M^2(X)}{\partial X^\mu}\frac{\partial}{\partial p_\mu}\right] F(X,p), \tag{2.297}$$

with $p^0 \equiv \omega$.

We can similarly transform the right-hand sides of (2.289) and (2.290) using the above prescription. To derive these expressions, we consider for functions $f(x,y)$ and $g(x,y)$ the integral $\int d^{d+1}z\, f(x,z)g(z,y)$. For

$$f(x,z) \equiv f\left(X_{xz} + \tfrac{1}{2}s_{xz}, X_{xz} - \tfrac{1}{2}s_{xz}\right), \tag{2.298}$$

we introduce the Fourier transform with respect to the relative coordinate:

$$f(X_{xz}, p) = \int d^{d+1}s_{xz}\, \exp(ip_\mu s_{xy}^\mu) f\left(X_{xz} + \tfrac{1}{2}s_{xz}, X_{xz} - \tfrac{1}{2}s_{xz}\right). \tag{2.299}$$

This quantity may also be written as

$$f(X_{xz}, p) \equiv f\left(X_{xy} + \tfrac{1}{2}s_{zy}, p\right) = \exp\left(\frac{s_{zy}^\mu}{2}\frac{\partial}{\partial X_{xy}^\mu}\right) f(X_{xy}, p), \tag{2.300}$$

where we have Taylor expanded to obtain the last equality. Similar steps for $g(z,y)$ lead to

$$g(X_{zy}, p) = \int d^{d+1}s_{zy}\, \exp\left(ip_\mu s_{zy}^\mu\right) g\left(X_{zy} + \tfrac{1}{2}s_{zy}, X_{zy} - \tfrac{1}{2}s_{zy}\right)$$

$$= g\left(X_{xy} - \tfrac{1}{2}s_{xz}, p\right)$$

$$= \exp\left(-\frac{s_{xz}^\mu}{2}\frac{\partial}{\partial X_{xy}^\mu}\right) g(X_{xy}, p). \tag{2.301}$$

---

[13] The expression is actually correct to NNLO since the first correction is $\mathcal{O}((s^\mu \partial_{X^\mu})^3)$.

Inverting (2.299) gives, with (2.300),

$$f(x,z) = \int \frac{\mathrm{d}^{d+1}p}{(2\pi)^{d+1}} \exp(-ip_\mu s_{xz}^\mu) \exp\left(\frac{s_{zy}^\mu}{2} \frac{\partial}{\partial X_{xy}^\mu}\right) f(X_{xy}, p),$$

$$g(z,y) = \int \frac{\mathrm{d}^{d+1}p'}{(2\pi)^{d+1}} \exp(-ip'_\mu s_{zy}^\mu) \exp\left(-\frac{s_{xz}^\mu}{2} \frac{\partial}{\partial X_{xy}^\mu}\right) g(X_{xy}, p'),$$

$$(2.302)$$

using also the equivalent expressions for $g(z,y)$. With this, we can write, employing a partial integration,

$$\int \mathrm{d}^{d+1}s_{xy} \exp(ip_\mu s_{xy}^\mu) \int \mathrm{d}^{d+1}z\, f(x,z)g(z,y)$$

$$= \exp\left[\frac{i}{2}\left(\frac{\partial}{\partial p_\mu}\frac{\partial}{\partial X_{xy}^{\mu\prime}} - \frac{\partial}{\partial p'_\mu}\frac{\partial}{\partial X_{xy}^{\mu}}\right)\right] f(X_{xy},p)g(X'_{xy},p')|_{X_{xy}=X'_{xy},p=p'}$$

$$\simeq f(X_{xy},p)g(X_{xy},p) + \frac{i}{2}\{f(X_{xy},p);g(X_{xy},p)\}_{\mathrm{PB}},$$

$$(2.303)$$

where the Poisson bracket reads

$$\{f(X,p);g(X,p)\}_{\mathrm{PB}} = \frac{\partial f(X,p)}{\partial p_\mu}\frac{\partial g(X,p)}{\partial X^\mu} - \frac{\partial f(X,p)}{\partial X^\mu}\frac{\partial g(X,p)}{\partial p_\mu}. \qquad (2.304)$$

The lowest-order equations are obtained by neglecting $\mathcal{O}\left(\partial_{X^\mu}\partial_{p_\mu}\right)$ and higher contributions in the gradient expansion. To this order, the transport equations then read

$$2p^\mu \frac{\partial F(X,p)}{\partial X^\mu} = \tilde{\Sigma}_\rho(X,p)\,F(X,p) - \Sigma_F(X,p)\,\tilde{\rho}(X,p), \qquad (2.305)$$

$$2p^\mu \frac{\partial \tilde{\rho}(X,p)}{\partial X^\mu} = 0. \qquad (2.306)$$

We note that the compact form of the gradient expanded equation to lowest order is very similar to the exact equation for the statistical function (2.145). The main technical difference is that there are no integrals over the time history. Furthermore, to this order in the expansion, the evolution equation for the spectral function $\tilde{\rho}(X,p)$ becomes trivial. Accordingly, this approximation describes changes in the occupation number while neglecting its impact on the spectrum and vice versa. Moreover, we see that in thermal equilibrium the fluctuation–dissipation relations (2.192) and (2.193) yield $\partial_X F(X,p) = 0$.

Vertex-resummed kinetic equation

In the following, we consider the lowest-order expressions (2.305) and (2.306) of the gradient expansion, where the self-energies are obtained from the 2PI $1/N$ expansion to NLO. It is convenient to introduce a suitable "occupation number distribution" $f(X,p)$. For the real scalar field theory, we have

$$F(X,-p) = F(X,p), \qquad \tilde{\rho}(X,-p) = -\tilde{\rho}(X,p). \tag{2.307}$$

Without loss of generality, we can write

$$F(X,p) = \left[f(X,p) + \tfrac{1}{2}\right]\tilde{\rho}(X,p), \tag{2.308}$$

which defines the function $f(X,p)$ for any given $F(X,p)$ and $\tilde{\rho}(X,p)$. We emphasize that without additional assumptions, (2.308) does not represent a fluctuation–dissipation relation (2.192), which holds only if $f(X,p)$ is replaced by a thermal distribution function. We will not assume this in the following and will keep $f(X,p)$ general at this stage. In particular, (2.307) then implies the identity

$$f(X,-p) = -[f(X,p) + 1]. \tag{2.309}$$

For spatially homogeneous ensembles, (2.306) implies a constant $\tilde{\rho}(p)$ that does not depend on time. In contrast, the statistical function $F(t,p)$ to this order can depend on time $t \equiv X^0$, and, using (2.308), we can write

$$\int_0^\infty \frac{dp^0}{2\pi} 2p^0 \frac{\partial}{\partial t} F(t,p) = \int_0^\infty \frac{dp^0}{2\pi} 2p^0 \tilde{\rho}(p) \frac{\partial f(t,p)}{\partial t} = C[f](t,\mathbf{p}). \tag{2.310}$$

Here $C[f]$ describes the effects of interactions to lowest order in the gradient expansion. It is instructive to consider for a moment a free spectral function given by

$$\tilde{\rho}^{(0)}(p) = 2\pi\, \mathrm{sgn}(p^0)\, \delta\big((p^0)^2 - \omega_{\mathbf{p}}^2\big) \tag{2.311}$$

for a relativistic scalar field theory with particle energy $\omega_{\mathbf{p}}$. Choosing a free field theory type of spectral function with zero "width" leads to the characterization of the dynamics in terms of a "gas" of particles, where

$$f(t,\mathbf{p}) = f(t,p^0 = \omega_{\mathbf{p}},\mathbf{p}) = \int_0^\infty \frac{dp^0}{2\pi} 2p^0 \tilde{\rho}^{(0)}(p) f(t,p) \tag{2.312}$$

denotes the "on-shell" number distribution. In the following, we will consider the resummed $1/N$ expansion of the 2PI effective action to NLO. If the spectral function $\tilde{\rho}(p)$ for such an approximation is taken, it is not of the free field form (2.311) but acquires a nonzero "width" as pointed out in Section 2.3. We may nevertheless use the same procedure to define for spatially homogeneous systems an effective number distribution

$$f(t,\mathbf{p}) \equiv \int_0^\infty \frac{dp^0}{2\pi} 2p^0 \tilde{\rho}(p) f(t,p), \tag{2.313}$$

which depends on time and spatial momenta only. This will allow us to formulate the quantum field theory in a way reminiscent of a Boltzmann equation. The time evolution of that distribution is, according to (2.305), given by

$$\frac{\partial f(t,\mathbf{p})}{\partial t} = C[f](t,\mathbf{p}) = \int_0^\infty \frac{dp^0}{2\pi}\left[\tilde{\Sigma}_\rho(t,p)F(t,p) - \Sigma_F(t,p)\tilde{\rho}(p)\right]. \quad (2.314)$$

For the following, it is useful to note that the last expression can be rewritten with

$$\tilde{\Sigma}_\rho(t,p)F(t,p) - \Sigma_F(t,p)\tilde{\rho}(p) \equiv \left[\Sigma_F(t,p) + \tfrac{1}{2}\tilde{\Sigma}_\rho(t,p)\right]\left[F(t,p) - \tfrac{1}{2}\tilde{\rho}(p)\right]$$
$$- \left[\Sigma_F(t,p) - \tfrac{1}{2}\tilde{\Sigma}_\rho(t,p)\right]\left[F(t,p) + \tfrac{1}{2}\tilde{\rho}(p)\right]. \quad (2.315)$$

With (2.308), we also have

$$F(t,p) - \tfrac{1}{2}\tilde{\rho}(p) = f(t,p)\tilde{\rho}(p),$$
$$F(t,p) + \tfrac{1}{2}\tilde{\rho}(p) = [f(t,p) + 1]\tilde{\rho}(p), \quad (2.316)$$

which will allow us to conveniently express everything in terms of factors of $f(t,p)$ or, including "Bose enhancement," $f(t,p) + 1$ and $\tilde{\rho}(p)$.

In the following, we consider the collision term $C[f]$ at NLO in the $1/N$ expansion of the 2PI effective action as described in Section 2.3.2. To this end, we introduce retarded and advanced quantities as $I_R(x,y) = \Theta(x^0 - y^0)I_\rho(x,y)$ and $I_A(x,y) = -\Theta(y^0 - x^0)I_\rho(x,y)$, and equivalently for $\Pi_R(x,y)$ and $\Pi_A(x,y)$, using the expressions (2.267) and (2.268) with $t_0 \to -\infty$. Then we Fourier transform with respect to the relative coordinates to obtain the lowest-order gradient expansion result. To ease the notation, we suppress the dependence on the global central coordinate and only write the momentum dependences, with $\int_q \equiv \int d^{d+1}q/(2\pi)^{d+1}$. For the retarded and advanced summation functions, we find, from (2.267),

$$I_R(t,p) = \frac{\Pi_R(t,p)}{1 + \Pi_R(t,p)}, \qquad I_A(t,p) = \frac{\Pi_A(t,p)}{1 + \Pi_A(t,p)}. \quad (2.317)$$

Proceeding in the same way for I_F, we obtain, with $I_\rho = I_R - I_A$,

$$I_\rho(t,p) = v_{\text{eff}}(t,p)\Pi_\rho(t,p), \qquad I_F(t,p) = v_{\text{eff}}(t,p)\Pi_F(t,p). \quad (2.318)$$

The vertex correction $v_{\text{eff}}(t,p)$ sums the infinite chain of "ring" diagrams appearing at NLO in the $1/N$ expansion and is given by

$$v_{\text{eff}}(t,p) = \frac{1}{|1 + \Pi_R(t,p)|^2}, \quad (2.319)$$

with $|1 + \Pi_R|^2 \equiv (1 + \Pi_R)(1 + \Pi_A)$ and the one-loop retarded self-energy

$$\Pi_R(t,p) = \frac{\lambda}{3}\int_q F(t, p-q)G_R(q). \quad (2.320)$$

At NLO in the 2PI $1/N$ expansion, the above linear combinations of self-energies are then given by

$$\Sigma_F(t,p) \pm \tfrac{1}{2}\tilde{\Sigma}_\rho(t,p) = -\frac{\lambda^2}{18N} \int_{ql} v_{\text{eff}}(t,p-q)$$
$$\times \left[F(t,p-q-l) \pm \tfrac{1}{2}\tilde{\rho}(p-q-l) \right]$$
$$\times \left[F(t,q) \pm \tfrac{1}{2}\tilde{\rho}(q) \right]$$
$$\times \left[F(t,l) \pm \tfrac{1}{2}\tilde{\rho}(l) \right], \tag{2.321}$$

which has the structure of a two-loop self-energy, but with a time- and momentum-dependent "effective coupling" entering via $v_{\text{eff}}(t,p)$. In terms of $f(t,p)$, these self-energy combinations read

$$\Sigma_F(t,p) + \tfrac{1}{2}\tilde{\Sigma}_\rho(t,p) = -\frac{\lambda^2}{18N} \int_{ql} v_{\text{eff}}(t,p-q)$$
$$\times [f(t,p-q-l)+1]\tilde{\rho}(p-q-l)$$
$$\times [f(t,q)+1]\tilde{\rho}(q)$$
$$\times [f(t,l)+1]\tilde{\rho}(l), \tag{2.322}$$

$$\Sigma_F(t,p) - \tfrac{1}{2}\tilde{\Sigma}_\rho(t,p) = -\frac{\lambda^2}{18N} \int_{ql} v_{\text{eff}}(t,p-q)$$
$$\times f(t,p-q-l)\tilde{\rho}(p-q-l)$$
$$\times f(t,q)\tilde{\rho}(q)$$
$$\times f(t,l)\tilde{\rho}(l). \tag{2.323}$$

Putting everything together, we obtain

$$(\tilde{\Sigma}_\rho F - \Sigma_F \tilde{\rho})(t,p) = -\frac{\lambda^2}{18N} \int_{qlr} (2\pi)^{d+1} \delta(p-q-l-r) v_{\text{eff}}(t,p-q)$$
$$\times \left\{ [f(t,q)+1][f(t,l)+1][f(t,r)+1]f(t,p) \right.$$
$$\left. - f(t,q)f(t,l)f(t,r)[f(t,p)+1] \right\}$$
$$\times \tilde{\rho}(q)\tilde{\rho}(l)\tilde{\rho}(r)\tilde{\rho}(p). \tag{2.324}$$

To bring this expression into a form that can be directly compared with kinetic or Boltzmann descriptions, we map onto positive frequencies. For this, we split the frequency integrals $\int_{-\infty}^{\infty} dq^0 [\ldots] = \int_{-\infty}^{0} dq^0 [\ldots] + \int_0^{\infty} dq^0 [\ldots]$, with $q^0 \to -q^0$ for the negative-frequency part, and employ $f(t,-q^0,\mathbf{q}) = -[f(t,q^0,\mathbf{q})+1]$.

Collecting the $2^3 = 8$ contributions from the different orthants in frequency space, we find a "collision integral" (2.314) that is accurate to NLO in the large-N expansion, including processes to all orders in the coupling constant. We obtain with $f_p \equiv f(t,p)$, suppressing the global time dependence,

$$
C^{\mathrm{NLO}}[f](\mathbf{p}) =
$$

$$
\int d\Omega^{2\leftrightarrow 2}[f](p,l,q,r)\left[(f_p+1)(f_l+1)f_q f_r - f_p f_l (f_q+1)(f_r+1)\right]
$$

$$
+ \int d\Omega^{1\leftrightarrow 3}_{(a)}[f](p,l,q,r)\left[(f_p+1)(f_l+1)(f_q+1)f_r - f_p f_l f_q (f_r+1)\right]
$$

$$
+ \int d\Omega^{1\leftrightarrow 3}_{(b)}[f](p,l,q,r)\left[(f_p+1)f_l f_q f_r - f_p (f_l+1)(f_q+1)(f_r+1)\right]
$$

$$
+ \int d\Omega^{0\leftrightarrow 4}[f](p,l,q,r)\left[(f_p+1)(f_l+1)(f_q+1)(f_r+1) - f_p f_l f_q f_r\right].
$$

$$(2.325)$$

Here

$$
\int d\Omega^{2\leftrightarrow 2}[f](p,l,q,r) = \frac{\lambda^2}{18N}\int_0^\infty \frac{dp^0\, dl^0\, dq^0\, dr^0}{(2\pi)^{4-(d+1)}} \int_{lqr} \delta(p+l-q-r)
$$
$$
\times \tilde\rho_p \tilde\rho_l \tilde\rho_q \tilde\rho_r \left[v_{\mathrm{eff}}(p+l) + v_{\mathrm{eff}}(p-q) + v_{\mathrm{eff}}(p-r)\right],
$$

$$
\int d\Omega^{1\leftrightarrow 3}_{(a)}[f](p,l,q,r) = \frac{\lambda^2}{18N}\int_0^\infty \frac{dp^0\, dl^0\, dq^0\, dr^0}{(2\pi)^{4-(d+1)}} \int_{lqr} \delta(p+l+q-r)
$$
$$
\times \tilde\rho_p \tilde\rho_l \tilde\rho_q \tilde\rho_r \left[v_{\mathrm{eff}}(p+l) + v_{\mathrm{eff}}(p+q) + v_{\mathrm{eff}}(p-r)\right],
$$

$$(2.326)$$

$$
\int d\Omega^{1\leftrightarrow 3}_{(b)}[f](p,l,q,r) = \frac{\lambda^2}{18N}\int_0^\infty \frac{dp^0\, dl^0\, dq^0\, dr^0}{(2\pi)^{4-(d+1)}} \int_{lqr} \delta(p-l-q-r)
$$
$$
\times \tilde\rho_p \tilde\rho_l \tilde\rho_q \tilde\rho_r\, v_{\mathrm{eff}}(p-l),
$$

$$
\int d\Omega^{0\leftrightarrow 4}[f](p,l,q,r) = \frac{\lambda^2}{18N}\int_0^\infty \frac{dp^0\, dl^0\, dq^0\, dr^0}{(2\pi)^{4-(d+1)}} \int_{lqr} \delta(p+l+q+r)
$$
$$
\times \tilde\rho_p \tilde\rho_l \tilde\rho_q \tilde\rho_r\, v_{\mathrm{eff}}(p+l),
$$

with $\int_q \equiv \int d^d q/(2\pi)^d$. We emphasize that the above expressions still contain the integrations over frequencies and spectral functions. No quasiparticle assumptions using a free field form of the spectral function have been employed yet and the only approximations are the $1/N$ expansion to NLO and the gradient expansion underlying (2.310).

We can see that for sufficiently large p, for which $\Pi^R(p) \ll 1$, the vertex function (2.319) approaches 1. In this case, the $2 \leftrightarrow 2$ contribution of the first line in (2.325) is reminiscent of the two-to-two scattering process in a Boltzmann equation. The main difference is that the latter assumes a δ-like spectral function such that all momenta are on shell. Therefore, in the perturbative expression (2.446), off-shell processes involving the decay of one into three particles or corresponding $3 \rightarrow 1$ annihilation processes or even $0 \leftrightarrow 4$ processes are absent. They can occur in principle at NLO in the 2PI $1/N$ expansion, which leads to the different terms contributing to the right-hand sides of (2.325), but they are typically small. At sufficiently high momenta, these off-shell contributions should be suppressed along with all quantum statistical corrections such that the spectral function approaches a δ-like behavior. In this case, one recovers the standard Boltzmann equation for elastic two-to-two scattering. In the infrared, $v_{\text{eff}}(p)$ may have a nontrivial momentum dependence, which incorporates important vertex corrections for the $2 \leftrightarrow 2$ scattering term that are discussed, in particular, in Section 2.6.

2.3.4 Bibliography

- Figure 2.5 is taken from J. Berges and J. Cox. Thermalization of quantum fields from time-reversal invariant evolution equations. *Phys. Lett. B* **517** (2001) 369, where thermalization in relativistic quantum field theory has been shown for $1+1$ dimensions. Similar studies in $2+1$ dimensions include S. Juchem, W. Cassing, and C. Greiner. Quantum dynamics and thermalization for out-of-equilibrium ϕ^4-theory. *Phys. Rev. D* **69** (2004) 025006. For thermalization in $3+1$ dimensions, including fermions and bosons, see J. Berges, S. Borsanyi and J. Serreau, Thermalization of fermionic quantum fields. *Nucl. Phys. B* **660** (2003) 51. For a related study of thermalization in the context of ultracold quantum gases, see J. Berges and T. Gasenzer. Quantum versus classical statistical dynamics of an ultracold Bose gas. *Phys. Rev. A* **76** (2007) 033604.
- Figure 2.6 is taken with permission from G. Aarts and J. Berges. Nonequilibrium time evolution of the spectral function in quantum field theory. *Phys. Rev. D* **64** (2001) 105010 (Copyright (2001) by the American Physical Society).
- Figures 2.8–2.10 are taken from J. Berges. Controlled nonperturbative dynamics of quantum fields out of equilibrium. *Nucl. Phys. A* **699** (2002) 847, which compares thermalization dynamics from the $1/N$ expansion at LO and NLO. Our presentation is also based on G. Aarts, D. Ahrensmeier, R. Baier, J. Berges, and J. Serreau. Far-from-equilibrium dynamics with broken symmetries from the $1/N$ expansion of the 2PI effective action. *Phys. Rev. D* **66** (2002) 045008.
- The phenomenon of prethermalization has been pointed out in J. Berges, S. Borsanyi, and C. Wetterich. Prethermalization. *Phys. Rev. Lett.* **93** (2004) 142002. Applications to condensed matter systems include M. Moeckel and S. Kehrein. Interaction quench in the Hubbard model. *Phys. Rev. Lett.* **100** (2008) 175702, and for experimental investigations with ultracold atoms, see M. Gring et al. Relaxation and prethermalization in an isolated quantum system. *Science* **337** (2012) 6100.
- The presentation of the gradient expansion in Section 2.3.3 follows J. Berges and S. Borsanyi. Range of validity of transport equations. *Phys. Rev. D* **74** (2006) 045022. The discussion of the vertex-resummed kinetic theory in Section 2.3.3 is based on J. Berges and D. Sexty. Strong versus weak wave-turbulence in relativistic field theory. *Phys. Rev. D* **83** (2011) 085004. For a discussion of transport in quantum field theory, see also S. Juchem, W. Cassing, and C. Greiner.

Nonequilibrium quantum field dynamics and off-shell transport for ϕ^4 theory in (2+1)-dimensions. *Nucl. Phys. A* **743** (2004) 92. For a corresponding nonrelativistic study, see A. Branschadel and T. Gasenzer. 2PI nonequilibrium versus transport equations for an ultracold Bose gas. *J. Phys. B* **41** (2008) 135302.

2.4   Classical aspects of nonequilibrium quantum fields

The extent to which nonequilibrium quantum field theory can be approximated by classical statistical field theory is an important question. A frequently employed strategy in the literature is to consider nonequilibrium classical dynamics instead of quantum dynamics, since the former can be simulated numerically up to controlled statistical errors. Classical statistical field theory indeed gives important insights when the number of bosonic field quanta per mode is sufficiently large that quantum fluctuations are suppressed compared with statistical fluctuations. We will derive below a sufficient condition for the validity of classical approximations to nonequilibrium quantum dynamics. The description in terms of spectral and statistical correlation functions as introduced in Section 2.2.6 is particularly suitable for comparisons, since these correlation functions possess well-defined classical counterparts.

Classical Rayleigh-Jeans divergences and the lack of genuine quantum effects—such as the approach to quantum thermal equilibrium characterized by Bose–Einstein statistics—limit the use of classical statistical field theory. To find out its use and its limitations we perform below direct comparisons of nonequilibrium classical and quantum evolutions for same initial conditions. It is found that classical methods can give an accurate description of quantum dynamics for the case of sufficiently large characteristic occupation numbers. This typically limits their application to not too late times, before the approach to quantum thermal equilibrium sets in.

Classical methods have been extensively used in the past to rule out "candidates" for approximation schemes applied to nonequilibrium quantum field theory. Approximations that fail to describe classical nonequilibrium dynamics should be typically discarded also for the quantum case. If the dynamics is formulated in terms of correlation functions, then approximation schemes for the quantum evolution can be straightforwardly implemented for the respective classical theory as well. For instance, the 2PI $1/N$ expansion introduced in Section 2.3.2 can be equally well implemented in the classical as in the quantum theory. Therefore, in the classical statistical approach, one can compare NLO results with results including all orders in $1/N$. This gives a rigorous answer to the question of what happens at NNLO or beyond in this case. In particular, for increasing occupation numbers per mode, the classical and quantum evolutions can be shown to approach each other if the same initial conditions are applied and for not too late times. For sufficiently high particle number densities, one can therefore strictly verify how rapidly the $1/N$ series converges for far-from-equilibrium dynamics.

Although we restrict ourselves here to scalar fields, it is important to note that crucial applications arise also by coupling classical statistical bosonic fields, including lattice gauge fields, to fermions. For instance, this is done to describe the nonequilibrium real-time evolution in quantum electrodynamics for ultrastrong laser fields or quantum chromodynamics related to collision experiments of heavy nuclei. Fermions are never classical in the sense that they cannot be strongly occupied because of the Pauli exclusion principle. However, since the fermions occur quadratically in the respective actions, their functional integral can be treated on the lattice without losing their genuine quantum nature. We refer to the reference list in Section 2.4.5 for further reading on these extensions of the classical statistical lattice simulation approach.

2.4.1   Nonequilibrium quantum field theory revisited

Functional integral

In order to discuss the different origins of quantum and of classical statistical fluctuations, it is convenient to rewrite the nonequilibrium generating functional of Section 2.2.3. We start by employing the notation introduced in Section 2.2.2, where the superscripts "+" and "−" indicate that the fields are taken on the forward branch $(\mathcal{C}^+)$ starting at t_0 and backward $(\mathcal{C}^-)$ along the closed time path, respectively. To be specific, we consider again the N-component scalar field theory for which the classical action (2.93) can be written as

$$
S[\varphi^+, \varphi^-] =
$$
$$
\int_{x,t_0} \left\{ \frac{1}{2} \partial^\mu \varphi_a^+(x) \partial_\mu \varphi_a^+(x) - \frac{m^2}{2} \varphi_a^+(x) \varphi_a^+(x) - \frac{\lambda}{4!N} [\varphi_a^+(x) \varphi_a^+(x)]^2 \right.
$$
$$
\left. - \frac{1}{2} \partial^\mu \varphi_a^-(x) \partial_\mu \varphi_a^-(x) + \frac{m^2}{2} \varphi_a^-(x) \varphi_a^-(x) + \frac{\lambda}{4!N} [\varphi_a^-(x) \varphi_a^-(x)]^2 \right\}. \quad (2.327)
$$

Here the minus sign in front of the "−" terms accounts for the reversed time integration of the closed time contour and $\int_{x,t_0} \equiv \int_{t_0} dx^0 \int d^d x$. The corresponding generating functional for correlation functions then reads in this notation

$$
Z[J^+, J^-, R^{++}, R^{+-}, R^{-+}, R^{--}; \varrho_0] = \int [d\varphi_0^+][d\varphi_0^-] \, \varrho_0 \, [\varphi_0^+, \varphi_0^-]
$$
$$
\times \int_{\varphi_0^+}^{\varphi_0^-} \mathscr{D}' \varphi^+ \, \mathscr{D}' \varphi^- \, \exp\left\{ i \left[S'[\varphi^+, \varphi^-] + \int_x \left(\varphi_a^+(x), \varphi_a^-(x) \right) \begin{pmatrix} J_a^+(x) \\ -J_a^-(x) \end{pmatrix} \right. \right.
$$
$$
\left. \left. + \frac{1}{2} \int_{xy} \left(\varphi_a^+(x), \varphi_a^-(x) \right) \begin{pmatrix} R_{ab}^{++}(x,y) & -R_{ab}^{+-}(x,y) \\ -R_{ab}^{-+}(x,y) & R_{ab}^{--}(x,y) \end{pmatrix} \begin{pmatrix} \varphi_b^+(y) \\ \varphi_b^-(y) \end{pmatrix} \right] \right\}. \quad (2.328)
$$

Again the superscripts "+" and "−" indicate that the respective time arguments of the sources are taken on the forward or backward branch of the closed time path, and the prime on the measure means that the integration over the fields at initial time t_0 is excluded.

In order to simplify the comparison with the classical statistical field theory, a standard linear transformation A of the fields is introduced as

$$\begin{pmatrix} \varphi \\ \tilde{\varphi} \end{pmatrix} \equiv A \begin{pmatrix} \varphi^+ \\ \varphi^- \end{pmatrix}, \tag{2.329}$$

where

$$A = \begin{pmatrix} \frac{1}{2} & \frac{1}{2} \\ 1 & -1 \end{pmatrix}, \qquad A^{-1} = \begin{pmatrix} 1 & \frac{1}{2} \\ 1 & -\frac{1}{2} \end{pmatrix}, \tag{2.330}$$

such that $\varphi = \frac{1}{2}(\varphi^+ + \varphi^-)$ and $\tilde{\varphi} = \varphi^+ - \varphi^-$, or $\varphi^\pm = \varphi \pm \frac{1}{2}\tilde{\varphi}$, respectively. To avoid a proliferation of symbols, we have used here φ, which agrees with the defining field in (2.93) only for $\varphi^+ = \varphi^-$. Since this will be the case for expectation values in the absence of sources, where physical observables are obtained, and since there is no danger of confusion in the following, we keep this notation.

Correspondingly, we write for the source terms

$$\begin{pmatrix} J \\ \tilde{J} \end{pmatrix} \equiv A \begin{pmatrix} J^+ \\ J^- \end{pmatrix}, \tag{2.331}$$

$$\begin{pmatrix} R^F & R^R \\ R^A & R^{\tilde{F}} \end{pmatrix} \equiv A \begin{pmatrix} R^{++} & R^{+-} \\ R^{-+} & R^{--} \end{pmatrix} A^T. \tag{2.332}$$

Inserting these definitions into the functional integral (2.328) and using the fact that (2.331) and (2.332) can be equivalently written as

$$\begin{pmatrix} \tilde{J} \\ J \end{pmatrix} \equiv (A^{-1})^T \begin{pmatrix} J^+ \\ -J^- \end{pmatrix}, \tag{2.333}$$

$$\begin{pmatrix} R^{\tilde{F}} & R^A \\ R^R & R^F \end{pmatrix} \equiv (A^{-1})^T \begin{pmatrix} R^{++} & -R^{+-} \\ -R^{-+} & R^{--} \end{pmatrix} A^{-1}, \tag{2.334}$$

we find

$$Z[J, \tilde{J}, R^F, R^R, R^A, R^{\tilde{F}}; \varrho_0] = \int [d\varphi_0]\,[d\tilde{\varphi}_0]\,\varrho_0 \left[\varphi_0 + \tfrac{1}{2}\tilde{\varphi}_0, \varphi_0 - \tfrac{1}{2}\tilde{\varphi}_0\right]$$

$$\times \int_{\varphi_0, \tilde{\varphi}_0} \mathscr{D}'\varphi\, \mathscr{D}'\tilde{\varphi} \exp\left\{i\left[S[\varphi, \tilde{\varphi}] + \int_{x, t_0} (\varphi_a(x), \tilde{\varphi}_a(x)) \begin{pmatrix} \tilde{J}_a(x) \\ J_a(x) \end{pmatrix}\right.\right.$$

$$\left.\left. + \frac{1}{2} \int_{xy, t_0} (\varphi_a(x), \tilde{\varphi}_a(x)) \begin{pmatrix} R^{\tilde{F}}_{ab}(x, y) & R^A_{ab}(x, y) \\ R^R_{ab}(x, y) & R^F_{ab}(x, y) \end{pmatrix} \begin{pmatrix} \varphi_b(y) \\ \tilde{\varphi}_b(y) \end{pmatrix}\right]\right\}. \tag{2.335}$$

Fig. 2.12   Classical vertex (left) and quantum vertex (right) in the scalar field theory.

Here $S[\varphi, \tilde{\varphi}] = S_0[\varphi, \tilde{\varphi}] + S_{\text{int}}[\varphi, \tilde{\varphi}]$ consists of the action for the free field theory

$$S_0[\varphi, \tilde{\varphi}] = \int_{x, t_0} \left[\partial^\mu \tilde{\varphi}_a(x) \partial_\mu \varphi_a(x) - m^2 \tilde{\varphi}_a(x) \varphi_a(x) \right] \tag{2.336}$$

and the interaction part

$$S_{\text{int}}[\varphi, \tilde{\varphi}] = -\frac{\lambda}{6N} \int_{x, t_0} \tilde{\varphi}_a(x) \varphi_a(x) \varphi_b(x) \varphi_b(x) - \frac{\lambda}{24N} \int_{x, t_0} \tilde{\varphi}_a(x) \tilde{\varphi}_a(x) \tilde{\varphi}_b(x) \varphi_b(x). \tag{2.337}$$

The two types of vertices appearing in the interaction part (2.337) are illustrated in Fig. 2.12. To understand their role for the dynamics, it is important to note that one can also write down a functional integral for the corresponding nonequilibrium classical statistical field theory. This is done in Section 2.4.2, where we will observe that the generating functionals for correlation functions are very similar in the quantum and the classical statistical theories. A crucial difference is that the quantum theory is characterized by an additional vertex: the interaction term $\sim \tilde{\varphi}^3$, appearing with (2.337) in the functional integral of the quantum theory, does not occur for the classical theory.

Connected one- and two-point functions

For later use, we define the following correlation functions in the above basis. From the generating functional for connected correlation functions, $W = -i \ln Z$, we define the macroscopic field ϕ_a and $\tilde{\phi}_a$ by

$$\frac{\delta W}{\delta \tilde{J}_a(x)} = \phi_a(x), \qquad \frac{\delta W}{\delta J_a(x)} = \tilde{\phi}_a(x). \tag{2.338}$$

The connected statistical correlation function $F_{ab}(x, y)$, the retarded/advanced propagators $G_{ab}^{\text{R/A}}(x, y)$, and the "anomalous" propagator $\tilde{F}_{ab}(x, y)$ are defined by

$$\frac{\delta W}{\delta R_{ab}^{\tilde{F}}(x, y)} = \tfrac{1}{2} \left[\phi_a(x) \phi_b(y) + F_{ab}(x, y) \right],$$

$$\frac{\delta W}{\delta R_{ab}^{\text{A}}(x, y)} = \tfrac{1}{2} \left[\phi_a(x) \tilde{\phi}_b(y) - i G_{ab}^{\text{R}}(x, y) \right],$$

$$\frac{\delta W}{\delta R_{ab}^{\text{R}}(x, y)} = \tfrac{1}{2} \left[\tilde{\phi}_a(x) \phi_b(y) - i G_{ab}^{\text{A}}(x, y) \right], \tag{2.339}$$

$$\frac{\delta W}{\delta R_{ab}^{F}(x, y)} = \tfrac{1}{2} \left[\tilde{\phi}_a(x) \tilde{\phi}_b(y) + \tilde{F}_{ab}(x, y) \right].$$

Equivalently, we can define the same connected two-point correlation functions by the second functional derivatives

$$\frac{\delta^2 W}{\delta \tilde{J}_a(x)\delta J_b(y)} = G^{\mathrm{R}}_{ab}(x,y), \qquad \frac{\delta^2 W}{\delta J_a(x)\delta \tilde{J}_b(y)} = G^{\mathrm{A}}_{ab}(x,y), \qquad (2.340)$$

$$\frac{\delta^2 W}{\delta \tilde{J}_a(x)\delta \tilde{J}_b(y)} = \mathrm{i}F_{ab}(x,y), \qquad \frac{\delta^2 W}{\delta J_a(x)\delta J_b(y)} = \mathrm{i}\tilde{F}_{ab}(x,y). \qquad (2.341)$$

We note that the propagators satisfy the symmetry properties $G^{\mathrm{A}}_{ab}(x,y) = G^{\mathrm{R}}_{ba}(y,x)$, $F_{ab}(x,y) = F_{ba}(y,x)$, and $\tilde{F}_{ab}(x,y) = \tilde{F}_{ba}(y,x)$. These properties follow directly from the definition of the propagators in terms of the second functional derivatives with respect to J and $\tilde{J}$. The spectral function ρ is given by the difference of the retarded and advanced propagators $\rho_{ab}(x,y) = G^{\mathrm{R}}_{ab}(x,y) - G^{\mathrm{A}}_{ab}(x,y)$ and $G^{\mathrm{R}}_{ab}(x,y) = \rho_{ab}(x,y)\theta(x^0 - y^0)$.

It is important to note that the anomalous propagator $\tilde{F}$ vanishes in the limit where the external sources are set to zero. This is a consequence of the algebraic identity (2.46), since $\tilde{F} = G^{++} + G^{--} - G^{-+} - G^{+-}$, as one may readily check by changing the basis:

$$\begin{pmatrix} F & -\mathrm{i}G^{\mathrm{R}} \\ -\mathrm{i}G^{\mathrm{A}} & \tilde{F} \end{pmatrix} = A \begin{pmatrix} G^{++} & G^{+-} \\ G^{-+} & G^{--} \end{pmatrix} A^{\mathrm{T}}$$

$$= \begin{pmatrix} \frac{1}{4}(G^{++} + G^{--} + G^{+-} + G^{-+}) & \frac{1}{2}(G^{++} - G^{--} + G^{-+} - G^{+-}) \\ \frac{1}{2}(G^{++} - G^{--} - G^{-+} + G^{+-}) & G^{++} + G^{--} - G^{-+} - G^{+-} \end{pmatrix}, \qquad (2.342)$$

More generally, in the absence of sources, we have

$$\frac{\delta W}{\delta J_a(x)}\bigg|_{J,\tilde{J},R^{\mathrm{R},\mathrm{A}},F,\tilde{F}=0} = \tilde{\phi}_a(x) = 0, \qquad (2.343)$$

$$\frac{\delta^2 W}{\delta J_a(x)\,\delta J_b(y)}\bigg|_{J,\tilde{J},R^{\mathrm{R},\mathrm{A}},F,\tilde{F}=0} = \mathrm{i}\tilde{F}_{ab}(x,y) = 0, \qquad (2.344)$$

and, correspondingly, arbitrary functional derivatives of the generating functional with respect to J vanish in the absence of sources.

We finally mention that the inverse of the two-point function matrix (2.342) in the absence of sources, where $\tilde{F} = 0$, reads

$$\begin{pmatrix} 0 & \mathrm{i}(G^{\mathrm{A}})^{-1} \\ \mathrm{i}(G^{\mathrm{R}})^{-1} & (G^{\mathrm{R}})^{-1} \cdot F \cdot (G^{\mathrm{A}})^{-1} \end{pmatrix}. \qquad (2.345)$$

Here we have employed the compact matrix notation

$$\left[(G^{\mathrm{R}})^{-1} \cdot F \cdot (G^{\mathrm{A}})^{-1}\right]_{ab}(x,y) \equiv \int_{zw} (G^{\mathrm{R}})^{-1}_{ac}(x,z) F_{cd}(z,w)(G^{\mathrm{A}})^{-1}_{db}(w,y). \qquad (2.346)$$

For vanishing sources, the exact inverse two-point function (2.345) can then be written as

$$
\begin{pmatrix} 0 & i(G^A)^{-1} \\ i(G^R)^{-1} & (G^R)^{-1} \cdot F \cdot (G^A)^{-1} \end{pmatrix} = \begin{pmatrix} 0 & G_0^{-1} \\ G_0^{-1} & 0 \end{pmatrix} - \begin{pmatrix} 0 & -i\Sigma^A \\ -i\Sigma^R & \Sigma^F \end{pmatrix},
$$

(2.347)

where the retarded, advanced and statistical self-energies are obtained as

$$
\begin{pmatrix} 0 & -i\Sigma^A \\ -i\Sigma^R & \Sigma^F \end{pmatrix} = (A^{-1})^{\mathrm{T}} \begin{pmatrix} \Sigma^{++} & -\Sigma^{+-} \\ -\Sigma^{-+} & \Sigma^{--} \end{pmatrix} A^{-1}
$$

(2.348)

from the self-energies written in the "±" basis. The retarded/advanced self-energy $\Sigma^{R/A}$ and the statistical self-energy Σ^F receive contributes only from graphs with propagator lines associated with $G^{R,A}$ and F, which can be obtained from closed 2PI graphs by opening one propagator line.

2PI effective action

The 2PI effective action written in terms of the rotated variables becomes a functional of the field expectation values $\phi, \tilde{\phi}$ and the propagators $G^R, G^A, F, \tilde{F}$:

$$
\begin{aligned}
\Gamma = W &- \int_x \left[\phi_a(x)\tilde{J}_a(x) + \tilde{\phi}_a(x)J_a(x)\right] \\
&- \frac{1}{2}\int_{xy} \Big\{ R_{ab}^{\tilde{F}}(x,y)\left[\phi_a(x)\phi_b(y) + F_{ab}(x,y)\right] \\
&\qquad + R_{ab}^A(x,y)\left[\phi_a(x)\tilde{\phi}_b(y) - iG_{ab}^R(x,y)\right] \\
&\qquad + R_{ab}^R(x,y)\left[\tilde{\phi}_a(x)\phi_b(y) - iG_{ab}^A(x,y)\right] \\
&\qquad + R_{ab}^F(x,y)\left[\tilde{\phi}_a(x)\tilde{\phi}_b(y) + \tilde{F}_{ab}(x,y)\right] \Big\}.
\end{aligned}
$$

(2.349)

From (2.349), we find the equations of motion for the fields,

$$
\frac{\delta\Gamma}{\delta\phi_a(x)} = -\tilde{J}_a(x) - \int_y \left[R_{ab}^{\tilde{F}}(x,y)\phi_b(y) + \tfrac{1}{2}R_{ab}^A(x,y)\tilde{\phi}_b(y) + \tfrac{1}{2}\tilde{\phi}_b(y)R_{ba}^R(y,x)\right],
$$

(2.350)

$$
\frac{\delta\Gamma}{\delta\tilde{\phi}_a(x)} = -J_a(x) - \int_y \left[R_{ab}^F(x,y)\tilde{\phi}_b(y) + \tfrac{1}{2}R_{ab}^R(x,y)\phi_b(y) + \tfrac{1}{2}\phi_b(y)R_{ba}^A(y,x)\right],
$$

(2.351)

as well as for the two-point functions,

$$
\frac{\delta\Gamma}{\delta F_{ab}(x,y)} = -\tfrac{1}{2}R_{ab}^{\tilde{F}}(x,y), \quad i\frac{\delta\Gamma}{\delta G_{ab}^R(x,y)} = -\tfrac{1}{2}R_{ab}^A(x,y),
$$

$$
i\frac{\delta\Gamma}{\delta G_{ab}^A(x,y)} = -\tfrac{1}{2}R_{ab}^R(x,y), \quad \frac{\delta\Gamma}{\delta\tilde{F}_{ab}(x,y)} = -\tfrac{1}{2}R_{ab}^F(x,y).
$$

(2.352)

In the presence of a nonvanishing field value $\phi \neq 0$ but $\tilde{\phi} = 0$, the interaction vertices are obtained from (2.337) by shifting in $S[\varphi, \tilde{\varphi}]$ the field $\varphi \to \phi + \varphi$, and collecting all cubic and quartic terms in the fluctuating fields φ and $\tilde{\varphi}$, i.e.,

$$S_{\text{int}}[\varphi, \tilde{\varphi}; \phi] = -\frac{\lambda}{6N} \int_x \tilde{\varphi}_a(x)\varphi_a(x)\varphi_b(x)\varphi_b(x)$$

$$-\frac{\lambda}{24N} \int_x \tilde{\varphi}_a(x)\tilde{\varphi}_a(x)\tilde{\varphi}_b(x)\varphi_b(x)$$

$$-\frac{\lambda}{3N} \int_x \tilde{\varphi}_a(x)\varphi_a(x)\varphi_b(x)\phi_b(x)$$

$$-\frac{\lambda}{6N} \int_x \tilde{\varphi}_a(x)\phi_a(x)\varphi_b(x)\varphi_b(x)$$

$$-\frac{\lambda}{24N} \int_x \tilde{\varphi}_a(x)\tilde{\varphi}_a(x)\tilde{\varphi}_b(x)\phi_b(x). \qquad (2.353)$$

The quadratic terms in the fluctuating fields are taken into account in the classical inverse propagator corresponding to a field-dependent $iG_0^{-1}(x, y; \phi)$ given by (2.102). For the classical statistical field theory, only those terms of (2.353) appear that are linear in $\tilde{\varphi}$. This is shown in the following.

2.4.2   Functional integral for the classical statistical theory

The classical field equation of motion for the N-component scalar field $\varphi_a(x)$ with action (2.93) is

$$\left[-\Box_x - m^2 - \frac{\lambda}{6N}\varphi_b(x)\varphi_b(x) \right] \varphi_a(x) = 0. \qquad (2.354)$$

Its solution, $\varphi_a^{\text{cl}}(x)$, requires specification of the initial conditions $\varphi_a^{\text{cl}}(t_0, \mathbf{x}) = \varphi_{0,a}(\mathbf{x})$ and $\pi_a^{\text{cl}}(t_0, \mathbf{x}) = \pi_{0,a}(\mathbf{x})$, with $\pi_a^{\text{cl}}(x) = \partial_{x^0}\varphi_a^{\text{cl}}(x)$ for the considered scalar field theory. We define the *macroscopic or average classical field* by

$$\phi_a^{\text{cl}}(x) = \langle \varphi_a(x) \rangle_{\text{cl}} = \int [\mathrm{d}\pi_0] [\mathrm{d}\varphi_0] \, W^{\text{cl}}[\varphi_0, \pi_0] \, \varphi_a^{\text{cl}}(x). \qquad (2.355)$$

Here $W^{\text{cl}}[\varphi_0, \pi_0]$ denotes the normalized probability functional at the initial time. The measure indicates integration over classical phase space,

$$\int [\mathrm{d}\pi_0] [\mathrm{d}\varphi_0] = \int \prod_{a=1}^{N} \prod_{\mathbf{x}} \mathrm{d}\pi_{0,a}(\mathbf{x}) \, \mathrm{d}\varphi_{0,a}(\mathbf{x}), \qquad (2.356)$$

and the theory may be defined on a spatial lattice. Similarly, the connected *classical statistical propagator* $F_{ab}^{\text{cl}}(x, y)$ is defined by

$$F_{ab}^{\text{cl}}(x, y) + \phi_a^{\text{cl}}(x)\phi_b^{\text{cl}}(y) = \langle \varphi_a(x)\varphi_b(y) \rangle_{\text{cl}} \equiv \int [\mathrm{d}\pi_0] [\mathrm{d}\varphi_0] \, W^{\text{cl}}[\varphi_0, \pi_0]\varphi_a^{\text{cl}}(x)\varphi_b^{\text{cl}}(y).$$
$$(2.357)$$

The classical equivalent of the quantum spectral function is obtained by replacing $-i$ times the commutator by the Poisson bracket:

$$\rho_{ab}^{\text{cl}}(x, y) = -\langle\{\varphi_a(x), \varphi_b(y)\}_{\text{PB}}\rangle_{\text{cl}}, \tag{2.358}$$

where we recall that the Poisson bracket with respect to the initial fields is

$$\{A(x), B(y)\}_{\text{PB}} = \sum_{a=1}^{N} \int_{\mathbf{z}} \left[\frac{\delta A(x)}{\delta \varphi_{0,a}(\mathbf{z})} \frac{\delta B(y)}{\delta \pi_{0,a}(\mathbf{z})} - \frac{\delta A(x)}{\delta \pi_{0,a}(\mathbf{z})} \frac{\delta B(y)}{\delta \varphi_{0,a}(\mathbf{z})} \right]. \tag{2.359}$$

As a consequence, we find the *equal-time relations for the classical spectral function*

$$\rho_{ab}^{\text{cl}}(x, y)|_{x^0 = y^0} = 0, \qquad \partial_{x^0} \rho_{ab}^{\text{cl}}(x, y)|_{x^0 = y^0} = \delta_{ab}\delta(\mathbf{x} - \mathbf{y}). \tag{2.360}$$

Although their origin is different, we note that they are in complete correspondence with the respective quantum relations (2.130).

In general, classical statistical correlation functions are obtained as phase-space averages over trajectories given by solutions of the classical field equation (2.354). Such averages, for an arbitrary functional of the field $f[\varphi_a]$, are defined as

$$\langle f[\varphi_a]\rangle_{\text{cl}} = \int [\text{d}\varphi_0] \, [\text{d}\pi_0] \, W^{\text{cl}}[\varphi_0, \pi_0] f[\varphi_a^{\text{cl}}]. \tag{2.361}$$

This expression will be the starting point for constructing a functional integral for the classical statistical field theory similar to the expression (2.335) for the quantum theory.

We define

$$S^{\text{cl}}[\varphi, \tilde{\varphi}] \equiv S_0[\varphi, \tilde{\varphi}] + S_{\text{int}}^{\text{cl}}[\varphi, \tilde{\varphi}]. \tag{2.362}$$

The free part, $S_0[\varphi, \tilde{\varphi}]$, is given by (2.336). Integrating by parts, we have to take care of the finite initial-time boundary values $\tilde{\varphi}_{0,a}(\mathbf{x}) = \tilde{\varphi}(t_0, \mathbf{x})$ and $\pi_{0,a}(\mathbf{x}) = \partial_{x^0}\varphi(x)|_{x^0 = t_0}$, and we can write

$$S_0[\varphi, \tilde{\varphi}] = -\int_{\mathbf{x}} \pi_{0,a}(\mathbf{x})\tilde{\varphi}_{0,a}(\mathbf{x}) + \int_{x,t_0} \tilde{\varphi}_a(x) \left(-\Box_x - m^2\right) \varphi_a(x). \tag{2.363}$$

The interaction part, $S_{\text{int}}^{\text{cl}}[\varphi, \tilde{\varphi}]$, reads

$$S_{\text{int}}^{\text{cl}}[\varphi, \tilde{\varphi}] = -\frac{\lambda}{6N} \int_x \tilde{\varphi}_a(x)\varphi_a(x)\varphi_b(x)\varphi_b(x). \tag{2.364}$$

This interaction part differs from (2.337) in that it contains fewer vertices. The absence of vertices beyond those that are linear in $\tilde{\varphi}$ turns out to be a crucial difference between a classical statistical and a quantum field theory, as is shown in the following. From the definition of $S^{\text{cl}}[\varphi, \tilde{\varphi}]$, the classical equation of motion (2.354) for the field $\varphi_a(x)$ can be obtained as

$$\frac{\delta S^{\text{cl}}[\varphi, \tilde{\varphi}]}{\delta \tilde{\varphi}_a(x)} = \left[-\Box_x - m^2 - \frac{\lambda}{6N}\varphi_b(x)\varphi_b(x) \right] \varphi_a(x) = 0. \tag{2.365}$$

We now rewrite this equation of motion as a δ-constraint in a functional integral using the Fourier transform representation

$$\delta\left[\frac{\delta S^{\mathrm{cl}}[\varphi,\tilde\varphi]}{\delta\tilde\varphi}\right] = \int \mathcal{D}\tilde\varphi \exp\left\{\mathrm{i}\int_{x,t_0}\tilde\varphi_a(x)\left[-\Box_x - m^2 - \frac{\lambda}{6N}\varphi_b(x)\varphi_b(x)\right]\varphi_a(x)\right\}$$

$$= \int \mathcal{D}\tilde\varphi \exp\left\{\mathrm{i}S^{\mathrm{cl}}[\varphi,\tilde\varphi] + \mathrm{i}\int_{\mathbf{x}}\pi_{0,a}(\mathbf{x})\tilde\varphi_{0,a}(\mathbf{x})\right\}, \qquad (2.366)$$

where we have used (2.363) for the last equality.

In order to complete the construction of the functional integral for classical statistical correlation functions, we note that any functional $f[\varphi^{\mathrm{cl}}]$ of the classical field solution can be written as

$$f[\varphi^{\mathrm{cl}}] = \int_{\varphi_0} \mathcal{D}'\varphi\, f[\varphi]\delta[\varphi - \varphi^{\mathrm{cl}}]$$

$$= \int_{\varphi_0} \mathcal{D}'\varphi\, f[\varphi]\,\delta\left[\frac{\delta S^{\mathrm{cl}}[\varphi,\tilde\varphi]}{\delta\tilde\varphi}\right]\mathcal{J}[\varphi]$$

$$= \int_{\varphi_0} \mathcal{D}'\varphi\,\mathcal{D}\tilde\varphi\, f[\varphi]\exp\left\{\mathrm{i}S^{\mathrm{cl}}[\varphi,\tilde\varphi] + \mathrm{i}\int_{\mathbf{x}}\pi_{0,a}(\mathbf{x})\tilde\varphi_{0,a}(\mathbf{x})\right\}\mathcal{J}[\varphi]. \quad (2.367)$$

The prime on the measure again signifies that no integration over the initial time is implied, since this is fixed by the initial condition. What we have employed in the second equality of (2.367) is the generalization of $\delta\big(g(x)\big) = \sum_i \delta(x - x_i)/|g'(x_i)|$ for zeros x_i of an ordinary function $g(x)$ to functionals. The Jacobian reads

$$\mathcal{J}[\varphi] = \left|\det\left(\frac{\delta^2 S^{\mathrm{cl}}[\varphi,\tilde\varphi]}{\delta\varphi\,\delta\tilde\varphi}\right)\right|. \qquad (2.368)$$

Here it turns out that the Jacobian plays the role of an irrelevant normalization constant. Putting things together, we can write

$$\langle f[\varphi_a]\rangle_{\mathrm{cl}} = \int [\mathrm{d}\varphi_0]\,[\mathrm{d}\pi_0]\,W^{\mathrm{cl}}[\varphi_0,\pi_0]$$

$$\times \int_{\varphi_0} \mathcal{D}'\varphi\,\mathcal{D}\tilde\varphi\, f[\varphi]\exp\left\{\mathrm{i}S^{\mathrm{cl}}[\varphi,\tilde\varphi] + \mathrm{i}\int_{\mathbf{x}}\pi_{0,a}(\mathbf{x})\tilde\varphi_{0,a}(\mathbf{x})\right\}\mathcal{J}[\varphi]. \quad (2.369)$$

It is also possible to specify the initial conditions by a classical density matrix $\varrho_0^{\mathrm{cl}}[\varphi_0 + \frac{1}{2}\tilde\varphi_0, \varphi_0 - \frac{1}{2}\tilde\varphi_0]$, which is characterized by the Fourier transform of the phase-space probability distribution $W^{\mathrm{cl}}[\varphi_0,\pi_0]$:

$$\varrho_0^{\mathrm{cl}}\left[\varphi_0 + \tfrac{1}{2}\tilde\varphi_0, \varphi_0 - \tfrac{1}{2}\tilde\varphi_0\right] = \int [\mathrm{d}\pi_0]\,W^{\mathrm{cl}}[\varphi_0,\pi_0]\exp\left\{\mathrm{i}\int_{\mathbf{x}}\pi_{0,a}(\mathbf{x})\tilde\varphi_{0,a}(\mathbf{x})\right\}. \quad (2.370)$$

Adding also sources, we obtain the generating functional for classical statistical correlation functions:

$$
Z^{\text{cl}}[J, \tilde{J}, R^F, R^{\text{R}}, R^{\text{A}}, R^{\tilde{F}}; \varrho_0] = \int [\mathrm{d}\varphi_0]\, [\mathrm{d}\tilde{\varphi}_0]\, \varrho_0^{\text{cl}}\left[\varphi_0 + \tfrac{1}{2}\tilde{\varphi}_0, \varphi_0 - \tfrac{1}{2}\tilde{\varphi}_0\right]
$$

$$
\times \int_{\varphi_0, \tilde{\varphi}_0} \mathscr{D}'\varphi\, \mathscr{D}'\tilde{\varphi} \exp\left\{ \mathrm{i}\left[S^{\text{cl}}[\varphi, \tilde{\varphi}] + \int_{x, t_0} (\varphi_a(x), \tilde{\varphi}_a(x)) \begin{pmatrix} \tilde{J}_a(x) \\ J_a(x) \end{pmatrix} \right.\right.
$$

$$
\left.\left. + \frac{1}{2}\int_{xy, t_0} (\varphi_a(x), \tilde{\varphi}_a(x)) \begin{pmatrix} R_{ab}^{\tilde{F}}(x, y) & R_{ab}^{\text{A}}(x, y) \\ R_{ab}^{\text{R}}(x, y) & R_{ab}^{F}(x, y) \end{pmatrix} \begin{pmatrix} \varphi_b(y) \\ \tilde{\varphi}_b(y) \end{pmatrix} \right] \right\} \mathscr{J}[\varphi]. \qquad (2.371)
$$

Comparing with the quantum generating functional in (2.335), and using the fact that the Jacobian $\mathscr{J}[\varphi]$ plays the role of an irrelevant normalization constant, we find that the generating functionals for correlation functions are very similar in the quantum and the classical statistical theories. The main difference is that the quantum theory is characterized by more vertices.

In particular, all definitions of correlation functions given in Section 2.4.1 for the quantum theory apply as well for the respective classical correlators. For instance, from (2.341), we infer that the classical spectral function in the absence of sources is

$$
\rho_{ab}^{\text{cl}}(x, y) = \mathrm{i} \int [\mathrm{d}\varphi_0]\, [\mathrm{d}\tilde{\varphi}_0]\, \varrho_0^{\text{cl}}\left[\varphi_0 + \tfrac{1}{2}\tilde{\varphi}_0, \varphi_0 - \tfrac{1}{2}\tilde{\varphi}_0\right]
$$

$$
\times \int_{\varphi_0, \tilde{\varphi}_0} \mathscr{D}'\varphi\, \mathscr{D}'\tilde{\varphi}\, \left[\varphi_a(x)\tilde{\varphi}_b(y) - \tilde{\varphi}_a(x)\varphi_b(y)\right] \exp\left\{ \mathrm{i} S^{\text{cl}}[\varphi, \tilde{\varphi}] \right\}. \qquad (2.372)
$$

In order to understand the equivalence with the definition as a phase-space average of the Poisson bracket in (2.358), we may consider the retarded case $x^0 < y^0$ first and take $x^0 = t_0$. With the help of (2.370), we can then write

$$
-\mathrm{i} \int [\mathrm{d}\varphi_0]\, [\mathrm{d}\tilde{\varphi}_0]\, \varrho_0^{\text{cl}}\left[\varphi_0 + \tfrac{1}{2}\tilde{\varphi}_0, \varphi_0 - \tfrac{1}{2}\tilde{\varphi}_0\right] \int_{\varphi_0, \tilde{\varphi}_0} \mathscr{D}'\varphi\, \mathscr{D}'\tilde{\varphi}\, \tilde{\varphi}_a(x)\varphi_b(y) \exp\left\{ \mathrm{i} S^{\text{cl}}[\varphi, \tilde{\varphi}] \right\}
$$

$$
= -\mathrm{i} \int [\mathrm{d}\varphi_0]\, [\mathrm{d}\pi_0]\, W^{\text{cl}}[\varphi_0, \pi_0] \int_{\varphi_0, \tilde{\varphi}_0} \mathscr{D}'\varphi\, \mathscr{D}\tilde{\varphi}\, \tilde{\varphi}_a(x)\varphi_b(y)
$$

$$
\times \exp\left\{ \mathrm{i} S^{\text{cl}}[\varphi, \tilde{\varphi}] + \mathrm{i} \int_{\mathbf{x}} \pi_{0,a}\tilde{\varphi}_{0,a} \right\} \mathscr{J}[\varphi]
$$

$$
= -\int [\mathrm{d}\varphi_0]\, [\mathrm{d}\pi_0]\, W^{\text{cl}}[\varphi_0, \pi_0] \frac{\delta}{\delta\pi_{0,a}(\mathbf{x})} \int_{\varphi_0, \tilde{\varphi}_0} \mathscr{D}'\varphi\, \mathscr{D}\tilde{\varphi}\, \varphi_b(y)
$$

$$
\times \exp\left\{ \mathrm{i} S^{\text{cl}}[\varphi, \tilde{\varphi}] + \mathrm{i} \int_{\mathbf{x}} \pi_{0,a}\tilde{\varphi}_{0,a} \right\} \mathscr{J}[\varphi]
$$

$$= -\int [d\varphi_0]\,[d\pi_0]\,W[\varphi_0, \pi_0]\frac{\delta\varphi_b^{cl}(y)}{\delta\pi_{0,a}(\mathbf{x})}$$

$$= -\int [d\varphi_0]\,[d\pi_0]\,W[\varphi_0, \pi_0]\int_{\mathbf{z}}\frac{\delta\varphi_a^{cl}(t_0, \mathbf{x})}{\delta\varphi_{0,c}(\mathbf{z})}\frac{\delta\varphi_b^{cl}(y)}{\delta\pi_{0,c}(\mathbf{z})}, \qquad (2.373)$$

where for the last line we have employed (2.367) and summation over repeated field indices is implied. Doing the corresponding steps for the retarded case $x^0 > y^0$ taking $y^0 = t_0$ shows the equivalence.

2.4.3 Classicality condition

Since the generating functionals for correlation functions are very similar in the classical and quantum theories, the same techniques can be used to derive time-evolution equations of classical correlation functions. In particular, the classical dynamic equations have the very same form (2.145) and (2.151) as their quantum analogues with the replacements $\phi_a(x) \to \phi_a^{cl}$, $F_{ab}(x, y) \to F_{ab}^{cl}(x, y)$, and $\rho_{ab}(x, y) \to \rho_{ab}^{cl}(x, y)$. The corresponding classical statistical self-energies $\Sigma_{ab}^{F,cl}(x, y)$ and $\Sigma_{ab}^{\rho,cl}(x, y)$ have in general the same diagrammatic contributions, but lack certain terms because of the reduced number of vertices.

In order to compare classical and quantum corrections to self-energies, we note that the classical statistical generating functional (2.371) exhibits an important reparametrization property: if the fluctuating fields are rescaled according to

$$\varphi_a(x) \to \varphi_a'(x) = \sqrt{\lambda}\,\varphi_a(x), \qquad \tilde{\varphi}_a(x) \to \tilde{\varphi}_a'(x) = \frac{1}{\sqrt{\lambda}}\,\tilde{\varphi}_a(x), \qquad (2.374)$$

then the coupling λ drops out of $S^{cl}[\varphi, \tilde{\varphi}] = S_0[\varphi, \tilde{\varphi}] + S_{int}^{cl}[\varphi, \tilde{\varphi}]$ defined in (2.336) and (2.364). The free part $S_0[\varphi, \tilde{\varphi}]$ remains unchanged and the interaction part becomes

$$S_{int}^{cl}[\varphi', \tilde{\varphi}'] = -\frac{1}{6N}\int_x \tilde{\varphi}_a'(x)\varphi_a'(x)\varphi_b'(x)\varphi_b'(x). \qquad (2.375)$$

Moreover, the functional measure in (2.371) is invariant under the rescaling (2.374), and the sources can be redefined accordingly. Therefore, the classical statistical generating functional becomes independent of λ, except for the coupling dependence entering the probability distribution fixing the initial conditions. Accordingly, the coupling does not enter the classical dynamic equations for correlation functions. All the λ dependence enters the initial conditions that are required to solve the dynamic equations.

In contrast to the classical case, this reparametrization property is absent in the quantum theory: after the rescaling (2.374), one is left with $S[\varphi', \tilde{\varphi}']$, whose coupling dependence is given by the interaction part

$$S_{int}[\varphi', \tilde{\varphi}'] = -\frac{1}{6N}\int_x \tilde{\varphi}_a'(x)\varphi_a'(x)\varphi_b'(x)\varphi_b'(x) - \frac{\lambda^2}{24N}\int_x \tilde{\varphi}_a'(x)\tilde{\varphi}_a'(x)\tilde{\varphi}_b'(x)\varphi_b'(x),$$

$$(2.376)$$

according to (2.337). Comparing with (2.375), we can see that the quantum vertex, which is absent in the classical statistical theory, encodes all the λ dependence of the dynamics.

The comparison of quantum versus classical dynamics becomes particularly transparent using the above rescaling. The rescaled macroscopic field and statistical correlation function are given by

$$\phi'_a(x) = \sqrt{\lambda}\,\phi_a(x), \qquad F'_{ab}(x,y) = \lambda F_{ab}(x,y), \tag{2.377}$$

while the spectral function $\rho_{ab}(x,y)$ remains unchanged according to (2.372). Similarly, we define for the statistical self-energy $\Sigma^{F\prime}_{ab}(x,y) = \lambda \Sigma^{F}_{ab}(x,y)$. For instance, for the two-loop self-energies for vanishing macroscopic field given in (2.378) and (2.379), we obtain

$$\Sigma'_F(x,y) = -\frac{N+2}{18N^2}F'(x,y)\left[F'^2(x,y) - \frac{3}{4}\lambda^2\rho^2(x,y)\right], \tag{2.378}$$

$$\Sigma_\rho(x,y) = -\frac{N+2}{6N^2}\rho(x,y)\left[F'^2(x,y) - \frac{1}{12}\lambda^2\rho^2(x,y)\right]. \tag{2.379}$$

The corresponding classical statistical self-energies $\Sigma^{\prime\text{cl}}_F(x,y)$ and $\Sigma^{\text{cl}}_\rho(x,y)$ are obtained from the same expressions by dropping the λ-dependent terms, which are proportional to ρ^2. More precisely, one observes that the quantum evolution equations would be accurately described by the classical ones if the *classicality condition*

$$\boxed{F^2(x,y) \gg \rho^2(x,y)} \tag{2.380}$$

in terms of the non-rescaled correlation functions holds. A similar analysis of the loop corrections in the presence of a nonzero macroscopic field given in Section 2.3.1, or of the NLO $1/N$ expansion of Section 2.3.2, yields the same condition.

However, the requirement that (2.380) be satisfied for all space–time arguments is too restrictive and it can typically be achieved only for a limited range of time and relevant momenta. One expects the classical description to become a reliable approximation for the quantum theory if the number of field quanta in each mode is sufficiently high. The classicality condition (2.380) entails the justification of this expectation. In order to illustrate the condition in terms of a more intuitive picture of occupation numbers, we employ the free field theory type form of the spectral function and statistical propagator with mode frequency $\omega_{\mathbf{p}}$ as given in (2.249). From this, we obtain the following estimates for the time-averaged correlators:

$$\overline{F^2}(t,t';\mathbf{p}) \equiv \frac{\omega_{\mathbf{p}}}{2\pi}\int_{t-2\pi/\omega_{\mathbf{p}}}^{t} dt'\, F^2(t,t';\mathbf{p}) \rightarrow \frac{\left[f_{\mathbf{p}}(t)+\frac{1}{2}\right]^2}{2\omega_{\mathbf{p}}^2(t)},$$

$$\overline{\rho^2}(t,t';\mathbf{p}) \rightarrow \frac{1}{2\omega_{\mathbf{p}}^2(t)}. \tag{2.381}$$

Inserting these estimates into (2.380) for equal momenta yields

$$\left[f_{\mathbf{p}}(t) + \tfrac{1}{2} \right]^2 \gg 1, \qquad \text{or} \qquad f_{\mathbf{p}}(t) \gg \tfrac{1}{2}. \tag{2.382}$$

If the dominant momentum modes have occupancies much larger than the quantum half, then a classical statistical description can be an accurate approximation of the quantum dynamics. It is important to note that this is in general *not* the case in thermal equilibrium, where at temperature T the typical momenta $p \sim T$ have an occupancy of order one. Consequently, the late-time approach to thermal equilibrium in quantum theories is beyond the range of applicability of classical statistical approximations. Important examples where a classical statistical description is accurate include continuous thermal phase transitions, since the relevant momenta for scaling behavior have $p \ll T$ with occupancy $T/p \gg 1$. Other examples are nonequilibrium instabilities, which yield high occupation numbers of characteristic modes, such as described in Section 2.5, or wave turbulence and nonthermal scaling phenomena, such as described in Section 2.6.

2.4.4 Precision tests of quantum versus classical statistical dynamics

When the nonequilibrium quantum dynamics of a highly occupied system can be accurately mapped onto a classical statistical field theory evolution, one can solve it without further approximations using lattice simulation techniques. This mapping is valid as long as the classicality condition (2.380) is fulfilled for typical momenta. The lattice field theory is then defined on a spatial grid with spacing a_{s} and side length $N_{\mathrm{s}} a_{\mathrm{s}}$ in a box with periodic boundary conditions, as explained around (2.248) at the end of Section 2.3.1. Classical statistical simulations consist of numerically solving the classical field equations of motion and Monte Carlo sampling of initial conditions according to (2.361). Thus, observables are obtained by averaging over the different classical trajectories that arise from the different initial field configurations.

For the relativistic second-order differential equation of the scalar field (2.354), the numerical integrations on a d-dimensional grid can be efficiently done using a standard leapfrog algorithm. For the nonrelativistic first-order equation (2.30) a conventional split-step method may be applied. These classical equations are then supplemented by suitable quantum initial conditions. For instance, typical Gaussian initial conditions represented by spatially homogeneous ensembles for a relativistic field equation as explained in Section 2.2.4 can be specified by a macroscopic field $\phi(t_0) = \phi_0$, its derivative $\dot{\phi}(t_0) = \dot{\phi}_0$, and the distribution function $f_{\mathbf{p}}(t_0)$ as

$$\varphi(t_0, \mathbf{x}) = \phi_0 + \int \frac{d^d p}{(2\pi)^d} \sqrt{\frac{f_{\mathbf{p}}(t_0) + \tfrac{1}{2}}{\omega_{\mathbf{p}}(t_0)}} \, c_{\mathbf{p}} e^{i \mathbf{p} \cdot \mathbf{x}}, \tag{2.383}$$

with the initial frequency $\omega_{\mathbf{p}}(t_0) = \sqrt{\mathbf{p}^2 + M^2}$. For Gaussian initial conditions, the coefficients $c_{\mathbf{p}}$ have to satisfy the relations

$$\langle c_{\mathbf{p}} c_{\mathbf{p}'}^* \rangle_{\mathrm{cl}} = (2\pi)^d \delta(\mathbf{p} - \mathbf{p}'),$$

$$\langle c_{\mathbf{p}} c_{\mathbf{p}'} \rangle_{\mathrm{cl}} = \langle c_{\mathbf{p}}^* c_{\mathbf{p}'}^* \rangle_{\mathrm{cl}} = 0. \tag{2.384}$$

They can be realized by taking $c_{\mathbf{p}}$ as Gaussian random numbers multiplied by complex random phase factors. This can be formulated as $c_{\mathbf{p}} = A(\mathbf{p})e^{\mathrm{i}2\pi\alpha(\mathbf{p})}$, with a Gaussian-distributed amplitude $A(\mathbf{p})$ and uniformly distributed phase $\alpha(\mathbf{p})$ between 0 and 1. The random numbers $c_{\mathbf{p}}$ additionally have to satisfy $c^*_{-\mathbf{p}} = c_{\mathbf{p}}$ to ensure that $\varphi(t_0, \mathbf{x})$ is real-valued.

The conjugate momentum field $\pi(t_0, \mathbf{x}) = \dot{\varphi}(t_0, \mathbf{x})$ is initialized in a very similar way:

$$\pi(t_0, \mathbf{x}) = \dot{\phi}_0 + \int \frac{\mathrm{d}^d p}{(2\pi)^d} \sqrt{\left[f_{\mathbf{p}}(t_0) + \tfrac{1}{2}\right] \omega_{\mathbf{p}}(t_0)} \, \tilde{c}_{\mathbf{p}} e^{\mathrm{i}\mathbf{p}\cdot\mathbf{x}}, \qquad (2.385)$$

with complex Gaussian random numbers $\tilde{c}_{\mathbf{p}}$ satisfying the same relations as $c_{\mathbf{p}}$ such as (2.384). Since $c_{\mathbf{p}}$ and $\tilde{c}_{\mathbf{p}}$ are independent random numbers, one also has $\langle c_{\mathbf{p}}\tilde{c}_{\mathbf{p}'}\rangle_{\mathrm{cl}} = 0$. Therefore, $\langle \varphi(t_0, \mathbf{x})\pi(t_0, \mathbf{y}) + \pi(t_0, \mathbf{x})\varphi(t_0, \mathbf{y})\rangle_{\mathrm{cl}}$ vanishes automatically at initial time $t = t_0$ for the initial conditions discussed here. For instance, the macroscopic field or the statistical correlation function at times larger than t_0 are then given by the averages (2.355) or (2.357), and equivalently for higher statistical n-point correlation functions. Since $\rho^{\mathrm{cl}}(t, t', \mathbf{x} - \mathbf{x}')$ is given by the Poisson bracket (2.358), spectral properties cannot be obtained from simple products of classical fields. However, in contrast to solving the equations (2.145) for correlation functions, knowledge of the spectral function is not a prerequisite for the computation of correlation functions from classical statistical simulations.

In the following, we apply these simulation techniques to the scalar N-component field theory. Since the nonequilibrium evolution of classical statistical correlation functions can be obtained numerically up to controlled statistical errors, the results include all orders in $1/N$. Consequently, they can be used for a precision test of approximation schemes such as the 2PI $1/N$ expansion implemented in classical statistical field theory. We emphasize that this compares two very different calculational procedures: the results from the simulation involve thousands of individual runs from which the correlators are constructed, while the corresponding results employing the 2PI $1/N$ expansion involve only a single run directly solving the evolution equation for the correlators. The accuracy of the simulations manifests itself also in the fact that the time-reversal-invariant dynamics can be explicitly reversed in practice for not too late times. We then compare to the corresponding NLO results of the quantum theory and check the classicality condition of Section 2.4.3.

Here, we consider a system in $d = 1$ that is invariant under space translations and work in momentum space. Similar calculations have been performed also for $d = 3$ and for theories including fermions, as described in in the literature cited in Section 2.4.5. At $t_0 = 0$, we choose the same initial conditions for the classical and the corresponding quantum theory: a Gaussian initial state with zero macroscopic field and a statistical propagator $F(0, 0; p) = [f_p(0) + \tfrac{1}{2}]/\omega_p(0)$, where the initial $f_p(0)$ represents a peaked distribution around the momentum $p = p_{\mathrm{ts}}$ as in Section 2.3.1. The initial mode energy is given by $\omega_p(0) = \sqrt{p^2 + M^2}$, where M is the one-loop renormalized mass in the presence of the nonequilibrium medium, determined from the corresponding one-loop gap equation as in (2.281). As a renormalization condition we choose the one-loop

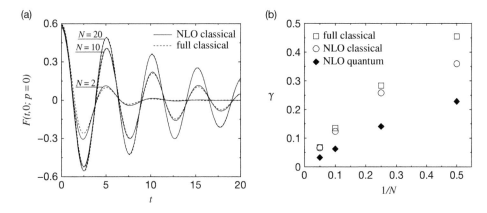

Fig. 2.13 (a) Unequal-time two-point function $F(t, 0; p = 0)$ at zero momentum for $N = 2$, 10, 20. The solid lines show results from the NLO classical evolution and the dashed lines results from the full classical statistical simulation. A convergence of classical NLO and full results can be seen already for moderate values of N. (b) Damping rate extracted from $F(t, 0; p = 0)$ as a function of $1/N$. Open symbols represent NLO and full classical evolution. The quantum NLO results are shown with black symbols for comparison. The initial conditions are characterized by low occupation numbers, so that quantum effects become sizable. It can be seen that in the quantum theory, the damping rate is reduced compared with the classical theory. (All in units of m_{R}.)

renormalized mass in vacuum $m_{\mathrm{R}} \equiv M|_{f_p(0) = 0}$ as our scale. The results shown below are obtained using a fixed coupling constant $\lambda/m_{\mathrm{R}}^2 = 30$.

Figure 2.13(a) shows the classical statistical propagator $F(t, 0; p = 0)$ for three values of N. All other parameters are kept constant. The figure compares the time evolution using the 2PI $1/N$ expansion to NLO and the classical statistical simulation.[14] We can see that the approximate time evolution of the correlation function shows a rather good agreement with the full result even for small values of N. For $N = 20$, the exact and NLO evolution can hardly be distinguished. A very sensitive quantity for comparisons is the damping rate γ, which is obtained from an exponential fit to the envelope of $F(t, 0; p = 0)$. The systematic convergence of the NLO and the Monte Carlo result as a function of $1/N$ can be observed from Fig. 2.13(b). The accuracy of the description of far-from-equilibrium processes within the classical statistical NLO approximation of the 2PI effective action is remarkable.

Figure 2.13(b) also shows the damping rate from the quantum evolution, using the same initial conditions and parameters. We can see that the damping in the quantum theory differs from and, in particular, is reduced compared with the classical result. The effective loss of details about the initial conditions takes more time for

---

[14] The results presented here have been obtained from sampling 50 000–80 000 independent initial conditions to approximate the exact evolution of correlation functions. In general, far fewer configurations are required for $d > 1$, because of self-averaging.

the quantum system than for the corresponding classical one. In the limit $N \to \infty$, damping of the unequal-time correlation function goes to zero since the nonlocal part of the self-energies vanishes identically at LO large-N and scattering is absent. In this limit, there is no difference between evolution in a quantum and a classical statistical field theory for the same initial conditions.

For finite N, scattering is present and the quantum and classical evolutions differ in general. However, as discussed in Section 2.4.3, the classical field approximation may be expected to become a reliable description of the quantum theory if the number of field quanta in each field mode is sufficiently high. We observe that increasing the initial particle number density leads to a convergence of quantum and classical time evolutions at not too late times. Fig. 2.14(a) shows the time evolution of the equal-time correlation function $F(t, t; p)$ for several momenta p and $N = 10$. Here the initial integrated particle density $\int [dp/(2\pi)] f_p(0)/M = 1.2$ is six times as high as in Fig. 2.13. At $p = 2p_{\text{ts}}$, one finds $f_{2p_{\text{ts}}}(0) \simeq 0.35$ and a slightly larger value at this momentum of about $\simeq 0.5$ at later times shown. For these evolutions, the classicality condition (2.382) is therefore approximately fulfilled up to momenta $p \simeq 2p_{\text{ts}}$, and we can indeed observe from Fig. 2.14(a) a rather good agreement of quantum and classical evolution in this range. For an estimate of the NLO truncation error, we also give the full Monte Carlo result for $N = 10$, showing a quantitative agreement with the classical NLO evolution during these times.

From Fig. 2.14(a), we can see that the initially highly occupied modes "decay" as time proceeds and the low-momentum modes become more and more populated. At late times, the classical and quantum theories approach their respective equilibrium

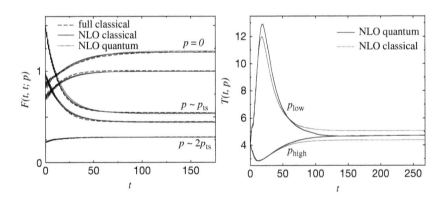

Fig. 2.14 (a) Nonequilibrium evolution of the equal-time two-point function $F(t, t; p)$ for $N = 10$ for various momenta p. A good agreement can be seen between the full simulation (dashed) and the NLO classical result (full). The quantum evolution is shown by dotted lines. The integrated initial particle density is six times as high as in Fig. 2.13. (b) A very sensitive quantity to study deviations is the time-dependent inverse slope $T(t, p)$ defined in the text. When a Bose–Einstein-distributed occupation number is approached, all modes become equal, $T(t, p) = T_{\text{eq}}$, as can be observed to high accuracy for the quantum evolution. For classical thermal equilibrium, the defined inverse slope remains momentum-dependent.

distributions. Since classical and quantum thermal equilibria are distinct, the classical and quantum time evolutions have to deviate at sufficiently late times. Figure 2.14(b) shows the time-dependent inverse slope parameter

$$T(t,p) \equiv -f_p(t)[f_p(t)+1]\left(\frac{df_p}{d\epsilon_p}\right)^{-1}, \qquad (2.386)$$

which was introduced in Section 2.3.2 to study the approach to a Bose–Einstein distribution.[15] It employs the effective particle number $f_p(t)$ defined in (2.283) and mode energy $\epsilon_p(t)$ given by (2.285). Initially, we can see very different behavior of $T(t,p)$ for the low- and high-momentum modes, indicating that the system is far from equilibrium. The quantum evolution approaches a Bose–Einstein-distributed occupation number with a momentum-independent inverse slope $T_{\text{eq}} = 4.7m_{\text{R}}$ to very good accuracy. In contrast, in the classical theory, the slope parameter remains momentum-dependent since the classical dynamics does, of course, not reach a Bose–Einstein distribution.

To see this in more detail, we note that for a Bose–Einstein distribution $f_\beta(\varepsilon_p) = \{[\exp(\epsilon_p/T_{\text{eq}})-1]\}^{-1}$, the inverse slope (2.386) is independent of the mode energies and equal to the temperature T_{eq}. During the nonequilibrium evolution, effective thermalization can therefore be observed if $T(t,p)$ becomes time- and momentum-independent, $T(t,p) \to T_{\text{eq}}$. This is indeed seen in Fig. 2.14(a) for the quantum system. If the system approaches classical equilibrium at some temperature T_{cl} and is not too strongly coupled, the following behavior is expected. From the definition (2.283) of $f_p(t)$ in terms of two-point functions, we expect to find approximately

$$T(t,p) \to T_{\text{cl}}\left(1 - \varepsilon_p^2/T_{\text{cl}}^2\right), \qquad (2.387)$$

i.e. a remaining momentum dependence with $T(t,p) < T(t,p')$ if $\varepsilon_p > \varepsilon_{p'}'$. Indeed, this is what one observes for the classical field theory result in Fig. 2.14.

For a classical theory, a very simple test for effective equilibration is available. An exact criterion can be obtained from the classical counterpart of the KMS condition for thermal equilibrium discussed in Section 2.2.7. In coordinate space, the classical equilibrium KMS condition reads

$$\frac{1}{T_{\text{cl}}}\frac{\partial}{\partial x^0}F_{\text{cl}}^{(\text{eq})}(x-y) = -\rho_{\text{cl}}^{(\text{eq})}(x-y), \qquad (2.388)$$

and, in momentum space,

$$F_{\text{cl}}^{(\text{eq})}(k) = -if_{\text{cl}}(k^0)\rho_{\text{cl}}^{(\text{eq})}(k), \qquad f_{\text{cl}}(k^0) = \frac{T_{\text{cl}}}{k^0}. \qquad (2.389)$$

Differentiating (2.388) with respect to y^0 at $x^0 = y^0 = t$ gives

$$\frac{1}{T_{\text{cl}}}\frac{\partial}{\partial y^0}\frac{\partial}{\partial x^0}F_{\text{cl}}^{(\text{eq})}(x-y)\bigg|_{x^0=y^0=t} = -\frac{\partial}{\partial y^0}\rho_{\text{cl}}^{(\text{eq})}(x-y)\bigg|_{x^0=y^0=t}. \qquad (2.390)$$

---

[15] Note that $d\ln[f_p^{-1}(t)+1]/d\epsilon_p(t) = T^{-1}(t,p)$.

Combining this KMS relation with the equal-time condition (2.360) for the spectral function leads to

$$\partial_t \partial_{t'} F_{\mathrm{cl}}^{(\mathrm{eq})}(t,t';\mathbf{x}-\mathbf{y})|_{t=t'} = T_{\mathrm{cl}}\delta(\mathbf{x}-\mathbf{y}). \qquad (2.391)$$

In terms of the classical conjugate momentum fields $\pi_a(x) \equiv \partial_{x^0}\varphi_a(x)$, this represents the well-known equilibrium relation $\langle \pi_a(t,\mathbf{x})\pi_b(t,\mathbf{y})\rangle_{\mathrm{cl}}^{(\mathrm{eq})} = T_{\mathrm{cl}}\delta(\mathbf{x}-\mathbf{y})\delta_{ab}$. Out of equilibrium, one can define an effective classical mode temperature

$$T_{\mathrm{cl}}(t,p) = \partial_t \partial_{t'} F_{\mathrm{cl}}(t,t';p)|_{t=t'}. \qquad (2.392)$$

Effective classical equilibration is observed if $T_{\mathrm{cl}}(t,p)$ becomes time- and momentum-independent, $T_{\mathrm{cl}}(t,p) \to T_{\mathrm{cl}}$. This is indeed the case for the results presented at sufficiently late times for a given lattice regularization.

Apart from the 2PI $1/N$ expansion, the late-time behavior can also be studied from the 2PI loop expansion. For the quantum theory, this has been demonstrated in Section 2.3.1 using two-loop self-energy corrections. We show that this can also be employed for the classical statistical theory and use the corresponding loop approximation in the following to demonstrate the above statements about classical equilibration.

In Fig. 2.15, the nonequilibrium evolution of the classical mode temperature $T_{\mathrm{cl}}(t,p)$ is shown for various momentum modes in the $(1+1)$-dimensional classical scalar field theory for $N=1$. The equations are solved by a lattice discretization with spatial lattice spacing $m_{\mathrm{R}}a_{\mathrm{s}}=0.4$, time step $a_t/a_{\mathrm{s}}=0.2$, and $N_{\mathrm{s}}=24$ sites for $\lambda/m_{\mathrm{R}}^2=1$. For the initial ensemble, we take $F_{\mathrm{cl}}(0,0;p)=T_0/(p^2+m_{\mathrm{R}}^2)$ and $\partial_t \partial_{t'} F_{\mathrm{cl}}(t,t';p)|_{t=t'=0}=T_0$, with $T_0/m_{\mathrm{R}}=5$. We have observed that at sufficiently late times, the contributions from early times to the dynamics are effectively

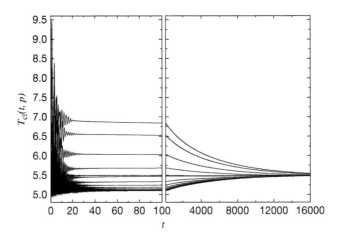

Fig. 2.15 Nonequilibrium time evolution in $1+1$ dimensions from the three-loop approximation of the classical statistical 2PI effective action for one scalar field, $N=1$. The effective mode temperature $T_{\mathrm{cl}}(t,p)$ is shown. The approach to classical equilibrium, $T_{\mathrm{cl}}(t,p) \to T_{\mathrm{cl}}$, can be seen at sufficiently late times for a given lattice regularization.

suppressed. This fact has been employed in Fig. 2.15 to reach the very late times. We can see that at sufficiently late times, the system relaxes toward classical equilibrium with a final temperature $T_{\text{cl}}/m_{\text{R}} \approx 5.5$. Although the typical classical equilibration times observed are substantially larger than the times required to approach thermal equilibrium in the respective quantum theory, this is not a regularization-independent statement for the classical theory. In contrast to the quantum theory, statements about equilibration times are sensitive to the lattice regularization employed for the classical theory, because of the Rayleigh–Jeans divergence.

2.4.5 Bibliography

- Figures 2.13–2.15 are taken from G. Aarts and J. Berges. Classical aspects of quantum fields far from equilibrium. *Phys. Rev. Lett.* **88** (2002) 0416039 and J. Berges. Nonequilibrium quantum fields and the classical field theory limit. *Nucl. Phys. A* **702** (2002) 351.
- The presentation given here follows to a large extent J. Berges and T. Gasenzer. Quantum versus classical statistical dynamics of an ultracold Bose gas. *Phys. Rev. A* **76** (2007) 033604. See also S. Jeon. The Boltzmann equation in classical and quantum field theory. *Phys. Rev. C* **72** (2005) 014907.
- Many of these ideas can be traced back to the original work of P. C. Martin, E. D. Siggia, and H. A. Rose. Statistical dynamics of classical systems. *Phys. Rev. A* **8** (1973) 423.
- The use of classical statistical field theory for the relativistic reheating problem was pointed out in D. T. Son. Classical preheating and decoherence. arXiv: hep-ph/9601377 and S. Y. Khlebnikov and I. I. Tkachev. Classical decay of inflaton. *Phys. Rev. Lett.* **77** (1996) 219.
- Classical statistical field theory studies related to approximation schemes in QFT can be found in G. Aarts, G. F. Bonini, and C. Wetterich. Exact and truncated dynamics in nonequilibrium field theory. *Phys. Rev. D* **63** (2001) 025012; G. Aarts, G. F. Bonini, C. Wetterich. On thermalization in classical scalar field theory. *Nucl. Phys. B* **587** (2000) 403; F. Cooper, A. Khare, and H. Rose. Classical limit of time-dependent quantum field theory: a Schwinger–Dyson approach. *Phys. Lett. B* **515** (2001) 463; and A. Arrizabalaga, J. Smit, and A. Tranberg. Tachyonic preheating using 2PI-1/N dynamics and the classical approximation. *JHEP* **0410** (2004) 017.
- Classical statistical simulations including fermions were pioneered in G. Aarts and J. Smit. Real time dynamics with fermions on a lattice. *Nucl. Phys. B* **555** (1999) 355 for $d = 1$, and became possible for $d = 3$ in J. Berges, D. Gelfand and J. Pruschke. Quantum theory of fermion production after inflation. *Phys. Rev. Lett.* **107** (2011) 061301. For a review of real-time lattice QED and QCD, see V. Kasper, F. Hebenstreit, and J. Berges. Fermion production from real-time lattice gauge theory in the classical statistical regime. *Phys. Rev. D* **90** (2014) 025016.
- For diagrammatics in classical field theory, see G. Aarts and J. Smit. Classical approximation for time-dependent quantum field theory: diagrammatic analysis for hot scalar fields. *Nucl. Phys. B* **511** (1998) 451.

2.5   Nonequilibrium instabilities

2.5.1   Parametric resonance

In classical mechanics, parametric resonance is the phenomenon of resonant amplification of the amplitude of an oscillator having a time-dependent periodic frequency. In the context of quantum field theory, a similar phenomenon describes the amplification of quantum fluctuations, which can be interpreted as particle production. It provides an important building block for our understanding of the (pre)heating of the early universe at the end of an inflationary period, and may also be operative at certain stages in relativistic heavy-ion collision experiments. The example of parametric resonance can lead to nonperturbative phenomena such as strong turbulence and Bose condensation far from equilibrium, even in the presence of arbitrarily small couplings, which will be discussed in Section 2.6. Here we will consider the phenomenon as a "paradigm" for far-from-equilibrium dynamics following nonequilibrium instabilities. Much of the nonlinear physics turns out to be universal and thus independent of the details of the underlying mechanism that triggers the instability.

We recall first the classical mechanics example of resonant amplitude growth for an oscillator with time-dependent periodic frequency. A physical realization of this situation is a pendulum with a periodically changing length as shown here:

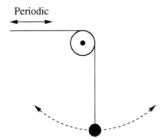

Periodic

In the linear regime, the amplitude $y(t)$ is described by the second-order differential equation

$$\ddot{y}(t) + \omega^2(t)\, y(t) = 0 \tag{2.393}$$

with periodic $\omega(t + \Delta T) = \omega(t)$ of period ΔT. Since the equation is invariant under $t \to t + \Delta T$, one expects periodic solutions

$$y(t + \Delta T) = A y(t). \tag{2.394}$$

Writing the time-independent amplitude as

$$A = e^{\alpha\, \Delta T} \tag{2.395}$$

in terms of the so-called Floquet index α, these solutions can be expressed as

$$y(t) = e^{\alpha t}\Pi(t), \tag{2.396}$$

with periodic $\Pi(t + \Delta T) = \Pi(t)$. This can be directly verified since

$$y(t + \Delta T) = e^{\alpha(t + \Delta T)}\Pi(t + \Delta T) = A \underbrace{e^{\alpha t}\Pi(t)}_{y(t)}. \qquad (2.397)$$

With $y(t)$, $y(-t)$ is also a solution, and we conclude from (2.396) that, for real Floquet index $\alpha \neq 0$, there is an instability characterized by exponential growth.

In contrast to this mechanics example, in closed systems described by quantum field theory, there will be no external periodic source. We will see below that a large coherent field amplitude coupled to its own quantum fluctuations can trigger the phenomenon of parametric resonance. Mathematically, however, important aspects are very similar to the above classical example for sufficiently early times. There is even a precise mapping: the mechanical oscillator amplitude y plays the role of the statistical two-point function F in quantum field theory, and the periodic $\omega^2(t)$ plays the role of an effective mass term $M^2(\phi(t))$ whose time dependence is induced by an oscillating macroscopic field $\phi(t)$. Simple linear approximations to the problem turn out to be mathematically equivalent to the above mechanics example. Accordingly, well-known Lamé-type solutions of the mechanics problem will also play a role in the quantum field theory study. Substantial deviations do, however, quickly set in with important nonlinear effects.

Linearized classical analysis

We consider the $O(N)$-symmetric scalar field theory with classical action (2.93). Since the phenomenon of parametric resonance is essentially classical, for the purpose of comparison with the quantum treatment, we start with a linearized classical analysis. The N-component field is written as

$$\phi_a(x) = \phi \delta_{a1} + \delta\phi_a(x), \qquad (2.398)$$

where we will consider a spatially homogeneous but time-dependent "background field," i.e., $\phi = \phi(x^0)$. The classical equations of motion are obtained from the stationarity of the action, and in the following we will consider them in a linear approximation in the "fluctuations" $\delta\phi_a(x)$. These equations will also be seen to correspond to the evolution equations in the corresponding quantum field theory in the limit where all nonlinear or loop corrections are neglected.

We denote the fluctuations in the "longitudinal" field direction as $\delta\phi_{\parallel}(x) \equiv \delta\phi_1(x)$ and those in the "transverse" direction as $\delta\phi_{\perp}(x) \equiv \delta\phi_{a>1}(x)$. The square of the fields appearing in the action (2.93) then reads

$$\phi_a(x)\phi_a(x) = \phi^2 + 2\phi\,\delta\phi_{\parallel}(x) + \delta\phi_{\parallel}(x)^2 + (N-1)\,\delta\phi_{\perp}(x)^2, \qquad (2.399)$$

whereas the quartic interaction term involves

$$[\phi_a(x)\phi_a(x)]^2 = \phi^4 + 4\phi^3\,\delta\phi_{\parallel}(x) + 2\phi^2\left[3\,\delta\phi_{\parallel}(x)^2 + (N-1)\,\delta\phi_{\perp}(x)^2\right] + \mathcal{O}\left(\delta\phi^3\right).$$

$$(2.400)$$

Here we have expanded the quartic term, neglecting cubic and quartic powers of the fluctuations, since this is sufficient to obtain the evolution equations to linear order.

The classical field equation for the longitudinal component is given by the stationarity condition $\delta S[\phi + \delta\phi]/\delta\delta\phi_{\|}(x) = 0$. Using (2.399) and (2.400), we see that to lowest order in an expansion in powers of $\delta\phi_{\|}(x)$, it leads to the background field equation

$$\left[\frac{\partial^2}{\partial x_0^2} + m^2 + \frac{\lambda}{6N}\phi^2(x_0) \right] \phi(x_0) = 0. \tag{2.401}$$

To NLO in powers of $\delta\phi_{\|}(x)$, the same stationarity condition gives

$$\left[\Box_x + m^2 + \frac{\lambda}{2N}\phi^2(x_0) \right] \delta\phi_{\|}(x) = 0. \tag{2.402}$$

For the transverse components, we find from $\delta S[\phi + \delta\phi]/\delta\delta\phi_{\perp}(x) = 0$ the equation

$$\left[\Box_x + m^2 + \frac{\lambda}{6N}\phi^2(x_0) \right] \delta\phi_{\perp}(x) = 0. \tag{2.403}$$

In the following, we will choose $m = 0$ as motivated in Section 2.1.3 and denote $t \equiv x_0$. It is convenient to introduce the rescaled background field

$$\sigma(t) = \sqrt{\frac{\lambda}{6N}}\, \phi(t). \tag{2.404}$$

We will consider the relevant case of a parametrically large initial field amplitude $\phi(t=0) \sim \sqrt{6N/\lambda}$ with $\lambda \ll 1$, such that the initial $\sigma_0 \equiv \sigma(t=0)$ is of order one. In terms of the rescaled field, the evolution equation (2.401) reads

$$\ddot{\sigma}(t) + \sigma^3(t) = 0. \tag{2.405}$$

The solution of this equation can be given in terms of Jacobi elliptic functions. For an initial amplitude σ_0 and derivative $\dot{\sigma}(t=0) = 0$, we have

$$\sigma(t) = \sigma_0 \,\mathrm{cn}(\sigma_0 t). \tag{2.406}$$

Here the Jacobi cosine $\mathrm{cn}(z) \equiv \mathrm{cn}(z; n = \frac{1}{2})$ is a doubly periodic function in z with periods $4K_n$ and $4iK_{1-n}$, where $K_n = \int_0^{\pi/2} d\Theta/\sqrt{1 - n\sin^2\Theta}$ is the complete elliptic integral of the first kind. We denote the characteristic frequency of the oscillations of $\sigma(t)$ as

$$\omega_0 = \frac{\pi\sigma_0}{2K_{1/2}} \simeq 0.847\sigma_0, \tag{2.407}$$

where $K_{1/2} \simeq 1.854$. Since the function will enter quadratically the equations for fluctuations, it will also be important that it is a quasiperiodic function with period

$\mathrm{cn}(z + 2K_{1/2}) = -\mathrm{cn}(z)$. Averaged over one period, the square of the field amplitude gives $\overline{\sigma^2} = \int_0^{2K_{1/2}} \mathrm{d}t\, \sigma^2(t)/(2K_{1/2}) \simeq 0.457\sigma_0^2$.

In order to study the fluctuations, we will consider the products

$$F_{\|}(x, y) = \delta\phi_{\|}(x)\,\delta\phi_{\|}(y), \qquad F_{\perp}(x, y) = \delta\phi_{\perp}(x)\,\delta\phi_{\perp}(y), \qquad (2.408)$$

which will be the relevant quantities with which to make comparisons in the quantum treatment below. For spatially homogeneous systems, they only depend on the relative spatial coordinates, $F_{\|,\perp}(x, y) = F_{\|,\perp}(x^0, y^0; \mathbf{x} - \mathbf{y})$. According to (2.402) and (2.403), their spatial Fourier transforms $F_{\|,\perp}(t, t'; \mathbf{p}) = \int \mathrm{d}^d s\, e^{-i\mathbf{p}\cdot\mathbf{s}} F_{\|,\perp}(t, t'; \mathbf{s})$ obey, in this linearized classical approach,

$$\left[\partial_t^2 + \mathbf{p}^2 + 3\sigma_0^2\,\mathrm{cn}^2(\sigma_0 t)\right] F_{\|}(t, t'; \mathbf{p}) = 0, \qquad (2.409)$$

$$\left[\partial_t^2 + \mathbf{p}^2 + \sigma_0^2\,\mathrm{cn}^2(\sigma_0 t)\right] F_{\perp}(t, t'; \mathbf{p}) = 0. \qquad (2.410)$$

We will consider initial conditions characterizing a pure-state initial density matrix, where we take $\partial_t F_{\|,\perp}(t, 0; \mathbf{p})|_{t=0} = 0$ and $\partial_t \partial_{t'} F_{\|,\perp}(t, t'; \mathbf{p})|_{t=t'=0} \equiv \frac{1}{4} F_{\|,\perp}^{-1}(0, 0; \mathbf{p}) = \frac{1}{2}\omega_{\|,\perp}(\mathbf{p})$ in accordance with the results of Section 2.2.4. The frequency of the longitudinal modes is $\omega_{\|}(\mathbf{p}) = \sqrt{\mathbf{p}^2 + 3\sigma_0^2}$, and $\omega_{\perp}(\mathbf{p}) = \sqrt{\mathbf{p}^2 + \sigma_0^2}$ for the transverse modes.

Of course, in these linear equations the two-point functions can be factorized as products of momentum-dependent single-time functions $f_{\|,\perp}(t; \mathbf{p})$, with

$$F_{\|,\perp}(t, t'; \mathbf{p}) = \frac{1}{2}\left[f_{\|,\perp}(t; \mathbf{p})f_{\|,\perp}^*(t'; \mathbf{p}) + f_{\|,\perp}^*(t; \mathbf{p})f_{\|,\perp}(t'; \mathbf{p})\right], \qquad (2.411)$$

where $f_{\|,\perp}^*(t; \mathbf{p})$ denotes the complex conjugate of $f_{\|,\perp}(t; \mathbf{p})$. In terms of these so-called mode functions, the equations of motion read, for example, for the transverse modes,

$$\left[\partial_t^2 + \mathbf{p}^2 + \sigma_0^2\,\mathrm{cn}^2(\sigma_0 t)\right] f_{\perp}(t; \mathbf{p}) = 0. \qquad (2.412)$$

Up to an overall arbitrary phase, the above initial conditions for the two-point functions translate into

$$f_{\perp}(0, \mathbf{p}) = \frac{1}{\sqrt{2\omega_{\perp}(\mathbf{p})}}, \qquad \partial_t f_{\perp}(t, \mathbf{p})|_{t=0} = -i\sqrt{\frac{\omega_{\perp}(\mathbf{p})}{2}}, \qquad (2.413)$$

and equivalently for $f_{\|}$. For a given momentum $\mathbf{p}$, the equation (2.412) is mathematically equivalent to the classical mechanics oscillator described above. Before we summarize the analytical solution of the Lamé-type equation (2.412), we discuss some general properties. In order to be able to observe a significant resonance, the frequency ω_0 of the field $\sigma(t)$ and the characteristic frequency of the fluctuations $f_{\|,\perp}(t; \mathbf{p})$ should be similar. Therefore, for transverse modes, the dominant resonances should fulfill the approximate condition

$$\omega_{\perp}(t, \mathbf{p}) \equiv \sqrt{\mathbf{p}^2 + \sigma^2(t)} \simeq \sqrt{\mathbf{p}^2 + \tfrac{1}{2}\sigma_0^2} \stackrel{!}{\sim} \omega_0, \qquad (2.414)$$

where we have replaced the field square by its approximate time average. Since $\omega_0 \sim \sigma_0$, this condition could be met for not too high momenta $\mathbf{p}^2 \lesssim \frac{1}{2}\sigma_0^2$. In contrast, for longitudinal modes, the same analysis leads to the approximate condition $\sqrt{\mathbf{p}^2 + \frac{3}{2}\sigma_0^2} \overset{!}{\sim} \omega_0$, which, even for vanishing momenta, is not fulfilled, and therefore only subdominant resonances may be expected. The results of this qualitative analysis are indeed validated by the analytic solutions of the Lamé-type equation to which we turn next.

In the following, we concentrate on the dominant transverse modes to analyze the instability dynamics at early times. We choose $U_p(t)$ and $U_p(-t)$ as an independent set of solutions of (2.412) such that $f_\perp(t; \mathbf{p})$ is a linear combination of those. Here we will be interested in the unstable modes that show exponential amplification and we will state the corresponding growth rates. Similar to the above classical oscillator example, we consider a Floquet analysis of the periodic solution

$$U_p(t + \Delta T) = e^{\alpha_p \, \Delta T} U_p(t), \tag{2.415}$$

where the time-independent index α_p is a function of spatial momentum. Correspondingly, we can write

$$U_p(t) = e^{\alpha_p \, t} \Pi_p(t), \tag{2.416}$$

with the periodic function $\Pi_p(t + \Delta T) = \Pi_p(t)$ of period $\Delta T = 2K_{1/2}\sigma_0$. Again, this can be directly verified with $U_p(t + \Delta T) = e^{\alpha_p(t+\Delta T)}\Pi(t + \Delta T) = e^{\alpha_p \, \Delta T}U_p(t)$. The solutions of Lamé equations can be found in textbooks on differential equations, and we give here only the relevant properties of the Floquet index α_p for (2.412). In accordance with our qualitative discussion above, resonant amplification of modes can be found for not too large momenta. More precisely, for $\mathbf{p}^2 \geq \frac{1}{2}\sigma_0^2$, purely oscillating solutions are obtained. In contrast, for $0 < \mathbf{p}^2 < \frac{1}{2}\sigma_0^2$,the Floquet index

$$\alpha_p = i\omega_0 - \gamma_p \tag{2.417}$$

has a real part γ_p and an imaginary part ω_0, given by the characteristic frequency (2.407). The real part implies exponential solutions with a growth or decay rate for either $U_p(t)$ or $U_p(-t)$ given by

$$\gamma_p = \sigma_0 Z \left(\mathrm{cn}^{-1} \left(\sqrt{\frac{2\mathbf{p}^2}{\sigma_0^2}} \right) \right). \tag{2.418}$$

Here, the Jacobi zeta function $Z(x)$ has the series expansion

$$\begin{aligned} Z(x) &= \frac{2\pi}{K_{1/2}} \sum_{n=1}^{\infty} \frac{e^{-n\pi}}{1 - e^{-2n\pi}} \sin\left(\frac{n\pi x}{K_{1/2}}\right) \\ &\simeq \frac{2\pi}{K_{1/2}} e^{-\pi} \sin\left(\frac{\pi x}{K_{1/2}}\right), \end{aligned} \tag{2.419}$$

where the latter approximation gives an excellent description for our purposes. Taking into account that $\mathrm{cn}^{-1}(0) = K_{1/2}$ and $\mathrm{cn}^{-1}(1) = 0$, we note that the rate γ_p vanishes at the boundaries $\mathbf{p}^2 = 0$ and $\mathbf{p}^2 = \frac{1}{2}\sigma_0^2$ of the instability band.

In order to get a further analytic understanding, we may use the somewhat crude approximation

$$\mathrm{cn}^{-1}(x) \simeq \frac{2K_{1/2}}{\pi} \arccos(x), \tag{2.420}$$

which in our case will be good at the few percent level, and the identity

$$\sin\left[2\arccos\left(\sqrt{x}\,\right)\right] = 2\sqrt{2x(1-x)}. \tag{2.421}$$

Then the rate (2.418) can be expressed as

$$\gamma_p \simeq \frac{4\pi\sigma_0}{K_{1/2}} e^{-\pi} \sqrt{\frac{2\mathbf{p}^2}{\sigma_0^2}\left(1 - \frac{2\mathbf{p}^2}{\sigma_0^2}\right)}, \tag{2.422}$$

and the maximally amplified growth rate $\gamma_0 \equiv \gamma(\mathbf{p}^2 = \mathbf{p}_0^2)$ occurs for $\mathbf{p}_0^2 = \frac{1}{4}\sigma_0^2$, with

$$\gamma_0 \simeq \frac{2\pi\sigma_0}{K_{1/2}} e^{-\pi} \simeq 0.146\sigma_0. \tag{2.423}$$

Therefore, there is a separation of scales between the characteristic frequency of oscillations, ω_0, and the characteristic growth rate, γ_0:

$$\gamma_0 \ll \omega_0. \tag{2.424}$$

When analyzing the growth in the nonlinear regime below, we will often exploit this separation of scales by considering suitable time averages to smooth out the rapid oscillations.

2.5.2 Nonlinear regime: secondary instabilities

Dynamical power counting

Above we obtained the set of evolution equations (2.405) and (2.410) for the field $\phi(t) = \sigma(t)\sqrt{6N/\lambda}$ and the fluctuations $F_{\|,\perp}$ from linearized classical dynamics. According to Section 2.2.6, the very same set of equations can also be obtained starting from the quantum evolution of the rescaled field expectation value and the statistical two-point functions by neglecting all loop corrections: setting $\Sigma_{\|,\perp}^F$, $\Sigma_{\|,\perp}^\rho$, and $\Sigma_{\|,\perp}^{(0)}$ to zero in the field equation (2.151) and the equations for the two-point functions (2.145) leads precisely to (2.405) and (2.410). We note that the evolution equation for the spectral function $\rho_{\|,\perp}$ decouples from the rest in this linear approximation. It is important to point out that this linearized classical approximation is only valid at early times for $\lambda \ll 1$. For strong coupling, loop corrections cannot be neglected in this case.

In turn, we may use the quantum evolution equations (2.151) and (2.145) to determine the range of validity of the above classical approach. Parametrically the initial statistical propagators are of order one, i.e.,

$$F_{\|,\perp}(0,0;\mathbf{p}) \sim \mathcal{O}(N^0\lambda^0), \tag{2.425}$$

and the initial field is

$$\phi(0) \sim \mathcal{O}(N^{1/2}\lambda^{-1/2}). \tag{2.426}$$

As a consequence, all loop corrections are initially suppressed by powers of the coupling constant $\lambda \ll 1$ and the classical approximation represents an accurate description of the early quantum dynamics. However, parametric resonance leads to an exponential amplification of modes, where the dominant growth behavior is given parametrically by

$$F_\perp(t,t';\mathbf{p}) \sim e^{\gamma_p(t+t')}. \tag{2.427}$$

Accordingly, the maximally amplified mode $F_\perp(t,t;\mathbf{p}_0)$ grows with a rate $2\gamma_0$. As a consequence of the exponential amplification of the statistical propagator, there is a characteristic time when $F_\perp(t,t;\mathbf{p}_0)$ turns out to be no longer parametrically of order one. This is the time where the linearized classical approximation breaks down. To make further analytical progress, the nonlinear regime is most efficiently described using the evolution equations of Section 2.2.6. This allows us to study what happens when loop corrections start to become relevant. Indeed, we will find that there is a well-separated hierarchy of times, where different loop integrals start to become of order one at corresponding separate times. At some point, an infinite number of diagrams with an arbitrary number of loops becomes of order one and no power counting based on a small coupling can be performed. We will address this later stage in Section 2.6 using nonperturbative large-N techniques.

A general loop contribution to the evolution equations contains powers of λ, the field ϕ, the propagators $F_{\|,\perp}$, and the spectral functions $\rho_{\|,\perp}$. Here it is important to note that the "weight" of the spectral functions in loop integrals remains parametrically of order one at all times, as encoded in the equal-time commutation relations (2.130). It is also important to take into account the fact that transverse fluctuations ($F_\perp$) exhibit the dominant growth in the linear regime. Consequently, contributions containing more transverse propagators ($F_\perp$) can become important earlier than those diagrams containing longitudinal propagators ($F_\|$) instead. For instance, an expression containing powers $\lambda^n F_\perp^m \phi^{2l}$ with integers n, m, and l may be expected to give sizable corrections to the linearized evolution equations once $F_\perp \sim 1/\lambda^{(n-l)/m}$ for typical momenta. Here n yields the suppression factor from the coupling constant, whereas m introduces the enhancement due to large fluctuations for typical momenta and l those due to a large macroscopic field. The power counting can become more involved as time proceeds, and it is remarkable that one can indeed identify a sequence of characteristic timescales with corresponding growth rates.

Before we consider the calculations in more detail, the situation is schematically summarized in Fig. 2.16. The time when $F_\perp(t,t;\mathbf{p}_0) \sim \mathcal{O}(N^0\lambda^{-1/2})$ is denoted by

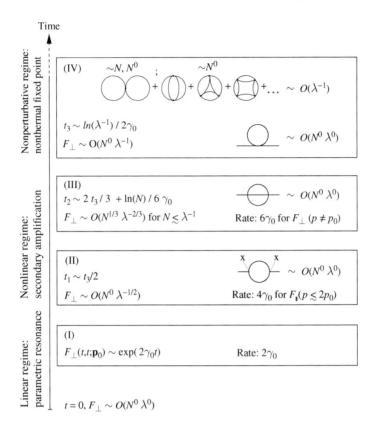

Fig. 2.16 Schematic overview of the characteristic timescales and the respective relevant diagrammatic contributions.

$t = t_1$. At this time, the one-loop diagram with two field insertions indicated by crosses (x) in Fig. 2.16 will make a contribution of order one to the evolution equation for $F_{\parallel}(t, t; \mathbf{p})$. For instance, the two powers of the coupling coming from the vertices of that diagram are canceled by the field amplitudes and by propagator lines associated with the amplified $F_{\perp}(t, t'; \mathbf{p}_0)$. Similarly, at the time $t = t_2$ the maximally amplified transverse propagator mode has grown to $F_{\perp}(t, t'; \mathbf{p}_0) \sim \mathcal{O}(N^{1/3}\lambda^{-2/3})$. As a consequence, the "setting sun" diagram in Fig. 2.16 becomes of order one and is therefore of the same order as the classical contributions. Although the loop corrections become of order one later than the initial time, they induce amplification rates that are multiples of the rate $2\gamma_0$, which lead to a very rapid growth of modes in a wide momentum range. Finally, when the fluctuations have grown nonperturbatively large with $F_{\perp}(t, t'; \mathbf{p}_0) \sim \mathcal{O}(N^0\lambda^{-1})$, any loop correction will no longer be suppressed by powers of the small coupling λ. In this case, the nonperturbative $1/N$ expansion of the 2PI effective action, which will be discussed later, becomes of crucial importance for a quantitative description of the dynamics.

Nonlinear amplification

The $\mathcal{O}(\lambda^0)$ approximation for longitudinal modes breaks down at the time

$$t \simeq t' = t_1: \qquad F_\perp(t, t'; \mathbf{p}_0) \sim \mathcal{O}(N^0 \lambda^{-1/2}). \tag{2.428}$$

This can be derived from the $\mathcal{O}(\lambda)$ evolution equations to which one-loop self-energies contribute, which are represented diagrammatically by

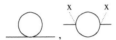

The approximate evolution equation reads

$$\left[\partial_t^2 + \mathbf{p}^2 + M^2(t) + 3\sigma^2(t)\right] F_\|(t, t'; \mathbf{p})$$

$$\simeq \frac{2\lambda(N-1)}{3N} \sigma(t) \left\{ \int_0^t dt'' \sigma(t'') \Pi_\perp^\rho(t, t''; \mathbf{p}) F_\|(t'', t'; \mathbf{p}) \right.$$

$$\left. - \frac{1}{2} \int_0^{t'} dt'' \sigma(t'') \Pi_\perp^F(t, t''; \mathbf{p}) \rho_\|(t'', t'; \mathbf{p}) \right\} = \text{RHS}, \tag{2.429}$$

where

$$M^2(t) = m^2 + \frac{\lambda}{6N} \left[3T_\|(t) + (N-1)T_\perp(t) \right], \tag{2.430}$$

$$T_{\|,\perp}(t) = \int^\Lambda \frac{d^3 p}{(2\pi)^3} F_{\|,\perp}(t, t; \mathbf{p}). \tag{2.431}$$

The "tadpole" contributions $T_\|$ and $T_\perp$ from the longitudinal and transverse propagator components, respectively, are regularized by some $\Lambda \gg p_0$, whose precise value is irrelevant as long as the integral is dominated by momenta much smaller than the cutoff. Here we have abbreviated

$$\Pi_\perp^X(t, t''; \mathbf{p}) = \int \frac{d^3 q}{(2\pi)^3} F_\perp(t, t''; \mathbf{p} - \mathbf{q}) X_\perp(t, t''; \mathbf{q}), \tag{2.432}$$

with $X = \{F, \rho\}$, and we have used the fact that $F_\perp^2 \gg \rho_\perp^2$.[16] We can see that indeed, for $F_\perp \sim \mathcal{O}(N^0 \lambda^{-1/2})$, the right-hand side of (2.429), RHS, becomes $\mathcal{O}(1)$ and cannot be neglected.

In order to make analytical progress, we have to evaluate the "memory integrals" in (2.429). This is dramatically simplified by the fact that the integral is at this

---

[16] This corresponds to the classicality condition of Section 2.4.3.

stage approximately local in time, since the exponential growth lets the latest-time contributions dominate the integral. For the approximate evaluation of the memory integrals, we consider

$$\int_0^t dt'' \longrightarrow \int_{t-c/\omega_0}^t dt'' \qquad (2.433)$$

with $c \sim \mathcal{O}(1)$. As long as we are only interested in exponential growth rates and characteristic timescales, the value for the constant c will turn out to be irrelevant to "leading-log" accuracy. We then perform a Taylor expansion around the latest time t (t'):

$$\rho_{\|,\perp}(t,t'';\mathbf{p}) \simeq \partial_{t''}\rho_{\|,\perp}(t,t'';\mathbf{p})|_{t=t''}(t''-t) \equiv (t-t''),$$
$$F_{\|,\perp}(t,t'';\mathbf{p}) \simeq F_{\|,\perp}(t,t;\mathbf{p}), \qquad (2.434)$$

where we have used the equal-time commutation relations (2.130). With these approximations, the right-hand side of (2.429) can be evaluated as

$$\text{RHS} \simeq \lambda\sigma^2(t)\frac{c^2}{\omega_0^2}\frac{(N-1)}{3N}T_\perp(t)F_\|(t,t';\mathbf{p}) \qquad \text{(mass term)} \qquad (2.435)$$

$$+ \lambda\sigma(t)\sigma(t')\frac{c^2}{\omega_0^2}\frac{(N-1)}{6N}\Pi_\perp^F(t,t';\mathbf{p}) \qquad \text{(source term)}. \qquad (2.436)$$

The first term is a contribution to the effective mass, whereas the second term represents a source. Note that both the "tadpole" mass term and the above correction to this mass are of the same order in λ, but with opposite signs. To evaluate the momentum integrals, we use a saddle-point approximation around the dominant $p \simeq p_0$, valid for $t,t' \gg \gamma_0^{-1}$, with $F_\perp(t,t',\mathbf{p}) \simeq F_\perp(t,t',\mathbf{p}_0)\exp\left[-\frac{1}{2}|\gamma_0''|(t+t')(p-p_0)^2\right]$.[17] From this, we obtain for the above mass term

$$\lambda T_\perp(t) \simeq \lambda\frac{p_0^2 F_\perp(t,t;\mathbf{p}_0)}{2(\pi^3|\gamma_0''|t)^{1/2}}. \qquad (2.437)$$

The result can be used to obtain an estimate at what time t this loop correction becomes an important contribution to the evolution equation. Note that to this order in λ it is correct to use $F_\perp(t,t';\mathbf{p}_0) \sim e^{\gamma_0(t+t')}$ on the right-hand side of (2.437). The condition $\lambda T_\perp(t=t_3) \sim \mathcal{O}(1)$ can then be written for $\lambda \ll 1$ as

$$\boxed{t_3 \simeq \frac{1}{2\gamma_0}\ln\lambda^{-1}.} \qquad (2.438)$$

---

[17] Here $\gamma(p) \simeq \gamma_0 + \frac{1}{2}\gamma_0''(p-p_0)^2$, with $\gamma_0'' < 0$.

The same saddle-point approximation can be performed to evaluate the source term (2.436):

$$\lambda \Pi_{\perp}^{F}(t, t'; 0) \simeq \lambda \, \frac{p_0^2 F_{\perp}^2(t, t'; \mathbf{p}_0)}{4 \left[\pi^3 |\gamma_0''| (t + t') \right]^{1/2}}. \tag{2.439}$$

Here we have only written the source term for $\mathbf{p} = 0$, where it has its maximum, although it affects all modes with $p \lesssim 2p_0$. Again this can be used to estimate the time $t = t_1$ at which $\lambda \Pi_{\perp}^{F} \sim \mathcal{O}(1)$:

$$\boxed{t_1 \simeq \tfrac{1}{2} t_3} \tag{2.440}$$

We arrive at the important conclusion that the source term (2.436) becomes of order one *earlier* than the mass term (2.435): for $t_1 \lesssim t \lesssim t_3$, the source term dominates the dynamics! Using these estimates in (2.429), we find that the longitudinal modes with $0 \lesssim p \lesssim 2p_0$ become amplified with twice the rate $2\gamma_0$:

$$F_{\|}(t, t; \mathbf{p}) \sim \lambda F_{\perp}^2(t, t; \mathbf{p}_0) \sim \lambda e^{4\gamma_0 t}. \tag{2.441}$$

Although the nonlinear contributions start later, they grow twice as fast. The analytical estimates for t_1 and rates agree well with the numerical solution[18] of the evolution equations without memory expansion as shown in Fig. 2.17, where we plot the effective particle number distribution

$$f_{\|, \perp}(t, \mathbf{p}) = \sqrt{F_{\|, \perp}(t, t; \mathbf{p}) \partial_t \partial_{t'} F_{\|, \perp}(t, t'; \mathbf{p})} \Big|_{t = t'} - \tfrac{1}{2} \tag{2.442}$$

and $M_0^2 \equiv M^2(t = 0)$.

A similar analysis can be performed for the transverse modes. Beyond the Lamé-type $\mathcal{O}(\lambda^0)$ description, the evolution equation for $F_{\perp}$ receives contributions from the feedback of the longitudinal modes at $\mathcal{O}(\lambda)$ as well as from the amplified transverse modes at $\mathcal{O}(\lambda^2)$. They represent *source terms* in the evolution equation for $F_{\perp}(t, t'; \mathbf{p})$, which are both parametrically of the form $\sim \lambda^2 F_{\perp}^3 / N$ and can be depicted as follows:

---

[18] The plots are obtained for the N-component field theory at NLO in the $1/N$ expansion of the 2PI effective action with nonzero mass parameter m. The latter is not relevant for our purposes.

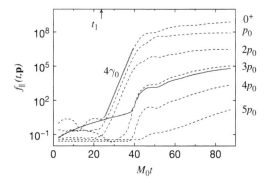

Fig. 2.17 Effective particle number for the longitudinal modes as a function of time for various momenta. For $t \gtrsim t_1$, the nonlinear corrections trigger an exponential growth with rate $4\gamma_0$ for $p \lesssim 2p_0$. The solid line corresponds to a mode in the parametric resonance band, whereas the dashed lines represent modes outside the band. The resonant amplification is quickly dominated by source-induced amplification.

Following along the lines of the above paragraph and using (2.438) this leads to the characteristic time $t = t_2$ at which these source terms become of order one:

$$t_2 \simeq \frac{2}{3}t_3 + \frac{\ln N}{6\gamma_0}. \tag{2.443}$$

For $t \simeq t' \simeq t_2$, the dominant transverse mode has grown to

$$F_\perp(t, t'; \mathbf{p}_0) \sim \mathcal{O}(N^{1/3}\lambda^{-2/3}). \tag{2.444}$$

Correspondingly, for $t_2 \lesssim t \lesssim t_3$, one finds a large growth rate $\sim 6\gamma_0$ in a momentum range $0 \lesssim p \lesssim 3p_0$, in agreement with the numerical results shown in Fig. 2.18. In this time range, the longitudinal modes exhibit an enhanced amplification as well (cf. Fig. 2.17). It is important to realize that the phenomenon of source-induced amplification repeats itself: the newly amplified modes, together with the primarily amplified ones, act as a source for other modes, and so on. In this way, even higher growth rates of multiples of γ_0 can be observed and the amplification rapidly propagates toward higher-momentum modes. We note that this regime is present if $t_2 \leq t_3$, i.e., as long as $N \lesssim \lambda^{-1}$.

Around $t \lesssim t_3$ is the earliest time when sizable corrections to the maximally amplified mode $F_\perp(t, t; \mathbf{p}_0)$ and to the field appear. Around t_3, there are corrections of order one that come from diagrams with an arbitrary number of loops. As a consequence, the dynamics is no longer characterized in terms of the small expansion parameter λ. This is the nonperturbative regime, where the $1/N$ expansion may be used to describe also the subsequent evolution governed by a nonthermal fixed point, as is explained in Section 2.6.

The accuracy of the above description based on the 2PI effective action can be tested using numerical simulations on a space–time lattice. This exploits the fact that

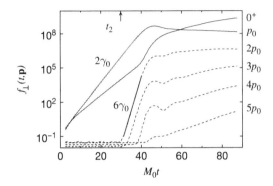

Fig. 2.18 Effective particle number for the transverse modes as a function of time for various momenta $p \leq 5p_0$. At early times, modes with $p \simeq p_0$ are exponentially amplified with a rate $2\gamma_0$. It can be seen that, owing to nonlinearities, there is subsequently an enhanced growth with rate $6\gamma_0$ for a broad momentum range.

classical statistical field theory descriptions of the dynamics have an overlapping range of validity with the underlying quantum field theory, as described in Section 2.4. As an example, Fig. 2.19 shows $F_\perp(t, t, \mathbf{p})$ for times $t = 10$ and 40 in the nonlinear regime and $t = 90$ in the nonperturbative regime. The solid lines give the results for the quantum evolution for $\lambda = 0.01$ and $N = 4$ by solving the NLO equations from the 2PI effective action of Section 2.3.2 in $d = 3$ numerically. For comparison, the dashed lines give the same quantity as obtained from simulations of the corresponding

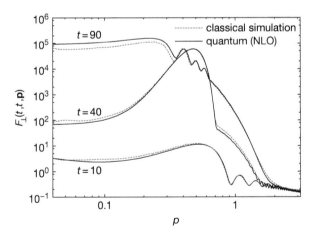

Fig. 2.19 Two-point function $F_\perp(t, t; \mathbf{p})$ as a function of momentum $|\mathbf{p}|$ for three different times. The quantum evolution (solid curves) from the 2PI $1/N$ expansion to NLO and the full classical statistical simulation using Monte Carlo sampling (dashed curves) agree remarkably well.

classical statistical field theory on a lattice with the same initial conditions. The level of agreement between the different results is remarkable. Quantum fluctuations are expected to be suppressed if the classicality condition $F^2 \gg \rho^2$ of Section 2.4.3 is fulfilled, which is very well realized for typical $F \sim 1/\lambda$ with ρ being of order unity. In turn, contributions beyond NLO in the $1/N$ expansion seem to play an inferior role even for the nonperturbative regime in this case.

2.5.3   Bibliography

- Figures 2.17 and 2.18 are taken with permission from J. Berges and J. Serreau. Parametric resonance in quantum field theory. *Phys. Rev. Lett.* **91** (2003) 111601 (Copyright (2003) by the American Physical Society). The presentation for the scalar field theory is based on that reference, where the 2PI $1/N$ expansion to NLO is used to describe the relevant nonlinear phenomena. Parts of the linearized analysis are based on D. Boyanovsky, H. J. de Vega, R. Holman, and J. F. Salgado. Analytic and numerical study of preheating dynamics. *Phys. Rev. D* **54** (1996) 7570, where a leading-order large-N approximation is employed. Figure 2.19 is taken with permission from J. Berges, A. Rothkopf, and J. Schmidt. Non-thermal fixed points: effective weak coupling for strongly correlated systems far from equilibrium. *Phys. Rev. Lett.* **101** (2008) 041603 (Copyright (2008) by the American Physical Society).
- The preheating phenomenon was first developed in L. Kofman, A. D. Linde, and A. A. Starobinsky. Reheating after inflation. *Phys. Rev. Lett.* **73** (1994) 3195. Its description in terms of classical statistical simulations was worked out in S. Yu. Khlebnikov and I. I. Tkachev. Classical decay of the inflaton. *Phys. Rev. Lett.* **77** (1996) 219.
- For a review about the inclusion of fermions, see J. Berges, D. Gelfand, and D. Sexty. Amplified fermion production from overpopulated Bose fields. *Phys. Rev. D* **89** (2014) 025001.

2.6   Nonthermal fixed points and turbulence

From the example of parametric resonance in Section 2.5, we have seen that once the exponential growth of fluctuations stops, the systems slows down considerably. At this stage, there is an infinite number of loop corrections of order unity. The coupling "drops out" of the problem and no power counting in terms of a small coupling parameter can be given. In this section, we want to study the physics of this slow evolution for the examples of relativistic N-component scalar field theory as well as nonrelativistic (Gross–Pitaevskii) field theory. They serve here as a paradigm to investigate universal behavior of isolated many-body systems far from equilibrium, which is relevant for a wide range of applications from high-energy particle physics to ultracold quantum gases, as pointed out in Section 2.1. The universality is based on the existence of *nonthermal fixed points*, which represent nonequilibrium attractor solutions with self-similar scaling behavior. The corresponding dynamic universality classes turn out to be remarkably large, encompassing both relativistic and nonrelativistic quantum and classical systems.

We start with a standard introduction to the phenomenon of *wave turbulence* in Section 2.6.1, where perturbative kinetic theory can describe the stationary transport of energy from a cascade toward higher momenta. Perturbative kinetic theory has also been employed in the literature to describe infrared phenomena such as Bose condensation; however, such an approach neglects important vertex corrections because the large occupancies at low momenta lead to strongly nonlinear dynamics. Here we apply the vertex-resummed kinetic theory based on the 2PI $1/N$ expansion to NLO, which is described in Section 2.3.3. This allows us to gain analytic understanding of the formation of a *dual cascade* and the phenomenon of *far-from-equilibrium Bose condensation*.

However, isolated systems out of equilibrium have no external driving forces that could realize stationary transport solutions. Instead, the transport in isolated systems is described in terms of the more general notion of a *self-similar* evolution, where the physics is described in terms of universal scaling exponents and scaling functions as outlined in Section 2.1. We can use the vertex-resummed kinetic theory for an analytic description of the phenomena, which we present in Section 2.6.2. Because of the typical high occupancies, this can be compared with results from classical statistical lattice simulations using the methods of Section 2.4.

2.6.1 Stationary transport of conserved charges

While many aspects of stationary transport of conserved charges associated with turbulence have long reached textbook level, there is still rather little known about turbulent behavior in nonperturbative regimes of quantum field theories. Here strong interest is also driven by related questions concerning the dynamics of relativistic heavy-ion collisions. In this subsection, we consider stationary solutions of transport equations with a net flux of energy and particles across momentum scales, thus violating detailed balance. We start with standard perturbative results relevant for high momenta in Section 2.6.1 and consider the nonperturbative infrared regime in Section 2.6.1.

Perturbative regime: weak wave turbulence

In perturbative kinetic theory, when two particles scatter into two particles, the time evolution of the distribution function $f_{\mathbf{p}}(t)$ for a spatially homogeneous system is given by

$$\frac{\partial f_{\mathbf{p}}(t)}{\partial t} = C^{2\leftrightarrow2}[f](t,\mathbf{p}). \tag{2.445}$$

The collision integral is of the form

$$C^{2\leftrightarrow2}[f](\mathbf{p}) = \int d\Omega^{2\leftrightarrow2}(\mathbf{p},\mathbf{l},\mathbf{q},\mathbf{r})\left[(f_{\mathbf{p}}+1)(f_{\mathbf{l}}+1)f_{\mathbf{q}}f_{\mathbf{r}} - f_{\mathbf{p}}f_{\mathbf{l}}(f_{\mathbf{q}}+1)(f_{\mathbf{r}}+1)\right], \tag{2.446}$$

where we have suppressed the global time dependence to ease the notation. The details of the model enter $\int d\Omega^{2\leftrightarrow 2}(\mathbf{p}, \mathbf{l}, \mathbf{q}, \mathbf{r})$, which, for the example of the relativistic N-component scalar field theory with quartic $\lambda/(4!N)$ interaction, reads

$$\int d\Omega^{2\leftrightarrow 2}(\mathbf{p}, \mathbf{l}, \mathbf{q}, \mathbf{r}) = \lambda^2 \frac{N+2}{6N^2} \int_{\mathbf{l}\mathbf{q}\mathbf{r}} (2\pi)^{d+1}\delta(\mathbf{p}+\mathbf{l}-\mathbf{q}-\mathbf{r})$$

$$\times\, \delta(\omega_{\mathbf{p}} + \omega_{\mathbf{l}} - \omega_{\mathbf{q}} - \omega_{\mathbf{r}}) \frac{1}{2\omega_{\mathbf{p}}2\omega_{\mathbf{l}}2\omega_{\mathbf{q}}2\omega_{\mathbf{r}}}, \qquad (2.447)$$

with $\omega_{\mathbf{p}} = \sqrt{\mathbf{p}^2 + m^2}$ and $\int_{\mathbf{p}} \equiv \int d^d p/(2\pi)^d$. This kinetic equation can be obtained from quantum field theory using the two-loop self-energies (2.240) and (2.241) for the lowest-order gradient expansion along the lines of Section 2.3.3 and taking the on-shell spectral function (2.252) of the free theory. The expression (2.446) with (2.447) for the collision integral represents a standard Boltzmann equation for a gas of relativistic particles.

Clearly, this approximation cannot be used if the occupation numbers per mode become so large that higher loop corrections become sizable. Parametrically, for a weak coupling λ, a necessary condition for its validity is $f_{\mathbf{p}} \ll 1/\lambda$, since otherwise any loop order contributes significantly, as explained in detail in Section 2.5. On the other hand, the phenomenon of weak wave turbulence is expected for not too small occupation numbers per mode, as is explained in the following. For the corresponding regime $1 \ll f_{\mathbf{p}} \ll 1/\lambda$, we may use the above Boltzmann equation, which becomes approximately

$$\frac{\partial f_{\mathbf{p}}(t)}{\partial t} \simeq \int d\Omega^{2\leftrightarrow 2}(\mathbf{p}, \mathbf{l}, \mathbf{q}, \mathbf{r}) \left[(f_{\mathbf{p}} + f_{\mathbf{l}})f_{\mathbf{q}}f_{\mathbf{r}} - f_{\mathbf{p}}f_{\mathbf{l}}(f_{\mathbf{q}} + f_{\mathbf{r}}) \right]. \qquad (2.448)$$

For number-conserving $2 \leftrightarrow 2$ scatterings, apart from the energy density ϵ, the total particle number density n is also conserved, with these being given by

$$\epsilon = \int_{\mathbf{p}} \omega_{\mathbf{p}} f_{\mathbf{p}}, \qquad n = \int_{\mathbf{p}} f_{\mathbf{p}}. \qquad (2.449)$$

The fact that they are conserved may be described by a continuity equation in momentum space, such as

$$\frac{\partial}{\partial t}(\omega_{\mathbf{p}} f_{\mathbf{p}}) + \nabla_{\mathbf{p}} \cdot \mathbf{j}_{\mathbf{p}} = 0 \qquad (2.450)$$

for energy conservation. Similarly, particle number conservation is described by formally replacing $\omega_{\mathbf{p}} \to 1$ in the above equation together with a corresponding substitution of the flux density. For the isotropic situation, we can consider the energy flux $A(k)$ through a momentum sphere of radius k. Then only the radial component of the flux density $\mathbf{j}_{\mathbf{p}}$ is nonvanishing and

$$\int_{\mathbf{p}}^{k} \nabla_{\mathbf{p}} \cdot \mathbf{j}_{\mathbf{p}} = \int_{\partial k} \mathbf{j}_{\mathbf{p}} \cdot d\mathbf{A}_{\mathbf{p}} \equiv (2\pi)^d A(k). \qquad (2.451)$$

Since in this approximation $\omega_{\mathbf{p}}$ is constant in time, we can write, with the help of (2.450) and the kinetic equation (2.445),

$$A(k) = -\frac{1}{2^d \pi^{d/2} \Gamma(d/2+1)} \int^k dp \, |\mathbf{p}|^{d-1} \omega_{\mathbf{p}} C^{2\leftrightarrow2}(\mathbf{p}). \qquad (2.452)$$

Stationary wave turbulence is characterized by a *scale-independent flux* $A(k)$, for which the respective integral does not depend on the integration limit k. To this end, we consider scaling solutions

$$f_{\mathbf{p}} = s^{\kappa} f_{s\mathbf{p}}, \qquad \omega_{\mathbf{p}} = s^{-1} \omega_{s\mathbf{p}}, \qquad (2.453)$$

with occupation number exponent κ and assuming a linear dispersion relation relevant for momenta $|\mathbf{p}| \gg m$. Since the physics is scale-invariant, we can choose $s = 1/|\mathbf{p}|$ such that $f_{\mathbf{p}} = |\mathbf{p}|^{-\kappa} f_1$ and $\omega_{\mathbf{p}} = |\mathbf{p}| \omega_1$.

Using these scaling properties, we obtain for the collision integral (2.447) and (2.448) of the theory with quartic self-interaction

$$C^{2\leftrightarrow2}(\mathbf{p}) = s^{-\mu_4} C^{2\leftrightarrow2}(s\mathbf{p}), \qquad (2.454)$$

where the scaling exponent is given by

$$\mu_4 = (3d-4) - (d+1) - 3\kappa = 2d - 5 - 3\kappa. \qquad (2.455)$$

The first term in parentheses comes from the scaling of the measure, the second from energy–momentum conservation for two-to-two scattering, and the third from the three factors of the distribution function appearing in (2.448). Apart from the 4-vertex interaction considered, it will be relevant to investigate also scattering in the presence of a nonvanishing field expectation value such that an effective 3-vertex appears according to Section 2.2.5.

To keep the discussion more general, we may write for the scaling behavior of a generic collision term

$$C(\mathbf{p}) = |\mathbf{p}|^{\mu_l} C(\mathbf{1}), \qquad (2.456)$$

in terms of the scaling exponent μ_l for l-vertex scattering processes. In the general case of an l-vertex, we obtain along these lines

$$\mu_l = (l-2)d - (l+1) - (l-1)\kappa. \qquad (2.457)$$

For the scaling properties of the energy flux, we can then write

$$A(k) = -\frac{1}{2^d \pi^{d/2} \Gamma(d/2+1)} \int^k dp \, |\mathbf{p}|^{d+\mu_l} \omega_1 C(\mathbf{1}). \qquad (2.458)$$

If the exponent in the integrand is nonvanishing, the integral gives

$$A(k) \sim \frac{k^{d+1+\mu_l}}{d+1+\mu_l} \omega_1 C(\mathbf{1}). \qquad (2.459)$$

Thus, scale invariance may be obtained for

$$d + 1 + \mu_l = 0. \tag{2.460}$$

This gives the scaling exponent

$$\kappa = d - \frac{l}{l-1} \qquad \text{for the perturbative } \textit{relativistic energy cascade.} \tag{2.461}$$

We can see that stationary turbulence requires in this case the existence of the limit

$$\lim_{d+1+\mu_l \to 0} \frac{C(1)}{d+1+\mu_l} = \text{const} \neq 0, \tag{2.462}$$

such that the collision integral must have a corresponding zero of first degree. Similarly, starting from the continuity equation for particle number, one can study stationary turbulence associated with particle number conservation. This leads to the scaling exponent

$$\kappa = d - \frac{l+1}{l-1} \qquad \text{for the perturbative } \textit{relativistic particle cascade.} \tag{2.463}$$

Accordingly, for quartic self-interactions, we get $\kappa = d - \frac{4}{3}$ for the energy cascade and $\kappa = d - \frac{5}{3}$ for the particle cascade. In the presence of a coherent field, when a 3-vertex can become relevant, the associated scaling exponents are $\kappa = d - \frac{3}{2}$ for the energy cascade and $\kappa = d - 2$ for the particle cascade.

Nonperturbative regime: strong turbulence

The above perturbative description is expected to become invalid at sufficiently low momenta. In particular, the occupation numbers $f_{\mathbf{p}} \sim |\mathbf{p}|^{-\kappa}$ for $\kappa > 0$ would grow nonperturbatively large in the infrared to the extent that the approximation (2.446) would become questionable. This concerns, for instance, the relevant case of $d = 3$ for the description of the early universe dynamics outlined in Section 2.1.3. To understand where the picture of weak wave turbulence breaks down and to compute the properties of the infrared regime, we have to consider nonperturbative approximations. For this purpose, we employ the vertex-resummed kinetic theory of Section 2.3.3 based on the expansion of the 2PI effective action in the number of field components to NLO.

For a relativistic theory with dispersion $\omega_{\mathbf{p}} = \sqrt{\mathbf{p}^2 + m^2}$, the scaling assumption (2.453) should be valid for sufficiently high momenta $|\mathbf{p}| \gg m$ such that the dispersion is approximately linear, with $\omega_{\mathbf{p}} \sim |\mathbf{p}|$. However, this is more involved at low momenta if a mass gap exists. An effective mass gap is typically expected because of medium effects even if the mass parameter in the Lagrangian is set to zero. In that case, the infrared modes behave effectively nonrelativistically, as explained in Section 2.2.8. In the following, we will analyze the two cases of a relativistic theory without a mass gap and a nonrelativistic theory separately for comparison.

Following similar lines as in Section 2.6.1, we first look for relativistic scaling solutions. To be able to cope with occupancies of order $\sim 1/\lambda$, we replace the perturbative collision term (2.448) by the vertex-resummed expression (2.325), which can be approximated for $f_p \gg 1$ accordingly by

$$C^{\mathrm{NLO}}[f](\mathbf{p}) \simeq \int d\Omega^{\mathrm{NLO}}[f](p,l,q,r) \left[(f_p + f_l) f_q f_r - f_p f_l (f_q + f_r) \right], \qquad (2.464)$$

with

$$\int d\Omega^{\mathrm{NLO}}(p,l,q,r) \simeq \frac{\lambda^2}{18N} \int_0^\infty \frac{dp^0 \, dl^0 \, dq^0 \, dr^0}{(2\pi)^{4-(d+1)}} \int_{\mathbf{lqr}} \delta(p + l - q - r)$$

$$\times \, \tilde{\rho}_p \tilde{\rho}_l \tilde{\rho}_q \tilde{\rho}_r \left[v_{\mathrm{eff}}(p+l) + v_{\mathrm{eff}}(p-q) + v_{\mathrm{eff}}(p-r) \right]. \qquad (2.465)$$

We emphasize that this still involves integration over frequencies as well as spatial momenta, since no free field form for the spectral function $\tilde{\rho}_p \equiv \tilde{\rho}(p^0, \mathbf{p})$ has been used so far, and $f_p \equiv f(p^0, \mathbf{p})$. A crucial difference from the perturbative kinetic equation is the appearance of the vertex function $v_{\mathrm{eff}}(p^0, \mathbf{p})$ given by (2.319), which encodes the emergence of a momentum-dependent effective coupling from the NLO corrections of the $1/N$ expansion. It should be emphasized that v_{eff} itself depends on the distribution function via the retarded self-energy (2.320). By writing down (2.464), we have neglected off-shell processes included in (2.325). However, their contributions are expected to be small for the scaling behavior considered, and we discard them in the following.

In principle, nonperturbative scaling phenomena may involve an anomalous scaling exponent for $\tilde{\rho}(p^0, \mathbf{p})$. Using isotropy, we write

$$\tilde{\rho}(p^0, \mathbf{p}) = s^{2-\eta} \tilde{\rho}(s^z p^0, s\mathbf{p}), \qquad (2.466)$$

with a nonequilibrium "anomalous dimension" η. The dynamical scaling exponent z appears since only spatial momenta are related by rotational symmetry and frequencies may scale differently because of the presence of medium effects. Accordingly, the scaling behavior of the statistical correlation function

$$F(p^0, \mathbf{p}) = s^{2+\kappa_{\mathrm{s}}} F(s^z p^0, s\mathbf{p}) \qquad (2.467)$$

will be described using a scaling exponent κ_{s}. This translates with the definition (2.308) for $f_p \gg 1$ into

$$f(p^0, \mathbf{p}) = s^{\kappa_{\mathrm{s}}+\eta} f(s^z p^0, s\mathbf{p}). \qquad (2.468)$$

Using these definitions, we can infer the scaling behavior of $v_{\mathrm{eff}}(p^0, \mathbf{p})$, which is given in terms of the "one-loop" retarded self-energy $\Pi_R(p^0, \mathbf{p})$ according to (2.319). From (2.320), it follows that

$$\Pi_R(p^0, \mathbf{p}) = s^\Delta \Pi_R(s^z p^0, s\mathbf{p}), \qquad (2.469)$$

with

$$\Delta = 4 - d - z + \kappa_{\mathrm{s}} - \eta. \tag{2.470}$$

If $\Delta > 0$, we find from (2.319) the infrared scaling behavior

$$v_{\mathrm{eff}}(p^0, \mathbf{p}) = s^{-2\Delta} v_{\mathrm{eff}}(s^z p^0, s\mathbf{p}). \tag{2.471}$$

(For $\Delta \leq 0$, the effective coupling would become trivial with $v_{\mathrm{eff}} \simeq 1$, on which we comment below.) Employing these scaling properties, (2.465) gives

$$\int d\Omega^{\mathrm{NLO}}(p, l, q, r) = s^{-2\kappa_{\mathrm{s}} - z - 2\eta} \int d\Omega^{\mathrm{NLO}}(s^z p^0, s^z l^0, s^z q^0, s^z r^0; s\mathbf{p}, s\mathbf{l}, s\mathbf{q}, s\mathbf{r}). \tag{2.472}$$

Following the procedure of Section 2.6.1, for any conserved quantity, we can compute the flux through a momentum sphere k. Stationary turbulence solutions then require that the respective integral does not depend on k. Energy conservation, expressed in terms of the effective particle number distribution (2.313), corresponds to the fact that

$$\epsilon = \int_0^\infty \frac{dp^0}{2\pi} \int \frac{d^d p}{(2\pi)^d} 2(p^0)^2 \tilde{\rho}_p f_p \tag{2.473}$$

is a constant of motion in this description. Also, the effective particle number density

$$n = \int_0^\infty \frac{dp^0}{2\pi} \int \frac{d^d p}{(2\pi)^d} 2p^0 \tilde{\rho}_p f_p \tag{2.474}$$

is constant for the collision integral (2.464). Similar to (2.452), the flux for this effective particle number reads

$$A(k) = -\frac{1}{2^d \pi^{d/2} \Gamma(d/2 + 1)} \int^k dp\, |\mathbf{p}|^{d-1} C^{\mathrm{NLO}}(\mathbf{p}). \tag{2.475}$$

The momentum integral can be evaluated along similar lines as before using the above scaling properties such that

$$A(k) \sim \frac{k^{d - \kappa_{\mathrm{s}} + z - \eta}}{d - \kappa_{\mathrm{s}} + z - \eta} C^{\mathrm{NLO}}(\mathbf{1}). \tag{2.476}$$

Therefore, scale invariance may be obtained for the *particle cascade*, with

$$\kappa_{\mathrm{s}} = d + z - \eta, \tag{2.477}$$

in the nonperturbative low-momentum regime. Similarly, for the scaling solution associated with energy conservation, we find, taking into account the additional power of p^0 in the integrand of (2.473), the exponent for the *energy cascade*,

$$\kappa_{\mathrm{s}} = d + 2z - \eta. \tag{2.478}$$

With these solutions, we can now reconsider the above assumption that $\Delta > 0$ by plugging (2.477) or (2.478) into (2.470). Indeed, it is fulfilled under the sufficient condition that $\eta < 2$. Taking into account that the anomalous dimension for scalar field theory is expected to be small for not too low dimension, with η maybe of the order of a few percent for $d = 3$, and employing a relativistic $z \simeq 1$, we find $\kappa_{\mathrm{s}} \simeq d + 1$ for the particle cascade and $\kappa_{\mathrm{s}} \simeq d + 2$ for the energy cascade solution.

The scaling of the effective occupation number distribution $f(\mathbf{p})$, which depends only on spatial momentum, can finally be obtained from the definition (2.313). This we write as

$$f(\mathbf{p}) + \tfrac{1}{2} = \int_0^\infty \frac{\mathrm{d}p^0}{2\pi}\, 2p^0\, \tilde{\rho}(p^0, \mathbf{p}) \left[f(p^0, \mathbf{p}) + \tfrac{1}{2} \right] = \int_0^\infty \frac{\mathrm{d}p^0}{2\pi}\, 2p^0 F(p^0, \mathbf{p}), \qquad (2.479)$$

using the fact that $\int_0^\infty [\mathrm{d}p^0/(2\pi)]\, p^0 \tilde{\rho}(p^0, \mathbf{p}) = \tfrac{1}{2}$ from the commutation relation (2.130) in Fourier space. Of course, the scaling behavior (2.467) for $F(p^0, \mathbf{p})$ may only be observed in a momentum regime with sufficiently high occupancies $f(\mathbf{p})$, as implied in the above derivation. Then (2.479) yields

$$f(\mathbf{p}) = s^{\kappa_{\mathrm{s}} + 2 - 2z} f(s\mathbf{p})$$
$$\sim \begin{cases} |\mathbf{p}|^{-(d+2-z-\eta)} & \text{relativistic particle cascade,} \\ |\mathbf{p}|^{-(d+2-\eta)} & \text{relativistic energy cascade.} \end{cases} \qquad (2.480)$$

These estimates show that vertex corrections can lead to a strongly modified infrared scaling behavior as compared with the perturbative treatment of Section 2.6.1.

Anticipating that there is a mass gap in the relativistic theory, and because of the very interesting applications such as those to the physics of ultracold atoms, we would like to compare this with the nonrelativistic case. The nonrelativistic limit is outlined in Section (2.2.8) for the complex Gross–Pitaevskii field theory, and we present here the relevant changes as compared with the relativistic case. To this end, we consider a nonrelativistic N-component complex scalar field theory and perform again the $1/N$ expansion to NLO. Since already the relativistic scaling exponents (2.480) indicate no explicit dependence on N at NLO—it can only enter indirectly via η and z—this seems to be a very good starting point to understand also the single complex field case of Gross–Pitaevskii.

To proceed with the analytic estimate, we first note that for the nonrelativistic theory (2.206) with quartic $(g/2)(\chi^*\chi)^2$ interaction, the right-hand side of the evolution equation (2.211) for $F^{(\mathrm{nr})}$ has the same functional dependence on $\rho^{(\mathrm{nr})}$ and $F^{(\mathrm{nr})}$ as its relativistic counterpart. Therefore, proceeding in the same way as for the relativistic theory, with the corresponding scaling ansatz (2.466) for $\rho^{(\mathrm{nr})}$ and (2.467) for $F^{(\mathrm{nr})}$, leads to the very same solutions (2.477) and (2.478). A crucial difference arises when the occupation number distribution $f_{\mathrm{nr}}(\mathbf{p})$ is determined. Using the nonrelativistic definition (2.33) for the distribution function, we have, with the notation (2.209) in the absence of a condensate,

$$f_{\mathrm{nr}}(\mathbf{p}) + \tfrac{1}{2} = \int_0^\infty \frac{\mathrm{d}p^0}{2\pi} \, F_{aa}^{(\mathrm{nr})}(p^0, \mathbf{p})$$

$$\equiv \frac{1}{2} \int \mathrm{d}^3 x \, \mathrm{e}^{-\mathrm{i}\mathbf{p}\cdot\mathbf{x}} \langle \chi(t, \mathbf{x})\chi^*(t, 0) + \chi(t, 0)\chi^*(t, \mathbf{x}) \rangle. \qquad (2.481)$$

Comparison with the relativistic case shows that a difference in the scaling behavior is caused by the additional factor of $\sim p^0$ in the integrand of (2.479). Therefore, we find that

$$f_{\mathrm{nr}}(\mathbf{p}) = s^{\kappa_{\mathrm{s}}+2-z} f_{\mathrm{nr}}(s\mathbf{p})$$

$$\sim \begin{cases} |\mathbf{p}|^{-(d+2-\eta)} & \text{nonrelativistic particle cascade,} \\ |\mathbf{p}|^{-(d+2+z-\eta)} & \text{nonrelativistic energy cascade} \end{cases} \qquad (2.482)$$

scales with one "z" difference than the relativistic solution (2.480). This can have important consequences, such as the fact that the scaling exponent for the nonrelativistic particle cascade is now independent of the dynamic exponent z describing the dispersion $\omega_{\mathbf{p}} \sim |\mathbf{p}|^z$. As a consequence, the same particle cascade scaling solution is found also in the presence of a condensate, where for the Gross–Pitaevskii theory the approximate (Bogoliubov) dispersion is given by

$$\omega_{\mathbf{p}} = \sqrt{\frac{\mathbf{p}^2}{2m}\left(\frac{\mathbf{p}^2}{2m} + 2g|\chi_0|^2\right)}. \qquad (2.483)$$

At large momenta, or in the absence of a condensate, one recovers $\omega_{\mathbf{p}} = \mathbf{p}^2/(2m)$, while for low momenta, one has $\omega_{\mathbf{p}} \sim |\mathbf{p}|$. The particular importance of the particle cascade and the associated phenomenon of Bose condensation far from equilibrium, as outlined in Section 2.1, will now be addressed in more detail.

2.6.2 Nonthermal fixed points

Self-similarity

In contrast to the stationary turbulence described above, we now turn to time evolution. In general, isolated systems out of equilibrium cannot realize stationary transport solutions. Instead, we have to consider the more general notion of a *self-similar* evolution, where the dynamics is described in terms of *time-independent scaling functions* and *scaling exponents*.

A self-similar evolution of the distribution function $f(t, \mathbf{p})$ for a spatially homogeneous and isotropic system is characterized as

$$f(t, \mathbf{p}) = s^{\alpha/\beta} f(s^{-1/\beta}t, s\mathbf{p}), \qquad (2.484)$$

with the real scaling exponents α and β. Again, all quantities are considered to be dimensionless by use of some suitable momentum scale. Choosing $s^{-1/\beta}t = 1$, we recover (2.25), i.e.,

$$f(t, \mathbf{p}) = t^\alpha f_S(t^\beta \mathbf{p}), \tag{2.485}$$

where the time-independent scaling function $f_S(t^\beta \mathbf{p}) \equiv f(1, t^\beta \mathbf{p})$ denotes the *fixed-point distribution*. This scaling form represents an enormous reduction of the possible dependence of the dynamics on variations in time and momenta, since $t^{-\alpha} f(t, \mathbf{p})$ depends on the product $t^\beta |\mathbf{p}|$ instead of depending separately on time and momenta. Therefore, an essential part of the time evolution is encoded in the momentum dependence of the fixed-point distribution $f_S(\mathbf{p})$. Moreover, the values for α and β determine the rate and direction of transport processes, since a given characteristic momentum scale $K(t_1) = K_1$ evolves as $K(t) = K_1(t/t_1)^{-\beta}$ with amplitude $f(t, K(t)) \sim t^\alpha$. This aspect is also further discussed in Section 2.1.

For the self-similar distribution (2.485), the scaling behavior of a generic collision integral is then given by

$$C[f](t, \mathbf{p}) = s^{-\tilde{\mu}} C[f](s^{-1/\beta}t, s\mathbf{p}) = t^{-\beta\tilde{\mu}} C[f_S](1, t^\beta \mathbf{p}), \tag{2.486}$$

where $\tilde{\mu}$ is a function of scaling exponents similar to the discussion above. Substituting this scaling form into the kinetic equation leads to the time-independent *fixed-point equation* for $f_S(\mathbf{p})$,

$$(\alpha + \beta \, \mathbf{p} \cdot \nabla_{\mathbf{p}}) \, f_S(\mathbf{p}) = C[f_S](1, \mathbf{p}), \tag{2.487}$$

and the *scaling relation*

$$\alpha - 1 = -\beta\tilde{\mu}. \tag{2.488}$$

This follows from comparing the left-hand side of the kinetic equation,

$$\frac{\partial}{\partial t} \left[t^\alpha f_S(t^\beta \mathbf{p}) \right] = t^{\alpha - 1} \left(\alpha + \beta \, \mathbf{q} \cdot \nabla_{\mathbf{q}} \right) f_S(\mathbf{q}) \big|_{\mathbf{q} = t^\beta \mathbf{p}}, \tag{2.489}$$

with its right-hand side given by (2.486).

Further relations can be obtained by imposing either energy conservation or particle number conservation if applicable. For constant

$$n = \int \frac{d^d p}{(2\pi)^d} f(t, \mathbf{p}) = t^{\alpha - \beta d} \int \frac{d^d q}{(2\pi)^d} f_S(\mathbf{q}), \tag{2.490}$$

we obtain the relation for *particle conservation*:

$$\alpha = \beta d. \tag{2.491}$$

Similarly, from *energy conservation*, we obtain

$$\alpha = \beta(d + z). \tag{2.492}$$

We can see that there is no single scaling solution conserving both energy and particle number. As outlined already in Section 2.1, in this case, a *dual cascade* is expected to emerge such that in a given inertial range of momenta only one conservation law governs the scaling behavior.

- *Perturbative scaling behavior.* We will start again with a perturbative analysis of the relativistic theory, which is relevant at momenta above a possible mass gap such that $\omega_{\mathbf{p}} = s^{-1}\omega_{s\mathbf{p}}$, after which we will consider the highly nonlinear infrared regime. For the self-similar distribution (2.485), the scaling behavior of a generic collision integral for l-vertex scattering is then given by

$$C(t, \mathbf{p}) = s^{(l-1)\alpha/\beta+l+1-(l-2)d}C(s^{-1/\beta}t, s\mathbf{p}). \tag{2.493}$$

The reasoning is the same as for (2.457), with the only difference being that now the distribution function scales according to (2.484) instead of (2.453). Consequently, the factor $(l-1)\alpha/\beta$ appears, replacing the term $(l-1)\kappa$ of (2.457). Using the scaling relation (2.488) and (2.491) from particle conservation gives the perturbative solution for *relativistic particle transport*:

$$\alpha = -\frac{d}{l+1}, \qquad \beta = -\frac{1}{l+1}. \tag{2.494}$$

Similarly, we find the perturbative solution for *relativistic energy transport*:

$$\alpha = -\frac{d+1}{2l-1}, \qquad \beta = -\frac{1}{2l-1}. \tag{2.495}$$

For instance, for quartic self-interactions, the perturbative energy transport is characterized by $\alpha = -\frac{1}{7}(d+1)$ and $\beta = -\frac{1}{7}$, where the latter is independent of the dimensionality of space d. Likewise, for the 3-vertex in the presence of a coherent field, one has for the energy transport $\alpha = -\frac{1}{5}(d+1)$ and $\beta = -\frac{1}{5}$. The latter is used for $d = 3$ to describe the direct energy cascade of the inflaton dynamics in Section 2.1.3.

- *Infrared regime.* In order to go beyond the perturbative estimates, which are not applicable in the infrared, we again consider the $1/N$ expansion to NLO. At the lowest-order gradient expansion employed, the spectral function is time-independent for a homogeneous system according to (2.306). Consequently, its scaling properties are still described by (2.466) for the self-similar evolution. However, for $f(t, p^0, \mathbf{p})$, we have to take into account the scaling with time:

$$f(t, p^0, \mathbf{p}) = s^{\alpha_s/\beta} f(s^{-1/\beta}t, s^z p^0, s\mathbf{p}), \tag{2.496}$$

which increases the number of scaling exponents as compared with (2.468) for the more restrictive case of stationary turbulence. With these definitions, we can analyze the scaling of the collision integral (2.464) and (2.465), which gives

$$C^{\mathrm{NLO}}(t, \mathbf{p}) = s^{\alpha_s/\beta - z} C^{\mathrm{NLO}}(s^{-1/\beta}t, s\mathbf{p}) = t^{\alpha_s - \beta z} C^{\mathrm{NLO}}(1, t^\beta \mathbf{p}), \tag{2.497}$$

following the corresponding steps taken in Section 2.6.1. Using the scaling relation (2.488), we infer

$$\alpha = \alpha_{\mathrm{s}} - \beta z + 1. \tag{2.498}$$

In particular, with the same reasoning as before, we note that (2.497) applies also to the corresponding nonrelativistic collision integral.

Again, the crucial difference emerges when the occupation number distributions $f(t, \mathbf{p})$ for the relativistic and $f_{\mathrm{nr}}(t, \mathbf{p})$ for the nonrelativistic theories are determined. For the former, we have, according to (2.479),

$$f(t, \mathbf{p}) = s^{\alpha_{\mathrm{s}}/\beta + 2 - 2z - \eta} \int_{0}^{\infty} \frac{dp^{0}\,s^{z}}{2\pi} 2(s^{z} p^{0}) \tilde{\rho}(s^{z} p^{0}, s\mathbf{p}) f(s^{-1/\beta} t, s^{z} p^{0}, s\mathbf{p})$$

$$\equiv s^{\alpha_{\mathrm{s}}/\beta + 2 - 2z - \eta} f(s^{-1/\beta} t, s\mathbf{p}) = t^{\alpha_{\mathrm{s}} + \beta(2 - 2z - \eta)} f_{S}(t^{\beta} \mathbf{p}), \tag{2.499}$$

so that, relativistically, $\alpha = \alpha_{\mathrm{s}} + \beta(2 - 2z - \eta)$, in comparison with (2.485). Together with (2.498), we find in the absence of a mass gap that for *relativistic transport of particles and energy*,

$$\beta = \frac{1}{2 - z - \eta} \tag{2.500}$$

and

$$\alpha = \begin{cases} \dfrac{d}{2 - z - \eta} & \text{for particles,} \\[2mm] \dfrac{d + z}{2 - z - \eta} & \text{for energy.} \end{cases} \tag{2.501}$$

Remarkably, the scaling exponent β is obtained without using in addition energy or particle conservation, whereas the different solutions for α arise from imposing (2.491) or (2.492), respectively.

Similarly, the nonrelativistic case yields, according to the definition (2.481),

$$f_{\mathrm{nr}}(t, \mathbf{p}) = t^{\alpha_{\mathrm{s}} + \beta(2 - z - \eta)} f_{S,\mathrm{nr}}(t^{\beta} \mathbf{p}). \tag{2.502}$$

Therefore, $\alpha = \alpha_{\mathrm{s}} + \beta(2 - z - \eta)$ by comparing with (2.485), which scales again with one "z" difference from the relativistic case. Correspondingly, we find that for *nonrelativistic transport of particles and energy*,

$$\beta = \frac{1}{2 - \eta} \tag{2.503}$$

and

$$\alpha = \begin{cases} \dfrac{d}{2 - \eta} & \text{for particles,} \\[2mm] \dfrac{d + z}{2 - \eta} & \text{for energy.} \end{cases} \tag{2.504}$$

The nonrelativistic particle transport solution for $\eta \to 0$ is referred to in (2.26).

Inverse particle cascade and condensate formation

In order to check the above analytic estimates, we may use the fact that the nonequilibrium quantum dynamics of the highly occupied system can be accurately mapped onto a classical statistical field theory evolution as described in Section 2.4. This mapping is valid for the system considered, as long as $f_{\mathrm{nr}}(t, \mathbf{p}) \gg 1$ for typical momenta $\mathbf{p}$ according to (2.382). We employ initial conditions with high occupation numbers as motivated in Section 2.1, where it is also described that the nonrelativistic and relativistic theories show the same universal behavior for low momenta.

In the following, we consider the nonrelativistic Bose gas described by (2.206) in the dilute regime ($\zeta \ll 1$). The initial distribution function at $t = 0$ is taken as

$$f_{\mathrm{nr}}(0, \mathbf{p}) \sim \frac{1}{\zeta} \Theta(Q - |\mathbf{p}|), \tag{2.505}$$

which describes overoccupation up to the characteristic momentum Q. The initial condensate fraction is taken to be zero, i.e., $|\psi_0|^2 (t = 0) = 0$ in the defining equation (2.33), with an initial $f_{\mathrm{nr}}(0, \mathbf{p}) = (50/2mgQ)\Theta(Q - |\mathbf{p}|)$. We always plot dimensionless quantities obtained by the rescalings $f_{\mathrm{nr}}(t, \mathbf{p}) \to 2mgQ f_{\mathrm{nr}}(t, \mathbf{p})$, $t \to (Q^2/2m)t$, and $\mathbf{p} \to \mathbf{p}/Q$. This reflects the classical statistical nature of the dynamics in the highly occupied regime, which has the important consequence that if we measure time in units of $2m/Q^2$ and momentum in units of Q, then the combination $2mgQ f_{\mathrm{nr}}(t, \mathbf{p})$ does not depend on the values of m, g, and Q.

The initial mode occupancies (2.505) are expected to be rapidly redistributed at the beginning of the nonequilibrium evolution, after which a slower behavior sets in. The latter reflects the dynamics near the nonthermal fixed point, where universality can be observed. We concentrate on the low-momentum part of the distribution and analyze its infrared scaling properties.

Figure 2.20 shows the results for the rescaled distribution $(t/t_{\mathrm{ref}})^{-\alpha} f_{\mathrm{nr}}(t, \mathbf{p})$ as a function of $(t/t_{\mathrm{ref}})^{\beta}|\mathbf{p}|$, where the reference time $t_{\mathrm{ref}} Q^2/2m = 300$ after which self-similarity is well developed is the earliest time shown. In contrast, the inset gives the curves at different times together with the initial distribution without rescaling for comparison. With an appropriate choice of the infrared scaling exponents α and β, all the curves at different times lie remarkably well on top of each other after rescaling. This is a striking manifestation of the self-similar dynamics (2.25) near the nonthermal fixed point. The numerical estimates for the scaling exponents obtained are

$$\alpha = 1.66 \pm 0.12,$$
$$\beta = 0.55 \pm 0.03. \tag{2.506}$$

Comparing these values with the analytic estimates for the particle cascade in (2.26) or in (2.503) and (2.504), we see that the numerical results (2.506) agree rather well with the NLO approximation for a vanishing anomalous dimension η. Furthermore, the simulation results confirm that $\alpha = 3\beta$ to very good accuracy, as expected for $d = 3$ from number conservation (2.491) in the infrared scaling regime.

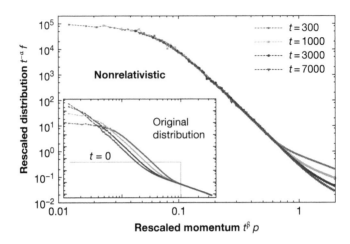

Fig. 2.20 Rescaled distribution function of the nonrelativistic theory as a function of the rescaled momentum for different times. The inset shows the original distribution without rescaling.

The positive values for the exponents imply that particles are transported toward low momenta, which has important consequences. For the initial conditions (2.505), there is no condensate present at $t = 0$. However, the inverse particle cascade continuously populates the zero mode, which will lead to the formation of a condensate far from equilibrium.

Of course, the time needed to fill the entire volume with a single condensate increases with the volume V in which the simulations are done. Using the fact that the parametrically slow power-law dynamics dominates the time until condensation is completed, we can use the scaling exponent α describing the growth in the infrared to estimate this condensation time. Taking the value of the zero-momentum correlator $V^{-1}F^{(\mathrm{nr})}(t, t, \mathbf{p} = 0)$ at the initial time t_0 of the self-similar regime as $V^{-1}f_{\mathrm{nr}}(t_0, 0)$ and its final value at the time t_f as $|\psi_0|^2(t_f)$, we can estimate from $V^{-1}F^{(\mathrm{nr})}(t, t, \mathbf{p} = 0) \sim t^\alpha$ the condensation time as

$$t_f \simeq t_0 \left(\frac{|\psi_0|^2(t_f)}{f_{\mathrm{nr}}(t_0, 0)} \right)^{1/\alpha} V^{1/\alpha}. \tag{2.507}$$

The time-dependent condensate fraction at zero momentum is given by

$$\frac{N_0(t)}{N_{\mathrm{total}}} = \frac{|\psi_0|^2(t)}{\displaystyle\int \frac{d^3 p}{(2\pi)^3} f_{\mathrm{nr}}(t, p) + |\psi_0|^2(t)}. \tag{2.508}$$

Accordingly, we may define the condensate fraction for the case of finite volume as $N_0/N_{\mathrm{total}} \to V^{-1}F^{(\mathrm{nr})}(t, t, \mathbf{p} = 0)/F^{(\mathrm{nr})}(t, t, \mathbf{x} = 0)$, using the fact that $N_{\mathrm{total}} = F^{(\mathrm{nr})}(t, t, \mathbf{x} = 0)$. Figure 2.21 shows the evolution of the condensate fraction for different volumes, with the time axis rescaled by $V^{-1/\alpha}$. Indeed, as predicted by (2.507), the

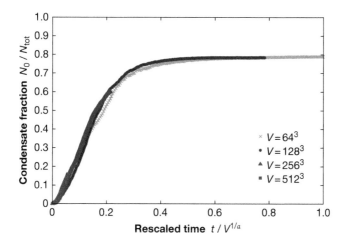

Fig. 2.21 Evolution of the condensate fraction for the nonrelativistic Bose gas for different volumes V. The different curves become approximately volume-independent after rescaling of time by $V^{-1/\alpha}$, in agreement with (2.507).

different curves are approximately volume-independent. We find that the condensate fraction saturates at $N_0/N_{\text{total}} \simeq 0.8$.

2.6.3 Bibliography

- Figures 2.20 and 2.21 are taken with permission from A. Piñeiro Orioli, K. Boguslavski, and J. Berges. Universal self-similar dynamics of relativistic and nonrelativistic field theories near nonthermal fixed points. *Phys. Rev. D* **92** (2015) 025041 (Copyright (2015) by the American Physical Society). Much of the discussion about self-similar evolution is taken from the same reference. The presentation of the perturbative scaling behavior mostly follows the review by R. Micha and I. I. Tkachev. Turbulent thermalization. *Phys. Rev. D* **70** (2004) 043538.
- The discussion of strong turbulence in relativistic theories follows partly that in J. Berges, A. Rothkopf, and J. Schmidt. Non-thermal fixed points: effective weak-coupling for strongly correlated systems far from equilibrium. *Phys. Rev. Lett.* **101** (2008) 041603 and in J. Berges and D. Sexty. Strong versus weak wave-turbulence in relativistic field theory. *Phys. Rev. D* **83** (2011) 085004. The presentation of nonrelativistic aspects partly follows C. Scheppach, J. Berges, and T. Gasenzer. Matter wave turbulence: beyond kinetic scaling. *Phys. Rev. A* **81** (2010) 033611 and B. Nowak, J. Schole, D. Sexty, and T. Gasenzer. Nonthermal fixed points, vortex statistics, and superfluid turbulence in an ultracold Bose gas. *Phys. Rev. A* **85** (2012) 043627.
- The discussion of Bose condensation out of equilibrium follows to some extent J. Berges and D. Sexty. Bose condensation far from equilibrium. *Phys. Rev. Lett.*

108 (2012) 161601 and B. Nowak, J. Schole, and T. Gasenzer. Universal dynamics on the way to thermalization. *New J. Phys.* **16** (2014) 093052.

- For a presentation of nonthermal fixed points in the language of the renormalization group, see J. Berges and G. Hoffmeister. Nonthermal fixed points and the functional renormalization group. *Nucl. Phys. B* **813** (2009) 383 and J. Berges and D. Mesterhazy. Introduction to the nonequilibrium functional renormalization group. *Nucl. Phys. B Proc. Suppl.* **228** (2012) 37.

- Nonthermal fixed points in the presence of fermions are discussed in J. Berges, D. Gelfand, and J. Pruschke. Quantum theory of fermion production after inflation. *Phys. Rev. Lett.* **107** (2011) 061301.

3

Slow relaxations and nonequilibrium dynamics in classical and quantum systems

Giulio BIROLI

IPhT, CEA/DSM-CNRS/URA 2306,
CEA Saclay, F-91191 Gif-sur-Yvette Cedex,
France

Strongly Interacting Quantum Systems out of Equilibrium. First Edition. Thierry Giamarchi et al.
© Oxford University Press 2016. Published in 2016 by Oxford University Press.

Chapter Contents

Preface

This chapter is based on a series of lectures I gave in 2012 at the Les Houches Summer School of Physics on Strongly Interacting Quantum Systems Out of Equilibrium. The aim of these lectures was to provide an introduction to several important and interesting facets of out-of-equilibrium dynamics. In recent years, there has been a boost in research on quantum systems out of equilibrium. If 15 years ago hard condensed matter and classical statistical physics remained somewhat separate research fields, the present focus on several kinds of out-of-equilibrium dynamics is bringing them closer and closer together. The aim of my lectures was to present to students the richness of this topic, insisting on common concepts and showing that there is much to gain in considering and learning out-of-equilibrium dynamics as a whole research field. This chapter is by no means self-contained. Each section is a door open toward a vast research area. I have endeavored to present the most striking or useful or important concepts and tools, often just in an informal and introductory way. I hope this will stimulate the interest of readers who will then turn to specialized reviews and books (cited in the chapter) to fully satisfy their curiosity and obtain more complete knowledge.

3.1 Coupling to the environment: what is a reservoir?

Depending on the problem one is facing and the fundamental questions one is addressing, a physical system can be considered *closed* or *open*, i.e., decoupled or not decoupled from the environment. Actually, no system is truly isolated; however, some can be considered so on a limited range of timescales over which the coupling to the environment does not take place. In the burgeoning field of nonequilibrium quantum dynamics, examples are provided by cold atoms and by electrons in solids. In both cases, there are situations in which the timescale over which dissipative effects take place is longer than that corresponding to relaxation, so that the systems can be considered isolated to some extent; see the contributions to this volume by Altman and Perfetti (Chapters 1 and 9).

In statistical physics or condensed matter, one generically deals with open systems that exchange energy and particles with the environment, which henceforth we shall call the *reservoir* (or bath). Just to cite a few examples of reservoirs, think of solvent molecules for colloidal particles or phonons for electron systems. Of course, any open system can always be considered as a subpart of a closed one, so actually the real decision to be made when considering whether a system is open or closed depends on the kind of questions one wants to address.

In this chapter, we will mainly focus on open systems, except when discussing some fundamental issues related to thermalization. When studying open systems, one declares all the rest of the world apart from the system as "the environment" and changes the fundamental dynamical (Newton or Schrödinger) laws to take into account the role of "the environment" on the system. This *reductionist* step from closed to open systems is a very subtle one. The aim of this section is to unravel the main physical ideas and technical steps underlying it.

3.1.1 What does a reservoir do to a system?

Let us start our analysis with some informal considerations. We shall focus on classical physics just to keep the discussion as easy as possible. The dynamics of an isolated system is governed by Newton's law:

$$m\frac{d^2x}{dt^2} = F, \tag{3.1}$$

where for simplicity we have written Newton's equation for a one-dimensional particle of mass m at position x and subject to an external force F. There are two main effects due to the environment:

- **Dissipation.** Throw a ball in the air and model your experiment with (3.1). What do you get? The ball accelerates (you exert a force on it) and then, once you have thrown it, it should go straight with a constant velocity ... forever. What really happens is quite different—even if you are a good baseball thrower! The ball inevitably falls down and stops.[1] The initial kinetic energy given to the ball is dissipated in the environment. This effect is general and often is taken into account by adding a friction term to the equation of motion, which then in the simplest case reads

$$m\frac{d^2x}{dt^2} + \eta\frac{dx}{dt} = F, \tag{3.2}$$

  where η is the strength of the dissipation.
- **Thermal noise.** There is something missing in the previous description. Statistical mechanics teaches us that the equilibrium distribution of a system is given by the Boltzmann probability law. In particular, for a particle in an external potential V, the Boltzmann law is $P(x) = Z^{-1}\exp[-V(x)/T]$ (henceforth the Boltzmann constant will be put equal to one). Consider for example the simplest case of a quadratic potential $V(x) = \frac{1}{2}x^2$, for which $F = -V'(x) = -x$. Equation (3.2) implies that at long time the particle will sit in the bottom of the potential not moving at all, no matter what the initial condition is. The Boltzmann law instead predicts that the particle position fluctuates around the bottom of the well over a distance of the order of $\sqrt{T}$. What is missing in (3.2) is a term corresponding to thermal noise: an environment at temperature T acts on the system with the net effect that energy is exchanged—sometimes the reservoir sucks energy out from the system and sometimes it releases energy to it. Langevin proposed to take this effect into account by adding a *random force* to the previous equation, which in the simplest case then reads

$$m\frac{d^2x}{dt^2} + \eta\frac{dx}{dt} = F + \xi(t), \tag{3.3}$$

---

[1] As a matter of fact, the longest ball throw was apparently by Glen Gorbous, a Canadian minor leaguer, who in 1957 threw a baseball 135.89 m.

where $\xi(t)$ is a Gaussian field with zero mean and with variance $\langle \xi(t)\xi(t') \rangle = 2T\eta\delta(t - t')$. This stochastic equation guarantees that the particle thermalizes at long times and that its probability distribution is given by the Boltzmann law for any initial condition. The proof is simple and will be shown in Section 3.3; as an exercise, you can verify this result by explicitly solving (3.3) in the quadratic case, $V(x) = \frac{1}{2}x^2$, and averaging over the thermal noise.

We have discussed the main effects of the environment in the case of a classical particle in one dimension, an arguably very simple case. It turns out that more general cases may be more difficult technically—the dissipative term can be represented by retarded friction, the noise can be colored and even multiplicative and non-Gaussian, and in the quantum case a formalism more involved than stochastic equations must be used—still, the main physical effects, namely, dissipation and thermal activation, are the same and play the same role.

An important outcome of the previous analysis is that by taking into account these two effects and by changing the dynamical laws, the particle naturally thermalizes at long times at the temperature T of the environment. The fact that the environment is at equilibrium at temperature T is actually encoded in the relationship between fluctuation and dissipation. A generic nonequilibrated environment also leads to fluctuation and dissipation. However, only for an equilibrated one are these tightly related: the fact that the variance of the noise is $2T\eta$ is an expression of this fundamental law that we shall discuss in more detail later.

3.1.2 The simplest reservoir: linearly coupled harmonic oscillators

We now go over the "reductionist step" discussed before, showing how one can obtain an open system from a closed one by integrating out the degrees of freedom that correspond to the reservoir. We do it for a simple, instructive, and also quite general model: a system coupled to a very large set of harmonic oscillators. In order to simplify the notation a bit, we take the system to be one-dimensional.

The total Hamiltonian reads

$$H = H_{\text{syst}} + H_I + H_R, \tag{3.4}$$

where

$$H_{\text{syst}} = \frac{p^2}{2m} + V(x)$$

is the Hamiltonian of the system on which we focus,

$$H_R = \sum_k^N \frac{p_k^2}{2} + \frac{\omega_k^2}{2} Q_k^2$$

is the Hamiltonian of the harmonic oscillators, and

$$H_I = -\sum_k^N \gamma_k Q_k$$

is the interaction between the system and the oscillators. The number of oscillators is very large: $N \gg 1$. Moreover, the coupling γ_k is such that $\gamma_k = \tilde{\gamma}_k / \sqrt{N}$, where $\tilde{\gamma}_k$ is of the order of one, i.e., all oscillators interact with the system, but very weakly. We will discuss the physics behind these two important assumptions later. For the moment, we just notice that these choices are those that come up naturally if one wants to study a system coupled to classical phonons.

Classical phonons

The phonon Hamiltonian is that of harmonic vibrations and reads

$$H_R = \sum_{\ell=-N/2+1}^{N/2} \frac{p_\ell^2}{2} + \sum_{\ell=-N/2+1}^{N/2} \frac{(q_\ell - q_{\ell+1})^2}{2},$$

where ℓ is the lattice site index, which runs over N different values (we assume periodic boundary conditions). The phonon–system interaction is local, and, without loss of generality, we consider the case in which the system is located at the origin:

$$H_I = -q_0 x.$$

We shall take the coupling to be linear, which is correct at low temperature (small fluctuations of both q_ℓ and x), but more complicated couplings can also be considered. It is easy to check that this model can be rewritten in the same form as the previous one: indeed, by going to Fourier space, i.e., defining $Q_k = (1/\sqrt{N}) \sum_{\ell=-N/2+1}^{N/2} e^{-ik\ell} q_\ell$, all phonons are decoupled and are characterized by the frequency $\omega_k^2 = 2(1 - \cos k)$; each of them interacts with the system via a coupling constant $\gamma_k = 1/\sqrt{N}$.

The equations of motion for the system read

$$m \frac{d^2 x}{dt^2} = -V'(x) + \sum_k \gamma_k Q_k, \tag{3.5}$$

$$\frac{d^2 Q_k}{dt^2} = -\omega_k^2 Q_k + \gamma_k x. \tag{3.6}$$

A reservoir that is formed by harmonic oscillators is simple to treat because it can be integrated out exactly; the generic solution of (3.6) can be readily found:

$$Q_k(t) = \frac{\gamma_k}{\omega_k^2} x(t) + \left[Q_k(0) - \frac{\gamma_k}{\omega_k^2} x(0) \right] \cos(\omega_k t)$$

$$+ \dot{Q}_k(0) \frac{\sin(\omega_k t)}{\omega_k} - \frac{\gamma_k}{\omega_k^2} \int_0^t \cos[\omega_k(t-s)] \, \dot{x}(s) \, ds. \tag{3.7}$$

By plugging this expression into (3.5), we obtain

$$
m\frac{d^2x}{dt^2} + \underbrace{\int_0^t \sum_k \frac{\gamma_k^2}{\omega_k^2}\cos[\omega_k(t-s)]\,\dot{x}(s)\,ds}_{\text{dissipation}} = -V'(x) + \underbrace{\sum_k \frac{\gamma_k^2}{\omega_k^2}x(t)}_{\text{effective potential}}
$$

$$
+ \underbrace{\sum_k \left\{ \dot{Q}_k(0)\gamma_k\frac{\sin(\omega_k t)}{\omega_k} + \gamma_k\left[Q_k(0) - \frac{\gamma_k}{\omega_k^2}x(0)\right]\cos(\omega_k t)\right\}}_{\text{noise}} \tag{3.8}
$$

We have found that once the reservoir is integrated out, new terms appear in the equation of motion for x: we have called them dissipation, effective potential, and noise for reasons that will be clear in the following. Before discussing them, let us rewrite the previous equation in a more appealing form:

$$
m\frac{d^2x}{dt^2} + \int_0^t K(t-s)\dot{x}(s)\,ds = -V'(x) - V'_{\text{eff}}(x) + \xi(t), \tag{3.9}
$$

where

$$
K(\tau) = \sum_k \frac{\gamma_k^2}{\omega_k^2}\cos(\omega_k\tau),
$$

$$
V_{\text{eff}} = \frac{1}{2}\sum_k \frac{\gamma_k^2}{\omega_k^2}x^2,
$$

$$
\xi(t) = \sum_k \left\{ \dot{Q}_k(0)\gamma_k\frac{\sin(\omega_k t)}{\omega_k} + \gamma_k\left[Q_k(0) - \frac{\gamma_k}{\omega_k^2}x(0)\right]\cos(\omega_k t)\right\}.
$$

Now the meaning of each new term should become clear. The term corresponding to dissipation has indeed the form of a retarded friction. This is expected on general grounds, since the system releases energy by interacting with the phonons. In our previous discussion of the Langevin equation, we did not add an effective force, $-V'_{\text{eff}}(x)$, due to the environment, but this of course has to be expected too—and is indeed what we find. The final term, the noise, is more problematic than the others. As its name makes clear, it should be characterized by some kind of randomness. However, its definition also makes clear that it is perfectly deterministic. Where is the subtlety? We postpone a complete discussion to Section 3.1.3; for the moment, we just notice that for typical values of $Q_K(0)$ and $\dot{Q}_k(0)$, i.e., values extracted from the equilibrium Boltzmann measure, the function $\xi(t)$ behaves so erratically that in practice it is indistinguishable from a random function if N is moderately large ($N > 20$); moreover, $\xi(t)$ indeed contains some randomness, since the $Q_K(0)$'s and $\dot{Q}_k(0)$'s are random.

In the following, we assume that the reservoir, i.e., the phonons, is at equilibrium at time $t = 0$ given the position $x(0)$ of the system. In consequence, the reservoir's probability measure reads

$$P\big[\{Q_k(0), \dot{Q}_k(0)\}\big|x(0)\big] =$$

$$\mathcal{N} \exp\left\{-\sum_k \left[\frac{1}{2T}\dot{Q}_k(0)^2 + \frac{\omega_k^2}{2T}Q_k(0)^2 - \frac{1}{T}\gamma_k Q_k(0)x(0)\right]\right\},$$

where $\mathcal{N}$ is the normalization constant. By redefining the normalization constant, this expression can be rewritten as

$$P\big[\{Q_k(0), \dot{Q}_k(0)\}\big|x(0)\big] =$$

$$\mathcal{N}' \exp\left\{-\sum_k \frac{1}{2T}\dot{Q}_k(0)^2 - \sum_k \frac{\omega_k^2}{2T}\left[Q_k(0) - \frac{\gamma_k}{\omega_k^2}x(0)\right]^2\right\}.$$

This simple rewriting reveals that $\dot{Q}_k(0)$ and $Q_k(0) - (\gamma_k/\omega_k^2)x(0)$ are independent Gaussian variables of zero mean and variance T and T/ω_k^2, respectively. It is now straightforward to show that $\xi(t)$ is a Gaussian random function, since it is a sum of Gaussian variables; its mean and variance are given by

$$\langle \xi(t) \rangle = 0, \qquad \langle \xi(t)\xi(t') \rangle = TK(t - t').$$

This completes the "reductionist" step from the closed to the open system. The main results that we have found are a first-principles derivation of the Langevin equation and that dissipation and fluctuations (or noise) are related: the memory function appearing in the retarded friction term is exactly equal to the variance of the noise times the temperature.[2]

3.1.3 Noise and irreversibility

Let us now discuss in more detail the noise term and the extent to which it is random. In a real experiment, the initial condition of the system and the environment is fixed; in particular, $\dot{Q}_k(0)$ and $Q_k(0)$ are fixed. Consequently, $\xi(t)$ is a completely deterministic function. Hence, one might wonder about the extent to which $\xi(t)$ does represent random noise. Well, it is known that a deterministic function can be identical for any practical purposes to a random one. Think, for example, of the random number generators used in a Monte Carlo simulation: these are deterministic functions producing numbers that one considers random for all practical purposes. In reality, however, these are not truly random, because the sequence one obtains repeats after

---

[2] The reader might be surprised, since we have got a factor-of-two reduction with respect to the Langevin equation introduced in Section 3.1.1. In fact, there is no contradiction here. The subtlety is that in the limit where K becomes a delta function, only half of the delta function contributes to the integrated friction. Thus, one has to take $K(\tau) \to 2\eta\delta(\tau)$ to get a friction term $\eta\dot{x}$. In this case, one recovers that the variance of the noise is indeed $2T\delta(\tau)$ as expected.

a certain number of trials. The better the random number generator, the longer is the period. The "noise" $\xi(t)$ behaves in a similar way; the larger the value of N, the more it resembles a random function. In the limit $N \to \infty$, it is indeed indistinguishable from a random function.[3] In order to better clarify this last statement, let us recall that there are two ways of doing averages:

- **Time average.** In this case, given an observable $O(x(t))$, we compute its average as

$$\langle O \rangle_T = \lim_{\tau \to \infty} \frac{1}{\tau} \int_0^\tau O(x(t')) \, dt'.$$

  This is what one does in an experiment assuming that the system has reached a steady state.
- **Ensemble average.** In this case, given an observable $O(x(t))$, we compute its average by averaging over all possible realizations belonging to the ensemble. In the case we are considering, this means averaging over all possible initial conditions for the oscillators, i.e., over the values of $\dot{Q}_k(0)$ and $Q_k(0)$. This leads to an average over $\xi(t)$, which then indeed becomes a random function:

$$\langle O \rangle_E = \lim_{\tau \to \infty} \int \prod dQ_k(0) \, d\dot{Q}_k(0) \, O(x(\tau)) P\big[\{Q_k(0), \dot{Q}_k(0)\} \big| x(0)\big].$$

The previous statement about the indistinguishability of $\xi(t)$ from a random function means that time averages and ensemble averages coincide. In particular,

$$\langle \xi(t)\xi(t') \rangle_T = \langle \xi(t)\xi(t') \rangle_E = TK(t - t').$$

As stated previously, the parameter controlling how much the pseudo-noise $\xi(t)$ behaves like a true noise is N, the number of degrees of freedom in the environment; see [49, 69] for more details.

We arrived at the conclusion that for large enough N, our deterministic open system is indistinguishable from a stochastic one based on the Langevin equation (3.9). It is easy to prove, as we shall show later on, that a system obeying such an equation thermalizes at long time, i.e., it goes to a steady state where the distribution is Maxwell–Boltzmann and all equilibrium relationships, like fluctuation–dissipation, are satisfied.

Just to highlight how subtle is this reduction from a closed to an open system, let us discuss an apparent paradox. Consider as a system a harmonic oscillator, with $V(x) = \frac{1}{2}kx^2$, and think about what the previous statement means: the oscillator thermalizes at long time for any initial condition. Is this not weird? The global closed system is just a collection of coupled harmonic oscillators—the easiest integrable system we can think of! And, as you know, integrable systems do not thermalize. Where is the catch? The subtle point is that we do not have to resort to any chaotic dynamics to get

---

[3] The frequencies ω_k should be incommensurate in order for the mapping to a random function to hold.

irreversibility and equilibration. There is nothing surprising here once we realize that the fact of having an environment with a very large number of degrees of freedom is the key ingredient: when $N = 4$, we get only quasiperiodic motion and no equilibration at all, but for $N \gg 1$ the quasiperiodicity is pushed up to times that are exponentially large in N; see [69].

It is instructive to disentangle two processes that take place when a system reaches equilibrium: irreversible behavior and steady-state properties given by a Maxwell–Boltzmann distribution. The former is in general related to reaching a steady state and to damping. An environment formed by a very large number of degrees of freedom leads to irreversible behavior and generically to a steady state, but only one that is at equilibrium leads to true thermalization (and this independently of its chaoticity: $N \gg 1$ is enough). See [23, 69] for a general discussion on the relationship between irreversibility, thermodynamic limit, and red herrings related to chaos.

3.1.4   The environment

By looking at the Langevin equation (3.9), we can see that the only characteristic of the environment that we need to know is $K(t)$. In order to describe $K(t)$, it is useful to introduce the spectral function

$$J(\omega) = \frac{1}{N} \sum_k \delta(\omega - \omega_k) \frac{\tilde{\gamma}_k^2}{\omega_k}$$

and rewrite

$$K(t) = \int_0^\infty d\omega \, \frac{J(\omega)}{\omega} \cos(\omega t).$$

Environments are classified on the basis of the low-frequency behavior of $J(\omega) \simeq \eta \omega^s$: an Ohmic bath corresponds to $s = 1$, whereas a sub(super)-Ohmic one corresponds to $s < 1 (s > 1)$ [65]. Note that a purely Ohmic bath leads to $K(t) = \eta \delta(t)$. Of course, in reality, there is always a change of behavior at high frequency, and hence the delta function acquires a finite width.

Before concluding this section on the role of the environment, let us stress that the previous procedure of integrating out the environment can be repeated in the quantum case as well. Indeed, it is at the root of first-principles descriptions of quantum open systems: examples are small quantum systems coupled to phonons, quantum dots coupled to noninteracting leads, etc.

We postpone a complete discussion of quantum reservoirs to the following sections. However, we anticipate that it is not possible in general to obtain a stochastic equation as one can do for classical systems. In the literature, there are quantum versions of the Langevin equation (3.9) that are identical to the classical one except for the fact that the dissipation and fluctuations are related by the *quantum* fluctuation–dissipation relation. These are approximations valid only for small quantum fluctuations, as we shall explain later on. For more discussion and details on open systems and the role of reservoirs, see [23, 32, 65, 69].

3.2   Field theory, time-reversal symmetry, and fluctuation theorems

In the following, we present a brief introduction to and derivation of the field theories used to study off-equilibrium dynamics in classical and quantum systems. We then go on to study a particularly important symmetry for equilibrium systems: time reversal. Out-of-equilibrium systems violate this symmetry, but, quite interestingly, this violation leads to a variety of interesting consequences that go under the name of fluctuation theorems and that have received a lot of attention recently. We shall discuss them and conclude the section by addressing generalizations to the quantum case.

3.2.1   Martin–Siggia–Rose–DeDominicis–Janssen field theory

We consider an open classical system whose dynamics is governed by a stochastic equation. For the sake of concreteness and for simplicity, we focus on a simplified version of the Langevin equation (3.9):

$$\eta \dot{x}(t) = -V'(x(t)) + \xi(t), \qquad \langle \xi(t)\xi(t') \rangle = 2T\eta\delta(t - t'). \tag{3.10}$$

We have neglected inertia and memory effects here. This is reasonable in certain physical systems such as colloids in which the particles of the system (colloidal particles) are much bigger and heavier (and hence much slower) than the particles forming the environment (solvent molecules). We present the derivation in this case, but also discuss the results we would have obtained in more general cases at the end. Note that we still focus on one-dimensional systems just to keep the notation as least involved as possible (to this end, we also set $\eta = 1$, i.e., we reabsorb η by redefining the unit of time). It is straightforward to generalize the results to higher dimensions.

Studying the dynamics of a system means computing (or measuring) averages, correlations, and response functions. Suppose that we want to compute the average value of the observable $O(x(t))$, where $O(x)$ is a generic function of x:

$$\langle O(x(t)) \rangle = \int D[\xi]\, dx_0\, O(x_\xi(t)) \exp\left[-\frac{1}{4T} \int_0^t \xi^2(t')\, dt' \right] P(x_0). \tag{3.11}$$

What the functional integration here means in words is that we have to solve (3.10) for a given $\xi(t)$ and initial condition x_0, compute $O(x(t))$, and then average the result over all possible Gaussian thermal noises and initial conditions.[4] Formally, we can insert in the integral in (3.11) a useful representation of unity by integrating over all paths $x(t)$ the Dirac δ-functional that imposes (3.10):

$$\int_{x(0)=x_0} D[x]\, \delta[\dot{x} + V'(x) - \xi] =$$

$$\int_{x(0)=x_0} D[x]\, D[\hat{x}] \exp\left\{ -\int_0^t dt'\, \hat{x}(t')\, [\dot{x}(t') + V'(x(t')) - \xi(t')] \right\} = 1.$$

---

[4] Note that the average in (3.11) is the ensemble average (cf. Section 3.1.3). Henceforth, we shall neglect the subscript E to lighten the notation.

Actually, a Jacobian should be also present, but since it can be proved to be a constant, we just absorb it in the normalization of the functional integral.[5] This insertion is useful because it allows us to integrate out the noise—just a functional Gaussian integral—and obtain the famous Martin–Siggia–Rose–DeDominicis–Janssen (MSRDJ) field theory:

$$\langle O(x(t)) \rangle = \int D[x]\, D[\hat{x}]\, dx_0\, O(x(t)) \exp(S)\, P(x_0), \tag{3.12}$$

where the action S reads

$$S = -\int_0^t dt'\, \hat{x}(t')\, [\dot{x}(t') + V'(x(t')) - T\hat{x}(t')]. \tag{3.13}$$

Using this field theory, we can also compute correlation functions $\langle O(x(t_1))O(x(t_2)) \rangle$ and response functions as discussed below.

3.2.2 Meaning of the fields

The MSRDJ field theory contains two fields: $x(t)$ and $\hat{x}(t)$. The meaning of the former is clear: it represents the fluctuating degrees of freedom of the system. The latter is called the response field (see [21]), because it intervenes in response functions. To see how this works, let us add an external field coupled to $x(t)$: $V(x) \to V(x) - hx$. By repeating the previous derivation, we find that the action gets an extra term:

$$S = -\int_0^t dt'\, \hat{x}(t')\, [\dot{x}(t') + V'(x(t')) - T\hat{x}(t')] + \int_0^t dt'\, \hat{x}(t')h(t'). \tag{3.14}$$

The response function measuring the variation of the average of $x(t)$ due to an infinitesimal and instantaneous external field acting at time t' can be expressed in terms of a correlation function of the fields x and $\hat{x}$:

$$R(t, t') = \left. \frac{\delta\langle x(t) \rangle}{\delta h(t')} \right|_{h=0} = \langle x(t)\hat{x}(t') \rangle.$$

Note that from this identity, we can immediately obtain that the average of $\hat{x}$ is zero, since the average of 1 does not change if we apply a field![6] Similarly, we can show that correlation functions with only $\hat{x}$ or in which the field with largest time is a response field vanish by causality.

---

[5] This is actually a tricky business; see, e.g., [21, 42, 67]. In order to compute the Jacobian, one has to specify the discretization of the Langevin equation. We choose the Itô discretization $x_{t+\Delta t} - x_t = [-V'(x_t) + \xi_t]\Delta t$. In this case, it is easy to show that the Jacobian is indeed a constant. Other discretizations lead to different results. As long as the noise is not multiplicative, different discretizations lead to different field theories, but physical averages are the same. In the case of multiplicative noise, instead different discretizations correspond to different physical processes.

[6] We have $\langle \hat{x} \rangle = \left. \frac{\delta\langle 1 \rangle}{\delta h(t')} \right|_{h=0} = 0.$

3.2.3 Generalizations

We now report the resulting MSRDJ field theory that would have been obtained had we started from the full Langevin equation (3.9). The derivation, which is a straightforward generalization of the previous one, is left as an exercise for the reader. We still end up with a field theory on $x, \hat{x}$; the difference is in the form of the action, which reads

$$
S = - \int_0^\infty dt' \, \hat{x}(t') \left[m\ddot{x}(t') + V'(x(t')) \right]
$$
$$
- \int_0^\infty dt \int_0^t dt' \, \hat{x}(t)\eta(t - t')\dot{x}(t') + \frac{1}{2} \int_0^\infty dt \int_0^t dt' \, \hat{x}(t)\nu(t - t')\hat{x}(t'), \quad (3.15)
$$

where $\eta(t - t') = K(t - t')$ is the retarded friction and $\nu(t - t') = TK(t - t')$ is the variance of the noise. Because of the inertia, we now have to specify both $x(0)$ and $\dot{x}(0)$ as initial conditions. For more details, see [3, 42, 45].

3.2.4 Relationship with Schwinger–Keldysh field theory

We now want to obtain a relationship between the MSRDJ field theory used to study the dynamics of classical open systems and the Schwinger–Keldysh field theory used for open quantum systems. As we have shown in Section 3.1, the Langevin equation that we used as a starting point to derive MSRDJ is obtained starting from a sub-system coupled to an environment, the combination of the two being a closed system characterized by the Hamiltonian (3.4). The starting point to derive the Schwinger–Keldysh field theory is to write down a path integral representation for the dynamics of the subsystem coupled to the reservoir. Then, at the level of the functional integral, we can integrate out the reservoir degrees of freedom and obtain an action for the subsystem only. This is shown by Berges in Chapter 2 of this volume, so we shall not repeat it here; see also [42]. We will only quote the main results.

In the closed-contour representation of the functional integral, we have two fields $x^+(t)$ and $x^-(t)$ corresponding to the two (forward and backward) time paths. In order to make the relationship with MSRDJ explicit, it is useful to make a change of variables and define

$$
x(t) = \frac{x^+(t) + x^-(t)}{2}, \qquad \hat{x}(t) = i\frac{x^+(t) - x^-(t)}{\hbar}
$$

In the literature, these are also called classical and quantum fields, respectively. In terms of these fields, the action of the Schwinger–Keldysh field theory can be written as

$$
S = \frac{i}{\hbar} \left(\mathcal{L}[x - i\hbar\hat{x}/2] - \mathcal{L}[x + i\hbar\hat{x}/2] \right)
$$
$$
+ \int_0^\infty dt \int_0^t dt' \, \hat{x}(t)\eta(t - t')\dot{x}(t') + \frac{1}{2} \int_0^\infty dt \int_0^t dt' \, \hat{x}(t)\nu(t - t')\hat{x}(t'), \quad (3.16)
$$

where

$$\mathcal{L}[x] = \int_0^t m \frac{\dot{x}^2}{2} - V(x)$$

is the Lagrangian of the subsystem and $\eta(t - t')$, which is equal to $K(t - t')$ as in MSRDJ, has the interpretation of a retarded friction; the other term, $\nu(t - t')$, plays the role of the variance of the thermal noise and is related to η by the quantum fluctuation–dissipation theorem [65]. In Fourier space, their relationship reads

$$\nu(\omega) = \hbar \coth\left(\frac{\hbar\omega}{2T}\right) \text{Im}\{\eta(\omega)\} \tag{3.17}$$

The properties of the environment are all encoded in the specific form of $\eta(\omega)$. It is easy to check that MSRDJ field theory defined by the action (3.9) is the classical limit of the Schwinger–Keldysh field theory. Indeed, by taking the limit $\hbar \to 0$, the quantum fluctuation–dissipation relation becomes the classical one, i.e., $\nu(\tau) \to TK(\tau)$, so that the second part of the action (3.16) coincides with the corresponding part of MSRDJ. Concerning the first part, by taking the limit $\hbar \to 0$, only the terms linear in $\hat{x}$ survive and we indeed find

$$\lim_{\hbar \to 0} \frac{i}{\hbar} \left(\mathcal{L}[x - i\hbar\hat{x}/2] - \mathcal{L}[x + i\hbar\hat{x}/2] \right) = -\int_0^\infty dt' \, \hat{x}(t') \left[m\ddot{x}(t') + V'(x(t')) \right].$$

Thus, the quantum nature of the Schwinger–Keldysh field theory is encoded in (1) the relationship between η and ν and (2) the existence of terms with possibly all-odd powers of $\hat{x}$. In general, we cannot interpret the action (3.16) as originating from a physical stochastic equation. However, if we disregard all odd terms beyond the linear one, then the corresponding field theory is MSRDJ-like for a Langevin equation (3.9) with a noise related to the friction via the quantum dissipation relation. This stochastic equation is the so-called quantum Langevin equation. It corresponds to the first term in the expansion in the quantum field $\hbar\hat{x}$ and hence is valid only in the regime of small quantum fluctuations.

3.2.5   Time-reversal symmetry

Time-reversal is *the symmetry* that characterizes an equilibrium system: being at thermodynamic equilibrium means being unable to determine the arrow of time. In the following, we shall show how this symmetry arises within the MSRDJ formalism. Consider a trajectory $x(t)$ starting from x_0 at time $t = 0$ and reaching x_f at time τ. The time-reversed trajectory of $x(t)$ is defined as $x_R(t) = x(\tau - t)$. Although the response field is not a physically measurable field, it is useful to also define its time-reversed counterpart

$$\hat{x}_R(t) = \hat{x}(\tau - t) + \frac{1}{T} \frac{dx(\tau - t)}{d(t)}.$$

Let us now study how the action transforms under time reversal. First, we shall rewrite it as

$$S[x(t), \hat{x}(t)] = -\int_0^t dt' \, \hat{x}(t') \left[\dot{x}(t') + V'(x(t')) - T\hat{x}(t') \right]$$

$$= -\int_0^t dt' \, \hat{x}(t') V'(x(t')) + T \left[\hat{x}(t') - \frac{\dot{x}^2}{2T} \right]^2 + \frac{\dot{x}^2}{4T}. \qquad (3.18)$$

The second and third terms are clearly invariant under time reversal, whereas the first term leads to an extra contribution:

$$S[x_R(t), \hat{x}_R(t)] = S[x(t), \hat{x}(t)] + \frac{1}{T} \int_0^T \dot{x}(t') V'(x(t')) \, dt'.$$

The probability to measure the time-reversed trajectory for the system, given the initial condition x_f at time 0, reads

$$P\big[x_R(t)\big|x_R(0) = x_f\big] = \int D[\hat{x}_R] \exp\{S[x_R(t), \hat{x}_R(t)]\}.$$

By applying the time-reversal transformation, which amounts just to a functional change of variables (whose Jacobian is one), we find

$$P\big[x_R(t)\big|x_R(0) = x_f\big] = \int D[\hat{x}] \exp\left\{ S[x(t), \hat{x}(t)] + \frac{1}{T} \int_0^T \dot{x}(t') V'(x(t')) \, dt' \right\}.$$

By separating the two terms in the exponential, performing the integration over time for the second, and formally integrating over $\hat{x}$, we find

$$P\big[x_R(t)\big|x_R(0) = x_f\big] = P\big[x(t)\big|x(0) = x_0\big] \exp\left[\frac{V(x_f) - V(x_0)}{T} \right]. \qquad (3.19)$$

By dividing both terms by the partition function $Z = \int dx \exp[V(x)/T]$, bringing $\exp[V(x_f)/T]$ onto the right-hand side, and recalling that the equilibrium probability measure is $P_{\text{eq}}(x) = (1/Z) \exp[-V(x)/T]$, we finally find the expression of time-reversal symmetry:

$$P\big[x_R(t)\big|x_R(0) = x_f\big] P_{\text{eq}}(x_f) = P\big[x(t)\big|x(0) = x_0\big] P_{\text{eq}}(x_0),$$

which simply means that the probability of starting from x_0 and following a given path to x_f coincides with the probability of starting from x_f in equilibrium and following the time-reversed path. This is a general property of equilibrium dynamics. We have shown it for Langevin dynamics, i.e., a classical system coupled to a bath, but it is generically valid for all physical equilibrium dynamics, in particular Newton and Schrödinger evolutions.

A general principle in physics is that the existence of symmetries implies non-trivial identities for correlation functions and observables, the most famous example being momentum and angular momentum conservations laws and invariance with respect to translations and rotations. We thus expect that time-reversal symmetry leads to important consequences. Indeed, three fundamental relationships that are generically taken as fundamental signature of equilibrium dynamics can be derived from it: time-reversal symmetry of correlation functions, fluctuation–dissipation relations, and Onsager reciprocity relations.

Consequences of time-reversal symmetry

- *Time-reversal symmetry of correlation functions*:

$$\langle O_1(t_1) \cdots O_n(t_n) \rangle = \langle O_1(-t_1) \cdots O_n(-t_n) \rangle,$$

  where $O_i(t_i)$ are observables at time t_i. This is a direct implication of the fact that it is not possible to identify microscopically an arrow of time: the direct and time-reversed paths have the same probability.
- *Fluctuation–dissipation relations*:

$$R(t - t') = \frac{1}{T} \partial_{t'} C(t - t'),$$

  where $C(t - t') = \langle x(t)x(t') \rangle$ and R is the response of $\langle x(t) \rangle$ to a field applied at time t'.
- *Onsager reciprocity relations*:

$$R_{AB}(t - t') = R_{BA}(t - t'),$$

  where R_{AB} denotes the response of $\langle A(t) \rangle$ when a field conjugated to the observable B is applied at time t'.

These three identities can be easily proved following the procedure outlined above and performing the change of path in the functional integral. They are identities between correlation functions implied by time-reversal symmetry (formally, the response function is a correlation function between the physical and response fields). Their proof is left as an exercise to the reader. Note that we also need to use the fact that the system is in a steady state, i.e., it is time-translation-invariant, which means that it is indeed at equilibrium (the only steady state possible for a finite system coupled to an equilibrium reservoir is equilibrium dynamics, as we shall discuss in the following sections).

3.2.6 Time-reversal symmetry breaking and fluctuation relations out of equilibrium

In the last 15 years it has been realized that studying the way in which time-reversal symmetry is broken out of equilibrium leads to new kinds of fluctuation relations. These are valid out of equilibrium and belong to what is now called the field of stochastic thermodynamics (see, e.g., [59]). This topic deserves a series of lectures on its own, and a detailed discussion is certainly outside the scope of this chapter. On the other hand, we cannot avoid mentioning it. With this in mind in the following, we shall just consider one of the most well-known relations, discuss its physical origin in a nutshell, and then point out complete reviews for interested readers. Like the identities discussed before at equilibrium, these out-of-equilibrium relations hold for all types of physical dynamics. Our analysis will be performed using Langevin dynamics.

As the setting for out-of-equilibrium dynamics, we choose

$$\dot{x}(t) = F(t) + \xi(t), \qquad \langle \xi(t)\xi(t') \rangle = 2T\delta(t - t'),$$

where $F(t)$ is a generic time-dependent force that we shall write in the form $F(t) = -V'(x(t), \lambda(t)) + f(t)$. This representation means that the system is out of equilibrium for two reasons: a control parameter of the external potential is changed during the dynamics and moreover there is a nonconservative force applied to the system.[7] This is a very simple model, but it contains the essence of the problem (treatments for more complicated models are straightforward generalizations). Repeating the derivation that lead to (3.19), we now find

$$P\big[x_R(t)\big|x_R(0) = x_f\big] = P\big[x(t)\big|x(0) = x_0\big] \exp\left(- \int_0^\tau dt \, \frac{\dot{x}F}{T} \right). \qquad (3.20)$$

The term $\int_0^\tau dt \, \dot{x}F$ has the interpretation of the heat exchanged between the system and the bath [59]. To see where this interpretation comes from, remember the first law of thermodynamics (applied in a small interval of time dt): $dw = dV + dq$, the work done on the system is equal to the change in energy plus the heat exchanged. The work done on the system reads, from the mechanical point of view, $dw = (\partial V/\partial\lambda)\dot{\lambda}\, dt + f\, dx$. Using the chain rule $dV = (\partial V/\partial\lambda)\dot{\lambda}\, dt + (\partial V/\partial x)\, dx$, we can eliminate dw from the two previous equations to get $dq = (-\partial V/\partial x + f)\, dx = F\, dx$.

From the relationship (3.20), all fluctuation relations valid out of equilibrium can be derived straightforwardly. In the following, we show how this works in a specific case and derive the Jarzynski–Crooks identities.

Consider the specific out-of-equilibrium protocol where the system is at equilibrium at $t = 0$ and is brought out of equilibrium by the changing of the external parameter $\lambda(t)$. We take $f = 0$, i.e., $F = -V'(x(t), \lambda(t))$. The term in the exponential of (3.20) can be rewritten in this specific case as

$$-\frac{1}{T} \int_0^\tau dt \, \dot{x}F = \int_0^\tau dt \left(\frac{dV}{dt} - \frac{\partial V}{\partial\lambda}\dot{\lambda} \right)$$

$$= \frac{V(x_f, \lambda(\tau)) - V(x_0, \lambda(0))}{T} - \frac{W}{T},$$

---

[7] The force is nonconservative because it depends explicitly on time and, moreover, it is generically assumed not to be derived from a potential (for this latter statement, we need to consider spatial dimensions higher than one, since in one dimension a generic function of x can always be written as the derivative of a potential).

where $W = \int_0^\tau (\partial V/\partial \lambda)\dot{\lambda}\, dt$ is the work performed on the system between times 0 and τ. Note that W is a stochastic quantity since it is a functional of the dynamical trajectory. Multiplying (3.20) by the Boltzmann measure that the system would have reached had it been equilibrated at τ, i.e., $(1/Z_f)\exp[-V(x_f, \lambda(\tau))/T]$, and multiplying the right-hand side by Z_i/Z_i, we finally get the equation

$$P\big[x_R(t)\big|x_R(0) = x_f\big]\,P_{\text{eq}}(x_f)$$
$$= P\big[x(t)\big|x(0) = x_0\big]\,P_{\text{eq}}(x_0)\exp\left(-\frac{W}{T} + \frac{\Delta F}{T}\right), \qquad (3.21)$$

where $\Delta F = -\log(Z_f/Z_i)/\beta$ is the difference in the free energies corresponding to the two equilibrium states characterized by values of λ equal to $\lambda(\tau)$ and $\lambda(0)$.

Let us now integrate (3.21) over all paths. This leads to the celebrated Jarzynski identity [40, 41]

$$\langle e^{-W/T}\rangle = e^{-\Delta F/T}, \qquad (3.22)$$

which is remarkable since it express the average over an out-of-equilibrium process (the left-hand side) in terms of a pure equilibrium quantity (the right-hand side). This is at first sight a surprising result, which even led to debates early on, since on the left-hand side there is an average over a completely out-of-equilibrium process, whereas on the right-hand side we find only the equilibrium free energies for $\lambda = \lambda(\tau)$ and $\lambda = \lambda(0)$. Using Jensen's inequality, we derive that $\langle W\rangle \geq \Delta F$, which is an expression of the well-known Clausius inequality of classical thermodynamics: the dissipated part of the work, $\langle W\rangle - \Delta F$, has to be positive or equal to zero. This is one example of a main result of stochastic thermodynamics: a microscopic statistical mechanics identity, (3.22), from which macroscopic ones characteristic of classical thermodynamic can be derived.

Another interesting relation can be obtained by integrating (3.21) over all paths that correspond to a given value of the work W. This leads to the Crooks identity [24]:

$$P_R(-W) = P(W)e^{-W/T + \Delta F/T}, \qquad (3.23)$$

where $P(W)$ is the probability to observe the work W during the nonequilibrium process and P_R is its time-reversed counterpart. It is interesting to consider a cyclic process, which means that $\lambda(0) = \lambda(\tau)$. In this case, the identity (3.23) is a constraint on the form of $P(W)$:

$$P(-W) = P(W)e^{-W/T}. \qquad (3.24)$$

If the process is adiabatic, then $\langle W\rangle = \Delta F = 0$, i.e., no work is done on the system globally. On the other hand, a typical out-of-equilibrium process leads to dissipation and to an average value $\langle W\rangle = W_{\text{diss}} > 0$. Equation (3.24) implies that even though $P(W)$ is centered around $W_{\text{diss}} > 0$, it necessarily has tails extending to negative values: the probability of measuring a *negative* work in a cycle—an apparent violation of thermodynamics—is finite and given by $\int_0^\infty dW\, P(W)e^{-W/T}$.

This example contains the essence of stochastic thermodynamics and its main consequences. First, it allows us to understand, and actually derive, the laws of thermodynamics as identities on averages. Second, it also shows that because for large systems quantities like $W, \Delta F, \ldots$ are extensive, i.e., they scale as the number of degrees of freedom, the exponential weights in these out-of-equilibrium identities are extremely small. Thus, for large systems, the probability of observing violations of the laws of thermodynamics is extremely small. It vanishes in the thermodynamic limit as expected, which is a very nice way to show the emergence of thermodynamics as a theory of macroscopic systems. The previous observation also implies that these fluctuation identities are in practice meaningful only for small systems, i.e., in situations where the terms in the exponential are not too negative and hence the probability of "violations" of thermodynamic laws is not too low. There are several experimental systems where this is the case, and these laws have been tested and/or usefully applied, examples include colloidal particles, proteins, among others. We refer to the review by Seifert [59] for a comprehensive introduction to this field.

3.2.7 Quantum systems

Time-reversal symmetry and its implications discussed in Sections 3.2.5 and 3.2.6 also hold for quantum systems, even though the derivations are sometimes more involved.

In the case of equilibrium quantum dynamics, time-reversal symmetry implies, as in the classical case, time-reversal symmetry of correlation functions, fluctuation–dissipation relations, and Onsager reciprocity relations, the only difference being the form of the fluctuation–dissipation relations.

In the case of out-of-equilibrium quantum dynamics, identities analogous to those obtained for classical system can be derived. Sometimes, issues related to process measurements in quantum mechanics arise. This makes the quantum version a bit more subtle.

We discuss, as an example, the Jarzynski–Crooks equalities, for which the issue of providing a correct definition of work arises. The easiest derivation is obtained for an isolated system, so we focus on this case to keep things as simple as possible. In the classical case, since the system is closed and does not exchange heat, the work is simply the difference between the final and the initial energy. In the quantum case, one has to take into account the measurement process in order to recover the Jarzynski–Crooks equalities [30, 45]. The correct measurement protocol is defined as follows: we start at $t = 0$ from the Boltzmann density matrix, measure the energy E_i, do the nonequilibrium process, and measure the energy again E_f at the end. The measurements lead to a collapse of the wavefunction, and hence at the beginning of the process the system is in an eigenstate of the initial Hamiltonian $H(\lambda(0))$ and at the end in an an eigenstate of the final Hamiltonian $H(\lambda(\tau))$. With this in mind, it is easy to write down the probability distribution of the work:

$$P(W) = \sum_{n,m} \frac{e^{-E_n/T}}{Z_i} \left| \langle m | \mathsf{T} \exp\left[-\frac{i}{\hbar} \int_0^\tau H(s)\, ds\right] | n \rangle \right|^2 \delta(W - (E_m - E_n)), \quad (3.25)$$

where T denotes time-ordering. The probability for the reversed process reads

$$P_R(W) = \sum_{n,m} \frac{e^{-E_m/T}}{Z_f} \left| \langle n | \tilde{\mathsf{T}} \exp\left[-\frac{i}{\hbar} \int_\tau^0 H(s)\,ds \right] |m\rangle \right|^2 \delta(W - (E_n - E_m)), \quad (3.26)$$

where $\tilde{\mathsf{T}}$ denotes reverse time-ordering. By interchanging the indices n, m in (3.26) and noticing that the amplitude-squared term is the same in (3.25) and (3.26), we recover the Crooks identity that we derived previously in the classical case.

As discussed for classical systems, fluctuation relations out of equilibrium are useful for small systems. In the context of quantum systems, the recent interest in nanodevices has provided a natural framework for further study and development of these new relations; see [30] for a comprehensive review on the subject.

3.3 Thermalization

In this section, we shall discuss in some detail the issue of thermalization, which is the process by which a system initially out of equilibrium relaxes to the thermal Maxwell–Boltzmann distribution, the cornerstone of equilibrium statistical mechanics. We shall discuss first systems coupled to a bath and then focus on the more difficult problem of isolated systems. Understanding how and why physical systems thermalize is a fundamental problem that is at the root of statistical physics. Note that not all systems do it, and actually several physical systems are found to be out of equilibrium. Some of them are driven out of equilibrium by shear, currents, dissipation, etc., and hence they cannot equilibrate; others instead just fail to equilibrate, at least on experimental timescales. Before discussing these latter cases in Section 3.4, we now explain why equilibrating is more the rule than the exception, which is also why avoiding equilibration is so interesting and puzzling.

3.3.1 The meaning of thermalization

A macroscopic system, classical or quantum, is said to have reached thermalization if all the time-averaged local observables and correlation functions coincide with those obtained by the ensemble average over the Maxwell–Boltzmann distribution:

$$\lim_{\tau \to \infty} \frac{1}{\tau} \int_0^\tau O(t)\,dt = \langle O \rangle_T, \quad (3.27)$$

where O denotes a local observable, which means an observable whose value depends on degrees of freedom that are locally close in real space: for example, products of spin operators contained in a finite subsystem or products of local densities in a liquid.[8] The average on the right-hand side of (3.27) is the usual ensemble average of statistical mechanics: it is performed using the Maxwell–Boltzmann distribution at temperature T. The value of T is fixed by the dynamical evolution: for an open system, it is

---

[8] In the quantum case, $O(t)$ is the expectation value under Schrödinger evolution.

given by the temperature of the bath, for an isolated system, it is the value of the microcanonical temperature leading to an average intensive energy equal to that of the system (which is conserved during the dynamics since the system is isolated).[9] The reason for insisting that observables have to be local will be discussed at the end of this section. It is important for macroscopic systems only.

What makes thermalization mind-boggling at first sight is that the left-hand side of (3.27) depends on a lot of things, in particular on the initial condition and on the bath evolution if the system is open. On the right-hand side, instead, the only parameter that matters is the temperature of the bath, or the initial energy if the system is isolated: all the extra knowledge (or information) contained on the left-hand side has to be lost irreversibly during the dynamical evolution. This is the reason why equilibrium statistical mechanics is so powerful. We need to know almost nothing about the past history of the system to predict its equilibrium properties.

As anticipated, thermalization is expected to be the rule for macroscopic systems. However, proving that a system does thermalize can be a tough problem, especially for isolated systems. In the following, we shall first discuss the case of open classical systems, which is the easiest case.

3.3.2   Classical systems coupled to a bath

Showing thermalization for classical systems coupled to a bath is the easiest case, and for this reason we shall discuss it in detail in the following. As we have shown in Section 3.1, the *deterministic* dynamics of a classical system coupled to a bath reduces to a *stochastic* damped dynamics for the system alone. Within this framework, time averages coincide with averages over the stochastic noise. Thus, proving thermalization means showing that at long times, averages over the stochastic noise converge to Maxwell–Boltzmann averages:

$$\lim_{t\to\infty} \langle O(t) \rangle = \langle O \rangle_T.$$

We shall consider finite systems, so there is no difference between local and nonlocal observables. For simplicity, we focus again on a one-dimensional system undergoing overdamped Langevin dynamics:

$$\frac{dx}{dt} = -V'(x) + \xi(t),$$

$$\langle \xi(t)\xi(t') \rangle = 2T\delta(t - t'),$$

(3.28)

where the friction η has been reabsorbed in a redefinition of the unit of time and $V(x)$ is an external potential. Although this is the simplest setting one can choose, it is very instructive and not so reductive after all, since all other cases can be treated technically

---

[9] We assume that there are no conserved quantities other than the energy. If this is not the case, then we have to add more control parameters conjugated to the extra conserved variables, e.g., the pressure for the volume.

as simple generalizations. Indeed, introducing inertia, considering nonwhite noise, and considering systems with more than one degree of freedom lead to more complicated stochastic equations that can always be rewritten as sets of coupled Langevin equations by introducing extra variables. For example, in the case of inertia, we introduce the momentum $p = \dot{x}$ and write the second-order stochastic equation as two first-order coupled Langevin equations of the type described above.

In the classical case, an observable is a function of the configuration of the system; in the simple case we are considering here, it is just a function of x. Introducing $P(x,t)$, the probability distribution of x at time t, thermalization can be expressed as follows:

$$\lim_{t\to\infty} \int dx\, P(x,t)O(x) = \int dx\, \frac{e^{-V(x)/T}}{Z} O(x) \qquad \forall O(x),$$

and since this has to hold for any function $O(x)$, it implies that

$$\lim_{t\to\infty} P(x,t) = \frac{e^{-V(x)/T}}{Z},$$

where Z is the partition function.

The way to proceed to show that this is indeed what happens consists in deriving an equation for the evolution of $P(x,t)$. This is quite standard and can be found in several textbooks (e.g., [42, 45]). The starting point is the identity

$$\frac{d}{dt}\langle O(x(t))\rangle = \int dx\, \partial_t P(x,t)O(x) \qquad \forall O(x).$$

The left-hand side can be rewritten in a discretized way as

$$\frac{1}{dt}\left[\left\langle \frac{dO}{dx}\dot{x}\, dt + \frac{1}{2}\frac{d^2 O}{dx^2}(\dot{x}\, dt)^2 \right\rangle + \cdots \right]. \tag{3.29}$$

In order to go further and understand why the second term has been retained, we have to specify the discrete form of the Langevin equation. We choose the so-called Itô prescription,[10] which corresponds to the discretization

$$\frac{x(t+dt) - x(t)}{dt} = -V'(x) + \xi(t),$$

$$\langle \xi(t)\xi(t')\rangle = 2T\frac{\delta_{t,t'}}{dt}, \tag{3.30}$$

---

[10] Another prescription would lead to the same results but a different derivation. Strictly speaking, the stochastic equation we have written is ill defined in the continuum limit, and the only correct mathematical way is to define it through its integrated version.

where $\delta_{t,t'}$ is the Kronecker delta. Now that everything is well defined, we can evaluate the terms obtained in (3.29):

$$\left\langle \frac{dO}{dx}\dot{x} \right\rangle = \left\langle \frac{dO}{dx}[-V'(x) + \xi(t)] \right\rangle$$

$$= \left\langle \frac{dO}{dx}[-V'(x)] \right\rangle$$

$$= -\int P(x,t)\frac{dO}{dx}V'(x)\,dx.$$

In the first equality, we have used the expression of the Langevin equation and in the second the fact that $x(t)$ is uncorrelated from the noise at the same time t (see (3.30)). For the second term in (3.29),

$$\frac{1}{2}\left\langle \frac{d^2O}{dx^2}(\dot{x}\,dt)^2 \right\rangle = \frac{1}{2}\left\langle \frac{d^2O}{dx^2}[-V'(x) + \xi(t)]^2 \right\rangle,$$

only the term containing the square of the thermal noise is of order dt. It is equal to

$$\frac{1}{2}\left\langle \frac{d^2O}{dx^2}\xi(t)^2 \right\rangle = T\,dt\left\langle \frac{d^2O}{dx^2} \right\rangle = T\,dt\int P(x,t)\frac{d^2O}{dx^2}\,dx,$$

where we have used the same tricks as before. Collecting all the pieces together and dividing by dt, we finally reach the equation

$$\int \partial_t P(x,t)O(x)\,dx = \int O(x)\frac{d}{dx}[V(x)P(x,t)]\,dx + \int O(x)T\frac{d^2P(x,t)}{dx^2}\,dx.$$

Since this has to be true for any function $O(x)$, the distribution $P(x,t)$ satisfies the so-called Fokker–Planck equation

$$\partial_t P(x,t) = \frac{d}{dx}\left[V'(x) + T\frac{d}{dx}\right]P(x,t). \qquad (3.31)$$

On examining (3.31), two observations immediately come to mind. First, it is clear that the Maxwell–Boltzmann distribution is stationary, since the right-hand side vanishes for $P(x,t) = e^{-V(x)/T}/Z$. Second, it looks very much like an imaginary-time Schrödinger equation

$$\partial_t P(x,t) = -H_{FP}P(x,t),$$

with

$$H_{FP} = -\frac{d}{dx}\left[V'(x) + T\frac{d}{dx}\right].$$

However, the analogy is not as straightforward as it seems at first sight, since thermal averages cannot be written as quantum averages. In fact, introducing the bra–ket notation, formally solving the Fokker–Planck equation, and denoting by $\langle 1|$ the bra corresponding to the function equal to one for all x, i.e., $\langle 1|x\rangle = 1$, we get

$$\langle O(x(t))\rangle = \langle 1|Oe^{-H_{FP}t}|P_0\rangle = \langle 1|e^{H_{FP}t}Oe^{-H_{FP}t}|P_0\rangle, \tag{3.32}$$

where we have explicitly used the fact that $\langle 1|H_{FP} = 0$ and $|P_0\rangle$ represents the initial probability distribution. For a fixed initial condition x_0, we have $\langle x|P_0\rangle = \delta(x - x_0)$. The expression (3.32) differs from its quantum mechanical counterpart for three main reasons: (1) the time is imaginary, (2) H_{FP} is not a Hermitian operator, and (3) the bra $\langle 1|$ is not the transpose of the ket $|P_0\rangle$. Nevertheless the analogy is useful, as we discuss below. First, let us list some general properties of H_{FP}. Their derivation is easy and is left as an exercise for the reader.

Properties of H_{FP}

- H_{FP} is not Hermitian, but the operator $H_S = e^{V/2T}H_{FP}e^{-V/2T}$ is Hermitian, as can be checked by simple algebra.
- H_S is a positive operator; i.e., all its eigenvalues $\lambda_i \geq 0$.
- The eigenvalues of H_{FP} coincide with those of H_S and hence are all positive.
- Under a very general hypothesis, namely, that the potential $V(x)$ grows fast enough when $|x| \to \infty$, all but one of the λ_i are positive. The single eigenvalue $\lambda_1 = 0$ has the Boltzmann distribution as right eigenvector and $\langle 1|$ as left eigenvector.

These properties imply that the Maxwell–Boltzmann distribution is the only stationary distribution. Moreover, by decomposing the initial distribution $|P_0\rangle$ and using right and left eigenvectors of H_{FP} (which form a basis of the Hilbert space), $|P_0\rangle = \sum_i c_i|i\rangle_R$, where $c_i =_L \langle i|P_0\rangle$, we find

$$P(x,t) = \sum_i e^{-\lambda_i t}c_i\langle x|i\rangle_R.$$

Consequently, since all but one of the exponentials vanish at long times and $c_1 = 1$ by normalization, we have proved the convergence to the Maxwell–Boltzmann distribution for any initial condition:

$$P(x,t) \xrightarrow{t\to\infty} \frac{e^{-V/T}}{Z}.$$

Along the way, we have also found that the the eigenvalues of H_{FP} can be interpreted as inverse relaxation times. The time to reach equilibrium is given by $1/\lambda_2$.

Several other interesting properties emerge from this analogy with quantum mechanics for both classical and quantum systems. From the classical point of view, using spectral analysis of H_{FP} and WKB methods, it is possible to show that the right and left eigenvectors allow one to define very precisely and quantitatively metastable states and their basins of attraction [9, 14–16, 33–35]. From the quantum point of view, the analogy was fruitful in the construction of variational wavefunctions à la Jastrow [47] and also in the analysis of quantum spin liquids. In this context, quantum dimer models were shown to exhibit special points in their phase diagram, so-called Rokhsar–Kivelson points, at which the quantum Hamiltonian maps exactly to a Fokker–Planck operator H_{FP} [57], thus allowing several results on the ground-state properties to be obtained from knowledge of the corresponding classic stochastic dynamics [39].

3.3.3 Quantum systems coupled to a bath

Showing thermalization is more difficult for open quantum systems; however, general results have been obtained recently. One has to show in full generality that the Schwinger–Keldysh field theory for a quantum system coupled to a bath leads to thermalization at long times. Although this has been done in the classical limit for the MSRDJ field theory [37], up to now it has been only performed on specific, albeit quite general, settings in the quantum case[11] [29]. The arguments that have been used are too involved to be presented here, and we refer to the original publications [29, 37].

Despite these technical difficulties, from the physical point of view, there is no reason to believe that a finite quantum system coupled to a bath does not thermalize. Actually, equilibration is even more likely for a quantum system than for a classical one, because of quantum fluctuations. An instructive case is provided by the double-well potential in Fig. 3.1, which models a situation in which there is a metastable state that can trap the dynamics for some time. A classical system starting at $t = 0$ in the left well with small kinetic energy needs to receive energy from the bath in order to

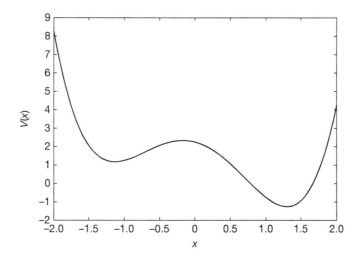

Fig. 3.1 Double-well potential.

---

[11] If one approximates the dynamics of a quantum open system using Lindblad operators, then proving thermalization is rather straightforward [32].

jump across the barrier and reach equilibrium. This is a less and less likely process the smaller is the temperature, and it leads to a lifetime of the metastable state that follows the Arrhenius law $e^{\Delta E/T}$, where ΔE is the height of the energy barrier that the system has to overcome (see, e.g., [45] for a derivation of this result). Strictly, at zero temperature, a classical system starting in the left well loses kinetic energy, which is absorbed by the bath through friction, and remains trapped forever in that well, never reaching the right well, which corresponds to the true ground state of the system at $T = 0$. Instead, in the quantum case, because of quantum tunneling, the system would never be trapped and eventually relax to equilibrium at $T = 0$.

Only if the two wells have exactly the same height could the quantum system avoid thermalization. Depending on the strength of the coupling to the bath and the spectral properties of the bath, localization in one well can occur at $T = 0$. This is the famous Caldeira–Leggett problem [46].

3.3.4   Isolated systems

The theory of thermalization of isolated systems is one of the most venerable parts of statistical mechanics. For classical systems, it is by now a textbook subject (good textbooks only though!), but for quantum systems, it is still a focus of current research. My aim in the following is certainly not to give a full presentation, which is well beyond the scope of this chapter. Instead, I will discuss some important points and to describe briefly the results on quantum systems.

3.3.5   Classical systems

In the case of classical systems, the theory of thermalization is intertwined with that of ergodic systems. The bottom line is that a system whose dynamics is chaotic enough (mixing) thermalizes. Roughly (very roughly), these two properties mean that if one starts with a distribution of initial conditions strongly peaked around a certain point $\mathbf{x}_0$ of configuration space, then, after a certain (equilibration) time, the distribution obtained by following the dynamical trajectories will be almost uniformly spread on the hypersurface corresponding to all points with an energy equal to $E(\mathbf{x}_0)$.

We shall not spend more time on ergodic theory and instead refer the interested reader to modern textbooks such as [23]. We would just like to mention two important points: first proving that a system is ergodic is extremely difficult and has been achieved in idealized cases only. On the other hand, it is thought that generically a macroscopic interacting system is ergodic. This actually is routinely observed in molecular dynamics simulations of interacting particle systems that thermalize rapidly. Integrable systems are an exception to this whole scenario. They are not ergodic, and, by Kolmogorov–Arnold–Moser (KAM) theory [23], perturbations of such systems are not ergodic if the perturbation is small enough. On the other hand, it is currently believed and has sometimes been shown that the KAM region corresponding to the values of the perturbation where the system remains nonergodic shrinks to zero in the thermodynamic limit; see, for example, the studies of the Fermi–Pasta–Ulam model [22]. In summary, all macroscopic interacting systems should thermalize, except for an ensemble of very finely tuned ones, say of measure zero in the space of models, which correspond to integrable systems. If a macroscopic system is found numerically or experimentally to avoid thermalization, this has to do with an underlying phase transition and the emergence of very long timescales, but certainly not with a loss of chaoticity.

One point that is often not stressed enough is that there is more to thermalization than just ergodic theory. The other very important aspect is to consider macroscopic systems and focus on special observables, in particular local observables because these are naturally what one looks for in simulations and experiments. Why is this important? Let us make an analogy with Monte Carlo dynamics. Imagine having N Ising spins that flip with a rate $\lambda = 1/\tau_0$ and starting the dynamics from the configuration with all spins up. This Monte Carlo dynamics is the one characteristic of $T = \infty$, since spins flip independently of the initial and final configurations. Hence, the Monte Carlo dynamics is just a random walk in the configuration space (which is a hypercube in N dimensions). It is easy to show (exercise!) that it thermalizes toward the flat distribution, which, as expected, corresponds to the Maxwell–Boltzmann distribution for $T = \infty$ ($e^{-\beta H} = 1$). Spins flip independently with a rate $1/\tau_0$, and hence the time averages of local correlation functions or observables converge to their thermalized values on a timescale τ_0, which is therefore the relaxation time. On the other hand, on the timescale τ_0, all spins have flipped a finite number of time in a given run, so the system has visited a number of configuration that is of the order of N only, i.e., much much smaller than the 2^N configurations available. How then is it possible that the system seems thermalized as if the measure were flat in configuration space and at the same time the number of configurations visited is extremely small? The way out, sketched pictorially in Fig. 3.2, is that typical dynamical trajectories visit the configuration space in such a way that for the observables on which we usually focus, time averages are essentially identical to those obtained by the Maxwell–Boltzmann measure: the random walk very quickly spreads over the configuration space. Observables local in real space cannot probe the granularity of this spreading, and, as far as they are concerned, the distribution obtained after a timescale τ_0 is identical to the flat one, even though this is clearly not the case. For instance, imagine that we focus on the observable equal to one if and only if the configuration is that with all spins

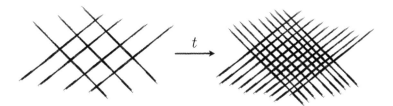

Fig. 3.2 Sketch of the progressive blurring of the probability measure over time. The usual physical observables are unable to probe the granularity of the measure, whereas unphysical ones such as the characteristic function corresponding to being in one given many-body configuration do.

down. This is clearly not a local observable: in fact, it corresponds to an N-point correlation function. Its thermalized value is $1/2^N$. In order to reach thermalization for this observable, one has to wait a time long enough to be sure that this configuration is encountered at least once during the dynamics. An estimate of this time—more a lower bound actually—is given by τ_0 times the number of configurations available divided by the number of configurations visited in a time τ_0. The resulting timescale is $\tau_0 2^N/N \gg \tau_0$, which is much larger than the relaxation time τ_0 and is actually enormous for realistic values of N. This example is very instructive, since it contains the essence of the issue we wanted to clarify. Although we pointed it out for Monte Carlo dynamics, the same ideas hold for for isolated systems. Also, in this case, it is crucial to take into account that we always focus on (1) macroscopic systems and (2) observables that are local. Nonlocal observables can avoid thermalization or display it on a timescale that is gigantic and anyhow completely irrelevant for all practical purposes.

3.3.6   Quantum systems

The problems mentioned above for classical systems are even more striking for quantum ones. In particular, it is clear that it does not make much sense to talk generically about thermalization of finite isolated quantum systems. In fact, many such systems display a discrete spectrum and, starting for a given initial condition, they show Rabi-like oscillation and do not even reach a steady state.

Let us then focus from the beginning on macroscopic systems. The initial condition for the dynamics is a given wavefunction $|\psi\rangle$. Thermalization means that

$$\lim_{\tau \to \infty} \frac{1}{\tau} \int_0^\tau \langle \psi | e^{iHt/\hbar} O e^{-iHt/\hbar} | \psi \rangle \, dt = \langle O \rangle_e,$$

where we have used on the right-hand side the microcanonical average at intensive energy e, which equals the average energy of the initial condition, $\langle \psi|H|\psi\rangle/N$. Decomposing $|\psi\rangle$ on the eigenvectors $|\alpha\rangle$ of H, we find

$$\lim_{\tau \to \infty} \sum_{\alpha \neq \beta} c_\beta^* c_\alpha \langle \beta|O|\alpha \rangle \frac{1 - e^{i(E_\beta - E_\alpha)\tau/\hbar}}{i(E_\alpha - E_\beta)\tau/\hbar} + \sum_\alpha |c_\alpha|^2 \langle \alpha|O|\alpha \rangle = \langle O \rangle_E.$$

Under very general hypotheses (see [5]), the first term on the left-hand side dies off in the large-τ limit. Hence, the formal expression of thermalization reads

$$\sum_\alpha |c_\alpha|^2 \langle \alpha|O|\alpha \rangle = \frac{1}{\mathcal{N}} \sum_{\alpha_e} \langle \alpha_e|O|\alpha_e \rangle,$$

where α_e are the eigenstates with intensive energy e and $\mathcal{N} = \sum_{\alpha_e} 1$ is the microcanonical normalization factor. Again, as at the beginning of this section, we face the riddle of thermalization: the left-hand side contains a lot of information on the initial condition, since the $c_\alpha = \langle \alpha|\psi\rangle$ depend on the initial wavefunction $|\psi\rangle$, whereas the right-hand side instead depends on e only. How is possible that by changing the initial condition but keeping the average intensive energy fixed, all the c_α change, but in such a way that the sum above does not? Clearly, this is generically impossible for a finite system. As stressed before, considering very large systems is mandatory. Only in the thermodynamic limit can the distribution of the c_α can reach a limit in which this becomes true.

The solution of this riddle started to be elucidated by Deutsch and Srednicki [28, 62] with the introduction of the so-called eigenstate thermalization hypothesis (ETH). The main idea is that a *given eigenstate is thermal*, i.e., the average of any local observable in an eigenstate must coincide with the one obtained by equilibrium thermal averaging: if $\langle \alpha | O | \alpha \rangle$ depends on α only through its intensive energy, then it is clearly equal to the microcanonical average, which is given by the flat sum over all eigenstates characterized by the same intensive energy. Again, this cannot be true in a finite system and can become true only as a limiting process for $N \to \infty$. Indeed, it is easy to show [11] that the distribution of the values of $\langle \alpha | O | \alpha \rangle$ for a given local observable becomes peaked around the microcanonical value in the thermodynamic limit. This, however, is not enough to provide an argument for thermalization, since even though the majority of eigenstates lead to the microcanonical average, if there exist rare ones for which this does not happen then thermalization could be avoided in principle provided that the initial conditions have a sufficiently large projection on them.

Two solutions to this impasse have been discussed in the literature:

1. The first, which we have called the strong version of the ETH [56], is sketched in Fig. 3.3: not only does the distribution of $\langle \alpha | O | \alpha \rangle$ become peaked, but also its support shrinks to zero so that effectively in the thermodynamic limit all states satisfy the ETH.

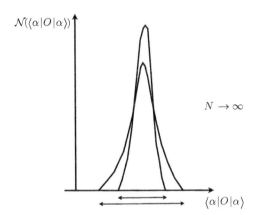

Fig. 3.3 Cartoon of how the distribution of $\langle \alpha | O | \alpha \rangle$ evolves with increasing N. In the strong version of the ETH, the support shrinks to zero so that in the thermodynamic limit, all states do indeed satisfy the ETH.

2. The second is that the support does not shrink to zero, but physical initial conditions are spread in configuration space in such a way that rare states never make a contribution, i.e., the $|c_\alpha|^2$ are rather uniform (see the discussion in [11] for more details).

The correct solution is not yet clear, since the values of N available in numerical diagonalization are not large enough to resolve this issue satisfactorily, nor are rigorous proofs available. What is becoming clear, however, is that for integrable systems, which indeed do not thermalize, the distribution of $\langle \alpha | O | \alpha \rangle$ has a finite support in the thermodynamic limit and the failure of thermalization is due to sampling of rare states.

In summary, macroscopic isolated quantum systems generically are expected to thermalize similarly to classical systems. There is, however, an important exception, which is a consequence of the presence of quenched disorder: quantum disordered systems can show many-body localization, i.e., localization in configuration space, and avoid thermalization [6, 54]. This is a pure quantum effect with no classical counterpart. In contrast to classical systems, in the quantum world, there could be a new kind of breaking of ergodicity that is not due to any underlying thermodynamic phase transition.

3.4 Quenches and coarsening

There are two ways to put a system off equilibrium: by an external drive (e.g., a shear or a voltage difference) or by changing a control parameter C (e.g., the magnetic field, the temperature, or the pressure). In the former case, the system is maintained out of equilibrium, whereas in the latter, it starts to evolve toward the new equilibrium state corresponding to the new value of C. This may take a long time—sometimes so long that the system remains out of equilibrium forever. In this section, we focus on this latter kind of slow relaxation and out-of-equilibrium dynamics. We shall consider systems coupled to a bath, since this is the common situation, and discuss at the end the case of isolated systems.

As we discussed in Section 3.3, the relaxation time of a an ensemble of interacting classical or quantum degrees of freedom coupled to a bath is generically finite (a possible exception is provided by zero-temperature baths; see Section 3.3). Thus, one expects that after a change in the control parameter C, a *finite* system reaches equilibrium in a *finite* time. Only in the thermodynamic limit can one observe neverending off-equilibrium relaxation, which is therefore, as a phase transition, a collective phenomenon with at least one growing correlation length. A simple argument to see why this must be the case is that if all correlation lengths were bounded during the relaxation, then one could roughly divide the system into independent *finite* subsystems, each with a *finite* relaxation time t_B. In consequence, the system would have to relax on timescales not larger than t_B, which contradicts the starting hypothesis of never reaching equilibrium on any finite timescales.

3.4.1 Thermal quenches

Let us consider a system that displays a second-order phase transition at a temperature T_c. We consider the following protocol, which is called a thermal quench: the system starts at equilibrium at a temperature T_i at time $t = 0$; see Fig. 3.4. The temperature of the heat bath is then suddenly changed at a value T_f. Theoretically, we shall study the limiting case where this happens instantaneously. In reality, the rate is finite, and actually often quite large, for example, a fraction of a kelvin per minute; see Fig. 3.5. In the literature, the time spent after the quench is called the waiting time and denoted by t_w. In our case, this coincides with t, since the quench occurs at time $t = 0$. Our argument is general, but for concreteness it is useful to have in mind an example, such as the ferromagnetic Ising model undergoing some kind of Monte Carlo dynamics. The Hamiltonian of the Ising model reads

$$H = -J \sum_{\langle i,j \rangle} \sigma_i \sigma_j, \qquad \sigma_i = \pm 1 \quad \forall i = 1, \ldots, N,$$

where the sum is over nearest neighbors on the three-dimensional cubic lattice. This model has a phase transition between a high-temperature paramagnetic phase and a low-temperature ferromagnetic phase characterized by long-range ferromagnetic order and a nonzero intensive magnetization m.

We consider first the protocol in which the final temperature T_f is larger than T_c (quench protocol 1 in Fig. 3.4). In this case, the system equilibrates after a

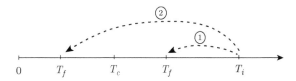

Fig. 3.4 The two different quench protocols discussed in the text: in protocol 1, the quench is toward a temperature $T_f > T_c$; in protocol 2, it is across the phase transition.

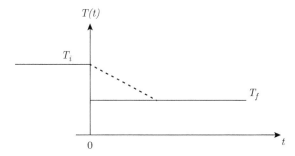

Fig. 3.5 Time dependence of temperature during a quench. The dashed line corresponds to realistic situations and the continuous line to the sudden quench studied theoretically.

characteristic timescale τ that depends on T_f (and possibly other control parameters and microscopic quantities). This means that for $t \gg \tau$, one-time quantities like the magnetization and the energy approach their equilibrium values and hence become independent of time, whereas, for example, the two-time correlations and response functions become invariant under time translation, i.e., they depend just on the time difference. Very often, this approach to equilibrium is exponential in time. See Fig. 3.6. Finally, typical equilibrium relations such as fluctuation–dissipation relations between correlation and response are satisfied.

In the second case, where $T_f < T_c$ (quench protocol 2 in Fig. 3.4), one-time quantities still approach their equilibrium values at long times, but slowly, with a power-law

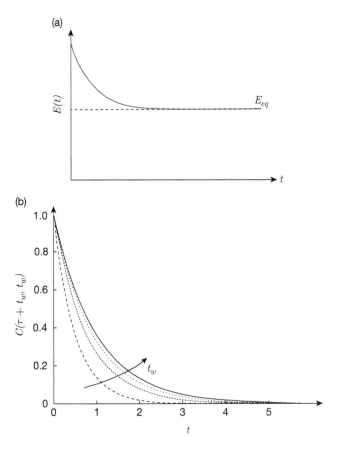

Fig. 3.6 (a) Time evolution of the energy after a quench with protocol 1 from Fig. 3.4. The relaxation to the equilibrium value is fast and often exponential in time. (b) Time evolution of the two-time correlation after a quench with protocol 1. Consider for instance the case of the spin–spin correlation function in the Ising model: $C(\tau + t_w, t_w) = N^{-1} \sum_i \langle \sigma_i(\tau + t_w) \sigma_i(t_w) \rangle$. Once t_w becomes of the order of the relaxation time, the different curves converge and the two-time correlation function becomes invariant under time translations.

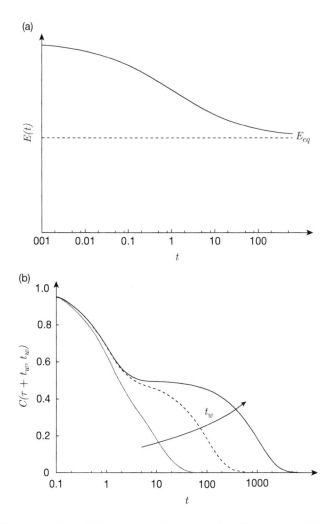

Fig. 3.7 (a) Time evolution of the energy after a quench with protocol 2 from Fig. 3.4. The relaxation to the equilibrium value is slow and often power-law in time. (b) Time evolution of the two-time correlation after a quench with protocol 2. Consider for instance the case of the spin–spin correlation function in the Ising model: $C(\tau + t_w, t_w) = N^{-1}\sum_i \langle \sigma_i(\tau + t_w)\sigma_i(t_w)\rangle$. The different curves never converge. The time to escape from the plateau becomes larger, the longer the time spent after the quench.

time dependence—a first signal that the dynamics changes crossing the phase transition; see Fig. 3.7(a). More striking behavior is found in two-time correlation functions, which display a first rapid relaxation to a plateau value over a timescale τ that depends on T_f and then a second relaxation from the plateau to zero, which takes place on an increasingly larger timescale; see Fig. 3.7(b). Two things are striking in this behavior. First, the first relaxation toward the plateau is the one characteristic of systems equilibrated at T_f, i.e., it coincides with what one would observe starting the dynamics from equilibrium at $T = T_f$. For example, in the case of the Ising model, the spin–spin correlation function converges to a plateau value equal to m^2. Second, no matter how long the time spent after the quench, the system does not equilibrate. Correlation functions do eventually escape from their plateau values and never become invariant under time translations. The timescale on which these nonequilibrium effects take place is not an intrinsic timescale fixed by T_f (or other parameters), but instead depends on the age of the system itself: it is larger the longer the time spent after the quench. For this reason, this behavior is called *aging* and emerges generally in quenches across phase transitions.

3.4.2 Coarsening

To understand what is going on and what is the physical origin of the out-of-equilibrium dynamics, the most useful thing is to look at the results of a numerical simulation—a picture actually. We shall focus on our preferred example: the aging (Monte Carlo) dynamics of the three-dimensional ferromagnetic Ising model after a quench from the disordered high-temperature phase to a temperature at which, in equilibrium, the system is ordered. Figure 3.8 shows two snapshots of a two-dimensional cut of the three-dimensional Ising model after a quench from the high- to the low-temperature phase (black represents minus spins and white plus spins). What do we see? There are four main features:

1. The system wants to break the symmetry and does it locally over a lengthscale $\xi(t)$ that grows with time.
2. The figure is self-similar; i.e., after rescaling the unit of length using $\xi(t)$, Figs. 3.8(a) and 3.8(b) would look the same.
3. The system reaches local equilibrium over domains of size $\xi(t)$.
4. The locally equilibrated domains do not persist forever: negatively magnetized regions become positively magnetized at larger times (and so on and so forth).

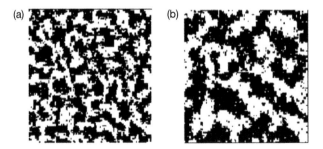

Fig. 3.8 Two snapshots of a two-dimensional cut of a three-dimensional ferromagnetic Ising model evolving with Glauber dynamics after a quench at time $t = 0$ from high temperature to $T < T_c$. Black represents minus spins and white plus spins. (a) Configuration reached after $t_w = 10^3$ Monte Carlo steps. (b) Configuration reached after $t_w = 10^5$ Monte Carlo steps. Statistically, the two configurations look the same after a rescaling of length.

These features are generically found in quenches across second-order phase transitions. The results of numerical simulations, experiments, and theoretical analyses lead to the emergence of a description of the resulting out-of-equilibrium dynamics that is called *coarsening* and is based on the following well-verified assumptions: the system breaks the symmetry locally and forms domains inside which a temporary local equilibrium is reached. Inside each domain, one of the possible low-temperature values of the order parameter is exhibited: in the case of the Ising model, these correspond to $\pm m$. The characteristic size of the domains, $\xi(t)$, grows with time generically in a power-law way. The out-of-equilibrium dynamics is characterized by scaling with respect to $\xi(t)$. For instance, two-time correlation functions can be expressed in terms of scaling functions and power laws:

$$\langle O_x(\tau + t)O_y(t)\rangle = \frac{1}{|x-y|^\eta}\, f\!\left(\frac{|x-y|}{\xi(t)}, \frac{\xi(\tau)}{\xi(t)}\right), \qquad \xi(t) \simeq \xi_0 t^z, \qquad (3.33)$$

where O_x is a generic observable evaluated at site x, such as σ_x for the Ising model. This scaling description holds only on large time- and lengthscales. At finite times and finite lengths, less than $\xi(t)$, the correlation functions have their equilibrium forms. This description is confirmed by all results obtained up to now and is considered to hold generically. What changes from one case to another, depending on the type of second-order phase transition that is crossed by quenching, is the growth law of $\xi(t)$ and the specific form of the scaling functions. These need case-by-case analyses. In the following, we shall explain in detail the analysis of the Ising model, or, more generically, quenches across phase transitions in which the order parameter acquires two different values.

Before doing that, we conclude this subsection by pointing out that only in the thermodynamic limit does the out-of-equilibrium dynamics go on forever. For a finite system, instead, when $\xi(t)$ reaches the size of the system, few domains remain, and, after a while, only one survives. The value of the order parameter to which it corresponds (e.g. plus or minus m for the Ising model) is a random event that depends on the previous (stochastic) dynamics. After the timescale $\xi(\tau_L) \simeq L$, where L is the linear system size, a different behavior sets in: for a very long time, much longer than τ_L, the system remains in one of the ordered states until a very rare random fluctuation creates a system-spanning domain corresponding to another ordered state that then takes over. In the case of the Ising model, this process corresponds to the creation of an interface between the plus and minus states and takes place on a timescale scaling as $\tau \simeq \exp(KL^2/T_f)$, where K is a constant.

3.4.3 Curvature-driven domain growth

As explained above, the out-of-equilibrium dynamics emerging after a quench across a phase transition is due to the evolution of the domain structure. In order to understand the main physical reason for this evolution, we shall focus on quenches across phase transitions in which the order parameter acquires two different values and on the simplest case of a single spherical domain.

We model the dynamics using a Langevin equation for the ordering field (the magnetization for ferromagnets):

$$\partial_t \phi(\mathbf{x}, t) = -\frac{\delta F}{\delta \phi(\mathbf{x}, t)} + \xi(\mathbf{x}, t),$$

where

$$F = \int d^3x \left\{ \tfrac{1}{2} [\nabla \phi(\mathbf{x}, t)]^2 + V(\phi(\mathbf{x}, t)) \right\}.$$

The potential $V(\phi(\mathbf{x}))$ has a double-well structure with two minima of equal height at ϕ_+ and ϕ_-. The thermal noise corresponds to a heat bath at temperature T, which at first we take equal to zero. We can of course wonder how much the results we will find depend on the large number of assumptions and approximations that we have already made. The answer is that, as for critical phenomena, the physics on large time- and lengthscales is independent of those that just determine a few constants, but not the scaling laws and the scaling functions.

The initial condition for the dynamics corresponds to a spherical domain of radius R, corresponding to $\phi(\mathbf{r}) = \phi_+$ for $r \ll R$, $\phi(\mathbf{r}) = \phi_-$ for $r \gg R$, and a sharp drop from ϕ_+ to ϕ_- at $r \simeq R$; see Fig. 3.9. Because of spherical symmetry, the zero-temperature Langevin equation reads

$$\partial_t \phi(\mathbf{r}, t) = \nabla^2 \phi(\mathbf{r}, t) - V'(\phi(\mathbf{r}, t))$$
$$= \frac{d^2\phi(r, t)}{dr^2} + \frac{d-1}{r}\phi(r, t) - V'(\phi(r, t)). \tag{3.34}$$

As we have discussed, the dynamics of large domains takes place on long timescales. Consequently, for $R \gg 1$, we seek a scaling solution of (3.34). It is natural—and, as we shall show, correct—to focus on a traveling-wave form, $\phi(r, t) = f(r - R(t))$. Since $\phi(r, t)$ is expected to maintain the shape corresponding to a domain, the function $f(x)$

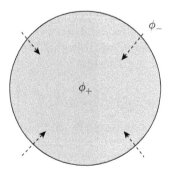

Fig. 3.9 Sketch of the spherical domain whose evolution is analyzed in the text.

equals ϕ_+ for $x \ll 0$, and ϕ_- for $x \gg 0$ and has a drop between these two values at $x = 0$. By plugging this ansatz into the evolution equation, we obtain

$$f''(r - R(t)) + \left[\frac{2}{r} + \dot{R}(t)\right] f'(r - R(t)) - V'(f(r - R(t))) = 0. \qquad (3.35)$$

Since $f'(r - R(t))$ vanishes for $|r - R(t)| \gg 1$, we can replace $2/r + \dot{R}(t)$ by $2/R(t) + \dot{R}(t)$ up to subleading corrections of order $1/R(t)$. The only way for (3.35) to be consistent is if the middle term, which violates the traveling-wave form, cancels out. This leads to two equations:

$$\dot{R}(t) = -\frac{2}{R(t)}, \qquad (3.36)$$

$$f''(x) - V'(f(x)) = 0. \qquad (3.37)$$

The solution of (3.36) is $R(t) = \sqrt{R^2(0) - 4t}$. Equation (3.37) is equivalent to the Newton equation for a particle moving in the inverted potential $-V(f)$ starting at time $-\infty$ at ϕ_+ and arriving at infinite time at ϕ_- (in this analogy, f and x denote the particle position and the time, respectively). Energy conservation imposes $f'(x)^2 + V(\phi_+) = V(f)$ at all times. From this equation, we obtain the solution in terms of the implicit equation

$$\int_0^{f(x)} \frac{df'}{\sqrt{V(f') - V(\phi_+)}} = x.$$

The final result is that a large spherical domain shrinks. The shape at the boundary of the interface is given by $f(x)$. The equation $\dot{R}(t) = -2/R(t)$ is a special form of the equation valid for nonspherical domains [18], which expresses the fact that the local speed of the interface of a domain is proportional to its curvature. Consequently, wavy domains tend to flatten out and small compact domains tend to shrink. In particular, a spherical domain of size R shrinks and disappears in a time $t \propto R^2$. This relationship between time and length, which follows from the equation for $R(t)$, is very general and implies that small domains disappear faster than large ones. The net result is that after a timescale t, domains of size less than $\sqrt{t}$ have typically disappeared. What remain are mainly domains of size $\sqrt{t}$ with a typical radius of curvature of the order of $\sqrt{t}$ [60, 61].

This is a way to justify the scaling assumption: indeed, we have just found that after a time, the characteristic lengthscale $\xi(t)$ is of the order of $\sqrt{t}$. To fully derive the scaling equation (3.33), we need to show also self-similarity, but this is beyond the scope of this chapter. For a thorough discussion and an exact derivation of the scaling laws, see [60, 61].

Let us now lift the the simplifying assumption of considering quenches to zero temperature. Qualitatively, the results remain the same for any $T_f < T_c$. Inside the domains, there is a local equilibration at temperature T_f and although there are now stochastic thermal fluctuations, the evolution of domains is on average still driven by the curvature. The only main change is in the prefactor of the growth law, which is renormalized by thermal fluctuations. We find that the typical lengthscale $\xi(t)$ grows as $\simeq D(T_f)\sqrt{t}$, where $D(T_f)$ vanishes when $T_f \rightarrow T_c$ as a power law (see Section 3.4.4 for a derivation of this result). The scaling functions are expected to remain unchanged.

Combining all previous results, we can now explain the behaviors sketched in Fig. 3.7. The energy approaches its equilibrium behavior as a power law because the only contribution comes from the boundary of the domains since the regions inside the domains, are equilibrated. Hence they do not make any extra contribution compared with the equilibrium case, whereas instead the boundaries of the domains cost extra energy. This extra surface energy is of the order of the surface of the domains $\xi(t)^{d-1}$. Therefore, the extra energy per unit volume scales as $1/\xi(t) \propto 1/\sqrt{t}$, which leads to a power-law approach to the asymptotic value for the energy as a function of time. With regard to the behavior of correlation functions shown in Fig. 3.7(b), the first rapid decrease to the plateau is related to the equilibration inside the domains (toward the value $m(T_f)^2$), whereas the secondary evolution is due to the slow motion of the domains and is described by the scaling theory.

A natural question is how general is this description of the out-of-equilibrium dynamics after a thermal quench across a phase transition. The answer is that it is very general. The main underlying mechanism is always the same: the system breaks the symmetry locally over a lengthscale $\xi(t)$. Below this length, the system is equilibrated in one of the possible symmetry-breaking states, whereas the slow out-of-equilibrium dynamics is due to the motion of the topological defects that break the symmetry. These are domain walls for the $\mathbb{Z}_2$ symmetry that we considered previously, but they can be more complicated objects, such as vortices for the XY model in two dimensions. In order to obtain a description of the out-of-equilibrium dynamics, one has to understand the evolution of the defects. This depends on the kind of symmetry breaking and the nature of the resulting topological defects: it is curvature-driven for domain walls, as we have explained, whereas it is (to a good approximation) diffusion for vortices. This actually again leads to a growth law $\xi(t) \propto \sqrt{t}$, but this has a very different physical origin. In general, the growth law for quenches across second-order phase transitions is a power law, $\xi(t) \propto t^{1/z}$, with the value of the exponent depending on the dynamics and the kind of symmetry breaking. As in critical phenomena, z depends on the dynamical conservation law; for example, a quench across the ferromagnetic transition, or equivalently the spinodal decomposition, is characterized by $z = 2$ if the dynamics does not conserve the order parameter, as we have already derived, and by $z = 3$ in the conserved case [18]. In contrast to critical phenomena, z is in general independent of dimension, and hence there is no upper critical dimension.

A final comment is called for here regarding quenches exactly at T_c, which are also called critical quenches. These are characterized by growth laws that are different from those considered previously and are related to the critical behavior characterizing the *equilibrium* phase transition [19]. In particular, the typical lengthscale over which the system is out of equilibrium, $\xi(t)$, grows as t^{1/z_d}, where z_d is the exponent relating time- and lengthscales in the equilibrium critical dynamics [19] (z_d and z are generically different; for example, for the two-dimensional Ising model, $z_d \simeq 2.17$ and $z = 2$

for nonconserved dynamics). There are no well-defined domains in this case; instead, the system should be considered a patchwork of equilibrated critically correlated regions of length $\xi(t)$. There are still scaling functions, but they are different from those obtained for $T_f < T_c$. In summary, critical quenches and thermal quenches below T_c both lead to out-of-equilibrium dynamics, although these are quite different.

3.4.4   Kibble–Zurek mechanism and slow thermal quenches

In Section 3.4.3, we studied very sudden quenches. Here we focus instead on very slow ones. This is interesting for several reasons. First, real quenches are generically slow: temperature can be decreased at rates of a fraction of kelvin per minute, i.e., on timescales that are very large compared with the typical microscopic ones (picoseconds or less). Moreover, theoretically, studying slow quenches is useful to understand the connection between critical and standard coarsening.

   Consider a situation in which the order parameter inducing the quench, for instance the temperature, is changed slowly over a timescale τ_Q:

$$T(t) = T_i + (T_f - T_i)\frac{t + \frac{1}{2}\tau_Q}{\tau_Q}, \qquad t \in \left(-\tfrac{1}{2}\tau_Q, \tfrac{1}{2}\tau_Q\right).$$

We consider a quench across a phase transition, i.e., $T_i > T_c$ and $T_f < T_c$. Clearly, if τ_Q is much larger than the characteristic microscopic timescales, then, at least at the beginning of the evolution, the system follows the temperature change adiabatically. Since, as T decreases, the phase transition is approached, the system develops long-range equilibrium correlations over a lengthscale $\xi(T(t)) \propto |T(t)) - T_c|^{-\nu}$. Because of this "stiffness" due to the spatial correlation, the characteristic timescale over which the equilibrium dynamics takes place also increases as $\tau(T(t)) \propto \xi_d^z \propto |T(t)) - T_c|^{-\nu z_d}$, where ν and z_d are the equilibrium critical exponents; see Fig. 3.10. This divergence implies that the adiabatic evolution inevitably breaks down when $T(t)$ approaches

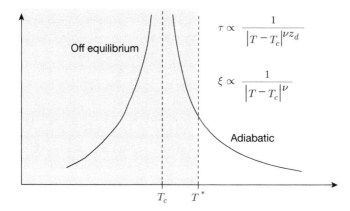

Fig. 3.10   Sketch of the phase diagram for slow quenches. Until T^*, the evolution is adiabatic and the system develops critical long-range correlations.

T_c too closely. In order to estimate when this happens, we consider a system that reaches the temperature $T(t)$ in equilibrium, (i.e., following the temperature change adiabatically) and compare the interval of time Δt that remains before crossing T_c. If this is smaller than the equilibrium relaxation time $\tau(T(t))$, then the system has to fall out of equilibrium. In this way, we obtain an equation for the temperature T^* at which adiabatic evolution breaks down:

$$\Delta t = \frac{T^* - T_c}{T_i - T_f} \tau_Q \simeq \tau(T^*) \quad \Rightarrow \quad (T^* - T_c) \propto \tau_Q^{-1/(\nu z_d + 1)}.$$

As expected, the smaller the cooling rate, i.e., the larger τ_Q, the closer can the system approach T_c while remaining in equilibrium. Nevertheless, for any large but finite value of τ_Q, the adiabatic evolution eventually breaks down. Our main aim is to find the growth of the characteristic lengthscale $R(t)$ over which the system is ordered as a function of time.

Physically, there are three regimes: first the system follows adiabatically the temperature change and develops critical correlations. When it goes out of equilibrium at T^*, it is a collection of critically correlated regions that start to grow by critical coarsening. There exist domains in this regime, but they are fractal and characterized by a magnetization that is not extensive. Eventually, a crossover from critical to standard coarsening takes place: domains become compact, the magnetization inside becomes extensive, and the evolution starts to be dominated by the curvature of the boundary of the domains. In the following, we assume scaling and use the fact that $R(t)$ tracks the equilibrium correlation length during the adiabatic regime, and at very long times we have to recover the coarsening growth law $R(t) \propto t^{1/z}$. Since the characteristic timescale governing the crossover from adiabatic to off-equilibrium evolution is $\tau(T(t))$, the scaling law reads

$$R(t) = \xi(T(t)) f\left(\frac{t}{\tau(T(t))}\right),$$

where the two regimes, adiabatic and standard coarsening, correspond respectively to the $-\infty$ and $+\infty$ limits of the scaling variable $x = t/T(t)$. In order to recover the expected growth law in these regimes, $f(x)$ has to tend to 1 for $x \to -\infty$ and to diverge as $x^{1/z}$ for $x \to +\infty$. The latter result allows us to obtain the prefactor of the coarsening growth law: we find $R(t) \propto [T(t) - T_c]^{\nu(-1+z_d/z)} t^{1/2z}$, and hence $D(T) \propto (T_f - T_c)^{\nu(-1+z_d/z)}$ if the temperature is kept constant at T_f at long times.

Kibble and Zurek originally studied this problem in cosmology, largely independently of coarsening in statistical physics. They were mainly interested in the density of topological defects created out of equilibrium when the temperature $T(t)$ is below T_c and symmetric with respect to T^* (the end of the so-called impulse regime). In the time interval between the adiabatic regime and the time at which $T(t)$ becomes equal to $T_c - (T^* - T_c)$, the length $R(t)$ grows following the critical coarsening law. As can be explicitly checked, this increase just leads to a value that is greater than that at the end of the adiabatic regime by a numerical factor greater than 1. Consequently, the density of topological defects, N, is proportional to $R(t^*)^{-d} \propto \tau_Q^{-d/z}$. Because of the original work by Kibble and Zurek, this result and some of the arguments explained above go under the name of the Kibble–Zurek mechanism for the formation of topological defects; see the review [68]. The explanation presented above is based on [12]. The Kibble–Zurek mechanism has recently been the focus of intense research activity in the context of quantum quenches for isolated quantum systems. On this point, see Altman's contribution to this volume (Chapter 1).

3.4.5 Quantum fluctuations

Let us now take quantum fluctuations into account. A generic phase diagram for a quantum system undergoing a second-order phase transition is shown in Fig. 3.11, in which Γ represents the strength of quantum fluctuations [58]. At zero temperature, a quantum critical point (QCP) occurs at Γ_c, from which originates the critical line separating the disordered and ordered phases. This line ends for $\Gamma = 0$ at the classical value of the critical temperature. Consider now quenches where the control parameter, T or Γ, is changed in order to cross the transition line; see Fig. 3.11. (Quenching T is experimentally feasible by changing the bath temperature; changing Γ is a more theoretically oriented protocol.) The resulting out-of-equilibrium dynamics is very similar to that described in the classical case: the system equilibrates locally over a lengthscale of size $\xi(t)$ that grows with time as a power law. In particular, for a quantum ferromagnet or generically for systems with only two competing low-temperature states, one again finds domain growth. Correlation functions exhibit a behavior similar to that discussed before. Equilibration seems to takes place on short timescales toward a plateau value that is related to the value of the order parameter at T_f, Γ_f. However, aging behavior, which is related to the motion of topological defects that break long-range order on the scale $\xi(t)$, eventually sets in. As is clear from Fig. 3.12, although

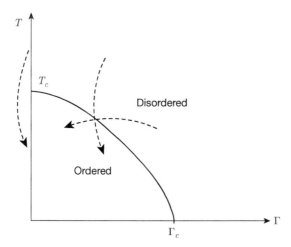

Fig. 3.11 Typical form of a phase diagram for a quantum system displaying a second-order phase transition. The dashed lines indicate different quenching protocols.

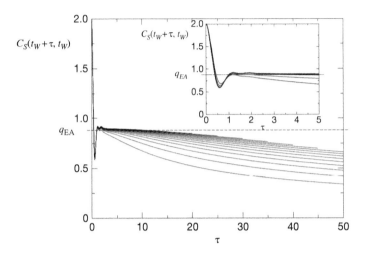

Fig. 3.12 Typical form of a correlation function after a quench in a quantum system [10]. The inset shows the evolution for times $\tau \ll t_w$: the system displays equilibrium dynamics. In the main panel, for $\tau \propto t_w$, aging behavior is manifest.

quantum fluctuations show up in the first part of the relaxation (for instance, the correlation function displays oscillating behavior due to quantum coherence), they seem to be absent in the out-of-equilibrium regime. This is indeed the case, since this regime appears at very long times. Hence, dissipation and decoherence inevitably kick in and quantum coherence is lost. Consequently, the motion of the topological defects is classical and, accordingly, the out-of-equilibrium behavior is governed by *classical coarsening laws*. Whether exceptions to this conclusion can be found is not clear. In particular, quenches across QCP, i.e., by changing Γ and keeping the bath temperature at $T = 0$, are still not well understood. In this case, depending on the bath density of states, the coherence effects can be of very long range in time, and so there could be room for a more important influence of quantum fluctuations on out-of-equilibrium aging dynamics.

Recently, there has been intense research activity on quantum quenches, i.e., protocols in which an isolated quantum system initially in the ground state is put out of equilibrium by suddenly changing a coupling in the Hamiltonian. A discussion of the resulting dynamics is outside the scope of this chapter and, moreover, is an issue still not completely settled. By quenching the coupling, one effectively injects energy into the system, since for the new Hamiltonian the old ground state is a wavefunction consisting of a mixture of excited states. Hence, we expect the rapid degrees of freedom to equilibrate and provide a bath for the slow degrees of freedom, which then undergo classical coarsening at long times. This is indeed what happens for classical systems in analogous situations [43]. Unexpected behavior could result from the nonthermal fixed point discussed in Berges' contribution to this volume (Chapter 2). Indeed, it has been found that, depending on the type of initial condition, isolated quantum systems could approach a quasistationary critical behavior reminiscent of turbulence. The competition between this behavior and coarsening is presently not understood. Other effects found in the literature are the so-called dynamical transitions; see, for example, [13] and references therein. These could actually be related to the nonthermal fixed point just discussed. As can readily be understood from this brief description, research on quantum quenches is still very much ongoing. Surprises and new explanations are to be expected.

3.5 Quenched disorder and slow dynamics

The presence of quenched disorder can induce very slow dynamics and lead to new kinds of out-of-equilibrium phenomena, some of which we present in this section. First we discuss the effect of disorder on equilibrium dynamics and then focus on out-of-equilibrium ones. This section is just an introduction, since a comprehensive presentation would deserve an entire series of lectures on its own—see, for instance, the 2002 Les Houches Lecture Notes [25]. Some topics will be left out for lack of space, in particular glasses and spin glasses, which would need quite a long discussion.

3.5.1 Broad distribution of relaxation times

One of the most relevant phenomenological consequences of quenched disorder is the presence of several relaxation times: the heterogeneity induced by the disorder leads to the simultaneous presence of regions that relax rapidly and others that relax slowly. Instead of having a characteristic timescale τ, disordered systems are characterized by a distribution of relaxation times, $P(\tau)$. A broad $P(\tau)$ can have an important impact on the dynamics, leading to rare regions that take a very long time to relax. This leads to a strong nonexponential relaxation, in contrast to nondisordered systems, which usually exhibit exponential relaxation.[12]

In the following, by focusing on the example of random ferromagnets, we shall present a general phenomenon taking place in disordered systems, which goes under the name of the Griffiths phase. The Hamiltonian of a simple model of a random ferromagnet reads

$$H = -\sum_{\langle i,j \rangle} J_{ij}\sigma_i^z \sigma_j^z,$$

where the couplings are independent and identically distributed random variables with the distribution $P(J_{ij}) = (1-p)\delta(J-J_1) + p\delta(J-J_2)$, and J_1 and J_2 are both positive $(J_2 > J_1)$. This system displays a ferromagnetic phase transition at a temperature T_c that depends on the values of p, J_1 and J_2. It is clear that T_c has to be a increasing function of p. In particular, $T_c(p, J_1, J_2)$ is bounded between the temperatures $T_c^{J_1}$ and $T_c^{J_2}$ at which a system with all couplings equal to J_1 or J_2 becomes ordered. Numerical simulations of the equilibrium dynamics show that at high temperature,

---

[12] When they have no quantities conserved by the dynamics.

roughly above $T_c^{J_2}$, correlation functions exhibit exponential relaxation, whereas at lower temperature, they become more and more stretched.

This result can be understood [17] (and actually proved rigorously [48]) by noticing that there are regions inside the system that have all couplings equal to J_2. Below $T_c^{J_2}$, they are "ordered," or, more correctly, they flip very slowly between the two low-temperature magnetized states and hence induce a very slow dynamics. The temperature regime below $T_c^{J_2}$ is called the Griffiths phase.[13]

Let us be precise and recall that for an ordered ferromagnet of size L, the time to flip between the two low-temperature magnetized states is of the order of $\tau_L \simeq \exp[2J_{\text{eff}}(T)L^{d-1}/T]$, where $J_{\text{eff}}(T)$ plays the role of a surface tension. It is equal to J at zero temperature, it is a decreasing function of T, and it vanishes at T_c. The spatially averaged equilibrium correlation function can be easily shown to be self-averaging, and hence

$$C(t) = \frac{1}{N} \sum_i \langle \sigma_i^z(t)\sigma_i^z(0) \rangle = \overline{\langle \sigma_0^z(t)\sigma_0^z(0) \rangle}.$$

If the origin belongs to a region of linear size L where all couplings are equal to J_2, then its relaxation is given by e^{-t/τ_L}. Therefore, considering only cubic regions centered around the origin, we can easily obtain the bound

$$C(t) \geq \sum_L p^{L^d}\left(1 - p^{2^d L^{d-1}}\right) \exp\left[-t\exp\left(-\frac{2J_{\text{eff}}L^{d-1}}{T}\right)\right],$$

where the first term is the probability that in the whole cubic region of size L couplings are equal to J_2 and on the boundary of the cube they are equal to J_1. This sum can be performed using the saddle-point approximation when t is large: one finds that the value $L(t)$ that dominates the sum is proportional to $[(T \log t)/J_{\text{eff}}]^{1/(d-1)}$ and that

$$C(t) \geq K_1 \exp\left[-K_2(\log t)^{d/(d-1)}\right],$$

where K_1 and K_2 are positive constants that depend on p, T, J_1, and J_2. This result shows that even in the high-temperature regime, before the phase transition takes place, equilibrium correlation functions relax much more slowly than exponentially. This is due to slow and rare Griffiths regions. We have focused on equilibrium dynamics here, but of course this effect is also present if the temperature is quenched in the high-temperature phase $(T_f > T_c)$.

3.5.2   Broad distribution of low-energy excitations

The Griffiths phase studied in Section 3.5.1 has a counterpart in quantum disordered systems. Analogous to the existence of rare slowly relaxing regions, quantum disordered systems contain rare regions that give rise to very low-energy excitations. In

---

[13] The name was chosen because of the relation to the singularities that Griffiths found when studying the thermodynamics of disordered systems [38].

particular, this leads to a density of states that vanishes as a power law at low frequency, in contrast to nondisordered quantum systems, which instead display a gap whose value is related to the typical energy scale. As before, let us describe this general phenomenon, which goes under the name of the quantum Griffiths phase [58], in a simple example, namely, the transverse random Ising ferromagnet, whose Hamiltonian reads

$$H = -\sum_{\langle i,j \rangle} J_{ij}\sigma_i^z\sigma_j^z - \Gamma\sum_i \sigma_i^x,$$

where σ^x and σ^z are Pauli matrices and J_{ij} are positive and independent random variables with the same distribution as in Section 3.5.1. At zero temperature, this model displays a phase transition between an ordered ferromagnetic phase for $\Gamma < \Gamma_c$ and a disordered paramagnetic phase for $\Gamma > \Gamma_c$ [58].

Γ_c plays the same role as T_c did before: it depends on p, J_1, and J_2 and is bounded between the values $\Gamma_c^{J_1}$ and $\Gamma_c^{J_2}$. The phase below $\Gamma_c^{J_2}$ is called the quantum Griffiths phase and is characterized by the existence of very low-energy excitations. The main argument parallels that presented in Section 3.5.1. The system contains regions of linear size L inside which all couplings are equal to J_2. Their density is of the order of p^{L^d}. For $\Gamma < \Gamma_c^{J_2}$, these regions are "ordered," and hence the lowest excitation energy scales as $\omega_L \simeq \exp(-KL^d)$. This corresponds to a multitude of virtual processes that allow one to go from one ordered state to the other and hence to lift the degeneracy between them.

The density of states associated with the spin–spin correlation function is given by

$$\rho(\omega) = \overline{\sum_\alpha |\langle\alpha|\sigma_0^z|GS\rangle|^2\delta(\omega - [E_\alpha - E_{GS}])}\,,$$

where the index α is associated with the many-body eigenstates and the subscript GS indicates the ground state. The contribution to the above sum due to Griffiths regions is of the form $\sum_L \exp[-(\log p)L^d]\,\delta(\omega - \exp(-K'L^d))$. Evaluating this sum for large L, one finds that the contribution of the Griffiths regions leads to a power law in the density of states, which, when $\omega \to 0$, behaves as

$$\rho(\omega) \simeq \frac{\omega^{K/K'-1}}{(\log \omega)^{d-1/d}}\,.$$

Thus, below $\Gamma_c^{J_2}$, the density of states is not gapped and instead displays a nonuniversal power law at small frequency. This in turn implies that the zero-temperature equilibrium dynamics is very slow below $\Gamma_c^{J_2}$.

3.5.3 Activated dynamic scaling and zero-temperature fixed points

In Sections 3.5.1 and 3.5.2, we have studied the effects of quenched disorder on the equilibrium dynamics and have shown that even if the system is not ordered and far from the critical point, the dynamics can become very slow. Here, we want instead to study the effect of the disorder on a critical point. From the point of view of the dynamical behavior, one of the most remarkable effects of quenched disorder is to produce in some cases a critical equilibrium dynamics that is ultraslow. In particular, there is a divergence of relaxation times that is super-Arrhenius instead of being power-law as in standard second-order phase transitions. This is due to a relationship between time- and lengthscales that is exponential, $\tau \simeq \tau_0 \exp[\xi(T)^\psi/T]$, instead of power-law, $\tau \propto \xi^z$. This phenomenon is called activated dynamic scaling [31, 64] and is due to the existence of a new kind of renormalization group fixed point that we shall now describe.

Let us first recall that for a standard second-order phase transition, close to the critical point, the system is formed of correlated regions of linear size ξ. Each region can be in one of the low-temperature phases, for example, positively or negatively magnetized in the case of a ferromagnet. In a (crude) real-space renormalization group sense, each correlated region can be represented by a coarse-grained degree of freedom, associated with the order parameter. The value of this degree of freedom indicates in which low-temperature phase the correlated region lies. The interaction between correlated regions is represented by one or more couplings between these degrees of freedom. For instance, in the ferromagnetic case, one associates with each region a spin and introduces a ferromagnetic coupling as an interaction between regions; see Fig. 3.13. This renormalization group picture emerges naturally in real-space renormalization group treatments such as the Migdal–Kadanoff one [1]. The fundamental property of second-order phase transitions is that the coarse-grained energy scale ΔF_ξ obtained by renormalization on the lengthscale ξ is of the order of the temperature.[14] In the previous approximate description, this means that $J_\xi \propto O(T)$. In a more formal treatment, one indeed finds that the renormalized theory on scales larger than ξ is perturbative and high-temperature-like. This has several important consequences, the most important of which is that correlated regions are "active," i.e., they flip rather easily, since $\tau \propto \exp(\Delta F_\xi/T)$. Actually, it is the pre-factor to the Arrhenius law that matters and gives the usual power-law scaling between time and length that is characteristic of critical dynamics: $\tau \propto \xi^z$. Another remarkable consequence is that the singular part of the free energy, which scales as $|T-T_c|^{2-\alpha}$, can be obtained as $\Delta F_\xi/\xi^d \propto |T-T_c|^{\nu d}$. This directly implies the hyperscaling relation $2 - \alpha = \nu d$.

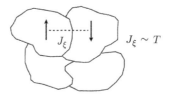

Fig. 3.13   Sketch of the renormalized effective model on the scale ξ for the Ising model.

---

[14] All these statements and those that follow are true only below the upper critical dimension where scaling holds.

In the presence of disorder, this picture can break down. A new kind of fixed point can emerge, for which the typical energy scale on the length ξ becomes much larger than the temperature. In this case, ΔF_ξ is found to diverge as ξ^θ approaching the phase transition. Consequently, the effective model on the scale ξ is now characterized by couplings much larger than the temperature.[15] Since what matters is the ratio between couplings and temperature, one can describe this situation as one of infinite-disorder couplings or a zero-temperature fixed point. One of the main consequences is that now the dynamics (e.g., the Monte Carlo dynamics) of the correlated regions becomes extremely slow. The relaxation time is given by the Arrhenius law applied to a correlated region: $\tau \propto \exp(\Delta F_\xi / T)$. This in turn implies a super-Arrhenius divergence for the characteristic timescale of critical dynamics:

$$\xi \propto \frac{1}{|T - T_c|^\nu} \quad \Rightarrow \quad \tau \propto \tau_0 \exp\left(\frac{K}{|T - T_c|^{\theta\nu}}\right).$$

An example of this behavior is provided by the random field Ising model, whose Hamiltonian reads

$$H = -J \sum_{\langle i,j \rangle} \sigma_i^z \sigma_j^z - \sum_i h_i \sigma_i^z,$$

where J is a ferromagnetic coupling and h_i are random fields corresponding to independent and identically distributed random variables of mean zero and variance Δ. A sketch of the phase diagram of this model is presented in Fig. 3.14. The transition line separates the paramagnetic and the ferromagnetic phases. Except for $\Delta = 0$, the critical behavior is governed by a zero-temperature fixed point. A sketchy renormalization group picture in terms of the effective model on the scale ξ is shown in Fig. 3.15: correlated regions are described by Ising spins interacting via random (ferromagnetic) couplings and coupled to random external fields. As found in the Migdal–Kadanoff treatment [20], which makes this renormalization group picture concrete, both the

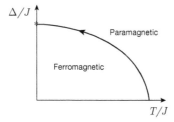

Fig. 3.14 Sketch of the phase diagram for the random field Ising model. The arrow on the transition line indicates that the renormalization group flow is controlled by the zero-temperature fixed point.

---

[15] In some analytical treatments where the energy scale is kept fixed under renormalization, it is the temperature scale that is renormalized: $T_\xi \propto T/\xi^\theta$.

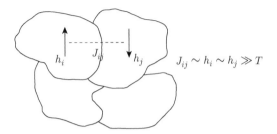

Fig. 3.15 Sketch of the renormalized effective model on the scale ξ for the random field Ising model.

coupling and the fields scale as $\xi^\theta \gg T$. More refined treatments, such as the nonperturbative functional renormalization group [63] and numerical simulations [50], confirm these results and make them more precise [63], giving numerical values for the exponents $\theta_{3D} \simeq 1.6$ and $\nu_{3D} \simeq 1.5$. As explained above, this implies an extremely rapid growth of the relaxation time as the transition is approached [4, 31]:

$$\tau \propto \tau_0 \exp\left(\frac{K}{|T - T_c|^{\theta_{3D}\nu_{3D}}}\right),$$

with $\theta_{3D}\nu_{3D} \simeq 2.4$.

3.5.4 Domain growth and coarsening in disordered systems

Let us now consider the effect of disorder on the off-equilibrium dynamics due to thermal quenches. We consider, as in Section 3.4, quenches across phase transitions. The physics is very similar to the nondisordered case: there is a tendency to local equilibration, which takes place over a lengthscale $\xi(t)$, but the system remains always out of equilibrium because of aging dynamics (because of the slow motion of topological defects). What changes is the dynamics of the topological defects, which now evolve in a disordered environment. Let us focus on the case of domain growth. Without quenched disorder, the motion of domains is curvature-driven, as we have explained. In the presence of disorder, instead, the motion of a domain boundary is governed by the competition between the tendency to reduce the interface energy and the gain obtained by wandering in profitable regions of the random environment [36, 66].

The study of domain boundaries in a random environment is another case in which the large-scale physics is governed by a zero-temperature fixed point. Theory and numerical simulations [36, 66] have established that in several cases, in particular in the case of random bond and random field ferromagnets, the domain boundary is rough: on a scale ℓ, it has transverse fluctuations of the order of ℓ^ζ; see Fig. 3.16. The free energy on the scale ℓ is given by two contributions,

$$\Delta F_\ell = \sigma \ell^{d-1} + \Upsilon \ell^\theta,$$

the first of which is the surface tension and is a nonfluctuating quantity, whereas the second, Υ, is a random variable with a universal distribution (in the renormalization

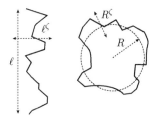

Fig. 3.16 Sketch of the wandering of domain walls in random ferromagnets.

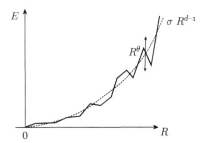

Fig. 3.17 Sketch of the energy as a function of the domain size: on top of the average contribution σR^{d-1}, there is a fluctuating part of the order of R^θ, which gives rise to barriers whose heights increase as R^θ.

group sense). Plotting the energy as a function of the domain radius gives a curve like that in Fig. 3.17, which on average is the same as for a nondisordered system but displays large-scale barriers. A spherical domain of size ℓ has a tendency to shrink as for nondisordered systems, since in this way the interface energy is decreased. However, in order to do so, it has to go over barriers that scale as ℓ^θ. It is assumed that domain growth for disordered system can still be described by a scaling theory in terms of a growing length $\xi(t)$. However, the relationship between time and length is no longer a power law, but rather an activated dynamic relationship:

$$\tau_\ell \propto \exp\left(\frac{\Upsilon \ell^{d-1}}{T}\right) \quad \Rightarrow \quad \xi(t) \propto (T\log t)^{1/\theta}.$$

In this case, the scaling relationship (3.11) reads

$$\langle O_x(t')O_y(t)\rangle = \frac{1}{|x-y|^\eta} f\left(\frac{|x-y|}{(T\log t)^{1/\theta}}, \frac{\log t'}{\log t}\right).$$

In practice, because of the logarithms in the scaling laws, domain growth in disordered systems is a particularly slow process. It takes a very long time for the length $\xi(t)$ to increase substantially.

3.5.5 Infinite-randomness fixed points

The effect of quenched disorder on quantum dynamics also gives rise to remarkable new phenomena. This is a topic that is currently attracting a lot of attention and on which a lot of work is ongoing. Instead of giving a comprehensive presentation, we shall briefly discuss what is one of the most influential concepts and frameworks—one that actually mirrors what has been explained in the previous sections for classical dynamics. The starting point is the study of random transverse Ising chains, for example that with the Hamiltonian

$$H = -\sum_i J_i \sigma_i^z \sigma_{i+1}^z - \tfrac{1}{2}\sum_i h_i \sigma_i^x,$$

where $J_i > 0$ and $h_i > 0$ are random quenched variables. This model displays a phase transition between a disordered and a ferromagnetically ordered phase when $\Delta = \overline{\log h}$ approaches a critical value Δ_c from above. D. S. Fisher showed that the corresponding critical point is governed by a new kind of renormalization fixed point, which is now called the infinite-randomness fixed point. By using the Dasgupta–Ma decimation procedure, a kind of real-space renormalization group where the strongest couplings in the system are decimated sequentially, he showed that at criticality, $\log J_i$ and $\log h_i$ grow as $\sqrt{\ell}$, where ℓ is the lengthscale over which the system is coarse-grained. This gives rise to several remarkable phenomena, such as strong differences between typical and average quantities and an intricate critical behavior.

The most important point for our discussion is that the renormalization group behavior actually resembles very much that described above for a zero-temperature fixed point. Indeed, in his seminal papers, Fisher proposed to call it "tunneling dynamics scaling," in analogy with the "activated dynamics scaling" discussed previously. Although for a long time these renormalization group techniques and this kind of fixed point were used to investigate equilibrium properties of the ground state, it was already pointed out in the earliest papers on this topic that this fixed point should have a very important influence on the dynamical properties and possibly give rise to very slow dynamics. Recently, this has been shown explicitly in the context of the many-body localization transition [6, 54]. This is a research theme that is likely to witness a great deal of progress in the next few years.

3.6 Effective temperature

The notion of effective temperature is a recurrent theme in out-of-equilibrium physics. The notion of temperature, a single quantity that determines the state of the system, is so useful that researchers keep trying to find out whether it somehow holds also out of equilibrium. In this section, we shall present several different notions of effective temperature that have been discussed in the literature.

3.6.1 Metastable equilibrium

The first notion of effective temperature that we wish to discuss is also the closest to the equilibrium temperature we are used to. Consider a system coupled to a thermal

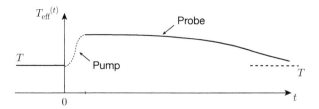

Fig. 3.18 Evolution of the effective temperature of electrons in a pump-and-probe experiment. During the pump, the system is truly out of equilibrium. After that, when is probed, electrons quickly reach a metastable equilibrium with a well-defined temperature that slowly evolves toward the bath temperature on much larger timescales.

bath that is at temperature T. There are two important timescales in the problem: one, τ_S, is the relaxation time of the degrees of freedom of the system in the absence of the coupling to the bath, while the other, τ_B, is the characteristic time over which the degrees of freedom of the bath evolve and exchange energy with the system. By perturbing a system that is initially at equilibrium at temperature T, one typically injects (or withdraws) energy from the system and puts it out of equilibrium. Subsequently, the system starts to evolve. If $\tau_S \ll \tau_B$, then, on a timescale $\tau_S \ll t \ll \tau_B$, the system reaches a metastable equilibrium as if it were decoupled from the bath. Since energy is conserved—it is not exchanged on these timescales—the system equilibrates to a temperature T_S different from that of the bath. This is the microcanonical temperature corresponding to the new value of intensive energy reached after the perturbation. The effective temperature T_S then slowly evolves and reaches T only on larger timescales, of the order of τ_B; see Fig. 3.18.

An example of this kind of situation is discussed by Perfetti in Chapter 9 of this volume: so-called pump-and-probe experiments that have recently been the focus of intense experimental research in the field of out-of-equilibrium hard condensed matter. In these cases, one pumps energy into an electron system coupled to a phonon bath and then probes the subsequent dynamics, for instance by using a femtosecond laser pulse. Since the characteristic timescales of electrons and phonons are widely separated—femtoseconds and picoseconds, respectively—electrons are indeed experimentally observed to equilibrate at temperatures $T_S \gg T$. An example in which electrons reach a metastable equilibrium state characterized by a Fermi–Dirac distribution at a temperature $T_S = 2080\,\mathrm{K}$ while the ambient temperature of the phonon bath is $T = 130\,\mathrm{K}$ is discussed in Chapter 9.

3.6.2 Driven and active systems

We focus now on systems that are permanently out of equilibrium. Some of them are generically coupled to an environment that provides dissipation and an external drive, for example magnetic systems through which an electronic current flows. Others constantly dissipate energy in order to evolve dynamically, like bacteria or other biological "degrees of freedom." These systems generically reach a steady state that cannot be described by equilibrium statistical mechanics. There are a few cases, however, where, despite the fact that the system is driven or actively withdraws and then dissipates energy through the bath, a pseudo-equilibrium state with an effective temperature is reached. We shall now explain how this comes about.

We take the general setting of quantum systems and consider the simplified case where the total effect of the environment and the driving can be condensed into additional self-energy terms, as we have already explained in Section 3.2; see in particular (3.16). In the absence of driving and if the environment is at equilibrium at temperature T, the additional self-energy terms must obey the quantum fluctuation–dissipation relation (3.17). For systems driven out of equilibrium, there is no reason for the ratio $\nu(\omega)/\text{Im}\{\eta(\omega)\}$ to acquire the quantum fluctuation–dissipation form and hence no way to define an effective temperature. However, if the dynamics of the system takes place on timescales much larger than that typical of the drive and the environment, τ_E, the only thing that matters is the behavior of the additional self-energy terms at low frequency. Recalling that $\eta(\omega)$ and $\nu(\omega)$ entering in the Schwinger–Keldysh action (3.16) have the interpretation of a generalized friction and amplitude fluctuations, respectively, it is not unreasonable that $\text{Im}\{\eta(\omega)\} \simeq \eta_0 \omega \tau_E$ and $\nu(\omega) \simeq \nu(0)$ for $\omega \ll 1/\tau_E$, as indeed happens in many cases. If these small-frequency behaviors hold, then, for $\omega \ll 1/\tau_E$,

$$\frac{\nu(\omega)}{\text{Im}\{\eta(\omega)\}} \rightarrow \frac{\nu(0)}{\eta_0 \omega \tau_E} = \frac{2T_{\text{eff}}}{\omega}.$$

This implies that the effect of the environment and the driving is *as if* the slow degrees of freedom of the system were only coupled to a classical thermal bath at temperature T_{eff}. As a theoretical example where this phenomenon happens, we consider the model of a two-dimensional itinerant magnet placed between two noninteracting leads studied by Mitra and Millis [51–53]; see Fig. 3.19(a). The schematic phase diagram for

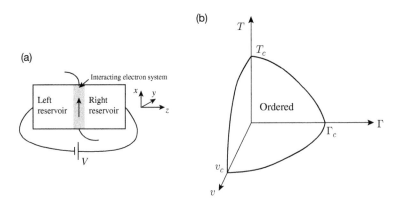

Fig. 3.19 (a) Schematic view of itinerant magnet placed between two noninteracting leads. (Reprinted with permission from [51]. Copyright (2006) by the American Physical Society.) (b) Schematic phase diagram.

this nonequilibrium system is shown in Fig. 3.19(b): it depicts a phase transition out
of equilibrium between a magnetically ordered state and a disordered state. A priori,
this transition should be unrelated to the equilibrium one. Instead, the mechanism explained above induces a classical pseudo-equilibrium state for the slow critical degrees
of freedom even when the quantum system is driven out of equilibrium by the voltage
and is applied to a zero-temperature bath: as far as universal quantities are concerned,
voltage acts as a temperature, as explained in [51–53]. Consequently, the critical behavior everywhere along the transition dome (except for $v = 0, T = 0$) corresponds
to classical critical slowing down, despite the fact that the system is strongly out of
equilibrium. Similarly, from a study of the slow out-of-equilibrium dynamics obtained
by suddenly changing the voltage from $v_i > v_c$ to $v_f < v_c$, it can be seen that the
system shows slow classical coarsening dynamics [2], even at zero temperature and in
the regime of strong quantum fluctuations. The reason is again that explained above:
the slow degrees of freedom—the domain walls in this case—evolve as if the total
effect of the environment, the driving, and the slow degrees of freedom is to provide a
thermal bath at a temperature T_{eff}.

Another interesting example is provided by active systems, a research topic that
ha recently received much attention in connection with the properties of living matter.
In these cases, elementary degrees of freedom move because they withdraw and then
dissipate energy from the environment. When the typical relaxation timescales of the
system become much longer than those of the bath, a phenomenon identical to that
discussed above takes place and a well-defined effective temperature emerges [8].

3.6.3 Glassy systems

We end this section by making an exception to what was stated in the introduction to
Section 3.5 and discussing briefly effective temperatures for glassy systems. Spin glasses
and structural glasses exhibit incredibly slow dynamics. After thermal quenches, they
display an aging dynamics richer than the coarsening dynamics discussed in the previous sections. As for coarsening, the aging dynamics is characterized by a separation of
timescales: short ones on which rapid degrees of freedom equilibrate and large ones on
which the slow degrees of freedom display out-of-equilibrium behavior. As we explained
in Section 3.2, one of the main signature of equilibrium dynamics is the fluctuation–
dissipation relation between correlation and response functions. The exact solution
of the aging dynamics of mean-field spin glasses [26] suggested that we look for generalizations of the fluctuation–dissipation relation out of equilibrium. Pragmatically,
one can define an effective temperature from the generalized fluctuation–dissipation
relation

$$-\frac{1}{T_{\text{eff}}}\partial_t C(t,t') = R(t,t'), \qquad (3.38)$$

where $C(t,t')$ and $R(t,t')$ are respectively the correlation and response functions for
a given observable. For an equilibrium system, this relation is satisfied with T_{eff} equal
to the equilibrium temperature and with C and R functions just of $t - t'$. In contrast,

during aging, these functions depend on both t and t' separately, and, using (3.38) to define T_{eff}, one generically finds an effective temperature that depends on t and t'.

It is remarkable that what can seem just a formal definition of temperature has been shown in the theoretical study of spin glasses and structural glasses to have a precise meaning at long times: T_{eff} is the effective temperature of the slow degrees of freedom, those that display aging dynamics [27]. Physically, this means that by probing the temperature of the system, one finds T or T_{eff}, depending on whether the timescales of the thermometer are short or long. This result was obtained first in the study of mean-field glass models and was then found in numerical simulations and a few experiments. This notion of effective temperature has been generalized also to (slowly) driven glassy systems, such as sheared supercooled liquids, where again it was found, first analytically and then in simulations, that slow degrees of freedom have an effective temperature different (generically higher) than the equilibrium one. Figure 3.20 shows the "classic" plot of the fluctuation–dissipation relation, in which the integrated response

$$\chi(t,t') = \int_t^{t'} ds\, R(s,t')$$

is plotted parametrically as a function of $C(t,t')$. The standard equilibrium fluctuation–dissipation relation would lead to

$$\chi = -\frac{1}{T}[C(t,t') - C(t,t)].$$

By normalizing the equal-time correlation function to 1, we would then find in equilibrium a straight line with slope $-1/T$ crossing the C axis at 1. Instead, the result found in simulations, theory, and a few experiments is that shown in Fig. 3.20: for large values of C, one indeed finds a slope $-1/T$, but below the value of C corresponding to

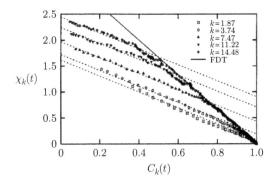

Fig. 3.20 Single-particle density fluctuations at wavevector k in a sheared supercooled liquid (from numerical simulations). Remarkably, all the different values of k are characterized by the same value of T_{eff}. (Reprinted with permission from [7]. Copyright (2002), AIP Publishing LLC.)

the plateau value in the plots of $C(t, t+\tau)$, one finds another slope equal to $-1/T_{\text{eff}}$. This indeed shows that the formal definition described above makes sense, since T_{eff} depends on t and t' only through C. The large values of C correspond to rapid degrees of freedom. These are the ones that relax from 1 to the plateau value in Fig. 3.7(b), and they are equilibrated at the bath temperature T as expected. Instead, the slow degrees of freedom, responsible for aging, which correspond to values of C lower than the plateau value, are characterized by an effective temperature[16] that is higher than T. It is not possible to give a brief theoretical explanation of this phenomenon, and so we refer the reader to a specific review on the subject [25].

We conclude by mentioning that although there is a theoretical framework that suggests and explains the existence of an effective temperature in glassy systems, and despite several simulations providing supporting evidence, numerical and experimental investigations are still ongoing to determine conclusively whether this notion of effective temperature holds for realistic systems [44]. The main issue is whether T_{eff} really is independent of observables, as a true temperature should be [55].

Acknowledgments

I wish to thank the organizers of the Les Houches Summer School of Physics on Strongly Interacting Quantum Systems Out of Equilibrium for their invitation to lecture. It was a great opportunity to think thoroughly about several topics in out-of-equilibrium dynamics.

Much of what I have learned about out-of-equilibrium dynamics, and which I have retransmitted to students, is due to collaboration and interaction with colleagues and friends. I will not list all their names here (too long a list would be needed ... I had a lot to learn), but I thank them all.

Last but not least, special thanks are due T. Giamarchi and L. F. Cugliandolo for their encouragement (and patience!).

References

[1] D. J. Amit and V. Martin-Mayor. *Field Theory, the Renormalization Group, and Critical Phenomena: Graphs To Computers* (3rd edn). World Scientific (2005).
[2] C. Aron, G. Biroli, and L. F. Cugliandolo. *Phys. Rev. Lett.* **102**, 050404 (2009).
[3] C. Aron, G. Biroli, and L.F. Cugliandolo. *J. Stat. Mech.* **1011**, P11018 (2010).
[4] I. Balog and G. Tarjus. *Phys. Rev. B* **91**, 214201 (2015).
[5] T. Barthel and U. Schollwöck. *Phys. Rev. Lett.* **100**, 100601 (2008).
[6] D. M. Basko, I. Aleiner, and B. A. Altshuler. *Ann. Phys. (NY)* **321**, 1126 (2006).
[7] L. Berthier and J.-L. Barrat. *J. Chem. Phys.* **116**, 6228 (2002).
[8] L. Berthier and J. Kurchan. *Nature Phys.* **9** 310 (2013).
[9] G. Biroli and J. Kurchan. *Phys. Rev. E*, **64**, 016101 (2001).

[16] Spin glasses are more complicated from this point of view. They are characterized by a hierarchy of T_{eff} values, one for each value of C [25].

[10] G. Biroli and O. Parcollet. *Phys. Rev. B* **65**, 094414 (2002).

[11] G. Biroli, C. Kollath, and A. M. Läuchli. *Phys. Rev. Lett.* **105**, 250401 (2010).

[12] G. Biroli, L. F. Cugliandolo, and A. Sicilia. *Phys. Rev. E* **81**, 050101(R) (2010).

[13] B. Sciolla and G. Biroli. *Phys. Rev. B* **88**, 201110R (2013).

[14] A. Bovier, M. Eckhoff, V. Gayrard, and M. Klein. *J. Phys. A. Math. Gen.* **33**, L447 (2000).

[15] A. Bovier, M. Eckhoff, V. Gayrard, and M. Klein. *Commun. Math. Phys.* **228** 219 (2002).

[16] A. Bovier, M. Eckhoff, V. Gayrard, and M. Klein. *J. Eur. Math Soc.* **6**, 399 (2004).

[17] A. J. Bray and G. J. Rodgers. *Phys. Rev. B* **38**, 9252 (1988).

[18] A. J. Bray. *Adv. Phys.* **51** 481 (2002).

[19] P. Calabrese and A. Gambassi. *J. Phys. A Math. Gen.***38**, R133 (2005).

[20] M. S. Cao and J. Machta. *Phys. Rev. B* **48**, 3177 (1993).

[21] J. Cardy. *Scaling and Renormalization in Statistical Physics*. Cambridge University Press (1996).

[22] L. Casetti, M. Cerruti-Sola, M. Pettini, and E. G. D. Cohen. *Phys. Rev. E* **55**, 6566 (1997).

[23] P. Castiglione, M. Falcioni, A. Lesne, and A. Vulpiani. *Chaos and Coarse Graining in Statistical Mechanics*. Cambridge University Press (2008).

[24] G. E. Crooks. *J. Stat. Phys.* **90**, 1481 (1998).

[25] L. F. Cugliandolo. Dynamics of glassy systems. In *Slow Relaxations and Nonequilibrium Dynamics in Condensed Matter. Les Houches, Session LXVII, July 2002* (ed. J.-L. Barrat et al.), p. 367. EDP Sciences/Springer-Verlag (2003).

[26] L. F. Cugliandolo and J. Kurchan. *J. Phys. A Math. Gen.* **27**, 5749 (1994).

[27] L. F. Cugliandolo, J. Kurchan, and L. Peliti. *Phys. Rev. E* **55**, 3898 (1997).

[28] J. M. Deutsch. *Phys. Rev. A* **43**, 2046 (1991).

[29] B. Doyon and N. Andrei. *Phys. Rev. B* **73** 245326 (2006).

[30] M. Esposito, U. Harbola, and S. Mukamel. *Rev. Mod. Phys.* **81** 1665 (2009).

[31] D. S. Fisher, *Phys. Rev. Lett.* **56**, 416 (1986).

[32] C. W. Gardiner and P. Zoller. *Quantum Noise*. Springer-Verlag (2000).

[33] B. Gaveau and L. S. Schulman *J. Math. Phys* **37**, 3897 (1996).

[34] B. Gaveau and L. S. Schulman. *Phys. Lett. A* **229**, 347 (1997).

[35] B. Gaveau and L. S. Schulman. *J. Math. Phys.* **39**, 1517 (1998).

[36] T. Giamarchi. Disordered elastic media. In *Encyclopedia of Complexity and Systems Science* (ed. R. A. Meyers et al.), p. 2019. Springer-Verlag (2009).

[37] E. Gozzi. *Phys.Lett. B* **143**, 183 (1984).

[38] R. B. Griffiths. *Phys. Rev. Lett.* **23**, 17 (1969).

[39] C. Henley. *J. Phys. Cond Matt.* **16**, S891 (2004).

[40] C. Jarzynski. *Phys. Rev. Lett.* **78**, 2690 (1997).

[41] C. Jarzynski. *Phys. Rev. E* **56**, 5018 (1997).

[42] A. Kamenev. *Field Theory of Non-Equilibrium Systems*. Cambridge University Press (2011).

[43] J. Kockelkoren and H. Chaté. *Physica D* **168** 80 (2002).

[44] J. Kurchan. *Nature* **433**, 222 (2005).

[45] J. Kurchan. Six out-of-equilibrium lectures. In *Long-Range Interacting Systems. Les Houches, Session XC, August 2008* (ed. T. Dauxois et al.), p. 67. Oxford University Press (2010).

[46] A. J. Leggett, S. Chakravarty, A. T. Dorsey, M. P. A. Fisher, A. Garg, and W. Zwerger. *Rev. Mod. Phys.* **59**, 1 (1987).

[47] G. D. Mahan. *Many Particle Physics* (3rd edn). Springer-Verlag (2000).

[48] F. Martinelli. *J. Stat. Phys.* **92** 337 (1998).

[49] P. Mazur and E. Montroll. *J. Math. Phys.* **1**, 70 (1960).

[50] A. A. Middleton and D. S. Fisher. *Phys. Rev. B* **65**, 134411 (2002).

[51] A. Mitra, S. Takei, Y. B. Kim, and A.J. Millis. *Phys. Rev. Lett.* **97**, 236808 (2006).

[52] A. Mitra and A. J. Millis. *Phys. Rev. B* **77** 220404 (2008).

[53] A. Mitra and A. J. Millis. *Phys. Rev. B* **84** 054458 (2011).

[54] R. Nandkishore and D. A. Huse. *Annu. Rev. Cond. Matt. Phys.* **6**, 15 (2015).

[55] A. S. Ninarello, N. Gnan, and F. Sciortino. *J. Chem. Phys.* **141**, 194507 (2014).

[56] M. Rigol, V. Dunjko, and M. Olshanii. *Nature* **452**, 854 (2008).

[57] D. Rokhsar and S. Kivelson. *Phys. Rev. Lett.* **61**, 2376 (1988).

[58] S. Sachdev. *Quantum Phase Transitions* (2nd edn). Cambridge University Press (2011).

[59] U. Seifert. *Eur. Phys. J. B* **64**, 423 (2008).

[60] A. Sicilia, J. J. Arenzon, A. J. Bray, and L. F. Cugliandolo. *Phys. Rev. Lett.* **98**, 145701 (2007).

[61] A. Sicilia, J. J. Arenzon, A. J. Bray, and L. F. Cugliandolo. *Phys. Rev. E* **76**, 061116 (2007).

[62] M. Srednicki. *Phys. Rev. E* **50**, 888 (1994).

[63] M. Tissier and G. Tarjus. *Phys. Rev. B* **85**, 104203 (2012).

[64] J. Villain. *Phys. Rev. Lett.* **52**, 1543 (1984).

[65] U. Weiss. *Quantum Dissipative Systems.* World Scientific (2008).

[66] K. J. Wiese and P. Le Doussal. *Markov Process. Relat. Fields* **13**, 777 (2007).

[67] J. Zinn-Justin. *Quantum Field Theory and Critical Phenomena* (4th edn). Oxford University Press (2002).

[68] W. H. Zurek. *Phys. Rep.* **276**, 177 (1996).

[69] R. Zwanzig. *Nonequilibrium Statistical Mechanics.* Oxford University Press (2001).

4

Numerical methods in the study of nonequilibrium strongly interacting quantum many-body physics

Ulrich SCHOLLWÖCK

Department of Physics,
University of Munich,
Theresienstrasse 37,
80333 Munich, Germany

Strongly Interacting Quantum Systems out of Equilibrium. First Edition. Thierry Giamarchi et al.
© Oxford University Press 2016. Published in 2016 by Oxford University Press.

Chapter Contents

4.1   Setting the problem

A lot of attention in quantum physics is paid to the properties of ground states of many-body systems—rightly so, because ground states form the basis for understanding a quantum system and often reveal interesting and unexpected phenomena at very low temperatures, where quantum effects dominate. However, it is quantum *dynamics* that is driving the unfolding history of our universe! Indeed, for most physicists, doing quantum physics in practice means dealing with the Schrödinger equation, which governs the nonrelativistic dynamics of quantum systems, although conceptually the choice of a linear state space, the superposition principle, the tensor product nature of many-body state spaces, and the resulting phenomenon of entanglement are much more fundamental.

Exactly solvable many-body dynamics is very rare, as it implies full knowledge of all eigenstates and eigenenergies. We can loosely distinguish between dynamical systems that are "close" to an exact solution, such as an exactly solvable model plus some perturbation or some interaction that is so weak that the system can be treated perturbatively, and systems where no such perturbative scheme can work, for example because of a strong interaction. This chapter will focus completely on the latter type of system, which has seen a dramatic surge in activity in recent years. On the one hand, progress in computing power and numerical methods has started to make such systems accessible to numerical analysis. On the other hand, very attractive experimental realizations have emerged.

The physics of *ultracold atoms* offers particularly rich theoretically and experimentally accessible dynamics. We shall therefore use it here to highlight a few of the questions that may be attacked by the numerical methods we are going to explore. Consider a gas of $O(10^6)$ ^{87}Rb atoms that have been confined in a magneto-optical trap at an ultralow temperature of about $100\,\mathrm{nK}$, many orders of magnitude below the cosmic microwave background radiation temperature of the universe, $2.7\,\mathrm{K}$. In practice, such a gas is extremely dilute and two-body collisions dominate. At the same time, the ultralow temperature ensures that kinetic energies are so small that scattering processes occur almost exclusively in the s-wave scattering channel: for such low energies, the internal structure of the atom with its fermionic constituents (protons, neutrons, and electrons) is not resolved, and the rubidium atoms act as bosons, since they consist of an even number of fermions, such that quantum statistics allows s-wave scattering. As the scattering length a_s is extremely short compared with average interatomic distances, the atoms are almost noninteracting, and form a Bose–Einstein condensate (BEC) at low enough temperatures, providing the cleanest experimental setup to date for this quantum phenomenon.

The observation of a purely statistical Bose–Einstein condensation in 1995 [3, 21] was a major experimental feat, duly awarded the Nobel Prize in 2001. The very weak repulsive interatomic interaction is nice to have, since the quantum phenomenon is minimally obscured by it—it just serves to ensure thermal equilibration—but on the other hand it limits the wealth of physical phenomena: in condensed matter physics, phenomena become particularly rich if interaction energy and kinetic energy compete, which is just the kind of situation in which we are interested in this chapter.

In the meantime, various ways of setting up such a competition for ultracold dilute atomic gases have emerged.

One of these is provided by the introduction of optical lattices: adding standing light-wave patterns by counter-propagating laser beams of wavelength λ (and wavenumber $k = 2\pi/\lambda$) in all three spatial directions creates a lattice-like structure where intensity maxima (or minima) are separated by $\lambda/2$, typically several hundred nanometers. If the wavelength is red-detuned compared with the single dominant absorption line of the alkali atom ^{87}Rb, the AC Stark effect attracts the atoms to the intensity maxima, which therefore act as lattice minima. Effectively, the atoms will assume positions as in a crystalline lattice, and their wavefunctions will form Bloch bands; let us assume that because of their very low energies, atoms will reside exclusively in the lowest Bloch band. As the scattering length is very short compared with the lattice spacing, the atoms will only interact appreciably if they are on the same lattice site. On the other hand, atomic motion is now confined to tunneling between neighboring lattice sites, which strongly suppresses kinetic energy. For suitable optical lattice intensities (or potential depths), kinetic and interaction energies turn out to be of comparable size and compete strongly. The use of Feshbach resonances provides another way of tuning the ratio of interaction to kinetic energy, which we will not pursue further here.

Theoretically, confined ^{87}Rb atoms in an optical lattice are modeled almost perfectly [40] by the single-band Bose–Hubbard model, which is closely related to its fermionic partner, the "original" single-band Hubbard model:

$$\hat{H}_{\mathrm{BH}} = -J \sum_{\langle i,j \rangle} (\hat{b}_i^\dagger \hat{b}_j + \hat{b}_j^\dagger \hat{b}_i) + U \sum_i \hat{n}_i(\hat{n}_i - 1)/2 + \sum_i \mu_i \hat{n}_i. \qquad (4.1)$$

Here, $\hat{b}_i^\dagger$ and $\hat{b}_i$ are the bosonic creation and annihilation operators on lattice site i; the first sum runs over pairs of nearest-neighbor lattice sites $\langle i,j \rangle$ and represents kinetic energy. We follow the convention of quantum optics of calling the tunneling amplitude J (as opposed to the t in condensed matter physics, where J is used to denote the magnetic exchange constant), since we want to reserve t for time. The second term models the interaction energy, which is taken to be $U > 0$ (repulsive) for every pair of bosons on the same lattice site i; $\hat{n}_i$ is just the number operator on site i. The last term looks like a chemical potential (with the suggestive μ_i), but in fact it models the site-dependent strength of the confining trap potential, which we take to be parabolic and, for simplicity, isotropic about the trap center, $\mu_i = \frac{1}{2}\omega_{\mathrm{trap}}^2 d_i^2$, where d_i is the distance between site i and the trap center. The number of atoms in a trap is conserved (except for undesired loss processes), so we are looking at a canonical, rather than a grand canonical, ensemble, and hence no conventional chemical potential appears.

Of course, many modifications are imaginable: the trap can be of oblong form; the lattice depth may vary with the spatial axis, introducing directionally dependent tunneling and allowing motion to be confined to one-dimensional tubes or two-dimensional sheets; and much more. For a glimpse of the enormous physical wealth provided by ultracold atoms in an optical lattice, see the review by Bloch et al. [11].

Let us now focus on nonequilibrium physics in these systems. A particularly attractive feature of dilute atomic gases in optical lattices is that Hamiltonian parameters can be calculated to high accuracy, as opposed to solids, where they are often very hard to determine. For the Bose–Hubbard Hamiltonian (4.1), the hopping parameter J would depend on the depth of the optical lattice $V(x) = V_{\text{lat}} \sin^2 kx$, scaled in units of the recoil energy $E_{\text{r}} = \hbar^2 k^2 / 2m$, where k is the wavenumber of the lattice lasers, as [86]

$$J(V_{\text{lat}}) \approx \sqrt{\frac{8}{\pi}} \left(\frac{V_{\text{lat}}}{E_{\text{r}}} \right)^{3/4} \text{e}^{-2\sqrt{V_{\text{lat}}/E_r}} \tag{4.2}$$

and the interaction U as

$$U(V_{\text{lat}}) \approx 4\sqrt{2\pi} \, \frac{a_{\text{s}}}{\lambda} \frac{V_{\text{lat}}}{E_{\text{r}}}. \tag{4.3}$$

(For simplicity, we have switched to one spatial dimension here.) Interestingly, the interaction depends only weakly on the lattice depth, which merely squeezes local atom wavefunctions, but the hopping depends exponentially on it! This allows to vary the ratio of kinetic to interaction energy by several orders of magnitude, paving the way to time-dependent out-of-equilibrium phenomena on a large scale.

If one changes the parameters of the Hamiltonian very slowly, one can observe adiabatic transformations of ground states, as was done in the seminal paper by Greiner et al. [30], where a superfluid condensate was transformed into a Mott insulating state; these two states are separated by a second-order quantum phase transition in the zero-temperature phase diagram. In many condensed matter systems, changing Hamiltonian parameters in a controlled way is very difficult, so this is a very appealing feature of cold-atom systems.

But further interesting physics emerges if one asks on which timescales the ratio of energies can be tuned. Ultracold atoms are characterized by very low energies and therefore very long timescales; typical timescales are of the order of milliseconds, which in quantum mechanics is almost an eternity compared with the phenomena of condensed matter physics. It is therefore now experimentally possible to change parameters much *faster* than typical quantum timescales, leading to a sudden and drastic change of the Hamiltonian, a so-called *quench*, which transforms a ground state (with respect to the old Hamiltonian) into a highly excited nonequilibrium state (with respect to the new Hamiltonian). This allows one to observe highly nontrivial quantum many-body dynamics, but poses an enormous theoretical challenge: the well-developed techniques of linear response, where the perturbation to the system (such as adding a particle, as in the calculation of a Green's function) is nonextensive and makes no energetic contribution in the thermodynamic limit, obviously do not apply. Exact analytical solutions for the full many-body state, as it explores the Hilbert space in time, are extremely rare.

One such exception is provided by the "collapse-and-revival" quench experiment of Greiner et al. [31]: a superfluid state ($U \sim J$) is quenched into the very deep Mott regime, i.e., $U \gg J$, by a sudden increase in the optical lattice depth. To a first approximation, tunneling between sites can then be neglected, and the full many-body Hamiltonian decouples into independent many-body single-site problems, with a Hamiltonian $\hat{H} = U\hat{n}(\hat{n}-1)/2$. Number states with n bosons are eigenstates of this Hamiltonian, with energies $E_n = Un(n-1)/2$. The quantum state, decomposed into these eigenstates, will evolve in time with different time-dependent phases for each term, namely, $e^{-iUn(n-1)t/(2\hbar)}$, destroying the superposition of the original quantum state ("collapse"). The very special level scheme, however, ensures that after multiples of times h/U, all time-dependent phases will be multiples of $2\pi i$, and hence as at $t = 0$, and the original wavefunction resurfaces ("revival"). This problem can still be solved analytically, but immediately becomes highly complicated if the small-hopping approximation breaks down. Then atoms will tunnel from site to site, nontrivial correlations between sites will arise, and we have to look at the full many-body Schrödinger equation. Can we solve it in general?

So far, the discussion has focused on pure states. Often this is good enough, but not always: if we want to cool fermionic atoms to ultralow temperatures, the Pauli principle prevents s-wave scattering, the only scattering process relevant at very low energy scales, making thermalization extremely ineffective. One way out is provided by sympathetic cooling: a background of bosonic atoms, who can collide both with each other and with the fermions in the s-wave channel, makes (mediated) thermalization for the fermions effective again. But the fermions can no longer be considered to be isolated! To describe the state of the fermionic atoms alone, a density operator description for a mixed state has to replace a pure-state description. As it turns out that cooling is not as effective as for bosons, we have to deal with the temperature (better: entropy) of the remaining fermions, if we remove the bosons after cooling by some experimental tricks. So, even then, the fermions cannot be described in the framework of a pure state. How can we calculate the dynamics of a mixed state? All states of Hilbert space may contribute, in contrast to the pure-state case. Is it still possible to come up with an effective scheme?

But this is not all: while these experiments are closed quantum systems, at least on short timescales, they will interact with the rest of the universe, the environment, in due course. For example, quite rarely there is a collision (interaction) not only between two atoms, where low energies and conservation laws prevent particle losses, but also between three particles, where ultimately the lattice may lose particles to the environment because they acquire energies that can no longer be confined. We therefore also have to cope with open quantum systems with dissipative dynamics. What can be done about those?

I claim that in all these cases at least partial answers can be found numerically. Let me outline the contents of this chapter. In a first step, we will establish the physical formulas that have to be evaluated numerically. Many of them should be quite familiar, perhaps with the exception of the simulation of open quantum systems. We will then move on to discuss the mathematical framework for all these problems, part of it being quite basic linear algebra, part of it an extensive discussion of the framework of Krylov spaces, on which all algorithms for the treatment of large sparse matrices (finding eigenvalues, solving equation systems) are based—and it turns out that our algorithms will make heavy use of them. In a third step, we will discuss quasi-exact simulations in the full Hilbert space, which of course will restrict us to relatively small system sizes, given the exponential explosion of Hilbert space with system size. They can deal with any type of system (provided it can be formulated meaningfully on a not-too-large lattice), and the results are exact and transparent. Therefore, these simulations give invaluable insights despite size constraints and provide the "mother algorithms" to be implemented in other approaches. In a fourth step, we will discuss whether it is possible to abandon the costly treatment of the full Hilbert space while identifying subspaces that can be handled numerically and still contain all of the essential physics. Related algorithms, which are essentially effective in one spatial dimension, will be discussed under the heading of "matrix-product state simulations" (or "density-matrix renormalization group simulations"). We will see that this strategy works very well on short timescales, but encounters fundamental issues of quantum physics as time evolves.

In 2005, I published an extensive review on the density-matrix renormalization group (DMRG) [61], which I still consider to be a good introduction to the DMRG in the language of statistical physics in which it was originally invented [78, 79] and which, given the usual training of a physics student, may be the easiest way of learning it. However, as is often the case in physics, the original path to a new idea is not necessarily the conceptually most straightforward one (just think about Planck's black-body radiation formula as the first step into quantum physics), so that, later on, a different approach or notation becomes the canonical one. I claim that in the last decade it has become very obvious that it is much more convenient to understand the DMRG (and its variants) as a method that does not work on the full many-body Hilbert space but in the subspace of so-called matrix-product states (MPS). MPS, as we will see, are a very special class of states that have peculiar physical properties, can be manipulated computationally at a speed that scales polynomially in their size (where we will discuss the meaning of size below), but definitely are a bit cumbersome and alien in their notation; however, the real conceptual core of the DMRG becomes much more visible. This is why the change to the MPS notation has sparked a lot of algorithmic development because it freed people's thinking from the inessential. In 2011, I published a review [62] on the DMRG in MPS formulation in which I hope to have established a consistent mathematical and graphical notation and in which I have provided a very detailed exposition of all major algorithms using MPS. As opposed to the first review, which aimed at giving a broad overview of algorithmic ideas and physical applications, the second review had a student in mind who wants to start programming and therefore is interested in the computational details (for less "computational" reviews, see [34, 73]). As I expect that many readers of this chapter, which does not have a specific focus on simulations using MPS but rather deals with nonequilibrium simulations for strongly correlated systems in general, will not aim at immediately implementing codes, I will summarize the algorithms, using the notation and presentation of [62], and refer interested readers to this easily accessible review. I will be more explicit about the calculation of spectral functions, as there have been interesting developments in the last few years that are not covered at all in [62].

4.2 Describing quantum many-body dynamics

4.2.1 Pure-state dynamics of closed quantum systems

For a pure state, we consider the *time-dependent Schrödinger equation*

$$\hat{H}|\psi(t)\rangle = i\partial_t|\psi(t)\rangle \tag{4.4}$$

with some initial condition $|\psi(t=0)\rangle$. As is conventional in the field, we have set $\hbar \equiv 1$. Let us assume for simplicity that $\hat{H}$ is time-independent; the generalization will be trivial for all numerical methods to come. Then the formal solution reads

$$|\psi(t)\rangle = e^{-i\hat{H}t}|\psi(0)\rangle. \tag{4.5}$$

If we knew the eigenbasis via $\hat{H}|\psi_n\rangle = E_n|\psi_n\rangle$, and the representation of $|\psi(0)\rangle = \sum_n c_n|\psi_n\rangle$ with $c_n = \langle\psi_n|\psi(0)\rangle$, the formal solution could be evaluated as

$$|\psi(t)\rangle = \sum_n c_n e^{-iE_n t}|\psi_n\rangle \equiv \sum_n c_n(t)|\psi_n\rangle. \tag{4.6}$$

Only relative phases change in this superposition, just as in the collapse-and-revival example above. The weight of each eigenstate $|\psi_n\rangle$ remains the same, $|c_n(t)|^2 = \text{constant}$. The crux is that for a general quantum problem, finding the *eigenbasis* is exponentially costly, and is impossible for almost all cases of interest. Usually, we calculate in some convenient *computational basis*, for example, for an assortment of L spin-$\frac{1}{2}$ objects, the basis $\{|\uparrow\rangle, |\downarrow\rangle\}^{\otimes L}$. If we call these states $|k\rangle$, then a double insertion of the identity, $\hat{I} = \sum_n |\psi_n\rangle\langle\psi_n|$ and $\hat{I} = \sum_k |k\rangle\langle k|$, gives

$$|\psi(t)\rangle = \sum_k \left(\sum_n e^{-iE_n t} c_n d_{kn} \right) |k\rangle, \tag{4.7}$$

where the c_n are as before and $d_{kn} = \langle k|\psi_n\rangle$. This implies no simplification, but we can see that the coefficients of $|k\rangle$ change not only phase, but also amplitude. This means that the state, unless it is an eigenstate, will very generally explore Hilbert space.

4.2.2 Mixed-state dynamics of closed quantum systems

$|\psi\rangle$ is not the most general description of a quantum state, which is provided by a density operator out of the set of Hermitian operators $\hat{\rho}$ with $\hat{\rho} \geq 0$, $\text{Tr}\,\hat{\rho} = 1$, and $\hat{\rho}^2 \leq \hat{\rho}$. Such density operators might occur as the description of a subsystem A of a composite system AB in a pure state, $\hat{\rho}_A = \text{Tr}_B |\psi\rangle_{AB}\,{}_{AB}\langle\psi|$, or as the statistical description of a thermal system, $\hat{\rho} = Z^{-1}e^{-\beta\hat{H}}$, with $Z = \text{Tr}\,e^{-\beta\hat{H}}$.

If we take A to be a closed quantum system, there is no interaction with the environment B, and hence $\hat{H}_{AB} = \hat{H}_A + \hat{H}_B$, where $[\hat{H}_A, \hat{H}_B] = 0$ since they act on different Hilbert spaces. Then the pure state of AB has dynamics

$|\psi(t)\rangle_{AB} = e^{-i\hat{H}_A t} e^{-i\hat{H}_B t} |\psi(0)\rangle_{AB}$. From this, we can find $\hat{\rho}_{AB}(t) = |\psi(t)\rangle_{AB}\,_{AB}\langle\psi(t)|$, and, by evaluating the trace in $\hat{\rho}_A(t) = \mathrm{Tr}_B\,\hat{\rho}_{AB}(t)$ in the eigenbasis of $\hat{H}_B$, we obtain

$$\hat{\rho}_A(t) = e^{-i\hat{H}_A t} \hat{\rho}_A(0) e^{+i\hat{H}_A t}. \tag{4.8}$$

Inserting this into $\hat{\rho}(t + dt) - \hat{\rho}(t)$ (we drop the subscript A for convenience) and expanding to first order in dt, we obtain a first-order differential equation, the *von Neumann equation*,

$$\frac{d\hat{\rho}}{dt} = -i[\hat{H}, \hat{\rho}]. \tag{4.9}$$

4.2.3   Dynamics of open quantum systems

Let us now assume that our quantum system A truly couples to the environment B. If taken to the extreme, B may comprise the entire rest of the universe. The Hamiltonian of the entire "universe" AB will now read $\hat{H}_{AB} = \hat{H}_A + \hat{H}_B + \hat{H}_{int}$, where $[\hat{H}_A, \hat{H}_B] = 0$, but $[\hat{H}_A, \hat{H}_{int}] \neq 0$ and $[\hat{H}_B, \hat{H}_{int}] \neq 0$. The new interaction term $\hat{H}_{int}$ couples A to B, which invalidates the above derivation of the von Neumann equation. What do we get instead?

Let us assume that at $t = 0$ the state of the universe is a product state given by

$$|\psi(0)\rangle_{AB} = |\psi(0)\rangle_A \otimes |0\rangle_B. \tag{4.10}$$

Assuming pure states on A and B is no loss of generality: as we will discuss later, a suitable enlargement of the state spaces of A and B allows any mixed state to be written as a pure state in the enlarged state space (purification). Assuming that the states of system and environment factorize at $t = 0$ is a restriction that we will discuss later on. While AB is again a closed quantum system, A alone is not—it interacts with B and is referred to as an *open* quantum system. The dynamics of AB is Hamiltonian, i.e., unitary, while that of A alone is non-Hamiltonian, nonunitary, or dissipative. Formally, this can be represented quite simply, and has been dealt with extensively in the literature. We closely follow Nielsen and Chuang's book [47] and Preskill's wonderful set of lecture notes [55]. For an excellent overview of the entire field of open quantum systems, see Breuer and Petruccione's book [12]. The dynamics of $|\psi(t)\rangle_{AB}$ is given by

$$|\psi(t)\rangle_{AB} = e^{-i\hat{H}_{AB} t}(|\psi(0)\rangle_A \otimes |0\rangle_B), \tag{4.11}$$

and hence that of $\hat{\rho}_{AB}(t) = |\psi(t)\rangle_{AB}\,_{AB}\langle\psi(t)|$ by

$$\hat{\rho}_{AB}(t) = e^{-i\hat{H}_{AB} t}(|0\rangle_B |\psi(0)\rangle_A\,_A\langle\psi(0)|\,_B\langle 0|) e^{+i\hat{H}_{AB} t}. \tag{4.12}$$

Now the typical physical situation is that we are interested only in A and that B remains a more or less elusive "bath" with its own inaccessible internal dynamics. We obtain the dynamics of A by tracing out the degrees of freedom of B, whose basis is given by an orthonormal set $\{|j\rangle_\mathrm{B}\}$, with $|0\rangle_\mathrm{B}$ the initial state of B in this basis. Then $\hat\rho_\mathrm{A}(t) = \mathrm{Tr}_\mathrm{B}\,\hat\rho_\mathrm{AB}(t)$, or

$$\hat\rho_\mathrm{A}(t) = \sum_j {}_\mathrm{B}\langle j|e^{-i\hat H_\mathrm{AB}t}|0\rangle_\mathrm{B} \left(|\psi(0)\rangle_\mathrm{A}\; {}_\mathrm{A}\langle\psi(0)|\right) {}_\mathrm{B}\langle 0|e^{+i\hat H_\mathrm{AB}t}|j\rangle_\mathrm{B}. \tag{4.13}$$

Identifying $\hat\rho_\mathrm{A}(0) = |\psi(0)\rangle_\mathrm{A}\; {}_\mathrm{A}\langle\psi(0)|$ as the reduced density operator of A at $t=0$ and introducing the *Kraus operators*

$$\hat E_\mathrm{A}^j(t) = {}_\mathrm{B}\langle j|e^{-i\hat H_\mathrm{AB}t}|0\rangle_\mathrm{B}, \tag{4.14}$$

we see that the time evolution of $\hat\rho_\mathrm{A}$ can be written in the *Kraus representation* as

$$\hat\rho_\mathrm{A}(t) = \sum_j \hat E_\mathrm{A}^j(t)\hat\rho_\mathrm{A}(0)\hat E_\mathrm{A}^{j\dagger}(t). \tag{4.15}$$

It is easy to see by insertion that the unitarity of the time evolution of AB implies that the Kraus operators meet the condition

$$\sum_j \hat E_\mathrm{A}^{j\dagger}(t)\hat E_\mathrm{A}^j(t) = \hat I_\mathrm{A}. \tag{4.16}$$

The Kraus operator $\hat E_\mathrm{A}^j(t)$ governs the time evolution of A provided that the environment was in state $|0\rangle_\mathrm{B}$ at time 0 and is in state $|j\rangle_\mathrm{B}$ at time t. Of course, this does not preclude it having been in states other than $|0\rangle_\mathrm{B}$ and $|j\rangle_\mathrm{B}$ in the meantime. For more generic situations, obtaining the Kraus operators in practice is not possible.

An important simplification occurs if the dynamics of A is without memory (or *Markovian*). Then it depends only on the density operator of A at an infinitesimally earlier time, and a first-order differential equation can be derived, which is called the master equation or, in this case, the *Lindblad equation*. When does physics allow this? In this discussion, we follow the arguments given by Haroche and Raimond [35]. To obtain a first-order differential equation, we linearize as

$$\hat\rho_\mathrm{A}(t+dt) - \hat\rho_\mathrm{A}(t) \equiv dt\,\frac{d\hat\rho_\mathrm{A}}{dt} \tag{4.17}$$

where dt is a small but finite time interval. For this to make sense, $dt \ll \tau_\mathrm{A}$, where τ_A is the characteristic timescale of A, in a sense that we will define below. At the same time, at some time t, the density operator of AB reads in fact

$$\hat\rho_\mathrm{AB}(t) = \hat\rho_\mathrm{A} \otimes [\bar\rho_\mathrm{B} + \delta\hat\rho_\mathrm{B}(t)] + \delta\hat\rho_\mathrm{AB}(t), \tag{4.18}$$

where $\bar\rho_\mathrm{B} = |0\rangle_\mathrm{B}\,{}_\mathrm{B}\langle 0|$ is the equilibrium state of B, which we assume right away to be a huge environment (otherwise Markovian dynamics will not apply for A; see below), $\delta\hat\rho_\mathrm{B}(t)$ are time-dependent fluctuations around the equilibrium state, and $\delta\hat\rho_\mathrm{AB}(t)$ is the entire entangled part of $\hat\rho_\mathrm{AB}(t)$. If we had to take the two "corrections" $\delta\hat\rho_\mathrm{B}(t)$ and $\delta\hat\rho_\mathrm{AB}(t)$ into account, the Kraus formalism would not emerge. But we do not have to do it: if B is huge, its energies span a huge range $\Delta E = \Delta\omega$ (since $\hbar \equiv 1$), which sets a very short relaxation timescale of $\tau_B = 1/\Delta\omega$. Memories (i.e., correlations between the fluctuations at different times, or the entangled part of the density operator $\hat{\rho}_{AB}$ at different times) die off on the scale τ_B. So, AB evolves coherently on a timescale τ_B, but on longer timescales, the evolution of $\hat{\rho}_A(t)$ can be understood as a random walk with (coherent) time steps $O(\tau_B)$. If $\|\hat{H}_{int}\| \sim V$, phases in A change as $V\tau_B$ in one coherent step. Over a time t, phase changes add quadratically (the direction of the change is random), and we have

$$[\Delta\phi(t)]^2 \sim V^2\tau_B^2 \frac{t}{\tau_B} \equiv \frac{t}{\tau_A}, \qquad \tau_A = \frac{1}{V^2\tau_B}, \tag{4.19}$$

which gives the characteristic timescale for A. For a Markovian assumption to be valid, we must have the characteristic timescale of B much shorter than that of A, $\tau_B \ll \tau_A$, or

$$V\tau_B \ll 1. \tag{4.20}$$

Our small finite time step dt must be longer than τ_B, otherwise memory effects would matter. So, the Markovian assumption is valid if we can choose

$$\tau_B \ll dt \ll \tau_A. \tag{4.21}$$

Let us now derive the Lindblad equation that arises under these assumptions, following the derivation by Preskill [55]. In the limit $dt \to 0$, the environment remains unchanged with probability $p_0 \to 1$ and changes (quantum jumps into one of its other states) with a probability linear in dt. If we associate the Kraus operator $\hat{E}_A^0(t)$ with the absence of change, a meaningful ansatz for the Kraus operators that scales out time is

$$\hat{E}_A^0(dt) = \hat{I}_A + O(dt), \qquad \hat{E}_A^j(dt) = \sqrt{dt}\,\hat{L}_A^j \quad (j > 0), \tag{4.22}$$

or, more precisely,

$$\hat{E}_A^0(dt) = \hat{I}_A + (\hat{K}_A - i\hat{H}_A)\,dt, \tag{4.23}$$

with two Hermitian operators $\hat{K}_A$ and $\hat{H}_A$. How can we understand this ansatz? Equation (4.23) is nothing other than the unique decomposition of the operator multiplying dt into an anti-Hermitian and a Hermitian part. The normalization condition of the Kraus operators, (4.16), entails that

$$\hat{K}_A = -\tfrac{1}{2}\sum_{j>0} \hat{L}_A^{j\dagger}\hat{L}_A^j. \tag{4.24}$$

If we assume that A has no interaction with B, then B stays in the state $|0\rangle_B$, all $\hat{L}_A^j = 0$, and hence $\hat{K}_A = 0$, and we can identify $\hat{H}_A$ with the Hamiltonian acting on A in order to recover von Neumann dynamics. In the general case, $\hat{H}_A$ will show level shifts compared with the Hamiltonian on A alone, since processes involving B can change energies; the most famous example is given by the interaction of an atom with vacuum fluctuations, which causes the Lamb shifts in its energy levels.

If we insert (4.22) and (4.23) into $\hat{\rho}_A(t + dt) - \hat{\rho}_A(t)$, we can derive a differential equation from (4.15), which is the Lindblad equation:

$$\frac{d\hat{\rho}}{dt} = -i[\hat{H}, \hat{\rho}] + \sum_{j>0} \left(\hat{L}^j \hat{\rho} \hat{L}^{j\dagger} - \tfrac{1}{2}\{\hat{L}^{j\dagger}\hat{L}^j, \hat{\rho}\} \right), \tag{4.25}$$

where the indices A have been dropped to simplify the notation. The last term is an anticommutator and the $\hat{L}^j$ are called quantum jump operators. In the absence of quantum jumps ($j = 0$ only) and no coupling to the environment, one recovers the von Neumann equation. At the price of non-Hermiticity, this equation can be simplified. If we introduce $\hat{H}_{\text{eff}} := \hat{H} + i\hat{K}$, then the last term is absorbed and we have

$$\frac{d\hat{\rho}}{dt} = -i[\hat{H}_{\text{eff}}\hat{\rho} - \hat{\rho}\hat{H}_{\text{eff}}^\dagger] + \sum_{j>0} \hat{L}^j \hat{\rho} \hat{L}^{j\dagger}. \tag{4.26}$$

The appearance of a prefactor $\sqrt{dt}$ in the $\hat{E}_A^j(dt)$ (typical for random walks or (memoryless) Brownian motion) can be understood in two ways: it is the only way to ensure the normalization condition of the Kraus operators; alternatively, it is needed to make short-time transition probabilities linear in time, which is physically reasonable. If we assume $|\psi(0)\rangle_{AB} = |\psi(0)\rangle_A \otimes |0\rangle_B$, then at time dt we have

$$|\psi(dt)\rangle_{AB} = \{[1 + (\hat{K}_A - i\hat{H})\,dt]\,|\psi(0)\rangle_A\} \otimes |0\rangle_B + \sqrt{dt} \sum_{j>0} (\hat{L}^j|\psi(0)\rangle_A) \otimes |j\rangle_B. \tag{4.27}$$

This can be viewed as the unread measurement of some operator acting on B with eigenstates $|0\rangle_B, \{|j\rangle_B\}$, which, owing to the entangling unitary dynamics of AB, also changes the state of A. If the measurement is read out, we find the associated state pairs in A and B with probabilities

$$p_0 = {}_A\langle\psi(0)|\hat{E}^{0\dagger}\hat{E}^0|\psi(0)\rangle_A = 1 - dt \sum_{j>0} {}_A\langle\psi(0)|\hat{L}^{j\dagger}\hat{L}^j|\psi(0)\rangle_A \tag{4.28}$$

and

$$p_{j>0} = dt\,{}_A\langle\psi(0)|\hat{L}^{j\dagger}\hat{L}^j|\psi(0)\rangle_A. \tag{4.29}$$

They sum to 1. After the measurement, the state of A is projected on either

$$|\psi_0(dt^+)\rangle_A = \frac{1 + (\hat{K} - i\hat{H})\,dt}{\sqrt{p_0}}\,|\psi(0)\rangle_A \tag{4.30}$$

or

$$|\psi_j(dt^+)\rangle_A = \frac{\hat{L}^j}{\sqrt{p_j/dt}}\,|\psi(0)\rangle_A, \tag{4.31}$$

taking proper normalization into account. Note that at the next time step, *we again assume the environment B to be in the state* $|0\rangle_B$, the equilibrium state, because it will relax from $|j\rangle_B$ to it on a timescale $\tau_B \ll dt$, so we can neglect the brief instant when it is not in this state.

Numerically, this interpretation of the Lindblad equation leads directly to the method of quantum trajectories, which has been widely applied in quantum optics [54]. Instead of solving the Lindblad equation directly, the quantum trajectory approach averages over many quantum trajectories that are sequences of coherent pure-state evolutions and random quantum jumps chosen to occur in such a way that a statistical average over many quantum trajectories reproduces the result of solving the Lindblad equation. The statistical averaging has to be carried out over both a distribution of initial states that reproduces an initial density operator $\hat{\rho}_A(0)$ and all sequences of coherent state evolutions and quantum jumps. Let us ignore the first average, which poses no particular problem, and assume that we start from a pure state $|\psi(0)\rangle_A$, such that this first average becomes irrelevant.

For each quantum trajectory, the algorithm proceeds as follows:

1. Set $t = 0$.
2. From $|\psi(t)\rangle_A$, calculate probabilities p_0 and $\{p_j\}$ from (4.28) and (4.29).
3. By choosing a uniformly distributed random number from $[0, 1]$, determine whether you carry out no jump (probability p_0) or a jump of type j (probability p_j) on the trajectory.
4. Depending on the outcome, calculate the state $|\psi(t+dt)\rangle_A$ from (4.30) and (4.31).
5. Set $t \leftarrow t + dt$ and continue the calculation of the next time step dt using steps 2–5.

For each quantum trajectory, evaluate observables at times of interest. Finally, for the observables calculated, average over all quantum trajectories thus calculated.

To see that this indeed reproduces the dynamics of the Lindblad equation, consider $\mathcal{N}$ quantum trajectories $|\psi^{(i)}(t)\rangle$ and define the average density operator on A as

$$\bar{\rho}_A = \mathcal{N}^{-1} \sum_i |\psi^{(i)}(t)\rangle_A \, _A\langle \psi^{(i)}(t)| \equiv \overline{|\psi(t)\rangle_A \, _A\langle \psi(t)|}. \tag{4.32}$$

Then

$$\frac{d\bar{\rho}_A}{dt} = \frac{1}{dt}[\bar{\rho}(t+dt) - \bar{\rho}(t)]$$

$$= \frac{1}{dt}\left[p_0 \frac{1 - i\hat{H}\,dt + \hat{K}\,dt}{\sqrt{p_0}} \overline{|\psi(t)\rangle_A \, _A\langle\psi(t)|} \frac{1 + i\hat{H}\,dt + \hat{K}\,dt}{\sqrt{p_0}} \right.$$

$$\left. + \sum_{j>0} p_j \frac{dt}{p_j} \hat{L}^j \overline{|\psi(t)\rangle_A \, _A\langle\psi(t)|} \hat{L}^{j\dagger} - \overline{|\psi(t)\rangle_A \, _A\langle\psi(t)|} \right]$$

$$= -i[\hat{H}, \bar{\rho}_A] + \hat{K}\bar{\rho}_A + \bar{\rho}_A\hat{K} + \sum_{j>0} \hat{L}^j \bar{\rho}_A \hat{L}^{j\dagger}.$$

But this is nothing other than the Lindblad equation, and with identical initial conditions $\hat{\rho}(0) = \bar{\rho}(0)$, we have $\hat{\rho}(t) = \bar{\rho}(t)$ for all $t > 0$.

4.3 Dynamics by full diagonalization

Let us try and follow through the "program" of doing dynamics suggested by the previous section. Let us start with the pure-state case, and assume that our Hamiltonian is represented by an $(n \times n)$-dimensional matrix. Then the first step is the full diagonalization of the Hamiltonian matrix $(H)_{kk'} = \langle k|\hat{H}|k'\rangle$ in some convenient orthonormal basis $\{|k\rangle\}$ (the "computational basis") to obtain eigenvalues (eigenenergies) E_i and eigenvectors (energy eigenstates) $|\psi_i\rangle$. Knowing those, it is easy to solve (4.4) by using (4.6) in the computationally useful form of (4.7). All remaining errors are due to machine precision, and time evolution is obtained exactly (or quasi-exactly, if you are a strict mathematician). Alas, it is easy to see that this procedure is extremely limited in practice.

Consider an even simpler Hamiltonian than the Bose–Hubbard model of (4.1), namely the isotropic Heisenberg Hamiltonian on a one-dimensional chain with nearest-neighbor exchange interaction J_{ex} and external magnetic field H on a chain of L lattice sites:

$$\hat{H}_{\mathrm{Hbg}} = J_{\mathrm{ex}} \sum_{i=1}^{L-1} \hat{\mathbf{S}}_i \cdot \hat{\mathbf{S}}_{i+1} - \sum_{i=1}^{L} H\hat{S}_i^z. \tag{4.33}$$

The spins are taken to be spin-$\frac{1}{2}$ with the site-local basis $\{|\uparrow\rangle, |\downarrow\rangle\}$. With a local state space of dimension $d = 2$, it is the simplest nontrivial quantum many-body system.

Which size of systems can we attack with our strategy? All numerical algorithms for the full diagonalization of an $n \times n$ matrix, including the calculation of the eigenvectors, are $O(n^3)$ in CPU time and $O(n^2)$ in memory. Depending on the type of computer, this becomes impractical for $\sim 10^5$ or so. Hilbert-space dimension grows exponentially as d^L, so for $S = \frac{1}{2}$ spins, this would correspond to roughly $L = 16$ spins ($n = 65\,536$), very far from the thermodynamic limit $L \to \infty$, which we usually want to understand. And there is no easy way out: imagine you had a computer that is 1000 times faster than your current one and with no memory limitation. Then matrix size can grow by a factor of 10, which is just a bit more than $8 = 2^3$. This means that your system size would grow from $L = 16$ to a meager $L = 19$. This in the simplest imaginable case!

Often we can do better by exploiting symmetries of the Hamiltonian. For example, (4.33) is invariant under rotations about the z axis selected by the magnetic field H, i.e., it is $U(1)$-invariant. This implies that the total magnetization $\hat{M} = \sum_i \hat{S}_i^z$ is conserved, $[\hat{H}_{\mathrm{Hbg}}, \hat{M}] = 0$. If we find a computational basis whose states have well-defined magnetization M (a "good quantum number"), we can diagonalize $\hat{H}_{\mathrm{Hbg}}$ independently in the subspaces of states of identical M. For $L = 16$, the largest such subspace has $M = 0$, i.e., $L_\uparrow = 8$ ↑-spins and $L_\downarrow = 8$ ↓-spins. They can be arranged in $16!/(8!8!) = 12\,870$ configurations, which is the dimension of the largest subspace. Its diagonalization will take only a fraction of $(12\,870/65\,536)^3 = 7.6 \times 10^{-3}$ of the original time needed. If the spins were on a ring, you could also exploit translational symmetry, but, given the exponential nature of the problem, even the impressive fraction just calculated only gives about two additional lattice sites. So symmetries help, and should be exploited, but one should not expect too much. Also, finding a suitable computational basis is not always easy: while the standard basis for spin $\frac{1}{2}$ of ↑- and ↓-spin states is perfectly adapted to conserved magnetization, it is much harder (albeit not impossible) to find a basis that would match the $SU(2)$ symmetry of the Heisenberg Hamiltonian for $H = 0$; here the reader is referred to the rich literature on the exact diagonalization of spin systems.

Can we do better? Let us take the following point of view: at each time t, the state at $t + \mathrm{d}t$ depends only on the state at time t itself, so knowing the full spectrum is not necessary for time evolution. The initial state sometimes is a "simple" state like $| \uparrow\downarrow\uparrow \ldots \uparrow\downarrow\rangle$, which in the computational basis has simple coefficients $(0, 0, \ldots, 0, 1, 0, \ldots, 0)$. Often, the initial state is the (usually much more complicated) ground state of some Hamiltonian $\hat{H}$, which is then quenched to some other Hamiltonian $\hat{H}_1$. We need methods to find this initial state, but again not the entire spectrum.

Indeed, such methods exist, and come under the heading of *exact diagonalization (ED)* in the literature. This is somewhat of a misnomer, because the Hamiltonian is no longer fully diagonalized, but its extreme (maximal and minimal) eigenvalues and associated eigenvectors are determined, which is, however, sufficient for our purposes. We will now consider this technique in some detail, for two reasons: (i) to allow us to get meaningful initial states and (ii) because many of the time-evolution algorithms we are going to discuss contain "auxiliary" problems to be solved that need such exact diagonalization methods. They all rely on the fact that most entries of the Hamiltonian matrices are known to be zero. Such matrices are called *sparse*, and they have many helpful properties. But first let us recall some useful basics of linear algebra and establish notation.

4.4   A reminder of linear algebra

In this section, we briefly revisit some concepts of linear algebra, most of which can be found in any book on linear algebra. Golub and van Loan [29] give a magisterial but somewhat more advanced overview of matrix computations. A beautifully concise overview of the field is given by Trefethen and Bau [70]. Parlett [51] gives a very accessible introduction to symmetric matrices and their eigensystems. From the point of view of practical implementation of algorithms, the books by Saad [58, 59] are very accessible, as are those by Barrett et al. [5] and Bai et al. [4]. Their arguments are used heavily throughout this text; but all of the following is standard fare in linear algebra and can be found very similarly in many other books, too.

We use Dirac's ket notation $|v\rangle$ for column vectors and the bra notation $\langle v|$ for the associated transposed complex-conjugated (!) row vectors. This will allow us to move seamlessly between mathematical and quantum physical notation. The inner (scalar) product, for example, is then written as

$$\langle v, w\rangle = \sum_i v_i^* w_i = \langle v|w\rangle.$$

4.4.1   Eigensystems

As the Hamiltonian matrix H is Hermitian, let us briefly review the properties of eigenvectors of a Hermitian matrix $H \in \mathbb{C}^{n \times n}$ and establish some notation. If $H|\lambda_i\rangle = \lambda_i|\lambda_i\rangle$, then the eigenvalues λ_i are real and the normalized eigenvectors $\{|\lambda_i\rangle\}$ form an orthonormal basis of $\mathbb{C}^n$, $\langle\lambda_i|\lambda_{i'}\rangle = \delta_{ii'}$, provided they are orthonormalized for degenerate eigenvalues λ_i. This statement can be expressed in alternative ways:

$$H = \sum_{i=1}^{n} \lambda_i|\lambda_i\rangle\langle\lambda_i| \tag{4.34}$$

and $Q^\dagger Q = QQ^\dagger = I$, $Q^\dagger = Q^{-1}$ for a unitary matrix Q whose columns are the eigenvectors of H, denoted as

$$Q = [|\lambda_1\rangle|\lambda_2\rangle \dots |\lambda_n\rangle]. \tag{4.35}$$

Introducing a matrix $\Lambda = \mathrm{diag}(\lambda_1, \lambda_2, \dots, \lambda_n)$, we obtain the compact matrix notation of the eigenvector–eigenvalue relationship:

$$HQ = Q\Lambda \quad \Leftrightarrow \quad Q^\dagger HQ = \Lambda \quad \Leftrightarrow \quad H = Q\Lambda Q^\dagger. \tag{4.36}$$

In the following, we will also assume that the eigenvalues are ordered as $\lambda_1 \geq \lambda_2 \geq \dots \geq \lambda_{n-1} \geq \lambda_n$.

4.4.2   The Rayleigh–Ritz quotient

We will frequently use the *Rayleigh–Ritz quotient* $\rho(H, |v\rangle)$ for $0 \neq |v\rangle \in \mathbb{C}^n$:

$$\rho(H, |v\rangle) := \frac{\langle v|H|v\rangle}{\langle v|v\rangle}, \tag{4.37}$$

which is bounded as $\lambda_1 \leq \rho(H, |v\rangle) \leq \lambda_n$. Its most important property is that it is stationary for $|v\rangle$ if and only if $|v\rangle$ is an eigenvector of H:

$$\delta\left(\frac{\langle v|H|v\rangle}{\langle v|v\rangle}\right) = 0, \quad \frac{\langle v|H|v\rangle}{\langle v|v\rangle} = \lambda \quad \Leftrightarrow \quad H|v\rangle = \lambda|v\rangle. \tag{4.38}$$

At this point, we can also introduce the *residual vector*

$$|r\rangle[|v\rangle] = [H - \rho(H, |v\rangle)]\,|v\rangle. \tag{4.39}$$

This allows us to quantify errors in eigendecompositions because it provides an orthogonal decomposition of $H|v\rangle$, as $|r\rangle[|v\rangle] \perp \rho(H, |v\rangle)|v\rangle$; see Fig. 4.1.

Assume we have found an approximate eigenvalue θ and associated approximate normalized eigenvector $|y\rangle$. Consider [51] the norm of the residual vector $\||r\rangle\|_2 = \|H|y\rangle - \theta|y\rangle\|_2$ and suppose that $\theta \neq \lambda_j$, where λ_j is the (unknown) eigenvalue of H closest to θ. Then $H - \theta I$ is invertible and for any $|v\rangle \neq 0$, we can write

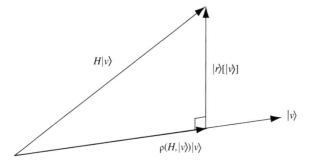

Fig. 4.1 Orthogonal decomposition of $H|v\rangle$ into components parallel and perpendicular to $|v\rangle$, which are specified by the Rayleigh–Ritz quotient $\rho(H, |v\rangle)$ and the residual vector $|r\rangle[|v\rangle]$.

$$\||v\rangle\|_2 = \|(H-\theta I)^{-1}(H-\theta I)|v\rangle\|_2 \le \|(H-\theta I)^{-1}\|_2\|(H-\theta I)|v\rangle\|_2 = |\lambda_j-\theta|^{-1}\||r\rangle\|_2.$$

In the last identity, we have used the fact that that the eigenvalue spectrum of $(H-\theta I)^{-1}$ is given by $\{(\lambda_i-\theta)^{-1}\}$ and hence its 2-norm is given by $|\lambda_j-\theta|^{-1}$. Therefore,

$$|\theta - \lambda_j| \le \||r\rangle\|_2. \tag{4.40}$$

If we know more about the spectrum of H, this bound can be improved. Let us define the *spread* of the spectrum by the difference between the largest and smallest eigenvalues,

$$\text{spread}(H) = \lambda_1(H) - \lambda_n(H), \tag{4.41}$$

and the *gap* between θ and the spectrum of H as the minimal distance between θ and any eigenvalue of H except that best approximated by θ, say λ_j,

$$\text{gap}(H, \theta) = \min_{i \ne j} |\lambda_i(H) - \theta|. \tag{4.42}$$

If $|\lambda_j\rangle$ is the eigenvector corresponding to λ_j, then we can bound both the eigenvalue and the angle between the true and approximated eigenvectors [51]:

$$\frac{\||r\rangle\|_2}{\text{spread}(H)} \le |\sin \angle(|y\rangle, |\lambda_j\rangle)| \le \frac{\||r\rangle\|_2}{\text{gap}(H, \theta)} \tag{4.43}$$

and

$$0 \le |\theta - \lambda_j| \le \frac{\||r\rangle\|_2^2}{\text{gap}(H, \theta)}. \tag{4.44}$$

For small residuals, this may be a stricter bound than (4.40).

4.4.3 Gram–Schmidt orthonormalization

Another reminder of standard linear algebra concerns the orthonormalization of a set of linearly independent vectors. Anticipating our interest in large sparse matrices, the most convenient procedure is the *Gram–Schmidt orthonormalization* scheme. Assuming a set of m linearly independent vectors $0 \neq |v_j\rangle \in \mathbb{C}^n$, we are looking for a set of m orthonormal vectors $|q_j\rangle \in \mathbb{C}^n$ with $\langle q_i|q_j\rangle = \delta_{ij}$ and with $\mathrm{span}(|v_1\rangle, \ldots, |v_m\rangle) = \mathrm{span}(|q_1\rangle, \ldots, |q_m\rangle)$. The scheme runs as follows:

1. Set $|q_1\rangle = |v_1\rangle/r_{11}$, where $r_{11} = \||v_1\rangle\|_2$.
2. Moving through $j = 2, \ldots, m$, calculate as follows:
   * Set $|\tilde{q}_j\rangle = |v_j\rangle - \sum_{i=1}^{j-1} r_{ij}|q_i\rangle$, where $r_{ij} = \langle q_i|v_j\rangle$.
   * Normalize $|q_j\rangle = |\tilde{q}_j\rangle/r_{jj}$, where $r_{jj} = \||\tilde{q}_j\rangle\|_2$.

It is very easy to show by induction that this procedure works. Linear independence of the $|v_j\rangle$ ensures that the normalization factors $r_{jj} \neq 0$ except for (in our context practically nonexistent) cases of numerical breakdown. A few remarks on the Gram–Schmidt orthonormalization scheme are in order:

1. If we form a matrix $R = [r_{ij}] \in \mathbb{C}^{m \times m}$, setting all r_{ij} not defined by the algorithm to be 0, it is upper-triangular. Defining $V = [|v_1\rangle|v_2\rangle \ldots |v_m\rangle] \in \mathbb{C}^{n \times m}$ and $Q = [|q_1\rangle|q_2\rangle \ldots |q_m\rangle] \in \mathbb{C}^{n \times m}$, we can introduce a matrix notation $V = QR$ and obtain, albeit not in the most efficient way, a so-called *QR-decomposition* of V, since $Q^\dagger Q = I$.
2. In order to calculate $|q_j\rangle$, we do not need $|v_i\rangle$ with $i > j$. We will use this to orthogonalize "along the way," since our later algorithms will produce the $|v_i\rangle$ iteratively. So $|v_i\rangle$ and $|q_i\rangle$ will be calculated at the same time.
3. The numerical accuracy of Gram–Schmidt orthonormalization can be significantly improved: in the calculation of $|\tilde{q}_i\rangle$ from $|v_i\rangle$, instead of calculating all $r_{ij} = \langle q_i|v_j\rangle$ first, it is better to proceed by calculating first $\langle q_1|v_j\rangle$, replace $|v_j\rangle \rightarrow |v_j\rangle - \langle q_1|v_j\rangle|q_1\rangle$, and calculate from this new $|v_j\rangle$ now $\langle q_2|v_j\rangle$, replace again $|v_j\rangle \rightarrow |v_j\rangle - \langle q_2|v_j\rangle|q_2\rangle$, and so forth. Mathematically, this is an equivalent procedure, because of the orthonormality of the $|q_i\rangle$. In numerical practice, this *modified Gram–Schmidt* scheme leads to smaller and hence more accurate subtractions—this form should always be used!

4.4.4 Tridiagonal matrices

A very important class of matrices appearing in our algorithms comprises tridiagonal matrices T where

$$|i - j| > 1 \Rightarrow (T)_{ij} = 0. \tag{4.45}$$

Let us define

$$\alpha_i = (T)_{ii}, \qquad \beta_i = (T)_{i+1,i}, \qquad \gamma_i = (T)_{i,i+1}. \tag{4.46}$$

In an explicit representation, we will be dealing with square tridiagonal matrices $T_m \in \mathbb{C}^{m \times m}$ given by

$$T_m = \begin{bmatrix} \alpha_1 & \gamma_1 & & & & & \\ \beta_1 & \alpha_2 & \gamma_2 & & & \Large 0 & \\ & \beta_2 & \alpha_3 & \gamma_3 & & & \\ & & \ddots & \ddots & \ddots & & \\ & \Large 0 & & \beta_{m-3} & \alpha_{m-2} & \gamma_{m-2} & \\ & & & & \beta_{m-2} & \alpha_{m-1} & \gamma_{m-1} \\ & & & & & \beta_{m-1} & \alpha_m \end{bmatrix} \tag{4.47}$$

and related tridiagonal matrices $\tilde{T}_m \in \mathbb{C}^{(m+1) \times m}$ with one more row than column,

$$\tilde{T}_m = \begin{bmatrix} \alpha_1 & \gamma_1 & & & & & \\ \beta_1 & \alpha_2 & \gamma_2 & & & \Large 0 & \\ & \beta_2 & \alpha_3 & \gamma_3 & & & \\ & & \ddots & \ddots & \ddots & & \\ & \Large 0 & & \beta_{m-3} & \alpha_{m-2} & \gamma_{m-2} & \\ & & & & \beta_{m-2} & \alpha_{m-1} & \gamma_{m-1} \\ & & & & & \beta_{m-1} & \alpha_m \\ & & & & & & \beta_m \end{bmatrix}. \tag{4.48}$$

Hermitian tridiagonal matrices ($\alpha_i \in \mathbb{R}$, $\gamma_i = \beta_i^*$) are at the core of the Hermitian eigenvalue problem for large sparse matrices, because any such matrix can be brought iteratively into tridiagonal form; the diagonalization of tridiagonal matrices is part of any standard linear algebra package, such as that provided by Section 11.4 in *Numerical Recipes* [56].

A particularly important class of Hermitian tridiagonal matrices is given by so-called *unreduced* tridiagonal matrices. A tridiagonal matrix is called unreduced if all side-diagonal entries $\beta_i \neq 0$. This will be a natural property of all Hermitian tridiagonal matrices produced by our numerical algorithms. A key result for unreduced tridiagonal matrices is that their eigenvalues are all *nondegenerate*, which will be seen to have important implications for the related algorithms (for a proof of this, see, e.g., [51]).

A further important property concerns the eigenvalues of tridiagonal submatrices. Consider some Hermitian tridiagonal matrix $T_n \in \mathbb{C}^{n \times n}$ with eigenvalues $\lambda_i^{(n)}$, $1 \leq i \leq n$, ordered in descending sequence. Let us now form all principal submatrices T_1, T_2, T_3, ..., T_m, by taking the $m \times m$ top-left elements of T_n. In descending order, the eigenvalues of T_m are $\lambda_i^{(m)}$, $1 \leq i \leq m$. Then they can be shown to obey

$$\lambda_i^{(m)} \geq \lambda_i^{(m-1)} \geq \lambda_{i+1}^{(m)}. \tag{4.49}$$

In the relevant case of unreduced tridiagonal matrices, even strict inequalities hold:

$$\lambda_i^{(m)} > \lambda_i^{(m-1)} > \lambda_{i+1}^{(m)} \tag{4.50}$$

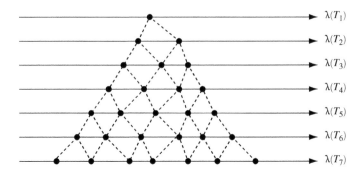

Fig. 4.2 Sturm chain formed by the eigenvalues $\lambda(T_m)$ of the principal submatrices T_m of an unreduced Hermitian tridiagonal matrix: the eigenvalues of T_{m-1} fit into the intervals defined by the eigenvalues of T_m.

The eigenvalues of the matrix sequence (T_m) form a so-called *Sturm chain*. This property is best visualized: assume that we depict the m eigenvalues of T_m as in Fig. 4.2. The smallest and largest eigenvalues of T_{m+1} are smaller and larger than the extreme eigenvalues of T_m, and all the others are inserted in the intervals formed by the previous eigenvalues. In the following, we will encounter large matrices H that are approximated in a new basis by increasingly large tridiagonal matrices forming a submatrix hierarchy $\{T_m\}$. This means that the biggest and smallest eigenvalues of the tridiagonal matrices form strict, monotonically improving lower and upper bounds to the biggest and smallest eigenvalues of H.

As we will see later, the calculation of dynamical correlators revolves around the expression $(E_0 + \omega + i\eta - H)^{-1}$, where $H \in \mathbb{C}^{n \times n}$ is the Hermitian Hamiltonian matrix and E_0 its ground-state energy. Let us assume that we have brought H into tridiagonal form T as $T = Q^\dagger H Q$, where $|q_1\rangle$ is the first column of Q. T is Hermitian, as was H. Hence, $\gamma_i = \beta_i^*$ in general; if H is a real matrix, then $\gamma_i = \beta_i$, which we will assume in the following. Our algorithms will achieve tridiagonalization iteratively, such that the choice of the normalized first column $|q_1\rangle$ is completely free and we will choose it to our advantage, since we can calculate the matrix element $\langle q_1 | (E_0 + \omega + i\eta - H)^{-1} | q_1 \rangle \equiv [(E_0 + \omega + i\eta - T)^{-1}]_{11}$ quite easily! For the argument worked out below and interesting applications, see [32].

We introduce the matrices $\tilde{T}_{ij}$, designating the tridiagonal matrix T with row i and column j eliminated, and $\overline{T}_{ij}$, again designating T but now with all rows $1, \ldots, i$ and columns $1, \ldots, j$ eliminated. $\overline{T}_{00} = T$ itself, and $E := E_0 + \omega$. Do not confuse $\overline{T}_{ij}$ and $\tilde{T}_{ij}$ with $(T)_{ij}$, a matrix element of T. Let us also slightly abuse notation and replace $\overline{(E + i\eta - T)}_{11}$ by $E + i\eta - \overline{T}_{11}$ (and so on, to make the intermediate steps more readable). Then

$$[(E + i\eta - T)^{-1}]_{11} = \frac{\det(E + i\eta - \overline{T}_{11})}{\det(E + i\eta - \overline{T}_{00})} = \frac{1}{\dfrac{\det(E + i\eta - \overline{T}_{00})}{\det(E + i\eta - \overline{T}_{11})}}. \tag{4.51}$$

The two determinants are related by

$$\det(E + i\eta - \overline{\overline{T}}_{00}) = (E + i\eta - \alpha_1)\det(E + i\eta - \overline{\overline{T}}_{11}) - \beta_1\det(E + i\eta - \tilde{T}_{21})$$
$$= (E + i\eta - \alpha_1)\det(E + i\eta - \overline{\overline{T}}_{11}) - \beta_1^2\det(E + i\eta - \overline{\overline{T}}_{22}),$$

or

$$\frac{\det(E + i\eta - \overline{\overline{T}}_{00})}{\det(E + i\eta - \overline{\overline{T}}_{11})} = E + i\eta - \alpha_1 - \frac{\beta_1^2}{\dfrac{\det(E + i\eta - \overline{\overline{T}}_{11})}{\det(E + i\eta - \overline{\overline{T}}_{22})}}. \qquad (4.52)$$

Similarly,

$$\frac{\det(E + i\eta - \overline{\overline{T}}_{ii})}{\det(E + i\eta - \overline{\overline{T}}_{i+1,i+1})} = E + i\eta - \alpha_{i+1} - \frac{\beta_{i+1}^2}{\dfrac{\det(E + i\eta - \overline{\overline{T}}_{i+1,i+1})}{\det(E + i\eta - \overline{\overline{T}}_{i+2,i+2})}}. \qquad (4.53)$$

Collecting all of these results, we obtain the *continued fraction representation of the resolvent*:

$$\langle q_1| \frac{1}{E + i\eta - H} |q_1\rangle = \cfrac{1}{E + i\eta - \alpha_1 - \cfrac{\beta_1^2}{E + i\eta - \alpha_2 - \cfrac{\beta_2^2}{E + i\eta - \alpha_3 - \cdots}}}. \qquad (4.54)$$

For a complex H, the only change would be to set $\beta_i^2 \to |\beta_i|^2$.

4.5 Handling large sparse matrices

4.5.1 Large sparse matrices

All remarks on linear algebra so far have been completely general for the respective types of matrices. In the following, we will heavily rely on matrices being additionally *large* and *sparse*. "Large" simply means that they cannot be treated by full diagonalization-type methods. "Sparse" means that most entries are strictly zero and that we can efficiently enumerate the remaining nonzero entries; otherwise, this property would be of little use.

Let us once again consider the Heisenberg Hamiltonian, which we make more explicit by introducing spin ladder operators:

$$\hat{H}_{\text{Hbg}} = J_{\text{ex}} \left\{ \sum_{i=1}^{L-1} \frac{1}{2}(\hat{S}_i^+ \hat{S}_{i+1}^- + \hat{S}_i^- \hat{S}_{i+1}^+) + \hat{S}_i^z \hat{S}_{i+1}^z \right\} - \sum_{i=1}^{L} H \hat{S}_i^z. \qquad (4.55)$$

Let us consider the case $L = 36$, way beyond what full diagonalization can achieve: the matrix dimension is $2^{36} = 6.87 \times 10^{10}$, 1 terabyte of storage *per column* for 16-byte complex numbers. If we wanted to store every matrix entry, we would need roughly

10^{11} terabytes—quite a whopper! Here, sparsity helps: each computational basis state $|i\rangle$ connects via the action of the Hamiltonian only (i) to itself through the diagonal terms containing $\hat{S}^z$ and (ii) to at most $L - 1 = 35$ other states through the spin-flip processes (many fewer in practice). We can directly enumerate these states and need at most 36 terabytes of storage, an impressive saving of about 10^9. In practice, no one would store even this matrix in its sparse form, but rather would provide an algorithm that can generate the matrix elements on demand efficiently.

What advantages can we draw in practice? Zeros in a matrix make matrix–vector multiplication very efficient, and for this an algorithm that generates matrix elements on demand is sufficient. For a general matrix in $\mathbb{C}^{n \times n}$, a matrix–vector multiplication is an operation of order n^2. In the case of a sparse matrix, where only a few entries per row are nonzero, a matrix–vector multiplication is of order cn, where c is some small constant, and hence many orders of magnitude faster for large n. What remains to be stored is the vector, but again savings are possible: if we consider the (largest) $M = 0$ magnetization subsector, it has dimension $36!/(18!18!) \approx 9.1 \times 10^9$, such that 16-byte complex coefficients can be stored in 135 gigabytes instead of 1 terabyte. Further symmetries help even more. Hence, in order to exploit sparsity in large matrices, we have to devise algorithms based on matrix–vector multiplications $H|v\rangle$ alone. There is a very elaborate mathematical framework for that, which we will discuss now.

4.5.2   Krylov spaces

The concept of *Krylov spaces* is highly useful for the manipulation of large sparse square matrices A in general, but even more so if they are Hermitian (then we call them H). Excellent accounts, which are being followed here, are given in [4, 29, 51, 58, 59, 70].

Consider a vector $0 \neq |v\rangle \in \mathbb{C}^n$ and a square matrix $A \in \mathbb{C}^{n \times n}$. Then we define *Krylov vectors* $|k_i\rangle$ as

$$|k_i\rangle := A^{i-1}|v\rangle \qquad (i = 1, 2, \ldots) \tag{4.56}$$

and the *Krylov space* $\mathbb{K}_m(A, |v\rangle)$ as

$$\begin{aligned} \mathbb{K}_m(A, |v\rangle) &:= \text{span}(|v\rangle, A|v\rangle, \ldots, A^{m-1}|v\rangle) \\ &\equiv \text{span}(|k_1\rangle, |k_2\rangle, \ldots, |k_m\rangle) \qquad (m = 1, 2, \ldots). \end{aligned} \tag{4.57}$$

Krylov spaces form a *hierarchy:*

$$\mathbb{K}_1(A, |v\rangle) \subset \mathbb{K}_2(A, |v\rangle) \subset \mathbb{K}_3(A, |v\rangle) \subset \mathbb{K}_4(A, |v\rangle) \subset \ldots \tag{4.58}$$

Moreover, as $A|k_1\rangle, A|k_2\rangle, \ldots, A|k_m\rangle = |k_2\rangle, |k_3\rangle, \ldots, |k_{m+1}\rangle$,

$$A\mathbb{K}_m(A, |v\rangle) \subset \mathbb{K}_{m+1}(A, |v\rangle). \tag{4.59}$$

Construction of orthonormal Krylov bases

So far, the entire discussion has been completely general. We are now going to make the (very strong) *full-rank assumption*, namely, that for $m \leq n$, all $|k_1\rangle, \ldots, |k_m\rangle$ are linearly independent,

$$m = \dim \operatorname{range} \mathbb{K}_m(A, |v\rangle) \qquad (m \leq n), \tag{4.60}$$

which of course must break down for $m > n$. If $|v\rangle$ is random, as will usually be the case in practice, it is overwhelmingly probable that the assumption will hold for $m \ll n$, which will be the case of interest for us; hence this strong assumption is a "good" one.

A first consequence is that

$$|k_{m+1}\rangle \notin \mathbb{K}_m(A, |v\rangle), \tag{4.61}$$

or $|k_{m+1}\rangle = |k_{m+1}^{\|}\rangle + |k_{m+1}^{\perp}\rangle$, with $|k_{m+1}^{\|}\rangle \in \mathbb{K}_m(A, |v\rangle)$ and $0 \neq |k_{m+1}^{\perp}\rangle \in \mathbb{K}_m^{\perp}(A, |v\rangle)$. This observation suggests that we form *preferential orthonormal bases* of dimension m for $\mathbb{K}_m(A, |v\rangle)$ with the additional property that the bases of the smaller Krylov spaces are subsets of the bases of the larger ones:

$$\begin{aligned} \mathbb{K}_1(A, |v\rangle) &= \operatorname{span}(|q_1\rangle), \\ \mathbb{K}_2(A, |v\rangle) &= \operatorname{span}(|q_1\rangle, |q_2\rangle), \\ &\vdots \\ \mathbb{K}_m(A, |v\rangle) &= \operatorname{span}(|q_1\rangle, |q_2\rangle, \ldots, |q_m\rangle), \end{aligned} \tag{4.62}$$

with $\langle q_i | q_j \rangle = \delta_{ij}$. Up to trivial factors this basis will be unique. We define orthonormal matrices

$$Q_m(A, |v\rangle) := [|q_1\rangle \, |q_2\rangle \, |q_3\rangle \, \cdots \, |q_m\rangle], \tag{4.63}$$

with $Q_m^{\dagger} Q_m = I_m$. Equations (4.62) suggest the use of Gram–Schmidt orthogonalization. We set $|q_1\rangle = |k_1\rangle / \||k_1\rangle\|_2$ and form

$$\begin{aligned} \beta_1 |q_2\rangle &= |k_2\rangle - \langle q_1 | k_2 \rangle |q_1\rangle, \\ \beta_2 |q_3\rangle &= |k_3\rangle - \langle q_1 | k_3 \rangle |q_1\rangle - \langle q_2 | k_3 \rangle |q_2\rangle, \\ &\vdots \\ \beta_m |q_{m+1}\rangle &= |k_{m+1}\rangle - \sum_{i=1}^{m} \langle q_i | k_{m+1} \rangle |q_i\rangle, \end{aligned}$$

where we choose $\beta_1, \beta_2, \ldots, \beta_m > 0$ such that $\||q_2\rangle\|_2 = \ldots = \||q_{m+1}\rangle\|_2 = 1$. Under the full-rank assumption, this is always possible.

In fact, $|k_{m+1}\rangle$ can be eliminated in favor of $A|q_m\rangle$ because

$$
\begin{aligned}
\mathbb{K}_{m+1}(A, |v\rangle) &= \mathrm{span}(|k_1\rangle, \ldots, |k_m\rangle, |k_{m+1}\rangle) = \mathrm{span}(|q_1\rangle, \ldots, |q_m\rangle, |k_{m+1}\rangle) \\
&= \mathrm{span}(|q_1\rangle, \ldots, |q_m\rangle, A|k_m\rangle) \\
&= \mathrm{span}(|q_1\rangle, \ldots, |q_m\rangle, A|q_1\rangle, \ldots, A|q_{m-1}\rangle, A|q_m\rangle) \\
&\overset{(4.59)}{=} \mathrm{span}(|q_1\rangle, \ldots, |q_m\rangle, A|q_m\rangle).
\end{aligned}
$$

Then the Gram–Schmidt procedure reads, for $m = 1, 2, \ldots, n-1$,

$$
\beta_m |q_{m+1}\rangle \equiv |w_m\rangle = A|q_m\rangle - \sum_{i=1}^{m} \langle q_i|A|q_m\rangle |q_i\rangle. \tag{4.64}
$$

Matrix restrictions and projections onto Krylov spaces

A maps $\mathbb{C}^n \to \mathbb{C}^n$. Consider some linear subspace $\mathcal{S} \subset \mathbb{C}^n$ and limit the domain of A to $\mathcal{S}$. If we limit the codomain to $\mathcal{S}$, we call this *projection* of A on $\mathcal{S}$; if we limit the codomain to range $A|_{\mathcal{S}}$, the range of A for domain $\mathcal{S}$, we call this *restriction* of A to $\mathcal{S}$.

The *restriction* of the matrix A to the Krylov space $\mathbb{K}_m(A, |v\rangle) = \mathrm{range}\, Q_m$ has a range given by $\mathrm{span}(\mathbb{K}_m(A, |v\rangle), |w_m\rangle) = \mathrm{range}\, Q_{m+1}$. Hence, the restriction of A in the preferential orthonormal basis is given by

$$
\tilde{A}_m = Q^\dagger_{m+1} A Q_m \in \mathbb{C}^{(m+1)\times m} \qquad (m < n), \tag{4.65}
$$

and the projection of A on the Krylov space $\mathbb{K}_m(A, |v\rangle)$ (Fig. 4.3) is given by

$$
A_m = Q^\dagger_m A Q_m \in \mathbb{C}^{m\times m} \qquad (m \le n). \tag{4.66}
$$

In matrix element notation,

$$
(\tilde{A}_m)_{ij} = \langle q_i|A|q_j\rangle \qquad (1 \le i \le m+1, 1 \le j \le m), \tag{4.67}
$$

$$
(A_m)_{ij} = \langle q_i|A|q_j\rangle \qquad (1 \le i, j \le m). \tag{4.68}
$$

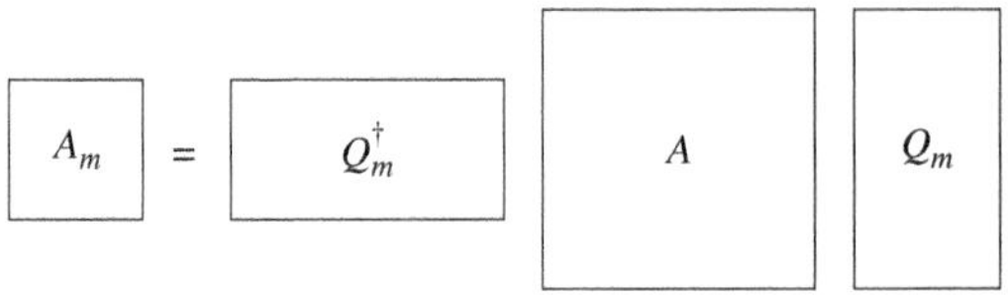

Fig. 4.3 Equation (4.66): A_m is the incomplete basis transformation (projection) of A to the basis formed by the orthonormal vectors in $Q_m = [|q_1\rangle \ldots |q_m\rangle]$.

We have $A_n = A$. The matrices $\tilde{A}_m$ and A_m are closely related by

$$\tilde{A}_m = \left[\begin{array}{c} A_m \\ \hline 0\,\ldots\,\ldots\,0\;\beta_m \end{array}\right].$$ (4.69)

Premultiplying (4.65) by $Q^\dagger_{m+1}Q_{m+1} = I_{m+1}$ on the left, we obtain

$$Q^\dagger_{m+1}\cdot Q_{m+1}\tilde{A}_m = Q^\dagger_{m+1}\cdot AQ_m,$$

and hence

$$Q_{m+1}\tilde{A}_m = AQ_m.$$ (4.70)

Applying $|q_{m+1}\rangle$ from the left to (4.64), orthonormality of the $|q_i\rangle$ gives

$$\beta_m = \langle q_{m+1}|A|q_m\rangle = \||w_m\rangle\|_2.$$ (4.71)

The β_i, $1 \le i \le m$, therefore form the first lower subdiagonal of $\tilde{A}_m$. (We have two mathematically equivalent equations for β_i: first, as a matrix element, second, as a vector norm; numerically, only the second is recommended.)
If we write $Q_{m+1} = [Q_m\,|q_{m+1}\rangle]$ and

$$\tilde{A}_m = \left[\begin{array}{c} A_m \\ \langle e_m|\beta_m \end{array}\right]$$ (4.72)

with $\langle e_m| = [0,\ldots,0,1]$ in (4.70), we obtain

$$Q_m A_m + \beta_m|q_{m+1}\rangle\langle e_m| = AQ_m,$$ (4.73)

or

$$Q_m A_m + |w_m\rangle\langle e_m| = AQ_m$$ (4.74)

as illustrated in Fig. 4.4. Equation (4.73) is of central importance in the application of Krylov spaces.
An important result is the relationship between eigenvalues and eigenvectors of A and A_m. Imagine that we have an eigenvalue $\lambda^{(m)}$ of A_m with associated eigenvector $|y\rangle \in \mathbb{C}^m$,

$$A_m|y\rangle = \lambda^{(m)}|y\rangle.$$ (4.75)

Then we use (4.66) and insert $Q^\dagger_m Q_m = I_m$ on the right to obtain

$$Q^\dagger_m \cdot AQ_m|y\rangle = Q^\dagger_m \cdot \lambda^{(m)}Q_m|y\rangle.$$ (4.76)

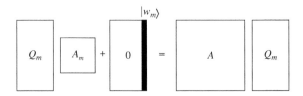

Fig. 4.4 Equation (4.73): the range of A restricted to the basis defined by Q_m exceeds the span of this basis by the vector $|w_m\rangle = \beta_m |q_{m+1}\rangle$.

We can read off that $\lambda^{(m)}$ is also an eigenvalue of A projected on $\mathrm{span}(|q_1\rangle, \dots, |q_m\rangle)$ and that the eigenvector in $\mathbb{C}^n$ of the projected A is given by

$$|x\rangle = Q_m |y\rangle \Leftrightarrow |x\rangle = \sum_{i=1}^{m} y_i |q_i\rangle. \tag{4.77}$$

As $A_n = A$, for $m = n$, the eigenvalues are exact eigenvalues of A and the eigenvectors are related by the transformation of (4.77).

The matrices A_m have a very special form:

$$(A_m)_{ij} = \langle q_i | A | q_j \rangle = 0 \quad \text{for } i > j + 1 \tag{4.78}$$

because $|q_i\rangle \perp \mathbb{K}_{i-1} \supset A\mathbb{K}_{i-2}$ and $|q_j\rangle \in \mathbb{K}_{i-2}$ for $j \leq i - 2$. So, only the top-right corner, the diagonal, and the first subdiagonal may be nonzero. Such a matrix is called an *upper-Hessenberg* matrix, and its entries are generated via (4.64) in the so-called *Arnoldi process*.

Reduction of Hermitian matrices to tridiagonal form

Many more properties of Krylov spaces can be derived for general square matrices A, but in view of what we will need in the following, *let us focus for the rest on Hermitian matrices H.*

For Hermitian H, the projection matrix A_m is Hermitian too: $(A_m)_{ij} = (A_m)^*_{ji}$. Then $\langle q_i | H | q_j \rangle = 0$ for $i > j + 1$ implies $\langle q_i | H | q_j \rangle = 0$ for $j > i + 1$; hence A_m is tridiagonal, and we relabel it T_m:

$$(T_m)_{ij} \equiv \langle q_i | H | q_j \rangle = 0 \quad (|i - j| > 1). \tag{4.79}$$

In view of (4.79), the Gram–Schmidt procedure in (4.64) simplifies to the three-term recursion

$$\beta_m |q_{m+1}\rangle = H |q_m\rangle - \langle q_m | H | q_m \rangle |q_m\rangle - \langle q_{m-1} | H | q_m \rangle |q_{m-1}\rangle, \tag{4.80}$$

first proposed by Lanczos in this form. We can formulate iterative tridiagonalization, sometimes called the *Lanczos process*, as in Algorithm 4.1, an embryonic version of the famous Lanczos algorithm to be discussed below.

Algorithm 4.1 Iterative reduction of a full-rank Hermitian matrix H to Hermitian tridiagonal form T

$|v\rangle \neq 0$ {random initial vector}
$\beta_0 = 0$
$|q_0\rangle = 0$
$|q_1\rangle = |v\rangle / \||v\rangle\|_2$
  for $i = 1 : n - 1$ **do**
    $|w_i\rangle = H|q_i\rangle$
    $\alpha_i = \langle q_i|w_i\rangle$
    $|w_i\rangle \leftarrow |w_i\rangle - \alpha_i|q_i\rangle - \beta_{i-1}|q_{i-1}\rangle$
    $\beta_i = \||w_i\rangle\|_2$
    $\gamma_i = \beta_i$
    $|q_{i+1}\rangle = |w_i\rangle / \beta_i$
  end for

We denote the entries of the diagonal and the two subdiagonals of the tridiagonal matrix T by α_i, β_i and γ_i as in (4.47). They are all real. It is in general not practical to bring H to tridiagonal form completely, n being very large. The interest is in the principal submatrices (top-left square matrices) T_m of dimension $m \times m$, the projections of H on $\mathbb{K}_m(H, |v\rangle)$. If we add a row $(0, 0, \ldots, 0, \beta_m)$ to T_m as in (4.48), we obtain $\tilde{T}_m$, the restriction of H to $\mathbb{K}_m(H, |v\rangle)$.

Why are these matrices of interest? While the exact relationship between eigenvalues of T and H only holds for $m = n$, even for $m \ll n$, T_m contains highly precise information on the extreme eigenvalues and associated eigenvectors of H; we will discuss this below.

Abandoning the full-rank assumption: invariant subspaces

If we abandon the full-rank assumption, there will be some m for which $H|q_m\rangle \in \mathbb{K}_m$ and the algorithm given above must break down as $|w_m\rangle = 0$, so we cannot construct $|q_{m+1}\rangle$. From the point of view of iterative reduction to tridiagonal form, this means that we have to restart the algorithm with an arbitrary normalized $|q_{m+1}\rangle$ that is orthogonal to all $|q_1\rangle, \ldots, |q_m\rangle$. In practice, orthonormalizing a random vector with respect to the previously constructed basis vectors will do. At the same time, we have constructed an m-dimensional invariant subspace $\mathcal{S}_m = \mathrm{span}(|q_1\rangle, \ldots, |q_m\rangle)$ with

$$H\mathcal{S}_m \subset \mathcal{S}_m. \tag{4.81}$$

In terms of the tridiagonal matrix being constructed, we have hit upon a reduced tridiagonal matrix. Diagonalizing it provides us with a subset of *exact* eigenvalues and eigenvectors of H because of (4.77). In practice, however, this will not occur: in our applications, the invariant subspaces will simply be too large.

4.6 Large sparse eigensystem solvers

In this section, we want to show how the theory of Krylov spaces allows us to design an algorithm that, merely based on matrix–vector multiplications, gives fast access to the eigenvalues at the ends of the spectrum of some large sparse Hermitian matrix H, in particular the lowest one, the ground-state energy if we think of a Hamiltonian, and the associated ground state. Again, this discussion leans heavily on [29], on the particularly accessible text [51], and on [4], which gives real algorithms in a way that can be easily adapted to one's own programs.

Before discussing this *Lanczos algorithm*, we consider the *power method*, a not very efficient, but reliable algorithm to find the largest eigenvalue, because it provides a good conceptual foundation for the Lanczos method.

4.6.1 The power method

Let us assume that the eigenvalues of a Hermitian matrix H are ordered by modulus, $|\lambda_1| \geq |\lambda_2| \geq \ldots \geq |\lambda_n|$; the eigenvectors $\{|v_i\rangle\}$ form an orthonormal basis of $\mathbb{C}^n$. Consider $|v\rangle = \sum_i \langle v_i|v\rangle |v_i\rangle$. Then $H|v\rangle = \sum_i \lambda_i \langle v_i|v\rangle |v_i\rangle$, and, after m multiplications by H,

$$H^m|v\rangle = \sum_i \lambda_i^m \langle v_i|v\rangle |v_i\rangle. \tag{4.82}$$

If $|\lambda_1|$ is strictly larger than any other $|\lambda_i|$ and if $\langle v_1|v\rangle \neq 0$, which is best ensured by taking $|v\rangle$ to be a normalized random vector, for $m \to \infty$ (4.82) will be dominated by the contribution $\lambda_1^m \langle v_1|v\rangle |v_1\rangle$, i.e., the eigenvector of the eigenvalue with the largest modulus, up to normalization. We may therefore formulate the *power algorithm* for obtaining the eigenpair $(\lambda_1, |\lambda_1\rangle)$ as in Algorithm 4.2. As termination criterion, we use that the eigenvalue equation is fulfilled up to a tolerance ϵ: in fact, it can be shown that this implies that an eigenvalue is approximated with precision better than $\sqrt{2}\,\epsilon$. If the largest eigenvalue is separated by more than $2\sqrt{2}\,\epsilon$ from all other ones, then the approximated eigenvalue is the largest eigenvalue.

How quickly will the power method approximate the largest eigenpair [51]? Let us consider the mth normalized vector $|k\rangle$ from Algorithm 4.2, which we label $|k_m\rangle$ and decompose as

$$|k_m\rangle = \cos\phi_m |\lambda_1\rangle + \sin\phi_m |\lambda_{1\perp}^{(m)}\rangle, \tag{4.83}$$

where $|\lambda_{1\perp}^{(m)}\rangle$ is a normalized vector perpendicular to the eigenvector $|\lambda_1\rangle$. ϕ_m is the error angle between the current vector and the eigenvector we are looking for. Up to normalization, $|k_{m+1}\rangle = |\tilde{k}\rangle$ is given by

$$H|k_m\rangle = \cos\phi_m \lambda_1 |\lambda_1\rangle + \sin\phi_m \frac{H|\lambda_{1\perp}^{(m)}\rangle}{\|H|\lambda_{1\perp}^{(m)}\rangle\|_2} \|H|\lambda_{1\perp}^{(m)}\rangle\|_2, \tag{4.84}$$

Algorithm 4.2 Power method determination of the largest absolute value eigenpair $(\lambda_1, |\lambda_1\rangle)$ of a Hermitian matrix H

---

   initialize $|k\rangle \neq 0$ {random initial vector}
   $|k\rangle \leftarrow |k\rangle / \||k\rangle\|_2$
   for $m = 1, 2, \ldots$ **do**
      $|\tilde{k}\rangle = H|k\rangle$
      $\theta_m = \langle k|\tilde{k}\rangle$
      if $\||\tilde{k}\rangle - \theta_m|k\rangle\|_2 < \epsilon$ **then**
         $\lambda_1 = \theta_m$
         $|\lambda_1\rangle = |k\rangle$
         terminate algorithm
      end if
      $|k\rangle = |\tilde{k}\rangle / \||\tilde{k}\rangle\|_2$
   end for

---

such that we can read off

$$\left|\lambda_{1\perp}^{(m+1)}\right\rangle = \frac{H\left|\lambda_{1\perp}^{(m)}\right\rangle}{\left\|H\left|\lambda_{1\perp}^{(m)}\right\rangle\right\|_2} \tag{4.85}$$

and

$$|\tan\phi_{m+1}| = \frac{\left\|H|\lambda_{1\perp}^{(m)}\rangle\right\|_2 |\sin\phi_m|}{|\lambda_1 \cos\phi_m|} \leq \left|\frac{\lambda_2}{\lambda_1}\right| |\tan\phi_m|. \tag{4.86}$$

We have bounded $\|H|\lambda_{1\perp}^{(m)}\rangle\|_2 \leq |\lambda_2|$ since the vector is perpendicular to the eigenvector of λ_1. In the mth iteration, the error angle is given by

$$|\tan\phi_m| \leq \left|\frac{\lambda_2}{\lambda_1}\right|^{m-1} |\tan\phi_1|, \tag{4.87}$$

where ϕ_1 is the angle between $|\lambda_1\rangle$ and the initial "guess" vector $|k_1\rangle$. Eigenvector convergence is said to be linear in m: each iteration adds one power of the convergence factor $|\lambda_2/\lambda_1|$. Convergence will mainly depend on the ratio of the two largest eigenvalues. Similarly, one can show [29]

$$\lambda_1 - \lambda_1^{(m)} \leq (\lambda_1 - \lambda_n)\tan^2\phi_1 \left(\frac{\lambda_2}{\lambda_1}\right)^{2m-2}, \tag{4.88}$$

where $\cos\phi_1 = \langle\lambda_1|k_1\rangle$, the overlap between the desired eigenvector and the normalized initial vector. The convergence of the eigenvalue is *quadratic* in m: each iteration gives a square of the convergence factor. We observe that convergence is faster if the overlap between the initial "guess" vector and the eigenvector is large. A good guess for $|\lambda_1\rangle$ should be used as starting vector, if available. The power algorithm will break down if the largest eigenvalue is not unique and critically slow down if it is almost degenerate. The fundamental problem is that the power method is slow even when it converges.

4.6.2   The Lanczos method

In the power method, we have iteratively calculated vectors $|v\rangle$, $H|v\rangle$, $H^2|v\rangle$, $H^3|v\rangle$, ..., always looking for a better eigenvector approximation in the last vector calculated, *neglecting* the fact that for variational reasons, there will usually be a much better approximation in the space spanned by all vectors calculated in the algorithm, which is nothing but the sequence of Krylov spaces $\mathbb{K}_m(H, |v\rangle)$. As matrix–vector multiplications is by far the most expensive operation in the power method, this is an enormous waste; but the Krylov sequence is optimally adapted for a search in increasingly larger subspaces of $\mathbb{C}^n$.

Let us look for optimal approximations to the eigenvectors of both the largest and smallest eigenvalues of H, λ_1 and λ_n, in the Krylov space

$$\mathbb{K}_m(H, |v\rangle) = \mathrm{span}(|v\rangle, H|v\rangle, H^2|v\rangle, \ldots, H^{m-1}|v\rangle), \tag{4.89}$$

by considering the Rayleigh–Ritz quotient $\rho(H, |x\rangle) = \langle x|H|x\rangle/\langle x|x\rangle$. Then the best approximation to the largest and smallest eigenvalues is given by

$$\lambda_1^{(m)} = \max_{|x\rangle \in \mathbb{K}_m(H,|v\rangle) \neq 0} \frac{\langle x|H|x\rangle}{\langle x|x\rangle}, \tag{4.90}$$

$$\lambda_m^{(m)} = \min_{|x\rangle \in \mathbb{K}_m(H,|v\rangle) \neq 0} \frac{\langle x|H|x\rangle}{\langle x|x\rangle}, \tag{4.91}$$

where

$$\lambda_1^{(m)} \leq \lambda_1, \qquad \lambda_m^{(m)} \geq \lambda_n. \tag{4.92}$$

To simplify the search, we introduce the preferential orthonormal basis of $\mathbb{K}_m(H, |v\rangle)$ as $|q_1\rangle, |q_2\rangle, \ldots, |q_m\rangle$ by iterative Gram–Schmidt orthonormalization. We assume for the moment that the Krylov subspace has full rank m. Then, for $|y\rangle \in \mathbb{C}^m$,

$$\lambda_1^{(m)} = \max_{|y\rangle \neq 0} \frac{\langle y|Q_m^\dagger H Q_m|y\rangle}{\langle y|y\rangle} = \max_{|y\rangle \neq 0} \frac{\langle y|T_m|y\rangle}{\langle y|y\rangle}, \tag{4.93}$$

$$\lambda_m^{(m)} = \min_{|y\rangle \neq 0} \frac{\langle y|Q_m^\dagger H Q_m|y\rangle}{\langle y|y\rangle} = \min_{|y\rangle \neq 0} \frac{\langle y|T_m|y\rangle}{\langle y|y\rangle}. \tag{4.94}$$

T_m allows a simple determination of $\lambda_1^{(m)}$, $\lambda_m^{(m)}$ and $|y_1^{(m)}\rangle$, $|y_m^{(m)}\rangle$, its eigenvalues and eigenvectors. By the properties of the Sturm chain, we have strict bounds $\lambda_1^{(m)} \leq \lambda_1$ and $\lambda_m^{(m)} \geq \lambda_n$.

Algorithm 4.3 Uncontrolled Lanczos algorithm for the iterative tridiagonalization of a Hermitian matrix H

$|w_1\rangle = |v\rangle \neq 0$ {random initial vector}
$\beta_0 = \||w_1\rangle\|_2$
$|q_0\rangle = 0$
$m = 1$
while $\beta_{m-1} \neq 0$ **do**
    $|q_m\rangle = |w_m\rangle/\beta_{m-1}$
    $|w_{m+1}\rangle = H|q_m\rangle$
    $\alpha_m = \langle q_m|w_{m+1}\rangle$
    $|w_{m+1}\rangle \leftarrow |w_{m+1}\rangle - \alpha_m|q_m\rangle - \beta_{m-1}|q_{m-1}\rangle$
    $\beta_m = \||w_{m+1}\rangle\|_2$
    $m = m + 1$
end while

How can we *optimally* enlarge our search space by another vector orthogonal to $\mathbb{K}_m(H,|v\rangle)$ in order to improve the eigenvalue approximation? The answer is to add the orthogonal component of the vector in whose direction the Rayleigh–Ritz quotient of the largest eigenvalue increases most strongly and the orthogonal component of the vector in whose direction the Rayleigh–Ritz quotient of the smallest eigenvalue decreases most strongly. The key observation is that while these two vectors will in general be different, their components orthogonal to $\mathbb{K}_m(H,|v\rangle)$ are such that both gradients are contained in the search space if we simply enlarge $\mathbb{K}_m(H,|\psi\rangle)$ to $\mathbb{K}_{m+1}(H,|\psi\rangle)$, suggesting the use of an iterative procedure for the eigenvalue approximation!

To see this, consider the gradient of the Rayleigh–Ritz quotient,

$$\nabla\rho(|x\rangle) = 2\frac{H|x\rangle - \rho(H,|x\rangle)|x\rangle}{\langle x|x\rangle}, \tag{4.95}$$

giving the direction of maximal increase and maximal decrease (with opposite signs). Inserting $Q_m|y_1^{(m)}\rangle$ and $Q_m|y_m^{(m)}\rangle$ will yield different gradients in general. However, in both cases, the gradient is a linear combination of a vector in $\mathbb{K}_m(H,|v\rangle)$ and $HQ_m|y_{\max,\min}^{(m)}\rangle \in \mathrm{span}(H|v\rangle,\dots,H^m|v\rangle)$. Thus, if we enlarge $\mathbb{K}_m(H,|v\rangle)$ to $\mathbb{K}_{m+1}(H,|v\rangle)$, both gradients, while different, will be contained, and the orthogonal vector we have been looking for is $H^m|v\rangle$ orthonormalized with respect to $\mathbb{K}_m(H,|\psi\rangle)$. The construction of that vector has already been achieved in the iterative construction of the orthonormal preferential basis using (4.80). We can therefore formulate Algorithm 4.3 for iterative tridiagonalization if we ignore issues of convergence for the time being. Numerical breakdown will be used as the termination criterion. This procedure is referred to as *Lanczos diagonalization*, after its inventor.

The Lanczos procedure provides us with a hierarchy of tridiagonal matrices T_1, T_2, ..., T_m, ..., where the smaller matrices are the top-left corners of the bigger matrices. If we further assume that no subdiagonal element β is zero, which would lead to termination of the algorithm, the eigenvalues of successive T_m obey the strict inequalities of a Sturm chain: if we denote the eigenvalues of T_m by $\lambda_i^{(m)}$, $1 \leq i \leq m$, assuming them to be ordered from the biggest to the smallest eigenvalue, then

$$\lambda_1^{(m)} > \lambda_1^{(m-1)} > \lambda_2^{(m)} > \lambda_2^{(m-1)} > \lambda_3^{(m)} > \ldots > \lambda_{m-1}^{(m)} > \lambda_{m-1}^{(m-1)} > \lambda_m^{(m)}. \quad (4.96)$$

We therefore approximate the largest eigenvalue from below and the smallest from above, with a monotonic improvement of the approximation.

The speed of convergence of eigenvalues is determined by the *Kaniel–Paige theorem* which we state without proof [29]: If we have a tridiagonal matrix $T_m \in \mathbb{R}^{m \times m}$ after m Lanczos steps with eigenvalues $\lambda_1^{(m)} \geq \lambda_2^{(m)} \geq \ldots \geq \lambda_m^{(m)}$, then the eigenvalues $\lambda_1^{(m)}$ and $\lambda_m^{(m)}$ approximate the largest and smallest eigenvalues of H as

$$\lambda_1^{(m)} \in \left[\lambda_1 - \frac{(\lambda_1 - \lambda_n) \tan^2 \phi_1}{[c_{m-1}(1 + 2s_1)]^2}, \lambda_1 \right],$$

$$\lambda_m^{(m)} \in \left[\lambda_n, \lambda_n + \frac{(\lambda_1 - \lambda_n) \tan^2 \phi_n}{[c_{m-1}(1 + 2s_n)]^2} \right], \quad (4.97)$$

where $\cos \phi_1 = \langle \lambda_1 | q_1 \rangle$, $\cos \phi_n = \langle \lambda_n | q_1 \rangle$, and

$$s_1 = \frac{\lambda_1 - \lambda_2}{\lambda_2 - \lambda_n}, \qquad s_n = \frac{\lambda_{n-1} - \lambda_n}{\lambda_1 - \lambda_{n-1}}.$$

The $c_{m-1}(x)$ are the *Chebyshev polynomials* of degree $m - 1$. Comparing with (4.88), the convergence of the Lanczos procedure will be essentially determined by the behavior of the Chebyshev polynomials for $x \geq 1$. The polynomials are given by $c_0(x) = 1$, $c_1(x) = x$, and the recursion $c_m(x) = 2x c_{m-1}(x) - c_{m-2}(x)$, so

$$c_2(x) = 2x^2 - 1, \qquad c_3(x) = 4x^3 - 3x, \qquad c_4(x) = 8x^4 - 8x^2 + 1, \quad \ldots \quad (4.98)$$

$c_m(1) = 1$, but for $x > 1$, which is the case for all nondegenerate extreme eigenvalues, $c_m(x)$ grows very rapidly as m increases. For small ϵ and large m, $c_m(1 + 2\epsilon) \approx \frac{1}{2} e^{2m\sqrt{\epsilon}}$. This shows that close to convergence, both the largest and smallest eigenvalues of T_m will converge *very rapidly* to the extreme eigenvalues of H. Other eigenvalues will converge too, but the further one moves away from the extreme ends of the spectrum, the slower convergence will be, as can be seen from the general form of the Kaniel–Paige theorem. The Lanczos method is therefore ideally suited to cases where we are interested in (a few) extreme eigenvalues and associated eigenvectors. As in the power method, the speed of convergence suffers a critical slowdown if the extreme eigenvalues are verging on degeneracy. Last but not least, it is important to emphasize that, like any other iterative method, the Lanczos method is strongly accelerated if we have a good initial guess for the solution! In a first-order approximate comparison between the Lanczos and power methods, for $\lambda_1/\lambda_2 = 1 + \epsilon$, the Lanczos method converges as $e^{-c_1 m \sqrt{\epsilon}}$ and the power method as $e^{-c_2 m \epsilon}$, which is much slower in the realistic case of not strongly separated dominant eigenvalues (i.e., ϵ small).

Convergence test

How do we terminate the algorithm? There are two reasons for termination, one good, one (seemingly) bad. The good reason is that we terminate because $|w_{m+1}\rangle = 0$ after orthogonalization with respect to $|q_1\rangle, \ldots, |q_m\rangle$. If we set

$$r = \operatorname{rank} \mathbb{K}_n(H, |v\rangle), \tag{4.99}$$

the dimension of the state space that can be reached from $|v\rangle$ by an arbitrary power H^i, then termination must occur for $m = r$. It can in fact be shown that—in a purely mathematical concept—it will not terminate beforehand. When this type of termination happens, we can determine all eigenvalues of H with eigenvectors in this state space exactly, since it is invariant under further applications of H.

For large sparse matrices, we will not encounter this type of termination in practice, since the spaces that can typically be reached from $|v\rangle$ have very high dimension such that the other reason for termination occurs first: the orthonormality of $|q_{m+1}\rangle$ with respect to $|q_1\rangle, \ldots, |q_{m-2}\rangle$ is ensured only on mathematical grounds. In fact, orthogonality between the first and last Lanczos vectors will be lost in any realistic implementation. This can be seen easily by calculating $\langle q_1 | q_m \rangle$ at each step; while it should be strictly zero for $m \geq 2$, it will acquire increasingly large nonzero values as m increases.

So we need a better test for the convergence of some eigenvalue $\lambda_i^{(m)}$ in the mth Lanczos iteration [29, 51]. For the Ritz pair $(\lambda_i^{(m)}, |y_i^{(m)}\rangle)$ obeying

$$T_m |y_i^{(m)}\rangle = \lambda_i^{(m)} |y_i^{(m)}\rangle, \tag{4.100}$$

the approximate eigenvector reads

$$|x_i^{(m)}\rangle = Q_m |y_i^{(m)}\rangle. \tag{4.101}$$

According to (4.40), the residual $\|H|x_i^{(m)}\rangle - \lambda_i^{(m)}|x_i^{(m)}\rangle\|_2$ is a good measure for the convergence of the Ritz pair to an eigenpair of H, because it bounds the distance of $\lambda_i^{(m)}$ to the closest true eigenvalue. It can be calculated as

$$
\begin{aligned}
\|H|x_i^{(m)}\rangle - \lambda_i^{(m)}|x_i^{(m)}\rangle\|_2 &= \|HQ_m|y_i^{(m)}\rangle - \lambda_i^{(m)}Q_m|y_i^{(m)}\rangle\|_2 \\
&= \|(HQ_m - Q_m T_m)|y_i^{(m)}\rangle\|_2 \\
&= \|\beta_m|q_{m+1}\rangle\langle e_m|y_i^{(m)}\rangle\|_2 \\
&= \beta_m |\langle e_m|y_i^{(m)}\rangle|.
\end{aligned}
\tag{4.102}
$$

It is convenient to abbreviate $\beta_{mi} = \beta_m |\langle e_m|y_i^{(m)}\rangle|$. Convergence is obtained when $\beta_{mi} \to 0$, or, more precisely,

$$\beta_{mi} \approx \sqrt{\epsilon} \, \|H\|_2, \tag{4.103}$$

with ϵ the machine precision. In practice, β_{mi} goes to zero because the bottom component of the Ritz vector $|y_i^{(m)}\rangle$ is vanishing.

Close to convergence, we can obtain reasonable estimates of the quality of the approximate eigenvalue and eigenvector by estimating $\mathrm{gap}(A, \lambda_i^{(m)})$ from the approximate spectrum of T_m,

$$\mathrm{gap}(A, \lambda_i^{(m)}) \approx \delta_i = \min_{j \neq i} |\lambda_i^{(m)} - \lambda_j^{(m)}|, \tag{4.104}$$

and using it in (4.43) and (4.44) to obtain

$$|\sin \angle(|x_i^{(m)}\rangle, |\lambda_i\rangle)| \approx \frac{\beta_{mi}}{\delta_i} \tag{4.105}$$

and

$$|\lambda_i^{(m)} - \lambda_i| \approx \frac{\beta_{mi}^2}{\delta_i}. \tag{4.106}$$

Observe that to judge convergence, it is not necessary to calculate the approximate eigenvector $|x_i^{(m)}\rangle$ explicitly.

Convergence and loss of orthogonality

In the Lanczos procedure, orthogonalization of a new basis vector $|q_{m+1}\rangle$ is carried out explicitly only with respect to the two previously added basis vectors $|q_m\rangle$ and $|q_{m-1}\rangle$. Orthogonality to all other basis vectors is an exact mathematical statement; see (4.79). Numerically, however, this orthogonality that is never explicitly enforced is in the danger of being lost. In fact, it turns out that orthogonality is well maintained until the first Ritz pair converges to a true eigenpair of H. Assuming that convergence happens at the mth iteration, $|q_{m+1}\rangle$ can be shown to have a large component in the direction of the converged eigenvector. But as the converged eigenvector is a linear combination of all previous $|q_i\rangle$, this also implies a generic loss of orthogonality of the new basis vectors to previous basis vectors.

If one continues to run the Lanczos algorithm without modification, the next Ritz pair will in general converge to the same eigenpair as before, not to the next eigenpair. This means that the Lanczos algorithm will suggest nonexistent multiplicities of eigenvalues. Therefore, if we are interested in more than just the lowest-lying eigenvalue for a ground-state calculation, we have to use more advanced orthogonalization techniques.

4.7 Large sparse equation solvers

Several important nonequilibrium algorithms, mainly for the calculation of spectral functions, ultimately rest on the solution of some large sparse equation system

$$A|x\rangle = |b\rangle, \tag{4.107}$$

where the matrix A is large and sparse, and not necessarily Hermitian. In the cases where Hermiticity applies, there is a very well-established and numerically very stable method, the *conjugate gradient method*. For all other cases, the current method of choice is the *generalized minimal residual method (GMRES)*. Very clear explicit algorithmic formulations of the methods can be found in [5, 29, 58].

By premultiplication with $A^\dagger$, any system in the form of (4.107) can be turned into a Hermitian one:

$$(A^\dagger A)|x\rangle = |c\rangle, \qquad |c\rangle = A^\dagger|b\rangle. \tag{4.108}$$

The reason why one should use this way out only as a last resort is that, as we have seen for the Lanczos method, convergence is governed strongly by ratios of eigenvalues, and similar results hold for solving large sparse equation systems. Upon "squaring" the matrix A, all these ratios are roughly squared, which dramatically slows down convergence.

4.7.1 Conjugate gradient method: solving Hermitian equation systems

Let us consider a large sparse equation system

$$H|x\rangle = |b\rangle, \tag{4.109}$$

where $|x\rangle, |b\rangle \in \mathbb{C}^n$ and $H \in \mathbb{C}^{n \times n}$ is Hermitian and full-rank (to ensure invertibility and the existence of a unique solution). H will also be *positive-definite* in the following, which we can assure to be the case in all our "real-world" problems later. The most naive approach to solving the problem using $H|v\rangle$ would be the *steepest-descent approach*. Let me assume for simplicity that all vectors and H are real (the complex case is an easy generalization, and not important for the following purpose). Then we can reformulate the linear system $H|x\rangle = |b\rangle$ as a minimization problem of

$$\phi(|x\rangle) = \tfrac{1}{2}\langle x|H|x\rangle - \langle x|b\rangle, \tag{4.110}$$

because the variation vanishes for

$$H|x\rangle = |b\rangle. \tag{4.111}$$

This is the analog variational problem to the Rayleigh–Ritz quotient.

Assume we have an initial guess $|x_0\rangle$ for the minimizing solution $|x\rangle$. It is a reasonable idea to look for a better solution in the direction of the *negative* gradient of $\phi(|x_0\rangle)$, which is given as

$$-\nabla\phi(|x_0\rangle) = |b\rangle - H|x_0\rangle = |r_0\rangle. \tag{4.112}$$

The *residual* $|r_0\rangle$ has a double function: it indicates by how much our initial guess is off the mark and it gives a direction in which to expect a better solution. How far we have to move along this direction is found by minimizing

$$\phi(|x_0\rangle + \alpha_1|r_0\rangle) = \phi(|x_0\rangle) + \tfrac{1}{2}\alpha_1^2\langle r_0|H|r_0\rangle - \alpha_1\langle r_0|r_0\rangle, \tag{4.113}$$

which gives

$$\alpha_1 = \frac{\langle r_0|r_0\rangle}{\langle r_0|H|r_0\rangle}. \tag{4.114}$$

Our next proposed solution is then given as $|x_1\rangle = |x_0\rangle + \alpha_1|r_0\rangle$, leading to a new residual $|r_1\rangle = |b\rangle - H|x_1\rangle$, in whose direction we will look for $|x_2\rangle$, and so forth.

Like the power method, this approach converges, but it may be excruciatingly slow, the reason for this being that at every step the new search direction is determined without accounting for the previous search directions, just like the power method discarding all information of the previously calculated approaches to the eigenvector to be determined. What may happen can be described in the following example appropriate to Les Houches: imagine you are in a mountain valley that slowly descends following a river, and goes up steeply on both sides. The lowest point of the valley is at its downriver exit. But when you use the steepest-descent method to find that point and start from a random point in the valley, the steepest-gradient method will basically take you from one steep side of the valley to the other and back, while only slowly zigzagging toward the exit of the valley. Excluding earlier search directions would have taken you out of this zigzag right away!

In analogy to the large sparse eigensolvers, we can instead search for the minimum of (4.110) in the space spanned by all search directions plus our initial guess, $\{|x_0\rangle +$ span$(|r_0\rangle, \dots, |r_{m-1}\rangle)\}$. Looking at the iterative generation of the $|r_i\rangle$, the span is just the Krylov space $\mathbb{K}_m(H, |r_0\rangle)$. We can therefore set up the following procedure: starting from $|r_0\rangle$, generate a sequence of tridiagonal matrices $T_m = Q_m^\dagger H Q_m$ by the Krlyov approach, iteratively generating the preferential orthonormal basis. Because of the link $|x\rangle = Q_m|y\rangle$, minimization of (4.110) amounts to minimizing for $|y\rangle \in \mathbb{C}^m$ the function

$$\phi(|x_0\rangle + Q_m|y\rangle) = \tfrac{1}{2}\left(\langle x_0|H|x_0\rangle + \langle y|Q_m^\dagger H|x_0\rangle + \langle x_0|HQ_m|y\rangle + \langle y|Q_m^\dagger HQ_m|y\rangle\right)$$
$$-\langle x_0|b\rangle - \langle y|Q_m^\dagger|b\rangle.$$

The minimum is obtained for $|y_m\rangle$ with

$$- Q_m^\dagger H|x_0\rangle + Q_m^\dagger|b\rangle = T_m|y_m\rangle, \tag{4.115}$$

or $Q_m^\dagger|r_0\rangle = T_m|y_m\rangle$. Given that $|r_0\rangle$ is the first basis vector of the preferential orthonormal basis up to its norm $\beta_1 := \||r_0\|_2$, we have to solve the problem $\beta_1|e_1\rangle = T_m|y_m\rangle$, which can be done by inverting the small tridiagonal matrix T_m; as the left-hand side is quite special, even more efficient techniques than full inversion exist. The iterative solution is then $|x\rangle = |x_0\rangle + Q_m|y_m\rangle$. This is the core idea of the *conjugate gradient method*, but in practice, one uses a mathematically equivalent, numerically better, but conceptually perhaps more obscure approach, which we will now discuss; see, for example, Section 2.7 in *Numerical Recipes* [56], whose variant of the conjugate gradient algorithm is adopted in the following.

The starting point in this approach to the conjugate gradient method is to break the issue of inefficient search directions by abandoning the identity of residual and search direction and by assuming that we generate a search direction $|p_1\rangle$ from $|r_0\rangle$ along which we search for $|x_1\rangle$ and so forth:

$$|x_0\rangle \to |r_0\rangle \to |p_1\rangle \to |x_1\rangle \to |r_1\rangle \to |p_2\rangle \to |x_2\rangle \to \dots \tag{4.116}$$

The good choice for the $|p_i\rangle$ is to still search along the steepest descent, but on the other hand also H-orthogonal (i.e., with respect to the scalar product induced by H,

$\langle v|H|w\rangle$) to all the previous search directions such that $\langle p_i|H|p_j\rangle \sim \delta_{ij}$; this condition is referred to as *H-conjugacy*. This set of vectors forms an (incomplete) basis and we are searching for the unique expansion $|x\rangle = |x_0\rangle + \sum_{i=1}^{n} \alpha_i|p_i\rangle$. *H*-conjugacy makes sense because we are essentially minimizing $\langle x|H|x\rangle$. We will then never search twice along the same direction. The two requirements suggest that we take as search direction $|p_m\rangle$ the last residual (i.e., gradient) $|r_{m-1}\rangle$, but orthogonalize it (with respect to the *H*-scalar product) to all previous search directions $|p_1\rangle, \ldots, |p_{m-1}\rangle$:

$$|p_m\rangle = |r_{m-1}\rangle - \frac{\langle p_{m-1}|H|r_{m-1}\rangle}{\langle p_{m-1}|H|p_{m-1}\rangle}|p_{m-1}\rangle - \frac{\langle p_{m-2}|H|r_{m-1}\rangle}{\langle p_{m-2}|H|p_{m-2}\rangle}|p_{m-2}\rangle - \ldots \quad (4.117)$$

This can be verified easily by premultiplying by $\langle p_i|H$ for $i < m$, assuming that all $|p_i\rangle$, $i < m$, are already *H*-conjugate among each other. For the new search direction, ϕ is given by

$$\phi(|x_{m-1}\rangle + \alpha_m|p_m\rangle) = \phi(|x_{m-1}\rangle) + \tfrac{1}{2}\alpha_m^2\langle p_m|H|p_m\rangle - \alpha_m\langle p_m|r_{m-1}\rangle \quad (4.118)$$

and minimized by

$$\alpha_m = \frac{\langle p_m|r_{m-1}\rangle}{\langle p_m|H|p_m\rangle}, \quad (4.119)$$

giving $|x_m\rangle = |x_{m-1}\rangle + \alpha_m|p_m\rangle$ and $|r_m\rangle = |r_{m-1}\rangle - \alpha_m H|p_m\rangle$. Two possible problems come to mind: (i) the *H*-orthogonalization will become very costly as m is increasing and (ii) even if it is obviously more efficient than the steepest-descent method, is it really fast?

Both issues are unproblematic, because it can be shown by induction that we are again just building a Krylov space from a Hermitian matrix. Then orthogonalization will shorten drastically from a m-term to a two-term expression (even less than in Lanczos), and, as before, one can show that an iterative search in Krylov space is the most efficient method to be obtained if we base the algorithm on subsequent calculations $H|v\rangle$.

If we iterate starting from some $|x_0\rangle$, associated $|r_0\rangle$, and $|p_0\rangle = 0$, then our algorithm, already sketched above, can be summarized as (without doing the induction explicitly)

$$|p_m\rangle = |r_{m-1}\rangle + \beta_m|p_{m-1}\rangle, \quad (4.120)$$

$$|x_m\rangle = |x_{m-1}\rangle + \alpha_m|p_m\rangle, \quad (4.121)$$

$$|r_m\rangle = |r_{m-1}\rangle - \alpha_m H|p_m\rangle, \quad (4.122)$$

with

$$\alpha_m = \frac{\langle p_m|r_{m-1}\rangle}{\langle p_m|H|p_m\rangle} = \frac{\langle r_{m-1}|r_{m-1}\rangle}{\langle p_m|H|p_m\rangle}, \quad (4.123)$$

$$\beta_m = -\frac{\langle p_{m-1}|H|r_{m-1}\rangle}{\langle p_{m-1}|H|p_{m-1}\rangle} = \frac{\langle r_{m-1}|r_{m-1}\rangle}{\langle r_{m-2}|r_{m-2}\rangle}. \quad (4.124)$$

Algorithm 4.4 Conjugate gradient solution of $H|x\rangle = |b\rangle$ for Hermitian positive-definite H of dimension $n \times n$

$|r_0\rangle = |b\rangle - H|x_0\rangle$ $\{|x_0\rangle$ initial guess$\}$
$m = 0$
while $|r_m\rangle \neq 0$ **do**
  $m = m + 1$
  if $m = 1$ **then**
    $|p_1\rangle = |r_0\rangle$
  else
    $\beta_m = \langle r_{m-1}|r_{m-1}\rangle / \langle r_{m-2}|r_{m-2}\rangle$
    $|p_m\rangle = |r_{m-1}\rangle + \beta_m |p_{m-1}\rangle$
  end if
  $\alpha_m = \langle r_{m-1}|r_{m-1}\rangle / \langle p_m|H|p_m\rangle$
  $|x_m\rangle = |x_{m-1}\rangle + \alpha_m |p_m\rangle$
  $|r_m\rangle = |r_{m-1}\rangle - \alpha_m H|p_m\rangle$
end while
$|x\rangle = |x_m\rangle$

This constructs an iterative solution

$$|x_m\rangle = |x_0\rangle + \sum_{k=1}^{m} \alpha_k |p_k\rangle, \tag{4.125}$$

where the sum is an element of

$$\text{span}(|p_1\rangle, \ldots, |p_m\rangle) = \text{span}(|r_0\rangle, \ldots, |r_{m-1}\rangle) = \mathbb{K}_m(H, |r_0\rangle). \tag{4.126}$$

This means that the spans of the search directions and the residuals thus generated are identical and in turn identical to the Krylov space generated by applying iteratively H to $|r_0\rangle$, with the residuals being (up to normalization) the Lanczos vectors. By construction, the search directions are H-conjugate, $\langle p_i|H|p_j\rangle \sim \delta_{ij}$, the residuals are mutually orthogonal, $\langle r_i|r_j\rangle \sim \delta_{ij}$, and orthogonality between a residual and all previous search directions holds: $\langle p_i|r_m\rangle = 0$ $(1 \leq i \leq m)$. What is remarkable is that there is no three-vector sequence as in the tridiagonalization of a Hermitian matrix H that occurs in the first form of the conjugate gradient method; only the results from the last iteration are involved. This follows immediately from the H-conjugacy of the $|p_i\rangle$.

This gives the *conjugate gradient* algorithm formulated as in Algorithm 4.4. Obviously, a good initial guess for $|x_0\rangle$ is crucial; for an appropriate *preconditioning* of H, the reader should consult the literature. As termination condition, we will in practice accept that the size of the residual drops below an acceptable accuracy.

4.7.2   The GMRES algorithm: solving non-Hermitian equation systems

If we are dealing with a general matrix A, we can again start from an initial solution $|x_0\rangle$ and search in the space $\{|x_0\rangle + \mathbb{K}_m(A, |r_0\rangle)\}$. Instead of tridiagonal matrices T_m, we will obtain upper-Hessenberg matrices $A_m = Q_m^\dagger A Q_m$. As A is not positive-definite and not Hermitian, we cannot formulate an equivalent variational problem to find $|x_m\rangle = |x_0\rangle + Q_m|y_m\rangle$ for some $|y_m\rangle \in \mathbb{C}^m$. Instead, we minimize the length of the residual

$$
\begin{aligned}
\||r_m\rangle\|_2 &= \||A|x_m\rangle - |b\rangle\|_2 \\
&= \|AQ_m|y_m\rangle + A|x_0\rangle - |b\rangle\|_2 \\
&= \|Q_{m+1}\tilde{A}_m|y_m\rangle - |r_0\rangle\|_2 \\
&= \|Q_{m+1}\tilde{A}_m|y_m\rangle - \beta_1 Q_{m+1}|e_1\rangle\|_2 \\
&= \|\tilde{A}_m|y_m\rangle - \beta_1|e_1\rangle\|_2.
\end{aligned}
\tag{4.127}
$$

β_1 is the norm of $|r_0\rangle$; $\tilde{A}_m$ is, as before, the upper-Hessenberg matrix A_m, augmented as in (4.69). The approximate solution at step m is then given by $|x_m\rangle = |x_0\rangle + Q_m|y_m\rangle$ as in the conjugate gradient method, where $|y_m\rangle$ minimizes the last expression. This is the core version of the *generalized minimal residual method (GMRES)*, first proposed by Saad and Schultz [60]; the above argument is the starting point of its exposition by Saad [58].

As convergence criterion, we will in practice accept that the achievable minimum of the residual has dropped below some acceptable accuracy ϵ. In the algorithm, H contains the growing upper-Hessenberg matrices, which are brought into upper-triangular form at the generation of new entries.

Many libraries for linear algebra can carry out a least-square minimization as in (4.127). Even if this is done from scratch every time, which makes it increasingly costly, this is comparatively fast given that the most expensive operation is still the matrix–vector multiplication $A|v\rangle$. Nevertheless, the most elegant solution is based on the following ideas. Any matrix A of dimension $m \times n$ can be *QR-decomposed* such that $A = QR$, where Q is $(m \times m)$-dimensional and unitary and R is $(m \times n)$-dimensional and an upper-triangular matrix. In our case, we QR-decompose as

$$
\tilde{A}_m = \tilde{Q}_m \tilde{R}_m,
\tag{4.128}
$$

where $\tilde{Q}_m$ has dimension $(m+1) \times (m+1)$ and $\tilde{R}_m$ has dimension $(m+1) \times m$. As $\tilde{R}_m$ is upper-triangular, its last row must be zero, and we can write it as

$$
\tilde{R}_m = \begin{bmatrix} R_m \\ 0 \end{bmatrix},
\tag{4.129}
$$

where R_m has dimension $m \times m$. We can then write

$$
\|\tilde{A}_m|y_m\rangle - \beta_1|e_1\rangle\|_2 = \|\tilde{Q}_m^\dagger(\tilde{A}_m|y_m\rangle - \beta_1|e_1\rangle)\|_2 = \|\tilde{R}_m|y_m\rangle - |\tilde{d}_m\rangle\|_2,
\tag{4.130}
$$

where $|\tilde{d}_m\rangle = \beta_1\tilde{Q}_m^\dagger|e_1\rangle$. We decompose this $(m+1)$-dimensional vector as

$$
|\tilde{d}_m\rangle = \begin{bmatrix} |d_m\rangle \\ d_{m+1} \end{bmatrix}.
\tag{4.131}
$$

As the last row of $\tilde{R}_m$ is zero, minimizing $\|\tilde{R}_m|y_m\rangle - |\tilde{d}_m\rangle\|_2$ is equivalent to solving the equation system

$$R_m|y_m\rangle = |d_m\rangle \tag{4.132}$$

for $|y_m\rangle$, and the minimum is $|d_{m+1}|$. As R_m is upper-triangular, it can be solved easily, working our way back upward from row m to row 1 by back-substitution. Back-substitution obviously runs into trouble if any $R_{ii} = 0$. We can ignore this issue in our calculation because it will only arise if we have already found the solution of our large sparse system.

The QR-decomposition is again part of any standard library of linear algebra; but here it is also possible to construct $\tilde{Q}_m$, R_m, and $|\tilde{d}_m\rangle$ iteratively: to transform an upper-Hessenberg matrix $\tilde{A}_m$ to an upper-triangular matrix $\tilde{R}_m$, we obviously have to eliminate the lower subdiagonal elements of $\tilde{A}_m$; if we can do this by suitable ortho-normal transformations, we have explicitly constructed $\tilde{Q}_m$ and $\tilde{R}_m$. The technique of choice is to use the so-called *Givens rotations* first proposed by Jacobi, which leads to the numerically most accurate and efficient implementations of GMRES; see, e.g., [58].

4.8   Exact time evolutions and spectral functions

Let us assume that we have a state $|\psi(t = 0)\rangle$, either a simple state we can construct by hand or the ground state of some Hamiltonian $\hat{H}_0$ that we have obtained by Lanczos diagonalization. This state is now evolved in time under some Hamiltonian $\hat{H}$ for which $|\psi(0)\rangle$ is not an eigenstate. In order to obtain $|\psi(t)\rangle$, we have to solve the time-dependent Schrödinger equation. There are two main strategies for doing this:

- We can apply a numerical integrator directly to the differential equation itself.
- We can write down the formal solution of the differential equation, $|\psi(t)\rangle = e^{-i\hat{H}t}|\psi(0)\rangle$ and evaluate it numerically.

4.8.1   Numerical integration

Let me briefly discuss strategies for direct numerical integration. Naively, we can replace infinitesimals dt by small finite time-differences Δt. Then the Schrödinger equation turns into

$$i\frac{|\psi(t + \Delta t)\rangle - |\psi(t)\rangle}{\Delta t} = \hat{H}|\psi(t)\rangle + O(\Delta t^2). \tag{4.133}$$

The order of the error can best be seen from an expansion of the formal solution. We can then propagate the wavefunction in time as

$$|\psi(t + \Delta t)\rangle = (1 - i\hat{H}\Delta t)|\psi(t)\rangle. \tag{4.134}$$

As we know how to calculate $\hat{H}|\psi(t)\rangle$, this scheme can be implemented easily. Equation (4.134) is simply applied repeatedly to reach times $t + \Delta t, t + 2\Delta t, t + 3\Delta t, t + 4\Delta t, \ldots$ The local error per time step is $O(\Delta t^2)$; to reach time T, $T/\Delta t$ evolution steps have to be carried out and the global error is $T \cdot O(\Delta t)$. We therefore speak of a *first-order method*, because the error is in the first power of Δt.

This scheme, originally due to Euler, is easily implemented, but it has the big disadvantage that it is not unitary. Let us introduce the time-evolution operator $\hat{U}(\Delta t) = 1 - i\hat{H}\Delta t$. In quantum mechanics, we should have $\hat{U}^\dagger(\Delta t)\hat{U}(\Delta t) = 1$. But here

$$\hat{U}^\dagger(\Delta t)\hat{U}(\Delta t) = (1 + i\hat{H}\Delta t)(1 - i\hat{H}\Delta t) = 1 + \hat{H}^2\Delta t^2, \qquad (4.135)$$

so there is a $O(\Delta t^2)$ violation of unitarity: norms are not conserved. This is undesirable.

Roughly speaking, there are two ways out, namely, to make time evolution strictly unitary or to push the violation of unitarity to a very high order in Δt such that the violation remains acceptable.

Making the numerical integrator strictly unitary always involves some more complicated calculation. Consider for example the following integration ansatz, the *Crank–Nicholson* scheme:

$$|\psi(t + \Delta t)\rangle = \frac{1 - i\hat{H}\Delta t/2}{1 + i\hat{H}\Delta t/2}|\psi(t)\rangle. \qquad (4.136)$$

If we expand the denominator using $(1 + \epsilon)^{-1} \approx 1 - \epsilon + O(\epsilon^2)$, we find that, up to $O(\Delta t^2)$, this agrees with the Euler scheme, which already made errors of $O(\Delta t^2)$. So the Crank–Nicholson approach is of the same order of approximation, but is unitary: identifying $\hat{U}(\Delta t) = (1 - i\hat{H}\Delta t/2)/(1 + i\hat{H}\Delta t/2)$, we find

$$\hat{U}^\dagger(\Delta t)\hat{U}(\Delta t) = \frac{1 + i\hat{H}\Delta t/2}{1 - i\hat{H}\Delta t/2}\frac{1 - i\hat{H}\Delta t/2}{1 + i\hat{H}\Delta t/2} = 1. \qquad (4.137)$$

This fortunate property, however, has a steep price tag: how do we calculate the intermediate result $|c\rangle = (1 + i\hat{H}\Delta t/2)^{-1}|\psi(t)\rangle$? As the state space is large, a full inversion is out of question. But we can rewrite this problem as a large sparse linear equation system,

$$(1 + i\hat{H}\Delta t/2)|c\rangle = |\psi(t)\rangle, \qquad (4.138)$$

with $|c_0\rangle = (1 - i\hat{H}\Delta t/2)|\psi(t)\rangle$ as an initial guess correct to $O(\Delta t^2)$. As we have seen, there are well-developed techniques for solving this large sparse equation system, namely, the GMRES method, since $1 + i\hat{H}\Delta t/2$ is non-Hermitian. We have to calculate $\hat{H}|v\rangle$ say five times or so to find the solution if we use the initial guess just introduced, which is quite a steep increase in computation time.

The other alternative is provided by higher-order integration schemes. Here the best-known workhorse of the field is the *fourth-order Runge–Kutta method (RK4)*. It is part of a huge class of similar methods (the Euler method would be a first-order Runge–Kutta method), which are routinely treated in numerical mathematics classes, so let us just state the method and its properties.

Starting from state $|\psi(t)\rangle$, calculate successively the following four vectors:

$$|k_1\rangle = -i\Delta t \hat{H} |\psi(t)\rangle,$$
$$|k_2\rangle = -i\Delta t \hat{H} (|\psi(t)\rangle + \tfrac{1}{2}|k_1\rangle),$$
$$|k_3\rangle = -i\Delta t \hat{H} (|\psi(t)\rangle + \tfrac{1}{2}|k_2\rangle),$$
$$|k_4\rangle = -i\Delta t \hat{H} (|\psi(t)\rangle + |k_3\rangle).$$

Then the new state vector at time $t + \Delta t$ is given as

$$|\psi(t + \Delta t)\rangle = |\psi(t)\rangle + \tfrac{1}{6}|k_1\rangle + \tfrac{1}{3}|k_2\rangle + \tfrac{1}{3}|k_3\rangle + \tfrac{1}{6}|k_4\rangle. \qquad (4.139)$$

The local error of this method is $O(\Delta t^5)$; the global error after $T/\Delta t$ time steps is fourth-order, $O(\Delta t^4)$, hence the name of the method. While it is easy to evaluate, one can show in a somewhat tedious calculation that the Runge–Kutta method is not norm-preserving (the evolution is not unitary), albeit with a much smaller error than in the Euler scheme (fifth-order).

For the Schrödinger equation, numerical evaluations of the formal solution are much more common, so let us turn to these approaches. There are two main approaches, one of which is based on *Trotter decompositions* and is very easy to implement, but works only for short-ranged Hamiltonians. The other is valid for any $\hat{H}$, but is a bit more complicated to implement—on the other hand, the errors are even smaller than the very small errors in the Trotter time evolution! As the approach uses Krylov-space concepts, let us call it *Krylov time evolution*.

4.8.2 Trotter time evolution

Let us assume that $\hat{H}$ connects nearest neighbors only. This is the case for many Hamiltonians in the field of strong-correlation physics: in the Heisenberg model, spin–spin interactions are only between nearest neighbors; in the Hubbard or Bose–Hubbard models, interactions are local, but the hopping term connects nearest neighbors. Then we can decompose $\hat{H}$ into bond Hamiltonians:

$$\hat{H} = \sum_i \hat{h}_i, \qquad (4.140)$$

where $\hat{h}_i$ contains the terms connecting sites i and $i + 1$. Local terms are often split equally between bond Hamiltonians. In the case of the Heisenberg model in an external field B, we would therefore have (omitting the special cases of bonds involving the first and last sites)

$$\hat{h}_i = \tfrac{1}{2} J_{\text{ex}} (\hat{S}_i^+ \hat{S}_{i+1}^- + \hat{S}_i^- \hat{S}_{i+1}^+) + J_{\text{ex}} \hat{S}_i^z \hat{S}_{i+1}^z - \tfrac{1}{2} H (\hat{S}_i^z + \hat{S}_{i+1}^z). \qquad (4.141)$$

While calculating $e^{-i\hat{H}T}|\psi(0)\rangle$ (a complicated L-site problem) is hard, calculating $e^{-i\hat{h}_iT}|\psi(0)\rangle$ (a 2-site problem) is not. It is therefore tempting to decompose

$$|\psi(T)\rangle = e^{-i\hat{H}T}|\psi(0)\rangle = e^{-i(\sum_i \hat{h}_i)T}|\psi(0)\rangle = \left(\prod_i e^{-i\hat{h}_i T}\right)|\psi(0)\rangle. \tag{4.142}$$

But this goes wrong in the last step, since in general $[\hat{h}_i, \hat{h}_{i+1}] \neq 0$; the exponential will then not simply factorize, in view of the Glauber formula

$$e^{\hat{A}+\hat{B}} = e^{\hat{A}}e^{\hat{B}}e^{\frac{1}{2}[\hat{A},\hat{B}]}, \tag{4.143}$$

which is valid if $[\hat{A},[\hat{A},\hat{B}]] = [\hat{B},[\hat{A},\hat{B}]] = 0$. (If the condition does not hold, even more terms invalidating the simple factorization will appear, but this is irrelevant to our argument.) But if we look more closely, the commutator appearing is

$$[-i\hat{h}_iT, -i\hat{h}_{i+1}T] = -[\hat{h}_i, \hat{h}_{i+1}]T^2. \tag{4.144}$$

If we let $T \to 0$, then it disappears as T^2. We therefore discretize time as $T = N\Delta t$ with $\Delta t \to 0$, $N \to \infty$ and (in the most naive approach) do a *first-order Trotter decomposition* for the evolution operator of a time step Δt, $e^{-i\hat{H}\Delta t}$, as

$$e^{-i\hat{H}\Delta t} = e^{-i\hat{h}_1\Delta t}e^{-i\hat{h}_2\Delta t}e^{-i\hat{h}_3\Delta t}\cdots e^{-i\hat{h}_{L-3}\Delta t}e^{-i\hat{h}_{L-2}\Delta t}e^{-i\hat{h}_{L-1}\Delta t} + O(\Delta t^2), \tag{4.145}$$

which contains an error $O(\Delta t^2)$ due to the noncommutativity of the bond Hamiltonians. Such a time step then has to be repeated N times.

As all time evolutions on odd and even bonds commute among each other, so we can factorize at will, it makes sense for the notation to group

$$\hat{H}_{odd} = \sum_k \hat{h}_{2k-1}, \qquad \hat{H}_{even} = \sum_k \hat{h}_{2k}, \tag{4.146}$$

so we can write the first-order Trotter decomposition as

$$e^{-i\hat{H}_{odd}\Delta t}e^{-i\hat{H}_{even}\Delta t}. \tag{4.147}$$

The error accumulates as in the Euler method: per time step, it is $O(\Delta t^2)$, after $T/\Delta t$ time steps it is $O(T\Delta t)$, and hence this Trotter decomposition is a first-order method.

One can construct more complicated decompositions of the evolution operator for a time step Δt with more than just two terms as in the last equation. The goal is to find decompositions such that more factorization errors appear but cancel each other, at least in the lowest orders of Δt. To make $O(\Delta t^2)$ disappear is still quite simple, but then formulas get increasingly complex.

The first improvement is to provided by the *second-order Trotter decomposition*

$$e^{-i\hat{H}\Delta t} = e^{-i\hat{H}_{odd}\Delta t/2}e^{-i\hat{H}_{even}\Delta t}e^{-i\hat{H}_{odd}\Delta t/2} + O(\Delta t^3), \tag{4.148}$$

where the error per time step is reduced by another order of Δt. At first sight, calculation time goes up by (only) 50% for gaining another order Δt, which is already quite
attractive (Δt often being quite small, say 0.001 or so). But very often we are not even
interested in the wavefunction at each time step, because the step is quite small, but
only every, say, 10 time steps or so. Then we can group two of the $\Delta t/2$-steps of two
"neighboring" time steps into one Δt-step for 9 out of 10 time steps, and we are more
or less back to the cost of the first-order decomposition, but have gained one order of
accuracy.

If this is not good enough, then higher-order decompositions are available. The
usual choice is a fourth-order Trotter decomposition that originates in quantum Monte
Carlo, due to Suzuki [68, 69]:

$$e^{-i\hat{H}\Delta t} = \hat{U}(\Delta t_1)\hat{U}(\Delta t_2)\hat{U}(\Delta t_3)\hat{U}(\Delta t_2)\hat{U}(\Delta t_1), \tag{4.149}$$

where

$$\hat{U}(\Delta t_i) = e^{-i\hat{H}_{\mathrm{odd}}\Delta t_i/2}e^{-i\hat{H}_{\mathrm{even}}\Delta t_i}e^{-i\hat{H}_{\mathrm{odd}}\Delta t_i/2} \tag{4.150}$$

and

$$\Delta t_1 = \Delta t_2 = \frac{1}{4 - 4^{1/3}}\Delta t, \qquad \Delta t_3 = \Delta t - 2\Delta t_1 - 2\Delta t_2. \tag{4.151}$$

Again, this can be contracted to save steps: instead of 15 exponentials, we just have 11
(one less if we do not evaluate at the time step). Even smaller errors can be achieved
at similar cost using less symmetric formulas [46].

Let us work out one example for a bond evolution operator by hand. Consider the
$S = \frac{1}{2}$ XY-model in an external field B, i.e., the local Hamiltonian of (4.141) without
the $\hat{S}_i^z \hat{S}_{i+1}^z$ term. In the $\{|\uparrow\uparrow\rangle, |\uparrow\downarrow\rangle, |\downarrow\uparrow\rangle, |\downarrow\downarrow\rangle\}$ basis, this local Hamiltonian can be
expressed as a 4×4 matrix, which we will call H:

$$H = \begin{bmatrix} -\frac{1}{2}B & 0 & 0 & 0 \\ 0 & 0 & +\frac{1}{2}J & 0 \\ 0 & +\frac{1}{2}J & 0 & 0 \\ 0 & 0 & 0 & +\frac{1}{2}B \end{bmatrix}. \tag{4.152}$$

If we diagonalize $HU = U\Lambda$ with

$$U = \begin{bmatrix} 1 & 0 & 0 & 0 \\ 0 & +\frac{1}{\sqrt{2}} & -\frac{1}{\sqrt{2}} & 0 \\ 0 & +\frac{1}{\sqrt{2}} & +\frac{1}{\sqrt{2}} & 0 \\ 0 & 0 & 0 & 1 \end{bmatrix}, \quad \Lambda = \begin{bmatrix} -\frac{1}{2}B & 0 & 0 & 0 \\ 0 & +\frac{1}{2}J & 0 & 0 \\ 0 & 0 & -\frac{1}{2}J & 0 \\ 0 & 0 & 0 & +\frac{1}{2}B \end{bmatrix}, \tag{4.153}$$

we have

$$e^{-iH\Delta t} = Ue^{-i\Lambda\Delta t}U^\dagger$$

$$= \begin{bmatrix} e^{+i(\frac{1}{2}B\Delta t)} & 0 & 0 & 0 \\ 0 & \cos(\frac{1}{2}J\Delta t) & -i\sin(\frac{1}{2}J\Delta t) & 0 \\ 0 & -i\sin(\frac{1}{2}J\Delta t) & \cos(\frac{1}{2}J\Delta t) & 0 \\ 0 & 0 & 0 & e^{-i(\frac{1}{2}B\Delta t)} \end{bmatrix}. \tag{4.154}$$

4.8.3 Krylov time evolution

It is not always possible to decompose the Hamiltonian using the Trotter approach, for example if we allow for longer-ranged hopping or longer-ranged interactions: even though the Coulomb interaction in a solid is strongly screened, in many cases it is unphysical to argue that electrons interact only on the same lattice site.

Alternatively, we can attempt to evaluate the exponential $e^{-i\hat{H}\Delta t}$ directly. Constructing this operator directly is essentially impossible, since $\hat{H}$ is large and sparse. If we consider a Taylor expansion

$$e^{-i\hat{H}\Delta t} = \sum_{k=0}^{\infty}(-i)^k\Delta t^k\frac{\hat{H}^k}{k!}, \tag{4.155}$$

it becomes clear that we have to carry out a sequence of matrix–matrix multiplications, which is very costly even in the case of sparse matrices; moreover, it is easy to see that upon increasing the order k, sparsity increasingly becomes lost. Thus, this approach is not viable.

Consider some matrix representation of $\hat{H}$, which we will call H. For a large sparse matrix H, the only time-efficient operation is a matrix–vector multiplication $H|v\rangle$, so any efficient method will be based on successive matrix–vector multiplications $H|v\rangle$, $H^2|v\rangle = H(H|v\rangle)$, and so on. The Taylor expansion for the evolution *of a given state* now reads

$$|\psi(t+\Delta t)\rangle = e^{-iH\Delta t}|\psi(t)\rangle = \sum_{k=0}^{\infty}(-i)^k\Delta t^k\frac{H^k}{k!}|\psi(t)\rangle, \tag{4.156}$$

and can be evaluated at reasonable cost. If we consider terms up to $k = m-1$, the error per time step will be $O(\Delta t^m)$, and globally $O(\Delta t^{m-1})$. But it turns out that using the concepts of Krylov space, we can do much (exponentially) better! The following is adapted from the guide to the EXPOKIT matrix exponentiator toolkit, available at http://www.maths.uq.edu.au/expokit/guide.html.

If we truncate the Taylor expansion at order $m - 1$, i.e., consider terms up to and including $[(-i\Delta t)^{m-1}H^{m-1}/(m-1)!]|\psi(t)\rangle$, then the resulting

$$|\tilde{\psi}_T(t+\Delta t)\rangle = \sum_{k=0}^{m-1}(-i\Delta t)^k\frac{H^k}{k!}|\psi(t)\rangle \tag{4.157}$$

will be an element of the Krylov subspace

$$\mathbb{K}_m(H, |\psi(t)\rangle) = \text{span}(|\psi(t)\rangle, H|\psi(t)\rangle, H^2|\psi(t)\rangle, \ldots, H^{m-1}|\psi(t)\rangle), \qquad (4.158)$$

i.e., a polynomial of order $m-1$ in H and some, not necessarily optimal, approximation to $|\psi(t + \Delta t)\rangle$. We may now ask what is the polynomial of the same order (i.e., calculable by the same number of matrix–vector multiplications) $|\tilde{\psi}\rangle \in \mathbb{K}_m(H, |\psi(t)\rangle)$ that yields the optimal approximation by minimizing

$$\||\tilde{\psi}\rangle - e^{-iH\Delta t}|\psi(t)\rangle\|_2, \qquad (4.159)$$

i.e., in the least-square sense. To answer this question efficiently, we generate the preferential orthonormal basis obtaining orthonormal vectors $|q_1\rangle, |q_2\rangle, \ldots$ and matrices

$$Q_m = [|q_1\rangle, \ldots, |q_m\rangle]. \qquad (4.160)$$

As before, we can construct tridiagonal matrices $\tilde{T}_m \in \mathbb{C}^{(m+1)\times m}$ and $T_m \in \mathbb{C}^{m\times m}$ with

$$(\tilde{T}_m)_{ij} = \langle q_i|H|q_j\rangle, \qquad (T_m)_{ij} = \langle q_i|H|q_j\rangle \qquad (4.161)$$

for the allowed indices. Hermitian $T_m = Q_m^\dagger H Q_m$ is just $\tilde{T}_m$ with the last line eliminated.

We now write the least-square optimal $|\tilde{\psi}\rangle = Q_m|y\rangle$ with $|y\rangle \in \mathbb{C}^m$. Hence, we are looking for the $|y\rangle$ minimizing

$$\|Q_m|y\rangle - e^{-iH\Delta t}|\psi(t)\rangle\|_2, \qquad (4.162)$$

which is solved by

$$|y\rangle = Q_m^+ e^{-iH\Delta t}|\psi(t)\rangle, \qquad (4.163)$$

where Q_m^+ is the Moore–Penrose pseudoinverse of Q_m. This can be shown in the framework of the singular value decomposition (SVD). Here, let us just define the pseudoinverse: SVD allows us to write any matrix $A = USV^\dagger$, where, in column notation, $U = [|u_1\rangle|u_2\rangle \ldots]$ and $V = [|v_1\rangle|v_2\rangle \ldots]$, with $\langle u_i|u_j\rangle = \langle v_i|v_j\rangle = \delta_{ij}$, and S is diagonal with non-negative real entries w_i. Then, with the sum only over strictly positive w_i,

$$A = \sum_i w_i|u_i\rangle\langle v_i|, \qquad A^+ = \sum_i w_i^{-1}|v_i\rangle\langle u_i|, \qquad A^\dagger = \sum_i w_i|v_i\rangle\langle u_i|. \qquad (4.164)$$

Because Q_m consists of orthonormal columns, it is U in its SVD and both S and $V^\dagger$ are identity matrices. Hence all (nonvanishing) $w_i = 1$, and therefore $Q_m^+ = Q_m^\dagger$, or

$$|y\rangle = Q_m^+ e^{-iH\Delta t}|\psi(t)\rangle = Q_m^\dagger e^{-iH\Delta t}|\psi(t)\rangle, \qquad (4.165)$$

which is convenient since $Q_m^\dagger$ is trivially obtained.

As $|\psi(t)\rangle = \||\psi(t)\rangle\|_2|q_1\rangle = \||\psi(t)\rangle\|_2 Q_m|e_1\rangle$ and $Q_m Q_m^\dagger = I$, we can write

$$|\tilde{\psi}\rangle = Q_m Q_m^\dagger e^{-iH\Delta t}|\psi(t)\rangle = \||\psi(t)\rangle\|_2 Q_m (Q_m^\dagger e^{-iH\Delta t} Q_m)|e_1\rangle \qquad (4.166)$$

for the optimal solution. Essentially, the matrix exponential has been projected onto the Krylov subspace basis. As this optimal solution still contains the unknown matrix exponential itself, we approximate it, replacing the projection of the exponential of H by the exponential of the projection of H, i.e.,

$$Q_m^\dagger \exp(-iH\Delta t)Q_m \to \exp(-iQ_m^\dagger HQ_m\Delta t) = \exp(-iT_m\Delta t). \qquad (4.167)$$

Collecting all results we have

$$|\tilde{\psi}\rangle = \||\psi(t)\rangle\|_2 Q_m \exp(-iT_m\Delta t)|e_1\rangle. \qquad (4.168)$$

It can be shown [37] that with increasing m the solution eventually becomes accurate *exponentially* fast (this is closely related to the exponential convergence of the Lanczos procedure for large m).

To calculate some exponential of H, we therefore calculate the exponential of T_m, which is a small dense matrix, for which there are many viable procedures, for example simple eigendecomposition,

$$T_m = \sum_i \lambda_i |\lambda_i\rangle\langle\lambda_i| \to e^{-iT_m\Delta t} = \sum_i e^{-i\lambda_i\Delta t}|\lambda_i\rangle\langle\lambda_i|. \qquad (4.169)$$

4.8.4 Spectral functions: the linear-response regime

In recent years, there has been so much emphasis on far-from-equilibrium dynamics that it has tended to be forgotten that a lot of dynamics happens close to equilibrium in the linear-response regime. Consider a thermodynamic-limit (or, numerically, sufficiently large) system in its ground state $|0\rangle$. Then any local operation (manipulation or measurement) is almost negligible, and the perturbed system will evolve "close to equilibrium," such that responses of the system will be linear in the perturbation.

A typical scenario would be to calculate

$$\langle 0|\hat{O}(t)\hat{P}(0)|0\rangle = \langle 0|e^{+i\hat{H}t}\hat{O}e^{-i\hat{H}t}\hat{P}|0\rangle, \qquad (4.170)$$

where we perturb the system by the operator $\hat{P}$ at time $t = 0$ and probe the response of the system by some operator $\hat{O}$ at some later time t. The first time evolution is just $e^{+iE_0 t}$, and, to complete the calculation, we need the time evolution of $\hat{P}|0\rangle$. Such a calculation can be carried out, but this is currently not the standard approach.

In condensed matter systems, we often have "translational invariance" for translations by lattice vectors in space and by some number in the time domain. It therefore makes sense to go to Fourier space, i.e., the momentum–frequency domain, obtaining potentially measurable structure functions. Let us just assume homogeneity in time. For some operator $\hat{A}$, one may define a (time-dependent) Green's function at $T = 0$ in the Heisenberg picture by

$$iG_A(t' - t) = \langle 0|\hat{A}^\dagger(t')\hat{A}(t)|0\rangle \qquad (4.171)$$

for a time-independent Hamiltonian $\hat{H}$ with $t' \geq t$ for causality. There are many slightly different, but essentially equivalent, definitions of Green's functions and spectral functions; the exposition and notation here follows that in [61]. Going to the frequency range by a Fourier transform

$$f(\omega) = \int_0^\infty f(t) e^{i\omega t} e^{-\eta t}\, dt, \tag{4.172}$$

the Green's function reads

$$G_A(\omega + i\eta) = \langle 0|\hat{A}^\dagger \frac{1}{E_0 + \omega + i\eta - \hat{H}} \hat{A}|0\rangle. \tag{4.173}$$

The integration interval is taken to be consistent with causality; η is some positive number to be taken to zero at the end in the purely mathematical sense. In practice, η ensures convergence of the Fourier transform; moreover, it introduces a finite lifetime $\Delta t \propto 1/\eta$ to excitations, such that in finite-size systems they can be prevented from inducing spurious finite-size effects.

We may also use the spectral or Lehmann representation of correlations in the eigenbasis of $\hat{H}$:

$$C_A(\omega) = \sum_n |\langle n|\hat{A}|0\rangle|^2 \delta(\omega + E_0 - E_n). \tag{4.174}$$

This frequency-dependent correlation function is related to $G_A(\omega + i\eta)$ by

$$C_A(\omega) = - \lim_{\eta \to 0^+} \frac{1}{\pi} \mathrm{Im}\, G_A(\omega + i\eta) \tag{4.175}$$

via

$$\delta(x) = \lim_{\eta \to 0^+} \frac{1}{\pi} \frac{\eta}{x^2 + \eta^2} = - \lim_{\eta \to 0^+} \frac{1}{\pi} \mathrm{Im}\, \frac{1}{x + i\eta}. \tag{4.176}$$

We shall also use $C_A(\omega + i\eta)$, where the limit $\eta \to 0^+$ will be assumed to have been omitted. η provides a Lorentzian broadening of $C_A(\omega)$,

$$C_A(\omega + i\eta) = \frac{1}{\pi} \int d\omega'\, C_A(\omega') \frac{\eta}{(\omega - \omega')^2 + \eta^2}, \tag{4.177}$$

which serves either to broaden the numerically obtained discrete spectrum of finite systems into some "thermodynamic-limit" behavior or to broaden analytical results for C_A for comparison with numerical spectra where $\eta > 0$.

We now highlight three ways of actually calculating the above expressions (cf. the exposition in [61], which will be followed here for the first two ways).

Continued fraction (or Lanczos vector) approach

As already seen in (4.54), the resolvent $1/(E_0 + \omega + i\eta - \hat{H})$ can be evaluated using *continued fractions* as first used by Gagliano and Balseiro [27] for a tridiagonal representation of the Hamiltonian, which we generate by a Krylov sequence. We only have to take into account that we use as starting vector of the Krylov sequence $\hat{A}|0\rangle$, which is not normalized: $\hat{A}|0\rangle = \||\hat{A}|0\rangle\| |q_1\rangle$. Then we have $(E = E_0 + \omega)$

$$G_A(\omega + i\eta) = \cfrac{\langle 0|\hat{A}^\dagger \hat{A}|0\rangle}{E + i\eta - \alpha_1 - \cfrac{\beta_1^2}{E + i\eta - \alpha_2 - \cfrac{\beta_2^2}{E + i\eta - \alpha_3 - \ldots}}}, \tag{4.178}$$

where the α_i and β_i are the diagonal and subdiagonal entries of the tridiagonal matrix representation of $\hat{H}$.

Sometimes one finds in the literature the completely equivalent prescription, where the vectors generated iteratively are not normalized:

$$|f_{n+1}\rangle = \hat{H}|f_n\rangle - a_n|f_n\rangle - b_n^2|f_{n-1}\rangle, \tag{4.179}$$

with

$$|f_0\rangle = \hat{A}|0\rangle, \tag{4.180}$$

$$a_n = \frac{\langle f_n|\hat{H}|f_n\rangle}{\langle f_n|f_n\rangle}, \tag{4.181}$$

$$b_n^2 = \frac{\langle f_{n-1}|\hat{H}|f_n\rangle}{\langle f_{n-1}|f_{n-1}\rangle} = \frac{\langle f_n|f_n\rangle}{\langle f_{n-1}|f_{n-1}\rangle}. \tag{4.182}$$

The global orthogonality of the states $|f_n\rangle$ (at least in a formal mathematical sense) and the tridiagonality of the new representation (i.e., $\langle f_i|\hat{H}|f_j\rangle = 0$ for $|i - j| > 1$) follow by induction. Again one obtains a continued fraction expansion for the Green's function G_A:

$$G_A(\omega + i\eta) = \cfrac{\langle 0|\hat{A}^\dagger \hat{A}|0\rangle}{E + i\eta - a_0 - \cfrac{b_1^2}{E + i\eta - a_1 - \cfrac{b_2^2}{E + i\eta - \ldots}}}, \tag{4.183}$$

where $E = E_0 + \omega$.

Instead of evaluating the continued fractions, one may also exploit the fact that in the Lanczos procedure the Hamiltonian is iteratively transformed into tridiagonal form in a new approximate orthonormal basis. Diagonalization of the tridiagonal Hamiltonian matrix gives us an approximate energy eigenbasis of $\hat{H}$, $\{|n\rangle\}$, with eigenenergies

E_n, so that, on double insertion of the decomposition of unity $1 = \sum_n |n\rangle\langle n|$, the Green's function can be written within this approximation as

$$G_A(\omega + i\eta) = \sum_n \frac{|\langle n|\hat{A}|0\rangle|^2}{E_0 + \omega + i\eta - E_n}, \tag{4.184}$$

where the sum runs over all approximate eigenstates. The dynamical correlation function is then given by

$$C_A(\omega + i\eta) = \frac{\eta}{\pi} \sum_n \frac{|\langle n|\hat{A}|0\rangle|^2}{(E_0 + \omega - E_n)^2 + \eta^2}, \tag{4.185}$$

where the matrix elements in the numerators are simply the $|q_1\rangle$ expansion coefficients of the approximate eigenstates $|n\rangle$, multiplied by the norm $\|\hat{A}|0\rangle\|_2$.

In practice, a severe limitation arises: the iterative generation of the continued fraction coefficients is a Lanczos diagonalization of $\hat{H}$ with starting vector $\hat{A}|0\rangle$. Typically, the convergence of the lowest eigenvalue of the transformed tridiagonal Hamiltonian to the ground-state eigenvalue of $\hat{H}$ will have occurred after $n \sim O(10^2)$ iteration steps for standard model Hamiltonians. Lanczos convergence, however, is accompanied by numerical loss of global orthogonality, which computationally is ensured only locally, invalidating the inversion procedure. The precision of this approach therefore depends on whether the continued fraction has sufficiently converged at termination. With $\hat{A}|0\rangle$ as starting vector, convergence will be rapid if $\hat{A}|0\rangle$ is a long-lived excitation (close to an eigenstate), such as would be the case if the excitation were part of an excitation band; this will typically not be the case if it is part of an excitation continuum.

Correction vector approach

A way to obtain more precise spectral functions has been proposed by Soos and Ramasesha [66]. After preselection of a fixed frequency ω, one may introduce a *correction vector*

$$|c(\omega + i\eta)\rangle = \frac{1}{E_0 + \omega + i\eta - \hat{H}}\hat{A}|0\rangle, \tag{4.186}$$

which, if known, allows for trivial calculation of the Green's function and hence the spectral function at this particular frequency:

$$G_A(\omega + i\eta) = \langle 0|\hat{A}^\dagger|c(\omega + i\eta)\rangle. \tag{4.187}$$

The correction vector itself is obtained by solving the large sparse linear equation system given by

$$(E_0 + \omega + i\eta - \hat{H})|c(\omega + i\eta)\rangle = \hat{A}|0\rangle. \tag{4.188}$$

To actually solve this non-Hermitian equation system, the most naive and brutal procedure is to split the correction vector into real and imaginary parts, to solve the Hermitian equation for the imaginary part and exploit the relationship to the real part:

$$[(E_0 + \omega - \hat{H})^2 + \eta^2] \operatorname{Im} |c(\omega + i\eta)\rangle = -\eta \hat{A}|0\rangle, \qquad (4.189)$$

$$\operatorname{Re} |c(\omega + i\eta)\rangle = \frac{\hat{H} - E_0 - \omega}{\eta} \operatorname{Im} |c(\omega + i\eta)\rangle. \qquad (4.190)$$

The standard method to solve such a large sparse Hermitian linear equation system is the previously discussed conjugate gradient method. But this slows convergence down drastically, because it amounts to "squaring" $(E_0 + \omega + i\eta - \hat{H})$. A somewhat more demanding, but computationally much more efficient, approach is to stay within the non-Hermitian equation system and use the GMRES method.

Chebyshev polynomial approach

Another method is the expansion of spectral functions using the orthonormal set of Chebyshev polynomials. In my view, this is the most powerful approach, combining accuracy with speed. Full details can be found in the review by Weiße et al. [77], whose notation will be adopted here. The key results will just be summarized here, following [39, 83], which combine the Chebyshev polynomial approach with matrix product states (see below).

Chebyshev polynomials of the first kind $T_n(x)$, $n = 0, 1, \ldots$ (not to be confused with the tridiagonal matrices frequently encountered above), can be defined in various different ways, recursively as

$$T_0(x) = 1, \qquad (4.191)$$
$$T_1(x) = x, \qquad (4.192)$$
$$T_{n+1}(x) = 2xT_n(x) - T_{n-1}(x). \qquad (4.193)$$

We will specifically focus on the interval $x \in [-1, 1]$, where an explicit representation is given by

$$T_n(x) = \cos(n \arccos x). \qquad (4.194)$$

On the interval $[-1, 1]$, $|T_n(x)| \le 1$, whereas outside this interval all Chebyshev polynomials except $T_0(x)$ grow without bound. The usefulness of the Chebyshev polynomials comes from the fact that on the interval $[-1, 1]$ they form an orthonormal function system with respect to a nontrivial weight function:

$$\int_{-1}^{+1} \frac{\mathrm{d}x}{\pi \sqrt{1 - x^2}} T_n(x) T_{n'}(x) = \delta_{nn'}. \qquad (4.195)$$

This implies that on this interval any function $f(x)$ can be expanded in terms of the $T_n(x)$, with one convention given by

$$f(x) = \frac{1}{\pi \sqrt{1 - x^2}} \left[\mu_0 + 2 \sum_{n=1}^{\infty} \mu_n T_n(x) \right], \qquad (4.196)$$

where the Chebyshev moments μ_n are given by

$$\mu_n = \int_{-1}^{+1} dx\, f(x) T_n(x). \tag{4.197}$$

This is in perfect analogy with the familiar Fourier expansion of a function. As with Fourier expansions, what is important is that one does not calculate infinitely many Fourier coefficients but only a finite, hopefully small, number that allows a good approximation of the function. As the convergence of the approximation to the exact function with the number of coefficients is in the L_2 norm, but not pointwise, the approximation may strongly deviate and oscillate locally, in particular in places where the original function has steps, kinks, etc. (Gibbs oscillations). Chebyshev polynomials have the great advantage that convergence is uniform (in the sense that a minimum quality of the approximation for any finite order is guaranteed for the entire interval $[-1, 1]$; of course, the approximation may be better in some parts of the interval than in others); but a weaker form of oscillations close to steps, kinks, etc. persists.

The conventional way of dealing with this runs as follows. If we keep N moments, a smoother approximation is given by

$$f_N(x) = \frac{1}{\pi\sqrt{1 - x^2}} \left[g_0 \mu_0 + 2 \sum_{n=1}^{\infty} g_n \mu_n T_n(x) \right], \tag{4.198}$$

with damping coefficients g_n that depend on N. For details, see [77]; there are various schemes for obtaining the damping factors: more frequently used ones seem to be the Jackson damping

$$g_n^J = \frac{(N - n + 1)\cos\dfrac{\pi n}{N + 1} + \sin\dfrac{\pi n}{N + 1}\cot\dfrac{\pi}{N + 1}}{N + 1} \tag{4.199}$$

and the Lorentz damping

$$g_{n,\lambda}^L = \frac{\sinh[\lambda(1 - n/N)]}{\sinh\lambda}, \tag{4.200}$$

where $\lambda = 4$ seems to be a good choice. All schemes share the property that they give a broadened version of the original function that resolves details on a scale $O(1/N)$.

More recently, it was found [28, 83, 84] that in certain cases a very attractive and superior alternative consists in the use of linear prediction (see Section 4.15.4); for a detailed analysis, see [83]. It is known that the Chebyshev moments μ_n have a convergence behavior dependent on $f(x)$: if $f(x)$ is smooth, they go to zero exponentially; if $f(x)$ is a step function, they go to zero algebraically, while their envelope remains constant for a delta function. In practice, exponential convergence holds already for functions that can only be differentiated a few times. Linear prediction is a technique that predicts further terms x_n in a series $x_0, x_1, \ldots$, where only the first N terms x_n are known, *if* the series can be written (or be very well approximated) as a superposition of exponential terms in the index n,

$$x_n = \sum_i \alpha_i e^{i\omega_i n} e^{-\eta_i n}. \tag{4.201}$$

Indeed, for suitable functions $f(x)$, the knowledge of, say, 100 or 200 moments can be used to calculate a few thousand with very high accuracy; the numerical cost is almost nil and the representation of the function almost exact.

If we now consider as $f(x)$ the spectral representation

$$C_A(\omega) = \langle 0 | \hat{A}^\dagger \delta(E_0 + \omega - \hat{H}) \hat{A} | 0 \rangle, \tag{4.202}$$

we obtain Chebyshev moments

$$\mu_n = \langle 0 | \hat{A}^\dagger \left[\int_{-1}^{+1} dx\, \delta(E_0 + x - \hat{H}) T_n(x) \right] \hat{A} | 0 \rangle$$

$$= \langle 0 | \hat{A}^\dagger T_n(\hat{H} - E_0) \hat{A} | 0 \rangle \tag{4.203}$$

to be inserted in (4.198). The calculation of the Chebyshev moments is achieved through Chebyshev vectors $|t_n\rangle$ as

$$\mu_n = \langle 0 | \hat{A}^\dagger | t_n \rangle, \tag{4.204}$$

with

$$|t_n\rangle = T_n(\hat{H} - E_0) \hat{A} | 0 \rangle. \tag{4.205}$$

Obviously, to make this computationally feasible (n may be up to the order of 10^3), $|t_n\rangle$ are calculated using the Chebyshev recursion relation

$$|t_0\rangle = \hat{A}|0\rangle, \quad |t_1\rangle = (\hat{H} - E_0)|t_0\rangle, \quad |t_{n+1}\rangle = 2(\hat{H} - E_0)|t_n\rangle - |t_{n-1}\rangle. \tag{4.206}$$

So, as so often in this field, the crucial point is carrying out a large sparse matrix–vector multiplication.

A few remarks are in order. The most important is that, so far, we have not yet ensured that the spectral function has support between $[-1, 1]$; we just have $\omega > 0$. If we look at a typical many-body Hamiltonian like the fermionic Hubbard model in one dimension,

$$\hat{H}_H = -J \sum_{i=1}^{L-1} (\hat{c}_i^\dagger \hat{c}_{i+1} + \hat{c}_{i+1}^\dagger \hat{c}_i) + U \sum_{i=1}^{L} \hat{n}_{i\uparrow} \hat{n}_{i\downarrow}, \tag{4.207}$$

the counterpart of (4.1), then both the minimal (ground-state) and maximal energies will grow extensively with system size L, so the support of the spectral function could be of a size of order L. Therefore, in principle, one would have to shift and rescale

$\hat{H}$ to achieve the desired support on $[-1, +1]$ with a scaling factor $1/W \sim O(1/L)$, where W is the bandwidth of the eigenspectrum of the Hamiltonian. Given that the resolution on that interval scales as $1/N$, the number of Chebyshev vectors, the resolution for the true spectral function would scale as $O(L/N)$, i.e., badly, as L increases. But the delta function of $C_A(\omega)$ is sandwiched between two operators acting on the ground state; usually, these operators are local creation or annihilation operators such that the excited states reached by the operators are quite low in energy, without any extensive change in energy. This means that effectively the spectral function is non-negligible only on an operator-dependent interval $[0, W_O]$, which can be shifted and rescaled to $[-1, +1]$; of course, this needs some numerical testing of W_O and "safety distances" by choosing W_O generously (but not too much so); for details, see [77] or the MPS-related papers [28, 39, 83, 84], which approach this issue from various perspectives. An interesting twist [83] is obtained by observing that many spectral functions have most of their "interesting" information in some narrow interval. While convergence of Chebyshev approximations is uniform, the convergence is about a factor 6 faster close to the ends of the interval $[-1, +1]$ than in its center. It would therefore be convenient to map the spectral function in a way that its key information sits close to the ends. A frequently encountered case would be the particle or hole part of a fermionic spectral function, where the relevant information then sits close to $\omega = 0$, which then is mapped to one end of the interval. On the other hand, decomposing the fermionic spectral function in this way introduces a step function into the particle and hole parts, making linear prediction inapplicable. However, one can modify these spectral functions in a small self-consistency loop such that the step disappears, and combine the advantage of higher Chebyshev resolution and linear prediction!

Last but not least, all our calculations are on finite systems, while we are usually interested in what the spectral functions would look like in the thermodynamic limit. The way the thermodynamic limit is extracted is somewhat different in the broadening and linear prediction approaches. In the broadening approach, the reasoning is as follows. What one obtains from N Chebyshev moments is a spectral function that resolves frequency details at a resolution $O(W_O/N)$. How does one choose N properly? To remove Gibbs oscillations, damping factors that depend on N were introduced, removing oscillations on a frequency scale $1/N$. If N becomes large, the artificial broadening of the spectral function will therefore be increasingly suppressed. It will at some point no longer suffice to smoothen the peaks in the spectral function that are due to the finite size L of the system. The finite system peaks are typically spread out by distances $O(1/L)$. Hence, N should be somewhat below L, or, more accurately, chosen such that the artificial broadening does not drop below what is required to smoothen out the finite system peaks. This is in analogy to the issue of the proper choice of η in the Lanczos and correction vector methods. In the linear prediction approach, the support of the spectral function is mapped much more loosely to $[-1, +1]$, such that the resolution of the Chebyshev approximation sees the finite system peaks only for much higher moments; the available lower moments are, however, amply sufficient to extract thermodynamic-limit behavior very precisely by linear prediction.

4.9 Quantum simulatability and matrix product states

The usual computational physics reasoning to explain why quantum simulations are difficult considers the size of the Hilbert space $\mathcal{H}$ on which the states are defined: in the case of a system with a local state-space dimension $d = 2$ (e.g., $S = \frac{1}{2}$) and size L, the Hilbert space $\mathcal{H}$ is of dimension d^L, which leads to states and operators with representations that are exponentially large in L and hence untreatable on a classical computer.

In the previous sections, we have indeed taken the point of view that we consider the full Hilbert space, enumerating the exponentially many coefficients of a quantum state. Roughly speaking, we can conceive of two ways out.

We could sample state-space coefficients stochastically, just as is done for the numerical evaluation of high-dimensional integrals. This is the core idea of the (quantum) Monte Carlo techniques, but they face the infamous negative-sign problem for important classes of problems such as frustrated magnets or fermionic many-body systems; moreover, additional analytic continuation problems arise for real-time evolutions.

The other way out is based on the (successful) identification of a subset of Hilbert space that allows for highly accurate approximation of the desired physical states. In the case of ground-state calculations, this is nothing but the fundamental concept of variational methods. What are the requirements for states in such a subset to achieve quantum simulatability (an awful word that means "feasibility of a quantum simulation on a classical computer")?

- *Efficient representation:* for states of physical interest, it must be possible to compress the state information contained in the exponentially many expansion coefficients into a much smaller number of coefficients determining an approximate state with minimal loss of information (high fidelity). "Small" in this context means that we are willing to accept polynomial growth of coefficients with system size L, because this is what a classical computer can do efficiently.
- *Efficient construction:* for states of physical interest, it is not sufficient that an efficient representation exists in principle; it must also be possible to express states in the new approximate representation in practice. This might be either by direct construction (Given a state $|\psi\rangle$, what is its approximate representation?) or by some physical algorithm (Given some Hamiltonian $\hat{H}$, what is its ground state in the efficient representation? Given $\hat{H}$ and some $|\psi(t = 0)\rangle$, what is the efficient representation of $|\psi(t)\rangle$?). Efficient construction means that the answers can be found in finite computation time, polynomial in system size (and/or time t).
- *Efficient evaluation:* for states of physical interest, we are interested in the values of observables, correlators, etc. Given an efficient representation, can the information of interest be extracted in polynomial time?

With hindsight, it is possible to define *without* motivation a special class of quantum states, the so-called *matrix product states (MPS)*, show under which circumstances they provide an efficient representation, present efficient construction methods where available, and discuss efficient evaluation strategies. This can be done in a very elegant mathematical framework, which will be stated later. At this point, to gain familiarity, we want to show that MPS emerge quite naturally if we identify entanglement as the key property of the quantum versus the classical world. Using the so-called AKLT state as a simple but powerful example, we will show how MPS can be evaluated very efficiently, highlighting their structural properties. We will then move on to discuss the circumstances under which MPS will provide an efficient representation of general quantum states. Algorithms for their construction—including time-evolution methods—will be deferred to later sections after a formal introduction to MPS. As mentioned before, we will lean heavily on [62], but leave out technicalities to a large extent and just reproduce the main ideas and results.

4.9.1 Starting point: product states as mean-field solutions

Consider a quantum system of L spin-$\frac{1}{2}$'s that do not interact but are exposed to a random magnetic field $\mathbf{H}_i$:

$$\hat{H} = -\sum_{i=1}^{L} \mathbf{H}_i \cdot \hat{\mathbf{S}}_i. \tag{4.208}$$

Assuming that $\mathbf{H}_i$ points in the direction $\mathbf{n}_i = (\sin\theta_i \cos\phi_i, \sin\theta_i \sin\phi_i, \cos\theta_i)$, the ground state for each spin is given by

$$|\psi_i\rangle = c^{\uparrow_i}|\uparrow_i\rangle + c^{\downarrow_i}|\downarrow_i\rangle, \tag{4.209}$$

where $c^{\uparrow_i} = e^{-i\phi_i/2}\cos(\theta_i/2)$ and $c^{\downarrow_i} = e^{+i\phi_i/2}\sin(\theta_i/2)$ are two complex numbers (owing to normalization and the irrelevance of a global phase, they contain as much information as the two real angles θ_i and ϕ_i) and the ground state of the overall system reads

$$|\psi\rangle = \prod_i |\psi_i\rangle, \tag{4.210}$$

a product state. The state is completely characterized by dL instead of d^L coefficients, $d = 2$, and therefore tractable for exponentially larger L compared with the case of a generic state. Indeed, such states arise as the result of *mean-field* calculations, where—to stay with the example of spins—the spins of the system do interact, but in such a way that it is justifiable to replace the effect of the surrounding spins on spin i by an effective additional field that acts on it, $\mathbf{H}_i \rightarrow \hat{\mathbf{H}}_i$, where the additional field has to be determined self-consistently from the orientations of the other spins.

A product state is the simplest quantum state treatable numerically. In a state representation

$$|\psi\rangle = \sum_{\sigma_1,\dots,\sigma_L} c^{\sigma_1\dots\sigma_L}|\sigma_1\dots\sigma_L\rangle, \qquad |\sigma_i\rangle \in \{|\uparrow_i\rangle, |\downarrow_i\rangle\}, \tag{4.211}$$

the scalar coefficients factorize as scalar products

$$c^{\sigma_1\dots\sigma_L} = c^{\sigma_1}c^{\sigma_2}\cdots c^{\sigma_L} \equiv c^{\sigma_1\dots\sigma_\ell}c^{\sigma_{\ell+1}\dots\sigma_L}, \tag{4.212}$$

where we have anticipated bipartitioning the universe of L spins into parts A (spins $1, \ldots, \ell$) and B (spins $\ell + 1, \ldots, L$). Note that product states do not form a linear subspace of the full Hilbert space (the sum of two product states is in general not a product state); this set of states is rather constrained by the multiplicative condition on the relationship between the scalar coefficients.

Alas, product states do not capture what is in many ways the key feature of quantum physics, namely, its nonlocality embodied in the notion of *quantum entanglement*.

4.9.2   Quantum entanglement and matrix product states

Ultimately, the intrinsic difference in complexity between the classical and quantum worlds rests on the different notion of a state of the world, which in quantum physics is a ray in Hilbert space, following the *superposition principle*, for example,

$$|\psi\rangle = c^\uparrow |\uparrow\rangle + c^\downarrow |\downarrow\rangle \tag{4.213}$$

as the most general state for a spin-$\frac{1}{2}$. The Hilbert space for two spin-$\frac{1}{2}$'s is the *tensor product* of the Hilbert spaces of the individual spin-$\frac{1}{2}$'s,

$$\mathcal{H}_{12} = \mathcal{H}_1 \otimes \mathcal{H}_2, \tag{4.214}$$

meaning that the most general pure state for two spin-$\frac{1}{2}$'s reads

$$|\psi\rangle_{12} = c^{\uparrow_1 \uparrow_2} |\uparrow_1 \uparrow_2\rangle + c^{\uparrow_1 \downarrow_2} |\uparrow_1 \downarrow_2\rangle + c^{\downarrow_1 \uparrow_2} |\downarrow_1 \uparrow_2\rangle + c^{\downarrow_1 \downarrow_2} |\downarrow_1 \downarrow_2\rangle. \tag{4.215}$$

There exist states $|\psi\rangle_{12}$ that are *not separable* (i.e., not product states) or *entangled*, where

$$|\psi\rangle_{12} \neq |\psi\rangle_1 |\psi\rangle_2 \tag{4.216}$$

for any states $|\psi\rangle_1$ and $|\psi\rangle_2$ living on sites 1 and 2, i.e., $c^{\sigma_1 \sigma_2} \neq c^{\sigma_1} c^{\sigma_2}$ for any c^{σ_1}, c^{σ_2}. Consider for example

$$|\psi_{\text{Bell}}\rangle_{12} = \frac{1}{\sqrt{2}} [| \uparrow_1 \uparrow_2\rangle + | \downarrow_1 \downarrow_2\rangle]. \tag{4.217}$$

There is no decomposition in the form of (4.216), which has important consequences for the foundations of physics, such as the nonlocality of quantum information. The above *Bell state* is quite iconic: any local measurement will give a completely random spin direction (hence there is no local information), but the outcome of the measurement of one spin completely determines the outcome of a measurement on the other spin. Information on the physical system here is therefore entirely nonlocal, an idea alien to classical physics. Complete nonlocality of the information corresponds to maximal entanglement between the two spins (we are going to quantify and sharpen this statement later on), entanglement being a measure of how much of the information in the system is distributed nonlocally across the system instead of residing with local parts of the system.

We may now ask whether we can amend the nonworking factorization $c^{\sigma_1\sigma_2} \neq c^{\sigma_1}c^{\sigma_2}$ for this state. As the state is a sum of two terms, one may wonder whether coefficients can be encoded as a sum over products, which mathematically leads naturally to the consideration of matrix–matrix multiplications. Let us therefore attempt a matrix-valued factorization

$$c^{\sigma_1\sigma_2} = M^{\sigma_1}M^{\sigma_2}, \tag{4.218}$$

where, in order to generate two terms and a resulting scalar, the (rectangular) matrices M^{σ_1} should be of dimension 1×2 and the matrices M^{σ_2} of dimension 2×1. It is easy to see (one way to do this is to think of "paths" that take us through spin configurations as we move through a chain of spins) that

$$M^{\uparrow_1} = \begin{bmatrix} 2^{-1/4} & 0 \end{bmatrix}, \qquad M^{\downarrow_1} = \begin{bmatrix} 0 & 2^{-1/4} \end{bmatrix},$$

$$M^{\uparrow_2} = \begin{bmatrix} 2^{-1/4} \\ 0 \end{bmatrix}, \qquad M^{\downarrow_2} = \begin{bmatrix} 0 \\ 2^{-1/4} \end{bmatrix}. \tag{4.219}$$

Check by multiplication!

Hence we obtain a state representation that does capture entanglement in the form

$$|\psi_{\text{Bell}}\rangle = \sum_{\sigma_1,\sigma_2} M^{\sigma_1}M^{\sigma_2}|\sigma_1\sigma_2\rangle, \tag{4.220}$$

which still conserves some of the local character of the product state in its mathematical form, but can encode nonlocal information because of its matrix-valued constituents. Such a state, here in its embryonic form, is called a *matrix product state (MPS)*; this type of ansatz goes back so many decades in many different, sometimes almost unrecognizable, disguises that no brief history can be given.

Can we generalize this idea to larger systems (in the end, we want to approach the thermodynamic limit)? To get some intuition, let us consider a chain of $L = 4$ spins and take the following two states:

$$|\psi_{\text{GHZ}}\rangle = \frac{1}{\sqrt{2}}[|\uparrow\uparrow\uparrow\uparrow\rangle + |\downarrow\downarrow\downarrow\downarrow\rangle],$$

$$|\psi_{\text{maxEnt}}\rangle = \frac{1}{2}\sum_{i=1}^{4}|i\rangle_A|i\rangle_B,$$

where A and B are subsystems formed from spins 1 and 2 (A) and 3 and 4 (B) respectively. The $|i\rangle$ are the four states of two spins, $|1\rangle = |\uparrow\uparrow\rangle$, $|2\rangle = |\uparrow\downarrow\rangle$, $|3\rangle = |\downarrow\uparrow\rangle$, and $|4\rangle = |\downarrow\downarrow\rangle$. We now work out their MPS representation

$$|\psi_*\rangle = \sum_{\sigma_1,\ldots,\sigma_4} M^{\sigma_1}M^{\sigma_2}M^{\sigma_3}M^{\sigma_4}|\sigma_1\ldots\sigma_4\rangle. \tag{4.221}$$

$|\psi_{\text{GHZ}}\rangle$ is a simple generalization of $|\psi_{\text{Bell}}\rangle$ above: the two "paths" through the chain (all up-spins and all down-spins, respectively) just have to be extended by diagonal 2×2 matrices in the "bulk" of the chain:

$$M^{\uparrow_1} = \begin{bmatrix} 2^{-1/4} & 0 \end{bmatrix}, \qquad\qquad M^{\downarrow_1} = \begin{bmatrix} 0 & 2^{-1/4} \end{bmatrix},$$

$$M^{\uparrow_2} = \begin{bmatrix} 1 & 0 \\ 0 & 0 \end{bmatrix}, \qquad\qquad M^{\downarrow_2} = \begin{bmatrix} 0 & 0 \\ 0 & 1 \end{bmatrix},$$

$$M^{\uparrow_3} = \begin{bmatrix} 1 & 0 \\ 0 & 0 \end{bmatrix}, \qquad\qquad M^{\downarrow_3} = \begin{bmatrix} 0 & 0 \\ 0 & 1 \end{bmatrix},$$

$$M^{\uparrow_4} = \begin{bmatrix} 2^{-1/4} \\ 0 \end{bmatrix}, \qquad\qquad M^{\downarrow_4} = \begin{bmatrix} 0 \\ 2^{-1/4} \end{bmatrix}.$$

Let us now consider $|\psi_{\mathrm{maxEnt}}\rangle$. There are now four terms in the sum, so we may expect matrices of dimension up to 4. As we do not yet have a procedure to generate the matrix decomposition, we have to guess. The solution is provided by matrices of dimensions 1×2, 2×4, 4×2, and 2×1 as

$$M^{\uparrow_1} = \begin{bmatrix} 2^{-1/2} & 0 \end{bmatrix}, \qquad\qquad M^{\downarrow_1} = \begin{bmatrix} 0 & 2^{-1/2} \end{bmatrix}$$

$$M^{\uparrow_2} = \begin{bmatrix} 1 & 0 & 0 & 0 \\ 0 & 0 & 1 & 0 \end{bmatrix}, \qquad\qquad M^{\downarrow_2} = \begin{bmatrix} 0 & 1 & 0 & 0 \\ 0 & 0 & 0 & 1 \end{bmatrix},$$

$$M^{\uparrow_3} = \begin{bmatrix} 1 & 0 \\ 0 & 1 \\ 0 & 0 \\ 0 & 0 \end{bmatrix}, \qquad\qquad M^{\downarrow_3} = \begin{bmatrix} 0 & 0 \\ 0 & 0 \\ 1 & 0 \\ 0 & 1 \end{bmatrix},$$

$$M^{\uparrow_4} = \begin{bmatrix} 2^{-1/2} \\ 0 \end{bmatrix}, \qquad\qquad M^{\downarrow_4} = \begin{bmatrix} 0 \\ 2^{-1/2} \end{bmatrix}.$$

One way to this solution is to consider the bipartitioning $A|B$. Then we can write the coefficients as $M^{\sigma_1\sigma_2}M^{\sigma_3\sigma_4}$, with matrix dimensions 1×4 and 4×1, where $M^{\uparrow_1\uparrow_2} = [2^{-1/2}\,0\,0\,0]$, $M^{\uparrow_1\downarrow_2} = [0\,2^{-1/2}\,0\,0]$, and so on; the $M^{\sigma_3\sigma_4}$ are simply the transposes. At the same time, it seems plausible from the structure of $|\psi_{\mathrm{maxEnt}}\rangle$ that the matrices for sites 1 and L are structurally as for $|\psi_{\mathrm{Bell}}\rangle$, and hence of dimensions 1×2 and 2×1, except for the different weights of states, and we can check whether we can construct matrices of dimensions 2×4 and 4×2 on sites 2 and 3, such that the products are just $M^{\sigma_1\sigma_2}$ and $M^{\sigma_3\sigma_4}$. This is indeed possible, and gives the above result. It works!

While the above states are also of fundamental interest in quantum physics, from the perspective of a genuine quantum-many body problem as in condensed matter or cold atoms, these states are a bit too trivial. To get more experience with MPS, let us now turn to a problem with real quantum many-body content, where I follow and summarize results discussed at greater length in [62].

4.9.3 The Affleck–Kennedy–Lieb–Tasaki (AKLT) model in a nutshell

The AKLT ground state and its parent Hamiltonian

A spin model that has attracted much attention for more than 25 years now is the so-called *Affleck–Kennedy–Lieb–Tasaki (AKLT)* model [1, 2]. It is a bilinear–biquadratic $S = 1$ spin chain,

$$\hat{H}_{\mathrm{AKLT}} = \sum_i \left[\hat{\mathbf{S}}_i \cdot \hat{\mathbf{S}}_{i+1} + \tfrac{1}{3}(\hat{\mathbf{S}}_i \cdot \hat{\mathbf{S}}_{i+1})^2 + \tfrac{2}{3} \right], \qquad (4.222)$$

where we will consider periodic boundary conditions. It looks like the Heisenberg antiferromagnet with a small correction by a biquadratic term, "small" here being justified in hindsight by the properties of the Hamiltonian: it turns out that it reproduces qualitatively many, if not most, of the properties of the ground state of the $S = 1$ isotropic Heisenberg antiferromagnet.

The crucial difference is that whereas the ground state of the isotropic Heisenberg antiferromagnet cannot be calculated exactly, the ground state of the AKLT model, which we will simply call the AKLT state, can be constructed explicitly as shown in Fig. 4.5, as was already realized in [1]: Each individual spin-1 is replaced by a pair of spin-$\frac{1}{2}$'s that are completely symmetrized; i.e., of the four possible states, we consider only the three triplet states naturally identified as $S = 1$ states:

$$|+\rangle = |\uparrow\uparrow\rangle,$$

$$|0\rangle = \frac{|\uparrow\downarrow\rangle + |\downarrow\uparrow\rangle}{\sqrt{2}}, \qquad (4.223)$$

$$|-\rangle = |\downarrow\downarrow\rangle.$$

On neighboring sites, adjacent pairs of spin-$\frac{1}{2}$'s are linked in a singlet state

$$\frac{|\uparrow\downarrow\rangle - |\downarrow\uparrow\rangle}{\sqrt{2}}. \qquad (4.224)$$

MPS representation of the AKLT ground state

It can now be shown (for a hopefully pedagogical explanation, see [62]) that the ground state of the AKLT Hamiltonian can be brought into a mathematically simple form. If

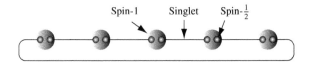

Fig. 4.5 The Affleck–Kennedy–Lieb–Tasaki (AKLT) ground state is built by expressing the local spin-1 as two totally symmetrized spin-$\frac{1}{2}$'s that are linked across sites by singlet states. The picture shows the case of PBC; for OBC, the "long" bond is cut, and two single spins-$\frac{1}{2}$'s appear at the ends.

we call the $S = 1$ basis $\{|\sigma_1\sigma_2\ldots\sigma_L\rangle\} \equiv \{|\{\sigma\}\rangle\}$, with the curly brackets serving as a shorthand for the spin states on the entire chain, then the ground state reads

$$|\psi\rangle = \sum_{\{\sigma\}} \mathrm{Tr}(A^{\sigma_1}A^{\sigma_2}\ldots A^{\sigma_L})|\{\sigma\}\rangle. \tag{4.225}$$

where we introduce three 2×2 matrices A^σ, i.e., one for each value σ can take:

$$A^+ = \begin{bmatrix} 0 & +\sqrt{\frac{2}{3}} \\ 0 & 0 \end{bmatrix}, \quad A^0 = \begin{bmatrix} -\sqrt{\frac{1}{3}} & 0 \\ 0 & +\sqrt{\frac{1}{3}} \end{bmatrix}, \quad A^- = \begin{bmatrix} 0 & 0 \\ -\sqrt{\frac{2}{3}} & 0 \end{bmatrix}. \tag{4.226}$$

Calculations in the AKLT ground state

In this subsection, we will carry out typical calculations for the AKLT state, following [62]. Let us start by checking the most fundamental property of a quantum state, its normalizability. We consider

$$\langle\psi|\psi\rangle = \sum_{\{\sigma\}} \mathrm{Tr}(A^{\sigma_1}\ldots A^{\sigma_L})^* \, \mathrm{Tr}(A^{\sigma_1}\ldots A^{\sigma_L}). \tag{4.227}$$

This expression can be simplified by using $\mathrm{Tr}(ABC\cdots)\,\mathrm{Tr}(FGH\cdots) = \mathrm{Tr}(A \otimes F)(B \otimes G)(C \otimes H)\cdots$, where the tensor product of two matrices is a matrix with multiplied row and column dimensions and entries $(A \otimes F)_{(ik),(jl)} = A_{i,j}F_{k,l}$. Hence

$$\langle\psi|\psi\rangle = \mathrm{Tr}\left[\left(\sum_{\sigma_1} A^{\sigma_1*} \otimes A^{\sigma_1}\right)\left(\sum_{\sigma_2} A^{\sigma_2*} \otimes A^{\sigma_2}\right)\cdots\left(\sum_{\sigma_L} A^{\sigma_L*} \otimes A^{\sigma_L}\right)\right], \tag{4.228}$$

where we have also broken up the multiple sums over the local states. As the A-matrices do not depend on the site i, each set of parentheses is identical, and we can write

$$\langle\psi|\psi\rangle = \mathrm{Tr}\,E^L = \sum_{i=1}^{4} \lambda_i^L, \tag{4.229}$$

where the λ_i are the four eigenvalues of the 4×4 matrix $E = \sum_\sigma A^{\sigma*} \otimes A^\sigma$, which reads

$$E = \sum_\sigma A^{\sigma*} \otimes A^\sigma = \begin{bmatrix} \frac{1}{3} & 0 & 0 & \frac{2}{3} \\ 0 & -\frac{1}{3} & 0 & 0 \\ 0 & 0 & -\frac{1}{3} & 0 \\ \frac{2}{3} & 0 & 0 & \frac{1}{3} \end{bmatrix}. \tag{4.230}$$

The four eigenvalues are $(1, -\frac{1}{3}, -\frac{1}{3}, -\frac{1}{3})$. Then

$$\langle\psi|\psi\rangle = 1^L + \left(-\frac{1}{3}\right)^L + \left(-\frac{1}{3}\right)^L + \left(-\frac{1}{3}\right)^L$$

$$= 1 + 3\left(-\frac{1}{3}\right)^L \to 1 \quad \text{for} \quad L \to \infty. \tag{4.231}$$

$|\psi\rangle$ is therefore normalizable for any length L and even properly normalized to 1 in the thermodynamic limit $L \to \infty$! This was, of course, achieved by trickery: if we rescaled the A-matrices by some factor γ, E would be rescaled by $|\gamma|^2$, and the largest eigenvalue 1 would be replaced by $|\gamma|^2$. Depending on γ, the norm would go to either zero or infinity and the state would not be normalizable in the thermodynamic limit. At least for a translationally invariant state, where E is not site-dependent, normalizability in the thermodynamic limit is given if no eigenvalue has modulus larger than 1, and if at least one has modulus 1. Early in the history of DMRG meets MPS, it was found by Östlund and Rommer [49] that this is ensured if the A-matrices have a special normalization property. To see this, let us write a (left-) eigenvector equation of E for eigenvalue 1, keeping multi-indices explicitly:

$$\sum_{ik} v_{ik} E_{(ik),(jl)} = \sum_{ik} v_{ik} \sum_{\sigma} A^{\sigma *}_{i,j} A^{\sigma}_{k,l} = v_{jl}. \tag{4.232}$$

Interestingly, this equation can be satisfied for the choice $v_{ik} = \delta_{ik}$, provided a certain normalization condition holds for the A-matrices:

$$\sum_{\sigma} \sum_{i} A^{\sigma *}_{i,j} A^{\sigma}_{i,l} = \sum_{\sigma} (A^{\sigma \dagger} A^{\sigma})_{j,l} = \delta_{jl}. \tag{4.233}$$

The normalization condition is therefore $\sum_{\sigma} A^{\sigma \dagger} A^{\sigma} = I$, which will later be referred to as the left-normalization condition. It ensures the existence of an eigenvalue 1, and thereby normalization in the thermodynamic limit, because one can also show [62] that then there is none with modulus larger than 1.

To calculate correlators such as $\langle \hat{S}^z_i \hat{S}^z_j \rangle$, we consider $\langle \psi | \hat{O} \hat{P} | \psi \rangle$, where $\hat{O}$ and $\hat{P}$ are local operators acting on sites i and j, respectively; $i < j$ in all that follows. Upon twice inserting the decomposition of unity, we obtain, following similar steps as for the calculation of the norm,

$$\langle \psi | \hat{O} \hat{P} | \psi \rangle = \sum_{\{\sigma\},\{\sigma'\}} \langle \psi | \{\sigma\} \rangle \langle \{\sigma\} | \hat{O} \hat{P} | \{\sigma'\} \rangle \langle \{\sigma'\} | \psi \rangle$$

$$= \sum_{\{\sigma\},\sigma'_i,\sigma'_j} \mathrm{Tr}(A^{\sigma_1} \cdots A^{\sigma_i} \cdots A^{\sigma_j} \cdots A^{\sigma_L})^* \langle \sigma_i | \hat{O} | \sigma'_i \rangle$$

$$\times \langle \sigma_j | \hat{P} | \sigma'_j \rangle \, \mathrm{Tr}(A^{\sigma_1} \cdots A^{\sigma'_i} \cdots A^{\sigma'_j} \cdots A^{\sigma_L})$$

$$= \mathrm{Tr}\Bigg[\left(\sum_{\sigma_1} A^{\sigma_1 *} \otimes A^{\sigma_1} \right) \cdots \left(\sum_{\sigma_i,\sigma'_i} A^{\sigma_i *} \otimes A^{\sigma'_i} \langle \sigma_i | \hat{O} | \sigma'_i \rangle \right) \cdots$$

$$\times \left(\sum_{\sigma_j,\sigma'_j} A^{\sigma_j *} \otimes A^{\sigma'_j} \langle \sigma_j | \hat{P} | \sigma'_j \rangle \right) \cdots \left(\sum_{\sigma_L} A^{\sigma_L *} \otimes A^{\sigma_L} \right) \Bigg].$$

Using E as defined above and introducing as a natural generalization

$$E_O = \sum_{\sigma\sigma'} A^{\sigma*} \otimes A^{\sigma'} \langle\sigma|\hat{O}|\sigma'\rangle, \tag{4.234}$$

we find

$$\langle\psi|\hat{O}\hat{P}|\psi\rangle = \mathrm{Tr}(E^{i-1}E_O E^{j-i-1}E_P E^{L-j}). \tag{4.235}$$

If we use the cyclicity of the trace and express the trace in the eigenbasis of E, $\{|m\rangle\}$, with eigenvalues $\{\lambda_m\}$, we obtain

$$\langle\psi|\hat{O}\hat{P}|\psi\rangle = \mathrm{Tr}(E^{i-1}E_O E^{j-i-1}E_P E^{L-j})$$

$$= \mathrm{Tr}(E_O E^{j-i-1}E_P E^{L-j+i-1})$$

$$= \sum_{m=1}^{4}\langle m|E_O E^{j-i-1}E_P E^{L-j+i-1}|m\rangle$$

$$= \sum_{m=1}^{4}\langle m|E_O E^{j-i-1}E_P \lambda_m^{L-j+i-1}|m\rangle. \tag{4.236}$$

This expression is valid for arbitrary L. As $\lambda_1=1$ and $|\lambda_{m>1}| < 1$, this reduces in the thermodynamic limit $L \to \infty$ further to

$$\langle\psi|\hat{O}\hat{P}|\psi\rangle = \langle 1|E_O E^{j-i-1}E_P|1\rangle$$

$$= \sum_{k=1}^{4}\langle 1|E_O|k\rangle\lambda_k^{j-i-1}\langle k|E_P|1\rangle, \tag{4.237}$$

where we have used the eigenrepresentation $E = \sum_{k=1}^{4}\lambda_k|k\rangle\langle k|$. As E is symmetric, the left and right eigenvectors of E are simple transposes of each other; the right eigenvectors read

$$|1\rangle = \begin{bmatrix} +\sqrt{\frac{1}{2}} \\ 0 \\ 0 \\ +\sqrt{\frac{1}{2}} \end{bmatrix}, \quad |2\rangle = \begin{bmatrix} 0 \\ 1 \\ 0 \\ 0 \end{bmatrix}, \quad |3\rangle = \begin{bmatrix} 0 \\ 0 \\ 1 \\ 0 \end{bmatrix}, \quad |4\rangle = \begin{bmatrix} +\sqrt{\frac{1}{2}} \\ 0 \\ 0 \\ -\sqrt{\frac{1}{2}} \end{bmatrix}. \tag{4.238}$$

Here eigenvector $|1\rangle$ belongs to eigenvalue $\lambda_1 = 1$.

Within this framework, we can work out any correlators. It is also quite easy to see how to calculate one-point or general n-point correlators in a similar way. The two most important correlators for the AKLT model are

- magnetic order $\langle \hat{S}_i^z \hat{S}_j^z \rangle$, $j > i$;
- hidden (or string) order [22]

$$\left\langle \hat{S}_i^z \left[\prod_{k=i+1}^{j-1} \exp(i\pi \hat{S}_k^z) \right] \hat{S}_j^z \right\rangle = \left\langle \hat{S}_i^z \left(\prod_{k=i+1}^{j-1} \hat{P}_k \right) \hat{S}_j^z \right\rangle, \tag{4.239}$$

where we have abbreviated the operators between i and j as $\hat{P}_k = \exp(i\pi \hat{S}_k^z)$.

We leave the explicit calculation as a simple exercise, and just state intermediate results. For the magnetic order, we need

$$E_{S^z} = A^{+*} \otimes A^+ - A^{-*} \otimes A^- = \begin{bmatrix} 0 & 0 & 0 & +\frac{2}{3} \\ 0 & 0 & 0 & 0 \\ 0 & 0 & 0 & 0 \\ -\frac{2}{3} & 0 & 0 & 0 \end{bmatrix}. \tag{4.240}$$

The only nonvanishing matrix elements relevant for (4.237) are

$$\langle 1|E_{S^z}|4\rangle = -\tfrac{2}{3}, \qquad \langle 4|E_{S^z}|1\rangle = +\tfrac{2}{3}, \tag{4.241}$$

and in the thermodynamic limit, we obtain

$$\langle \hat{S}_i^z \hat{S}_j^z \rangle = \langle \psi | \hat{S}_i^z \hat{S}_j^z | \psi \rangle = -\tfrac{2}{3} \cdot \tfrac{2}{3} \cdot \left(-\tfrac{1}{3}\right)^{j-i-1}$$
$$= \tfrac{4}{3} \left(-\tfrac{1}{3}\right)^{j-i} = \tfrac{4}{3}(-1)^{j-i} e^{-(j-i)/\xi}, \tag{4.242}$$

an antiferromagnetic correlation that decays exponentially with correlation length $\xi = 1/\ln 3 \approx 0.91$.

For the string order, we need also E_P for $\hat{P} = e^{i\pi \hat{S}^z}$:

$$E_P = -A^{+*} \otimes A^+ + A^{0*} \otimes A^0 - A^{-*} \otimes A^-$$

$$= \begin{bmatrix} +\frac{1}{3} & 0 & 0 & -\frac{2}{3} \\ 0 & -\frac{1}{3} & 0 & 0 \\ 0 & 0 & -\frac{1}{3} & 0 \\ -\frac{2}{3} & 0 & 0 & +\frac{1}{3} \end{bmatrix}. \tag{4.243}$$

In the thermodynamic limit, the expectation value is given by

$$\langle \psi | \hat{S}_i^z \prod_{i<k<j} \hat{P}_k \hat{S}_j^z | \psi \rangle = \sum_{m,m'=1}^{4} \langle 1|E_{S^z}|m\rangle \langle m|E_P^{j-i-1}|m'\rangle \langle m'|E_{S^z}|1\rangle$$
$$= \langle 1|E_{S^z}|4\rangle \langle 4|E_P^{j-i-1}|4\rangle \langle 4|E_{S^z}|1\rangle$$
$$= -\tfrac{4}{9} \langle 4|E_P^{j-i-1}|4\rangle.$$

Using the eigenrepresentation of E_P, the last expression simplifies to

$$\langle\psi|\hat{S}_i^z \prod_{i<k<j} \hat{P}_k\hat{S}_j^z|\psi\rangle = -\tfrac{4}{9}\cdot 1\cdot 1^{j-i-1}\cdot 1 = -\tfrac{4}{9}$$

for all $j - i \geq 2$, and hence string order is long-ranged.

For the two correlators that we have calculated explicitly, we have obtained two different types of distance dependence: exponential decay in the one case, long-range order in the other. Are these the only possibilities, or can we expect also, for example, critical behavior, a power-law decay of the type $(j-i)^{-\alpha}$? If we reconsider (4.237), the factor λ_k^{j-i-1} generates either a superposition of exponential decays (if only the $|\lambda_k| < 1$ contribute) or long-range order (if $|\lambda_k| = 1$ contributes, assuming for simplicity, and in agreement with what is usually observed in numerical practice, that there is only one eigenvalue of modulus 1). Which of the two cases occurs depends on which matrix elements $\langle 1|E_O|k\rangle$ and $\langle k|E_P|1\rangle$ are nonvanishing. In general, correlators will be a superposition of long-range and exponentially decaying contributions. If we generalize from 2×2 to $D \times D$ matrices, then the E-matrix will be $(D^2 \times D^2)$-dimensional. Upon proper normalization, it will have, without loss of generality, one $\lambda_1 = 1$, and all others $|\lambda_k| < 1$. Then

$$\frac{\langle\psi|\hat{O}^{[i]}\hat{O}^{[j]}|\psi\rangle}{\langle\psi|\psi\rangle} = c_1 + \sum_{k=2}^{D^2} c_k e^{i\phi_k r} e^{-r/\xi_k}, \qquad (4.244)$$

where $r = |j-i-1|$ and $c_k = \langle 1|E_O^{[i]}|k\rangle\langle k|E_O^{[j]}|1\rangle$ for $i < j$. Furthermore, $\lambda_k = |\lambda_k|e^{i\phi_k}$ and $\xi_k = -(\ln|\lambda_k|)^{-1}$.

It is therefore a generic result for matrix product states that they cannot contain critical (power-law) correlations, only long-range order or a superposition of exponential decays, maybe with oscillation factors. The former is not particularly relevant in one-dimensional quantum systems owing to the presence of strong quantum fluctuations: quantum fluctuations become more important as dimension is reduced. For short-ranged Hamiltonians with a continuous symmetry with at least two Lie generators, the Mermin–Wagner theorem forbids long-range order at any finite temperature; in one dimension, long-range order will usually be destroyed also at $T = 0$, the ferromagnetic ground state of the one-dimensional Heisenberg ferromagnet being an exception. More usually, correlators in one-dimensional quantum systems take either the Ornstein–Zernike form

$$\langle O_0 O_x\rangle \sim \frac{e^{-x/\xi}}{\sqrt{x}} \qquad (4.245)$$

or the critical power-law form (maybe with logarithmic corrections),

$$\langle O_0 O_x\rangle \sim x^{-\alpha}. \qquad (4.246)$$

In fact, the AKLT state with its purely exponential correlations is an exception [63]. Any finite-dimensional MPS will only approximate the true correlator by a superposition of exponentials. It turns out that this works very well on short distances, even for power laws. What one observes numerically is that the true correlation function will be represented accurately on increasingly long lengthscales as D is increased. Eventually, the slowest exponential decay will survive, turning the correlation into a pure exponential decay with $\xi = -1/\ln\lambda$, where λ is the largest eigenvalue of E that contributes to the correlator. Comparison of curves for various D is therefore an excellent tool to gauge the convergence of correlation functions and the lengthscale on which it has been achieved.

It is also important to observe that all operations have been reduced to matrix manipulations, which are polynomially expensive in matrix dimension D (if one follows the optimal implementations of these manipulations, for open boundary conditions, no operation is more expensive than $O(D^3)$, while for periodic boundary conditions, the upper limit is $O(D^5)$); without the benefit of translational invariance, we would have had to evaluate a sequence of matrix products scaling as L. Hence, evaluating matrix product states is only polynomially expensive in system size (unless it is necessary to choose D exponentially large in L to represent a state faithfully, an issue we are going to discuss below).

Let us pause for a moment. We have been able to represent a highly nontrivial quantum state very efficiently: instead of exponentially many coefficients, we are down to a handful of them, of course with the help of translational invariance, and the representation is even exact. It would be highly unrealistic to expect that this scenario generalizes to arbitrary states, and in fact it does not. But, once obtained, the state was an MPS like any other, which allowed us to get an idea of the efficiency of evaluating properties of this state.

4.10   Entanglement and MPS representability

Under what circumstances can a quantum state be represented efficiently, i.e., with little loss of information, as an MPS? Several observations in the last section suggest that there is some link to quantum entanglement:

- Unentangled states can be described by MPS of matrix dimension 1×1, i.e., product states. Our four-site states had MPS representations of different matrix sizes. $|\psi_{\mathrm{maxEnt}}\rangle$ has totally nonlocal information between A and B, and no local information within A and B, respectively, and is therefore maximally entangled for a bipartition AB. It also has the larger matrices. The other state has "less" (we do not yet have a measure!) entanglement. Is there a connection between the amount of entanglement in a quantum state and its MPS representation, for example the dimension of its matrices?
- How does the ratio between local and nonlocal information (i.e., entanglement) in a quantum many-body state scale with system size? How does this affect matrix dimensions? To give an example, for the maximally entangled state on two sites, the maximal matrix dimension was two, on four sites it was four. Given the supposed connection between matrix dimensions, and hence numerical complexity, and entanglement, the scaling behavior of entanglement and matrix dimensions will impact on our ability to simulate states in the thermodynamic limit.

4.10.1 Quantifying entanglement

Entanglement is not a direct physically measurable observable, but rather a tool of convenience to quantify the amount of nonclassicality of a quantum state. Motivating its measures by physical demands is therefore a rather complicated problem, which we will sidestep here and just give the most common measures in the field of many-body quantum states. In many ways, right down to the emerging formulas, this situation is as in classical information theory, where Shannon's measure of information was motivated by demands on a concept of "information" that ultimately allowed only for a single mathematical form. This analogy is not coincidental, since entanglement and information are deeply related to each other, as well as to the concept of entropy.

To quantify entanglement, the physical system (the "universe") is divided into subsystems A and B. We write a general quantum state of the universe AB as

$$|\psi\rangle_{AB} = \sum_{ij} \psi_{ij}|i\rangle_A |j\rangle_B, \qquad (4.247)$$

where we take $\{|i\rangle_A\}$ and $\{|j\rangle_B\}$ to be orthonormal bases of A and B, respectively. We can form the associated density operator $\hat{\rho}_{AB} = |\psi\rangle_{AB\ AB}\langle\psi|$. The subsystems A and B are then completely characterized by the *reduced density operators*

$$\hat{\rho}_A = \mathrm{Tr}_B\,\hat{\rho}_{AB}, \qquad \hat{\rho}_B = \mathrm{Tr}_A\,\hat{\rho}_{AB}, \qquad (4.248)$$

where the partial traces are taken over the bases of the other subsystem. Expressed in the orthonormal bases just introduced, the reduced density operator for A reads

$$\hat{\rho}_A = \sum_{ii'} \left(\sum_j \psi_{ij}\psi_{i'j}^* \right) |i\rangle_{A\ A}\langle i'| \qquad (4.249)$$

and the associated reduced density matrix elements $(\rho_A)_{i,i'} = \sum_j \psi_{ij}\psi_{i'j}^*$. Entanglement measures are functions $S(\{w_a\})$, where $\{w_a\}$ is the nonvanishing part of the eigenspectrum of $\hat{\rho}_A$. One might also consider $\hat{\rho}_B$, since the nonvanishing parts of the spectra are identical (as follows from the Schmidt decomposition; see below).

We define the *von Neumann entropy of entanglement* as

$$\mathcal{S}_1(|\psi\rangle_{AB}) \equiv \mathcal{S}_1(\hat{\rho}_A) = -\mathrm{Tr}_A\,\hat{\rho}_A \log_2 \hat{\rho}_A = -\sum_a w_a \log_2 w_a, \qquad (4.250)$$

through the quantum mechanical von Neumann entropy S_1 for the subsystem A. If we read the eigenvalues w_α as probabilities, this formula is also identical to Shannon's definition of information. As is usually the case in quantum information, we have chosen the binary logarithm here, but any other logarithm would do as well.

If we take A to be smaller than B (to avoid the technicality that a small B enforces vanishing eigenvalues in the reduced density operator of A), and assume that the local state space of the $|A|$ lattice sites in A is of dimension d, we can easily show that the von Neumann entanglement is bounded as

$$0 \le \mathcal{S}_1(|\psi\rangle_{AB}) \equiv \mathcal{S}_1(\hat{\rho}_A) \le |A| \log_2 d, \qquad (4.251)$$

where the two limits correspond to a pure state on A ($\hat{\rho}_A = |i\rangle_A \, _A\langle i|$) and to a maximally mixed state on A ($\hat{\rho}_A = d^{-|A|} \sum_i |i\rangle_A \, _A\langle i|$), which would be an infinite-temperature thermal state, therefore containing no local information whatsoever—all information is nonlocal and resides in quantum entanglement. It is important to observe that the maximal entanglement bound is extensive, i.e., proportional to $|A|$.

The von Neumann entropy is a special case of the *Renyi entropy* defined as

$$S_\alpha(\hat{\rho}_A) = \frac{1}{1-\alpha} \log_2 \mathrm{Tr}_A \hat{\rho}_A^\alpha \qquad (0 \le \alpha < \infty) \qquad (4.252)$$

for $\alpha \to 1$. The importance of the Renyi entropy stems from the fact that for exact mathematical statements on the usefulness of MPS in representing quantum states, von Neumann entropy is usually not enough; we will not use it in the following, since for our more hand-waving assessments, von Neumann entropy does the job, so we will drop the subscript 1. The relationship between simulatability and entropy is so complicated because the real quantity of interest is the complete spectrum $\{w_a\}$ of the reduced density operator $\hat{\rho}_A$, which allows for very strong statements on simulatability. Alas, we usually do not know it, so this is of little use. However, the various entropies for different α provide single-number information about the spectrum (knowing all of them would be equivalent to knowing the full spectrum), such that most, but not all, of the information of the spectrum is lost. This little bit is still enough to give reasonable, albeit not strict, assessments of simulatability and, as we will see, a lot is known about it, in particular for von Neumann entropy.

Let us nevertheless pretend that we are trying to calculate entanglement from the knowledge of the full spectrum, because it allows us to introduce a very "explicit" notation for a quantum state $|\psi\rangle_{AB}$ allowing us to read off entanglement directly, the *Schmidt decomposition*, which can be obtained from a *singular value decomposition* (SVD). The former is treated in all books on quantum information, the latter in all books on linear algebra, including all the references cited in earlier parts of this chapter; the summary here follows [62].

SVD guarantees for an arbitrary (rectangular) matrix M of dimension $N_A \times N_B$ the existence of a decomposition

$$M = USV^\dagger, \qquad (4.253)$$

where

- U is of dimension $N_A \times \min(N_A, N_B)$ and has orthonormal columns (the *left singular [column] vectors* $|u_a\rangle$), i.e., $U^\dagger U = I$; if $N_A \le N_B$, then this implies that it is square and unitary, and also $UU^\dagger = I$.
- S is of dimension $\min(N_A, N_B) \times \min(N_A, N_B)$, diagonal with non-negative entries $S_{aa} \equiv s_a$. These are the so-called *singular values*. The number r of nonzero singular values is the *(Schmidt) rank* of M. In the following, we assume descending order: $s_1 \ge s_2 \ge \ldots \ge s_r > 0$.
- $V^\dagger$ is of dimension $\min(N_A, N_B) \times N_B$ and has orthonormal rows (the *right singular [row] vectors* $\langle v_a|$), i.e., $V^\dagger V = I$. If $N_A \ge N_B$, then this implies that it is square and unitary, and also $VV^\dagger = I$.

As used previously, any pure state $|\psi\rangle_{AB}$ on a bipartite system AB can be written as

$$|\psi\rangle_{AB} = \sum_{ij} \Psi_{ij} |i\rangle_A |j\rangle_B, \tag{4.254}$$

where $\{|i\rangle_A\}$ and $\{|j\rangle_B\}$ are orthonormal bases of A and B with dimensions N_A and N_B respectively; we read the coefficients as entries of a matrix Ψ. If we carry out an SVD of the matrix Ψ in (4.254), we obtain

$$|\psi\rangle_{AB} = \sum_{ij} \sum_{a=1}^{\min(N_A,N_B)} U_{ia} S_{aa} V_{ja}^* |i\rangle_A |j\rangle_B$$

$$= \sum_{a=1}^{\min(N_A,N_B)} \left(\sum_i U_{ia} |i\rangle_A \right) s_a \left(\sum_j V_{ja}^* |j\rangle \right)$$

$$\equiv \sum_{a=1}^{\min(N_A,N_B)} s_a |a\rangle_A |a\rangle_B. \tag{4.255}$$

Owing to the orthonormality properties of U and $V^\dagger$, the sets $\{|a\rangle_A\}$ and $\{|a\rangle_B\}$ are orthonormal and can be extended to be orthonormal bases of A and B. If we restrict the sum to run only over the $r \le \min(N_A, N_B)$ positive nonzero singular values, we obtain the *Schmidt decomposition*

$$|\psi\rangle_{AB} = \sum_{a=1}^{r} s_a |a\rangle_A |a\rangle_B. \tag{4.256}$$

It is obvious that $r = 1$ corresponds to (classical) product or separable states and $r > 1$ to entangled (nonseparable) states.

The Schmidt decomposition allows us to read off the reduced density operators for A and B introduced above very conveniently: carrying out the partial traces, we find

$$\hat{\rho}_A = \sum_{a=1}^{r} s_a^2 |a\rangle_A\, _A\langle a|, \qquad \hat{\rho}_B = \sum_{a=1}^{r} s_a^2 |a\rangle_B\, _B\langle a|, \tag{4.257}$$

showing that they share the nonvanishing part of the spectrum, but not the eigenstates. The eigenvalues are the squares of the singular values, $w_a = s_a^2$, the respective eigenvectors are the left and right singular vectors. The von Neumann entropy of entanglement, for example, can therefore be read off directly from the Schmidt decomposition of $|\psi\rangle_{AB}$:

$$S(|\psi\rangle_{AB}) = -\operatorname{Tr} \hat{\rho}_A \log_2 \hat{\rho}_A = -\sum_{a=1}^{r} s_a^2 \log_2 s_a^2. \tag{4.258}$$

As only the nonvanishing part of the spectrum of $\hat{\rho}_A$ contributes, which the Schmidt decomposition reveals to be the same as for $\hat{\rho}_B$, entanglement can interchangeably be calculated from either $\hat{\rho}_A$ or $\hat{\rho}_B$, as already stated previously.

4.10.2   Entanglement entropy scaling for a pure quantum state: volume and area laws

Consider a universe AB, bipartitioned into A and B, with $|A|$ and $|B|$ the number of lattice sites in A and B; we take $|B| \to \infty$ and consider how entanglement depends on $|A|$. d is as always the number of local degrees of freedom per site. Then entanglement is bounded from above as

$$|A| \log_2 d \geq \mathcal{E}(\mathcal{S}(|\psi\rangle_{AB})), \tag{4.259}$$

provided that $|A| \leq |B|$, as we are assuming. At the same time, it can be shown that [25, 50, 65]

$$\mathcal{E}(\mathcal{S}(|\psi\rangle_{AB})) = \mathcal{E}(\mathcal{S}(\hat{\rho}_A)) > |A| \log_2 d - \frac{d^{|A|-|B|}}{2\log_2 2}, \tag{4.260}$$

where $\mathcal{E}(.)$ represents the expectation value taken over all states $|\psi\rangle_{AB}$ in the Hilbert space of AB with respect to the Haar measure. Therefore, for $|B| \to \infty$,

$$\mathcal{E}(\mathcal{S}(|\psi\rangle_{AB})) = |A| \log_2 d. \tag{4.261}$$

Entanglement entropy is therefore both extensive and maximal, except (possibly) for a subset of the Hilbert space of Haar measure 0. As we will see later, extensivity is a killer for classical simulation—hence the question arises whether any interesting quantum states are to be found in the measure 0 subset. Indeed there are! (Otherwise this would be the last sentence of this chapter!)

The origin of interest in quantum states whose entanglement is not extensive lies quite far from the physical systems we are looking at, such as spin chains and ultracold atomic gases. The topic initially arose in cosmology, where the fact that black holes extract matter and associated entropy (and information) from the rest of the universe would at first suggest a violation of the second law of thermodynamics. It was shown by Bekenstein [9] that this issue can be settled and the second law of thermodynamics can be retained if one assumes that a black hole has an entropy proportional to the *surface area* of its event horizon. Much research has focused on the quantum mechanism responsible for this entropy, and indeed it is related to entanglement due to quantum fluctuations connecting the inside and outside of the black hole. As the surface of a three-dimensional black hole is a two-dimensional area, this specific nonextensive dependence of entropy (quantum entanglement) is referred to as the *area law*, a name that has now migrated also to other fields of physics, namely, quantum many-body physics. For a detailed and up-to-date discussion of the actual state of knowledge, see the nice review by Eisert, Cramer, and Plenio [24].

The current state of knowledge as needed for our purposes can be summarized as follows:

- *One-dimensional quantum systems.* For ground states of local Hamiltonians (finite range of interactions) with a finite interaction strength and a gap ΔE between ground state and first excitation, the ground-state entanglement (or subsystem entropy) is bounded by a constant. As the surface of a subsystem in one dimension is just a point independent of L, one can therefore speak also of an area (surface) law in one dimension. This result has been observed numerically and analytically in many examples, and has been proved by Hastings [36].

For ground states of critical systems, conformal field theory gives [13, 38]

$$\mathcal{S}(|\psi\rangle_{\mathrm{AB}}) = \frac{c}{3} \log_2 \ell + O(1), \tag{4.262}$$

where c is the central charge of the field theory and ℓ the subsystem size measured in units of the lattice spacing. These results have also been found numerically [44, 76]. While entanglement no longer obeys a pure area law, it is not extensive either, but just a logarithmic correction to an area law.

Given the insight that ground states of local Hamiltonians are not maximally entangled, they can only occupy a tiny corner of Hilbert space (of measure 0); one might therefore say that the exponentially large Hilbert space is only an illusion as far as ground states are concerned.

- *Two-dimensional quantum systems.* In two dimensions, the situation is less well established in the sense that results are more empirical or quite special. It seems that for local Hamiltonians with a gapped spectrum, area laws hold (entanglement is therefore proportional to ℓ for subsystems of size $\ell \times \ell$); in the case of criticality, corrections may or may not appear. As it will turn out, for MPS, the area law in two dimensions is in itself already quite a killer for simulations, so the further complications due to criticality do not matter too much.

- *Dynamics in one dimension.* If we start time evolution from a state whose entanglement is already extensive and maximal—as is overwhelmingly likely if the state is random—it is also overwhelmingly likely that it will keep this property as time evolves. If, as we will show, such states are not classically simulatable, then no time evolution is feasible anyway. The more interesting and only practically relevant case is if we start time evolution from a state that obeys the area law, such as some simple product state or a ground state of some suitable Hamiltonian.

  Several typical scenarios may occur: we may change the Hamiltonian adiabatically with time, such that the state remains in the ground-state manifold, obeying the area law. In such a case, there is no restriction to simulatability in principle. For other changes, the original state of the system will start to explore Hilbert space in its generality, and therefore move into regions where the area law does not hold. In this case, simulatability will break down. As entanglement has to be built up physically by quantum fluctuations, this breakdown will not happen instantaneously, but only after some time. Two cases are particularly important:

1. A nonextensive change to the system, for example by inserting a particle locally, such that the system is no longer in the ground state, and evolving it under the action of the original Hamiltonian. Loosely, we refer to this as a *local quench*— it could also be engineered by retaining the original state, but changing the Hamiltonian locally, for example by changing a magnetic field at one site or so. In such cases, entanglement will grow in time, and is bounded as

$$\mathcal{S}(t) \leq \mathcal{S}(0) + c \log t. \tag{4.263}$$

2. An extensive change to the system, a so-called *global quench*, for example by starting from the ground state of some Hamiltonian $\hat{H}_0$, but evolving it under the action of some other $\hat{H}$, where it is no longer an eigenstate and where the energy change is extensive, $(\langle\psi|\hat{H}_0|\psi\rangle - \langle\psi|\hat{H}|\psi\rangle) \sim L$. It turns out that—following the Lieb–Robertson theorem—entanglement S can grow up to linearly in time for an out-of-equilibrium evolution of a quantum state:

$$S(t) \leq S(0) + ct, \tag{4.264}$$

where c is some constant related to the propagation speed of excitations in the lattice [14, 15]. This linear bound is actually reached for many global quantum quenches, where a Hamiltonian parameter is abruptly changed such that the global energy changes extensively.

4.10.3   When do matrix product states work in practice?

Let us assume (as is the case) that any quantum state on a lattice can be brought into MPS form exactly. It turns out, however, that the price to be paid is an exponentially divergent size of the matrices with lattice size L, rendering exact MPS representations of states useless except for a few special cases. This can already be seen by generalizing our maximally entangled states $|\psi_{\mathrm{Bell}}\rangle$ and $|\psi_{\mathrm{maxEnt}}\rangle$ to systems of length L, with L even,

$$|\psi_{\mathrm{maxEnt},L}\rangle = \frac{1}{d^{L/4}} \sum_{i=1}^{d^{L/2}} |i\rangle_{\mathrm{A}}|i\rangle_{\mathrm{B}}, \tag{4.265}$$

bipartitioning the system into two halves A and B, where the product of all matrices in A (in B) must yield a matrix of dimension $1 \times d^{L/2}$ (of dimension $(d^{L/2} \times 1)$ to encode the state. Hence, there must be exponentially growing matrix sizes involved (the detailed growth scheme is of no relevance here and will be discussed later).

In order to make numerical progress, let us assume that our computer can handle matrices up to a dimension $D \times D$. It then makes sense to assume we have an *exact* representation of some quantum state $|\psi\rangle$ by an MPS with matrix dimensions that are too large and to ask whether we can we find an optimal *approximate* representation of the state by another MPS $|\tilde{\psi}\rangle$ with smaller matrix dimensions (up to $D \times D$) minimizing $|||\psi\rangle - |\tilde{\psi}\rangle||_2^2$, where the approximation error is small.

Let us first observe that the optimal approximate representations for smaller matrix sizes will monotonically improve with matrix size. Even an approximate MPS is still a linear combination of all states of the Hilbert space, and no product basis state has been discarded; in that sense, MPS differ from the states generated by variational methods, which are based on restricting the underlying state space directly. Larger matrices simply give more variational parameters.

As we will see, any MPS can be brought into the form

$$|\psi\rangle = \sum_{\sigma_1,\ldots,\sigma_L} A^{\sigma_1}\ldots A^{\sigma_\ell}SB^{\sigma_{\ell+1}}\ldots B^{\sigma_L}|\sigma_1,\ldots,\sigma_L\rangle, \tag{4.266}$$

with S a diagonal matrix form with non-negative diagonal entries. At the same time, the states

$$|a\rangle_A = \sum_{\sigma_1,\ldots,\sigma_\ell} (A^{\sigma_1} \ldots A^{\sigma_\ell})_{1,a}, \qquad (4.267)$$

$$|a\rangle_B = \sum_{\sigma_{\ell+1},\ldots,\sigma_L} (B^{\sigma_{\ell+1}} \ldots B^{\sigma_L})_{a,1} \qquad (4.268)$$

form orthonormal sets respectively such that $(s_a = S_{a,a})$

$$|\psi\rangle = \sum_a s_a |a\rangle_A |a\rangle_B \qquad (4.269)$$

is a Schmidt decomposition. Assume that we have to shrink the column and row dimensions of the matrices A and B next to S as well as the size of S to D while minimizing the approximation (or *truncation*) error

$$\epsilon(D) = \||\psi\rangle - |\tilde\psi\rangle\|_2^2, \qquad (4.270)$$

where $|\tilde\psi\rangle$ is the new MPS with the reduced matrix sizes. It is quite easy to see that the optimal choice is to keep those D terms of the Schmidt decomposition that have the largest s_a and to truncate the matrices accordingly. The truncation error is then easily evaluated as

$$\epsilon(D) = \sum_{a=D+1} s_a^2 = 1 - \sum_{a=1}^{D} s_a^2. \qquad (4.271)$$

Here, we have assumed that the original state $|\psi\rangle$ was normalized, $\sum_a s_a^2 = 1$. The truncation error is given by the sum over the squares of the discarded Schmidt coefficients, which is the sum over the discarded eigenvalues of the reduced density operators $\hat\rho_A$ ($\hat\rho_B$). This is nothing but the discarded statistical weight of the reduced density operators. To minimize this makes immediate sense from a statistical physics point of view.

The above argument can be generalized from the approximation incurred by a single truncation to that incurred by $L-1$ truncations, one at each bond, to reveal that the error is at worst [71]

$$\||\psi\rangle - |\tilde\psi\rangle\|_2^2 \le 2 \sum_{i=1}^{L-1} \epsilon_i(D), \qquad (4.272)$$

where $\epsilon_i(D)$ is the truncation error at bond i incurred by truncating down to the leading D Schmidt coefficients. The problem of approximability remains related to the eigenvalue spectra of reduced density operators. For a *generic* state out of Hilbert space, a handwaving argument indicates that all eigenvalues of reduced density operators are $O(1/d^{|A|})$ if $|A|$ is the size of the subsystem, such that maximal entanglement arises (see the previously discussed results!). Then the eigenvalue distribution is very tail-heavy and an efficient MPS approximation impossible. Thus, MPS approximability emerges as a very peculiar property of a few (but important) states in one dimension. This has *important consequences for time evolutions that we are going to discuss later: the entire state space will be explored in general, such that time evolution must break down at some time.*

The problem now is that usually we do not know much about the eigenvalue spectrum; instead, we have quite good knowledge of the entanglement properties of quantum states, as discussed above. If we reconsider the MPS in the above form, and its Schmidt decomposition

$$|\psi\rangle = \sum_{a=1}^{D} s_a |a\rangle_\mathrm{A} |a\rangle_\mathrm{B}, \tag{4.273}$$

then we can see that the maximal (von Neumann) entanglement that can be encoded by this MPS is $\log_2 D$ in the case where all s_a are identical. To encode a state with entanglement $\mathcal{S}$ properly, one therefore needs an MPS of matrix dimension $D \geq 2^{\mathcal{S}}$. As the eigenvalue spectrum of the state will not normally be flat (the s_a all identical), in fact the MPS dimension might even have to be much larger, but from numerical practice, we can conclude that we may ignore this issue for the ballpark estimate done here. The loophole in this statement is (of course) rather the word "properly," because an exact encoding may still be exponentially expensive; what we am aiming at is that almost no entanglement is lost, and to quantify this properly is mathematically difficult (see the pathological states presented by [64], which show that this is even impossible—it just seems to work for "normal" states). If we ignore this and use the various insights we have obtained for $\mathcal{S}$, we can predict the following behavior of MPS:

- *Ground states of gapped systems in one dimension:* D will essentially not grow with system size.
- *Ground states of critical systems in one dimension:* $D \sim L^{c/3}$; even though this number diverges with L, it does so sufficiently slowly (c is usually $\frac{1}{3}$ or $\frac{1}{6}$ or so) that large system sizes ($L \sim O(10^3)$) are accessible.
- *Generic states in one dimension:* $D \sim 2^L$, owing to maximal and extensive entanglement; except for uselessly small L, not simulatable by MPS.
- *Ground states of gapped systems in two dimensions:* $D \sim 2^L$, as for generic states in one dimension less. This might therefore seem useless; however, if we consider long stripe systems $W \times L$, with $W \gg L$, we can set up MPS such that while L stays quite limited, W can be taken essentially to the thermodynamic limit and finite-size scaling can be attempted along the short direction (consider, for example, the recent very precise results for the difficult case of the Heisenberg model on the kagome lattice [23, 85]). This is a rapidly evolving field outside the scope of this chapter.
- *Time-evolving states in one dimension:* We have identified three cases of interest: adiabatic transformations, local quenches, and global quenches. If we stay adiabatically in the ground-state manifold, all the ground-state arguments continue

to hold. On the other hand, both from $D \sim 2^S$ and from a rigorous analysis [48], it follows that for quenches the matrix dimensions will have to go up with time: in the case of a local quench, we are looking at a power-law growth of $D(t) \sim t^\alpha$, and in the case of a global quench, at an exponential growth $D(t) \sim 2^t$. This is the worst case. Of course, we can also keep D fixed in a simulation; we will then look at results that deteriorate over time—in the worst case exponentially. Time-dependent MPS might therefore be set to fail for quenches, and indeed they will fail at some point. Fortunately, in many cases, the timescales reachable are long enough to observe the relevant physics, which is why this method we are going to explain in the following has attracted so much attention over the last 10 years.

4.11   Matrix product states (MPS)

We will now generalize the AKLT matrix product state, and see that not only is the manipulation of MPS computationally very efficient as already seen for the AKLT model, but also that there are efficient algorithms for actually constructing MPS, namely, finding ground states (this is nothing but the famous DMRG algorithm in different wording) and calculating their time evolution. First we have to establish the general notation.

4.11.1   Introduction of matrix product states

Consider a lattice of L sites with d-dimensional local state spaces with orthonormal bases $\{|\sigma_i\rangle\}$ on sites $i = 1, \ldots, L$. The most general pure quantum state on the lattice reads

$$|\psi\rangle = \sum_{\sigma_1, \ldots, \sigma_L} c_{\sigma_1 \ldots \sigma_L} |\sigma_1, \ldots, \sigma_L\rangle \equiv \sum_{\{\sigma\}} c_{\{\sigma\}} |\{\sigma\}\rangle, \qquad (4.274)$$

where we have exponentially many coefficients $c_{\sigma_1 \ldots \sigma_L}$. As before, we use the short-hand $\{\sigma\}$ for $\sigma_1, \ldots, \sigma_L$. Although lattice dimension is not explicit here, let us think about chains (one-dimensional quantum systems), because this is where MPS work best. Let us assume that $|\psi\rangle$ is normalized. It can be shown that through a sequence of SVDs *any* such state can be brought into MPS form; for details, see [62]. The most general form of a matrix product state is given by

$$|\psi\rangle = \sum_{\sigma_1, \ldots, \sigma_L} M^{\sigma_1} M^{\sigma_2} \cdots M^{\sigma_{L-1}} M^{\sigma_L} |\sigma_1, \ldots, \sigma_L\rangle. \qquad (4.275)$$

Here, we have introduced d matrices M^{σ_i} on each lattice site, one for each value of the local degree of freedom. Their row and column dimensions are arbitrary, but row and column dimensions of matrices on neighboring sites must match to allow for matrix multiplication. This is as in the AKLT case. The only difference is that we do not take a trace of the matrix product, but restrict the row dimensions of the first-site matrices and the column dimensions of the last-site matrices to 1, to obtain a scalar. This is adequate for open boundary conditions, and important for numerical efficiency.

All this emerges naturally from the explicit construction, but as this explicit construction of an MPS is exponentially complex in general, we leave it just as a mathematical observation and merely state the outcome. Let us remark right away that the outcome is *far from unique*: consider two neighboring sets of matrices M^{σ_i} and $M^{\sigma_{i+1}}$ of shared column/row dimension D. Then the MPS is invariant for any invertible matrix X of dimension $D \times D$ under

$$M^{\sigma_i} \to M^{\sigma_i} X, \qquad M^{\sigma_{i+1}} \to X^{-1} M^{\sigma_{i+1}}, \tag{4.276}$$

so we have essentially infinite "gauge" freedom on all sites. This can be exploited to simplify manipulations drastically. In fact, there are mainly three types of MPS that are useful in practice, and they even emerge naturally from the various ways of explicit construction: the *left-canonical*, *right-canonical*, and *mixed-canonical* MPS.

The difference between the three is that for the left-canonical MPS, one carries out a sequence of SVDs on $c_{\sigma_1 \dots \sigma_L}$ suitably reshaped into matrices from the left end of the chain to the right end, splitting off site after site. For the right-canonical MPS, one carries out the same sequence, now starting from the right; and for the mixed-canonical MPS, one starts from both ends, until all sites have been split off. The details do not matter (since they will not be used in any real algorithm), but the mathematical form of the three outcomes is essential:

Left-canonical matrix product state

This reads

$$|\psi\rangle = \sum_{\sigma_1,\dots,\sigma_L} A^{\sigma_1} A^{\sigma_2} \cdots A^{\sigma_{L-1}} A^{\sigma_L} |\sigma_1,\dots,\sigma_L\rangle, \tag{4.277}$$

where the maximal dimensions of the matrices are $1 \times d, d \times d^2, \dots, d^{L/2-1} \times d^{L/2}, d^{L/2} \times d^{L/2-1}, \dots, d^2 \times d, d \times 1$, going from the first to the last site (we have assumed L to be even for simplicity here), revealing the practical irrelevance of the exact decomposition in most circumstances. In some clever way, we will cut the matrix dimensions to something small and finite without losing too much information; if this is not possible, MPS are useless, but this we have already assessed in our discussion of simulatability. Of crucial importance and characteristic of left-canonical MPS is that the matrices obey some kind of normalization property,

$$\sum_{\sigma_\ell} A^{\sigma_\ell \dagger} A^{\sigma_\ell} = I. \tag{4.278}$$

This relationship has already been observed for the AKLT state. We will refer to matrices that obey this condition as *left-normalized*.

A very convenient aspect of MPS is that they find a simple graphical representation. In fact, once one has gained some familiarity with them, it will rarely be necessary to carry out calculations explicitly—the index load is simply too heavy. As with Feynman

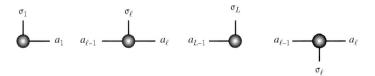

Fig. 4.6 Graphical representation of A-matrices at the ends and in the bulk of chains: the first diagram represents $A^{\sigma_1}_{1,a_1}$, the row vector at the left end, the third diagram represents $A^{\sigma_L}_{a_L,1}$, the column vector at the right end. The second diagram is $A^{\sigma_\ell}_{a_{\ell-1},a_\ell}$, and the fourth is its complex conjugate.

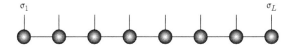

Fig. 4.7 Representation of an open-boundary-condition MPS.

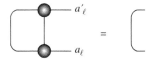

Fig. 4.8 If two *left*-normalized A-matrices are contracted over their *left* index and the physical indices, a $\delta_{a'_\ell,a_\ell}$ line results.

diagrams, everything can be done graphically, and be translated into formulas at the end. The graphical and translation rules are even much simpler.

The graphical rules for the A-matrices, which on the first and last sites are row and column vectors respectively, are summarized in Fig. 4.6: a site ℓ is represented by a solid circle, the physical index σ_ℓ by a vertical line, and the two matrix indices by horizontal lines; we drop the dummy index lines for sites 1 and L. We will also need the complex conjugates of matrices, A^*, which we represent by orienting the physical index line in the opposite direction.

The translation rule is simply that all legs that stick out freely are taken as index labels (like σ_ℓ) and that all connected legs are summed over their index after multiplying the connected objects, i.e., they are simply contracted. An open boundary condition MPS then looks like Fig. 4.7, and the graphical representation of (4.278) looks like Fig. 4.8.

Right-canonical matrix product state

Carrying out the decomposition procedure from the right (site L) instead of site 1, we obtain an MPS of the form

$$|\psi\rangle = \sum_{\sigma_1,\ldots,\sigma_L} B^{\sigma_1} B^{\sigma_2} \cdots B^{\sigma_{L-1}} B^{\sigma_L} |\sigma_1,\ldots,\sigma_L\rangle. \tag{4.279}$$

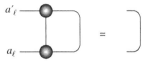

Fig. 4.9 If two *right*-normalized B-matrices are contracted over their *right* index and the physical indices, then a $\delta_{a'_\ell, a_\ell}$ line results.

This looks exactly as before: the B-matrices also have the same matrix dimension bounds as the A matrices. The crucial difference is that they obey a different kind of normalization condition,

$$\sum_{\sigma_\ell} B^{\sigma_\ell} B^{\sigma_\ell \dagger} = I, \tag{4.280}$$

and are referred to as *right-normalized* matrices; for a graphical representation of this property, see Fig. 4.9. We use the notation A and B to distinguish between the two properties; if none holds, we use the more general M for the MPS matrices.

Mathematically, all this is very smooth, but a seemingly less elegant notation does even better. This emerges if one tries to carry out a Schmidt decomposition on the MPS. Let us take the left-canonical MPS. If we view the L-site lattice as our universe, we can decompose it into subsystems (blocks) A and B, where A comprises sites $1, \ldots, \ell$ and B sites $\ell + 1, \ldots, L$, and introduce states living on A and B,

$$|a_\ell\rangle_\mathrm{A} = \sum_{\sigma_1, \ldots, \sigma_\ell} (A^{\sigma_1} A^{\sigma_2} \cdots A^{\sigma_\ell})_{1, a_\ell} |\sigma_1, \ldots, \sigma_\ell\rangle, \tag{4.281}$$

$$|a_\ell\rangle_\mathrm{B} = \sum_{\sigma_{\ell+1}, \ldots, \sigma_L} (A^{\sigma_{\ell+1}} A^{\sigma_{\ell+2}} \cdots A^{\sigma_L})_{a_\ell, 1} |\sigma_{\ell+1}, \ldots, \sigma_L\rangle \tag{4.282}$$

such that the MPS can be written as

$$|\psi\rangle = \sum_{a_\ell} |a_\ell\rangle_\mathrm{A} |a_\ell\rangle_\mathrm{B}. \tag{4.283}$$

This is not a Schmidt decomposition, because the $\{|a_\ell\rangle_\mathrm{A}\}$ form an orthonormal set, but the $\{|a_\ell\rangle_\mathrm{B}\}$ in general do not. To see the former, work out the overlap between states on A: the left-normalization condition ensures orthonormality, but the same calculation will fail on B. For right-canonical MPS, this naive attempt at a Schmidt decomposition also fails, but now on A, while on B orthonormality is achieved.

Mixed-canonical matrix product state

The last observation brings up the idea of proceeding with the decomposition of the state from left and right at the same time. Let us assume that we peel off the first ℓ sites from the left, and the rest from the right. This gives an MPS of the form

$$|\psi\rangle = \sum_{\sigma_1, \ldots, \sigma_L} A^{\sigma_1} \cdots A^{\sigma_\ell} S B^{\sigma_{\ell+1}} \cdots B^{\sigma_L} |\sigma_1, \ldots, \sigma_L\rangle. \tag{4.284}$$

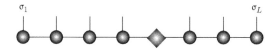

Fig. 4.10 Graphical representation of an exact MPS in the mixed-canonical representation. The diamond represents the diagonal matrix. Matrices to the left are left-normalized, while those to the right are right-normalized.

Obviously, there are A-matrices on the left and B-matrices on the right. What is new is the presence of S, which can be brought into diagonal form with non-negative diagonal entries. This state can be graphically represented as in Fig. 4.10. One may of course multiply S into either of the adjacent matrices, turning it into a general M-matrix.

The Schmidt decomposition into A and B, where A runs from sites 1 to ℓ and B from sites $\ell + 1$ to L, can now be read off immediately, as previously claimed. If we introduce vectors $|a_\ell\rangle_A$ and $|a_\ell\rangle_B$ as before, then the state takes the form ($s_a = S_{a,a}$)

$$|\psi\rangle = \sum_a s_a |a\rangle_A |a\rangle_B, \tag{4.285}$$

which is a valid Schmidt decomposition since the states on A *and* B are now orthonormal respectively.

4.11.2 Overlaps, expectation values, and matrix elements

Let us consider an overlap between states $|\psi\rangle$ and $|\phi\rangle$, described by matrices M and $\tilde{M}$, and focus on open boundary conditions. Then the overlap is given by

$$\langle\phi|\psi\rangle = \sum_{\{\sigma\}} \tilde{M}^{\sigma_1*} \cdots \tilde{M}^{\sigma_L*} M^{\sigma_1} \cdots M^{\sigma_L}, \tag{4.286}$$

or, equivalently,

$$\langle\phi|\psi\rangle = \sum_{\{\sigma\}} \tilde{M}^{\sigma_L\dagger} \cdots \tilde{M}^{\sigma_1\dagger} M^{\sigma_1} \cdots M^{\sigma_L}, \tag{4.287}$$

where we have used the fact that the transpose of a scalar is the scalar itself and $(AB)^T = B^T A^T$. Pictorially, the overlap looks like Fig. 4.11. The pictorial representation shows that there is an enormous amount of freedom in each sequence to carry out the many contractions. But they are hugely inequivalent in terms of efficiency (up to exponentially expensive—just contract all matrix indices before the physical indices). In this simple case, the optimal arrangement can still be found to be

$$\langle\phi|\psi\rangle = \sum_{\sigma_L} \tilde{M}^{\sigma_L\dagger} \left(\cdots \left(\sum_{\sigma_2} \tilde{M}^{\sigma_2\dagger} \left(\sum_{\sigma_1} \tilde{M}^{\sigma_1\dagger} M^{\sigma_1} \right) M^{\sigma_2} \right) \cdots \right) M^{\sigma_L}. \tag{4.288}$$

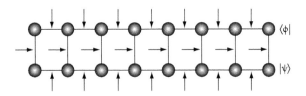

Fig. 4.11 Overlap between two states $|\phi\rangle$ and $|\psi\rangle$. All contractions (sums) over the same indices are indicated by arrows.

Evaluating the brackets from inside out, each bracket (except the special first step) contains a three-matrix multiplication, with the central matrix the result of the last bracket, and one d-term sum over a physical index. The three-matrix multiplication is of course done as a sequence of two standard matrix multiplications. If we assume for simplicity identical matrix dimensions D, each matrix multiplication is of complexity $O(D^3)$, and the total cost is $O(LD^3d)$. Pictorially, we run through the contractions like a zipper.

For $\langle\phi|\hat{O}^{[i]}\hat{O}^{[j]}|\psi\rangle$ with local operators

$$\hat{O}^{[\ell]} = \sum_{\sigma_\ell,\sigma'_\ell} O^{\sigma_\ell,\sigma'_\ell}|\sigma_\ell\rangle\langle\sigma'_\ell|,$$

we have to insert into the overlap expression operator matrix elements like $O^{\sigma_i,\sigma'_i}O^{\sigma_j,\sigma'_j}$, slightly complicating the onion-like overlap expression:

$$\langle\phi|\hat{O}^{[i]}\hat{O}^{[j]}|\psi\rangle = \sum_{\sigma_1,\ldots,\sigma_L,\sigma'_i,\sigma'_j} \tilde{M}^{\sigma_1*}\cdots\tilde{M}^{\sigma_L*}O^{\sigma_i,\sigma'_i}O^{\sigma_j,\sigma'_j}M^{\sigma'_1}\cdots M^{\sigma_L}$$

$$= \sum_{\sigma_L}\tilde{M}^{\sigma_L\dagger}\left(\cdots\left(\sum_{\sigma_j,\sigma'_j}O^{\sigma_j,\sigma'_j}\tilde{M}^{\sigma_j\dagger}\left(\cdots\left(\sum_{\sigma_i,\sigma'_i}O^{\sigma_i,\sigma'_i}\tilde{M}^{\sigma_i\dagger}\left(\cdots\left(\sum_{\sigma_1}\tilde{M}^{\sigma_1\dagger}M^{\sigma_1}\right)\right.\right.\right.\right.$$
$$\left.\left.\left.\left.\cdots\right)M^{\sigma'_i}\right)\cdots\right)M^{\sigma'_j}\right)\cdots\right)M^{\sigma_L}.$$

This amounts to the same calculation as for the overlap, except that on the operator sites, the single sum over the physical index turns into a double sum (Fig. 4.12), slightly adding to the numerical cost. But it is still proportional to LD^3.

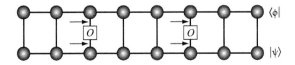

Fig. 4.12 Matrix elements between two states $|\phi\rangle$ and $|\psi\rangle$ are calculated like the overlap, with the operators inserted at the right places, generating a double sum of physical indices there, as indicated by the arrows.

4.11.3 Adding two matrix product states

For two generic MPS

$$|\psi\rangle = \sum_{\{\sigma\}} M^{\sigma_1} \ldots M^{\sigma_L} |\{\sigma\}\rangle,$$

$$|\phi\rangle = \sum_{\{\sigma\}} \tilde{M}^{\sigma_1} \ldots \tilde{M}^{\sigma_L} |\{\sigma\}\rangle,$$

(4.289)

the sum is given again by an MPS

$$|\psi\rangle + |\phi\rangle = \sum_{\{\sigma\}} N^{\sigma_1} \ldots N^{\sigma_L} |\{\sigma\}\rangle,$$

(4.290)

where

$$N^{\sigma_i} = M^{\sigma_i} \oplus \tilde{M}^{\sigma_i}$$

(4.291)

on sites $2, \ldots, L-1$ and, as special cases, $N^{\sigma_1} = [M^{\sigma_1} \tilde{M}^{\sigma_1}]$ and $N^{\sigma_L} = [M^{\sigma_L} \tilde{M}^{\sigma_L}]^{\mathrm{T}}$, as can be checked by evaluating the products.

Addition of MPS therefore leads to new matrices with dimension $D_N = D_M + D_{\tilde{M}}$, ignoring sites 1 and L, such that MPS of a certain dimension are not closed under addition, and dimensions start to grow. In order to keep algorithms that involve additions manageable, we therefore need compression algorithms that find the MPS of some smaller matrix dimension that best approximates the larger MPS.

4.11.4 Additional tools: canonization and compression

For the actual implementation of algorithms, two useful tools that go beyond simple manipulations of MPS are needed, canonization and compression. While both are not difficult, stating them explicitly is very index-heavy, so we will just outline the general ideas and refer to the details given in [62].

Canonization

As we have already seen, in an MPS, the matrices M^{σ_i} merely have to match in their dimensions. Nevertheless, orthonormality and decomposition properties make it highly desirable to have MPS in one of the three canonical forms. The procedure of achieving this we will call canonization. It is identical in all three cases except for the way one works through an MPS: for left-canonization, one starts from the left, for right-canonization from the right, and for mixed-canonization from both sides. In all three cases, the method rests on a sequence of SVDs on the M-matrices, because SVDs automatically generate new matrices with the desired orthonormality properties.

Compression of an MPS by SVD

All time-evolution algorithms lead to a substantial increase in matrix dimensions of the MPS under consideration. It is therefore a recurrent issue how to approximate a given

MPS with matrix dimensions $D'_i \times D'_{i+1}$ by another MPS with matrix dimensions $D_i \times D_{i+1}$, where $D_i < D'_i$, $D_{i+1} < D'_{i+1}$, as closely as possible. Fundamentally, two procedures are available, namely, *SVD compression* and *variational compression*. The former is usually fast, but not exactly optimal; the latter (see below) is optimal, but more complicated. For unitary time evolutions, the smallness of the time steps means that states do not change a lot, and the amount of information to be lost in compression is small anyway, and hence the SVD compression will do in practice.

The idea is that for an MPS in mixed-canonical representation, with S diagonal and non-negative,

$$|\psi\rangle = \sum_{\{\sigma\}} A^{\sigma_1} A^{\sigma_2} \cdots A^{\sigma_\ell} S B^{\sigma_{\ell+1}} \cdots B^{\sigma_{L-1}} B^{\sigma_L} |\{\sigma\}\rangle, \qquad (4.292)$$

the Schmidt decomposition can be read off directly. The Schmidt coefficients, the diagonal entries of S, tell us which contributions to the Schmidt decomposition are the most important ones, so the best approximation is given by keeping the D most important out of D' terms (we drop the indices here for simplicity). This provides the truncation prescription; in order to apply it, we have to be in the mixed-canonical representation, but this works only for the bond that connects parts A and B. To truncate therefore on all bonds (all matrix dimensions), we have to work our way through all mixed-canonical representations $ASBBBB$, $AASBBB$, $AAASBB$, and so on. The easiest way to do this is to bring the state into $BBBBB$ form, then to start left-canonizing it. This will take us through the configurations just listed, and we truncate at each step.

The small problem of the procedure is that—because the truncation is done in an asymmetric way from left to right—truncations further to the right depend on those previously done to the left, but not vice versa. Ideally, they would all be on the same footing. But if each truncation by itself is not losing much information, this asymmetric interdependence can be neglected, as is the case in time evolutions.

In practice, compression is often done by imposing some ϵ that at each truncation we accept as the maximum 2-norm distance between the original and the compressed state (given by the sum of the squares of the discarded singular values), implicitly defining D.

Compressing a matrix product state iteratively

The optimal approach is to start from an ansatz MPS $|\tilde\psi\rangle$ (matrices $\tilde M$) of the desired reduced dimension, and to minimize its distance to the MPS $|\psi\rangle$ (matrices M) to be approximated iteratively, i.e., by changing the $\tilde M^\sigma$ matrices of $|\tilde\psi\rangle$ iteratively. The matrices play the role of variational parameters.

The mathematically precise form of optimal compression of $|\psi\rangle$ from dimension D' to $|\tilde\psi\rangle$ with dimension D is to minimize $\||\psi\rangle - |\tilde\psi\rangle\|_2^2$, which means that we want to minimize $\langle\psi|\psi\rangle - \langle\tilde\psi|\psi\rangle - \langle\psi|\tilde\psi\rangle + \langle\tilde\psi|\tilde\psi\rangle$ with respect to $|\tilde\psi\rangle$.

Expressed in terms of the underlying matrices $\tilde M$, this is a highly multilinear optimization problem, but it can be solved iteratively as a sequence of linear optimization problems. We start with an initial guess for $|\tilde\psi\rangle$, which could be an SVD compression of $|\psi\rangle$, arguably not optimal, but a good starting point. Then we sweep through the set of $\tilde{M}^{\sigma_i}$ site by site, keeping all other matrices fixed and choosing the new $\tilde{M}^{\sigma_i}$, such that distance is minimized. The (usually justified hope) is that repeating this sweep through the matrices several times will lead to a converged optimal approximation.

4.12   Matrix product operators (MPO)

If we consider a single coefficient $\langle\{\sigma\}|\psi\rangle$ of an MPS,

$$\langle\{\sigma\}|\psi\rangle = M^{\sigma_1} M^{\sigma_2} \cdots M^{\sigma_{L-1}} M^{\sigma_L},$$

then it is a natural generalization to try to write coefficients $\langle\{\sigma\}|\hat{O}|\{\sigma'\}\rangle$ of operators as [17, 26, 45, 53, 72]

$$\langle\{\sigma\}|\hat{O}|\{\sigma'\}\rangle = W^{\sigma_1\sigma_1'} W^{\sigma_2\sigma_2'} \cdots W^{\sigma_{L-1}\sigma_{L-1}'} W^{\sigma_L\sigma_L'}, \qquad (4.293)$$

where the $W^{\sigma\sigma'}$ are matrices just like the M^σ, with the only difference being that as representations of operators they need both outgoing and ingoing physical states:

$$\hat{O} = \sum_{\{\sigma\},\{\sigma'\}} W^{\sigma_1\sigma_1'} W^{\sigma_2\sigma_2'} \cdots W^{\sigma_{L-1}\sigma_{L-1}'} W^{\sigma_L\sigma_L'} |\{\sigma\}\rangle\langle\{\sigma'\}|. \qquad (4.294)$$

The pictorial representation introduced for MPS can be extended in a straightforward fashion: instead of one vertical line for the physical state in the representation of M, we now have two vertical lines, one down, one up, for the ingoing and outgoing physical states in W (Fig. 4.13). The complete MPO itself then looks like Fig. 4.14.

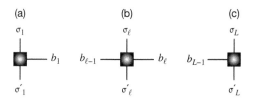

Fig. 4.13 Elements of an MPO: (a) a corner matrix operator $W^{\sigma_1\sigma_1'}_{1,b_1}$ at the left end of the chain; (b) a bulk matrix operator $W^{\sigma_\ell\sigma_\ell'}_{b_{\ell-1},b_\ell}$; (c) a corner operator $W^{\sigma_L\sigma_L'}_{b_{L-1},1}$ at the right end: the physical indices points up and down, the matrix indices are represented by horizontal lines.

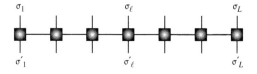

Fig. 4.14 An MPO acting on an entire chain: the horizontal matrix indices are contracted, and the MPO is ready to be applied to an MPS by simple contraction of vertical (physical) indices.

Following a similar procedure as for obtaining the MPS representation of quantum states, any operator can be brought into MPO form. The difference from states is that most operators of interest can be brought into MPO form very easily: this is immediately obvious for local operators or tensor products of local operators: then all matrices $W^{\sigma_\ell \sigma'_\ell}$ are scalars, and the scalars are given by matrix elements $(\hat{O})_{\sigma_\ell \sigma'_\ell}$ for an operator $\hat{O}$ on site ℓ. Even for seemingly complex operators like Hamiltonians, where the sum structure might seem to make the construction of the MPO very difficult, this is not complicated at all, and the emerging matrix dimensions are usually quite small (e.g., 5 for the Heisenberg model and 6 for the Hubbard model). This was discussed first explicitly by [45]; similar construction rules were given by [17]. For more complex Hamiltonians, see [26, 53] and, for a hopefully pedagogical introduction, [62].

The application of an MPO to an MPS runs as

$$\hat{O}|\psi\rangle = \sum_{\{\sigma\},\{\sigma'\}} (W^{\sigma_1,\sigma'_1} W^{\sigma_2,\sigma'_2} \cdots)(M^{\sigma'_1} M^{\sigma'_2} \cdots)|\{\sigma\}\rangle$$

$$= \sum_{\{\sigma\},\{\sigma'\}} \sum_{\{a\},\{b\}} (W^{\sigma_1,\sigma'_1}_{1,b_1} W^{\sigma_2,\sigma'_2}_{b_1,b_2} \cdots)(M^{\sigma'_1}_{1,a_1} M^{\sigma'_2}_{a_1,a_2} \cdots)|\{\sigma\}\rangle$$

$$= \sum_{\{\sigma\},\{\sigma'\}} \sum_{\{a\},\{b\}} (W^{\sigma_1,\sigma'_1}_{1,b_1} M^{\sigma'_1}_{1,a_1})(W^{\sigma_2,\sigma'_2}_{b_1,b_2} M^{\sigma'_2}_{a_1,a_2}) \cdots |\{\sigma\}\rangle$$

$$= \sum_{\{\sigma\}} \sum_{\{a\},\{b\}} N^{\sigma_1}_{(1,1),(b_1,a_1)} N^{\sigma_2}_{(b_1,a_1),(b_2,a_2)} \cdots |\{\sigma\}\rangle$$

$$= \sum_{\{\sigma\}} N^{\sigma_1} N^{\sigma_2} \cdots |\{\sigma\}\rangle,$$

or compactly as $|\phi\rangle = \hat{O}|\psi\rangle$, with $|\phi\rangle$ an MPS built from matrices N^{σ_i} with

$$N^{\sigma_i}_{(b_{i-1},a_{i-1}),(b_i,a_i)} = \sum_{\sigma'_i} W^{\sigma_i \sigma'_i}_{b_{i-1} b_i} M^{\sigma'_i}_{a_{i-1} a_i}. \qquad (4.295)$$

The MPS form therefore remains invariant, at a higher dimension given by the product of the original MPS and MPO dimensions (Fig. 4.15).

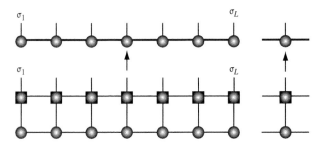

Fig. 4.15 An MPO acting on an MPS: matching physical (vertical) indices are contracted, and a new MPS emerges, with multiplied matrix dimensions and product structure in its matrices.

Once again, a seemingly exponentially complex operation (sum over exponentially many $\{\sigma\}$) is reduced to a low-cost operation: the operational count is of order $Ld^2 D_W^2 D^2$, D_W being the dimension of the MPO.

4.13 Ground-state calculations with MPS

Let us quickly discuss the calculation of ground states with MPS. Obviously, this is not strictly part of a discussion of nonequilibrium dynamics. Nevertheless, it is meaningful for two reasons: on the one hand, the ground-state calculation can be phrased as a special time evolution, so that understanding time evolution gives us ground states for free, albeit not in the numerically most efficient form. On the other hand, in many current quench experiments, nonequilibrium is generated by bringing the system into the (nontrivial) ground state of some Hamiltonian $\hat{H}_0$ and then changing $\hat{H}_0 \to \hat{H}$ at some time $t = 0$. So, setting up the nonequilibrium dynamics involves calculating a ground state first.

There are, loosely speaking, two ways of calculating the ground state of some Hamiltonian $\hat{H}$, which we briefly discuss in the following.

4.13.1 Imaginary-time evolution

Assume that the eigendecomposition of $\hat{H}$ reads

$$\hat{H} = \sum_n E_n |n\rangle\langle n| \tag{4.296}$$

and that the ground state is nondegenerate ($E_n > E_0$ for all $n > 0$). Let us also assume that $E_0 = 0$. Starting from some random state $|\psi\rangle = \sum_n c_n |n\rangle$, we calculate

$$\lim_{\beta \to \infty} e^{-\beta \hat{H}} |\psi\rangle = \lim_{\beta \to \infty} \sum_n e^{-\beta E_n} c_n |n\rangle = c_0 |0\rangle. \tag{4.297}$$

The result follows from the exponential suppression of all contributions from higher eigenstates. Therefore, if the initial state had some overlap with the ground state, we will obtain the ground state by an imaginary-time evolution $e^{-\beta \hat{H}} = e^{-i\hat{H}(-i\beta)}$ up to imaginary time $-i\beta$, which has to be taken to infinity. Of course, we do not know $E_0 = 0$, but this problem can be eliminated by keeping the time-evolving state normalized.

The appealing feature of this procedure is that if we have a real-time evolution algorithm, we can immediately adapt it to ground-state searches. It even has the nice feature that all errors are suppressed exponentially as imaginary time is growing, as opposed to unitary real-time evolution, where all errors are there to stay with us. The disadvantage of the method is that it is comparatively slow, because β has to be substantially larger than the inverse gap to have a dominant ground-state contribution in the time-evolved state.

4.13.2 Variational ground-state search

Alternatively, we can try to find the MPS $|\psi\rangle$ of some dimension D that minimizes

$$E = \frac{\langle\psi|\hat{H}|\psi\rangle}{\langle\psi|\psi\rangle}. \tag{4.298}$$

In order to solve this problem, we introduce a Lagrangian multiplier λ and extremize

$$\langle\psi|\hat{H}|\psi\rangle - \lambda\langle\psi|\psi\rangle; \tag{4.299}$$

in the end, $|\psi\rangle$ will be the desired ground state and λ the ground-state energy. The MPS network that represents (4.299) is shown in Fig. 4.16. Here, we have assumed that we know $\hat{H}$ in MPO form.

The problem with this approach is that the variables (the matrix elements of the MPS) appear in the form of products, making this a highly multi-linear optimization problem. But it can be done iteratively, too, and this is the idea that also drives DMRG ([61, 78, 79]—in fact, ground-state DMRG is exactly the same method, merely expressed in different terms): while keeping the matrices on all sites but one (ℓ) constant, consider only the matrix entries $M^{\sigma_\ell}_{a_{\ell-1}a_\ell}$ on site ℓ as variables. Then the variables appear in (4.299) only in quadratic form, for which the determination of the extremum is a benign linear algebra problem. This will lower the energy, and find a variationally better state, but of course not the optimal one. Now we continue to vary the matrix elements on another site to find a state again lower in energy, moving through all sites multiple times, until the energy no longer improves.

Graphically, if we take the extremum of (4.299) with respect to $M^{\sigma_\ell*}_{a_{\ell-1}a_\ell}$, we arrive at the equation represented in Fig. 4.17. The matrix taken as a variable appears twice. In both networks, if we take it as a large vector v, it is premultiplied by a matrix

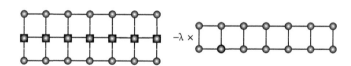

Fig. 4.16 Network to be contracted to obtain the functional to be extremized to find the ground state and its energy. The left-hand side represents the term $\langle\psi|\hat{H}|\psi\rangle$, the right-hand side the squared norm $\langle\psi|\psi\rangle$.

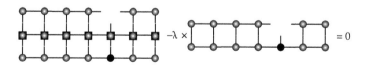

Fig. 4.17 Generalized eigenvalue problem for the optimization of $M^{\sigma_\ell}_{a_{\ell-1},a_\ell}$. The unknown matrix is solid black in both networks.

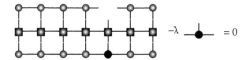

Fig. 4.18 Standard eigenvalue problem for the optimization of $M^{\sigma_\ell}_{a_{\ell-1},a_\ell}$. The unknown matrix is solid black.

(row indices are open legs; column indices are legs connecting to the matrix taken as a variable). We therefore have a generalized eigenvalue problem of the structure $Hv - \lambda Nv = 0$.

This simplifies if we take $|\psi\rangle$ to be in mixed-canonical representation, with all matrices of type A to the left of the variable matrix and all matrices of type B to its right. Then the left and right parts of the right network in the eigenequation contract to the identity, and we obtain the equation represented in Fig. 4.18, an *eigenvalue problem* of matrix dimension $dD^2 \times dD^2$:

$$Hv - \lambda v = 0. \tag{4.300}$$

Solving for the lowest eigenvalue λ_0 gives us a $v^0_{a_{\ell-1}\sigma_\ell a_\ell}$, which is reshaped back to $M^{\sigma_\ell}_{a_{\ell-1}a_\ell}$, λ_0 being the current ground-state energy estimate.

The dimension of v is dD^2, and in general dD^2 is too large for exact diagonalization, but as we are only interested in the lowest eigenvalue and eigenstate, an iterative eigensolver that aims for the ends of the spectrum will do. Here, we can use the Lanczos large sparse matrix solver discussed previously. The speed of convergence of such methods ultimately rests on the quality of the initial starting or guess vector. As this eigenproblem is part of an iterative approach to the ground state, the current M^{σ_ℓ} is a valid guess that will dramatically speed up calculations close to convergence, often reducing iteration numbers by more than an order of magnitude.

In order to work only with a standard eigenvalue problem, which is numerically to be preferred, we must maintain the current MPS of the algorithm always in mixed-canonical representation, such that all matrices to the left of the currently optimized M^{σ_ℓ} are in the A-form and all to the right in the B-form. By carrying out single steps of the iterative procedure that canonizes states, the position of the "central" matrix can be shifted to the left or right. This imposes that we "sweep" through the MPS from left to right to left to right, always shifting the position of the currently optimized matrix by one site, and moving the mixed-canonical representation along.

Technically, another issue is the calculation of the H-matrix, i.e., the contraction of the left part of the equation network. H is never calculated explicitly, but left as a product of the left part, the right part, and the MPO matrix directly connecting to the matrix currently being optimized. As for overlaps, the contraction of the left and right parts is best done column by column, and as we move systematically through the MPS either to the left or to the right, the contraction can also be built up iteratively during the algorithm or old contraction results be reused.

4.14   Time evolutions (real and imaginary) with MPS

Real-time evolutions $\mathrm{e}^{-\mathrm{i}\hat{H}t}$ and imaginary-time evolution operators $\mathrm{e}^{-\beta\hat{H}}$ (for ground-state calculations) find an easy encoding by MPS. This holds for both pure and mixed states, important at finite temperature. Both time-evolution approaches based on the evolution operator discussed previously (the Trotter and Krylov approaches) can be applied to MPS instead of a generic quantum state. In the following, we will first focus on time evolution based on a Trotter decomposition of the evolution operators [18, 72, 74, 75, 82], and then turn to the Krylov vector approach. We will then discuss the changes necessary for the simulation of the dynamics of mixed states. It should be mentioned here that other evolution schemes such as Runge–Kutta integration could be applied, but almost all applications nowadays use either the Trotter or Krylov schemes.

4.14.1   Time evolution of pure states

Trotter-based time evolution

Let us assume that $\hat{H}$ consists of nearest-neighbor interactions only. Then time evolutions separate, as seen previously, into evolutions on odd $(\mathrm{e}^{-\mathrm{i}\hat{H}_{\mathrm{odd}}\Delta t})$ and even $(\mathrm{e}^{-\mathrm{i}\hat{H}_{\mathrm{even}}\Delta t})$ bonds that respectively commute among each other. So we are looking for an MPO doing an infinitesimal time step on the odd bonds and for another MPO doing the same on the even bonds.

As any operator is guaranteed to be MPO-representable, the bond evolution operators $\mathrm{e}^{-\mathrm{i}\hat{h}\Delta t}$ can be written as an MPO, where we will show below that the dimension is (at most) d^2. The application of the MPO thus increases the bond dimensions from D to $d^2 D$, making MPS compression necessary.

This leads to the following algorithm: starting from $|\psi(t=0)\rangle$, repeat the following steps:

- Apply the MPO of the odd bonds to $|\psi(t)\rangle$.
- Apply the MPO of the even bonds to $\mathrm{e}^{-\mathrm{i}\hat{H}_{\mathrm{odd}}\Delta t}|\psi(t)\rangle$.
- Compress the MPS $|\psi(t+\Delta t)\rangle = \mathrm{e}^{-\mathrm{i}\hat{H}_{\mathrm{even}}\Delta t}\mathrm{e}^{-\mathrm{i}\hat{H}_{\mathrm{odd}}\Delta t}|\psi(t)\rangle$ from dimensions $d^2 D$ to D, monitoring the error. Obviously, one may also allow for some compression error (state distance) ϵ and choose a time-dependent D: it will typically grow strongly with time, limiting the reachable timescale, as discussed before. A good initial guess for the compressed state is given by $|\psi(t)\rangle$, since Δt is small. By analogy with the ground-state calculations, all results should be extrapolated in $D \to \infty$ or $\epsilon \to 0$.

What does the MPO look like explicitly? Let us consider the Trotter step for all odd bonds of a chain:

$$\mathrm{e}^{-\mathrm{i}\hat{h}_1\Delta t} \otimes \mathrm{e}^{-\mathrm{i}\hat{h}_3\Delta t} \otimes \ldots \otimes \mathrm{e}^{-\mathrm{i}\hat{h}_{L-1}\Delta t}|\psi\rangle; \tag{4.301}$$

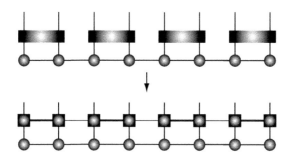

Fig. 4.19 A Trotter step. On all odd bonds, an (infinitesimal) bond time evolution is carried out. This merges two sites, such that the simple product form of MPS is lost at first sight, but the time evolution can be translated into MPOs. As the time evolution factorizes, the MPOs have dimension 1 on all even bonds (thin lines).

each bond-evolution operator like $e^{-i\hat{h}_1 \Delta t}$ takes the form

$$\sum_{\sigma_1 \sigma_2, \sigma_1' \sigma_2'} O^{\sigma_1 \sigma_2, \sigma_1' \sigma_2'} |\sigma_1 \sigma_2\rangle \langle \sigma_1' \sigma_2'|.$$

In both the pictorial and the explicit mathematical representations, it is obvious that this operator destroys the MPS form (Fig. 4.19).

It would therefore be desirable to have $O^{\sigma_1 \sigma_2, \sigma_1' \sigma_2'}$ in some form containing tensor products $O^{\sigma_1, \sigma_1'} \otimes O^{\sigma_2, \sigma_2'}$, to maintain the MPS form. To this purpose, we carry out the procedure for decomposing an arbitrary state into an MPS, adapted to an operator (two indices per site). It is very transparent here. We reorder O to group local indices and carry out an SVD:

$$\begin{aligned}
O^{\sigma_1 \sigma_2, \sigma_1' \sigma_2'} &= P_{(\sigma_1 \sigma_1'),(\sigma_2 \sigma_2')} \\
&= \sum_k U_{\sigma_1 \sigma_1', k} S_{k,k} (V^\dagger)_{k,(\sigma_2 \sigma_2')} \\
&= \sum_k U_k^{\sigma_1 \sigma_1'} \overline{U}_k^{\sigma_2 \sigma_2'} \\
&= \sum_k U_{1,k}^{\sigma_1 \sigma_1'} \overline{U}_{k,1}^{\sigma_2 \sigma_2'},
\end{aligned}$$

where

$$U_{1,k}^{\sigma_1 \sigma_1'} = U_{(\sigma_1 \sigma_1'),k} \sqrt{S_{k,k}}, \qquad \overline{U}_{k,1}^{\sigma_2 \sigma_2'} = \sqrt{S_{k,k}} (V^\dagger)_{k,(\sigma_2 \sigma_2')}.$$

In the very last step of the derivation, we have introduced a dummy index taking value 1 to arrive at the form of an MPO matrix. The index k may run up to d^2, giving the bond dimension D_W of the MPO.

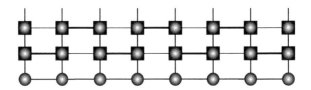

Fig. 4.20   A complete first-order Trotter time step (odd and even bonds). Thick and thin lines correspond to dimensions 1 and > 1 on MPO bonds. The MPOs in the top line on the first and last sites are trivial scalar identities 1.

The MPO representing the operator in (4.301), $U^{\sigma_1 \sigma'_1} \overline{U}^{\sigma_2 \sigma'_2} U^{\sigma_3 \sigma'_3} \overline{U}^{\sigma_4 \sigma'_4} \cdots$, factorizes on every second bond, as do the original unitaries. If one site does not participate in any bond evolution, we simply assign it the identity unitary as a 1×1 matrix: $I^{\sigma,\sigma'}_{1,1} = \delta_{\sigma,\sigma'}$. Then the global MPO can be formed trivially from local MPOs. The MPO for time evolution on all odd bonds would read $U\overline{U}U\overline{U}U\overline{U}\cdots$, whereas the even-bond time step reads $IU\overline{U}U\overline{U}U\overline{U}\cdots I$ (Fig. 4.20).

Let us conclude with a brief discussion of the sources of errors. On the one hand, there is the Trotter error due to the decomposition. As already discussed, this error grows linearly in time and is proportional to some power of Δt, depending on which order of the decomposition is being used. This error can be made very small and its growth with time is very benign. The more important source of error is the compression step. The compression itself would not generate any error of importance, if $|\psi(t)\rangle$ could be expressed with similar accuracy throughout time t for given matrix size D. As previously discussed, however, D may grow up to exponentially with t. This means that we either allow for a fixed compression (truncation) error (truncated weight) and increase D until we run out of memory or CPU, or we keep D fixed, such that the truncation error grows with t, which may be up to exponentially with t. Compared with this, the Trotter error is negligible.

To show an example, we consider a Heisenberg antiferromagnet with $J_{\text{ex}} = 1$ and various J^z_{ex} for the $\hat{S}^z_i \hat{S}^z_{i+1}$ term. The system size is $L = 128$, and we start with a Néel state $| \uparrow\downarrow\uparrow\downarrow\uparrow\downarrow\uparrow\downarrow \ldots\rangle$, an MPS of $D = 1$ since it is a product state. This state is not the ground state, not even an eigenstate, and has an energy with respect to the Heisenberg Hamiltonian that is extensively different from the ground-state energy. Hence, evolving the Néel state under the Heisenberg Hamiltonian corresponds to a global quench. For the time evolution, we use a fourth-order Trotter decomposition with time step 0.125, truncating reduced density matrix eigenvalues below 10^{-10}. This roughly corresponds to a truncated weight of 10^{-10} (in reality, it is slightly larger, but not much so).

Figure 4.21 shows how the magnetization on a site with initial state $| \uparrow\rangle$ decays from 0.5 in time for three different J^z_{ex} and might serve readers for a comparison with their own implementations. Similarly, Fig. 4.22 shows the time evolution of the two-point correlator $|\langle \hat{S}^+_0 \hat{S}^-_{\Delta x}\rangle|$. Here 0 is taken to be in the center of the chain.

Figure 4.23 shows resource usage in a log–linear plot of the bond dimension M_{64} at the center of the chain, the D of our CPU cost estimates. For all three J^z_{ex} considered

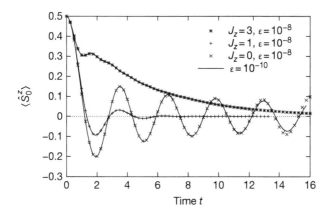

Fig. 4.21 Relaxation of the magnetization of a Néel state time-evolving under an antiferromagnetic Heisenberg Hamiltonian for various anisotropies. All calculations are shown for truncated weights $\epsilon = 10^{-8}$ per time step (symbols) and $\epsilon = 10^{-10}$ (solid lines). Results start to deviate only on very long timescales, gauging the high accuracy of the calculation.

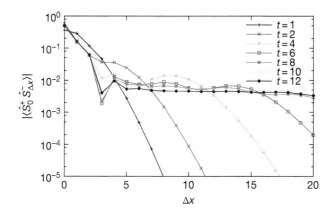

Fig. 4.22 Time evolution of two-point correlators $|\langle \hat{S}_0^+ \hat{S}_{\Delta x}^- \rangle|$ time-evolving under an antiferromagnetic isotropic Heisenberg Hamiltonian for various distances and times.

and two different truncated weights ϵ, roughly similar and almost linear curves emerge, indicating the almost exponential growth of resources needed for time evolution: even with matrices of dimension $O(5000)$, only times between 12 and 16 can be reached. But, as can be seen from the previous figures, for many purposes this may be enough.

Krylov-vector-based time evolution

Let us consider the case of a longer-ranged Hamiltonian, for example a frustrated spin chain or a spin ladder, where interactions become automatically long-ranged even if on the ladder interacting spins all sit next to each other. Then, if we give the sites

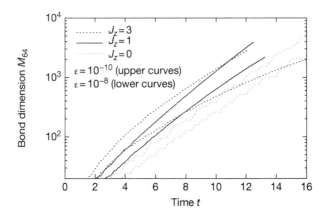

Fig. 4.23 Log–linear plot of the maximal matrix dimension of the Néel-state simulation at times t for different anisotropies J_{ex}^z and accepted truncated weights ϵ. All curves being almost linear, resource usage grows almost exponentially with time.

labels like $1, 3, 5, \ldots$ on the top leg and $2, 4, 6, \ldots$ on the bottom leg, interactions along the rungs of the ladder are nearest-neighbor, but those along the two legs are next-to-nearest-neighbor interactions. Such a ladder can still be simulated using a Trotter decomposition, but it becomes computationally heavy: as two spins have to be grouped in one, the local state-space dimension increases from d to d^2 and the dimension of the Trotter MPO matrices from $d^2 \times d^2$ to $d^4 \times d^4$. If $d = 2$, this is still feasible, but only barely so.

Alternatively, we can evaluate $e^{-i\hat{H}\Delta t}|\psi\rangle$ by the Krylov-exponentiation approach, and calculate $|\psi(t)\rangle \to |\psi(t + \Delta t)\rangle$, while maintaining the same matrix dimension D by subsequent compression. This immediately sets up a time-evolution procedure.

There are again two error sources, from the imperfect calculation of the exponential by the Krylov method, which always occurs, and from the truncations of the MPS matrices. As we have seen, the former can be made exponentially small; we control it by observing the convergence of the state $|\psi(t + \Delta t)\rangle$ with an increasing number of Krylov states. Mostly, if we choose a small Δt (which can nevertheless be bigger than in the Trotter approach), then say four or five Krylov states should be sufficient, but we have to check this for each problem individually and, of course, the number can be chosen adaptively to achieve a certain error bar.

More interesting for us is the MPS-based (compression) error, because an additional technical complication arises: for a Hermitian H, Krylov vectors $|q_i\rangle$ are determined by a three-vector recursion relation of the form

$$\beta_m |q_{m+1}\rangle = H|q_m\rangle - \langle q_m|H|q_m\rangle|q_m\rangle - \langle q_{m-1}|H|q_m\rangle|q_{m-1}\rangle. \qquad (4.302)$$

In the MPS language, this means that matrix dimensions increase as $D \to D_W D + 2D$, where we assume that the Krylov vectors $|q_m\rangle$ and $|q_{m-1}\rangle$ are MPS of dimension D and that the MPO of the Hamilton operator has dimension D_W, for example $D_W = 5$

for a Heisenberg chain. Then, in one step, $D \to 7D$, which means that truncation has to occur. This can be done by a compression algorithm. But the consequence of that is that the mathematical orthogonality of $|q_{m+1}\rangle$ with respect to all earlier Krylov vectors holds only approximately. Any scheme to fix that by re-orthogonalization will introduce new additions of states, etc., which leads to new compressions, and everything becomes messy. It is a possible procedure to take the new Krylov state at each iteration and calculate $\langle q_{m+1}|H|q_i\rangle$; then the small tridiagonal matrix becomes full (with the dominant entries still on the central three diagonals) and for exponentiation this new matrix is taken instead. The price is a certain loss of efficiency, since the Krylov states are no longer completely orthonormal and their span is smaller.

4.14.2 Time evolution of mixed states

Purification of mixed states

The simulation of mixed states seems to be much more complicated than that of pure states, since they are given by a potentially exponentially large sum of projectors on pure states; as we have seen, the efficient handling of pure states has already generated quite heavy numerics. In fact, the additional complexity cannot be completely removed, but it can be considerably alleviated; as we will see now, this can even be done in a fashion that allows for large-scale recycling of existing pure-state MPS code. The Schmidt decomposition of a state living on AB allowed us to read off, without further calculations, the reduced density operators on A and B, respectively. The "inversion" of this procedure is called *purification*: assume we have a density operator $\hat{\rho}_A$ on A. If we interpret it as the *reduced* density operator of a (pure or mixed) state on some larger system AB, B not being specified yet, then its nonvanishing eigenvalues s_a^2 must be identical to those of the reduced density operator on B. If B is at least as large as A, then its basis has at least as many basis states as A. If we match the orthonormal eigenstates of $\hat{\rho}_A$, $|a\rangle_A$, one by one with orthonormal basis states of B, $|a\rangle_B$, then

$$|\psi\rangle_{AB} = \sum_a s_a |a\rangle_A |a\rangle_B \qquad (4.303)$$

is a pure state on AB in Schmidt-decomposed form. Tracing out B gives us the original density operator on A. $|\psi\rangle_{AB}$ is called the *purification* of $\hat{\rho}_A$. Purification is far from unique: first of all, the choice of the so-called *auxiliary* system B is free up to the restriction that its dimension must not be less than that of A. The most economical choice conceptually and computationally is simply to take B to be a copy of A (not to be confused with a copy of the state living on A). Thus, if A is a spin chain, we simply choose B to be a second spin chain of the same length, so AB can be visualized as a spin ladder. Generically, therefore, we will look at ladder-like structures in one dimension. The problem is, of course, that for quantum many-body systems, we usually do not know $\hat{\rho}_A$, so purification would remain a purely conceptual tool. As was first shown by Verstraete, Garcia-Ripoll, and Cirac [72], purification can be made to work very elegantly for *thermal density operators*. In the following, we will first discuss

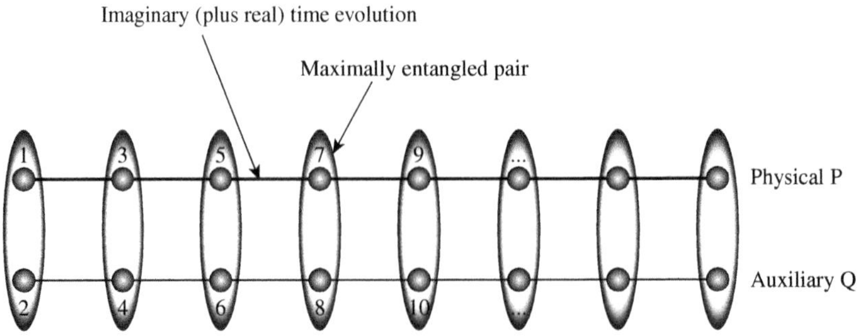

Fig. 4.24 Schematic representation of finite-temperature simulations: instead of a chain, we set up a ladder with an identical copy of the chain. Physical sites have odd labels and auxiliary sites even labels. Equivalent sites on the physical and auxiliary legs are linked by maximally entangled states. To reach inverse temperature β, an imaginary-time evolution is carried out up to "time" $-i\beta/2$.

the basic idea, as in [62], with a slightly updated notation, to prepare for the quite exciting recent algorithmic developments in this field [6, 7, 42], which will be outlined at the end.

To start with, we consider a physical system (state space) P and an *identical* auxiliary system (state space) Q (see Fig. 4.24). Purification then reads

$$\hat{\rho}_{\mathrm{P}} = \sum_{a=1}^{r} s_a^2 |a\rangle_{\mathrm{P}}\,_{\mathrm{P}}\langle a| \quad \rightarrow \quad |\psi\rangle = \sum_{a=1}^{r} s_a |a\rangle_{\mathrm{P}}|a\rangle_{\mathrm{Q}}, \qquad \hat{\rho}_{\mathrm{P}} = \mathrm{Tr}_{\mathrm{Q}}\,|\psi\rangle\langle\psi|. \quad (4.304)$$

We now consider a physical Hamiltonian $\hat{H}_{\mathrm{P}}$ acting on P. Consider the thermal density operator $\hat{\rho}_{\mathrm{P}}(\beta) = Z(\beta)^{-1}e^{-\beta\hat{H}_{\mathrm{P}}}$, with $Z(\beta) = \mathrm{Tr}_{\mathrm{P}}\,e^{-\beta\hat{H}_{\mathrm{P}}}$, and write

$$\hat{\rho}_{\mathrm{P}}(\beta) = Z(\beta)^{-1}e^{-\beta\hat{H}_{\mathrm{P}}} = Z(\beta)^{-1}e^{-\beta\hat{H}_{\mathrm{P}}/2} \cdot \hat{I}_{\mathrm{P}} \cdot e^{-\beta\hat{H}_{\mathrm{P}}/2}. \quad (4.305)$$

The identity $\hat{I}_{\mathrm{P}}$ is nothing other than $Z(0)\hat{\rho}_{\mathrm{P}}(0)$, the infinite-temperature density operator times the infinite-temperature partition function. Let us assume that we know the purification of $\hat{\rho}_{\mathrm{P}}(0)$ as an MPS on PQ, $|\psi(0)\rangle_{\mathrm{PQ}}$. Then

$$\hat{\rho}_{\mathrm{P}}(\beta) = \frac{Z(0)}{Z(\beta)}e^{-\beta\hat{H}_{\mathrm{P}}/2} \cdot \mathrm{Tr}_{\mathrm{Q}}\,|\psi(0)\rangle_{\mathrm{PQ}}\,_{\mathrm{PQ}}\langle\psi(0)| \cdot e^{-\beta\hat{H}_{\mathrm{P}}/2}$$

$$= \frac{Z(0)}{Z(\beta)}\,\mathrm{Tr}_{\mathrm{Q}}(e^{-\beta\hat{H}_{\mathrm{P}}/2} \otimes \hat{I}_{\mathrm{Q}})|\psi(0)\rangle_{\mathrm{PQ}}\,_{\mathrm{PQ}}\langle\psi(0)|(e^{-\beta\hat{H}_{\mathrm{P}}/2} \otimes \hat{I}_{\mathrm{Q}}).$$

The trace over Q can be pulled in front, since the Hamiltonian $\hat{H}_{\mathrm{P}}$ does not act on Q. But the result means that we just have to do an imaginary-time evolution

$$|\psi(\beta)\rangle_{\mathrm{PQ}} = (e^{-\beta\hat{H}_{\mathrm{P}}/2} \otimes \hat{I}_{\mathrm{Q}})|\psi(0)\rangle_{\mathrm{PQ}}, \quad (4.306)$$

which just acts on P. Expectation values are given by

$$\langle \hat{O} \rangle_\beta = \mathrm{Tr}_\mathrm{P}\, \hat{O} \hat{\rho}_\mathrm{P}(\beta)$$

$$= \frac{Z(0)}{Z(\beta)}\, \mathrm{Tr}_\mathrm{P}\, \hat{O}\, \mathrm{Tr}_\mathrm{Q}\, |\psi(\beta)\rangle_\mathrm{PQ}\, {}_\mathrm{PQ}\langle \psi(\beta)|$$

$$= \frac{Z(0)}{Z(\beta)}\, {}_\mathrm{PQ}\langle \psi(\beta)|\hat{O}|\psi(\beta)\rangle_\mathrm{PQ}. \tag{4.307}$$

The factor $Z(0)/Z(\beta) = d^L/Z(\beta)$ may seem difficult to obtain, but follows trivially from the expectation value of the identity,

$$1 = \langle \hat{I}_\mathrm{P} \rangle_\beta = \mathrm{Tr}_\mathrm{P}\, \hat{\rho}_\mathrm{P}(\beta) = \frac{Z(0)}{Z(\beta)}\, \mathrm{Tr}_\mathrm{P}\, \mathrm{Tr}_\mathrm{Q}\, |\psi(\beta)\rangle_\mathrm{PQ}\, {}_\mathrm{PQ}\langle \psi(\beta)|$$

$$= \frac{Z(0)}{Z(\beta)}\, {}_\mathrm{PQ}\langle \psi(\beta)|\psi(\beta)\rangle_\mathrm{PQ},$$

and hence $Z(\beta)/Z(0) = {}_\mathrm{PQ}\langle \psi(\beta)|\psi(\beta)\rangle_\mathrm{PQ}$, or, in complete agreement with standard quantum mechanics,

$$\langle \hat{O} \rangle_\beta = \frac{{}_\mathrm{PQ}\langle \psi(\beta)|\hat{O}|\psi(\beta)\rangle_\mathrm{PQ}}{{}_\mathrm{PQ}\langle \psi(\beta)|\psi(\beta)\rangle_\mathrm{PQ}}. \tag{4.308}$$

But this takes us right back to expressions that we know how to calculate. All we have to do is to find $|\psi(0)\rangle$ (we now drop the subscript), carry out imaginary-time evolution up to $-\mathrm{i}\beta/2$, and calculate expectation values as for a pure state. We can even subject the purified state $|\psi(\beta)\rangle$ to subsequent real-time evolutions, to treat time dependence at finite T, a topic we will discuss below.

We can also do thermodynamics quite simply, since $Z(\beta)/Z(0)$ is given by the square of the norm of $|\psi(\beta)\rangle$ and $Z(0) = d^L$. This means that we can obtain $Z(\beta)$ by keeping the purified state normalized at all temperatures and by accumulating normalization factors as temperature goes down and β increases. From $Z(\beta)$, we have $F(\beta) = -\beta^{-1}\ln Z(\beta)$. At the same time,

$$U(\beta) = \langle \hat{H}_\mathrm{P} \rangle_\beta = \frac{\langle \psi(\beta)|\hat{H}|\psi(\beta)\rangle}{\langle \psi(\beta)|\psi(\beta)\rangle}.$$

But this in turn gives us $S(\beta) = \beta[U(\beta) - F(\beta)]$. Further thermodynamic quantities follow similarly.

The purification of the infinite-temperature mixed state is a simple MPS of dimension 1, because it factorizes (if we take one ladder rung as a big site):

$$\hat{\rho}_\mathrm{P}(0) = \frac{1}{d^L}\hat{I}_\mathrm{P} = \left(\frac{1}{d}\hat{I}_{\mathrm{P, local}}\right)^{\otimes L}. \tag{4.309}$$

As $\hat{\rho}_P(0)$ factorizes, we can now purify the local mixed state on each physical site as a pure state on ladder rung i, to get $|\psi_i(0)\rangle$, and then

$$|\psi(0)\rangle_{PQ} = |\psi_1(0)\rangle|\psi_2(0)\rangle|\psi_3(0)\rangle\cdots,$$

a product state or an MPS of dimension 1. If we consider some rung i of the ladder, with states $|\sigma\rangle_P$ and $|\sigma\rangle_Q$ on the physical site $2i - 1$ and the auxiliary site $2i$, labeling the ladder sites 1 and 2 on the first rung, 3 and 4 on the second rung, and so on, we can purify as follows:

$$\frac{1}{d}\hat{I}_i = \sum_\sigma \frac{1}{d}|\sigma\rangle_P \ _P\langle\sigma|$$

$$= \mathrm{Tr}_Q \left[\left(\sum_\sigma \frac{1}{\sqrt{d}}|\sigma\rangle_P|\sigma\rangle_Q\right) \left(\sum_\sigma \frac{1}{\sqrt{d}} \ _P\langle\sigma| \ _Q\langle\sigma|\right) \right]. \tag{4.310}$$

Hence the purification is given by a maximally entangled state (the entanglement entropy is $\log_2 d$),

$$|\psi_i(0)\rangle = \sum_\sigma \frac{1}{\sqrt{d}}|\sigma\rangle_P|\sigma\rangle_Q. \tag{4.311}$$

If we introduce one set of A-matrices (here scalars) per rung, and consider as an example the case of a spin-$\frac{1}{2}$ chain, then the four "matrices" read

$$A^{\uparrow_P\uparrow_Q} = \sqrt{\tfrac{1}{2}}, \qquad A^{\uparrow_P\downarrow_Q} = 0, \qquad A^{\downarrow_P\uparrow_Q} = 0, \qquad A^{\downarrow_P\downarrow_Q} = \sqrt{\tfrac{1}{2}}. \tag{4.312}$$

These are local unitary transformations on both P and Q separately that leave maximal entanglement (and the induced maximal mixedness) invariant. This can be used to speed up calculations drastically: for example, for the purification of a spin-$\frac{1}{2}$ chain, the singlet state has good total spin and magnetization quantum numbers, and this allows to calculate in the subspace of total $S = 0$ and $S^z = 0$ only. It is also possible to go to A-matrices that are local to sites, not rungs, by SVD.

Beyond the calculation of thermodynamic quantities and static expectation values at finite temperature, we may also consider time-dependent expectation values, with the most important of these having the form

$$\langle \hat{B}_P(t)\hat{A}_P\rangle_\beta = Z(\beta)^{-1} \mathrm{Tr}\{e^{-\beta\hat{H}_P}\hat{B}_P(t)\hat{A}_P\}, \tag{4.313}$$

with operators in the Heisenberg picture, $\hat{B}_P(t) = e^{i\hat{H}_P t}\hat{B}_P e^{-i\hat{H}_P t}$. This might measure the response of the system to some perturbation $\hat{A}_P$ at time $t = 0$ through the observable $\hat{B}_P$ at some later time, or it might lead, after Fourier transformation, to some spectral or Green's function. In purified form, this reads

$$\langle \hat{B}_P(t)\hat{A}_P\rangle_\beta = Z(\beta)^{-1}\langle\psi(\beta)|e^{i\hat{H}_P t}\hat{B}_P e^{-i\hat{H}_P t}\hat{A}_P|\psi(\beta)\rangle$$

$$= Z(\beta)^{-1}\langle\psi(0)|e^{-\beta\hat{H}_P/2}e^{i\hat{H}_P t}\hat{B}_P e^{-i\hat{H}_P t}\hat{A}_P e^{-\beta\hat{H}_P/2}|\psi(0)\rangle. \tag{4.314}$$

In the first large-scale applications of this approach to finite-temperature dynamics [8], it was found that the timescales that could be reached were quite limited as in the $T = 0$ case.

An important observation is that there is now, as opposed to the case of $T = 0$, an additional freedom regarding the auxiliary system Q. Throughout, we have assumed a trivial time evolution on Q. However, we can take *any* nonsingular (invertible) operator $\hat{T}_\mathrm{Q}$ acting on Q only (a unitary time evolution on Q would be an important example) to obtain

$$\langle \hat{B}_\mathrm{P}(t)\hat{A}_\mathrm{P}\rangle_\beta = Z(\beta)^{-1}\langle\psi(0)|\hat{T}_\mathrm{Q}^{-1}\mathrm{e}^{-\beta\hat{H}_\mathrm{P}/2}\mathrm{e}^{\mathrm{i}\hat{H}_\mathrm{P}t}\hat{B}_\mathrm{P}\mathrm{e}^{-\mathrm{i}\hat{H}_\mathrm{P}t}\hat{A}_\mathrm{P}\mathrm{e}^{-\beta\hat{H}_\mathrm{P}/2}\hat{T}_\mathrm{Q}|\psi(0)\rangle. \quad (4.315)$$

As all other operators act on P, the positions of $\hat{T}_\mathrm{Q}$ and its inverse can be chosen freely. This additional freedom is of interest if we consider the time evolution of $\hat{\rho}_\mathrm{P}(\beta)$ under the von Neumann equation: because the thermal density operator commutes with $\hat{H}_\mathrm{P}$, it is constant in time. However, its purification, $|\psi(\beta)\rangle_\mathrm{PQ}$ is *not*, because it is not an eigenstate of $\hat{H}_\mathrm{P}$. Physically speaking, the real-time evolution builds up entanglement along the ladder, making the purification representation of $\hat{\rho}_\mathrm{P}$ computationally more and more inefficient with time even though the purified object does not change in time. The question is whether we can find operators $\hat{T}_\mathrm{Q}$ that counteract this entanglement build-up. As we will see, for the simple example just presented, this can be done exactly; in the more general case of the time evolution of a density operator after an operator has acted on it, finding the optimal $\hat{T}_\mathrm{Q}$ that limits resource usage most strongly is impossible. However, physical arguments lead to simple choices for $\hat{T}_\mathrm{Q}$ that are probably very close to the optimum.

Karrasch, Bardarson, and Moore [42] pointed out that a strong suppression of resource usage occurs if one chooses a time-dependent

$$\hat{T}_\mathrm{Q}(t) = \mathrm{e}^{\mathrm{i}\hat{H}_\mathrm{Q}t}, \quad (4.316)$$

where $\hat{H}_\mathrm{Q}$ is just the Hamiltonian $\hat{H}_\mathrm{P}$ acting on Q instead of P. Shifting the positions of $\hat{T}_\mathrm{Q}$ and its inverse next to the real-time evolutions on P, we see that this choice simply means that during the real-time evolution on P, we set up the same evolution on Q *with reversed time*:

$$\langle \hat{B}_\mathrm{P}(t)\hat{A}_\mathrm{P}\rangle_\beta = Z(\beta)^{-1}\langle\psi(0)|\mathrm{e}^{-\beta\hat{H}_\mathrm{P}/2}\mathrm{e}^{\mathrm{i}\hat{H}_\mathrm{P}t}\mathrm{e}^{-\mathrm{i}\hat{H}_\mathrm{Q}t}\hat{B}_\mathrm{P}\mathrm{e}^{-\mathrm{i}\hat{H}_\mathrm{P}t}\mathrm{e}^{\mathrm{i}\hat{H}_\mathrm{Q}t}\hat{A}_\mathrm{P}\mathrm{e}^{-\beta\hat{H}_\mathrm{P}/2}|\psi(0)\rangle. \quad (4.317)$$

Why does this improve the computational load and can we do even better? These questions can be answered most transparently by moving to a different representation of the purified state introduced in [6, 7]. As P and Q are chosen to be identical, $\mathcal{H}_\mathrm{P} = \mathcal{H}_\mathrm{Q} \equiv \mathcal{H}$. Then $|\psi(\beta)\rangle$ is defined on $\mathcal{H} \otimes \mathcal{H}$. There is an isomorphism between states $|\psi\rangle \in \mathcal{H}\otimes\mathcal{H}$ and linear maps (linear operators) $\hat{\Psi} \in \mathcal{B}(\mathcal{H}) : \mathcal{H} \mapsto \mathcal{H}$. With $|\{\sigma\}\rangle$ and $|\{\sigma'\}\rangle$ states in the tensored Hilbert spaces,

$$\langle\{\sigma\},\{\sigma'\}|\psi\rangle \equiv \langle\{\sigma\}|\hat{\Psi}|\{\sigma'\}\rangle. \quad (4.318)$$

For MPS and MPO, the isomorphism is explicitly given by the expressions

$$|\psi\rangle = \sum_{\{\sigma\},\{\sigma'\}} A^{\sigma_1,\sigma_1'} \dots A^{\sigma_L,\sigma_L'} |\{\sigma\},\{\sigma'\}\rangle, \tag{4.319}$$

$$\hat{\Psi} = \sum_{\{\sigma\},\{\sigma'\}} A^{\sigma_1,\sigma_1'} \dots A^{\sigma_L,\sigma_L'} |\{\sigma\}\rangle\langle\{\sigma'\}|, \tag{4.320}$$

where we interpret the same $A^{\sigma_\ell,\sigma_\ell'}$ as either the matrices for the physical and auxiliary local states on rung ℓ or the ingoing and outgoing physical local states on site ℓ. Pictorially, this can be represented as in Fig. 4.25.

Important entries in the dictionary are that

$$|\psi(0)\rangle \propto \sum_{\{\sigma\},\{\sigma'\}} |\{\sigma\},\{\sigma'\}\rangle \equiv |I\rangle \text{ is isomorphic to } \hat{I}, \text{ the identity operator.}$$

Then $|\psi(\beta)\rangle \propto e^{-\beta\hat{H}}|I\rangle$ is isomorphic to $e^{-\beta\hat{H}}$. Most important for us is

$$(\hat{P}\otimes\hat{Q})|\psi\rangle \quad \leftrightarrow \quad \hat{P}\hat{\Psi}\hat{Q}^{\mathrm{T}}, \tag{4.321}$$

which can be worked out analytically or observed pictorially: if we unfold the auxiliary legs of $|\psi\rangle$ into ingoing legs of the MPO of $\hat{\Psi}$, their direction changes from up to down: the in- and outgoing legs of $\hat{Q}$ have to be exchanged—hence the transpose.

In the new notation (where we can drop the superfluous index P), the conventional scheme

$$\langle\hat{B}(t)\hat{A}\rangle_\beta = Z(\beta)^{-1}\langle I|e^{-\beta\hat{H}/2}e^{\mathrm{i}\hat{H}t}\hat{B}e^{-\mathrm{i}\hat{H}t}\hat{A}e^{-\beta\hat{H}/2}|I\rangle \tag{4.322}$$

reads

$$\langle\hat{B}(t)\hat{A}\rangle_\beta = Z(\beta)^{-1}\,\mathrm{Tr}\big\{\big[e^{-\beta\hat{H}/2}e^{+\mathrm{i}\hat{H}t}\big]\hat{B}\big[e^{-\mathrm{i}\hat{H}t}\hat{A}e^{-\beta\hat{H}/2}\big]\big\}, \tag{4.323}$$

where the brackets $[\cdot]$ denote an MPO to be calculated (instead of a purified MPS). The trace appears because the bra $\langle I|$, translated into an MPO, closes the in- and

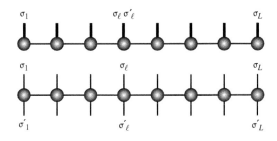

Fig. 4.25 Schematic representation of the isomorphism between states and linear maps. The thick vertical (physical) legs for the MPS correspond to pairs of local states that are "unfolded" into the ingoing and outgoing local states of an MPO.

outgoing legs of the MPO that arises from translating the remaining expression like a trace. Similarly, the scheme of [42] reads, in the new notation,

$$\langle \hat{B}(t)\hat{A}\rangle_\beta = Z(\beta)^{-1} \operatorname{Tr}\left\{ \left[e^{-i\hat{H}t} e^{-\beta\hat{H}/2} e^{+i\hat{H}t} \right] \hat{B} \left[e^{-i\hat{H}t} \hat{A} e^{-\beta\hat{H}/2} e^{i\hat{H}t} \right] \right\}. \qquad (4.324)$$

This representation is obtained by the translation rules, noting that the transpose of $e^{-i\hat{H}t}$ is again $e^{-i\hat{H}t}$, since for real $\hat{H}$, $\hat{H}^{\mathrm{T}} = \hat{H}$, and by commuting the order of evolution operators. It is now easy to see the advantages of this scheme: assume that we want to calculate $\langle \hat{B}(t)\rangle_\beta$, which will be a constant in time (since $\hat{B}$ is not explicitly time-dependent). For this, set $\hat{A} = \hat{I}$; then both MPOs collapse to $[e^{-\beta\hat{H}/2}]$, and no resources are wasted with time. This is optimal. More relevant, of course, is the case where $\hat{A}$ is nontrivial. The first MPO collapses to $[e^{-\beta\hat{H}/2}]$, and no time evolution has to be calculated. The second MPO has a nontrivial time dependence. If we rearrange it as $[e^{-i\hat{H}t}\hat{A}e^{i\hat{H}t}e^{-\beta\hat{H}/2}]$, we see that it contains a time evolution of $\hat{A}$. In most cases, $\hat{A}$ is local, and therefore the evolution is trivial and shows no growth of resources outside the light cone, i.e., those sites that "see" the effects of $\hat{A}$ being applied at $t = 0$ at some later time t. This is a far better use of resources than in the conventional scheme.

But we can do even better. Most recently, Barthel, Schollwöck, and Sachdev [7] (for details, see [6]) have found the following as the near-optimal scheme for the frequently occurring case $\hat{B}^\dagger = \hat{A}$:

$$\langle \hat{B}(2t)\hat{A}\rangle_\beta = Z(\beta)^{-1} \operatorname{Tr}\left\{ \left[e^{i\hat{H}t} e^{-\beta\hat{H}/2} \hat{B} e^{-i\hat{H}t} \right] \left[e^{-i\hat{H}t} \hat{A} e^{-\beta\hat{H}/2} e^{i\hat{H}t} \right] \right\}, \qquad (4.325)$$

where again the brackets delimit the MPOs to be calculated. Note the argument in $\hat{B}(2t)$: this scheme reaches twice the time as that in [42] for the same numerical resources. Only at very low temperatures does the conventional scheme seem to have a numerical advantage, as discussed in [6]. As opposed to the original purification scheme, we are now looking at the calculation of an MPO that is achieved by "Trotterizing" the exponentials with respect to imaginary or real time, contracting the time slices one by one, and compressing the resulting MPO, just as for an MPS, with the matrix 2-norm replacing the vector 2-norm in the calculation of the truncation error.

To get an idea of the power of the various schemes, consider Fig. 4.26, which shows which times can be reached at which temperature for the same amount of computational cost or maximally allowed MPO matrix dimension.

As an example of the current state of the art, we show in Fig. 4.27 the real part of the time-dependent correlator $\langle \hat{S}_k^z(t)\hat{S}_k^z(0)\rangle_\beta$ for an isotropic Heisenberg antiferromagnetic chain of length $L = 129$ at $\beta = 4$; this correlator is understood to be the momentum-space Fourier transform of $\langle \hat{S}_x^z(t)\hat{S}_0^z(0)\rangle_\beta$. Both real- and imaginary time evolutions were done using a fourth-order Trotter decomposition with time steps 0.125; results for acceptable truncated weights of 10^{-10} and 10^{-12} are shown. For such small truncated weights, the results are essentially exact and agree for both truncated weights on the timescales reached by each one. The larger truncated weight can be maintained by the algorithm somewhat longer, extending the maximally reached time.

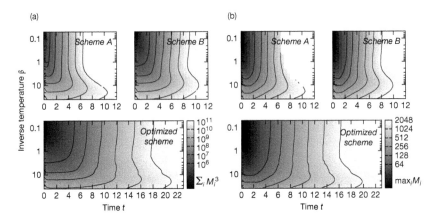

Fig. 4.26 Reachable times for the isotropic Heisenberg model with $\hat{B}$, $\hat{A}$, and $\hat{S}^{\pm}$ at the chain center for the three calculation schemes presented. Scheme A is the conventional scheme and scheme B is that of Karrasch, Bardarson, and Moore [42]. The plots in (a) show computational cost (proportional to the sum of the third power of the matrix dimensions), and those in (b) the maximal matrix dimension. The superiority of the optimized scheme is clearly visible. (Reprinted from [6].)

4.14.3   Minimally entangled typical thermal states

The purification method is now very powerful, given recent algorithmic advances, but nevertheless simulations become difficult at very low temperatures $T \to 0$. In this limit, the mixed state living on the physical system P will evolve toward the pure-state projector on the ground state, $\hat{\rho}^{\mathrm{P}}(\infty) = |\psi(\infty)\rangle_{\mathrm{P}\,\mathrm{P}}\langle\psi(\infty)|$ (here we use $|\psi(\infty)\rangle_{\mathrm{P}}$ as the ground state of the physical Hamiltonian $\hat{H}$, to refer to $\beta = \infty$ as in the purification section). Using the isomorphism of the last section, the purification is given by the product state

$$|\psi(\infty)\rangle = |\psi(\infty)\rangle_{\mathrm{P}}|\psi(\infty)\rangle_{\mathrm{Q}}, \qquad (4.326)$$

where the latter state is just the physical ground state defined on the auxiliary state space. Assuming that it can be described with sufficient precision using matrix dimension D, the product will be described by matrices of dimension D^2. Effectively, this means that our algorithm scales with the sixth instead of the third power of the characteristic matrix dimension for the problem under consideration. So there is lively interest in alternatives to the purification approach.

As White [80] has pointed out, one can avoid the purification approach entirely by sampling over a cleverly chosen set of thermal states, the so-called *minimally entangled typical thermal states* (METTS). This approach has already been shown to alleviate strongly the $T \to 0$ problem, although not much is yet known regarding how the best purification approaches compete with METTS for $T \to \infty$. We will limit the discussion here to an explanation of the conceptual basics, following [62]; for a very good account of the technical details as well as the concepts, see [67].

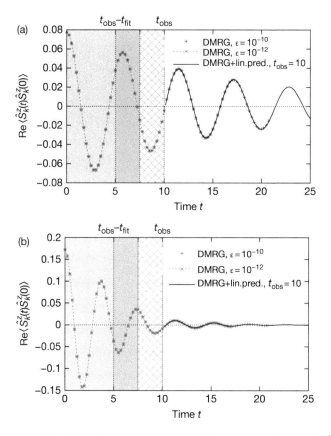

Fig. 4.27 Real parts of the $k = 0.25\pi$ (a) and 0.5π (b) Fourier transforms of $\langle \hat{S}_x^z(t)\hat{S}_0^z(0)\rangle_\beta$ at $\beta = 4$. The crosses correspond to different acceptable truncated weights ϵ per time step, leading to different maximally reachable times. The solid line is the result of the prediction technique explained further in the text: long-time data were predicted from data in the dark gray and cross-hatched time ranges, discarding data in the light gray range. Note that predicted and calculated data agree perfectly.

A thermal average is given by

$$\langle \hat{A}\rangle = \frac{1}{Z}\mathrm{Tr}\, \mathrm{e}^{-\beta\hat{H}}\,\hat{A} = \frac{1}{Z}\sum_n \mathrm{e}^{-\beta E_n}\langle n|\hat{A}|n\rangle, \qquad (4.327)$$

where we have chosen, like all textbooks do, the energy representation of the thermal density operator. As pointed out by Schrödinger many decades ago, this is mathematically correct, but unphysical in the sense that real systems at finite temperature will usually not be in a statistical mixtures of eigenstates, since eigenstates are highly fragile under coupling to an environment. But the choice of the basis for taking the trace is arbitrary, and we may also write

$$\langle A \rangle = \frac{1}{Z} \sum_i \langle i | e^{-\beta \hat{H}/2} \hat{A} e^{-\beta \hat{H}/2} | i \rangle = \frac{1}{Z} \sum_i P(i) \langle \phi(i) | \hat{A} | \phi(i) \rangle \tag{4.328}$$

where $\{|i\rangle\}$ is an arbitrary orthonormal basis and $|\phi(i)\rangle = P(i)^{-1/2} e^{-\beta \hat{H}/2} |i\rangle$. With $P(i) = \langle i | e^{-\beta \hat{H}} | i \rangle$, we recognize $|\phi_i\rangle$ to be normalized. It is easy to see that $\sum_i P(i) = Z$, and hence the $P(i)/Z$ are probabilities. We can therefore statistically estimate $\langle \hat{A} \rangle$ by *sampling* $|\phi(i)\rangle$ with probabilities $P(i)/Z$ and average over $\langle \phi(i) | \hat{A} | \phi(i) \rangle$.

Several questions arise before this can be turned into a practical algorithm. How can we sample correctly, given that we do not know the complicated probability distribution? Can we choose a set of states such that averages converge most rapidly? Given that an imaginary-time evolution will be part of the algorithm, can we find a low-entanglement basis $\{|i\rangle\}$ such that time evolution can be done with modest D, i.e., run fast?

To address these issues, White chooses as orthonormal basis the computational basis formed from product states,

$$|i\rangle = |i_1\rangle |i_2\rangle |i_3\rangle \cdots |i_L\rangle, \tag{4.329}$$

classical (unentangled) product states (CPS) that can be represented exactly as an MPS with $D = 1$. The corresponding states

$$|\phi(i)\rangle = \frac{1}{\sqrt{P(i)}} e^{-\beta \hat{H}/2} |i\rangle \tag{4.330}$$

are METTS: the hope is that while the imaginary-time evolution introduces entanglement through the action of the Hamiltonian, it is a reasonable expectation that the final entanglement will be lower than for similar evolutions of already entangled states. While this is not totally true in a strict mathematical sense, in a practical sense it seems to be! Compared with purification, this will be a much faster computation, in particular because the factorization issue of purification will not appear.

In order to sample the $|\phi(i)\rangle$ with the correct probability distribution, which we cannot calculate, we use the same trick as in Monte Carlo and generate a Markov chain of states, $|i_1\rangle \to |i_2\rangle \to |i_3\rangle \to \ldots$ such that the correct probability distribution is reproduced. From this distribution, we can generate $|\phi(i_1)\rangle, |\phi(i_2)\rangle, |\phi(i_3)\rangle, \ldots$ for calculating the average.

The algorithm runs as follows. We start with a random CPS $|i\rangle$. From this, we repeat the following three steps until the statistics of the result is good enough:

1. Calculate $e^{-\beta \hat{H}/2} |i\rangle$ by imaginary-time evolution and normalize the state (the squared norm is $P(i)$, but we will not need it in the algorithm).
2. Evaluate the desired quantities as $\langle \phi(i) | \hat{A} | \phi(i) \rangle$ for averaging.
3. Collapse the state $|\phi(i)\rangle$ to a new CPS $|i'\rangle$ by quantum measurements with probability $p(i \to i') = |\langle i' | \phi(i) \rangle|^2$, and restart with this new state.

Let us convince ourselves that this gives the correct sampling, following [67]. As the $|\phi(i)\rangle$ follow the same distribution as the $|i\rangle$, we only have to show that the latter are sampled correctly. Asking for the probability with which we collapse into some $|j\rangle$ provided the previous CPS $|i\rangle$ was chosen with the right probability $P(i)/Z$, we find

$$\sum_i \frac{P(i)}{Z} p(i \to j) = \sum_i \frac{P(i)}{Z} |\langle j|\phi(i)\rangle|^2$$

$$= \sum_i \frac{1}{Z} |\langle j|e^{-\beta\hat{H}/2}|i\rangle|^2$$

$$= \frac{1}{Z} \langle j|e^{-\beta\hat{H}}|j\rangle$$

$$= \frac{P(j)}{Z}. \tag{4.331}$$

This shows that the desired distribution is a fixpoint of the update procedure. It is therefore valid, but it is of course sensible to discard, as in Monte Carlo, a number of early data points, to eliminate the bias due to the initial CPS. It turns out that—after discarding the first few METTS, to eliminate effects of the initial choice—averaging quantities over only a hundred or so allows to calculate local static quantities (magnetizations and bond energies) with high accuracy.

While we already know how to do an imaginary-time evolution, we still have to discuss the collapse procedure. As it turns out, the structure of MPS can be exploited to make this part of the algorithm extremely fast compared with the imaginary-time evolution.

For each site i, we choose an *arbitrary* d-dimensional orthonormal basis $\{|\tilde{\sigma}_i\rangle\}$, to be distinguished from the computational basis $\{|\sigma_i\rangle\}$. From this, we can form projectors $\hat{P}^{\tilde{\sigma}_i} = |\tilde{\sigma}_i\rangle\langle\tilde{\sigma}_i|$, with the standard quantum mechanical probability of a local collapse into the state $|\tilde{\sigma}_i\rangle$ being given by $p_{\tilde{\sigma}_i} = \langle\psi|\hat{P}^{\tilde{\sigma}_i}|\psi\rangle$. If we collapse $|\psi\rangle$ into the CPS $|\psi'\rangle = |\tilde{\sigma}_1\rangle|\tilde{\sigma}_2\rangle\cdots|\tilde{\sigma}_L\rangle$, then the probability is given by $\langle\psi|\hat{P}^{\tilde{\sigma}_1}\cdots\hat{P}^{\tilde{\sigma}_L}|\psi\rangle = |\langle\psi'|\psi\rangle|^2$, as demanded by the algorithm. After a single-site collapse, the wavefunction reads

$$|\psi\rangle \to p_{\tilde{\sigma}_i}^{-1/2} \hat{P}^{\tilde{\sigma}_i}|\psi\rangle, \tag{4.332}$$

where the prefactor ensures proper normalization of the collapsed state as in elementary quantum mechanics. To give an example, for $S = \frac{1}{2}$ spins measured along an arbitrary axis $\mathbf{n}$, the projectors would read

$$\hat{P}^{\uparrow_n,\downarrow_n} = \frac{1}{2} \pm \mathbf{n}\cdot\hat{\mathbf{S}}_i. \tag{4.333}$$

Such a sequence of local measurements and collapses on all sites can be done very efficiently in the MPS formalism, as detailed in [67]. An important conceptual point is that at each site, the measurement basis can be chosen randomly. In order to obtain short autocorrelation "times" of the Markov chain, i.e., high quality of the sampling, this is certainly excellent, but also much more costly than always collapsing

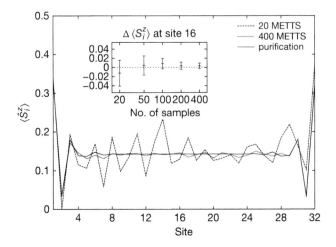

Fig. 4.28 Average local magnetization $\langle \hat{S}_i^z \rangle$ for an isotropic antiferromagnetic Heisenberg chain of $L = 32$ in a magnetic field $H = J = 1$ at temperature $\beta = 4$. Results for 20 and 400 METTS are compared with a high-precision purification calculation. The inset shows an error estimate for various numbers of METTS obtained from the statistics of bin averages.

into the same basis, which, however, generates ergodicity problems. It is very efficient to switch in an alternating manner between two bases where, for each basis, projectors are maximally mixed in the other basis (e.g., if we measure spins in an alternating manner along the x and z (or y) axis). Autocorrelation times then may go down to five steps or so [67]. For estimating the statistical error, in the simplest cases, it is enough to calculate averages over bins larger than the autocorrelation time, and to look at the statistical distribution of these bin averages to get an error bar.

Let us conclude this section by looking at an application of METTS. We consider an isotropic Heisenberg antiferromagnet on a chain of length $L = 32$ in a magnetic field $H = J = 1$ at an inverse temperature $\beta = 4$. The imaginary-time evolution is done in such a way that any errors from there are negligible. The remaining error is entirely due to the METTS sampling. In Fig. 4.28, METTS results for local magnetization on each site are compared with a highly accurate purification-based calculation. We can see how the method approaches the purification result (essentially exact here) rapidly with the number of METTS. At the moment, however, it seems that in most cases purification calculations are more efficient than METTS calculations [10].

4.14.4 Dissipative dynamics: quantum jumps

The simulation of Lindblad equations is quite easy in the MPS formalism, in particular using MPOs [72], but also in the form of a superoperator formalism [87]. A very attractive alternative, which allows maximal reuse of available pure-state codes, has been proposed by Daley et al. [19]. This combines pure-state time evolution with the method of quantum trajectories, which we have described earlier in this chapter.

In the context of MPS-based pure-state evolutions, it is convenient to replace the dt time steps of the original quantum trajectory algorithm by longer coherent evolutions, ended by quantum jumps, whose duration is chosen statistically such as to reproduce the original statistics. The algorithm then proceeds by generating $\mathcal{N}$ quantum trajectories in a time interval $[0, T]$ (where T is the final time of the simulation) as follows:

1. Generate a starting state $|\psi(0)\rangle$; it either samples the $t = 0$ density operator correctly or is simply always the same, depending on the physical problem.
2. Choose a uniformly distributed random number p_0 in $[0, 1]$.
3. Carry out, using a pure-state time evolution method, the time evolution of $|\psi(0)\rangle$ under $\hat{H}_{\text{eff}}$. As the effective Hamiltonian is non-Hermitian, the norm of the state will decrease over time. Stop the time evolution at time t_1, which is defined by $\langle \psi(t_1)|\psi(t_1)\rangle = 1 - p_0$; this is the time of the first quantum jump. Note that if $T < t_1$, our simulation stops at T and we have a trajectory without jump, and we normalize the final state.
4. To carry out the quantum jump at t_1, calculate

$$\tilde{p}_j = \langle \psi(t_1)|\hat{L}^{j\dagger}\hat{L}^j|\psi(t_1)\rangle, \qquad p_j = \frac{\tilde{p}_j}{\sum_{j>0} \tilde{p}_j} \qquad (j > 0) \qquad (4.334)$$

and choose a j according to the normalized probability distribution $\{p_j\}$.
5. Carry out this jump and normalize the state:

$$|\psi(t_1^+)\rangle = \frac{\hat{L}^j|\psi(t_1)\rangle}{\|\hat{L}^j|\psi(t_1)\rangle\|}. \qquad (4.335)$$

6. Continue with finding a new p_0, from which time evolution of $|\psi(t_1^+)\rangle$ with $\hat{H}_{\text{eff}}$ generates t_2, the location of the second quantum jump, and so on, until T is exceeded.

Physical expectation values

$$\frac{\langle \psi(t)|\hat{O}|\psi(t)\rangle}{\langle \psi(t)|\psi(t)\rangle}$$

(be careful: states are not always normalized here!) up to time T are now averaged over the $\mathcal{N}$ quantum trajectories that have been generated. Obviously, a careful analysis of convergence in $\mathcal{N} \to \infty$ has to be carried out, but it seems that for a small number of jump operators, even a few hundred trajectories may give highly reliable results [19]. If we use a Trotter decomposition, the local Hamiltonian to be diagonalized is a small non-Hermitian matrix, but this presents no practical problem. If we use a Krylov-based approach, tridiagonalization breaks down because of non-Hermiticity. Technically, this means that instead of a three-state recursion, the full Gram–Schmidt orthogonalization procedure has to be carried out at each step. Instead of a tridiagonal matrix (and ignoring the additional compression issues), we then obtain an upper-Hessenberg matrix (top-right half, diagonal, and first lower subdiagonal full), for which also very efficient techniques of diagonalization exist (the corresponding large sparse matrix methods are referred to as Arnoldi instead of Lanczos methods).

4.15 Linear response in the MPS formalism

The calculation of Green's and spectral functions in the MPS formalism essentially rephrases the algorithms we have already described (Lanczos vector dynamics, correction vector dynamics, and Chebyshev vector dynamics) in MPS language. This section will discuss the essential modifications. At the moment, Chebyshev vector dynamics as presented in [83] is the best combination of efficiency and precision. A fourth method, which we have only alluded to so far, is to calculate real-time correlators that are then transformed into frequency space by a Fourier transformation to yield frequency-dependent spectral functions. This is particularly useful in MPS, since the system sizes that are treatable are big enough for small finite-size effects. However, finite timescales are a nuisance, although the prediction method, discussed at the end, can help in many cases of interest. Currently, Chebyshev vector dynamics and real-time dynamics seem to be similarly efficient, but the issue is somewhat open (see the discussion below).

As with full Hilbert-space methods, finite system size is an issue. While system sizes are typically much larger (up to a few 100 sites are possible), MPS are most efficient for open boundary conditions, which imply $O(1/L)$ corrections that are much harder to remove than finite-size effects on rings, which often vanish exponentially fast. This problem is particularly pressing for dynamical structure functions $S(k, \omega)$ with a momentum dependence. The underlying perturbing operator is then spatially delocalized. The open boundary conditions introduce hard cuts in real space and and therefore large spreads in momentum space, leading to quite imprecise results. Kühner and White [43] alleviate this problem by filtering the perturbing operator through a filter function that dampens out contributions from the edges. Ideally, a filter function should exclude as few sites as possible while still being narrow in momentum space. In the more benign case of a local perturbing operator (as in conductivity calculations), we have to make sure that excitations do not see the chain ends. This can be achieved by the previously discussed η-damping, where we set the real frequency ω to be $\omega + i\eta$, where η is a small positive number. To approximate thermodynamic limit results, we have to take the limit $L \to \infty$ first and $\eta \to 0$ second. In our context, it was pointed out by Jeckelmann [41] that if one knows the typical propagation velocity c of the system, one may take a single limit by choosing

$$\eta(L) \geq c/L, \tag{4.336}$$

so that the finiteness of the system is not seen. For more detailed expositions, see the original paper [41] or the review [61].

4.15.1 Lanczos vector dynamics

The continued fraction method due to Gagliano and Balsiero [27] was first implemented in the DMRG context by Hallberg [33]. We can evaluate the continued fraction directly or exploit the fact that the generation of Lanczos vectors iteratively brings the

Hamiltonian into tridiagonal form in a new (incomplete) orthonormal basis. We may as well diagonalize this effective Hamiltonian to obtain an approximate energy eigenbasis of $\hat{H}$, $\{|n\rangle\}$ with eigenenergies E_n. The Green's function can be written within this approximation as

$$G_A(\omega + i\eta) = \sum_n \langle 0|\hat{A}^\dagger|n\rangle\langle n|\frac{1}{E_0 + \omega + i\eta - E_n}|n\rangle\langle n|\hat{A}|0\rangle, \qquad (4.337)$$

where the sum runs over all approximate eigenstates. The dynamical correlation function is then given by

$$C_A(\omega + i\eta) = \frac{\eta}{\pi}\sum_n \frac{|\langle n|\hat{A}|0\rangle|^2}{(E_0 + \omega - E_n)^2 + \eta^2}, \qquad (4.338)$$

where the matrix elements in the numerator are simply the expansion coefficients of the approximate eigenstates $|n\rangle$ in the first new basis state $|q_1\rangle = \hat{A}|0\rangle/\|\hat{A}|0\rangle\|$ times the norm $\|\hat{A}|0\rangle\|$.

As already discussed, the limitation of this method is mainly due to the loss of orthogonality of the Lanczos vectors, which is only ensured on a mathematical, not a numerical, level. In the DMRG context, Kühner and White [43] have proposed to monitor, for *normalized* Lanczos vectors $|q_i\rangle$, $\langle q_1|q_n\rangle > \epsilon$ as termination criterion in the nth iteration. So, again, the method works best for quite well-defined excitation bands, not continua.

It is crucial that beyond the above complications, for exact diagonalization with an exact $\hat{H}$, additional approximations are introduced in the original form of DMRG, since the Hamiltonian is not exact there. Moreover, the Lanczos states themselves are approximated, since DMRG cannot approximate them individually but only as a group of states: as the DMRG version of the Lanczos vector approach is in my view obsolete, let me just state that DMRG finds itself in the quandary of either describing the first few Lanczos vectors very accurately and all others quite badly, or all of them at the same intermediate quality. The first approach is fine if in the spectral functions the first few states carry the highest weight. This is usually only the case for very pronounced excitation bands, but not for excitation continua such as in the spin-$\frac{1}{2}$ Heisenberg antiferromagnetic chain.

In the MPS formulation, the same algorithm becomes much more efficient and reliable [20]. The Hamiltonian is represented exactly in MPO form. Subsequent applications of $\hat{H}$ increase the MPS dimension such that compression has to occur. Of course, compressions for higher Lanczos vectors depend on the earlier ones, but now each vector is at least approximated individually within the MPS framework and the available resources. The price to pay is that compression definitely kills orthonormality, so the full matrix $\langle q_i|H|q_j\rangle$ has to be calculated and diagonalized. But this is a small price to pay for a surprising increase in accuracy over the original formulation of the method for DMRG (for details, see [20]); for the first time (Fig. 4.29), it was possible to obtain the spinon continuum of the spin-$\frac{1}{2}$ Heisenberg antiferromagnetic chain not only without strong artifacts, but also in very good quantitative agreement with Bethe ansatz results, within the Lanczos vector method.

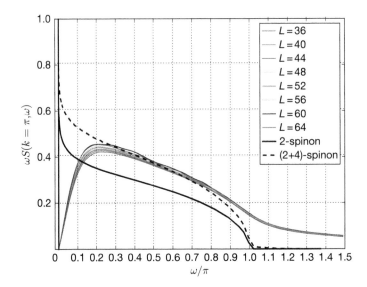

Fig. 4.29 Isotropic spectral function $\omega S(q = \pi, \omega)$ of the $S = \frac{1}{2}$ Heisenberg antiferromagnetic chain from Lanczos vector dynamics in the MPS formalism for various system sizes. The 2-spinon and $(2 + 4)$-spinon lines are from Bethe ansatz calculations in the thermodynamic limit; the latter should essentially represent the exact result for the spectral function. (Reprinted with permission from [20]. Copyright (2012) by the American Physical Society.)

4.15.2 Correction vector dynamics

The correction vector approach was first applied in the DMRG framework in various degrees of refinement by Ramasesha et al. [57], Kühner and White [43], and Jeckelmann [41]. Today, we use it in the MPS formalism, where the joint-basis issue of DMRG briefly hinted at simply does not exist. The Hamiltonian is exactly expressed, and if states become too large, they are compressed, so their approximation is individual.

As has been shown by Kühner and White [43], it is not necessary to calculate a very dense set of correction vectors in ω-space to obtain the spectral function for an entire frequency interval, mainly because of the finite broadening η. Comparing the results for some ω obtained by starting from neighboring anchoring frequencies allows for excellent convergence checks. This method is, for example, able to provide a high-precision result for the spinon continuum in the $S = \frac{1}{2}$ Heisenberg chain.

Reducing the broadening factor η, it is also possible to resolve finite-system peaks in the spectral function, obtaining to some approximation both location and weight of the Green's function's poles.

4.15.3 Chebyshev vector dynamics

Holzner et al. [39] were the first to use Chebyshev vector dynamics in the MPS context. Major parts of the algorithm remain unchanged, but the use of MPS introduces new issues. Here we will just highlight the major conceptual issues; for the technicalities, see

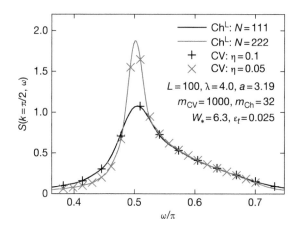

Fig. 4.30 Isotropic spectral function $\omega S(q = \pi/2, \omega)$ of the $S = \frac{1}{2}$ Heisenberg antiferromagnetic chain from Chebyshev vector dynamics in the MPS formalism versus correction vector dynamics (CV) for two different dampings η and corresponding numbers of Chebyshev vectors ($N = 111\,222$) for identical broadening. Agreement is excellent, despite the fact that matrix dimension in the correction vector method (1000) is much larger than that in the Chebyshev vector method (32). (Reprinted from [39]. Copyright (2011) by the American Physical Society.)

the original paper [39]. The calculation of the Chebyshev vectors $|t_n\rangle$ rests essentially on a recursion relations where $\hat{H}$ is applied to a state and another state is added. The result therefore has a much higher matrix dimension and has to be compressed without losing too much accuracy. This could potentially imply a runaway growth of the matrix dimension just as for time evolution. It turns out that this is not the case, for reasons that are not completely understood at present. A handwaving argument would point to the fact that the spectral function is encoded in a single MPS in , for example, the correction vector method, whereas the information is now "spread" over many states $|t_n\rangle$ such that each of them can be approximated in a satisfying manner by very small D.

At the same time, even if obviously not too much accuracy is lost, state compression may lead to leakage of the spectral function: in order to achieve reasonable frequency resolution, we do not map the entire, extensive energy bandwidth of the Hamiltonian into the interval $[-1, 1]$, but only the much smaller finite region where the spectral function is nonvanishing. State compression may introduce artificial high-energy components into the state, leading eventually to diverging Chebyshev vectors owing to the strong growth of the Chebyshev polynomials outside the interval $[-1, 1]$. This problem is remedied by a so-called energy truncation that forces the Chebyshev vector to have contributions only from states in the allowed energy range to a sufficiently good approximation.

Figure 4.30 shows a practical application to the isotropic Heisenberg model done in [39]. The accuracy matches that of the very reliable correction vector method, but at a much lower cost: the entire spectral function was calculated 25 times faster than a single point in the correction vector method.

The Chebyshev vector approach exploiting linear prediction pioneered in [28, 83, 84] does not resort to energy truncation, but chooses the interval into which the spectral function is mapped to be smaller to avoid leakage, and calculates as many Chebyshev vectors and moments as possible while keeping a fixed accuracy in the MPS representation of the Chebyshev vectors and the moments. The moments are then extrapolated using linear prediction (see Section 4.15.4), and a highly accurate spectral function is obtained. Overall, another couple of orders of magnitude in speed can be gained and many fewer moments have to be calculated, and this approach is in my view the most accurate, conceptually cleanest, and fastest way of calculating spectral functions in frequency space; it should be favored over all other competing approaches, including that of Holzner et al. [39]; for details, see [83]. In this new approach to Chebyshev polynomials in the framework of MPS, it was for the first time possible to carry out two-band dynamical mean-field theory (DMFT) calculations with DMRG as an impurity solver, both in the case of half-filling [28] and in the more demanding case of small doping [84]; this would have been impossible using any of the previous methods within competitive calculation times. It gives pause for thought, however, that Wolf et al. [83] also observed that Chebyshev polynomials can be used very efficiently to set up a real-time evolution directly from the Chebyshev moments, although this is not as efficient (but not by a wide margin) as a conventional time evolution with MPS. Given that the latter can be combined with linear prediction, too, and yields spectral functions after Fourier transformations, it is not quite clear at present whether it is not the most efficient technique after all to use real-time evolution plus Fourier transformation, when it comes to the calculation of spectral functions! Time will tell.

4.15.4   Linear prediction and spectral functions

The advent of powerful MPS-based time evolution methods (namely, time-dependent DMRG) has made it possible to calculate real-time real-space correlators such as $\langle \hat{S}_i^+(t)\hat{S}_j^-(0)\rangle$, and to carry out a double Fourier transform to momentum and frequency space. This approach, pioneered in [82], has the advantage that it can extend to finite T seamlessly, but it suffers from the limitations of reachable length and timescales. Of these, the limitations in time are much more serious, because of the rapid growth of entanglement in time. The timescales reachable are mostly so limited that a naive Fourier transform gives strong aliasing or that one has to introduce a windowing of the raw data that smears out spectral information quite strongly. This limitation can be circumvented, however, at very low numerical cost by a linear prediction technique both at $T = 0$ [52, 81] and $T > 0$ [8] that extends reachable t and thereby greatly refines results in the frequency domain.

For a time series of complex data $x_0, x_1, \ldots, x_n, \ldots, x_N$ at equidistant points in time $t_n = n\Delta t$ (and maximal time $t_{\text{obs}} := N\Delta t$) obtained by DMRG, one makes a prediction of $x_{N+1}, x_{N+2}, \ldots$ For the data points beyond $t = t_{\text{obs}}$, linear prediction gives the ansatz

$$\tilde{x}_n = -\sum_{i=1}^{p} a_i x_{n-i}. \tag{4.339}$$

The (predicted) value $\tilde{x}_n$ at time step n is assumed to be a linear combination of p previous values $\{x_{n-1}, \ldots, x_{n-p}\}$. Once the a_i have been determined from known data, they are used to calculate (an approximation of) all x_n with $n > N$.

The coefficients a_i are determined by minimizing the least-square error in the predictions over a subinterval of data points x_n, with $n \in \Omega$, such that data points are considered up to the maximum time reached, excluding short times (say, the first half of the entire time interval) for spurious effects, giving an effective fitting window. In the simplest approach, we minimize $E \equiv \sum_{n\in\Omega} |\tilde{x}_n - x_n|^2$. Minimization of E with respect to a_i yields the linear system

$$R\vec{a} = -\vec{r}, \tag{4.340}$$

where R and $\vec{r}$ are the autocorrelations,

$$R_{ji} = \sum_{n\in\Omega} x^*_{n-j} x_{n-i}, \qquad r_j = \sum_{n\in\Omega} x^*_{n-j} x_n.$$

Equation (4.340) is solved by $\vec{a} = -R^{-1}\vec{r}$.

One may wonder why general extrapolation toward infinite time is possible in this fashion. As demonstrated below, linear prediction generates a superposition of oscillating and exponentially decaying (or growing) terms, a type of time dependence that emerges naturally in many-body physics: Green's functions of the typical form $G(k,\omega) = [\omega - \epsilon_k - \Sigma(k,\omega)]^{-1}$ are in time–momentum representation dominated by the poles; for example, for a single simple pole at $\omega = \omega_1 - i\eta_1$ with residue c_1, the Green's function will read $G(k,t) = c_1 e^{-i\omega_1 t - \eta_1 t}$, and similarly it will be a superposition of such terms for more complicated pole structures. Often only few poles matter, and the ansatz of the linear prediction is well suited for the typical properties of the response quantities in which we are interested. Where such an ansatz does not hold, the method is probably inadequate: this is the case, for example, for the Green's function in real space, which does not have a suitable analytical structure, or the Chebyshev expansions of step functions as discussed previously. This is the point where we have secretly introduced additional analytical information.

To see the special form of time series generated by the prediction, we introduce vectors $\vec{x}_n := [x_n, \ldots, x_{n-p+1}]^{\mathrm{T}}$ such that (4.339) takes the form

$$\tilde{\vec{x}}_{n+1} = A\vec{x}_n, \tag{4.341}$$

with

$$A \equiv \begin{bmatrix} -a_1 & -a_2 & -a_3 & \cdots & -a_p \\ 1 & 0 & 0 & \cdots & 0 \\ 0 & 1 & 0 & \cdots & 0 \\ \vdots & \ddots & \ddots & \ddots & \vdots \\ 0 & \cdots & 0 & 1 & 0 \end{bmatrix}, \tag{4.342}$$

with the a_i as the elements of the vector $\vec{a}$ found above. Prediction therefore corresponds to applying powers of A to the initial vector $\vec{x}_N$. A (non-Hermitian) eigenvector decomposition of A with eigenvalues α_i leads to

$$\tilde{x}_{N+m} = [A^m \vec{x}_N]_1 = \sum_{i=1}^{p} c_i \alpha_i^m, \qquad (4.343)$$

where the coefficients c_i are determined from $\vec{x}_N$ and the eigenvectors of A. The eigenvalues α_i encode the physical resonance frequencies and dampings. The connection to exponentials is given by expressing $\alpha_i = e^{i\omega_i \Delta t - \eta_i \Delta t}$, revealing that linear prediction *models a superposition of exponential decays and oscillations.* Spurious $|\alpha_i| \geq 1$ may appear, but can be dealt with [8].

As a demanding example without a simple pole structure of the self-energy, we consider the spinon continuum of an isotropic antiferromagnetic $S = \frac{1}{2}$ Heisenberg chain; see Fig. 4.31. In the zero-temperature limit, the results agree extremely well with Bethe ansatz results [16]. At finite temperatures, simulations at different precisions indicate that results are fully converged and essentially exact. Note that for a good prediction we do not extrapolate $S(x,t) = \langle \hat{S}_x^z(t) \hat{S}_0^z(0) \rangle_\beta$, which does not work properly, but its Fourier transform in momentum space, $S(k,t)$, because the former does not fit into the analytical framework of prediction.

Figure 4.32 shows the consequences of the choice of the fitting window. The chain is of length $L = 129$ at $\beta = 4$; both real- and imaginary-time evolutions were calculated using a fourth-order Trotter decomposition with time steps 0.125 and truncated weight of eigenvalues 10^{-10} (checked by shorter time runs at truncated weight 10^{-12}) using

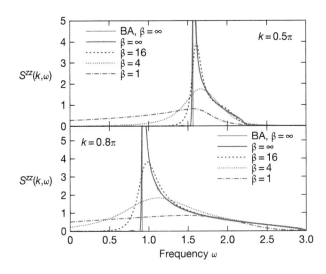

Fig. 4.31 Spectral functions of the isotropic Heisenberg chain at two momenta and four different temperatures. At $T = 0$, Bethe ansatz (BA) and numerics agree extremely well. (Adapted with permission from [8]. Copyrighted by the American Physical Society.)

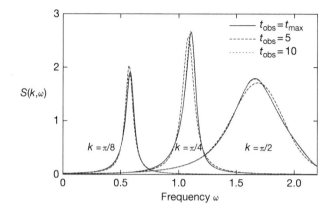

Fig. 4.32 Spectral functions of the isotropic Heisenberg antiferromagnet at various momenta k for three different fitting intervals $[t_{obs}/2, t_{obs}]$, where $t_{max} = 17$. Note that results obtained for the longer fitting intervals are independent of the chosen fitting interval, and are quasi-exact.

the most advanced scheme from [6]. The prediction ansatz was evaluated using various fitting intervals $[t_{obs}/2, t_{obs}]$. All three give essentially identical, and hence robust, results; only the one for the shortest t_{obs} deviates slightly, owing to the fewer available data points and short-time effects.

References

[1] Affleck, I., Kennedy, T., Lieb, E. H., and Tasaki, H. (1987). *Phys. Rev. Lett.*, **59**, 799.

[2] Affleck, I., Kennedy, T., Lieb, E. H., and Tasaki, H. (1988). *Commun. Math. Phys.*, **115**, 477.

[3] Anderson, M. H., Ensher, J. R., Matthews, M. R., Wieman, C. E., and Cornell, E. A. (1995). *Science*, **269**, 198.

[4] Bai, Z., Demmel, J., Dongarra, J., Ruhe, A., and van der Vorst, H. (2000). *Templates for the Solution of Algebraic Eigenvalue Problems: A Practical Guide.* SIAM.

[5] Barrett, R., Berry, M., Chan, T. F., Demmel, J., Donato, J., Dongarra, J., Eijkhout, V., Pozo, R., Romine, C., and van der Vorst, H. (1990). *Templates for the Solution of Linear Systems: Building Blocks for Iterative Methods* (2nd edn). SIAM.

[6] Barthel, T. (2013). *New J. Phys.*, **15**, 073010.

[7] Barthel, T., Schollwöck, U., and Sachdev, S. (2012). arXiv:1212.3570 [cond-mat.str-el].

[8] Barthel, T., Schollwöck, U., and White, S. R. (2009). *Phys. Rev. B*, **79**, 245101.

[9] Bekenstein, J. D. (1973). *Phys. Rev. D*, **7**, 2333.

[10] Binder, M. and Barthel, T. (2015). *Phys. Rev. B*, **92**, 125119.

[11] Bloch, I., Dalibard, J., and Zwerger, W. (2008). *Rev. Mod. Phys.*, **80**, 885.

[12] Breuer, H. P. and Petruccione, F. (2002). *The Theory of Open Quantum Systems*. Oxford University Press.

[13] Calabrese, P. and Cardy, J. (2004). *J. Stat. Mech.*, **2004**, P06002.

[14] Calabrese, P. and Cardy, J. (2006). *Int. J. Quantum Inf.*, **4**, 429.

[15] Calabrese, P. and Cardy, J. (2006). *Phys. Rev. Lett.*, **96**, 136801.

[16] Caux, J.-S. and Hagemans, R. (2006). *J. Stat. Mech.*, **2006**, P12013.

[17] Crosswhite, G. M. and Bacon, D. (2008). *Phys. Rev. A*, **78**, 012356.

[18] Daley, A. J., Kollath, C., Schollwöck, U., and Vidal, G. (2004). *J. Stat. Mech.: Theor. Exp.*, P04005.

[19] Daley, A. J., Taylor, J. M., Diehl, S., Baranov, M., and Zoller, P. (2009). *Phys. Rev. Lett.*, **102**, 040402.

[20] Dargel, P. E., Wöllert, A., Honecker, A., McCulloch, I. P., Schollwöck, U., and Pruschke, T. (2012). *Phys. Rev. B*, **85**, 205119.

[21] Davis, K. B., Mewes, M.-O., Andrews, M. R., van Druten, N. J., Durfee, D. S., Kurn, D. M., and Ketterle, W. (1995). *Phys. Rev. Lett.*, **75**, 3969.

[22] den Nijs, M. and Rommelse, K. (1989). *Phys. Rev. B*, **40**, 4709.

[23] Depenbrock, S., McCulloch, I. P., and Schollwöck, U. (2012). *Phys. Rev. Lett.*, **109**, 067201.

[24] Eisert, J., Cramer, M., and Plenio, M. B. (2010). *Rev. Mod. Phys.*, **82**, 277.

[25] Foong, S. K. and Kanno, S. (1994). *Phys. Rev. Lett.*, **72**, 1148.

[26] Fröwis, F., Nebendahl, V., and Dür, W. (2010). *Phys. Rev. A*, **81**, 062337.

[27] Gagliano, E. R. and Balseiro, C. A. (1987). *Phys. Rev. Lett.*, **59**, 2999.

[28] Ganahl, M., Thunström, P., Verstraete, F., Held, K., and Evertz, H. G. (2014). *Phys. Rev. B*, **90**, 045144.

[29] Golub, G. and van Loan, C. F. (1996). *Matrix Computations*. Johns Hopkins University Press.

[30] Greiner, M., Mandel, O., Esslinger, T., Hänsch, T. W., and Bloch, I. (2002). *Nature*, **415**, 39.

[31] Greiner, M., Mandel, O., Hänsch, T. W., and Bloch, I. (2002). *Nature*, **419**, 51.

[32] Grosso, G. and Parravicini, G. P. (2000). *Solid State Physics*. Academic Press.

[33] Hallberg, K. (1995). *Phys. Rev. B*, **52**, 9827.

[34] Hallberg, K. (2006). *Adv. Phys.*, **55**, 477.

[35] Haroche, S. and Raimond, J.-M. (2006). *Exploring the Quantum*. Oxford University Press.

[36] Hastings, M. (2007). *Phys. Rev. B*, **76**, 035114.

[37] Hochbruck, M. and Lubich, C. (1997). *SIAM J. Numer. Anal.*, **34**, 1911.

[38] Holzhey, C., Larsen, F., and Wilczek, F. (1994). *Nucl. Phys. B*, **424**, 443.

[39] Holzner, A., Weichselbaum, A., McCulloch, I. P., Schollwöck, U., and von Delft, J. (2011). *Phys. Rev. B*, **83**, 195115.

[40] Jaksch, D., Bruder, C., Cirac, J. I., Gardiner, C. W., and Zoller, P. (1998). *Phys. Rev. Lett.*, **81**, 3108.

[41] Jeckelmann, E. (2002). *Phys. Rev. B*, **66**, 045114.

[42] Karrasch, C., Bardarson, J. H., and Moore, J. E. (2012). *Phys. Rev. Lett.*, **108**, 227206.

[43] Kühner, T. D. and White, S. R. (1999). *Phys. Rev. B*, **60**, 335.

[44] Latorre, J. I., Rico, E., and Vidal, G. (2004). *Quantum Inf. Comput.*, **4**, 48.

[45] McCulloch, I. P. (2007). *J. Stat. Mech.*, **2007**, P10014.

[46] McLachlan, R. I. (1995). *SIAM J. Sci. Comput.*, **16**, 151.

[47] Nielsen, M. A. and Chuang, I. L. (2000). *Quantum Computation and Quantum Information*. Cambridge University Press.

[48] Osborne, T. J. (2006). *Phys. Rev. Lett.*, **97**, 157202.

[49] Östlund, S. and Rommer, S. (1995). *Phys. Rev. Lett.*, **75**, 3537.

[50] Page, D. N. (1993). *Phys. Rev. Lett.*, **71**, 1291.

[51] Parlett, B. N. (1998). *The Symmetric Eigenvalue Problem*. SIAM Classics in Applied Mathematics.

[52] Pereira, R. G., White, S. R., and Affleck, I. (2008). *Phys. Rev. Lett.*, **100**, 027206.

[53] Pirvu, B., Murg, V., Cirac, J. I., and Verstraete, F. (2010). *New J. Phys.*, **12**, 025012.

[54] Plenio, M. B. and Knight, P. L. (1998). *Rev. Mod. Phys.*, **70**, 101.

[55] Preskill, J. (1998). Lecture Notes on Quantum Information and Computation. California Institute of Technology. http://www.theory.caltech.edu/ preskill/ ph219/index.html#lecture.

[56] Press, W. H., Teukolsky, S. A., Vetterling, W. T., and Flannery, B. P. (2007). *Numerical Recipes: The Art of Scientific Computing* (3rd edn). Cambridge University Press.

[57] Ramasesha, S., Pati, S. K., Krishnamurthy, H. R., Shuai, Z., and Brédas, J. L. (1997). *Synth. Met.*, **85**, 1019.

[58] Saad, Y. (2003). *Iterative Methods for Sparse Linear Systems* (2nd edn). SIAM.

[59] Saad, Y. (2011). *Numerical Methods for Large Eigenvalue Problems* (rev. edn). SIAM.

[60] Saad, Y. and Schultz, M. (1986). *SIAM J. Sci. Stat. Comput.*, **7**, 856.

[61] Schollwöck, U. (2005). *Rev. Mod. Phys.*, **77**, 259.

[62] Schollwöck, U. (2011). *Ann. Phys. (NY)*, **326**, 96.

[63] Schollwöck, U., Jolicœur, T., and Garel, T. (1996). *Phys. Rev. B*, **53**, 3304.

[64] Schuch, N., Wolf, M. M., Verstraete, F., and Cirac, J. I. (2008). *Phys. Rev. Lett.*, **100**, 030504.

[65] Sen, S. (1996). *Phys. Rev. Lett.*, **77**, 1.

[66] Soos, Z. G. and Ramasesha, S. (1989). *J. Chem. Phys.*, **90**, 1067.

[67] Stoudenmire, E. M. and White, S. R. (2010). *New J. Phys.*, **12**, 055026.

[68] Suzuki, M. (1976). *Prog. Theor. Phys.*, **56**, 1454.

[69] Suzuki, M. (1991). *J. Math. Phys.*, **32**, 400.

[70] Trefethen, L. N. and Bau, D. (1997). *Numerical Linear Algebra*. SIAM.

[71] Verstraete, F. and Cirac, J. I. (2006). *Phys. Rev. B*, **73**, 094423.

[72] Verstraete, F., Garcia-Ripoll, J. J., and Cirac, J. I. (2004). *Phys. Rev. Lett.*, **93**, 207204.

[73] Verstraete, F., Murg, V., and Cirac, J. I. (2008). *Adv. Phys.*, **57**, 143.

[74] Vidal, G. (2003). *Phys. Rev. Lett.*, **91**, 147902.

[75] Vidal, G. (2004). *Phys. Rev. Lett.*, **93**, 040502.

[76] Vidal, G., Latorre, J. I., Rico, E., and Kitaev, A. (2003). *Phys. Rev. Lett.*, **90**, 227902.

[77] Weiße, A., Wellein, G., Alvermann, A., and Fehske, H. (2006). *Rev. Mod. Phys.*, **78**, 275.

[78] White, S. R. (1992). *Phys. Rev. Lett.*, **69**, 2863.

[79] White, S. R. (1993). *Phys. Rev. B*, **48**, 10345.

[80] White, S. R. (2009). *Phys. Rev. Lett.*, **102**, 190601.

[81] White, S. R. and Affleck, I. (2008). *Phys. Rev. B*, **77**, 134437.

[82] White, S. R. and Feiguin, A. E. (2004). *Phys. Rev. Lett.*, **93**, 076401.

[83] Wolf, F. A., Justiniano, J. A., McCulloch, I. P., and Schollwöck, U. (2015). *Phys. Rev. B*, **91**, 115144.

[84] Wolf, F. A., McCulloch, I. P., Parcollet, O., and Schollwöck, U. (2014). *Phys. Rev. B*, **90**, 115124.

[85] Yan, S., Huse, D. A., and White, S. R. (2011). *Science*, **332**, 1173.

[86] Zwerger, W. (2003). *J. Opt. B*, **5**, S9.

[87] Zwolak, M. and Vidal, G. (2004). *Phys. Rev. Lett.*, **93**, 207205.

Part II

Short lectures

5

Quench dynamics in integrable systems

Natan ANDREI, Deepak IYER, and Huijie GUAN

Department of Physics and Astronomy
Rutgers University
Piscataway, New Jersey 08854
USA

Strongly Interacting Quantum Systems out of Equilibrium. First Edition. Thierry Giamarchi et al.
© Oxford University Press 2016. Published in 2016 by Oxford University Press.

Chapter Contents

Preface

This chapter covers in some detail lectures that I gave at the 2012 Les Houches Summer School on Strongly Interacting Quantum Systems Out of Equilibrium. I describe work done with Deepak Iyer, with important contributions from Hujie Guan. I discuss some aspects of the physics revealed by quantum quenches and present a formalism for studying the quench dynamics of integrable systems. The formalism presented here generalizes an approach by Yudson and is applied to the Lieb–Liniger model, which describes a gas of N interacting bosons moving on the continuous infinite line while interacting via a short-range potential. We carry out the quench from several initial states and for any number of particles and compute the evolution of the density and the noise correlations. In the long-time limit, the system dilutes and we find that for any value of repulsive coupling, independently of the initial state, the system asymptotes toward a strongly repulsive gas, while for any value of attractive coupling, the system forms a maximal bound state that dominates at longer times. In either case, the system equilibrates but does not thermalize, an effect that is consistent with prethermalization. These results can be confronted with experiments.

For much greater detail, see our *Physical Review A* paper "Exact formalism for the quench dynamics of integrable models" [1], on which this chapter is based.

Further applications of the approach to the Heisenberg and Anderson models are presented in [2].

5.1   Introduction

In recent years, the study of the nonequilibrium dynamics of isolated quantum systems has seen significant advances in both theory and experiment. Although nonequilibrium processes abound in all fields of science, in particular in condensed matter physics, their detailed study has been hampered by the very short relaxation times that characterize, in conventional systems, the response to an external perturbation, as well as our inability to thermally isolate such systems from contact with phonon baths or other sources of decoherence, all of which leads to the blurring of nonequilibrium effects. The recent advances have followed the appearance of diverse experimental systems ranging from nano-devices and molecular electronic devices to optically trapped cold-atom gases where some of these limitations have been overcome, thus allowing systematic study of various aspects of nonequilibrium responses [3, 4]. In parallel with the experimental effort, much progress has also been made on the theoretical front, and advances in conceptual ideas and numerical tools have provided further insight into this new and burgeoning field [5–8]. This chapter is devoted to reviewing a particular direction, namely, the study of *quench dynamics in low-dimensional quantum systems described by integrable Hamiltonians*. Both terms require some discussion. We shall first discuss "quantum quenches" and then "integrable Hamiltonians."

5.1.1   Quantum quenches

A quantum quench is a nonequilibrium process where a system, initially in some given quantum state, is induced to time-evolve under the influence of an applied

Hamiltonian. Thus, one prepares the system in some initial state, $|\psi_0\rangle$, which may in practice be the ground state of an initial Hamiltonian H_{in}. The state will have certain characteristic properties and correlations inherited from the initial Hamiltonian. At time $t = 0$, one effects some change, and henceforth a new Hamiltonian, H, is applied. The state $|\psi_0\rangle$, not being an eigenstate of the new Hamiltonian, will evolve nontrivially under its influence. The response of the system to the quench reveals its intrinsic timescales, the mechanisms underlying its evolution, and the phase transitions it may cross as it evolves. Many questions arise: the fate of the initial correlations in the new regime, the development of new correlations, and whether the system thermalizes, among many others. The changes induced in the Hamiltonian and leading to the quench can be of a great variety: one may change an interaction coupling constant (e.g., via a Feshbach resonance in a cold atomic gas), one may release the interacting gas from a harmonic or periodic trap and allow it to expand in open space, or one may couple a quantum dot to a Fermi sea and observe the evolution of a Kondo peak. Further, the quench can be sudden, i.e., over a time window much shorter than other timescales in the system, or it can be driven at a constant rate or with a time-dependent ramp. Here we shall concentrate on sudden quenches

$$|\psi, t\rangle = e^{-iHt}|\psi_0\rangle, \tag{5.1}$$

where the initial state $|\Psi_0\rangle$ describes a system in a finite region of space with a particular density profile: lattice-like or condensate-like (see Fig. 5.1).

The quench dynamics can be studied theoretically and experimentally by observing the time evolution of physical quantities, which may be local operators, multiparticle observables, correlation functions, local currents, or also global quantities such as entanglements. Consider an observable $\hat{A}$, which will evolve under H according to

$$\langle \hat{A}(t) \rangle = \langle \psi_0 | e^{iHt} \hat{A} e^{-iHt} | \psi_0 \rangle. \tag{5.2}$$

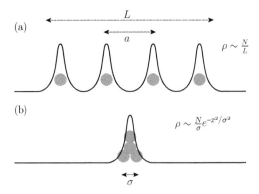

Fig. 5.1 Initial states. (a) For $a/\sigma \gg 1$, we have a lattice-like state, $|\Psi_{\mathrm{latt}}\rangle$. (b) For $a = 0$, we have a condensate-like state $|\Psi_{\mathrm{cond}}\rangle$. σ determines the spread. (Reprinted with permission from [1]. Copyright (2013) by the American Physical Society.)

To compute the evolution, it is convenient to expand the initial state in the eigenbasis of the evolution Hamiltonian,

$$|\Psi_0\rangle = \sum_{\{n\}} C_n |n\rangle, \tag{5.3}$$

where $|n\rangle$ are the eigenstates of H and $C_n = \langle n|\Psi_0\rangle$ are the overlaps with the initial state, determining the weights with which different eigenstates contribute to the time evolution:

$$|\Psi_0, t\rangle = \sum_{\{n\}} e^{-i\epsilon_n t} C_n |n\rangle. \tag{5.4}$$

The evolution of observables is then given by

$$\langle \hat{A}(t)\rangle_{\Psi_0} = \langle \Psi_0, t|\hat{A}|\Psi_0, t\rangle$$

$$= \sum_{\{n,m\}} e^{-i(\epsilon_m - \epsilon_n)t} C_n^* C_m \langle n|\hat{A}|m\rangle. \tag{5.5}$$

Many issues in dynamics and quantum kinetics revolve around expressions of this type, some already mentioned: Does the system thermalize [9–11], equilibrate, cross phase transitions in time, generate defects as it evolves? What entropy is best suited to describe the evolution? What is the quantum work done in the quench? Some of these questions are discussed in other chapters in this volume, and we shall confine our attention here to issues related to the dynamics and how to compute the evolution itself.

The quench is typically not a low-energy process. A finite amount of energy (or energy per unit volume) is injected into the system,

$$\epsilon_{\text{quench}} = \langle \Psi_0|H|\Psi_0\rangle$$

$$= \sum_{\{n\}} \epsilon_n |C_n|^2, \tag{5.6}$$

and is conserved throughout the evolution. It specifies the energy surface on which the system moves (and the temperature, if the system thermalizes). This surface is determined by the initial state through the overlaps C_n. Unlike the situation in thermodynamics, where the ground state and low-lying excitations play a central role, this is not the case when the system is out of equilibrium. A quench puts energy into the isolated system that it cannot dissipate and thereby relax to its ground state. Rather, the eigenstates that contribute to the dynamics depend strongly on the initial state via the overlaps C_n (see Fig. 5.2).

In the experiments that we seek to describe in this chapter, a system of N bosons is initially confined to a region of space of size L and then allowed to evolve on the infinite line while interacting via short-range interactions. The quantum Hamiltonian describing the system in one dimension is called the Lieb–Liniger Hamiltonian

$$H = \int dx\, \partial b^\dagger(x)\partial b(x) + c \int dx\, b^\dagger(x)b(x)b^\dagger(x)b(x), \tag{5.7}$$

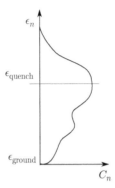

Fig. 5.2 Difference between quench dynamics and thermodynamics. After a quench, the system probes high-energy states and does not necessarily relax to the ground state. In thermodynamics, we minimize the energy (or free energy) of a system and probe the region near the ground state. (Reprinted with permission from [1]. Copyright (2013) by the American Physical Society.)

where the field $b^\dagger(x)$ creates a boson at point x. The strength and sign of the coupling constant c can be controlled in the experiment. When $c > 0$, the system is repulsive, whereas when $c < 0$, it is attractive and bound states appear. The choice of short-range potential $V(x - y) = c\delta(x - y)$ renders the model integrable, as we shall discuss below.

Several timescales underlie the phenomena we describe. One is determined by the initial condition (spatial extent, overlap of nearby wavefunctions), and the other is determined by the parameters of the quenched system (mass and interaction strength). For an extended system where we start with a locally uniform density (see Fig. 5.1(a)), we expect the dynamics to be in the constant-density regime as long as $t \ll L/v$, v being the characteristic velocity of propagation. Although the low-energy thermodynamics of a constant-density Bose gas can be described by a Luttinger liquid [12], we expect the collective excitations of the quenched system to behave as a highly excited liquid since the initial state is far from the ground state. It is also possible that, depending on the energy density $\epsilon_{\text{quench}}/L$, the Luttinger liquid description may break down altogether.

The other timescale that enters the description of nonequilibrium dynamics is the interaction timescale τ, a measure of the time it takes the interactions to develop fully: $\tau \sim 1/c^2$ for the Lieb–Liniger model.[1] Assuming that L is large enough that $\tau \ll L/v$, we expect a fully interacting regime to be operative at times beyond the interactions scale until $t \sim L/v$, when the density of the system can no longer be considered constant. In the low-density regime into which the expanding Lieb–Liniger gas enters

---

[1] The estimate for the interaction time can be refined by setting $\tau \sim 1/\delta E$, with $\delta E = \langle \Phi_0 | H_I | \Phi_0 \rangle$. Also, if we start from a lattice-like state, τ will include a short timescale $\tau_a \sim a/v$ before which the system only expands as a noninteracting gas, until neighboring wavefunctions overlap sufficiently.

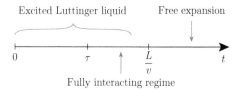

Fig. 5.3 Timescales involved in quench dynamics. τ is an intrinsic timescale that depends on the interaction strength. L/v is a characteristic time at which the system sees the finite extent. (Reprinted with permission from [1]. Copyright (2013) by the American Physical Society.)

when $t \gg L/v$, the effective strong coupling manifests itself as fermionization for repulsive interactions and as bound-state correlations for attractive interactions. Thus, the main operation of the interaction occurs in the time range $\tau \lesssim t \lesssim L/v$ over which the wavefunction rearranges and after which the system is dilute and expands in space while interacting. In this inhomogeneous low-density limit, one can no longer make contact with thermodynamic ensembles, in particular, with the generalized Gibbs ensemble [13–16]. When $L = \infty$, the system is homogeneous and free-space expansion is not present. Figure 5.3 summarizes the different timescales involved in a dynamical situation. We shall also consider initial conditions where the bosons are "condensed in space," occupying the same single-particle state characterized by some scale σ (see Fig. 5.1(b)). In this case, the short-time dynamics is not present, the timescales at which we can measure the system are typically much larger than σ/N, and we expect the dynamics to be in the strongly interacting and expanding regime.

5.1.2   Integrable Hamiltonians

We now turn to the topic of "integrable Hamiltonians," which is vast and will be addressed here from just one particular point of view, namely, the construction of eigenstates by means of the Bethe ansatz. As we have seen, to carry out the computation of the quench dynamics, we need to know the eigenstates of the propagating Hamiltonian. The Bethe ansatz approach is helpful in this respect as it provides us with the eigenstates of a large class of interacting one-dimensional Hamiltonians. A partial list includes the Heisenberg chain (magnetism), the Hubbard model (strong correlations), the Lieb–Liniger and sine–Gordon models (cold atoms in optical traps), the Kondo and Anderson models (impurities in metals, quantum dots) [12, 17–22]. Many of the Hamiltonians that can be solved in this way are of fundamental importance in condensed matter physics and have been proposed to describe various experimental situations. Integrable models not only pervade condensed matter physics but also appear in low-dimensional quantum and classical field theory and statistical mechanics. They can be realized experimentally in a variety of ways, and many were extensively studied long before their integrability became manifest.

For a Hamiltonian to possess Bethe ansatz eigenstates, it must have the property that multiparticle interactions can be consistently factorized into series of two-particle interactions, all of these being equivalent. Put differently, one must be able to express *in a consistent way* multiparticle wavefunctions in terms of single-particle wave-functions and two-particle S-matrices. We shall now do this for the Lieb–Liniger Hamiltonian. The eigenstates that we construct will subsequently be used to carry out quench evolution calculations in the model. The derivation of the model and its role in the description of cold atoms experiments are presented in other chapters in this volume.

The Bethe ansatz eigenstates

As the number of bosons is conserved, we can consider the Hilbert space of N bosons spanned by states of the form

$$|F\rangle = \int \prod_{j=1}^{N} dx_j \, F(x_1, \ldots, x_N) \prod_j b^\dagger(x_j)|0\rangle, \tag{5.8}$$

with $|0\rangle$ being the vacuum state with no bosons, $b(x)|0\rangle = 0$. In the N-boson sector, when acting on the symmetric wavefunction $F(x_1, \ldots, x_N)$, the Hamiltonian takes the form

$$H_N = -\sum_{j=1}^{N} \frac{\partial^2}{\partial x_j^2} + 2c \sum_{j<l} \delta(x_j - x_l). \tag{5.9}$$

The condition that the state $|F\rangle$ be an eigenstate of H is that the wavefunction $F(x_1, \ldots, x_N)$ satisfy $H_N F = EF$.

We proceed to obtain the single-particle wavefunctions and two-particle S-matrices in terms of which we shall express the multiparticle wavefunctions. In the one-boson sector,

$$H_1 = -\frac{\partial^2}{\partial x^2}, \tag{5.10}$$

and the eigenstates are plane waves $F^\lambda(x) = e^{i\lambda x}$ with energy $E = \lambda^2$. In the two-boson sector, an interaction term is present,

$$H_2 = -\frac{\partial^2}{\partial x_1^2} - \frac{\partial^2}{\partial x_2^2} + 2c\delta(x_1 - c_2). \tag{5.11}$$

As the bosons interact only along the boundary, $x_1 = x_2$, we can divide the configuration space into two regions, $x_1 < x_2$ and $x_2 < x_1$, in the interior of which the particles are free and where the wavefunctions take the forms $Ae^{i(\lambda_1 x_1 + \lambda_2 x_2)}$ and $Be^{i(\lambda_1 x_1 + \lambda_2 x_2)}$, respectively, with energy $E = \lambda_1^2 + \lambda_2^2$. The relation between A and B is determined by the interaction that becomes operative on the boundary between the the two regions. The fully symmetrized wavefunction becomes

$$
\begin{aligned}
F(x_1, x_2) &= \mathcal{S}e^{i(\lambda_1 x_1 + \lambda_2 x_2)}[A\theta(x_2 - x_1) + B\theta(x_1 - x_2)] \\
&= \tfrac{1}{2}e^{i(\lambda_1 x_1 + \lambda_2 x_2)}[A\theta(x_2 - x_1) + B\theta(x_1 - x_2)] \\
&\quad + \tfrac{1}{2}e^{i(\lambda_1 x_2 + \lambda_2 x_1)}[A\theta(x_1 - x_2) + B\theta(x_2 - x_1)].
\end{aligned} \tag{5.12}
$$

Here the step function $\theta(x)$ is defined as follows: $\theta(x) = 1$ for $x > 0$, $\theta(x) = 0$ for $x < 0$, and $\theta(x) = \frac{1}{2}$ for $x = 0$. Hence, $\theta(x) + \theta(-x) = 1 \forall x$ and $\partial_x \theta(x) = \delta(x)$. Note that the wavefunction is continuous (as it should be) on the boundary: $F(x, x^+) = F(x, x^-) = \frac{1}{2} e^{i(k_1+k_2)x}(A + B)$.

Now applying the Hamiltonian to the wavefunction, $H_2 F(x_1, x_2) = EF(x_1, x_2)$ we find that the condition for F to be an eigenfunction is

$$- e^{i(\lambda_1+\lambda_2)x_1} \delta(x_1 - x_2)[2i\lambda_1(A - B) + 2i\lambda_2(B - A) + 2i\lambda_1(A - B) + 2i\lambda_2(B - A)]$$

$$- e^{i(\lambda_1+\lambda_2)x_1} \delta'(x_1 - x_2)[(A - B) + (B - A) + (A - B) + (B - A)]$$

$$+ 2ce^{i(\lambda_1+\lambda_2)x_1} \delta(x_1 - x_2) \frac{A + B}{2} = 0. \tag{5.13}$$

The $\delta'(x)$ term cancels automatically as result of the continuity of the wavefunction on the boundary. The requirement that the $\delta(x)$-terms cancel leads to the condition

$$B = \frac{(\lambda_1 - \lambda_2) + ic}{(\lambda_1 - \lambda_2) - ic} A, \tag{5.14}$$

from which we deduce the two particle S-matrix

$$S^{12} = \frac{B}{A} = \frac{(\lambda_1 - \lambda_2) + ic}{(\lambda_1 - \lambda_2) - ic}, \tag{5.15}$$

completing the determination of the two-particle eigenstate:

$$|\lambda_1 \lambda_2\rangle = \mathcal{N}_{12} \int_x Z_{12}^x(\lambda_1 - \lambda_2) e^{i(\lambda_1 x_1 + \lambda_2 x_2)} b^\dagger(x_1) b^\dagger(x_2) |0\rangle. \tag{5.16}$$

Here $\mathcal{N}_{12}$ is a normalization factor determined by a particular solution, and

$$Z_{12}^x(z) = \frac{z - ic\,\mathrm{sgn}(x_1 - x_2)}{z - ic} \tag{5.17}$$

is the dynamic factor, which includes the S-matrix.

The generalization to any number of particles follows a similar argument. One divides the configuration space of N bosons into $N!$ regions conveniently labeled by elements Q of the permutation group S_N that specify their ordering. Thus, the region where $x_{Q1} < x_{Q2} < \cdots < x_{QN}$ corresponds to the permutation Q. The ordering of the particles in region Q can be obtained from the ordering in some standard region, say in region I, where $x_1 < x_2 < \cdots < x_N$, by a series of transpositions P^{ij} where the positions of two adjacent particles $Qm = i, Q(m + 1) = j$ are exchanged. The amplitude A_Q in region Q is then given as

$$A_Q = \left(\prod S \right) A_I \tag{5.18}$$

where the product of S-matrices is taken along a path of transpositions leading from I to Q. That path is not unique, and hence for the construction to be consistent, the S-matrices must satisfy the condition (the Yang–Baxter equation)

$$S^{ij} S^{ik} S^{jk} = S^{jk} S^{ik} S^{ij} \tag{5.19}$$

guaranteeing path independence. This condition is very stringent in general, when particles carry internal degrees of freedom and the S-matrices do not commute. Here the condition is satisfied because the bosons have equal mass and hence the individual particle momenta are conserved under scattering and the S-matrices are commuting phases.

The eigenstates thus take the form

$$|\vec{\lambda}\rangle = \int_x e^{i\vec{\lambda}\cdot\vec{x}} \sum_Q A_Q \theta(x_Q) \prod_{j=1}^{N} b^{\dagger}(x_j)|0\rangle, \tag{5.20}$$

where $\int_x \equiv \int \prod_j dx_j$ and we have omitted the symmetrizer since the wavefunction is automatically symmetrized under the integration. The eigenstates may be recast here into a more convenient form

$$|\vec{\lambda}\rangle = \mathcal{N}(\vec{\lambda}) \int_x \prod_{i<j} Z^x_{ij}(\lambda_i - \lambda_j) \prod_j e^{i\lambda_j x_j} b^{\dagger}(x_j)|0\rangle, \tag{5.21}$$

where $\mathcal{N}(\lambda)$ is a normalization factor determined by a particular solution. The energy eigenvalue corresponding to the eigenstate $|\vec{\lambda}\rangle$ is

$$E(\vec{\lambda}) = \sum_j \lambda_j^2. \tag{5.22}$$

The roles played by the eigenstates and eigenvalues thus obtained differ depending on whether one is studying the equilibrium or nonequilibrium properties of a given quantum system. To study the thermodynamic properties, one must be able to enumerate and classify all eigenstates in order to construct the partition function. To achieve this, some finite-volume boundary conditions are typically imposed. One may impose periodic boundary conditions to maintain translation invariance or open boundary conditions when the system has physical ends. Imposing periodic boundary conditions on the Bethe ansatz wavefunctions of the Lieb–Liniger model leads to N coupled equations

$$e^{i\lambda_j L} = \prod_l \frac{(\lambda_j - \lambda_l) + ic}{(\lambda_j - \lambda_l) - ic}, \qquad j = 1, \ldots, N, \tag{5.23}$$

whose solutions are the allowed momentum configurations $\{\vec{\lambda}\}$ from which the full spectrum is determined via (5.22). One can then identify the ground state and the low-lying excitations that dominate the low-temperature physics. The analysis and classification of the solutions of the model was given by Lieb and Liniger [19]. The thermodynamic partition function was then derived by summing over all energy eigenvalues with their appropriate degeneracies by Yang and Yang [23].

In the study of nonequilibrium quench dynamics, on the other hand, the main issue is the determination of the eigenstates of the propagating Hamiltonian that enter in the time evolution of a given initial state $|\psi_0\rangle$. This information is encoded by the overlaps $C_n = \langle n|\psi_0\rangle$, with the ground state and low-lying excitations playing no special role. Given the eigenstate expansion of the initial state, one needs then to resum it with the time phases, (5.4), to obtain the evolved state. The overlaps, however, are not easy to evaluate owing to the complicated nature of the Bethe eigenstates and their normalization. The problem is more pronounced in the far-from-equilibrium quench when the state with which we start suddenly finds itself far away from the eigenstates of the new Hamiltonian (see (5.4)) and all the eigenstates have nontrivial weights in the time evolution. In all but the simplest cases, the problem is nonperturbative and existing analytical techniques are not suited for a direct application to such a situation.

When quench calculations are carried out in finite volume, all the steps mentioned are involved: (i) solving the Bethe ansatz equation for the spectrum and the eigenstates; (ii) calculation of overlaps; (iii) resummation of the evolution series. If, however, one is interested in the physics in the infinite-volume limit, then one need not (unlike in thermodynamics) pass through the finite-volume calculation. Instead, one can carry out the quench directly in the infinite-volume limit, allowing the the overlaps to pick out the relevant contributions. Working in the infinite-volume limit allows us to replace the discrete sum in (5.5) by integrations over continuous momentum variables, transforming the difficult talk of computing overlaps to the simpler one of calculating residues in Cauchy-type integrals, as we discuss below. This approach, due to Yudson, circumvents some of the difficulties mentioned and leads to an efficient calculational formalism and to transparent physical results. Its application to the Lieb–Liniger model is the main topic of this chapter. We have applied this approach to other models too, and results are presented in [2].

5.2 Time evolution on the infinite line

In 1985, Yudson presented a new approach to time-evolve the Dicke model (a model for superradiance in quantum optics [24]) considered on an infinite line [25]. The dynamics in certain cases was extracted in closed form with much less work than previously required, and in some cases where it was even impossible with earlier methods. The core of the method is to bypass the laborious sum over momenta using an appropriately chosen set of contours and integrating over momentum variables in the complex plane. It is applicable in its original form to models with a particular pole structure in the two-particle S-matrix and a linear spectrum. We generalize the approach to the case of the quadratic spectrum and apply it to the study of quantum quenches.

As discussed earlier, in order to carry out the quench of a system given at $t = 0$ in a state $|\Psi_0\rangle$, one naturally proceeds by introducing a "unity" in terms of a complete set of eigenstates,

$$|\Psi_0\rangle = \sum_{\{\lambda\}} |\vec{\lambda}\rangle\langle\vec{\lambda}|\Psi_0\rangle, \qquad (5.24)$$

and then applies the evolution operator. The Yudson representation overcomes the difficulties in computing overlaps and carrying out this sum by using an integral representation for the (over-)complete basis directly in the infinite-volume limit. The argument consists of two independent parts:

1. Since the wavefunction for bosons, $F(x_1, \ldots, x_N)$, is fully symmetric, it suffices to compute overlaps in a region $(x_1 < \cdots < x_N)$.
2. If no boundary conditions are imposed, then the Schrödinger equation for N bosons, $H|\vec{\lambda}\rangle = \epsilon(\vec{\lambda})|\vec{\lambda}\rangle$, is satisfied for any value of the momenta $\{\lambda_j, j = 1, \ldots, N\}$.

The initial state can then be written as

$$|\Psi_0\rangle = \int_\gamma d^N\lambda \, C_{\vec{\lambda}} |\vec{\lambda}\rangle, \tag{5.25}$$

with the integration over $\vec{\lambda}$ replacing the summation over states. This is akin to summing over an overcomplete basis, the relevant elements in the sum being automatically picked up by the overlap with the physical initial state. The integration over momenta will be carried out over contours $\{\gamma\}$, fixed by the pole structure of the S-matrices, chosen so that (5.25) holds.

We proceed to carry out the evolution for the repulsive and attractive models for several initial states. The contours of integration, as we shall see, depend on the sign of the coupling constant c. We will notice that in the repulsive case, it is sufficient to integrate over the real line. The attractive case will require the use of contours separated out in the imaginary direction (to be qualified below). Those separated contours are consistent with the fact that the spectrum consists of "strings" with momenta taking values as complex conjugates. Actually, we shall find that the existence of string solutions (bound states) and their spectrum follow elegantly and naturally from the contour representation. They need not be known a priori, nor need they be computed from the Bethe ansatz equations.

5.2.1 Repulsive interactions

We begin by discussing the repulsive case, $c > 0$. For this case, a similar approach was independently developed by Tracy and Widom [26] and used by Lamacraft [27] to calculate noise correlations in the repulsive model.

A generic N-boson initial state

$$|\Psi_0\rangle = \int_{\vec{x}} \Phi_s(\vec{x}) \prod_j b^\dagger(x_j)|0\rangle, \tag{5.26}$$

with Φ_s symmetrized, can be rewritten, using the symmetry of the boson operators, in terms of basis states:

$$|\Psi_0\rangle = N! \int_{\vec{x}} \Phi_s(\vec{x})|\vec{x}\rangle, \tag{5.27}$$

where

$$|\vec{x}\rangle = \theta(\vec{x}) \prod_j b^\dagger(x_j)|0\rangle, \tag{5.28}$$

with $\theta(\vec{x}) = \theta(x_1 > x_2 > \cdots > x_N)$. It therefore suffices to show that we can express any coordinate basis state as an integral over the Bethe ansatz eigenstates,

$$|\vec{x}\rangle = \theta(\vec{x}) \int_\gamma \prod_j \frac{d\lambda_j}{2\pi} A(\vec{\lambda}, \vec{x})|\vec{\lambda}\rangle, \tag{5.29}$$

with appropriately chosen contours of integration $\{\gamma_j\}$ and $A(\lambda, \vec{x})$, which plays a role similar to the overlap of the eigenstates and the initial state. Establishing an identity of the type (5.29) corresponds to identifying the complete set of eigenstates. That is the reason why the contours for the attractive case, where bound states appear, differ from those for the repulsive case, where they are absent.

We claim that in the repulsive case, (5.29) is realized with

$$A(\vec{\lambda}, \vec{x}) = \prod_j e^{-i\lambda_j x_j} \tag{5.30}$$

and with the contours γ_j running along the real axis from minus to plus infinity. In other words, (5.29) takes the form

$$|\vec{x}\rangle = \theta(\vec{x}) \int_{\vec{y}} \int \prod_j \frac{d\lambda_j}{2\pi} \prod_{i<j} Z_{ij}^y(\lambda_i - \lambda_j) \prod_j e^{i\lambda_j(y_j - x_j)} b^\dagger(y_j)|0\rangle. \tag{5.31}$$

Equivalently, we claim that the λ integration above produces $\prod_j \delta(y_j - x_j)$.

We shall prove this in two stages. Consider first $y_N > x_N$. To carry out the integral using the residue theorem, we have to close the integration contour in λ_N in the upper half-plane. The poles in λ_N are at $\lambda_j - ic$, $j < N$. These are all below γ_N, and so the result is zero. This implies that any nonzero contribution comes from $y_N \leq x_N$. Let us now consider $y_{N-1} > x_{N-1}$. The only pole above the contour is $\lambda_{N-1}^* = \lambda_N + ic$. However, we also have $y_{N-1} > x_{N-1} > x_N \geq y_N \Rightarrow y_{N-1} > y_N$. This causes the only contributing pole to get canceled. The integral is again zero unless $y_{N-1} \leq x_{N-1}$. We can proceed is this fashion for the remaining variables, thus showing that the integral is nonzero only for $y_j \leq x_j$.

Now consider $y_1 < x_1$. We have to close the contour for λ_1 below. There are no poles in that region, and the residue is zero. Thus, the integral is nonzero only for $y_1 = x_1$. Consider $y_2 < x_2$. The only pole below, at $\lambda_2^* = \lambda_1 - ic$, is canceled as before, since we have $y_2 < x_2 < x_1 = y_1$. Again we find that the integral is nonzero only for $y_2 = x_2$. Carrying this on, we end up with

$$\theta(\vec{x}) \int_{\vec{y}} \prod_j \delta(y_j - x_j) b^\dagger(y_j)|0\rangle = |\vec{x}\rangle. \tag{5.32}$$

In order to time-evolve this state, we act on it with the unitary time-evolution operator. Since the integrals are well defined, we can move the operator inside the integral signs to obtain

$$|\vec{x}, t\rangle = \theta(\vec{x}) \int \prod_j \frac{d\lambda_j}{2\pi} e^{-i\epsilon(\vec{\lambda})t} A(\vec{\lambda}, \vec{x}) |\vec{\lambda}\rangle. \qquad (5.33)$$

5.2.2   Attractive interactions

We now consider the attractive case, $c < 0$. As mentioned earlier, the spectrum of the Hamiltonian now contains complex, so-called *string solutions*, which correspond to many-body bound states. In fact, the ground state at $T = 0$ consists of one N-particle bound state. We will see that the Yudson integral representation requires in this case another set of contours in order to establish (5.29). Similar properties are seen to emerge in [28], where Prolhac and Spohn obtain a propagator for the attractive Lieb–Liniger model by analytically continuing the results obtained by Tracy and Widom [26] for the repulsive model.

It can immediately be noted that for the attractive interaction the structure of the S-matrix is altered. This change, $c \to -c$, prevents the proof of Section 5.2.1 from working. In particular, the poles in the variable λ_N are at $\lambda_j + i|c|$ for $j < N$, and the residue within the contour closed in the upper half-plane is no longer zero. We need to choose a contour to avoid this pole. This can be achieved by separating the contours in the imaginary direction such that adjacent $\mathrm{Im}(\lambda_j - \lambda_{j-1}) > |c|$. At first sight, this seems to pose a problem since the quadratic term in the exponent diverges at large positive λ and positive imaginary part. There are two ways around this. We can tilt the contours as shown in Fig. 5.4 so that they lie in the convergent region of the Gaussian integral. The pieces toward the ends, which join the real axis, although though essential for the proof to work at $t = 0$, where we evaluate the integrals using

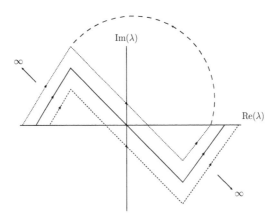

Fig. 5.4 Contours for the λ integration. Three contours are shown here, with the closing of the Nth (here the third) contour as discussed in the proof. (Reprinted with permission from [1]. Copyright (2013) by the American Physical Society.)

the residue theorem, do not contribute at finite time, because the integrand vanishes on them as they are taken to infinity. Another more natural means of doing this is to use the finite spatial support of the initial state. The overlaps of the eigenstates with the initial state effectively restrict the support for the λ integrals, making them convergent.

The proof of (5.29) now proceeds as in the repulsive case. We start by assuming that $y_N > x_N$, requiring us to close the contour in λ_N in the upper half-plane. This encloses no poles owing to the choice of contours, and the integral is zero unless $y_N \leq x_N$. Now we assume that $y_{N-1} > x_{N-1}$. Closing the contour above encloses one pole at $\lambda_{N-1}^* = \lambda_N - i|c|$; however, since $y_{N-1} > x_{N-1} > x_N \geq y_N$, this pole is canceled by the numerator and again we have $y_{N-1} \leq x_{N-1}$. We proceed in this fashion and then backward to show that the integral is nonzero only when all the poles cancel, giving us $\prod_j \delta(y_j - x_j)$, as required.

5.2.3   Two-particle dynamics

We begin with a detailed discussion of the quench dynamics of two bosons. As we have seen, it is convenient to express any initial state in terms of an ordered coordinate basis, $|\vec{x}\rangle = \theta(x_1 > x_2 > \cdots > x_N) \prod_j b^\dagger(x_j)|0\rangle$. At finite time, the wavefunction of bosons initially localized at x_1 and x_2 and subsequently evolved by a repulsive Lieb–Liniger Hamiltonian is given by

$$|\vec{x}, t\rangle_2 = e^{-iHt}\theta(x_1 - x_2)b^\dagger(x_1)b^\dagger(x_2)|0\rangle$$

$$= \int_{y,\lambda} Z_{12}^y(\lambda_1 - \lambda_2)e^{-i\lambda_1^2 t - i\lambda_2^2 t + i\lambda_1(y_1-x_1)+i\lambda_2(y_2-x_2)}b^\dagger(y_1)b^\dagger(y_2)|0\rangle$$

$$= \int_y \frac{e^{i(y_1-x_1)^2/4t+i(y_2-x_2)^2/4t}}{4\pi it}$$

$$\times \left[1 - c\sqrt{\pi it}\,\theta(y_2 - y_1)e^{i\alpha^2/8t}\,\mathrm{erfc}\left(\frac{i-1}{4}\frac{i\alpha}{\sqrt{t}}\right)\right] b^\dagger(y_1)b^\dagger(y_2)|0\rangle, \quad (5.34)$$

where $\alpha = \alpha(x_1 - x_2, y_1 - y_2) = 2ct + i(y_1 - x_1) - i(y_2 - x_2)$. This expression retains the Bethe form of wavefunctions defined in different configuration sectors. The only scales in the problem are the interaction strength c and $x_1 - x_2$, the initial separation between the particles.

In order to get physically meaningful results, we need to start from a physical initial state. We choose the state $|\Psi(0)_{\mathrm{latt}}\rangle$ where bosons are trapped in a periodic trap, forming initially a lattice-like state (see Fig. 5.1(a)):

$$|\Psi(0)_{\mathrm{latt}}\rangle = \prod_j \left[\frac{1}{(\pi\sigma^2)^{1/4}} \int_{\vec{x}} e^{-[x_j+(j-1)a]^2/2\sigma^2} b^\dagger(x_j)\right]|0\rangle. \quad (5.35)$$

We shall consider two limits. The first, which we will continue to refer to as $|\Psi(0)_{\mathrm{latt}}\rangle$, is $\sigma \ll a$, with wavefunctions of neighboring bosons not overlapping significantly, i.e., $e^{-a^2/\sigma^2} \ll 1$. In this case, the ordering of the initial particles needed for the Yudson representation is induced by the nonoverlapping support and it becomes possible to carry out the integral analytically. The other limit, $a \ll \sigma$, when there is maximal initial overlap, will be referred to as the condensate (in position space) wavefunction $|\Psi(0)_{\mathrm{cond}}\rangle$.

Evolution of the density

We consider the evolution of the density $\hat{\rho}(x) = b^{\dagger}(x)b(x)$ at $x = 0$ in the state $|\Psi(0)_{\mathrm{latt}}\rangle$. Figure 5.5 shows $\langle \Psi_{\mathrm{latt}}, t|\rho(0)|\Psi_{\mathrm{latt}}, t\rangle$ for repulsive, attractive, and noninteracting bosons. No difference is discernible between the three cases. The reason is obvious: the local interaction is operative only when the wavefunctions of the particles overlap. As we have taken $\sigma \ll a$, this will occur only after a long time when the wavefunction is spread out and overlap is negligible.

When the separation a is set to zero, with maximal initial overlap between the bosons, $|\Psi(0)_{\mathrm{cond}}\rangle$, we expect the interaction to be operative from the start. Figure 5.6 shows the density evolution for attractive, repulsive, and no interaction. The decay of the density for the attractive model is slower than the decay for the noninteracting model, which in turn is slower than that for the repulsive model. Still, the density does not show much difference between repulsive and attractive interactions in this case. A dramatic difference will appear when we study the noise correlations $\langle \Psi_0, t|\rho(x_1)\rho(x_2)|\Psi_0, t\rangle$, as will be shown below.

The appearance of bound states

Actually, the apparent similarity in the behavior observed in Fig. 5.6 is somewhat misleading. We shall show here that for the attractive interaction, bound states appear

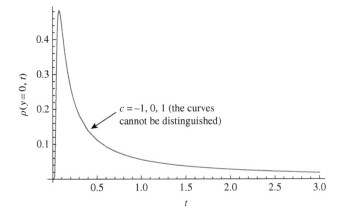

Fig. 5.5 $\langle \rho(x = 0, t)\rangle$ versus t after the quench from $|\Psi_{\mathrm{latt}}\rangle$. $\sigma/a \sim 0.1$. The curves appear indistinguishable (i.e., lie on top of each other) since the particles start out with insignificant overlap. The interaction effects would show up only when they have propagated long enough to have spread sufficiently to reach a significant overlap, at which time the density is too low. (Reprinted with permission from [1]. Copyright (2013) by the American Physical Society.)

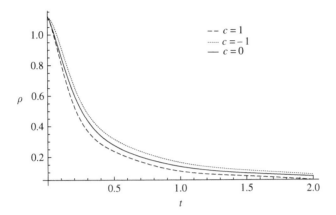

Fig. 5.6 $\langle \rho(x = 0, t)\rangle$ versus t after the quench from $|\Psi_{\text{cond}}\rangle$. $\sigma \sim 0.5$, $a = 0$. As the bosons overlap, interaction effects show up immediately. The upper curve is for $c = -1$, the middle curve for $c = 0$, and the lower curve for $c = 1$. (Reprinted with permission from [1]. Copyright (2013) by the American Physical Society.)

in the spectrum, although their effect on the evolution—in the case of two bosons—can be absorbed into the same mathematical expression as for the repulsive case.

Recall that the contours of integration are separated in the imaginary direction. In order to carry out the integration over λ, we shift the contour for λ_2 to the real axis and add the residue at the pole at $\lambda_2 = \lambda_1 + i|c|$. The two-particle finite-time state can be written as

$$|\vec{x}, t\rangle_2 = \int_y \int_{\gamma_c} \prod_{i<j=1,2} Z_{ij}^y(\lambda_i - \lambda_j) \prod_{j=1,2} e^{-i\lambda_j^2 t + i\lambda_j(y_j - x_j)} b^\dagger(y_j)|0\rangle$$

$$= \int_y \left[\int_{\gamma_r} \prod_{i<j=1,2} Z_{ij}^y(\lambda_i - \lambda_j) \prod_{j=1,2} e^{-i\lambda_j^2 t + i\lambda_j(y_j - x_j)} b^\dagger(y_j)|0\rangle \right.$$

$$\left. + \theta(y_2 - y_1) I(\lambda_2 = \lambda_1 + i|c|, t) b^\dagger(y_1) b^\dagger(y_2)|0\rangle \right]. \tag{5.36}$$

γ_c refers to contours that are separated in imaginary direction, while γ_r refers to all λ integrated along the real axis. $I(\lambda_2 = \lambda_1 + i|c|, t)$ is the residue at the pole at $\lambda_2 = \lambda_1 + i|c|$ obtained by shifting the λ_2 contour to the real axis. It is given by

$$I(\lambda_2 = \lambda_1 + i|c|, t) = -2c \int_y \int_{\lambda_1} e^{i\lambda_1(y_1 - x_1) + i(\lambda_1 + i|c|)(y_2 - x_2) - i\lambda_1^2 t - i(\lambda_1 + i|c|)^2 t}$$

$$= -2c \int_y \int_{\lambda_1} e^{i\lambda_1(y_1 - x_1 + y_2 - x_2) - \frac{1}{2}|c|(y_2 - y_1) - \frac{1}{2}|c|(x_1 - x_2)} e^{-i(\lambda_1 - i\frac{1}{2}|c|)^2 t - i(\lambda_1 + i\frac{1}{2}|c|)^2 t}$$

$$= -c \int_y \int_{\lambda_1} e^{i\lambda_1(y_1 - x_1 + y_2 - x_2) - \frac{1}{2}|c||y_2 - y_1| - \frac{1}{2}|c|(x_1 - x_2)} e^{-2i\lambda_1^2 t + i\frac{1}{2}|c|^2 t}. \tag{5.37}$$

This second term corresponds to the two-particles propagating as a bound state with the wavefunction $e^{-\frac{1}{2}|c||y_2-y_1|}$ and with center-of-mass position and momentum $e^{i\lambda_1(y_1+y_2-x_2-x_1)}$. It has a kinetic energy $2\lambda_1^2$ and a binding energy of $-c^2$, as reported in [29]. Also, the overlap of the bound state with the initial state follows immediately from the representation. It is given by $-ce^{-\frac{1}{2}|c|(x_1-x_2)}$. Such bound states appear for any number of particles involved. For instance, for three particles, the Yudson representation with complex λ's automatically produces multiple bound states coming from the poles, i.e., $I(\lambda_2 = \lambda_1 + i|c|)$, $I(\lambda_3 = \lambda_1 + i|c|)$, etc. They give rise to two- and three-particle bound states of the form $e^{-\frac{1}{2}|c|(|y_1-y_2|+|y_1-y_3|+|y_2-y_3|)}$. The contribution to the above integral coming from the real axis corresponds to the nonbound state.

Finally, putting everything together,

$$
|\vec{x}, t\rangle_2 = \int_y \frac{e^{i(y_1-x_1)^2/4t+i(y_2-x_2)^2/4t}}{4\pi i t}
$$

$$
\times \left[1 + |c|\sqrt{\pi i t}\, \theta(y_2 - y_1) e^{i\tilde{\alpha}^2/8t}\, \mathrm{erfc}\left(\frac{i-1}{4} \frac{i\tilde{\alpha}}{\sqrt{t}} \right) \right] \times b^\dagger(y_1) b^\dagger(y_2)|0\rangle, \qquad (5.38)
$$

where $\tilde{\alpha} = -2|c|t + i(y_1 - x_1) - i(y_2 - x_2)$. Surprisingly, the wavefunction maintains its form and we only need to replace $c \to -c$. This simple result is not valid for more than two particles.

Evolution of noise correlations

The density–density correlation function,

$$
\langle \Psi_0, t | \rho(x_1)\rho(x_2) | \Psi_0, t \rangle,
$$

involves the interaction in an essential way because the geometry of the experimental set-up measures the interference of "direct" and "crossed" propagating waves; see Fig. 5.7(a). The interaction among bosons is therefore expected to have a significant effect in noise measurements—more so than in density measurements, which do not directly involve scattering; see Fig. 5.7(b).

The set-up is the famous Hanbury Brown–Twiss experiment [30], where, for free bosons or fermions, the crossing produces a phase of ± 1 and causes destructive or constructive interference. Originally designed to study the photons arriving from two stars providing constant sources of light, in our case the set-up is generalized to multiple time-dependent sources with the phase given by the two-particle S-matrix capturing the interactions between the particles. In Fig. 5.8, we present the density–density correlation matrix $\langle \rho(x_1)\rho(x_2) \rangle$ for the repulsive gas, attractive gas, and noninteracting gas, shown at different times, starting with the lattice initial state of two bosons. Figure 5.9 shows the same for the condensate initial state. In both initial states, we note that the repulsive gas develops strong antibunching, repulsive correlations at long times akin to fermionic correlations, while the attractive gas shows strong bunching correlations, enhanced bosonic in nature. We shall discuss these in detail below. We expect the results to be qualitatively similar for higher particle number, although

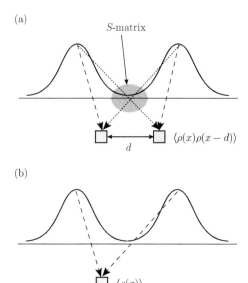

Fig. 5.7 (a) The Hanbury Brown–Twiss effect, where two detectors are used to measure the interference of the direct (long dashes) and the crossed waves (short dashes). The S-matrix enters explicitly. (b) The density measurement is not directly sensitive to the S-matrix. The thick black line shows the wavefunction amplitude and the dashed lines show time propagation. (Reprinted with permission from [1]. Copyright (2013) by the American Physical Society.)

beyond two particles the integrations cannot be carried out exactly. However, we can extract the asymptotic behavior of the wavefunctions analytically, as we show below.

5.2.4 Multiparticle dynamics at long times

Here we derive expressions for multiparticle wavefunction evolution at long times. The number of particles, N, is kept fixed in the limiting process, and hence, as discussed in Section 5.1, in the long-time limit we are in the low-density limit where interactions are expected to be dominant. The other regime, where N is sent to infinity first, is discussed elsewhere [31]. We first deal with the repulsive model, for which no bound states exist and the momentum integrations can be carried out over the real line, and then proceed to the attractive model. In a separate subsection, we examine the effect of starting with a condensate-like initial state.

Repulsive interactions: quench asymptotics in the lattice state

At large time, we use the stationary-phase approximation to carry out the λ integrations. The phase oscillations come primarily from the exponent $e^{-i\lambda_j^2 t + i\lambda_j(y_j - x_j)}$. At large t (i.e., $t \gg 1/c^2$), the oscillations are rapid, and the stationary point is obtained by solving

$$\frac{d}{d\lambda_j}[-i\lambda_j^2 t + i\lambda_j(y_j - x_j)] = 0. \tag{5.39}$$

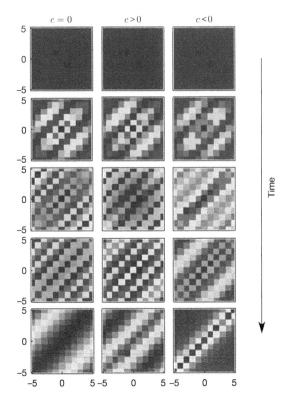

Fig. 5.8 Time evolution of the density–density correlation matrix $\langle\rho(x)\rho(y)\rangle$ for the $|\Psi_{\mathrm{latt}}\rangle$ initial state. In the color version of the figure, available in [1] and arXiv:1304.0506, blue is zero and red is positive. The repulsive model shows antibunching, i.e., fermionization at long times, while the attractive model shows bunching. (Reprinted with permission from [1]. Copyright (2013) by the American Physical Society.)

Note that typically one would ignore the second term above since it does not oscillate faster with increasing t, but here we cannot do that here, since the integral over y produces a nonzero contribution for $y \sim t$ at large time. Doing the Gaussian integral around this point (and fixing the S-matrix prefactor to its stationary value), we obtain for the repulsive case,

$$|\vec{x},t\rangle \rightarrow \int_y \prod_{i<j} Z^y_{ij}\left(\frac{y_i - y_j - x_i + x_j}{2t}\right)$$

$$\times \prod_j \frac{1}{\sqrt{4\pi i t}}\, e^{-i(y_j - x_j)^2/4t + i(y_j - x_j)^2/2t} b^\dagger(y_j)|0\rangle. \qquad (5.40)$$

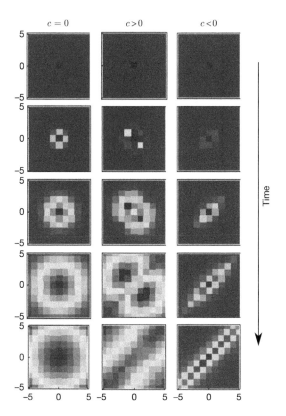

Fig. 5.9 Time evolution of the density–density correlation matrix $\langle \rho(x)\rho(y) \rangle$ for the $|\Psi_{\mathrm{cond}}\rangle$ initial state. In the color version of the figure, available in [1] and arXiv:1304.0506, blue is zero and red is positive. The repulsive model shows antibunching, i.e., fermionization at long times, while the attractive model shows bunching. (Reprinted with permission from [1]. Copyright (2013) by the American Physical Society.)

In this expression, the wavefunction has support mainly from regions where y_j/t is of order 1. In an experimental setup, one typically starts with a local finite-density gas, i.e., a finite number of particles localized over a finite length. With this condition, at long time, we can neglect x_j/t in comparison with y_j/t, giving

$$|\vec{x}, t\rangle \rightarrow \int_y \prod_{i<j} Z_{ij}(\xi_i - \xi_j) \prod_j \frac{1}{\sqrt{4\pi i t}} \, e^{it\xi_j^2 - i\xi_j x_j} b^\dagger(y_j)|0\rangle, \qquad (5.41)$$

where $\xi = y/2t$.

We turn now to calculating the asymptotic evolution of some observables. To compute the expectation value of the density, we start from the coordinate basis states $\langle \vec{x}', t | \rho(z) | \vec{x}, t \rangle$, which we then integrate with the chosen initial state:

$$\langle \vec{x}', t | \rho(z) | \vec{x}, t \rangle = \sum_{\{P\}} \int_y \sum_j \delta(y_j - z) \prod_{i<j} Z_{ij}(\xi_i - \xi_j) Z^*_{P_i P_j}(\xi_{P_i} - \xi_{P_j})$$

$$\times \prod_j \frac{1}{4\pi t} e^{-i(\xi_j x_j - \xi_{P_j} x'_j)}. \tag{5.42}$$

Note that the above product of S-matrices is actually independent of the ordering of the y. First, only those terms appear in the product for which the permutation P has an inversion. For example, say for three particles, if $P = 312$, then the inversions are 13 and 23. It is only these terms that give a nontrivial S-matrix contribution. For the noninverted terms, here 12, we get

$$\frac{\xi_1 - \xi_2 - ic\,\mathrm{sgn}(y_1 - y_2)}{\xi_1 - \xi_2 - ic} \frac{\xi_1 - \xi_2 + ic\,\mathrm{sgn}(y_1 - y_2)}{\xi_1 - \xi_2 + ic}, \tag{5.43}$$

which is always unity irrespective of the ordering of y_1, y_2. For a term with an inversion, say 23, we get

$$\frac{\xi_2 - \xi_3 - ic\,\mathrm{sgn}(y_2 - y_3)}{\xi_2 - \xi_3 - ic} \frac{\xi_3 - \xi_2 + ic\,\mathrm{sgn}(y_3 - y_2)}{\xi_3 - \xi_2 + ic}, \tag{5.44}$$

which is always equal to

$$\frac{\xi_2 - \xi_3 + ic}{\xi_2 - \xi_3 - ic} \equiv S(\xi_2 - \xi_3), \tag{5.45}$$

irrespective of the sign of $y_2 - y_3$. This allows us to carry out the integration over the y_j.

In order to calculate physical observables, we have to choose initial states. We treat here the lattice state with N particles distributed uniformly in a series of harmonic traps given by

$$|\Psi_{\mathrm{latt}}\rangle = \int_x \prod_{j=1}^{N} \frac{1}{(\pi\sigma^2)^{1/4}} e^{-[x_j + (j-1)a]^2/2\sigma^2} b^\dagger(x_j)|0\rangle, \tag{5.46}$$

such that the overlap between the wavefunctions of two neighboring particles is negligible. In this particular case, the ordering of the particles is induced by the limited nonoverlapping support of the wavefunction.

In this lattice-like state, the initial wavefunction starts out with the neighboring particles having negligible overlap. At short time (as seen from (5.34)), the particles repel each other and never cross, owing to the repulsive interaction. So, at long time, the interaction does not play a role since the wavefunctions are sufficiently non-overlapping. It is only the $P = 1$ contribution then that survives, and we get for the density

$$\langle \vec{x}', t | \rho(z) | \vec{x}, t \rangle = \sum_j \prod_{k \neq j} \frac{\delta(x_k - x'_k) e^{-i(z/2t)(x_j - x'_j)}}{4\pi t}. \tag{5.47}$$

We need to integrate the position basis vectors $|\vec{x}\rangle$ over some initial condition. We do this here for the lattice state (5.46). This gives

$$\rho_{\text{latt}}(\xi_z) = \langle \Psi_{\text{latt}}, t | \rho(z) | \Psi_{\text{latt}}, t \rangle = \frac{N\sigma}{2\sqrt{\pi}\, t} e^{-\xi_z^2/\sigma^2}. \tag{5.48}$$

Mathematically, any S-matrix factor that appears will necessarily have zero contribution from the pole—this is easy to see from the pole structure and the ordering of the coordinates. In order to get a nonzero result, we need to fix at least two integration variables (i.e., the y_j). Thus, the first nontrivial contribution comes from the two-point correlation function.

We now proceed to calculate the evolution of the noise, i.e., the two-body correlation function:

$$\rho_2(z, z'; t)_{\text{latt}} = \langle \Psi_{\text{latt}}, t | \rho(z) \rho(z') | \Psi_{\text{latt}}, t \rangle \tag{5.49}$$

The contributions can be grouped in terms of number of crossings, which corresponds to a grouping in terms of the coefficient e^{-ca} [27]. The leading-order term can be explicitly evaluated, and we show below which terms contribute. In general, we have

$$\rho_{2\,\text{latt}}(z, z'; t) = \sum_{\{P\}} \int_y \left[\sum_{j,K} \delta(y_j - z) \delta(y_k - z') \right]$$
$$\times \prod_{\substack{i<j \\ (ij) \in P}} S(\xi_i - \xi_j) \prod_j \frac{1}{4\pi t} e^{-i(\xi_j x_j - \xi_{P_j} x'_j)}. \tag{5.50}$$

The above shorthand in the S-matrix product means that only the (ij) that belong to the inversions in P are included. A detailed discussion of how to carry out the summation is given in [1]. Here we just quote the result:

$$\rho_{2\,\text{latt}}(z, z') = \frac{N^2 \sigma^2}{4\pi t^2} e^{-(\xi_z^2 + \xi_{z'}^2)\sigma^2}$$

$$\times \left\{ 1 + \frac{2}{N^2} \text{Re}[S(\xi_z - \xi_{z'})]\, e^{ia(z-z')} \frac{N(1 - e^{ia(z-z')}g) + e^{iaN(z-z')}g_{zz'}^N - 1}{(1 - g_{zz'} e^{ia(z-z')})^2} \right\}, \tag{5.51}$$

where

$$g_{zz'} =$$
$$1 - 2c\sqrt{\pi}\sigma S(\xi_z - \xi_{z'} - ic)\{e^{(c+i\xi_z)^2\sigma^2}\operatorname{erfc}[(c+i\xi_z)\sigma] + e^{(c-i\xi_{z'})^2\sigma^2}\operatorname{erfc}[(c-i\xi_{z'})\sigma]\}.$$

$$(5.52)$$

To compare with the Hanbury Brown–Twiss result, we calculate the normalized spatial noise correlations, given by

$$C_2(z, z') \equiv \frac{\rho_2(z, z')}{\rho(z)\rho(z')} - 1.$$

$$(5.53)$$

In the noninteracting case, i.e., $c = 0$, $S(\xi) = 1$ and $g_{zz'} = 0$, and we recover the Hanbury Brown–Twiss result for $N = 2$:

$$C_2^0(\xi_z, \xi_{z'}) = \tfrac{1}{2}\cos[a(\xi_z - \xi_z')].$$

$$(5.54)$$

One can also check that the limit $c \to \infty$ gives the expected answer for free fermions, namely,

$$C_2^\infty(\xi_z, \xi_{z'}) = -\tfrac{1}{2}\cos[a(\xi_z - \xi_z')].$$

$$(5.55)$$

At finite c, we can see a sharp fermionic character appear that broadens with increasing c as shown in Fig. 5.10. The large-time behavior is captured in a small window around $\xi = 0$. It can be seen that at any finite c, the region near zero develops a strong fermionic character, indicating that, irrespective of the value of the coupling with which we start, the model flows toward an infinitely repulsive model at large time,

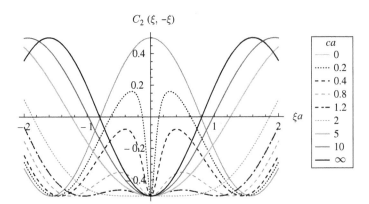

Fig. 5.10 Normalized noise correlation function $C_2(\xi, -\xi)$. Fermionic correlations develop on a timescale $\tau \sim c^{-2}$, so, for any c, we get a sharp fermionic peak near $\xi = 0$, i.e., at large time. (Reprinted with permission from [32]. Copyright (2012) by the American Physical Society.)

which can be described in terms of free fermions. We also obtained this result "at" $t = \infty$ at the beginning of this section.

For higher particle number, we see "interference fringes" that get narrower and more numerous, retaining memory of the initial lattice state. However, the asymptotic fermionic character does not disappear. Figures 5.11 and 5.12 show the noise correlation function for five and ten particles, respectively. The large peaks are interspersed with smaller peaks, and so on. This reflects the character of the initial state.

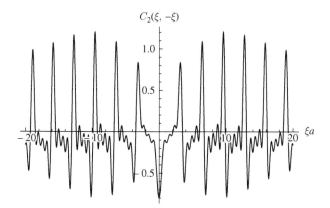

Fig. 5.11 Normalized noise correlation function for five particles released for a Mott-like state for $c > 0$. (Reprinted with permission from [32]. Copyright (2012) by the American Physical Society.)

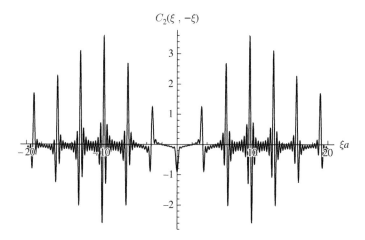

Fig. 5.12 Normalized noise correlation function for ten particles released for a Mott-like state for $c > 0$. (Reprinted with permission from [1]. Copyright (2013) by the American Physical Society.)

Attractive interactions: quench asymptotics in the lattice state

For the attractive case, since the contours of integration are spread out in the imaginary direction, we have the contributions from the poles in addition to the stationary-phase contribution at large time. The stationary-phase contribution is picked up on the real line, but as we move the contour, it stays pinned above the poles and we need to include the residue obtained from going around them, leading to a sum over several terms. Figure 5.13 shows an example of how this works.

In [32], a formula was provided for the asymptotic state. Here we give a more careful treatment by taking into account that the fixed point of the approximation moves for terms that come from a pole of the S-matrix. It is therefore necessary to first shift the contours of integration, and then carry out the integral at long time. We do this below.

Shifting a contour over a pole leads to an additional term from the residue:

$$\int_{\gamma_2} \frac{d\lambda_2}{2\pi} \to \int_{\gamma_2^R} \frac{d\lambda_2}{2\pi} - i\mathcal{R}(\lambda_2 \to \lambda_1 + i|c|), \tag{5.56}$$

where $\mathcal{R}(x)$ indicates that we evaluate the residue given by the pole x. γ_j indicates the original contour of integration and γ_j^R indicates that integration is carried out over the real axis. Proceeding with the other variables, we end up with the following expression:

$$\int_{\gamma_1, \gamma_2, \ldots, \gamma_N} \to \int_{\gamma_1^R} \left[\int_{\gamma_2^R} i\mathcal{R}(\lambda_2 \to \lambda_1 + i|c|) \right]$$

$$\times \left[\int_{\gamma_3^R} i\mathcal{R}(\lambda_3 \to \lambda_1 + i|c|) + i\mathcal{R}(\lambda_3 \to \lambda_2 + i|c|) \right] \cdots$$

$$\times \left[\int_{\gamma_N^R} +i\mathcal{R}(\lambda_N \to \lambda_1 + i|c|) + i\mathcal{R}(\lambda_N \to \lambda_2 + i|c|) \right.$$

$$\left. + \ldots + i\mathcal{R}(\lambda_N \to \lambda_{N-1} + i|c|) \right]. \tag{5.57}$$

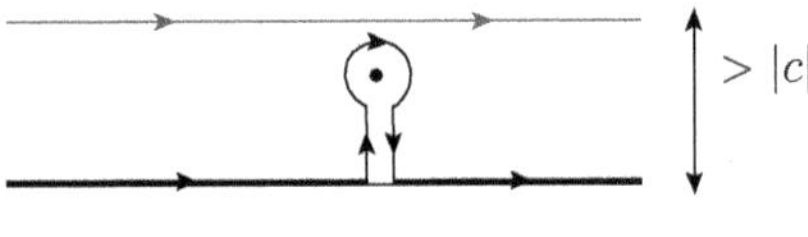

Fig. 5.13 Contribution from stationary phase and pole at large time in the attractive model. The lower contour represents the shifted contour. (Reprinted with permission from [1]. Copyright (2013) by the American Physical Society.)

The integrals can now be evaluated using the stationary-phase approximation. The correction produced by the above procedure does not affect the qualitative features observed in [32].

We now calculate the evolution of the density and the two-body correlation function in order to compare with the repulsive case. We will first study the two-particle case. Although we have a finite-time expression for this case from which we can directly take a long-time limit, we will study the asymptotics using the above scheme for an N-particle state, since we have an analytical expression to go with. We get two terms, the first of which is the stationary-phase contribution and is just like the repulsive case with $c \to -c$. The second is the contribution from the pole. It contains the bound-state contribution, which brings about another interesting feature of the attractive case. While the asymptotic dynamics of the repulsive model is solely dictated by the new variables $\xi_j \equiv y_j/2t$, and all the time dependence of the wavefunction enters through this "velocity" variable, this is not the case in the attractive model. While it is true that the system is naturally described in terms of ξ variables, there still exists a nontrivial time dependence.

First, we integrate out the x dependence, assuming an initial lattice-like state. This gives

$$|\Psi_{\text{latt}}(t)\rangle = \int_y \sum_{\substack{\xi_j^* = \xi_j \\ \xi_i^* + ic, i<j}} \prod_{i<j} S_{ij}(\xi_i^* - \xi_j^*) \prod_j \frac{(4\pi\sigma^2)^{1/4}}{\sqrt{4\pi it}}$$

$$\times e^{-(\sigma^2/2 + it)(\xi_j^*)^2 + i\xi_j^*[2t\xi_j + a(j-1)]} b^\dagger(y_j)|0\rangle. \tag{5.58}$$

Defining $\phi(\xi, t)$ from

$$|\Psi_{\text{latt}}(t)\rangle = \int_y \phi(\xi, t) \prod_j b^\dagger(y_j)|0\rangle, \tag{5.59}$$

we have for the density evolution under attractive interactions, $c < 0$,

$$\bar{\rho}_{\text{latt}}(z; t) = \sum_{\{P\},j} \int_y \delta(y_j - z)\phi^*(\xi_P, t)\phi(\xi, t). \tag{5.60}$$

We can show numerically (the expressions are a bit unwieldy to write here) that, asymptotically, the density shows the same Gaussian profile that we expect from a uniformly diffusing gas, namely, $e^{-\xi^2 \sigma^2}$.

With this, we can proceed to compute the noise correlation function. The two-particle case is easy, since there are no more integrations to carry out. We get

$$\bar{\rho}_{2\,\text{latt}}(z, z'; t) = \sum_{\{P\},j,k} \int_y \delta(y_j - z)\delta(y_k - z')\phi^*(\xi_P, t)\phi(\xi, t)$$

$$= |\phi_s(\xi_z, \xi_z')|^2, \tag{5.61}$$

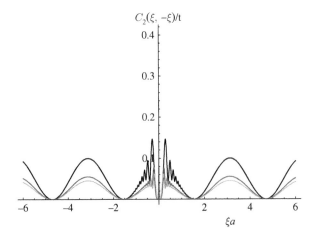

Fig. 5.14 Noise correlation $C_2(\xi, -\xi)$ for two particles in the attractive case plotted for three different times. Note the growth of the central peak. At larger times, the correlations away from zero fall off. The top, middle, and bottom curves are for $ta^2 = 20$, 40, and 60, respectively. (Reprinted with permission from [1]. Copyright (2013) by the American Physical Society.)

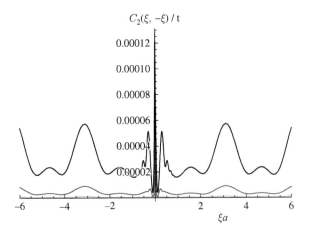

Fig. 5.15 Noise correlation $C_2(\xi, -\xi)$ for three particles in the attractive case plotted for three different times. At larger times, the correlations away from zero fall off. The top, middle, and bottom curves are for $ta^2 = 20$, 40, and 60, respectively. (Reprinted with permission from [32]. Copyright (2012) by the American Physical Society.)

where ϕ_s is the symmetrized wavefunction. Figures 5.14 and 5.15 show the normalized noise correlations for different values of t for two and three particles, respectively.

For more particles, we see interference fringes similar to the repulsive case. We note that the central peak increases and sharpens with time, indicating an increasing contribution from bound states to the correlations (see Fig. 5.15 for an example).

The condensate: attractive and repulsive interactions

We study the evolution of the Bose gas after a quench from an initial state where all the bosons are in a single level of a harmonic trap. For $t < 0$, the state is described by

$$|\Psi_{\text{cond}}\rangle = \int_x \mathcal{S}_x \prod_j \frac{e^{-x_j^2/\sigma^2}}{(\pi\sigma^2)^{1/4}} b^\dagger(x_j)|0\rangle. \tag{5.62}$$

Recall that in order to use the Yudson representation, the initial state needs to be ordered. We can rewrite the above state as

$$|\Psi_{\text{cond}}\rangle = \int_x \mathcal{S}_x \theta(x_1 > \cdots > x_N) \prod_j \frac{e^{-x_j^2/\sigma^2}}{(\pi\sigma^2)^{1/4}} b^\dagger(x_j)|0\rangle, \tag{5.63}$$

where $\mathcal{S}$ is a symmetrizer. The time evolution can be carried out via the Yudson representation, and again we concentrate on the asymptotics. For the repulsive model, the stationary-phase contribution is all that appears, and we obtain the following expression:

$$|\vec{x}\rangle = \int_y \prod_{i<j} S_{ij}^y \left(\frac{y_i - y_j - x_i + x_j}{2t} \right)$$
$$\times \prod_j \frac{1}{\sqrt{2\pi it}} e^{-i(y_j - x_j)^2/4t + i(y_j - x_j)^2/2t} b^\dagger(y_j)|0\rangle. \tag{5.64}$$

At large time t, we therefore have

$$|\Psi_{\text{cond}}(t)\rangle = \int_{x,y} \theta(x_1 > \cdots > x_N)\phi_2(x)I(y,x,t) \prod_j b^\dagger(y_j)|0\rangle. \tag{5.65}$$

$\phi_2(x)$ is symmetric in the x, while $I(y,x,t)$ is symmetric in the y but not in the x. Therefore, we have to carry out the x integration over the wedge $x_1 > \cdots > x_N$. This is not straightforward. If $I(y,x,t)$ were also symmetric in x, then we could add the other wedges to rebuild the full space in x. However, owing to the presence of the S-matrix factors, symmetrizing in y does not automatically symmetrize in x. The exponential factors, on the other hand, are automatically symmetric in both variables if one of them is symmetrized, because their functional dependence is of the form $f(y_j - x_j)$. It is, however, possible to make the S-matrix factors approximately symmetric in x (we will define what we mean by approximately shortly). What is important is to obtain a $y_j - x_j$ dependence. As of now, the S-matrix that appears in (5.64) is

$$S_{ij}^y \left(\frac{y_i - y_j - x_i + x_j}{2t} \right) = \frac{\dfrac{y_i - y_j - x_i + x_j}{2t} - ic\,\text{sgn}(y_i - y_j)}{\dfrac{y_i - y_j - x_i + x_j}{2t} - ic}. \tag{5.66}$$

First, we can change $\text{sgn}(y_i - y_j)$ to $\text{sgn}((y_i - y_j)/2t)$, since $t > 0$. Next, we note that, asymptotically in time, the stationary-phase contribution comes from $y/2t \sim \mathcal{O}(1)$.

However, since x has finite extent, at large enough time, $x/2t \sim 0$. We are therefore justified in writing

$$
\mathrm{sgn}\left(\frac{y_i - x_i}{2t} - \frac{y_j - x_j}{2t} \right).
$$

The only problem could arise when $y_i \sim y_j$. However, if this occurs, then the S-matrix is approximately $\mathrm{sgn}(y_i - y_j)$, which is antisymmetric in ij. With this prefactor, the particles are effectively fermions, and therefore at $y_i \sim y_j$, the wavefunction has an approximate node. At large time, therefore, we do not have to be concerned with the possibility of particles overlapping, and our inclusion of the x_i inside the sgn function is valid. With this change, the S-matrix also becomes a function of $y_j - x_j$, and when symmetrizing over y, one automatically symmetrizes over x.

In short, we have established that asymptotically in time, the wavefunction can be made symmetric in x. This allows us to rebuild the full space. We get

$$
|\Psi_{\mathrm{cond}}(t)\rangle = \int_{x,y} \sum_P \theta(x_P)\phi_2(x)I^s(y,x,t) \prod_j b^\dagger(y_j)|0\rangle
$$

$$
= \int_{x,y} \phi_2(x)I^s(y,x,t) \prod_j b^\dagger(y_j)|0\rangle, \tag{5.67}
$$

where the superscript s indicates that we have established that $I(y,x,t)$ is also symmetric in x. With this in mind, we can do away with the ordering when we are integrating over the x if we symmetrize the initial-state wavefunction and the final wavefunction. Note that when we calculate the expectation value of a physical observable, the symmetry of the wavefunction is automatically enforced, and thus taken care of automatically.

Recall that when we calculated the noise correlations of the repulsive gas, in order to get an analytic expression for N particles, we considered the leading-order term, i.e., the Hanbury Brown–Twiss term. We did this by showing that higher-order crossings produced terms of higher order in e^{-2ca}, which we claimed was a small number. Now, however, $a = 0$, and although the calculation is essentially the same with our approximate symmetrization, this simplification does not occur. The two- and three-particle results remain analytically calculable, but for higher numbers of particles, we have to resort to numerical integration. Figure 5.16 shows the noise correlation for two and three repulsive bosons starting from a condensate. For noninteracting particles, we expect a straight line $C_2 = \frac{1}{2}$. When repulsive interactions are turned on, we see the characteristic fermionic dip develop. The plots for the attractive Bose gas are shown in Figs. 5.17 and 5.18. As expected from the noninteracting case, the oscillations arising from the interference of particles separated spatially do not appear. The attractive case, however, does show the oscillations near the central peak that are also visible in the case when we start from a lattice-like state. It is interesting to note that for three particles we do not see any additional structure develop in the attractive case.

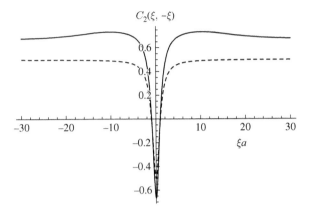

Fig. 5.16 Noise correlation $C_2(\xi, -\xi)$ for two (dashed, bottom, curve) and three (full, upper, curve) repulsive bosons starting from a condensate. Unlike the attractive case, there is no explicit time dependence asymptotically. $ca = 3$. (Reprinted with permission from [32]. Copyright (2012) by the American Physical Society.)

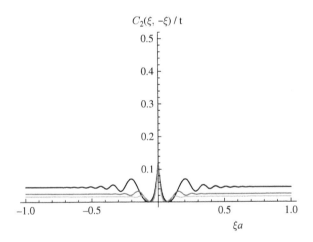

Fig. 5.17 Noise correlation $C_2(\xi, -\xi)$ for two attractive bosons starting from a condensate. The top, middle, and bottom curves are for $tc^2 = 20$, 40, and 60, respectively. As time increases, the central peak dominates. (Reprinted with permission from [1]. Copyright (2013) by the American Physical Society.)

Quenching from a bound state

In this brief subsection, our initial state is the ground state of the attractive Lieb–Liniger Hamiltonian (with interaction strength $-c_0 < 0$. For two bosons, this takes the form [19]

$$|\Psi_{\text{bound}}\rangle = \int_{\vec{x}} e^{-c_0|x_1-x_2|-x_1^2/2\sigma^2-x_2^2/2\sigma^2} \, b^\dagger(x_1)b^\dagger(x_2)|0\rangle, \qquad (5.68)$$

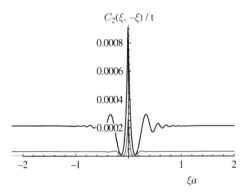

Fig. 5.18 Noise correlation $C_2(\xi, -\xi)$ for three attractive bosons starting from a condensate. Note that the side peak structure found in Fig. 5.15 is missing because of the initial condition. The top, middle, and bottom curves are for $tc^2 = 20$, 40, and 60, respectively. As time increases, the oscillations near the central peak die out. (Reprinted with permission from [32]. Copyright (2012) by the American Physical Society.)

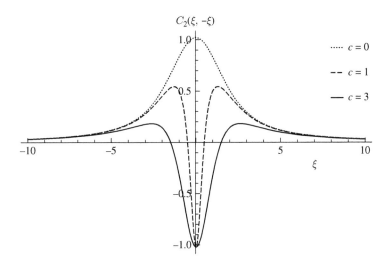

Fig. 5.19 Normalized noise correlation function for two particles quenched from a bound state into the repulsive regime. The legend indicates the values of c into which the state is quenched. We start with $c_0\sigma^2 = 3$, $\sigma = 1$, c_0 being the interaction strength of the initial-state Hamiltonian. Again, we see the fermionic dip, but the rest of the structure is determined by the initial state. (Reprinted with permission from [1]. Copyright (2013) by the American Physical Society.)

and we quench it with a repulsive Hamiltonian. The long-time noise correlations are shown in Fig. 5.19. We see that while the initial-state correlations are preserved over most of the evolution, in the asymptotic long-time limit the characteristic fermionic dip appears. We expect similar effects for any number of bosons.

A scaling argument

We noticed that as time evolved, the system developed strong correlations, enhancing the effective coupling in both the attractive and repulsive cases. This was shown through detailed calculations of the asymptotics carried out by means of saddle-point arguments. Here we proceed to show it via a simpler, though less powerful, reasoning. We begin by considering repulsive interactions. From (5.29), we see that the scaling $\lambda \to \lambda\sqrt{t}$ gives $Z_{ij}^y(\lambda_i - \lambda_j) \to \text{sgn}(y_i - y_j) + O(1/\sqrt{t})$, yielding, to leading order,

$$
\begin{aligned}
|\Psi_0, t\rangle &\to \int_x \int_y \int_\lambda \theta(\vec{x})\Psi_0(\vec{x}) \prod_j \frac{1}{\sqrt{t}} e^{-i\lambda_j^2 + i\lambda_j(y_j - x_j)/\sqrt{t}} \prod_{i<j} \text{sgn}(y_i - y_j) b^\dagger(y_j)|0\rangle \\
&= \int_{x,y,\lambda,k} \theta(\vec{x})\Psi_0(\vec{x}) \prod_j e^{-i\lambda_j^2 t + i\lambda_j(y_j - x_j)} e^{-ik_j y_j} c_{k_j}^\dagger |0\rangle \\
&= \int_{x,k} \theta(\vec{x})\Psi_0(\vec{x}) \prod_j e^{-ik_j^2 t - ik_j x_j} c_{k_j}^\dagger |0\rangle \\
&= e^{-iH_0^f t} \int_x \mathcal{A}_x \theta(\vec{x})\Psi_0(\vec{x}) \prod_j c^\dagger(x_j)|0\rangle,
\end{aligned}
\tag{5.69}
$$

with $c^\dagger(y)$ being fermionic creation operators replacing the hard-core bosonic operators, $\prod_j c^\dagger(y_j) = \prod_{i<j} \text{sgn}(y_i - y_j) b^\dagger(y_j)$. We denote by $H_0^f = \int_x \partial c^\dagger(x)\partial c(x)$ the free fermionic Hamiltonian, and $\mathcal{A}_y$ is an antisymmetrizer acting on the y variables. Thus, the repulsive Bose gas, for *any* value of $c > 0$, is governed at long time by the $c = \infty$ hard-core boson limit (or its fermionic equivalent) [33, 34], and the system equilibrates (but does not thermalize), with antisymmetric wavefunction $\tilde{\Psi}_0(\vec{y}) = \mathcal{A}_y \theta(\vec{y})\Psi_0(\vec{y})$ and total energy $E_{\Psi_0} = \langle\Psi_0|H|\Psi_0\rangle$. Hence the "fermionic" dip we observed in the noise correlation function.

For attractive interactions, the argument needs to be refined, since the contributions of the poles have to be taken into account. These dominate in the long-time limit, again corresponding to an effective increase in the attractive interaction strength as time evolves.

5.3   Conclusions and the dynamic renormalization group hypothesis

We have shown that the Yudson contour integral representation for arbitrary states can indeed be used to understand aspects of the quench dynamics of the Lieb–Liniger model and to obtain the asymptotic wavefunctions exactly. This representation overcomes some of the major difficulties involved in using the Bethe ansatz to study the dynamics of some integrable systems by automatically accounting for complicated states in the spectrum.

We see some interesting dynamical effects at long times. The infinite-time limit of the repulsive model corresponds to particles evolving with a free fermionic Hamiltonian. It retains, however, a memory of the initial state and therefore is not a thermal state. The correlation functions approach that of hard-core bosons at long time, indicating a dynamical increase in interaction strength. The attractive model also shows a dynamical strengthening of the interaction, and the long-time limit is dominated by a multiparticle bound state. This of course does not mean that it condenses. In fact, the state diffuses over time, but remains strongly correlated.

We may interpret our results in terms of a "dynamic renormalization group" in time. The asymptotic evolutions of the model for both $c > 0$ and $c < 0$ are given by the Hamiltonians $H_{\pm}^{*}$ with $c \to \pm\infty$, respectively. Accepting the renormalization group logic behind the conjecture, one would expect there to be basins of attraction around the Lieb–Liniger Hamiltonian with models whose long-time evolution would bring them close to the "dynamic fixed points" $H_{\pm}^{*}$. One such Hamiltonian would have short-range potentials replacing the δ-function interaction that renders the Lieb–Liniger model integrable. Perhaps lattice models could also be found in this basin.[2]

We have to emphasize, however, that as these models are not integrable, we do not expect that they would actually flow to $H_{\pm}^{*}$. Instead, starting close enough in the "basin," they would flow close to $H_{\pm}^{*}$ and spend much time in its neighborhood, eventually evolving into another, thermal state. We thus conjecture that away from integrability, a system would approach the corresponding nonthermal equilibrium, where the dynamics will slow down, leading to a "prethermal" state [35]. Figure 5.20 shows a schematic of this. Such prethermalization behavior has indeed been observed in lattice models [36]. The system is expected to eventually find a thermal state. It is therefore of interest to characterize different ways of breaking integrability to see when a system is "too far" from integrability to see this effect and to determine in what regimes a system can be considered to be close to integrability. For a review and background on this subject, see [5].

Further, the flow diagram in Fig. 5.20 might have another axis that represents initial states. Study of the Bose–Hubbard model [1] shows that it has an interesting initial-state dependence. Whereas the sign of the interaction does not affect the quench dynamics, the asymptotic state depends strongly on the initial state, with a lattice-like state leading to fermionization and a condensate-like state retaining bosonic correlations. The strong dependence on the initial state in the quench dynamics is evident from (5.4) and is the subject of much debate, in particular with regard to the eigenstate thermalization hypothesis [9–11].

---

[2] In spite of its similarity to the Lieb–Liniger model, the Bose–Hubbard model is not such a model. The Hamiltonian as defined on the bipartite lattice has a symmetry that leads attractive and repulsive interactions to evolve identically [1] Such a symmetry is not present in the Lieb–Liniger model. Adding to the lattice model such terms as the next-nearest hopping or interactions that break this symmetry may lead to Hamiltonians in the same basin of attraction

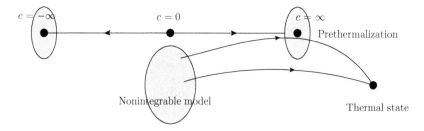

Fig. 5.20 Schematic diagram showing pre-thermalization of states in a nonintegrable model. (Reprinted with permission from [1]. Copyright (2013) by the American Physical Society.)

We have shown here how to compute quantum quenches for the Lieb–Liniger model and have provided predictions for experiments that can be carried out in the context of continuum cold-atom systems. Applications of the approach to other models are underway. This work also opens up several new questions. Although the representation is provable mathematically, further investigation is required to understand physically how it achieves the tedious sum over eigenstates, while automatically accounting for the details of the spectrum. It would also be useful to tie this approach to other means of calculating overlaps in the algebraic Bethe ansatz, i.e., the form-factor approach. The representation can essentially be thought of as a different way of writing the identity operator. From that standpoint, it could serve as a new way of evaluating correlation functions using the Bethe ansatz.

Acknowledgments

We are grateful to G. Goldstein for very useful discussions. This work was supported by NSF Grant DMR 1410583.

References

[1] D. Iyer, H. Guan, and N. Andrei. *Phys. Rev. A* **87**, 053628 (2013).
[2] W. Liu and N. Andrei. *Phys. Rev. Lett.* **112**, 257204 (2014).
[3] I. Bloch, J. Dalibard, and W. Zwerger. *Rev. Mod. Phys.* **80**, 885 (2008).
[4] H. Moritz, T. Stöferle, M. Köhl, and T. Esslinger. *Phys. Rev. Lett.* **91**, 250402 (2003).
[5] A. Polkovnikov, K. Sengupta, A. Silva, and M. Vengalattore. *Rev. Mod. Phys.* **83**, 863 (2011).
[6] M. A. Cazalilla, R. Citro, T. Giamarchi, E. Orignac, and M. Rigol. *Rev. Mod. Phys.* **83**, 1405 (2011).
[7] X.-W. Guan, M. T. Batchelor, and C. Lee. arXiv:cond-mat/1301.6446.
[8] L. Vidmar, S. Langer, I. McCulloch, U. Schneider, U. Schollwoeck, and F. Heidrich-Meisner. arXiv:cond-mat/1305.5496.
[9] M. Rigol, V. Dunjko, and M. Olshanii. *Nature* **451**, 854 (2008).
[10] M. Srednicki. *Phys. Rev. E* **50**, 888 (1994).

[11] J. M. Deutsch. *Phys. Rev. A* **43**, 2046 (1991).

[12] T. Giamarchi. *Quantum Physics in One Dimension* (Oxford University Press, 2004).

[13] M. Rigol, V. Dunjko, V. Yurovsky, and M. Olshanii. *Phys. Rev. Lett.* **98**, 050405 (2007).

[14] P. Calabrese and J. Cardy. *J. Stat. Mech.* **2007**, P10004 (2007).

[15] J.-S. Caux and F. H. L. Essler. arXiv:1301.3806.

[16] J.-S. Caux and R. M. Konik. *Phys. Rev. Lett.* **109**, 175301 (2012).

[17] H. Bethe. *Z. Phys.* **71**, 205 (1931).

[18] E. H. Lieb and F. Y. Wu. *Phys. Rev. Lett.* **20**, 1445 (1968).

[19] E. H. Lieb and W. Liniger. *Phys. Rev.* **130**, 1605 (1963).

[20] N. Andrei, K. Furuya, and J. H. Lowenstein. *Rev. Mod. Phys.* **55**, 331 (1983).

[21] N. Andrei. In *Low-Dimensional Quantum Field Theories for Condensed Matter Physicists: Lecture Notes of ICTP Summer Course, Trieste, Italy, September 1992*, ed. S. Lundqvist, G. Morandi, and Yu Lu (World Scientific, 1995).

[22] A. Tsvelik and P. Wiegmann. *Adv. Phys.* **32**, 453 (1983).

[23] C. N. Yang and C. P. Yang. *J. Math. Phys.* **10**, 1115 (1969).

[24] R. H. Dicke. *Phys. Rev.* **93**, 99 (1954).

[25] V. I. Yudson. *Sov. Phys. JETP* **61**, 1043 (1985).

[26] C. A. Tracy and H. Widom. *J. Phys. A: Math. Theor.* **41**, 485204 (2008).

[27] A. Lamacraft. *Phys. Rev. A* **84**, 043632 (2011).

[28] S. Prolhac and H. Spohn. *J. Math. Phys.* **52**, 122106 (2011).

[29] C. N. Yang. *Phys. Rev.* **168**, 1920 (1968).

[30] R. Hanbury Brown and R. Q Twiss. *Nature* **177**, 27 (1956).

[31] G. Goldstein and N. Andrei. arXiv:1309:3471 [cond-mat.quant-gas].

[32] D. Iyer and N. Andrei. *Phys. Rev. Lett.* **109**, 115304 (2012).

[33] D. Jukić, R. Pezer, T. Gasenzer, and H. Buljan. *Phys. Rev. A* **78**, 053602 (2008).

[34] M. Girardeau. *J. Math. Phys.* **1**, 516 (1960).

[35] J. Berges, S. Borsányi, and C. Wetterich. *Phys. Rev. Lett.*, **93**, 142002 (2004).

[36] C. Kollath, A. M. Läuchli, and E. Altman. *Phys. Rev. Lett.* **98**, 180601 (2007).

6

Nonequilibrium density matrix for thermal transport in quantum field theory

Benjamin DOYON

Department of Mathematics
King's College
London, UK

Strongly Interacting Quantum Systems out of Equilibrium. First Edition. Thierry Giamarchi et al.
© Oxford University Press 2016. Published in 2016 by Oxford University Press.

Chapter Contents

Preface

In this chapter, I explain how to describe one-dimensional quantum systems that are simultaneously near to, but not exactly at, a critical point, and in a far-from-equilibrium steady state. This description uses a density matrix on scattering states (of the type of Hershfield's density matrix), or equivalently a Gibbs-like ensemble of scattering states. The steady state I am considering is one where there is a steady flow of energy along the chain, coming from the steady draining/filling of two faraway reservoirs put at different temperatures. The context I am using is that of massive relativistic quantum field theory, which is the framework for describing the region near quantum critical points in any universality class with translation invariance and with dynamical exponent z equal to 1. I show, in this completely general setup, that a particular steady-state density matrix occurs naturally from the physically motivated Keldysh formulation of nonequilibrium steady states. In this formulation, the steady state occurs as a result of a large-time evolution from an initial state where two halves of the system are separately thermalized. I also show how this suggests a particular dependence (a "factorization") of the average current on the left and right temperatures. The idea of this density matrix has already been proposed in a recent publication with my collaborator Denis Bernard, where we studied it in the context of conformal field theory.

6.1   Introduction

This chapter will explain how to describe quantum systems that are simultaneously near to (but not exactly at) a quantum critical point, and in certain steady states away from equilibrium. This situation is particularly intriguing because it mixes in an essential fashion both rich collective quantum effects and nonequilibrium physics. Also, steady states, being time-independent, are as near as possible to the usual states of equilibrium physics while still being out of equilibrium, and we can provide general descriptions in parallel to parts of equilibrium statistical physics.

We have in mind quantum systems composed of a macroscopic number of microscopic constituents with local (few-neighbors) interactions on some lattice. In certain situations, for instance when a parameter (a magnetic field, an interaction strength, etc.) is adjusted appropriately, the system becomes critical. Near to such a quantum critical point, quantum effects dominate the low-energy and large-distance physics. Indeed, on the one hand, the ground state encodes quantum correlations between local observables that exist much beyond few lattice sites. This is nontrivial, because with local interactions, one expects that constituents separated by many sites do not "feel" each other very much, which is indeed what happens away from critical points. On the other hand, at least intuitively, the low-lying excitations can be understood as large-scale collective behaviors of the underlying constituents that emerge thanks to quantum fluctuations—as if large groups started to act together forming new quantum objects. Critical points arguably give some of the most interesting physics: they are where many nontrivial physical effects arise (one of the most well-known being the Kondo effect); and their large-distance / low-energy physics is actually universal, because most of the intricacies of the local interactions have been lost in the averaging out over many sites that arises in large-distance correlations or collective objects.

The physical theory that describes what is observed "near" critical points is quantum field theory (QFT)—more precisely, when not exactly at the critical point, massive QFT. By "near" we mean in the limit where observation lengths, lengths related to the system's geometry, and correlation lengths are all large in lattice spacings, these three kind of lengths being taken to infinity in fixed proportions.[1] This is the scaling limit. QFT is the physical theory that will provide predictions for what happens in the scaling limit (for instance, how correlations decay to zero, giving both the exponents and universal amplitudes). Of course, in reality, one never reaches exactly the scaling limit. But when criticality is approached in macroscopic systems, all large-distance and low-energy observations are effectively predicted by QFT.

Note that if the correlation lengths are much larger than everything else, then we are exactly "at" the critical point. In these cases, QFT possesses scale invariance, and, often with this, invariance under the full global conformal group. In $1 + 1$ dimensions (the scaling limit of quantum chains), this very often comes with local conformal invariance, and we have a conformal field theory (CFT). We will not describe these situations here, and refer the reader to [1] for the CFT counterpart of the nonequilibrium steady states we will study here.

The equilibrium physics of critical and near-critical points has been very well studied over the years, and there are by now a wide range of techniques to obtain exact and approximate results. Not as much can be said of the physics away from equilibrium. One reason is, of course, the large variety of out-of-equilibrium situations (instead of saying what the situation is, we say what it is not ...). But also, even in the relatively simple cases where a steady state exists, there is a lack of a framework as powerful as and as widely applicable as that of Gibbs ensembles. Hence, one often must revert to other formulations, usually rather adapted to the type of physical model at hand. Here, we will start, in the massive QFT context, with an "explicit," real-time construction of the steady state, where we implement theoretically the idealized experimental set-up that should produce the steady state.[2] This will lead us to rather general statements, paralleling Gibbs ensembles.

6.2 Basic notions

Consider a quantum chain of length L with local (nearest-neighbor, say) interactions. We want to describe a state where energy flows along the chain owing to a difference of temperatures at its extremities. There are various ways to do this: one could use Lindblad's theory of open quantum systems where the thermal baths have been traced out and the time evolution is nonunitary [2], or, in a somewhat related fashion, one

---

[1] The concept of observation length is kept slightly nebulous here, but, for example, this could be the distance between two points in a correlation function or, translating into energy–momentum language by the usual Fourier-transform inversion, the temperature.

[2] Similar real-time constructions, where unitary evolution is performed from a state that is not a Hamiltonian eigenstate, are usually referred to, in general, as "quenches."

could explicitly connect Caldeira–Leggett baths [3] with different temperatures on the extremities of the chain (as is done, for instance, in [4, 5]). One also could consider an infinite chain, instead of a finite one, and try to describe the flow using asymptotic states in a Gibbs-like ensemble, through what is usually referred to in the condensed matter community as Hershfield's density matrix (Hershfield considered this for general charge transport [6]; Ruelle obtained rigorous general results in the context of C^*-algebras with similar underlying ideas for heat transport [7], see also [8]); this of course requires a clear principle to fix this density matrix. Finally, also with boundaries sent to infinity, one could explicitly construct the baths with different temperatures by using parts of the system itself as baths: the infinite chain is initially cut into two semi-infinite parts, each thermalized at different temperatures, which are then connected and evolved unitarily for a long time. This latter method is in a sense the most "physical," since the steady state is explicitly constructed in real time (it is sometimes referred to as the "partitioning approach" of Caroli et al. [9], and sometimes as the Keldysh formulation, used for instance in the context of mesoscopic systems [10], and it is related to quantum quenches). Note that in this method, thanks to the unitary evolution, no part of the system is kept at a fixed temperature during time evolution; the information on the temperature is only in the initial mixed state. When combined with ideas of perturbation theory and path integrals, the Keldysh formulation leads to the powerful Keldysh perturbation theory (in fact, Keldysh perturbation theory can be used as well with the method of Caldeira–Leggett baths).

In the present chapter, we will start with the Keldysh formulation in the context of general, nonperturbative massive QFT, and we will show, using nonrigorous but well-motivated QFT arguments, that it leads to a Hershfield formulation, giving the precise Gibbs-like ensemble of asymptotic states describing the steady state where there is an energy flow. This has already been proposed in [1], where it was studied in the context of CFT (there, the heat current and its full counting statistics were found). See also [1, 11–18] for similar constructions concerning charge or heat transport in a variety of other situations and at various levels of rigor.

To be more specific, we consider the Hamiltonians H^l and H^r of the two semi-infinite, separate halves (left and right) of a quantum chain, and the initial density matrix

$$\rho_0 = e^{-\beta_l H^l - \beta_r H^r}, \tag{6.1}$$

where the halves are at different temperatures $T_{l,r} = \beta_{l,r}^{-1}$. We then evolve this density matrix with the full Hamiltonian $H = H^l + H^r + \Delta H$, where ΔH represents the energy term for the single link connecting both halves:

$$\rho(t) = e^{-iHt} \rho_0 e^{iHt}. \tag{6.2}$$

The steady-state average of observables is obtained upon evaluating

$$\lim_{t \to \infty} \frac{\text{Tr}[\rho(t)\mathcal{O}]}{\text{Tr}[\rho(t)]}. \tag{6.3}$$

More precisely, if v is a typical velocity of quantum propagation in the chain (which we will take in the following to be 1), then the limit is that vt must be much greater than the observation length (which could be taken as the length of the support of the observable $\mathcal{O}$), and much smaller than the length of the system itself (which we have implicitly taken as infinity). This is so that both halves of the system are big enough to play the role of thermal baths, absorbing and emitting energy as necessary, so that a nonequilibrium steady state may arise. See [12, 14, 17] for discussions about such steady-state limits.

Remark 6.1 *One may worry about the physical meaning, in the Hershfield and Keldysh formulations, of sending the extremities of the chain to infinity. Indeed, according to Fourier's law, the heat current (which is, in many situations, the same as the energy current) is proportional to the gradient of the "local temperature": $j = -c\nabla T$, with heat conductivity c. But if the extremities are infinitely far from each other, and no part of the system is actually kept at a fixed temperature during the evolution, surely the local temperature gradient, after infinite time, will be zero, and hence there should be no heat current, unless the heat conductivity c is infinite. Yet one of course finds that the heat current is nonzero (although we will not evaluate any current in the present chapter). We must realize that in low dimensions, Fourier's law is known to be broken: heat conduction is anomalous [19–21], and receives nonlocal contributions. In particular, it was shown in [28] that a nonzero Drude weight plays an important role. An underlying principle in the present context may be that of quantum coherence. Returning to Fourier's law for the sake of the argument, it should be understood in a coarse-grained fashion: one must take the limit of $T(x+\Delta x)/\Delta x$, where Δx is much smaller than the distance at which T changes significantly, but much greater than the distance at which things start to mix and scatter (so that, in particular, the concept of a local temperature is clear). But quantum coherence precludes scattering and mixing, so Δx must be much greater than the quantum coherence length. Near to quantum criticality, the correlation length is very large (in lattice spacings), and hence one may indeed expect the quantum transport behavior to be very different from that predicted by Fourier's law.*

Remark 6.2 *We will consider the case where there is a quantum critical point, and take the scaling limit, described by QFT. See [17, 22] for a discussion of the interplay between the scaling limit and the steady-state limit. In particular, the scaling limit describes the low-energy behaviors, and one may worry that some low-energy states of $H^l + H^r$ are high-energy states of H and are not captured by a QFT description of the steady-state limit (i.e., lie far outside the low-energy, universal range). The basic explanation [17, 22] is that if the renormalization of the observable with respect to the $H^l + H^r$ ground state is also finite with respect to the H ground state, then the correct steady-state average of this observable is captured by QFT.*

Remark 6.3 *Instead of connecting a single link, we could also think of connecting a whole finite-size (in lattice spacing) subpart of the chain (i.e., ΔH corresponds to a finite number of sites). Near to critical points, this will lead to the same low-energy* *results, because the correlation length is infinite in lattice spacings. In fact, we also expect that with ΔH corresponding to a number of site being a finite proportion of the correlation length, we still find the same steady state.*

6.3   Energy-flow steady states in massive $(1 + 1)$-dimensional quantum field theory

We restrict our attention to translation-invariant quantum chains in the thermo-dynamic limit and assume that there is a quantum critical point with dynamical exponent $z = 1$. Averages near to, or at, quantum critical points have asymptotic be-haviors described by QFT. With $z = 1$, the near-critical (not exactly critical) behavior of the quantum chain is described by massive relativistic QFT, which is the context that we consider. See, for instance, the standard reference [23] for a description of quantum critical points. In essence, if m is the gap between the lowest-energy state of the quantum chain and the next state, and J is some microscopic energy scale (like the intersite coupling), then the scaling limit leading to massive QFT is that where $T_l, T_r <, \simeq m \ll J$.

6.3.1   Asymptotic states and field configurations

One way of thinking about QFT is as a theory of quantized fields. Although this is exactly what the name relates to, it is not entirely accurate—in particular because of the need for renormalization. However, it does provide a good intuitive picture, and it gives rise to precise results when used correctly. In this picture, QFT is a quantum theory whose Hilbert space $\mathcal{C}$ has for basis the set $\{|\phi\rangle\}$ parametrized by all field configurations ϕ on real space, with overlap $\langle\phi|\phi'\rangle = \delta[\phi, \phi']$, the delta-functional supported on field configurations that are equal.[3] In the $(1 + 1)$-dimensional case, these are field configurations $\phi : \mathbb{R} \to \mathcal{S}$ on the line, where the field takes values in some space $\mathcal{S}$. There is the associated x-dependent operator $\hat{\phi}(x)$ (which, following tradition, we refer to as a field) defined by $\hat{\phi}(x)|\phi\rangle = \phi(x)|\phi\rangle$, and there is an energy density $\hat{h}(x)$, formed out of $\hat{\phi}(x)$ and its derivatives, using which the Hamiltonian is

$$H = \int_{-\infty}^{\infty} dx\, \hat{h}(x). \tag{6.4}$$

We can think intuitively of a field configuration as being the conglomeration of the individual basis states that the microscopic components of the underlying quantum model are in, but we must realize that on any finite length on $\mathbb{R}$, there are infinitely many such components, and that in the effective description of QFT, the space $\mathcal{S}$ is not necessarily the set of such microscopic-component basis states.

A somewhat surprising consequence of the principles of QFT (locality and Poincaré invariance) is that the Hamiltonian eigenfunctions and eigenvalues are completely

---

[3] It is not clear that this is actually a Hilbert space—restriction to the set of field configurations may be necessary, beside other subtleties—but it is sufficient for our intuitive description.

characterized by few parameters: the particle spectrum. Indeed, given a set of quantum numbers $\mathcal{N}$ and associated particle masses $\mathcal{M} = \{m_\alpha : \alpha \in \mathcal{N}\}$, a set of orthonormal basis states in the Hilbert space is given by

$$\{|\theta_1, \ldots, \theta_k\rangle_{\alpha_1, \ldots, \alpha_k} : \theta_1 > \ldots > \theta_k, \ \alpha_j \in \mathcal{N}, \ k = 0, 1, 2, \ldots\} \tag{6.5}$$

on which the Hamiltonian has eigenvalues

$$\sum_{j=1}^{k} m_{\alpha_j} \cosh \theta_j. \tag{6.6}$$

Here we are using the rapidity θ, in terms of which the relativistic (on-shell) energy and momentum are given by $E = m \cosh \theta$ and $p = m \sinh \theta$. When $k = 0$, we denote the state by $|\mathrm{vac}\rangle$. The sets $\mathcal{N}$ and $\mathcal{M}$ are characteristics of the QFT model, although they do not completely define it.

Remark 6.4 *Hence, in massive relativistic QFT, the problem of diagonalizing the Hamiltonian is almost solved: one only needs to determine a few parameters! Determining the masses can be hard, of course, and even when this is done, the QFT model is not yet solved, nor is it even fully defined. One more piece of information is necessary: that having to do with locality. One needs to say what the stress–energy tensor is, as a (tensorial) operator. This provides a definition not only of space via the momentum operator, but also of what operators are local: one can write the quantum locality condition of vanishing commutators with the energy density at space-like distances. This fully defines the QFT model, and to solve it (according to a certain definition), one needs to find all local operators and calculate their correlation functions. This in principle provides the measurable physical information encoded in the QFT model.*

The basis (6.5) is a basis of asymptotic states. Naturally, asymptotic states are particular linear combinations of field configurations, formally

$$|\theta_1, \ldots, \theta_k\rangle_{\alpha_1, \ldots, \alpha_k} = \sum_{\phi} c_{\alpha_1, \ldots, \alpha_k}(\theta_1, \ldots, \theta_k \,|\, \phi) \,|\phi\rangle. \tag{6.7}$$

Intuitively, they are linear combinations such that if we were to evolve the state, say, back in time for a very long time, we would find large wavepackets widely separated and propagating at rapidities $\theta_1, \ldots, \theta_k$. The wavepackets would be associated with the various particle types $\alpha_1, \ldots, \alpha_k$, so they would be wavepackets of the various "fundamental fields" related to the various particle types. These various fundamental fields can be components of $\hat{\phi}(x)$ (remember that ϕ takes values in $\mathcal{S}$, which may be $\mathbb{R}^n$ for an n-component field for instance), or some composite fields (like $\hat{\phi}(x)^2$, appropriately renormalized), depending on the model.

Of course, this last paragraph cannot be entirely true, because asymptotic states diagonalize the Hamiltonian; hence they evolve trivially (only getting a phase). A more correct statement is that one must construct linear combinations with the properties of the above paragraph, and then take the limit of infinite time, simultaneously infinitely large and infinitely separated wavepackets. In order to represent scattering states, this limit must be taken in such a way that the "free" relativistic evolution of wavepackets would produce space–time trajectories (for the centers of the wavepackets) that are found, at time 0, in a finite region of space. Where exactly they are found (so-called impact factors) is up to us, and just affects our choice of basis elements.

Remark 6.5 *What are described here are in-states. There is a similar construction where the condition is obtained by evolving forward in time; the resulting asymptotic states are out-states. These then provide two bases for the Hilbert space that diagonalize the energy operator; they have the same description and lead to the same energies, but are different bases. The overlaps between in-states and out-states forms the scattering matrix of the model. It is expected that giving the scattering matrix is equivalent to specifying the stress–energy tensor: it provides the locality information of the QFT model. This is rather explicit in the context of integrable QFT, with the so-called form factor program, but not obvious out of integrability.*

Remark 6.6 *Not all possible linear combinations $\sum_\phi c_\phi |\phi\rangle$ of field configurations occur in the basis (6.5) (or in finite linear combinations of basis elements); in particular, the states $|\phi\rangle$ themselves do not occur. But we expect that by taking the completion, involving all (possibly infinite, and continuously so) linear combinations that converge, one recovers an appropriate set of field configurations $|\phi\rangle$ and their (again possibly infinite, and continuously so) linear combinations $\sum_\phi c_\phi |\phi\rangle$ with well-behaved coefficients c_ϕ. In fact, we expect to obtain all square-integrable field configurations ϕ—this is natural, since ϕ is supposed to represent a wavefunction in the "second-quantization" viewpoint on QFT. This is similar to the situation in ordinary quantum mechanics, where the wavefunctions for energy eigenstates may have certain continuity and differentiability properties, and appropriate infinite linear combinations need to be taken in order to obtain all possible square-integrable functions, including, for instance, the discontinuous ones. In the present case, however, this is much more complicated, because the basis (6.5) has continuous components (it is not a countable basis); hence the full Hilbert space structure (including completeness) is much more subtle. We will avoid these tricky details here, and only discuss basic properties in due course.*

It is extremely important to realize that asymptotic states are just particular linear combinations of field configurations. The term "asymptotic" refers here to the way these linear combinations are chosen: by an asymptotic condition on a hypothetical time evolution. It does not refer to the states "living only in the far past" (or the far future). It turns out that this construction produces linear combinations of field configurations that diagonalize the Hamiltonian.

We can verify the latter sentence and make the intuitive construction above much more precise as follows. Recall first that the vacuum in quantum field theory is itself a nontrivial linear combination of field configurations:

$$|\mathrm{vac}\rangle = \sum_\phi c(\mathrm{vac}|\phi) \, |\phi\rangle. \tag{6.8}$$

One standard way of defining this linear combination is via the path integral (appropriately renormalized). Assume that the Hamiltonian is associated with the action functional $S[\varphi]$, for $\varphi : \mathbb{R}^2 \to \mathcal{S}$, where $\varphi(x,t)$ is the trajectory in time t of the field configuration ϕ. From this, it is of course possible to write down the action $S_R[\varphi]$ on any open region R of $\mathbb{R}^2$. One defines the coefficients $c(\mathrm{vac}|\phi)$ by

$$c(\mathrm{vac}|\phi) = \int_{\substack{\varphi(x,0^-) = \phi(x)\,:\,x \in \mathbb{R} \\ \varphi(x,t) \to 0\,:\,x \in \mathbb{R},\, t \to -(1-i0^+)\infty}} \mathcal{D}_{\mathbb{L}}\varphi \, e^{iS_{\mathbb{L}}[\varphi]}, \tag{6.9}$$

where $\mathbb{L} = \{(x,t) \in \mathbb{R}^2 : t < 0\}$ and $\mathcal{D}_{\mathbb{L}}\varphi$ is an integration measure over all configurations of $\varphi : \mathbb{L} \to \mathcal{S}$. That is, one performs Feynman's path integral on field trajectories over half of space–time, the half with negative times, with the boundary condition in the limit of zero time $t \to 0^-$ given by the field configuration ϕ, and with infinite-past asymptotic condition (under Feynman's imaginary-time prescription) equal to 0 (here we assume that $\phi = 0$ is a minimum of the potential representing the interaction).

The wavepackets are constructed on top of this vacuum. Let us denote by $\hat{\phi}_\alpha(x)$ a fundamental field associated with the particle α. This means that $\hat{\phi}_\alpha(x)$ has the quantum numbers of the particle α, and that the Fourier transform of the two-point correlation function

$$\langle \mathrm{vac}|\, \mathcal{T}\big(\hat{\phi}_\alpha(x,t)\hat{\phi}_\alpha(0,0)\big)|\mathrm{vac}\rangle \tag{6.10}$$

(where $\mathcal{T}$ is the time-ordering operation, bringing the earliest operator to the right, and $\hat{\phi}_\alpha(x,t) = e^{iHt}\hat{\phi}_\alpha(x)e^{-iHt}$) has a pole, as a function of the square two-momentum $E^2 - p^2$, at m_α^2 and no other singularities at lower values. We assume for simplicity of the discussion that $\hat{\phi}_\alpha$ is a real bosonic field, transforming under the Poincaré group in the spin-0 representation (generalizing to other representations and statistics and to complex fields is straightforward, if somewhat tedious). The statement above implies that the correlation functions of $\hat{\phi}_\alpha(x,t)$ with any other local fields satisfy the Klein–Gordon equation asymptotically in time (other equations occur for other representations):

$$\left(m^2 - \frac{\partial^2}{\partial x^2} - \frac{\partial^2}{\partial t^2}\right)\langle\mathcal{T}\big(\hat{\phi}_\alpha(x,t)\cdots\big)\rangle = O\left(e^{-m'\sqrt{t^2 - x^2}}\right) \qquad \text{as} \quad t^2 - x^2 \to \infty. \tag{6.11}$$

Here m' is the lowest mass, greater than m, created by $\hat{\phi}_\alpha$, and $\cdots$ represents other fields at fixed positions.

Define the following operators:

$$A_\alpha(\theta) = i \lim_{\substack{t \to -\infty \\ L \to \infty}} \int_{-\infty}^{\infty} dx \left[f_\theta(x,t)\frac{\partial}{\partial t}\hat{\phi}_\alpha(x,t) - \frac{\partial}{\partial t}f_\theta(x,t)\hat{\phi}_\alpha(x,t) \right], \tag{6.12}$$

with

$$f_\theta(x,t) = \exp\left\{ im\cosh(\theta)\,t - im\sinh(\theta)\,x - \frac{[x - \coth(\theta)t]^2}{L^2} \right\}, \tag{6.13}$$

as well as their Hermitian conjugates $A_\alpha^\dagger(\theta)$. Here the limits on t and L are such that $t \gg L$, so that the wavepackets are well separated (without going into more details). Using (6.11) and taking into account the limits taken in (6.12), one can show that these are eigenoperators of the Hamiltonian; in particular,

$$[H, A_\alpha^\dagger(\theta)] = m_\alpha \cosh\theta\, A_\alpha^\dagger(\theta). \tag{6.14}$$

The vacuum having zero energy, this means that we can construct eigenstates of the Hamiltonian by multiple actions of $A_\alpha^\dagger(\theta)$:

$$|\theta_1, \dots, \theta_k\rangle_{\alpha_1,\dots,\alpha_k} = A_{\alpha_1}^\dagger(\theta_1) \cdots A_{\alpha_k}^\dagger(\theta_k)|\text{vac}\rangle. \tag{6.15}$$

These are the asymptotic (in-)states.

One may wonder why we did not use the operators $A_\alpha(\theta)$ instead to create different Hamiltonian eigenstates out of the vacuum. This is because

$$A_\alpha(\theta)|\text{vac}\rangle = 0 \tag{6.16}$$

and

$$[A_\alpha(\theta), A_{\alpha'}^\dagger(\theta')] = 4\pi\delta_{\alpha,\alpha'}\delta(\theta - \theta'). \tag{6.17}$$

The former is a (slightly subtle) consequence of (6.8), (6.9) and (6.12), (6.13). The latter is more involved, and a consequence of locality and of the pole structure of the Fourier transforms of the two-point functions $\langle\text{vac}|\mathcal{T}(\hat\phi_\alpha(x,t)\hat\phi_{\alpha'}(0,0))|\text{vac}\rangle$. The relations (6.16) and (6.17) mean that the vacuum (6.8) can alternatively be defined as the vacuum of the Fock space of the canonical algebra generated by $A_\alpha(\theta)$ and $A_\alpha^\dagger(\theta)$; this is a definition that does not rely on the existence of an action functional.

Remark 6.7 *Note that the canonical commutation relations (6.17) do not imply that the model is free. They only relate to the fact that the asymptotic states have a particularly simple description, which can be obtained as the Fock space of a canonical algebra. But for interacting models, the local fields are not simple Fourier transforms of the operators $A_\alpha(\theta)$, $A_\alpha^\dagger(\theta)$, and taking the limit $t \to \infty$ instead of $t \to -\infty$ in (6.12) gives rise to operators creating the out-states, which are in general different from the in-states.*

Out of the states (6.15), we may now construct the Hilbert space of the model by completion. Without going into any details of this rather subtle mathematical process (see Remark 6.6), we will denote by $\mathcal{H}$ the resulting Hilbert space; this is expected to be a proper subspace of the space of field configurations $\mathcal{C}$.

6.3.2 Asymptotic states in the presence of a cut-impurity

Impurities and boundaries on the quantum chain give rise to QFT models with inhomogeneities running, in space–time, along the time direction (hence there is a natural, preferred time direction). These can be treated in similar ways as above. The case of interest below is that where there is a cut at one point along the chain (i.e., we have two separate half-infinite chains). In this case, the field configurations ϕ are on $\mathbb{R}^0 := \mathbb{R} \setminus \{0\}$ (the basis $|\phi\rangle$ forms the space $\mathcal{C}^0$), and the Hamiltonian H^0 is (slightly) different from H, in its expression in terms of fields $\hat{\phi}$, having a discontinuity at $x = 0$:

$$H^0 = \int_{\mathbb{R}^0} dx \, \hat{h}(x). \tag{6.18}$$

Obviously, we can write $H^0 = H^l + H^r$ as a sum of a term on $(-\infty, 0)$ and a term on $(0, \infty)$, respectively. These terms commute with each other, $[H^l, H^r] = 0$, and hence can be diagonalized simultaneously.

The vacuum can be described as

$$|\text{vac}\rangle^0 = \sum_{\phi} c^0(\text{vac}|\phi) \, |\phi\rangle, \tag{6.19}$$

where

$$c^0(\text{vac}|\phi) = \int_{\substack{\varphi(x,0^-) = \phi(x) \, : \, x \in \mathbb{R}^0 \\ \varphi(x,t) \to 0 \, : \, x \in \mathbb{R}^0, \, t \to -(1-i0^+)\infty}} \mathcal{D}_{\mathbb{L}^0}\varphi \, e^{iS_{\mathbb{L}^0}[\varphi]}, \tag{6.20}$$

with $\mathbb{L}^0 = \{(x,t) \in \mathbb{R}^2 : x \neq 0, t < 0\}$. Here we choose the free boundary conditions as $x \to 0^-$ and $x \to 0^+$, and hence we do not constrain the limit value of the field. This is clearly a factorized vacuum, $|\text{vac}\rangle^0 = |\text{vac}\rangle^l \otimes |\text{vac}\rangle^r$.

Then, asymptotic states are constructed in the same way as for the model on the whole line. We define A_α^0 by

$$A_\alpha^0(\theta) = i \lim_{\substack{t \to -\infty \\ L \to \infty}} \int_{-\infty}^{\infty} dx \left[f_\theta(x,t) \frac{\partial}{\partial t} \hat{\phi}_\alpha^0(x,t) - \frac{\partial}{\partial t} f_\theta(x,t) \hat{\phi}_\alpha^0(x,t) \right], \tag{6.21}$$

with $\hat{\phi}_\alpha^0(x,t) = e^{iH^0 t} \hat{\phi}_\alpha(x) e^{-iH^0 t}$. Then we simply act on the vacuum with the operators $(A_\alpha^0)^\dagger(\theta)$:

$$|\theta_1, \ldots, \theta_k\rangle^0_{\alpha_1, \ldots, \alpha_k} = (A_{\alpha_1}^0)^\dagger(\theta_1) \cdots (A_{\alpha_k}^0)^\dagger(\theta_k) |\text{vac}\rangle^0. \tag{6.22}$$

The l–r factorization also occurs for asymptotic states. Note that for positive rapidities, the wavepackets (6.13) are supported on $(-\infty, 0)$, and for negative rapidities, they are supported on $(0, \infty)$ (apart from exponentially decaying tails that vanish in the limit $t \to -\infty$). Since asymptotic state operators commute for rapidities of different signs, we can then write

$$|\theta_1, \ldots, \theta_k\rangle^0_{\alpha_1, \ldots, \alpha_k} = \prod_{i: \theta_i > 0} (A_{\alpha_i}^0)^\dagger(\theta_i) |\text{vac}\rangle^l \otimes \prod_{j: \theta_j < 0} (A_{\alpha_j}^0)^\dagger(\theta_j) |\text{vac}\rangle^r. \tag{6.23}$$

The factors are the separate eigenstates of H^l and H^r.

Again, it is important to recall that the asymptotic states (6.22) are just particular linear combinations of field configurations on $\mathbb{R}^0$. In particular, the field configurations involved, seen as functions on $\mathbb{R}$, are generically discontinuous at $x = 0$ (i.e., the left and right limits generically do not agree). Despite the similar formulation, the asymptotic states (6.22) are very different from the asymptotic states (6.15), when seen as linear combinations of field configurations (in particular, there is reflection on the point $x = 0$ in the case of the theory on $\mathbb{R}^0$).

Remark 6.8 *A model on $\mathbb{R}^0$ is equivalent to two independent copies of the model on a half-line. In each copy, the set of asymptotic states is half the size of that for the model on the line, and one can describe them by choosing a sign of the rapidities. This is clear from the asymptotic state construction above: for instance, only states where particles have positive momenta have wavepackets supported inside $(-\infty, 0)$, as noted. Putting two copies together, we get a set of asymptotic states isomorphic to that of the model on the line. There are different ways of describing the resulting tensor product states. But if one copy is interpreted as being on $(-\infty, 0)$ and the other on $(0, \infty)$ (so that we have a model on $\mathbb{R}^0$, as we did above), then we have both signs of rapidities involved, and the asymptotic state construction that we have gives the same description (6.5) both for the set of asymptotic states on $\mathbb{R}$ and for that on $\mathbb{R}^0$. Here we neglect states with particles of zero rapidity, because they contribute only infinitesimally to averages of local operators.*

Remark 6.9 *Impurities and boundaries give rise to more intricate scattering phenomena, which include in general reflection and transmission phenomena. This is not seen in the construction of asymptotic in-states or of asymptotic out-states; it is rather seen in the structure of the scattering matrix, formed by the overlaps between in- and out-states. The scattering matrix will have additional components corresponding to reflections and transmissions. In the case of the cut-impurity, of course, only reflection phenomena will occur at $x = 0$, without transmission. In general, one could also have additional bound states to the impurity, but this is not expected to occur with a cut-impurity.*

Out of the states (6.22), we may now construct the Hilbert space $\mathcal{H}^0 = \mathcal{H}^l \otimes \mathcal{H}^r$ of the model by completion. We omit any discussion of the subtleties of how $\mathcal{H}^0$ compares with $\mathcal{H}$ (e.g., how it is embedded into $\mathcal{H}$), but note that the two Hilbert spaces are closely related, since only one point on the line is missing in the construction of $\mathcal{H}^0$. From the viewpoint of the asymptotic state basis, the only difference between $\mathcal{H}$ and $\mathcal{H}^0$ is of course the states with zero-rapidity particles, which, as mentioned, are expected to contribute only infinitesimally to averages of local (or finite-extent) observables.

Remark 6.10 *It is not essential to single out the point $x = 0$ as we have done above, since in QFT, the value of the field at one point in space does not affect physical results (unless infinite potentials exist at that point). From this viewpoint, we may expect $\mathcal{H}^0 \cong \mathcal{H}$, and we could simply add the point at $x = 0$ from a configuration* *on $\mathbb{R}^0$ in some more or less arbitrary fashion (e.g., the average of the right and left limits). However, singling out $x=0$ clarifies the discussion and avoids these technical details. We may also think of the point $x=0$ as representing the scaling limit of some finite, contiguous-sites subsystem of the underlying quantum chain that separates the half-infinite left and right parts. Our results apply to this general situation. In fact, we may replace the point $x=0$ by any finite interval $[a,b]$, $a,b \in \mathbb{R}$. On the left/right of the interval, we may use the description of $\mathcal{H}^l \otimes \mathcal{H}^r$ as above, and we would have an additional space $\mathcal{L}$ to describe the configurations on the interval itself. Our main results would still hold in this case.*

6.3.3   Steady-state density matrix for energy flows

Statement of the result

According to the steady-state construction in Section 6.2, and taking the scaling limit, we need to evaluate the average of observables H-evolved to time t, in the mixed state with density matrix ρ_0 given by (6.1). The left half of the system, with energy H^l, is at temperature β_l^{-1}, and the right half, with energy H^r, is at temperature β_r^{-1}. The steady state is obtained in the limit where $t \to \infty$. For our observable, we will take any operator $\mathcal{O}$ that is *finitely supported*. Formally, this is an operator that commutes with the energy density $\hat{h}(x)$ whenever $|x|$ is large enough; informally, it is composed of local fields (in products, integrals) at points that all lie on some finite closed interval (the support). This guarantees that far enough in space, the observable has no effect.

We will show that

$$\lim_{t\to\infty} \frac{\mathrm{Tr}_{\mathcal{H}^0}\left[\rho_0\, e^{iHt}\mathcal{O}e^{-iHt}\right]}{\mathrm{Tr}_{\mathcal{H}^0}[\rho_0]} = \frac{\mathrm{Tr}_{\mathcal{H}}\left[\rho_{\text{stat}}\mathcal{O}\right]}{\mathrm{Tr}_{\mathcal{H}}[\rho_{\text{stat}}]} \tag{6.24}$$

for every finitely supported operator $\mathcal{O}$. Here, ρ_0 is diagonal on the asymptotic-state basis (6.22) of $\mathcal{H}^0$ with eigenvalues

$$e^{-\beta_l E^l - \beta_r E^r}, \tag{6.25}$$

where E^l and E^r are the eigenvalues of H^l and H^r, respectively, and ρ_{stat} is diagonal on the asymptotic-state basis (6.15) of $\mathcal{H}$ with eigenvalues

$$e^{-\beta_l E^+ - \beta_r E^-}, \tag{6.26}$$

where

$$E^+ = E^l = \sum_{i:\theta_i>0} m_{\alpha_i} \cosh\theta_i,$$
$$E^- = E^r = \sum_{j:\theta_j<0} m_{\alpha_j} \cosh\theta_j. \tag{6.27}$$

Note that E^+ and E^- have the same forms as E^l and E^r, respectively, when expressed in terms of the rapidities of asymptotic states. However, we emphasize that the asymptotic states associated with the values $E^{\pm}$ are very different from those associated with the values $E^{l,r}$.

Using the density operators $n_\alpha(\theta) = A_\alpha^\dagger(\theta)A_\alpha(\theta)$, which measure the density of asymptotic particles of type α at rapidity θ, we can write explicitly ρ_{stat} as

$$\rho_{\text{stat}} = \exp\left[-\beta_l \int_0^\infty d\theta \sum_\alpha n_\alpha(\theta) - \beta_r \int_{-\infty}^0 d\theta \sum_\alpha n_\alpha(\theta)\right]. \tag{6.28}$$

Hence we have a description of the nonequilibrium steady state via a density matrix—a "quasi-equilibrium" description.

Remark 6.11 *Note that $E^+ + E^-$ is the eigenvalue of H on the state (6.15), but that $\beta_l E^+ + \beta_r E^-$ is not, in general, the eigenvalue of a local charge (the integration of a local density). This is in contrast to the situation on $\mathcal{H}^0$: not only is $E^l + E^r$ the eigenvalue of H^0, but also $\beta_l E^+ + \beta_r E^-$ is the eigenvalue of $\beta_l H^l + \beta_r H^r$, a local charge. Nonlocality of the charge involved in the nonequilibrium steady-state density matrix is a quite general phenomenon, observed also in the case of charge currents in impurity models [12, 24].*

Remark 6.12 *Although we discuss density matrices and traces over them, there are a great many subtleties involved in a proper definition of such objects. One way to define them is to regularize to a finite-length quantum chain in the same universality class and then take the limit after evaluating the trace. This is correct but in general rather messy, and the difficulty in taking the scaling limit makes it hard to extract the universal structures associated with QFT from the particularities of the quantum chain chosen. Another way is to consider the operator algebra of QFT independently of the states (C^*-algebra), and to see the trace as a particular linear functional on this algebra. This is the algebraic-QFT way, is much more elegant, and can be directly applied to the infinite-length limit, which is useful given that this limit must be taken first in the construction of the steady state. Unfortunately, it is hard to do anything other than the free theory (and aspects of CFT) in this completely rigorous picture. Taking ideas from both perspectives may be the most fruitful way.*

The derivation of (6.24) is based on the statement that the operator $e^{-iHt}e^{iH^0t}$, in the large-t limit, is a map $\mathcal{H}^0 \to \mathcal{H}$ that takes the state (6.22), times any state on $\mathcal{L}$, to the state (6.15):

$$\lim_{t\to\infty} e^{-iHt}e^{iH^0t}|\theta_1,\ldots,\theta_k\rangle^0_{\alpha_1,\ldots,\alpha_k} = |\theta_1,\ldots,\theta_k\rangle_{\alpha_1,\ldots,\alpha_k}. \tag{6.29}$$

This statement is true in the context of evaluating matrix elements of finitely supported operators (i.e., in the "weak" sense); in particular, there is no problem with a possible loss of invertibility after the limit process (see Remark 6.10). This statement shows (6.24), because, H^0 commuting with ρ_0, we have

$$\frac{\text{Tr}_{\mathcal{H}^0}\left[\rho_0\, e^{iHt}\mathcal{O}e^{-iHt}\right]}{\text{Tr}_{\mathcal{H}^0}[\rho_0]} = \frac{\text{Tr}_{\mathcal{H}^0}\left[\rho_0\, e^{-iH^0t}e^{iHt}\mathcal{O}e^{-iHt}e^{iH^0t}\right]}{\text{Tr}_{\mathcal{H}^0}[\rho_0]}, \tag{6.30}$$

and we just have to take the trace by summing over the states (6.22), and use the statement on every term (see Remark 6.12 for the subtleties in principle involved in taking the trace).

Remark 6.13 *The result derived below can be seen as naturally related to the so-called Gell-Mann–Low theorem of QFT [25]. This theorem essentially implies that if the limit on the left-hand side of (6.29) exists, then it must give the right-hand side. There are, however, many subtleties in the existence of this limit, and of the limit on the trace (6.24) itself; in particular, the theorem itself involves an adiabatic process, whereas here we have a quench. The arguments we present are based on the particular setup of the Keldysh formulation, with the particular order of limits involved, different from what is used in standard QFT constructions. Hence we believe it is worth going through the derivation without the explicit use of the Gell-Mann–Low theorem.*

Remark 6.14 *In integrable systems, one expects that time evolutions from states that are not Hamiltonian eigenstates (quenches) give rise, after a large time, to generalized Gibbs ensembles [26]. These ensembles are described by density matrices of the form $\rho_{\text{quench}} = e^{-\sum_j \beta_j Q_j}$, where Q_j are the local conserved charges (including the Hamiltonian) and where the generalized (inverse) temperatures β_j can be determined by requiring that the averages of the Q_j before the start of the time evolution be reproduced by ρ_{quench}. In the case of integrable QFT, it is tempting to view our density matrix ρ_{stat} in this context. However, our result holds beyond integrability. Further, contrary to the usual quench situation, the density matrix ρ_{stat} does not reproduce the conserved averages of the charges Q_j. This is because ρ_{stat} reproduces averages of local fields and describes the steady-state region only, while conserved charges Q_j are integration of local densities over the whole length of space, including the faraway baths. The subtlety lies in the order of the limits that we must take: infinite length first, then infinite time.*

Existence of steady state

As a first step, we will show that the steady state exists, and that, in particular, we can replace the limit $t \to \infty$ by the Wick-rotated limit,

$$\lim_{\substack{t=-i\tau \\ \tau \to \infty}}, \tag{6.31}$$

for the operator $e^{-iHt}e^{iH^0 t}$ that is on the right of $\mathcal{O}$ in (6.30) (and $t = i\tau$ for the one on the left). In order to show this, we write

$$e^{-iHt}e^{iH^0 t} = \mathcal{P}\exp\left[i\int_0^{-t} d\kappa\, \delta H(\kappa)\right], \tag{6.32}$$

where $\delta H = H - H^0 = dx\, h(0)$ and $\delta H(\kappa) = e^{iH^0\kappa}\, \delta H\, e^{-iH^0\kappa}$. Then, we have

$$
\frac{\mathrm{Tr}\left[\rho_0\, e^{-iH^0 t} e^{iHt} \mathcal{O} e^{-iHt} e^{iH^0 t}\right]}{\mathrm{Tr}\left[\rho_0\, e^{-iH^0 t} e^{iHt} e^{-iHt} e^{iH^0 t}\right]}
$$

$$
= \frac{\mathrm{Tr}\left[\rho_0\, \mathcal{P}\exp\left[i\int_{-t}^{0} d\kappa\, \delta H(\kappa)\right] \mathcal{O}\, \mathcal{P}\exp\left[i\int_{0}^{-t} d\kappa\, \delta H(\kappa)\right]\right]}{\mathrm{Tr}\left[\rho_0\, \mathcal{P}\exp\left[i\int_{-t}^{0} d\kappa\, \delta H(\kappa)\right] \mathcal{P}\exp\left[i\int_{0}^{-t} d\kappa\, \delta H(\kappa)\right]\right]}
$$

$$
= \left\langle \mathcal{P}\exp\left[i\int_{-t}^{0} d\kappa\, \delta H(\kappa)\right] \mathcal{O}\, \mathcal{P}\exp\left[i\int_{0}^{-t} d\kappa\, \delta H(\kappa)\right] \right\rangle_{0}^{\mathrm{connected}} . \tag{6.33}
$$

Then, we use the fact that for local fields in QFT, we have the form

$$
\langle \mathcal{O}_1(0)\mathcal{O}_2(\kappa)\rangle_0^{\mathrm{connected}} \sim \kappa^{\#} e^{-\#T|\kappa|} e^{im\kappa} \quad \text{as} \quad |\kappa| \to \infty, \tag{6.34}
$$

and similar equations for time separation of groups of fields. We use this, with the local fields being $\mathcal{O}$ and various products of $\delta H(\kappa)$ at various times κ. This implies that the integrals in (6.33) all converge as $t \to \infty$, and hence the existence of the steady state. Further, in the path-ordered exponential on the right, we can replace $t \mapsto -i\tau$ (and then in the integral $\kappa = ik$), and we get convergent integrals still (opposite shift on the left), with the same value; hence we can use (6.31).

Vacuum mapping

Since the asymptotic states (6.22) are constructed from the vacuum $|\mathrm{vac}\rangle^0$, we start by describing the effect of $e^{-iHt} e^{iH^0 t}$ on $|\mathrm{vac}\rangle^0$. We employ a standard argument from elementary QFT, used in textbooks in order to obtain, for instance, the perturbative series from the operator formalism.

Let us write $|\mathrm{vac}\rangle^0$ as a linear combination of eigenstates of H:

$$
|\mathrm{vac}\rangle^0 = B_{\mathrm{vac}}|\mathrm{vac}\rangle + \underbrace{\sum B_s|s\rangle}_{\substack{\text{excited} \\ \text{states } s}} = B_{\mathrm{vac}}|\mathrm{vac}\rangle + \int d\theta \sum_{\alpha} B_{\alpha}(\theta)|\theta\rangle_{\alpha} + \dots \tag{6.35}
$$

Then

$$
e^{-iHt} e^{iH^0 t}|\mathrm{vac}\rangle^0 = e^{-iHt}|\mathrm{vac}\rangle^0 = B_{\mathrm{vac}}|\mathrm{vac}\rangle + \underbrace{\sum B_s e^{-iE_s t}|s\rangle}_{\substack{\text{excited} \\ \text{states } s}}. \tag{6.36}
$$

We have shown that the large-t limit can be taken by doing (6.31). Then all contributions from the excited states vanish, and we get

$$
\lim_{t\to\infty} e^{-iHt} e^{iH^0 t}|\mathrm{vac}\rangle^0 = B_{\mathrm{vac}}|\mathrm{vac}\rangle. \tag{6.37}
$$

Hence, in the large-time limit, the vacuum $|\text{vac}\rangle^0$ collapses to $|\text{vac}\rangle$, as long as the overlap B_{vac} is nonzero (in which case it will be normalized away by taking the ratio of traces in (6.24)).

In order to justify $B_{\text{vac}} \neq 0$, we use arguments similar to those of standard QFT textbooks, except for some changes due to the construction of the steady state requiring us to take the length of the system to be infinite first; the arguments then use in an essential way the locality of $\delta H = H - H^0$.

In standard QFT, in the context of perturbation theory, one compares a "true" vacuum $|\text{vac}\rangle$ with the vacuum of a free theory $|\text{vac}\rangle^0$. The differences between the Hamiltonian is a term integrated over the whole volume—here the length of the system ℓ—and hence the overlap $\langle \text{vac}|\text{vac}\rangle^0 \sim e^{-\ell}$ decreases exponentially with ℓ. Hence one usually assumes that the volume is finite and takes the limit of infinite volume afterward; this gives rise to the usual perturbation theory. If we had a similar situation here, this would be very bad, since for the construction of the steady state, we need to take the limit $\ell \to \infty$ before $t \to \infty$. Happily, the difference between H^0 and H is a local term (in fact, it is an infinitesimal term, but this is not the main point). This implies that the overlap $B_{\text{vac}} = \langle \text{vac}|\text{vac}\rangle^0$ is stable in the limit $\ell \to \infty$. Indeed, in the language of path integrals, we can represent it as

$$B_{\text{vac}} = \frac{\langle \text{vac}|\text{vac}\rangle^0}{\sqrt{\langle \text{vac}|\text{vac}\rangle^{\,0}\langle \text{vac}|\text{vac}\rangle^0}}$$

$$= \frac{\int \mathcal{D}_{\mathbb{L}^0 \cup \mathbb{H}} \varphi\, e^{iS}}{\sqrt{\int \mathcal{D}_{\mathbb{R}^2} \varphi\, e^{iS} \int \mathcal{D}_{\mathbb{R}^0 \times \mathbb{R}} \varphi\, e^{iS}}}, \tag{6.38}$$

where $\mathbb{H} = \{(x,t) : x \in \mathbb{R}, t \geq 0\}$. Let us divide space–time into cells of extent much larger than the inverse of the smallest mass, m^{-1} (but still finite). Then the large-volume asymptotics of the path integral is obtained by factorizing it into such cells. Using time-reversal symmetry, we see that all the cells, except three, cancel out in the ratio; in particular, for every cell in the numerator containing a space–time region ($t < 0$, $x \approx 0$) cut into two pieces, there are two equal cells in the denominator, under the square root sign, containing a similar space–time region ($t < 0$, $x \approx 0$ and $t > 0$, $x \approx 0$). Those not canceled are those containing the space–time point $(0,0)$, because they are different in the three overlaps involved. Since only three cells remain, the result has a finite infinite-volume limit. Here we do not discuss UV divergences: these are present, but one can UV-regularize the theory and take the full scaling limit only at the end of the calculation of the nonequilibrium average (see Remarks 6.2 and 6.12).

Excited-states mapping

Finally, in order to map the full excited states as in (6.29), we only need to show that (see (6.12) and (6.21))

$$\lim_{t \to \infty} e^{-iHt} e^{iH^0 t} A_\alpha^0(\theta) e^{-iH^0 t} e^{iHt} = A_\alpha(\theta). \tag{6.39}$$

The main idea is to equate the "asymptotic-state" time $t = t_{\text{asymp}}$, used in (6.12) and (6.21) to construct the asymptotic states, with the negative $-t_{\text{steady}}$ of the "steady-state" time $t = t_{\text{steady}}$, used, for instance in (6.39), to construct the steady state. This should be fine by the fact that the observable $\mathcal{O}$ is of finite extent. Then (6.39) follows simply from

$$e^{-iHt}e^{iH^0t}\hat{\phi}^0_\alpha(x,-t)e^{-iH^0t}e^{iHt} = \hat{\phi}_\alpha(x,-t). \tag{6.40}$$

In order to justify equating $t_{\text{asymp}} = -t_{\text{steady}}$, consider the trace (6.30). We may use, instead of the asymptotic states, a set of states created as in (6.22), but with A^0_α replaced by an expression like (6.21) where the t limit has not yet been taken (hence neither has the L limit), fixing $t = t_{\text{asymp}}$. This set is not a basis of states, but it is likely that such states, at least for t_{asymp} sufficiently negative, span (a dense subset of) $\mathcal{H}^0$. Hence the trace can be written as a sum over such almost-asymptotic states, with appropriate normalization coefficients. Denoting by $|s\rangle^0_{t_{\text{asymp}}}$ the almost-asymptotic state corresponding to the asymptotic state $|s\rangle^0$, and by $C^0_s(t_{\text{asymp}})$ the associated coefficients (satisfying $\lim_{t_{\text{asymp}}\to-\infty} C^0_s(t_{\text{asymp}}) = 1$), we have formally

$$\text{Tr}_{\mathcal{H}^0}[\rho_0 \cdot] \approx \sum_s w^0_s\, C^0_s(t_{\text{asymp}})\, {}_{t_{\text{asymp}}}{}^0\langle s| \cdot |s\rangle^0_{t_{\text{asymp}}} \tag{6.41}$$

where w^0_s is the weight (6.25). Denoting by $|s\rangle_{t_{\text{asymp}}}$ the similar almost-asymptotic states corresponding to (6.12), and by $C_s(t_{\text{asymp}})$ the associated coefficients, we have similarly

$$\text{Tr}_{\mathcal{H}}[\rho_{\text{stat}} \cdot] \approx \sum_s w_s\, C_s(t_{\text{asymp}})\, {}_{t_{\text{asymp}}}\langle s| \cdot |s\rangle_{t_{\text{asymp}}}, \tag{6.42}$$

where w_s is the weight (6.26). Now choosing $t_{\text{steady}} = -t_{\text{asymp}}$, the relation (6.40) and the equality (6.27) for the weights imply that we obtain, for the trace (6.30),

$$\sum_s w_s\, C^0_s(-t_{\text{steady}})\, {}_{-t_{\text{steady}}}\langle s|\mathcal{O}|s\rangle_{-t_{\text{steady}}}. \tag{6.43}$$

For general $\mathcal{O}$, not necessarily of finite extent, we could not expect the limit $t_{\text{steady}} \to \infty$ of this expression to give the trace on the right-hand side of (6.24), because it is the different coefficient $C_s(t_{\text{steady}})$ that would be required. However, for observables of finite extent, we expect

$$\lim_{t_{\text{asymp}}\to-\infty} {}_{t_{\text{asymp}}}\langle s|\mathcal{O}|s\rangle_{t_{\text{asymp}}} = \langle s|\mathcal{O}|s\rangle. \tag{6.44}$$

Since the coefficient $C^0_s(-t_{\text{steady}})$ itself tends to 1, the large steady-state time can be taken, and we obtain (6.24).

Remark 6.15 *Note that it is crucial to take the limit of infinite system length first, since otherwise we clearly cannot replace the asymptotic-state time by the (negative of the) steady-state time (with finite lengths, the large time limit will not give asymptotic states).*

6.3.4 A consequence for the average current

The statement (6.24) with (6.26) does not provide an immediate way of evaluating the average current or its fluctuations in general (in CFT, it does—see the exact results in [1]). Yet, an intuitive argument leads to the nice observation of left and right "factorization" of the average current. Clearly, the average energy current is

$$J = \frac{\text{Tr}_{\mathcal{H}}[\rho_{\text{stat}}p(0)]}{\text{Tr}_{\mathcal{H}}[\rho_{\text{stat}}]}, \tag{6.45}$$

where $p(x)$ is the momentum density. In cases where there is a sense in the notion of quasiparticles and quasimomenta even in finite volume (for instance, in integrable models via the Bethe ansatz; see, e.g., [27]), there is a natural finite-length L extrapolation ρ_{stat}^L of ρ_{stat}. There is of course a finite-L notion of p, and, by translation invariance, we have $J = \lim_{L\to\infty} J_L$, with

$$J_L = L^{-1}\frac{\text{Tr}_{\mathcal{H}_L}[\rho_{\text{stat}}^L P]}{\text{Tr}_{\mathcal{H}_L}[\rho_{\text{stat}}^L]}, \tag{6.46}$$

where $P = \int_{-L/2}^{L/2} dx\, p(x)$ is the total momentum. Clearly, for every state $|v\rangle$ with n quasiparticles of momenta p_j, $j = 1, \ldots, n$, we have

$$P|v\rangle = \left(\sum_{j=1}^{n} p_j\right)|v\rangle. \tag{6.47}$$

This means that we can write $P = P_+ + P_-$, where $P_\pm$ are operators that measure the momenta of positive- and negative-momentum particles:

$$P_\pm|v\rangle = \left(\sum_{j:\, p_j \gtrless 0} p_j\right)|v\rangle. \tag{6.48}$$

Since ρ_{stat} (and similarly ρ_{stat}^L) factorizes into factors $\rho_{\text{stat},+}$ and $\rho_{\text{stat},-}$ acting on left- and right-moving particles, $\rho_{\text{stat}} = \rho_{\text{stat},+}\rho_{\text{stat},-}$, similar to each other but depending on β_l and β_r respectively, we obtain

$$J = f(\beta_l) - f(\beta_r). \tag{6.49}$$

The above derivation holds even if the quasiparticle/quasimomentum description is not exact at finite L, as long as it becomes "more and more exact" as $L \to \infty$, which should be the case for any near-critical or critical system.

The "factorization" relation (6.49), as well as an equivalent factorization for the moment-generating function, was established in the context of CFT in [1]; further, interestingly, (6.49) was observed numerically in [28] in spin chain models—hence beyond QFT.[4] As was remarked in [28], (6.49) immediately implies a simple relation between the linear conductance and the nonequilibrium current:

$$J = J(T_l, T_r) = \int_{T_r}^{T_l} dT\, G(T),$$

(6.50)

where

$$G(T) = \left.\frac{dJ(T', T)}{dT'}\right|_{T'=T}.$$

(6.51)

6.4 Concluding remarks

We have shown how the Keldysh formulation leads, by QFT arguments, to a Gibbs-like ensemble formulation for nonequilibrium steady states of energy flow in quantum systems near to critical points.

We note that all the characteristics of the constructions were essential in order to obtain the proof: that the system's length be infinite, otherwise we cannot identify asymptotic-state and steady-state times, and that the quench be local (the term $\delta H = H - H^0$ is local), or at least of finite extent, otherwise B_{vac} vanishes. The existence proof of the steady state presented also uses both the infinite length and the locality (or finite extent) of δH.

Trying to generalize this construction to higher dimensions lead to interesting problems. If we have one additional dimension of finite length, then everything goes through essentially unchanged (except for subtleties in defining asymptotic states). However, as can be expected physically, an additional dimension of infinite length will cause problems. There, one would expect, immediately after connection of the two halves, to have an infinite total flow of energy, while in the steady state itself, one would expect to have a finite total flow of energy and hence a vanishing energy current density. This slightly "anomalous" situation is translated into the vanishing of the overlap B_{vac}: in the cell-cancellation argument presented, there remain infinitely many cells, leading to a vanishing proportional to $e^{-\ell}$, where ℓ is the extent of the additional dimension.

As we mentioned, the idea of certain Gibbs-like ensembles of asymptotic states representing a nonequilibrium steady state, and its link with the Keldysh formulation, has been used many times in the past. Hershfield's density matrix [6] was developed for charge transport, and was studied and used afterward in the context of quantum impurities [12–15, 17, 24]. Charge transport through impurities has also been studied

---

[4] To my knowledge, the relation (6.49) was first presented in my talk about the work [1] with Denis Bernard at the Workshop on Non-Equilibrium Physics and Asymmetric Exclusion Processes, ICMS, Edinburgh, 8 December 2011. There it was proposed in the context of general $(1 + 1)$-dimensional QFT. It was also presented in my talk at the Galileo Galilei Institute workshop "New Quantum States of Matter in and out of Equilibrium" on 23 April 2012, and appeared in the first version (arXiv:1202.0239v1) of the preprint of [1].

using the asymptotic states of integrable QFT [29–32], and using Bethe ansatz asymptotic states [33], without explicit reference to a density matrix (although it is in some sense there implicitly). Nonequilibrium Gibbs ensembles of asymptotic states in spin chains were discussed with mathematical precision in [7, 8, 11] for heat transport, and were also studied in [18] in free fermion and boson models. Energy transport, including exact results for the current and the so-called full counting statistics, was studied in the context of CFT in [1], where a nonequilibrium density matrix was obtained in terms of the Virasoro algebra. The present chapter gives the first general study of energy transport in the context of massive relativistic QFT, providing the density matrix for nonequilibrium heat flows (which was proposed in [1]) and a direct connection with the Keldysh formulation.

Much still needs to be done, of course. The evaluation of the energy current and its fluctuations is foremost: can concepts of integrability, combined with the present Gibbs-like ensemble description, be fruitfully used? (I have work in progress with my collaborators on this subject.) Could one develop an efficient perturbation theory, including all the subtleties of renormalization, paralleling that for finite-temperature QFT? Deeper questions about the thermodynamic of nonequilibrium steady states may also be accessible from the present formulation. I hope that this chapter will open the way to further developments in these directions.

Note added: after this chapter was originally written, the statement of equation (6.49) was conjectured to be weakly broken in interacting integrable models in [O. Castro-Alvaredo, Y. Chen, B. Doyon and M. Hoogeveen. Thermodynamic Bethe ansatz for non-equilibrium steady states: exact energy current and fluctuations in integrable QFT, *J. Stat. Mech.* **2014**, P03011 (2014)], due to the lack of factorization of the density of states.

Acknowledgments

I am very much indebted to my collaborator Denis Bernard for many (still ongoing) discussions about the subject of nonequilibrium steady states, which ultimately led to my understanding of the situation in massive QFT. I also thank the participants in the Galileo Galilei Institute Workshop "New Quantum States of Matter in and out of Equilibrium" for questions raised on the subject that eventually led to this work, and I acknowledge in particular discussions with M. J. Bhaseen and F. Essler. I am of course very grateful to the students and lecturers at this Les Houches Summer School for their insightful questions and comments, and in particular to Natan Andrei for comments and encouragement to write this chapter.

References

[1] D. Bernard and B. Doyon. Heat flow in non-equilibrium conformal field theory. *J. Phys. A: Math. Theor.* **45**, 362001 (2012) [arXiv:1202.0239].

[2] G. Lindblad. On the generators of quantum dynamical semigroups. *Commun. Math. Phys.* **48**, 119–130 (1976).

[3] A.O. Caldeira and A.J. Leggett. Influence of dissipation on quantum tunneling in macroscopic systems. *Phys. Rev. Lett.* **46**, 211 (1981).

[4] K. Saito and A. Dhar. Fluctuation theorem in quantum heat conduction. *Phys. Rev. Lett.* **99**, 180601 (2007) [arXiv:cond-mat/0703777].

[5] K. Saito and A. Dhar. Generating function formula of heat transfer in harmonic networks. *Phys. Rev. E* **83**, 041121 (2011) [arXiv:1012.0622].

[6] S. Hershfield. Reformulation of steady state nonequilibrium quantum statistical mechanics. *Phys. Rev. Lett.* **70**, 2134 (1993).

[7] D. Ruelle. Natural nonequilibrium states in quantum statistical mechanics. *J. Stat. Phys.* **98**, 57 (2000) [arXiv:math-ph/9906005].

[8] W. H. Aschbacher, V. Jaksic, Y. Pautrat, and C.-A. Pillet. Topics in non-equilibrium statistical mechanics. In *Open Quantum Systems III* (ed. S. Attal, A. Joye, and C.-A. Pillet), pp. 1–66. Lecture Notes in Mathematics, Vol. 1882. Springer-Verlag (2006).

[9] C. Caroli, R. Combescot, P. Nozières and D. Saint-James. Direct calculation of the tunneling current. *J. Phys. C* **4**, 916 (1971).

[10] A.-P. Jauho, N. S. Wingreen and Y. Meir. Time-dependent transport in interacting and noninteracting resonant-tunneling systems. *Phys. Rev. B* **50**, 5528 (1994) [arXiv:cond-mat/9404027].

[11] W. H. Aschbacher and C.-A. Pillet. Non-equilibrium steady states of the *XY* chain. *J. Stat. Phys.* **112**, 1153 (2003).

[12] B. Doyon and N. Andrei. Universal aspects of non-equilibrium currents in a quantum dot. *Phys. Rev. B* **73**, 245326 (2006) [arXiv:cond-mat/0506235].

[13] J.E. Han. Mapping of strongly correlated steady-state nonequilibrium system to an effective equilibrium. *Phys. Rev. B* **75**, 125122 (2007) [arXiv:cond-mat/0604583].

[14] B. Doyon. The density matrix for quantum impurities out of equilibrium. Lecture Notes for the Fifth Capri Spring School on Transport in Nanostructures, 29 March–5 April 2009. http://benjamindoyon.weebly.com/lecture-notes.html.

[15] P. Dutt, J. Koch, J. Han, and K. Le Hur. Effective equilibrium theory of nonequilibrium quantum transport. *Ann. Phys. (NY)* **326** 2963–2999 (2011) [arXiv:1101.1526].

[16] V. Moldoveanu, H. D. Cornean, and C.-A. Pillet, Non-equilibrium steady-states for interacting open systems: exact results. *Phys. Rev. B* 84, 075464 (2011) [arXiv:1104.5399].

[17] D. Bernard and B. Doyon. Full counting statistics in the resonant-level model. *J. Math. Phys.* **53**, 122302 (2012) [arXiv:1105.1695].

[18] A. Dhar, K. Saito, and P. Hanggi. Nonequilibrium density matrix description of steady state quantum transport. *Phys. Rev. E* 85, 011126 (2012) [arXiv:1106.3207].

[19] F. Bonetto, J. L. Lebowitz, and L. Rey-Bellet. Fourier's law: a challenge for theorists. In *Mathematical Physics 2000* (ed. A. Fokas et al.), p. 128. Imperial College Press (2000) [arXiv:math-ph/0002052].

[20] S. Lepri, R. Livi, and A. Politi, Thermal conduction in classical low-dimensional lattices. *Phys. Rep.* **377**, 1 (2003) [arXiv:cond-mat/0112193].

[21] A. Dhar. Heat transport in low-dimensional systems. *Adv. Phys.* **57**, 457 (2008) [arXiv:0808.3256].

[22] D. Bernard and B. Doyon. Non-equilibrium steady states in conformal field theory. *Ann. Henri Poincaré* **16**, 113–161 (2015) [arXiv:1302.2105].

[23] S. Sachdev. *Quantum Phase Transitions* (2nd edn). Cambridge University Press (2011).

[24] B. Doyon. New method for studying steady states in quantum impurity problems: the interacting resonant level model. *Phys. Rev. Lett.* **99**, 076806 (2007) [arXiv:cond-mat/0703249].

[25] M. Gell-Mann and F. Low. Bound states in quantum field theory. *Phys. Rev.* **84**, 350 (1951).

[26] M. Rigol, V. Dunjko, V. Yurovsky, and M. Olshanii. Relaxation in a completely integrable many-body quantum system: an ab initio study of the dynamics of the highly excited states of 1D lattice hard-core bosons. *Phys. Rev. Lett.* **98**, 050405 (2007) [arXiv:cond-mat/0604476].

[27] L. D. Faddeev. How the algebraic Bethe ansatz works for integrable models. In *Quantum Symmetries. Les Houches Session LXIV, 1995* (ed. A. Connes, K. Gawędzki, and J. Zinn-Justin), pp. 149–219. Elsevier (1998) [arXiv:hep-th/9605187].

[28] C. Karrasch, R. Ilan, and J. E. Moore. Nonequilibrium thermal transport and its relation to linear response. *Phys. Rev. B* **88**, 195129 (2013) [arXiv:1211.2236].

[29] P. Fendley, A. W. W. Ludwig, and H. Saleur. Exact non-equilibrium transport through point contacts in quantum wires and fractional quantum Hall devices. *Phys. Rev. B* **52**, 8934 (1995) [arXiv:cond-mat/9503172].

[30] R. Konik, H. Saleur, and A. Ludwig. Transport through quantum dots: analytic results from integrability. *Phys. Rev. Lett.* **87**, 236801 (2001) [arXiv:cond-mat/0010270].

[31] R. Konik, H. Saleur, and A. Ludwig. Transport in quantum dots from the integrability of the Anderson model. *Phys. Rev. B* **66**, 125304 (2002) [arXiv:cond-mat/0103044].

[32] E. Boulat and H. Saleur. Exact low temperature results for transport properties of the interacting resonant level model. *Phys. Rev. B* **77**, 033409 (2008) [arXiv:cond-mat/0703545].

[33] P. Mehta and N. Andrei. Nonequilibrium transport in quantum impurity models: the Bethe ansatz for open systems. *Phys. Rev. Lett.* **96**, 216802 (2006) [arXiv:cond-mat/0508026]; see also P. Mehta, S.-P. Chao, and N. Andrei. Erratum. arXiv:cond-mat/0703426.

7

Nonthermal fixed points: universality, topology, and turbulence in Bose gases

Boris NOWAK, Sebastian ERNE, Markus KARL,
Jan SCHOLE, Dénes SEXTY,
and Thomas GASENZER

Institut für Theoretische Physik der Ruprecht-Karls-Universität Heidelberg,
Philosophenweg 16, 69120 Heidelberg, Germany, and
ExtreMe Matter Institute EMMI, GSI Helmholtzzentrum für Schwerionenforschung
GmbH, Planckstraße 1, 64291 Darmstadt, Germany

Strongly Interacting Quantum Systems out of Equilibrium. First Edition. Thierry Giamarchi et al.
© Oxford University Press 2016. Published in 2016 by Oxford University Press.

Chapter Contents

7.1   Introduction

At a second-order phase transition, different types of order of a physical system meet one another, giving rise to universal critical properties that are independent of the microscopic details of the system. This kind of universality is an extremely successful concept in characterizing equilibrium states of matter and classifying different phenomena in terms of just a few classes governed by the same critical properties. The appearance of an ordered state that does not possess the full symmetry deriving from the conservation laws obeyed by the system's dynamics is called spontaneous symmetry breaking. The order induced by this symmetry breaking has long been known to allow for the appearance of (quasi-)topological defects such as solitons or vortices, objects we are particularly interested in here.

In this chapter, we leave equilibrium systems aside and aim at sketching a picture of related concepts far from thermal equilibrium. We consider the example of nearly coherent Bose gases brought far out of equilibrium and discuss their behavior in view of connections between universal properties, (quasi-)topological field configurations, and turbulent dynamics. We demonstrate that the isolated Bose gas, on its way back to thermal equilibrium, can approach metastable nonequilibrium configurations and spend a long time in their vicinity. In such configurations, which have been termed *nonthermal fixed points*, the system shows universal long-range properties manifest through scaling, i.e., self-similar correlations. The time evolution near such fixed points is demonstrated to undergo critical slowing down. The spatial field pattern, at the same time, is characterized by the appearance of defects and domain formation, whose geometry gives rise to the particular scaling laws seen in the correlation functions. We obtain an overall picture that connects well-known concepts for describing universal dynamics such as wave turbulence, superfluid turbulence, and (quasi-)topological excitations. This allows us to bring together an excitingly wide range of concepts and methods with an excitingly wide spectrum of applications. Beyond the immediate implications for simple low-energy degenerate quantum gases, phenomena such as topological configurations in solids and in soft matter, the dynamics of the quark–gluon plasma created in heavy-ion collisions, and the reheating of the post-inflationary universe come into view. Vice versa, an ultracold quantum gas offers itself as an example of the often-discussed "quantum simulator" for universal dynamics of systems that are more difficult to access experimentally, like some of the above. Not least, the extension to fermionic and gauge fields should provide many new interesting aspects.

The time evolution far from equilibrium of a system with many degrees of freedom is often characterized by the appearance of widely different scales. Usually, a period of fast motion is followed by a much longer period of much slower motion. In the same way, a separation of spatial scales allows descriptions in terms of statistical concepts such as hydrodynamics or renormalization group theory. Taking into account the well-developed concepts and concrete results for equilibrium and near-equilibrium systems (see, e.g., [66]), it has to be expected that far-from-equilibrium time evolution will show universal behavior, with fixed points or partial fixed points being just a small subclass of the possible phenomena. Recently, there has been increased discussion of potential further discoveries, in particular in the context of possible "prethermalization" phenomena [2, 9, 15, 16, 29, 53, 60, 82, 86]. Such dynamics could be considered to hint at a wider class of universal phenomena that will allow classification and deeper understanding of nonequilibrium physics.

7.2 Strong wave turbulence and vortical flow

In analogy to equilibrium phase transitions and criticality in driven systems, it has been proposed that transient stationarity arising in the time evolution of an initially strongly perturbed quantum many-body system can reflect the existence of *nonthermal fixed points* [19] (see also Chapter 2 by Berges in this volume). At the same time, these fixed points have been numerically demonstrated to occur in the field evolution imposed by an $O(N)$-symmetric, relativistic, nonlinear scalar model. By means of parametrically resonant oscillations of the field expectation value and nonlinear amplification, a broad range of modes can be excited initially. While falling back to equilibrium, their occupation number spectra show scaling behavior. Remarkably, the respective power-law exponents have confirmed analytical predictions in the infrared domain of long-wavelength excitations, where standard descriptions in terms of Boltzmann-type kinetic equations break down. Initially found in searches for stationary scaling solutions of the nonperturbative dynamic Dyson equations, with the help of the Zakharov integral transformations from wave-turbulence theory [106, 165], these infrared power laws are interpreted as constituting the previously mysterious strong wave turbulence that had been considered out of reach of kinetic theory [140].

In the following, considering the example of a nonrelativistic superfluid Bose gas in two spatial dimensions, we identify this strong wave turbulence with the appearance of *quantized vortices*. On scales considerably smaller than the mean distance between vortices, the velocity field $v(r)$ associated with the rotational flow decays as $1/r$ with growing distance r from the nearest vortex core. As a result, the angle-averaged kinetic energy distribution $\sim mv^2/2$ of bosons gives rise to a single-particle momentum spectrum $n(k) \sim k^{-4}$ identical to that predicted for strong wave turbulence by Scheppach et al. [140]. Hence, the nonthermal fixed point appears to correspond to a configuration bearing a dilute ensemble of vortices. In this section, we give evidence for this interpretation, which allows us to conjecture a deep link between the extended kinetic-theory picture of wave turbulence on the one hand and the theory of nonlinear (quasi-)topological field configurations and superfluid turbulence on the other.

7.2.1 Nonthermal fixed points and weak wave turbulence

Most generally, a nonthermal fixed point can be defined as a metastable state of a many-body system [2, 19, 29]. While one may include open systems in the discussion, for example, driven ones and those experiencing dissipation [45], we will restrict ourselves in the following discussion to closed systems. In analogy to fixed points in scaling flows, we furthermore consider in particular solutions with power-law behavior of correlation functions [17, 19, 140]. Although, precisely at the fixed point, the system is metastable and characterized by scaling correlation functions in the infrared limit of infinitesimally slow modes, it will vary in time away from the fixed point and show power-law correlations within a finite scaling regime. From the physics of turbulence, it is well known that such states occur as a consequence of local conservation laws in momentum space. Such unintuitive locally conserved currents in momentum space instead of position space imply the possibility of gain- and loss-less transport processes between different scales, giving rise to so-called cascades. The most prominent example of this occurs in fully developed classical fluid turbulence, where there is a quasistationary flow of kinetic energy from large to small spatial scales, i.e., from low to high momenta [133]. The energy is fed in, for example by a stirrer, at a large scale and finally dissipated into heat at the microscopic scale defined by the fluid's viscosity. The corresponding energy spectrum exhibits the famous Kolmogorov–Obukhov five-thirds scaling of the radial energy distribution: $E(k) \sim k^{-5/3}$ [87, 119].

Considering a dilute Bose gas, one has to take into account its compressibility, allowing for collective sound excitations of the particles. In addition to the density of an incompressible fluid, the system is thus characterized by the dispersion relation between the momentum and energy of its excitations. This allows the techniques of wave turbulence to be invoked when looking into turbulence phenomena of a Bose gas [106, 165]. The mathematically best-controlled case is that of weak wave turbulence, which rests on the analysis of stationary solutions of Boltzmann-type kinetic equations. Within a certain range of momenta k and times t, for not too strongly excited systems, the quantum Boltzmann equation (QBE)

$$\partial_t n(\mathbf{k}, t) = I(\mathbf{k}, t), \tag{7.1}$$

with

$$I(\mathbf{k}, t) = \int d^d p\, d^d q\, d^d r\, |T_{\mathbf{kpqr}}|^2 \delta(\mathbf{k} + \mathbf{p} - \mathbf{q} - \mathbf{r})\, \delta(\omega_\mathbf{k} + \omega_\mathbf{p} - \omega_\mathbf{q} - \omega_\mathbf{r})$$
$$\times [(n_\mathbf{k} + 1)(n_\mathbf{p} + 1)n_\mathbf{q} n_\mathbf{r} - n_\mathbf{k} n_\mathbf{p}(n_\mathbf{q} + 1)(n_\mathbf{r} + 1)], \tag{7.2}$$

provides a good description of the time evolution of the momentum-mode occupation numbers $n_\mathbf{k} \equiv n(\mathbf{k}, t) = \langle \Phi^\dagger(\mathbf{k}, t)\Phi(\mathbf{k}, t)\rangle$ of an interacting degenerate Bose gas. Here, Φ denotes the quantum field operator describing the Bose system, with $[\Phi(\mathbf{x}, t), \Phi^\dagger(\mathbf{y}, t)] = \delta(\mathbf{x} - \mathbf{y})$, while all other equal-time commutators vanish. We consider only two-to-two elastic collisions quantified by the T-matrix elements, which for dilute, weakly interacting atomic gases reduce to a single quantity, the s-wave scattering length a, i.e., $|T_{\mathbf{kpqr}}| \equiv g = \text{const} \times a$.

Zeros of the scattering integral $I(\mathbf{k})$ correspond to fixed points of the time evolution within the regime of applicability of the QBE [165]. Most prominent among these are the thermal fixed point corresponding to the system in thermal equilibrium, $n_\mathbf{k} = (e^{\beta\omega(\mathbf{k})} - 1)^{-1}$, and the trivial fixed point $n_\mathbf{k} = \text{const}$. At both fixed points, the scattering integral vanishes and $n(\mathbf{k}, t)$ becomes independent of t. Note that both the trivial and the Bose–Einstein distribution (in the Rayleigh–Jeans regime), taking $\omega(\mathbf{k}) \sim k^2$, show a power-law behavior of the form $n_\mathbf{k} \sim k^{-\zeta}$, with $\zeta = 0$ and $\zeta = 2$, respectively.

The theory of weak wave turbulence [165] allows the analytical derivation of further, *nonthermal*, fixed points at which the occupation number $n_{\mathbf{k}}$ obeys a scaling law of the form $n_{\mathbf{k}} \sim k^{-\zeta}$, with, in general, $\zeta \neq 2$. As in classical turbulence, one expects that universal scaling appears within a certain regime of momenta, the inertial range. According to this picture, outside the scaling regime, excitation quanta enter the system from an external or internal source and/or leave it into a sink, whereas there are no sources and sinks within the inertial interval, where the quanta are transported from momentum shell to momentum shell. This process is described by a continuity equation in momentum space, with a momentum-independent, radially oriented current vector.[1] A central aspect of weak-wave-turbulence theory is that the quantum Boltzmann equation can be cast into different such equations [165], for the radial densities of particles, $N(k) = (2k)^{d-1}\pi n(k)$, and energy, $E(k) = (2k)^{d-1}\pi\varepsilon(k)$, with $\varepsilon(k) = \omega(k)n(k)$:

$$\partial_t N(k,t) = -\partial_k Q(k), \tag{7.3}$$

$$\partial_t E(k,t) = -\partial_k P(k). \tag{7.4}$$

Taking either the radial particle current $Q(k) = (2k)^{d-1}\pi Q_k(k)$ or the energy current $P(k) = (2k)^{d-1}\pi P_k(k)$ to be independent of k, one derives different scaling exponents. The resulting exponents[2] are

$$\zeta_Q^{\mathrm{UV}} = d - \tfrac{2}{3}, \qquad \zeta_P^{\mathrm{UV}} = d. \tag{7.5}$$

These exponents can be obtained by simple power counting. Combining (7.1) and (7.3) gives the radial relation $\partial_k Q(k) \sim k^{d-1}I(k)$, which implies that stationarity requires $k^d I(k)$ to become k-independent, i.e., to scale as k^0. Counting all powers of k in $I(k)$, (7.2), in the wave-kinetic regime where the terms of third order in the occupation numbers dominate the scattering integral, this requires $n(k) \sim k^{-d+2/3}$. Analogously, one infers the exponent ζ_P^{UV} from the balance equation (7.4) for the energy density $\varepsilon(k) \sim k^2 n(k)$. Despite this simple procedure, the existence of the respective scaling solutions has to, and can, be derived rigorously from the QBE by means of Zakharov conformal integral transforms [165].

As we will illustrate for our case in Section 7.3.2, the energy flux generically constitutes a direct cascade to larger k, whereas the particle flux corresponds to an inverse cascade. While the character of the fluxes is entirely determined by the properties of the system, it turns out that there must be at least two sinks to which the particles and the energy can flow [165]. Let us assume that an external source introduces particles with energy ω_s at some scale k_s and that energy and particles leave at scales $k = 0$ and $k = K$. Denoting by Γ_s, Γ_0, and Γ_K the injection/ejection rates of particle

---

[1] Note that this assumption, i.e., locality of the transport, needs to be justified for each particular wave-turbulent solution (see, e.g., [165]).

[2] The superscript UV (ultraviolet) in (7.5) refers to the regime of large momenta where the description in terms of a kinetic equation is expected to be accurate.

number at the respective scales, number and energy conservation imply $\Gamma_s = \Gamma_0 + \Gamma_K$ and $\omega_s \Gamma_s = \omega_0 \Gamma_0 + \omega_K \Gamma_K$, respectively. Inverting these conditions,

$$\Gamma_0 = \frac{\omega_K - \omega_s}{\omega_K - \omega_0} \Gamma_s, \qquad \Gamma_K = \frac{\omega_s - \omega_0}{\omega_K - \omega_0} \Gamma_s, \tag{7.6}$$

shows that for $\omega_0 \ll \omega_s \ll \omega_K$, the particles are ejected at ω_0, whereas energy is dissipated at ω_K [61].

7.2.2 Infrared scaling as strong wave turbulence

Given a positive scaling exponent ζ, momentum occupation numbers $n(k) \sim k^{-\zeta}$ grow large in the IR regime of small k. Keeping the coupling g fixed, the QBE fails for $n_k \gtrsim g^{-1}$, where perturbative contributions to the scattering integral $I(k)$ of order higher than g^2 are no longer negligible. To find scaling solutions in the IR, an approach beyond this perturbative approximation is required.[3] This is available through quantum field dynamic equations derived from the two-particle irreducible (2PI) effective action or Φ-functional [10, 41, 98], which can be expanded in terms of 2PI closed loop diagrams— the lowest-order ones are sketched in Fig. 7.1(a). The solid lines denote the time-ordered Green's function $G(x, y) = \langle T \Phi^\dagger(x) \Phi(y) \rangle_c$, which, in turn, is a solution of the real-time Dyson equation. The Dyson equation is derived from the action by use of Hamilton's variational principle. It contains a time-evolution equation (7.1) for the momentum-mode occupation numbers $n_\mathbf{k}$. As before, one considers zeros of the resulting scattering integral, which can be expressed, within the 2PI approach, in terms of the self-consistently determined Green's function G, connected by the bare scattering vertices of the theory. The scattering integral (7.2) of the QBE (7.1) is recovered within the expansion of the action up to the two-loop diagrams in Fig. 7.1(a).

   To describe the IR kinetics, one needs to go beyond this approximation. Resumming an infinite set of loop diagrams contributing to the 2PI effective action [1, 12] (see also, e.g., [13, 52, 53]) leads to a nonperturbative, effectively renormalized coupling $g_{\text{eff}}(k)$ in the dynamic equations [17, 19, 140]; see Fig. 7.1(b, c). In particular, this coupling

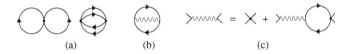

(a)   (b)   (c)

Fig. 7.1 2PI diagrams of the loop expansion of $\Gamma_2[G]$. (a) The two lowest-order diagrams of the loop expansion, which lead to the quantum Boltzmann equation. Black dots represent the bare vertex $\sim g\delta(x - y)$, solid lines the full propagator $G(x, y)$. (b) Diagram representing the resummation approximation, which, in the IR, replaces the diagrams in (a) and gives rise to the scaling of the T-matrix in the IR regime. (c) The wiggly line is the two-point resummed vertex function, which is represented as a sum of bubble-chain diagrams.

---

[3] We recommend that the reader consult [140] for more details of the procedure summarized in the following (see also Chapter 2 by Berges in this volume).

becomes suppressed in the IR to below its bare value g, which, in effect, leads to an even steeper rise of the particle spectrum $n_{\mathbf{k}}$ [19]. The IR scaling exponents for the radial particle and energy flows of a Bose gas in d dimensions that constitute the resulting *strong wave turbulence* were derived by Scheppach et al. [140] to be

$$\zeta_Q^{\mathrm{IR}} = d + 2, \qquad \zeta_P^{\mathrm{IR}} = d + 2 + z, \tag{7.7}$$

where z is the dynamical scaling exponent accounting for the scaling of the dispersion $\omega(sk) = s^z \omega(k)$. From the point of view of its scaling, the T-matrix in the scattering integral of the QBE can be replaced by an effective many-body T-matrix, $T_{\mathbf{kpqr}}^{\mathrm{eff}} \equiv T_{\mathbf{k+p,q+r}}^{\mathrm{eff}}$. This effective T-matrix is more complex than, but scales like,

$$|T_{\mathbf{k}}^{\mathrm{eff}}| \equiv |T_{\mathbf{k,k}}^{\mathrm{eff}}| \sim \left| \frac{gCk^{z-2}}{1 + C'gk^{d-2}n_{\mathbf{k}}} \right|, \tag{7.8}$$

$k = |\mathbf{k}|$, where C' is some constant that fine tunes the position of the transition from UV to IR scaling. The second term in the denominator can be related to the validity criterion of the kinetic equation (7.1) (see, e.g., [150]) in d dimensions,

$$g \int_0^k \mathrm{d}^d k' \, n(\mathbf{k}') \ll \frac{k^2}{2m}. \tag{7.9}$$

For a scaling distribution $n(k) \sim k^{-\zeta}$, this translates into $2\Omega_d m g k^{d-2} n(k) \ll 1$, with Ω_d the surface of a unit sphere in d dimensions. For small $n_{\mathbf{k}}$ and $z = 2$, one recovers the weak-wave-turbulence case discussed in Section 7.2.1. For large $n_{\mathbf{k}}$, the second term in the denominator dominates, which implies a power-law behavior $|T_{\mathbf{k}}^{\mathrm{eff}}|^2 \sim k^{2(\zeta - d + z)}$ and, as a consequence, the modified scaling (7.7) of $n_{\mathbf{k}}$ in the infrared regime of small wavenumbers. Moreover, the coupling becomes universal in the sense that it is now independent of g, which is canceled out by the leading denominator term in (7.8).

In Fig. 7.2, we summarize the nonthermal fixed point scaling predicted with the loop-resummed 2PI effective action for a dilute Bose gas in d dimensions. Direct confirmation of the scaling (7.7) by integration of the 2PI dynamic equation is complicated by the required computational effort. However, as we will show in Section 7.2.3, this challenge can be met by the use of semiclassical simulations of the field equations of motion.

Before we proceed, let us comment on the possibility of understanding the turbulent scaling by taking a renormalization group viewpoint [17, 18, 122]. Since the seminal work of Kolmogorov [49, 87, 119], turbulence has served as one of the first phenomena for which to develop renormalization group techniques out of equilibrium. The effectively local transport processes in momentum space that are at the basis of turbulent cascades immediately suggest themselves for a renormalization group analysis. Building on functional renormalization group techniques [24, 58, 121, 164], specifically out of equilibrium [17, 34, 36, 54, 56, 57, 67, 68, 72, 78, 88, 99, 105, 143, 167], more refined scaling analyses are being developed [35, 101]. The scaling exponents given in (7.7) result as canonical exponents, from power counting of flow equations for G and higher-order vertices. Additional flow equations that, in particular, account for

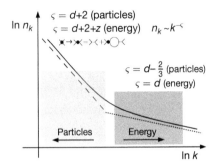

Fig. 7.2 (Color online: see [114]) Sketch of the single-particle mode occupation number $n(k)$ of a Bose gas in d dimensions, as a function of the radial momentum k of the system at the nonthermal fixed point, as predicted by Scheppach et al. [140]. At large momenta, weak-wave-turbulence scaling is recovered, with the Kolmogorov–Zakharov spectra $\zeta = d - \frac{2}{3}$ (particle flux) and $\zeta = d$ (energy flux). At low momenta, new scaling solutions are predicted, with $\zeta = d + 2$ (particle flux) and $\zeta = d + 2 + z$ (energy flux), termed strong turbulence. The different shadings (colors online) indicate the expected character of an IR particle flux and a UV energy flux.

the interaction effects in the spectral functions are expected to fix potentially relevant anomalous dimensions. We remark that the vortex picture developed in the following leads to the conjecture that such anomalous scaling will become important in driven systems, where interaction effects between finitely spaced defects become relevant.

7.2.3 Vortices in a superfluid

Near degeneracy, the strongly occupied low-momentum modes of an ideal Bose gas, $n_{\mathbf{k}} \gg 1$, constitute the Rayleigh–Jeans distribution $n_{\mathrm{RJ}}(k) = 2mT/k^2$ of classical waves.[4] One can show that, also out of equilibrium, strongly occupied modes can be described by the dynamics of classical waves [16, 28, 33, 126]. The state of the gas can be defined in terms of the Wigner quasi-probability distribution $W[\phi, \phi^*]$ for the complex field $\phi(\mathbf{x})$ and its conjugate momentum $\phi^*(\mathbf{x})$ at each point in space. In the considered wave-classical limit, W is positive-definite. Since quantum effects arising from coupling to sparsely occupied modes are small, the dynamics of the Wigner function follows a classical Liouville equation. In this semiclassical limit, the so-called truncated Wigner approximation allows us to follow the evolution *exactly*, within computational errors, by evaluating many trajectories evolving according to the classical field equation

$$i\partial_t \phi(\mathbf{x}, t) = \left[-\frac{\nabla^2}{2m} + V(\mathbf{x}) + g|\phi(\mathbf{x}, t)|^2 \right] \phi(\mathbf{x}, t). \qquad (7.10)$$

While (7.10) is equivalent in form to the Gross–Pitaevskii equation (GPE) for the quantum field expectation value $\langle \Phi \rangle$, the statistical sampling procedure leads to a

---

[4] In the following, we use units where $\hbar = k_{\mathrm{B}} = 1$.

quasi-exact result for the full many-body evolution. Correlation functions are obtained by averaging over many trajectories. The nonlinear classical field equation (7.10) has some interesting properties relevant for our purposes. It can be mapped to an Euler-type hydrodynamic equation, implying the interpretation of the gas dynamics in terms of (superfluid) flow. Among the possible solutions of this equation, those resembling eddy flow and shock waves lead to topologically nontrivial configurations. These include vortices, solitons, and related nonlinear stationary states if more than two field components couple to each other (see, e.g., [123, 125]).

Hydrodynamic representation

The polar representation $\phi = \sqrt{n}\, e^{i\varphi}$ allows us to express the particle current $\mathbf{j} = i(\phi^*\nabla\phi - \phi\nabla\phi^*)/2m = n\mathbf{v}$ in terms of the velocity field $\mathbf{v} = m^{-1}\nabla\varphi$ and the particle density $n = |\phi|^2$. With this, the GPE (7.10) can be rewritten as the continuity equation and an effective Euler equation for a compressible inviscid (i.e., nonviscous) fluid with modified pressure $m^{-1}\tilde{\mu}$:

$$\partial_t n + \nabla \cdot (n\mathbf{v}) = 0, \tag{7.11}$$

$$\partial_t \mathbf{v} + (\mathbf{v} \cdot \nabla)\mathbf{v} = -m^{-1}\nabla\tilde{\mu}, \qquad \tilde{\mu} \equiv gn - \frac{1}{2\sqrt{n}}\triangle\sqrt{n}. \tag{7.12}$$

As the velocity is a potential field, it is irrotational wherever the density is nonvanishing, and therefore the phase φ of the field ϕ is well defined. Small excitations of ϕ that give rise to small density fluctuations and small velocities are described by the linearized version of the hydrodynamic equations (7.12). At wavelengths longer than the healing length $\xi = (2mng)^{-1/2}$, these are collective sound-wave excitations.

Solitons

These are quasi-topological one-dimensional solutions of (7.10) that travel with a fixed velocity but are nondispersive, i.e., stationary in shape (see, e.g., [79, 166]). For positive coupling constant $g > 0$, the solitons are "dark," i.e., characterized by an exponentially localized density depression in the surrounding bulk matter. This and the corresponding phase shift in complex field are given by

$$\phi_\nu(x,t) = \sqrt{n}\left\{\gamma^{-1}\tanh\left[\frac{x - x_{\mathrm{sol}}(t)}{\sqrt{2}\,\gamma\xi}\right] + i\nu\right\}, \tag{7.13}$$

where $x_{\mathrm{sol}}(t) = x_0 + \nu t$ is the position of the soliton at time t. Depending on the depth of the density depression, the dark soliton is called either gray or, at maximum depression, black. $\gamma = 1/\sqrt{1 - \nu^2}$ is the "Lorentz factor" corresponding to the velocity v of the gray soliton in units of the speed of sound, $\nu = v/c_{\mathrm{sound}} = |\phi_\nu(vt,t)|/\sqrt{n}$. Being related to the density minimum, ν measures the "grayness" of the soliton, ranging between 0 (black soliton, $|\phi_\nu(vt,t)| = \nu\sqrt{n} = 0$) and 1 (no soliton, $|\phi_\nu(vt,t)| = \sqrt{n}$). Owing to the interaction with sound, solitons can continuously vanish, which means that they are not topologically stable. In $d > 1$ dimensions, solitons decay into vortices [5, 31]. The energy to create a soliton on top of a uniform background is $E_{\text{sol}} = \left(1 - \nu^2\right)^{3/2} 4nc_{\text{sound}}/3$ [125]. For small velocities $\nu \ll 1$, this gives $E_{\text{sol}} \simeq 4nc_{\text{sound}}/3 - 2nc_{\text{sound}}\nu^2$, reminiscent of a classical point particle with negative mass $-4n/c_{\text{sound}}$. The energy of a soliton monotonically decreases with increasing velocity, which hints at a dynamical instability.

Vortices

These are topologically stable solutions of (7.10) in $d > 1$ dimensions that form the superfluid analogies of eddy flows in classical fluids. We recall that the ground-state manifold given by the minimum of the effective potential $V(\phi) = \mu|\phi|^2 + g|\phi|^4$ of the classical field $\phi(\mathbf{x}) = \sqrt{n(\mathbf{x})}\, e^{i\varphi(\mathbf{x})}$ requires constant density $n(\mathbf{x}) = n$, but is degenerate in the phase $\varphi(\mathbf{x})$. The true ground state has a constant phase, and is therefore called topologically trivial. On the other hand, we can consider field configurations that have constant density on the boundary of, for example, a two-dimensional volume, but a varying phase. If we use this freedom to evolve the phase angle $\varphi(\mathbf{x})$ from 0 to 2π when going around the boundary, we arrive at a topologically nontrivial state. Configurations are topologically distinct, because one cannot define a continuous function that transforms one into the other. As a consequence of the phase winding, the phase cannot be well defined at some point inside the volume, and hence the density has to go to zero at that point. The stationary state of (7.10) that exhibits these properties is called a vortex [124]. Following the phase angle along a closed path around the vortex core, it continuously varies from 0 up to $\varphi = 2\pi\kappa$, where the integer κ is called the winding number or circulation. Only a singly quantized vortex with $\kappa = \pm 1$ is stable. It is described, in polar coordinates centered at the vortex core, by the field $\phi(r, \varphi) = f(r)\, e^{i\kappa\varphi}$, where $f(r)$ can be chosen real and approaches the square root of the bulk density n for large distances r from the vortex core. At $r \to 0$, $f(r) \sim r$ rises linearly. $\phi(r, \varphi)$ is a stationary solution of (7.10), evolving as $\phi(\mathbf{r}, t) = \phi(\mathbf{r}, 0)\, e^{-i\mu t}$, with $\mu = gn$.

We remark that the irrotational nature of the velocity field defined in the hydrodynamic formulation of the GPE is restricted to those points where the density is nonvanishing. The velocity field of the vortex is, in the polar coordinates used before, $\tilde{\mathbf{v}}(r, \varphi) = \kappa \mathbf{e}_\varphi/mr$. Its curl is concentrated locally at the vortex core. The compact phase $\varphi \in [0, 2\pi)$, becomes a noncompact velocity potential. As a consequence, vortex creation or annihilation is not described by the effective Euler equation. In fact, because of the Thomson circulation theorem, vorticity is locally conserved in an inviscid flow [95].

The energy associated with the vortex is extremely nonlocal. For a singly quantized vortex in a two-dimensional homogeneous gas, the energy within the volume $V = \pi R^2$ grows logarithmically with the radius, $E_{\text{v}} = \pi nm^{-1} \ln(1.46R/\xi)$; see, e.g., [125]. In $d = 3$ dimensions, point vortices extend to vortex lines around which the fluid rotates [123, 124]. In the simplest case, a vortex line of length L goes straight from one end of the volume to the other. This requires the excitation energy LE_{v}. Vortex lines cannot end inside the medium, but can form closed loops of all shapes, like rings, ellipses, and also knots. The GPE (7.10), moreover, supports linear wave excitations of the position of the vortex lines, so-called Kelvin waves [83, 92, 149, 160]. Higher-dimensional vortices exist in which the dimensionality of the vortex-core geometry is always $d - 2$, for example leading to vortex surfaces in four dimensions.

Decomposition of the flow field

A nonequilibrium flow features the presence of multiple types of excitations. In order to distinguish longitudinal excitations (sound waves) from rotational excitations (vortices), we close this section by discussing a decomposition of the kinetic energy density proposed in [112]. The total kinetic energy

$$E_{\text{kin}} = \frac{m}{2} \int d^d x \, \langle |\nabla\phi(\mathbf{x}, t)|^2 \rangle \tag{7.14}$$

can be split, $E_{\text{kin}} = E_{\text{v}} + E_{\text{q}}$, into a "classical" part

$$E_{\text{v}} = \frac{m}{2} \int d^d x \, \langle |\sqrt{n}\mathbf{v}|^2 \rangle \tag{7.15}$$

and a "quantum-pressure" component

$$E_{\text{q}} = \frac{1}{2m} \int d^d x \, \langle |\nabla\sqrt{n}|^2 \rangle. \tag{7.16}$$

The radial energy spectra for these fractions involve the Fourier transforms of the generalized velocities $\mathbf{w}_{\text{v}} = \sqrt{n}\,\mathbf{v}$ and $\mathbf{w}_{\text{q}} = \nabla\sqrt{n}/m$:

$$E_\delta(k) = \frac{m}{2} \int d\Omega_d \, \langle |\mathbf{w}_\delta(\mathbf{k})|^2 \rangle, \qquad \delta = \text{v}, \text{q}. \tag{7.17}$$

which we cast further into occupation numbers $n_\delta(k) = k^{-2} E_\delta(k)$, $\delta = \text{v}, \text{q}$. Since the superfluid velocity $\mathbf{v} = \nabla\varphi$ is a potential field, it does not exhibit a transverse flow component, $\nabla \times \tilde{\mathbf{v}} = 0$ (outside vortex cores). In contrast, $\mathbf{w}_{\text{v}}$ is not a potential field and the divergence of $\mathbf{v}$ at $r \to 0$ is regularized by the vanishing $\sqrt{n(r)}$. Following [112], the regularized velocity $\mathbf{w}_{\text{v}}$ can be further decomposed into "incompressible" (divergence-free) and "compressible" (solenoidal) parts, $\mathbf{w}_{\text{v}} = \mathbf{w}_{\text{i}} + \mathbf{w}_{\text{c}}$, with $\nabla\cdot\mathbf{w}_{\text{i}} = 0$, $\nabla \times \mathbf{w}_{\text{c}} = 0$, to distinguish vortical superfluid and irrotational motion of the fluid. By construction, the generalized velocity $\tilde{\mathbf{w}}_{\text{v}} = f(r)\tilde{\mathbf{v}}(r, \varphi)$ of a vortex has only an incompressible component, since

$$\nabla \cdot \tilde{\mathbf{w}}_{\text{v}} = \nabla f(r) \cdot \tilde{\mathbf{v}}(r, \varphi) + f(r)\nabla \cdot \tilde{\mathbf{v}}(r, \varphi) = 0. \tag{7.18}$$

The first term vanishes owing to the transverse nature of the vortex velocity field, and the second one equals zero because the superfluid velocity is a potential field. The density of incompressible energy $|\tilde{\mathbf{w}}_{\text{i}}|$ of a vortex is constant up to about one healing length distance from the core and then falls off as $1/r$. Sound waves are purely compressible excitations. In our simulations, their oscillating density and phase profiles will be visible as maxima and minima in the compressible energy density in position space.

7.2.4 Vortex statistics

To understand the implications of vortex defects appearing in the dynamical evolution of degenerate Bose gases, in particular to make contact with the observables studied in the context of wave turbulence, we turn to a statistical viewpoint. The point-vortex model studied in the following was introduced by Onsager [120]. It describes the complex flow pattern in terms of the statistics of classical point objects with Coulomb-type interactions. However, owing to the absence of a kinetic energy term in the Hamiltonian, there is no kinematic transfer of potential energy into energy of motion. The model is constructed as a discrete-vorticity approximation of classical fluid turbulence, but it is even more suitable to describe superfluid turbulence consisting of quantized vortices.

We restrict our discussion to the example of $d = 2$ dimensions. We have seen that an isolated, singly quantized vortex is described by the complex field $\phi(r, \varphi) \equiv \sqrt{n(r)}\, e^{i\varphi}$. As the r dependence of the density $n(r)$ only becomes important at small scales of the order of the healing length $\xi = (2mgn)^{-1/2}$ at which in practice thermal excitations dominate, we assume n to be uniform. A set of M vortices can be described by $\phi(\mathbf{x}) = \Pi_i^M \phi_i(\mathbf{x})$, where $\phi_i(\mathbf{x}) = \phi(\mathbf{x} - \mathbf{x}_i)$ is the single-vortex field centered around $\mathbf{x}_i$. Let us derive the corresponding bosonic single-particle spectrum by considering the velocity field $\mathbf{v} = \nabla\varphi/m$. We can express the mean classical kinetic energy density of the velocity field $\tilde{\mathbf{v}}(\mathbf{x})$ of a single vortex as

$$E_{\mathrm{v}}(\mathbf{x}) = \frac{m}{2}\langle|\mathbf{v}(\mathbf{x})|^2\rangle = \frac{m}{2}\left\langle\left|\int d^2x'\,\tilde{\mathbf{v}}(\mathbf{x} - \mathbf{x}')\rho(\mathbf{x}')\right|^2\right\rangle, \tag{7.19}$$

where $\rho(\mathbf{x}) = \sum_{i=1}^M \kappa_i\delta(\mathbf{x} - \mathbf{x}_i)$ defines the spatial distribution of vortices with winding number $\kappa_i = \pm 1$. Here and in the following, $\langle\cdot\rangle$ denotes an ensemble average over different realizations of the classical field $\phi(\mathbf{x})$. We derive the low-k scaling of $n(k)$ from the kinetic energy spectrum $E_{\mathbf{v}}(k)$, the angle-averaged Fourier transform of $E_{\mathbf{v}}(\mathbf{x})$, taking into account that at low k, the velocity field $\mathbf{v}$ dominates the dynamics:

$$n(k) \simeq 2mk^{-2}E_{\mathbf{v}}(k). \tag{7.20}$$

We have, from (7.19),

$$E_{\mathrm{v}}(\mathbf{k}) \sim \langle|\mathbf{v}(\mathbf{k})|^2\rangle = \langle|\rho(\mathbf{k})|^2\,|\tilde{\mathbf{v}}(\mathbf{k})|^2\rangle, \tag{7.21}$$

with

$$|\rho(\mathbf{k})|^2 = \sum_{i,j}^M \kappa_i\kappa_j\, e^{i\mathbf{k}\cdot(\mathbf{x}_i - \mathbf{x}_j)}. \tag{7.22}$$

Below the healing lengthscale $k_\xi = 2\pi/\xi$, the modulus of the velocity field of a single vortex scales as $|\tilde{\mathbf{v}}| \sim k^{-1}$ and is radially symmetric.

To distinguish contributions from vortex–vortex and vortex–antivortex correlations, we write the distribution $\rho(\mathbf{x}) = \rho^{\mathrm{V}}(\mathbf{x}) - \rho^{\mathrm{A}}(\mathbf{x})$ as the sum of distributions

$\rho^{\mathrm{V}}(\mathbf{x}) = \sum_{i=1}^{M} \delta(\mathbf{x} - \mathbf{x}_i^{\mathrm{V}})$ of M vortices and $\rho^{\mathrm{A}}(\mathbf{x}) = \sum_{i=1}^{M} \delta(\mathbf{x} - \mathbf{x}_i^{\mathrm{A}})$ of M antivortices. Hence,

$$\langle |\rho(\mathbf{k})|^2 \rangle = \int \mathrm{d}^2 x \, \mathrm{d}^2 x' \, \epsilon^{\mathrm{i}\mathbf{k}\cdot(\mathbf{x}-\mathbf{x}')} C(\mathbf{x}, \mathbf{x}'), \qquad (7.23)$$

with $C(\mathbf{x}, \mathbf{x}') = \langle \rho_{\mathbf{x}}^{\mathrm{V}} \rho_{\mathbf{x}'}^{\mathrm{V}} \rangle - \langle \rho_{\mathbf{x}}^{\mathrm{V}} \rho_{\mathbf{x}'}^{\mathrm{A}} \rangle - \langle \rho_{\mathbf{x}}^{\mathrm{A}} \rho_{\mathbf{x}'}^{\mathrm{V}} \rangle + \langle \rho_{\mathbf{x}}^{\mathrm{A}} \rho_{\mathbf{x}'}^{\mathrm{A}} \rangle$. This allows a derivation of the kinetic energy distribution in terms of correlation functions of vortex positions.

We can now model pairing by the density–density correlation functions

$$\langle \rho_{\mathbf{x}}^{\mathrm{V(A)}} \rho_{\mathbf{x}'}^{\mathrm{V(A)}} \rangle = \frac{M}{V_R} \delta(\mathbf{x} - \mathbf{x}') + P_{\mathbf{x}, \mathbf{x}'}, \qquad (7.24)$$

$$\langle \rho_{\mathbf{x}}^{\mathrm{V(A)}} \rho_{\mathbf{x}'}^{\mathrm{A(V)}} \rangle = \frac{M}{V_R V_\lambda} \theta(\lambda - |\mathbf{x} - \mathbf{x}'|) + P_{\mathbf{x}, \mathbf{x}'} \qquad (7.25)$$

where V_R is the volume in which we take averages, and $V_\lambda = \pi \lambda^2$ is the area where the theta function equals one, measuring the correlation regime of vortices and antivortices. The contributions

$$P_{\mathbf{x}, \mathbf{x}'} = \frac{M(M-1)}{V_R(V_R - V_\Lambda)} \theta(|\mathbf{x} - \mathbf{x}'| - \Lambda) \qquad (7.26)$$

take into account that, besides pairing, vortices and antivortices keep a minimum distance Λ in the dilute gas. This is due to vortex–vortex repulsion and fast vortex–antivortex annihilation at small distances. The functions $P_{\mathbf{x}, \mathbf{x}'}$ cancel out in (7.23).[5]

From this ansatz, two scaling regimes can be found [116]. In the case of pairing, the flow field far away from the cores is given by the field of a vortex pair, which decays as r^{-2}, and the low-momentum power law is dominated by the flow of random vortex pairs $n_k \sim k^{-2}$. Above $k_{\mathrm{pair}} \simeq \pi/\lambda$, the distribution exhibits the scaling of an ensemble of independent vortices, $n_k \sim k^{-4}$, up to the healing lengthscale k_ξ, above which one can observe the vortex-core scaling $\sim k^{-6}$. The above results show that in a vortex-dominated flow, particles with low momenta are found far away from the vortex cores. Particles closer to the vortex cores pick up a higher momentum.

For conciseness, we remark that, as was shown in [113], one can obtain Kolmogorov $5/3$ scaling from the statistics of point vortices, by choosing the density–density correlation functions to decay as

$$\langle \rho_{\mathbf{x}}^{\mathrm{V(A)}} \rho_{\mathbf{x}'}^{\mathrm{V(A)}} \rangle \sim |\mathbf{x} - \mathbf{x}'|^{-\alpha} + P_{\mathbf{x}, \mathbf{x}'}, \qquad (7.27)$$

$$\langle \rho_{\mathbf{x}}^{\mathrm{V(A)}} \rho_{\mathbf{x}'}^{\mathrm{A(V)}} \rangle = P_{\mathbf{x}, \mathbf{x}'}, \qquad (7.28)$$

where the contributions $P_{\mathbf{x}, \mathbf{x}'}$ are assumed to be equal. The integral in (7.23) is convergent for $\frac{1}{2} < \alpha < 2$. This includes $\alpha = 4/3$, which gives $n(k) \sim k^{-4.66}$ and thus Kolmogorov scaling $E(k) \sim k^{-5/3}$. Note that the presence of vortex–antivortex correlations destroys the $5/3$ scaling, as discussed by Bradley and Anderson [30].

---

[5] If different avoidance scales Λ apply for vortices and antivortices, the terms do not cancel, but the remaining term does not alter the results for pair scaling derived here.

Infrared cutoff

Consider a Bose gas with density n_V containing a random distribution of vortices of either sign. We expect a decay of the coherence $\langle \phi^*(\mathbf{x})\phi(\mathbf{x}')\rangle$ over a distance $|\mathbf{x}-\mathbf{x}'|$ of the order of the mean vortex distance $l_V = n_V^{-1/2}$, corresponding to a momentum scale $k_V = \pi/l_V$. That is because vortices appear on average at this distance and induce a rapid change of the phase angle φ. For momenta $k < k_V$, the momentum distribution $n(k)$ needs to be sufficiently flat in order to ensure convergence of the integral that gives the total number of particles. This restriction is not imposed upon the particle numbers defined by the hydrodynamic decomposition in terms of n_i, n_c, and n_q. Hence, we expect a deviation of $n(k)$ from the incompressible momentum distribution n_i below k_V. In Fig. 7.3, we present numerical evidence for our reasoning. The plot shows the single-particle distribution (dots) as well as the incompressible momentum distribution (lines) for three different vortex numbers N_V. We can see the vortex density dependence of the IR cutoff in the single-particle momentum distribution. The spectrum of the incompressible velocity field does not show this feature. Instead, the k^{-4} scaling persists all the way to the lowest momenta.

Onsager model: nonthermal fixed point as a maximum-entropy state

We close with the picture that Onsager [120] developed of thermodynamic equilibrium states of a fixed number of vortices and antivortices in two dimensions. He used the Hamiltonian of vortical flow [97],

$$H = -\frac{1}{2\pi} \sum_{i>j}^{M} \kappa_i \kappa_j \ln(|\mathbf{r}_i - \mathbf{r}_j|), \tag{7.29}$$

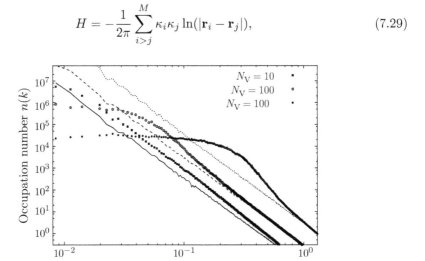

Fig. 7.3 Radial momentum distributions for a set of infinitely thin ($\xi = 0$) randomly distributed vortices. Note the log–log scale. At the highest momenta, the total number spectrum $n(k)$ (dots) and the incompressible component $n_i(k)$ (lines) follow the k^{-4} scaling of independent vortices. However, at momenta $k < k_V$, with $k_V(N_V = 10) \simeq 10^{-2}$, $k_V(N_V = 100) \simeq 3 \times 10^{-2}$, and $k_V(N_V = 1000) \simeq 10^{-1}$, $n(k)$ becomes flat, while $n_i(k)$ continues to rise.

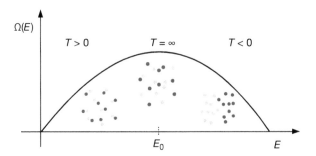

Fig. 7.4 Illustration of Onsager's picture of thermodynamic equilibrium states of the balanced two-dimensional Coulomb-type vortex gas at fixed defect number. We plot the qualitative form of the number of available states $\Omega(E)$ as a function of energy E as well as snapshots of typical microscopic configurations of vortices and antivortices.

to describe the dynamics of a system of M vortices in a superfluid which hence interact like charge carriers in a Coulomb gas. Here, the position of the ith vortex is denoted by $\mathbf{r}_i = (x_i, y_i)$. Because the x and y coordinates of each vortex are canonical conjugates of each other, phase space is identical with position space. Hence, for M vortices moving in a two-dimensional volume V, the total phase space is V^M. The Hamiltonian (7.29) implies that low-energy configurations feature vortices of opposite sign close to each other, whereas high-energy configurations require vortices of equal sign to group. Owing to these constraints, the number of configurations, $\Omega(E)$, available for the system at a given energy E decreases toward high and low energies, with a maximum at some intermediate $E = E_0$. This concept is illustrated in Fig. 7.4. According to Boltzmann, the entropy is

$$S(E) = \ln[\Omega(E)], \tag{7.30}$$

and the inverse temperature $1/T = \partial S/\partial E$ is positive for $E < E_0$ and negative for $E > E_0$. It follows that positive-temperature states are characterized by vortex–antivortex pairing, while negative-temperature states feature clustering of vortices of the same circulation. At the point of maximum entropy $S(E_0)$ and infinite temperature, Onsager expected a state of uncorrelated vortices and antivortices.

Three-dimensional systems

We briefly comment on the case of vortex lines and loops in three dimensions. A formulation similar to the Onsager point-vortex model is possible [109, 110, 156].

For taking into account the most general case of squeezed vortex loops, it is helpful to consider elliptical filaments, characterized by major radius r_a and minor radius r_b. Three scaling regimes can be distinguished. For the lowest momenta, one has $n(k) \sim k^{-2}$, which equals the IR scaling in the presence of a vortex ring. For momenta $k_a \ll k \ll k_b$, one finds $n(k) \sim k^{-3}$, which coincides with the IR scaling of two anticirculating vortex lines. For the ellipse, above k_b, the momentum distribution scales like $n(k) \sim k^{-5}$. This is the scaling of a single vortex line and can also be found as the high-momentum scaling of a vortex ring or a pair of straight vortex lines. For more details, we refer to [116].

7.3 Nonthermal fixed point of a vortex gas

In extending the concept of universality to time evolution far from thermal equilibrium, one expects that also, away from the thermal limit, the character of dynamical evolution can become independent of the microscopic details. Looking at closed systems, this implies that in approaching critical configurations, the evolution must become independent of the particular initial state from which the system has started, and critical slowing down in the actual time evolution will be observed. Time evolution near the fixed point becomes equivalent to a coarsening transformation. It should look like the self-similar pattern that is observed through a microscope as one continuously turns the magnification of the lens to smaller focal lengths. Precisely at the fixed point, the system becomes stationary owing to its self-similarity under time translations. A generic isolated system can evolve into the vicinity of a nonthermal fixed point and stay there over long times, before it eventually undergoes thermalization. This picture is supported by the derivation of the strong-turbulence scaling laws in the framework of renormalization group theory for correlation functions [17].

In this section, we discuss mostly numerical results obtained for the time evolution of an isolated two-dimensional gas. This is characterized, from the turbulence point of view, by direct and inverse cascades, with fluxes determined by local conservation laws in Fourier space. From the perspective of defect formation, a diluting ensemble of vortices and antivortices marks the approach of the nonthermal fixed point. We identify the mechanism for this dilution process and demonstrate that it is consistent with the coarsening transformation picture of universal dynamics. By reducing the characterization of the momentaneous configuration to a few length parameters, we can illustrate the slowing of the evolution near the fixed point. We show that how closely the critical point is approached depends on the chosen initial state.

7.3.1 Time evolution of vortex patterns and momentum spectra

In this subsection, we have a closer look at the process of vortex formation and of the Bose gas approaching the nonthermal fixed point, concentrating again on the two-dimensional case. Vortical excitations can be created in large numbers, for example within shock waves forming during the nonlinear evolution of a coherent matter-wave field. We follow the example of evolution of phase and density profiles illustrated in Fig. 7.5. Six snapshots are shown, taken at dimensionless times $\bar{t}$ as indicated in the caption. The initial field configurations were prepared by macroscopically populating a few of the lowest-momentum modes in the computation such that the resulting condensate density in position space varied between zero and some maximum value. We can see strong phase gradients forming owing to the nonlinear evolution. At around $t \simeq 100$, these gradients produce shock fronts delimited by phase defects that in the following collapse into vortex trains. Scattering processes between vortices quickly isotropize phase and density fluctuations.

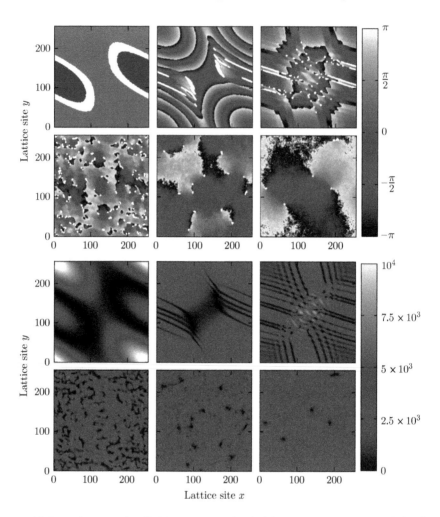

Fig. 7.5 (Color online: see [114]) Phase angle $\varphi(\mathbf{x}, t)$ (a) and spatial density $n(\mathbf{x})$ (b) at six times during a single run of the simulations in $d = 2$ on a space–time lattice with side lengths $L = N_s a_s$, lattice spacing a_s, with periodic boundary conditions imposed. The GPE is written in terms of the dimensionless variables $\bar{g} = 2mg a_s^{2-d}$, $\bar{t} = t/(2m a_s^2)$, and $\bar{\phi}(\bar{t}) = \phi\sqrt{a_s^d}\,e^{2i\bar{t}}$. Parameters are $\bar{g} = 3 \times 10^{-5}$, $N = 10^8$, and $N_s = 256$. The times shown are as follows (by line, each from left to right): $\bar{t} = 0$: ordered phase at initial preparation. $\bar{t} = 104$: during build-up of strong density and phase gradients. $\bar{t} = 210$: formation of vortices and antivortices. $\bar{t} = 420$: unbinding of vortex–antivortex pairs. $\bar{t} = 6550$: slowing down of vortex dynamics near the nonthermal fixed point. $\bar{t} = 10^5$: a few separated vortex–antivortex pairs close to the nonthermal fixed point.

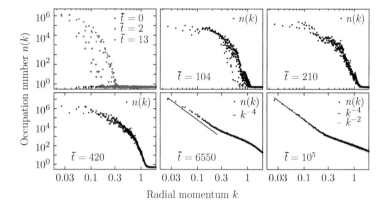

Fig. 7.6 Single-particle angle- and ensemble-averaged mode occupation numbers $n(k) = \int d^{d-1}\Omega_k \langle \phi^*(\mathbf{k})\phi(\mathbf{k})\rangle_{\text{ensemble}}$ as functions of the radial momentum k, at the same evolution times as in Fig. 7.5. The dimensionless lattice momenta are $k = \left[\sum_{i=1}^{d} 4\sin^2(k_i/2)\right]^{1/2}$, $k_i = 2\pi n_i/N_s$, $n_i = -N_s/2, \ldots, N_s/2$. The parameters are the same as in Fig. 7.5. Note the double-logarithmic scale. At early times (top left), scattering between macroscopically occupied modes leads to excitation transfer to high momenta. Vortex formation ($\bar{t} \simeq 210$) sets in when this process has reached the healing lengthscale $k_\xi \simeq 1.26$. At $\bar{t} \gtrsim 10^4$, a bimodal power law emerges, characteristic of the nonthermal fixed point.

In Fig. 7.6, we show the corresponding time evolution of the angle- and ensemble-averaged radial momentum spectrum. The early times (top left) are characterized by scattering between macroscopically occupied modes. Once excitations of the order of the healing length ξ have been created, vortex formation sets in. Shortly after vortices have been created, the spectrum exhibits a power-law behavior $2.85 \lesssim \zeta \lesssim 3.0$ within a range of momenta $k \in [0.04, 0.4]$, see the lower left panel of Fig. 7.6. Subsequently, the evolution slows down and a quasistationary period is entered. During an intermediate stage (bottom center and right panels of Fig. 7.6) of the vortex-bearing phase, two distinct power laws develop that are in excellent agreement with the analytical prediction in (7.5) and (7.7). In the UV, the exponent $\zeta_P^{\text{UV}} = d$ exhibits weak wave turbulence, (7.5), whereas in the IR, the exponent confirms the field-theory prediction $\zeta_Q^{\text{IR}} = d+2$; see (7.7). During the ensuing evolution, the weak-wave-turbulence scaling decays toward $\zeta = 2$, reflecting a thermal UV tail. Note that in $d = 2$, the weak-turbulence exponent $\zeta_P^{\text{UV}} = 2$ is identical to that in thermal equilibrium in the Rayleigh–Jeans regime, $n(k) \sim T/k^2$ [165]. In $d = 3$, we observe, at late times, a change of the IR scaling behavior from $\zeta = d + 2 = 5$ to $\zeta = 3$, pointing to the development of pairing correlations [116].

At late times, after the last vortical excitations have disappeared, we observe the entire spectrum to become thermal, i.e., exhibit Rayleigh–Jeans scaling with $\zeta = 2$ (not shown). We emphasize that thermal scaling of the single-particle occupation number has $\zeta = 2$ despite the fact that quasiparticles with a linear dispersion are expected to thermalize in the regime of wavenumbers smaller than the inverse healing length.

7.3.2   Local transport in momentum space and inverse particle cascade

From the above findings, the question arises as to why the system selects the particular exponents $\zeta_P^{\mathrm{UV}} = d$ and $\zeta_Q^{\mathrm{IR}} = d + 2$ from the set of four possible exponents given in (7.5) and (7.7). For this, the fluxes underlying the stationary but nonequilibrium distributions are relevant [22]. The timeline of distributions shown in Fig. 7.6 suggests that the evolution of the gas involves transport of particles originating from the intermediate-momentum regime $k \simeq 0.05$–0.2, which during the initial evolution becomes strongly *overpopulated*. The particles drift toward both lower and higher wavenumbers, building up a bimodal power-law distribution. To describe the character of this bidirectional flux, we plot in Fig. 7.7 the radial particle and kinetic energy flux distributions Q_k and P_k, respectively, at $\bar{t} = 6550$, corresponding to the bottom center panel of Fig. 7.6. Note that the radial particle flux density Q_k is multiplied by gn to have the same units as the energy flux density $P(k)$. These flux densities are defined through the balance equations (7.3) and (7.4), respectively, with kinetic energy density $\varepsilon_k = n_k k^2/2m$. They are determined by integrating the numerically obtained particle and energy spectra $N(k)$ and $E(k)$ up to the scale k.

The graph supports the interpretation of the transport in terms of an inverse particle cascade in the IR and a direct energy cascade in the UV [116], in accordance with the appearance of the bimodal momentum distributions in Fig. 7.6. Although the derivation of the IR exponents requires the full dynamical theory with nonperturbatively resummed self-energies, the signs of the fluxes correspond to the respective scaling exponents, i.e., ζ_Q^{IR} in the IR and ζ_P^{UV} in the UV. Moreover, at late times, the kinetic energy flux P almost vanishes owing to a thermalized UV momentum distribution, but Q still reshuffles particles and therefore energy, with the zero mode acting as a sink, keeping the system out of equilibrium close to the nonthermal fixed point.

We finally remark that a necessary condition for a nonequilibrium stationary distribution is energy damping at large k [165]. Moreover, energy and particle number conservation in the interaction of different momentum modes can be shown to imply

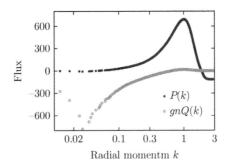

Fig. 7.7   Direct kinetic energy and inverse particle fluxes in $d = 2$ at $\bar{t} = 6550$. The parameters are as in Fig. 7.6. Note that the k axis has a logarithmic scale. A positive kinetic energy flux is seen in the UV and a negative particle flux in the IR. Units are $[P] = [gnQ] = (4m^2 a_s^{d+4})^{-1}$. (Reprinted with permission from [116]. Copyright (2012) by the American Physical Society.)

the existence of at least one more sink, i.e., a region where the right-hand side of (7.3) effectively has an additional damping term $\sim \Gamma(k)n(k)$, with negative Γ. In between these sinks, a source region supplies the input to the bidirectional flux pattern toward the UV and IR. Using kinetic theory, one can show that under certain conditions, a positive k-independent flux transports energy, $P > 0$, while a negative flux transfers particles, $Q < 0$ [165], Remarkably, this pattern remains valid in our case, beside the UV weak-wave-turbulence regime also in the IR region where the exponent emerges from a fixed point of the nonperturbative dynamic equations for Green's functions. As already pointed out by Scheppach et al. [140], the derivation of the IR exponent $\zeta_Q^{\text{IR}} = d+2$ requires sufficiently well-defined quasiparticles, suggesting that a treatment in terms of the QBE with a momentum-dependent scattering matrix element would be applicable. From this point of view, the negative flux Q and scaling in the IR and the positive flux P and weak wave turbulence in the UV, as observed in the numerics, emerge as a necessary consequence of conservation laws and transport processes described by wave-kinetic transport equations with nontrivial interactions.

7.3.3 Approach of the nonthermal fixed point and critical slowing down

Let us study in greater detail the evolution of the system toward and away from the nonthermal fixed point and focus on universal aspects of the dynamics of vortices. For details, we refer to [144]. Figure 7.8(a) shows the time evolution of the vortex density $\rho(t) = \langle N^V(t) + N^A(t)\rangle/V$, where $N^{V(A)}(t)$ is the mean number of vortices (antivortices) in the volume V at time t, for different specific choices of the initial state; see [144]. In all runs, vortex formation occurs around $t_V \simeq 10^3$, as is apparent from the steep increase in vortex density around this time. For $t \gtrsim t_V$, two distinct stages in the decay of vortex density are observed: a rapid early stage and a slow late stage. We have repeated our simulations on various grid sizes, $N_s \in [256, \dots, 4096]$, and have found that decay exponents saturate for and above $N_s = 512$. We attribute deviations on smaller grids to effects from regular (integrable) dynamics of few-vortex systems [6]. We remark that the onset of the slow decay coincides with the development of a particular scaling behavior in the single-particle momentum distribution $n(k) \sim k^{-4}$, which was shown by Nowak et al. [116, 117] to signal the approach of the nonthermal fixed point and the formation of a set of randomly distributed vortices. In this context, the reduction of the vortex density decay exponent, compared with the early stage of rapid decay, is interpreted as being due to (critical) slowing down of the nonlinear dynamics near the nonthermal fixed point.

We can discuss the dynamical transition in the vortex annihilation dynamics in terms of characteristic features of the vortex–antivortex correlation function

$$g_{VA}(\mathbf{x}, \mathbf{x}', t) = \frac{\langle \rho^V(\mathbf{x}, t)\rho^A(\mathbf{x}', t)\rangle}{\langle \rho^V(\mathbf{x}, t)\rangle\langle \rho^A(\mathbf{x}', t)\rangle}, \tag{7.31}$$

where $\rho^{V(A)}(\mathbf{x}, t) = \sum_i \delta(\mathbf{x} - \mathbf{x}_i(t))$ is the distribution of (anti)vortices in a single run at time t. For sufficiently large ensembles, g_{VA} is a function of $r = |\mathbf{x}-\mathbf{x}'|$ only. At early times, one finds a strong pairing peak in $g_{VA}(r, t)$ near $r = 0$; see Fig. 7.9(a). This peak quickly becomes smaller, and a hole is "burned" into the correlation function near the

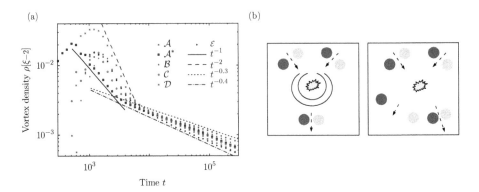

Fig. 7.8 (a) Vortex density ρ as a function of time $\bar{t}$: evolution for various initial conditions, averaged over 20 runs on a grid of size $N_s = 1024$. The lines show different power-law evolutions. The vortex density follows power laws $\rho(t) \sim t^{-\alpha_i}$ with two different exponents α_i, $i = 1, 2$. The exponent during the early stage depends considerably on initial conditions $1 \lesssim \alpha_1 \lesssim 2$, whereas the late stage features a decay exponent in a narrow interval $0.3 \lesssim \alpha_2 \lesssim 0.4$. The closest approach to the nonthermal fixed point is at $t \simeq (5 \ldots 10) \times 10^5$. (b) Sketches of different scattering events between two vortex–antivortex pairs in $d = 2$ dimensions. The left panel shows the scattering of two vortex pairs resulting in mutual annihilation of two vortices and the emission of density waves. The right panel shows the scattering of two vortex pairs leading to changes in the vortex–antivortex distance l_D and in the pair velocity. We remark that once $l_D \sim \xi$, a vortex pair decays rapidly under the emission of density waves. (Reprinted with permission from [144]. Copyright (2012) by the American Physical Society.)

origin; see Fig. 7.9(b). Following the time evolution of the spatial vortex distribution, we observe that this involves qualitatively different processes. Mutual annihilations of closely positioned vortices and antivortices occur under the emission of sound waves. Further separated vortices can approach each other in different ways, as illustrated in Fig. 7.8(b). The scattering of two pairs can lead directly to the annihilation of one pair under the emission of sound waves. This includes events where the dipole length decreases below a certain threshold, implying a density dip rather than a vortex pair. This density dip can still interact with other vortices, but will quickly vanish. Alternatively, the scattering reduces the vortex–antivortex separation within one pair while increasing it within the other, in accordance with the Onsager point-vortex model [120]. We refer to this characteristic change in $g_{VA}(r)$ as a vortex-unbinding process. Recall that the Onsager model does not contain a kinetic term for the point defects, so their interaction potential energy cannot induce relative acceleration of the vortices. Changes in their separation must occur dynamically via collisions.

At around the time $t_3 \lesssim t \lesssim t_4$, the power-law exponent of the vortex density decay changes to about a third of its previous value; see the inset in Fig. 7.9(a). Computing the mean vortex–antivortex pair distance l_D, by averaging over distances between each vortex and its nearest antivortex, we find that, in accordance with the previous discussion, l_D grows continuously, exhibiting two characteristic stages; see

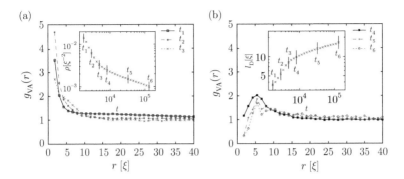

Fig. 7.9 Normalized pairing correlation function g_{VA}, (7.31), as a function of the vortex–antivortex distance r, for six different times t_i (in lattice units). (a) $g_{VA}(r)$ during the rapid-decay stage, averaged over 174 runs on a grid of $N_s = 1024$. Inset: Time evolution of the vortex density ρ, from Fig. 7.8(a). (b) $g_{VA}(r)$ during the slow-decay stage, with the same averaging as in (a). Inset: Evolution of the mean vortex–antivortex pair distance l_D. (Reprinted with permission from [144]. Copyright (2012) by the American Physical Society.)

the inset in Fig. 7.9(b). At times $t \gtrsim 10^4$, $l_D(t)$ approaches the power-law solution $l_D \sim \rho^{-1/2}$, as expected for uncorrelated vortices.

We close by focusing on the growth of long-range coherence, associated with the annihilation of topological defects [25, 42, 69, 70, 90, 96, 107, 151]. From this point of view, freely decaying superfluid turbulence is a particular example of phase-ordering dynamics after a quench into the ordered phase [32]. Whereas in three dimensions, a second-order phase transition connects a normal-fluid phase and a superfluid phase, a Bose gas in two dimensions experiences a Berezinskii–Kosterlitz–Thouless (BKT) transition [11, 89]. For the two-dimensional ultracold Bose gas, experimental and theoretical results support the understanding of the phase transition in terms of vortices undergoing an unbinding–binding transition [26, 48, 59, 63, 145, 147, 162].

In this context, we are interested in a comparison between correlation properties observed in the nonequilibrium dynamics near a nonthermal fixed point and those known from equilibrium studies. We compute the dynamical trajectory of the vortex gas in the space of inverse coherence length and inverse mean vortex–antivortex pair distance. We compare our results with simulations of a thermal two-dimensional Bose gas specifically for our system parameters. We define a coherence length l_C^* in terms of the integral over the angle-averaged first-order coherence function:

$$l_C^* = \int dr\, g^{(1)}(r), \qquad g^{(1)}(r) = \int d\theta\, \frac{\langle \phi^*(\mathbf{x})\phi(\mathbf{x}+\mathbf{r})\rangle}{\sqrt{\langle n(\mathbf{x})\rangle\langle n(\mathbf{x}+\mathbf{r})\rangle}}. \qquad (7.32)$$

l_C^* measures the spatial extension of the first-order coherence function. In contrast to $r_{coh} = \int dr\, r^2 g^{(1)}(r) / \int dr\, r\, g^{(1)}(r)$, the quantity l_C^* does not sum up values of $g^{(1)}(r)$ weighted by the distance, which would enlarge insignificant contributions at large r. In addition, it does not overestimate the coherence of flat distributions. In equilibrium,

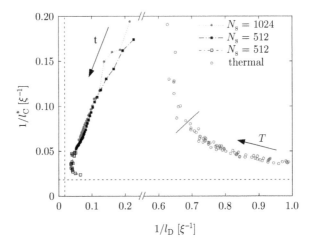

Fig. 7.10 Trajectories of multivortex states in the space of inverse coherence length $1/l_C$ and inverse mean vortex–antivortex distance $1/l_D$, starting from $t = t_V$. Comparison of the thermal line $(l_D^{-1}(T), l_C^{*-1}(T))$ for a range of temperatures T (right) with the corresponding dynamical evolution (left). Dashed lines mark the minimal values $2/L = 0.018\xi^{-1}$ available on a grid of size $N_s = 512$. Note that the $(1/l_D)$-axis interval $[0.25, 0.55]$ has been cut out. (Reprinted with permission from [144]. Copyright (2012) by the American Physical Society.)

this quantity smoothly interpolates between the regime of exponential decay of $g^{(1)} \sim \epsilon^{-r/\xi_C}$ above the BKT transition and its power-law decay below.

In Fig. 7.10, we follow the time evolution of the gas for $t > t_V$. We can see that a state of low coherence and small mean vortex–antivortex pair distance evolves toward larger coherence and larger vortex–antivortex separation. As discussed above, this is due to vortex annihilations and vortex–antivortex unbinding. For times $t > 10^4$, the coherence length grows as $l_C^* \sim \rho^{-1/2}$, in the same way as l_D shown in Fig. 7.9(b). The evolution slows down considerably for $1/l_C^* \sim 1/l_D \to 0$, when the gas starts to show the characteristic scaling $n(k) \sim k^{-4}$. After spending a long time near the nonthermal fixed point, l_D declines because the last remaining vortex–antivortex pairs reduce their size before their annihilation and the equilibration of the system. Our understanding of the nonthermal fixed point as a configuration with a few maximally separated pairs on an otherwise maximally coherent background implies that it is located near the crossing of the dashed lines. Hence, the nonthermal fixed point is approached most closely between $t \simeq 5 \times 10^5$ and $t \simeq 10^6 = t_3$. To set the above evolution in relation to equilibrium configurations, we also show the thermal line $\{l_D^{-1}(T), l_C^{*-1}(T)\}$ for a range of temperatures T for which the zero-mode population does not vanish.

Our results allow a picture to be drawn of the evolutionary path toward and away from the fixed point. The nonthermal fixed point is characterized by a few pairs—in the extreme case one pair—of far-separated anticirculating vortices and bears similarities with the equilibrium BKT fixed point. However, while the phase transition also features unbinding of vortices, finite temperature implies the simultaneous excitation of many rotons, i.e., strongly bound vortex–antivortex pairs. The nonthermal fixed point is clearly identified by strong-wave-turbulent scaling in the infrared limit, $n(k) \sim k^{-4}$. The high-energy modes are much more weakly populated, for example far below the BKT critical temperature, or remain out of equilibrium. The details of the UV-mode populations are determined by the way in which the nonthermal fixed point is being approached.

The way in which the system is forced, here, to approach the nonthermal fixed point generalizes the protocol of Kibble [81] and Zurek [168]. A strong sudden quench replaces the near-adiabatic approach of the BKT transition. The dynamical evolution in the vicinity of the BKT critical point was studied by Mathey and Polkovnikov [100] in terms of a perturbative renormalization group analysis. The route to a non-perturbative analysis in the strong-coupling regime is provided by out-of-equilibrium functional techniques; see the references at the end of Section 7.2.2.

7.3.4   Dependence on driving and a new route to Bose condensation

Let us finally give a taste of the relevance of the strength with which the system is driven away from thermal equilibrium for the approach of the nonthermal fixed point and, as a by-product, find new aspects of how superfluid turbulence affects the process of Bose condensation [115]. In this subsection, we restrict ourselves to a gas in $d = 3$ dimensions. We choose the overpopulated momentum distribution of the gas directly, eliminating the early instability phase of our previous simulations during which overpopulation was induced through nonlinear scattering of wave modes. The initial field in momentum space, $\phi(\mathbf{k}, 0) = \sqrt{n(\mathbf{k}, 0)} \, e^{i\varphi(\mathbf{k},0)}$, is parametrized in terms of a randomly chosen phase $\varphi(\mathbf{k}, 0) \in [0, 2\pi)$ and a density $n(\mathbf{k}, 0) = f(k)\nu_{\mathbf{k}}$, with $\nu_{\mathbf{k}} \geq 0$ drawn from an exponential distribution $P(\nu_{\mathbf{k}}) = e^{-\nu_{\mathbf{k}}}$ for each $\mathbf{k}$. The spectrum is flat at low k and falls off according to the function $f(k) = f_{\alpha}/(k_0^{\alpha} + k^{\alpha})$, where α controls the deviation from a thermal decay with $\alpha = 2$. We choose a cutoff $(k_0\xi)^{\alpha} = 0.2/0.44^{\alpha}$ and normalization $f_{\alpha} = 400/0.44^{\alpha}$. We compare results for a range of different cooling quenches defined by the power laws $\alpha = 2.5, \ldots, 10.0$, varying the total number between $N = 10^9$ ($\alpha = 2.5$) and $N = 4.3 \times 10^8$ ($\alpha = 10$).

In Fig. 7.11, we show $n(k, t)$ over the radial momenta $k = |\mathbf{k}|$ at the initial time as well as at a moment when the system is closest to the nonthermal fixed point, for the different choices of α. During the initial evolution ($t \lesssim 10^2\tau$), the mode occupations gradually spread to lower wavenumbers, at the same time depositing energy into the high-momentum tail. We emphasize that cutting away sufficiently much population at high momenta initially is necessary if the system is supposed to approach the nonthermal fixed point during its rethermalization. As described in Sec. 7.3.2, the approach to the fixed point is characterized by a dual cascade in which the energy of the intermediate-k overpopulation is deposited in the high-k tail, carried there by a few particles, while the majority of overpopulation particles move toward the IR, conserving overall energy and particle number. Hence, only strong cooling quenches allow for the build-up of a steep population far into the IR.

At late times, the spectra developing from the different initial α differ strongly. For $\alpha \gtrsim 3$, the distribution develops a bimodal structure, with power-law behavior

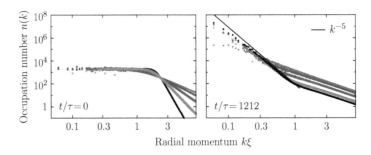

Fig. 7.11 (Color online: see [114]) Momentum distributions of a three-dimensional gas for different initial conditions. (a) Initial conditions with varying high momentum decay $k^{-\alpha}$, characterized by the parameter $\alpha = 2.5, 3, 4, 6, 10$, representing different evaporative cooling quenches. (b) At intermediate times, when the system is closest to the nonthermal fixed point, showing the steep IR power laws $\sim k^{-5}$ for initial conditions representing a strong cooling quench, $\alpha \gtrsim 3$. (Reprinted with permission from [115], © IOP Publishing & Deutsche Physikalische Gesellschaft. CC BY http://creativecommons.org/licenses/by/3.0/.)

$n(k) \sim k^{-5}$ in the IR and $n(k) \sim k^{-2}$ in the UV. At very long times, this bimodal structure decays toward a global $n(k) \sim k^{-2}$ (not shown). For $\alpha \lesssim 3$, the distribution directly reaches a thermal Rayleigh–Jeans scaling $n(k) \sim T/k^2$. As is shown in Fig. 6 of [115], the trajectories, when plotted as in Fig. 7.10, approach the nonthermal fixed point in the lower left corner more closely, the larger the value of α that is chosen. While $\alpha = 2.33$ leads to a trajectory near the thermal states, $\alpha = 10$ induces a motion similar to that of the black points.

Much experimental effort is currently being undertaken to study the dynamics of condensation under conditions of rapid evaporative cooling [125, 136, 148, 162]. Moreover, the many-body dynamics of coherent bosonic excitations is being intensively studied in solid state systems consisting of magnons [43, 44, 76, 118] or polaritons [4, 8, 75, 77, 93, 138]. Recently, condensation has been discussed for the case of gluons as an intermediate stage of heavy-ion collisions [14, 27] and, for relativistic scalars, has been found to rely on a nonperturbative inverse particle cascade [23].

Our results confirm that Bose–Einstein condensation in a nonequilibrated and undercooled gas can have the characteristics of a turbulent inverse cascade [25, 69, 70, 150], corresponding to a quasilocal transport process in *momentum* space, into the low-energy modes of the Bose field. The main new finding is that the superfluid turbulence period can appear in two different forms [115]. The two possible paths to Bose–Einstein condensation are shown schematically in Fig. 7.12. If a sufficiently small amount of energy is removed, a thermal Rayleigh–Jeans distribution forms in a quasi-adiabatic way. The chemical potential increases, and a fraction of particles is deposited in the lowest mode, forming a Bose–Einstein condensate. In the second scenario, after a sufficiently strong cooling quench, the system develops transient scaling behavior in the momentum distribution prior to condensate formation, i.e., it approaches the nonthermal fixed point. These scenarios differ qualitatively in how the condensate mode builds up as a function of time [115].

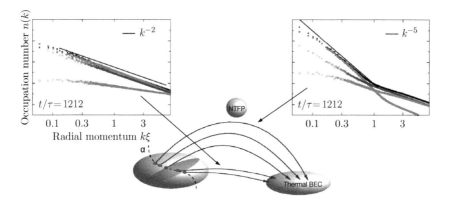

Fig. 7.12 (Color online: see [114]) Depending on the strength α of the initial cooling quench, the gas can thermalize in a near-adiabatic manner to a Bose–Einstein condensate. Alternatively, it can first approach and be critically slowed down near a nonthermal fixed point, where it is characterized by a scaling spectrum $n(k) \sim k^{-5}$ in the IR. Furthermore, dynamical scale separation of the incompressible (red points, second from above in the IR) and incompressible (blue points, third from above in the IR) components of the gas (see [116] and the last paragraph of Section 7.2.3) occurs. The incompressible component corresponds to transverse vortical flow. The gray points (fourth from above in the IR) indicate the quantum-pressure component. (Reprinted with permission from [115], © IOP Publishing & Deutsche Physikalische Gesellschaft. CC BY http://creativecommons.org/licenses/by/3.0/.)

7.4 Other systems

7.4.1 Soliton ensembles

In the preceding sections, it was discussed how strong-wave-turbulent scaling of the momentum spectrum of a degenerate Bose gas can be understood from the statistics of vortices. In this context, correlations between vortices and antivortices play a crucial role. In three spatial dimensions, similar relations exist between scaling and the creation of vortex lines and rings [116]. Looking at systems in one spatial dimension, we find that solitons, in particular dark solitons, play a crucial role in realizing strong IR wave turbulence. Solitons interact with other defects as well as sound excitations and show characteristics of localized quasiparticle excitations. In the following, we present a brief summary of the extent to which random ensembles of solitons can be seen as characterizing a nonthermal fixed point in one-dimensional bosonic systems.

A model of randomly positioned gray solitons that were solutions of the Gross–Pitaevskii equation (7.10) in the vicinity of each soliton density dip was discussed by Schmidt et al. [142]. In the limit of soliton separations large compared with the healing length, analytic expressions for the momentum spectra were derived in a homogeneous system as well as under the constraint of a trapping potential. A central result is depicted in Fig. 7.13 for the case of randomly distributed black solitons in a homogeneous background. The momentum distributions show a k^{-2} power-law behavior between the scale k_{n_s} corresponding to the mean soliton separation and k_{kink}, which marks

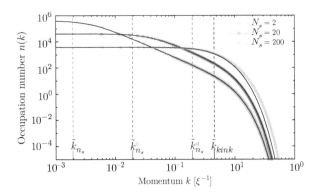

Fig. 7.13 Momentum distributions for N_s randomly distributed black, i.e., static solitons (symbols), showing a k^{-2} power-law behavior between the scale k_{n_s} given by the mean soliton separation and k_{kink} given by the healing length. The spectra are found to be in excellent agreement with the analytically predicted curves [142].

the healing lengthscale. At higher momenta, the distribution features an exponential decay following from the characteristic way in which the spatial density drops inside the soliton cores, while at low momenta, a flat distribution marks the long-range exponential decay of the coherence. The soliton configurations in one dimension resemble the turbulent phenomena in $d = 2$ and 3 dimensions. Their presence is accompanied by a quasistationary power-law momentum distribution that marks the self-similarity of a random distribution of sharp phase kinks: looking at the system within a window whose size is below the mean soliton separation and well above the core width, the system looks the same irrespective of the window size. There is either a kink seen within the window or not.

Transport in momentum space, in the form of an inverse cascade toward the IR, corresponds to soliton decay, which increases the inter-soliton distance and extends the self-similar regime toward smaller momenta (see Fig. 7.13). We remark that the predicted strong-wave-turbulent scaling $\sim k^{-d-2}$, which was found to be consistent with the vortex scaling in $d = 2$ and 3, does not give the scaling $\sim k^{-2}$ for the solitons in $d = 1$. The reason for this is expected to be similar to that in the case of domain walls in $d = 2, 3$, in a multicomponent system as described in Section 7.4.2: new transport equations must be set up in which a different conserved current, related to transport of spin-wave excitations, leads to a different IR scaling. A soliton in $d = 1$ does not exhibit transverse (incompressible) superfluid flow like that around a vortex core, which decays as $|\mathbf{v}(r)| \sim 1/r$. In this context, we remark that the longitudinal (compressible) component in $d = 2$ and 3 dimensions (Fig. 7.12) shows an IR exponent weaker by 1.

The formation and evolution of soliton excitations in trapped one-dimensional Bose gases can be studied by means of the classical field equation. A possible scenario for far-from-equilibrium dynamics involves an initially noninteracting thermal gas that is quenched by a sudden ramp of the interaction, as studied by Schmidt et al. [142].

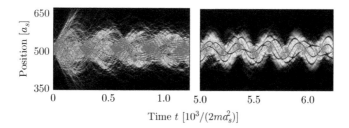

Time t $[10^3/(2ma_s^2)]$

Fig. 7.14 (Color online: see [114]) Time intervals of a single run of the classical field equation (7.10), showing solitons that oscillate inside a trapped one-dimensional ultracold Bose gas. The gas is initially noninteracting and thermalized, with $T = 360\omega_{\text{ho}}$, in a trap with oscillator length $l_{\text{ho}} = 8.5$ (in grid units). At time $t = 0$, the interaction is switched to $g_{1\text{D}} = 7.3 \times 10^{-3}$, and a cooling period using a high-energy knife is applied. The panels show the one-dimensional color-encoded density distribution as a function of time. Left panel: The initially imposed interaction quench causes strong breathing-like oscillations and the creation of many solitons. Right panel: Breathing oscillations have damped out, leaving a dipolar oscillation of the bulk distribution, with clearly distinct solitons in the trap. For details, see [142]. (Reprinted with permission from [142], © IOP Publishing & Deutsche Physikalische Gesellschaft. CC BY-NC-SA http://creativecommons.org/licenses/by-nc-sa/3.0/.)

To allow the emerging collective excitations to form solitons at a desired density, evaporative cooling helps to achieve the required reduction of the UV-mode populations. An example of position-space evolution is depicted in Fig. 7.14.

7.4.2 Domains and defects in two-component systems

As a further example, we briefly comment on what is already known in the context of strong wave turbulence and defect formation in multicomponent (e.g., spinor) gases [71]; see Fig. 7.15. A two-component Bose gas with contact interactions is well known to possess two different ground states depending on the value of the parameter $\alpha = g_{11}g_{22}/g_{12}$ that parametrizes the relative strength of intra-species couplings g_{11} and g_{22} as compared with inter-species interaction g_{12} [73, 153] and is the subject of an increasing number of experimental investigations [62, 111, 139, 159]. For $\alpha > 1$, which is called the immiscible regime, the inter-species interaction energy is greater than that of the interactions within one species. Hence, in the ground state of the system, the spatial overlap of the components is minimized through domain formation and spatial separation of the two components. In contrast, for $\alpha < 1$, the two components become miscible and uniformly distributed when in the ground state. One can use the parameter α to change the properties of the system in the as-yet unexplored region of nonequilibrium quasistationary states.

Dynamical instabilities serve to drive the system far from equilibrium. This leads to momentum distributions of the different components that are characterized by a strong overpopulation at intermediate momenta as compared with thermal equilibrium distributions with the same energy and particle numbers. Similar to our finding for the case of a one-component Bose gas in $d = 1, 2$, and 3 dimensions, such a far-from-equilibrium

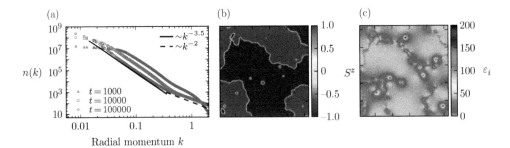

Fig. 7.15 (Color online: see [114]) Nonthermal fixed point and domain formation in an immiscible two-component Bose gas. (a) Occupation number spectrum in $d = 2$ spatial dimensions at different times (note the double-log scale). The spectrum at late times shows a bimodal structure with a thermal scaling $n(k) \sim k^{-2}$ at high momenta and an IR power law $n(k) \sim k^{-3.5}$, similar to the scaling seen when charge separation is found in the relativistic simulations (see Fig. 7.16). (b) Polarization $\langle S_z(\mathbf{x}) \rangle$, i.e., density difference of the two gas components, $S^z(\mathbf{x}) = (n_1 - n_2)/(n_1 + n_2)$. It demonstrates the formation of spin domains as a result of dynamic demixing. (c) Spatial distribution of incompressible kinetic energy $\varepsilon_i(\mathbf{x})$, showing high amounts of transverse hydrodynamic, i.e., vortical, flow around spin domains and especially around point-like defects.

distribution of particle momenta induces a redistribution of particles in momentum space toward both lower and higher momenta. Subsequently, the system encounters long-lived transient states with nontopological and quasi-topological defects, including domain walls, vortices in a single species, and skyrmions in the coupled spin system. Distinguishable types of defects are produced for different values of the external parameter, thereby allowing the induction of a transition between metastable nonequilibrium ordered states. In summary, one obtains an example of how to extend the concept of a phase transition into the realm of far-from-equilibrium time evolution.

7.4.3 Charge separation in reheating after cosmological inflation

In cosmological models, reheating describes the epoch starting at the end of inflation [3]. During this period, the potential energy of the inflaton field is redistributed into a homogeneous and isotropic hot plasma of particle excitations. These become a substantial part of the further expanding universe. Simple models describing reheating after inflation invoke self-interacting scalar fields. One of the popular scenarios involves the parametrically resonant amplification of quantum fluctuations of the macroscopically oscillating inflaton field. The amplified modes represent the emerging matter content of the universe [84, 155]. Various theoretical approaches have been proposed to model reheating. As both the inflaton and the amplified modes are strongly populated, classical field simulations can be applied to describe their evolution [80, 128, 154].

We focus on a scenario of parametric resonance in a globally $O(2)$ or, equivalently, $U(1)$ symmetric relativistic scalar field theory (mass parameter $m = 0$) in $d = 2$ and 3 dimensions [55]. Shortly after the resonant excitations have set in, a

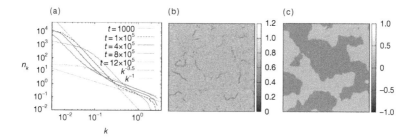

Fig. 7.16 (Color online: see [114]) Nonthermal fixed point and charge separation in the non-linear Klein–Gordon equation for a complex field. (a) Occupation number spectrum in $d = 2$ spatial dimensions at different times. The double-log scale reveals the bimodal power-law, with $n(k) \sim k^{-3.5}$ at low momenta. (b) Modulus of the field, $\phi(\mathbf{x})$, showing worm-like regions (dark gray) of near-zero field. (c) Charge distribution of the field, $j_0(\mathbf{x})$, showing homogeneous regions of opposite charge.

spatial separation of charges occurs as shown in Fig. 7.16(c). Both charge and anti-charge overdensities become uniformly distributed within slowly varying regions that are separated by sharp boundary walls of grossly invariant thickness. These walls have a character similar to topological defects and appear for initial conditions cor-responding to the parametric reheating scenario. We see that the presence of charge domains coincides with the visibility of nonthermal stationary scaling solutions—see Fig. 7.16(a)—as discussed earlier in the context of ultracold atomic gases. In this way, a link is established between wave turbulence phenomena as discussed, for example, in [17, 19, 22, 38, 102, 103, 140], and long-lived quasi-topological structures in the inflaton field. Strong nonthermal stationary scaling solutions have also been observed for the case of $O(4)$- and $O(10)$-symmetric scalar fields in $d = 3$ and 4 dimensions [19, 22]. The latter work raises the interesting question of the corresponding spatial configuration at the fixed point.

7.5 Outlook

In this chapter, we have pointed out the possibility of a universal duality between de-caying topological defects and a nonperturbative inverse wave-turbulent cascade. This cascade requires the generation of (quasi-)topological configurations far from thermal equilibrium and their slow decay, together with increased coherence and defect sep-aration. Under these conditions, we expect power-law scaling in a regime between the scales $1/\xi$, ξ being the microscopic extent of the defect core, and $1/l_{\text{COH}}$, where l_{COH} is the coherence length approximated by the mean distance between defects. In this setting, an inverse particle cascade is generated by defect dilution, and the associated power laws can be found from the scaling properties of the respective sin-gle defect. We have shown this mechanism to exist in soliton- and vortex-dominated single-component Bose gases [115–117, 142], decaying domain walls and vortices in

Table 7.1 Strong-wave-turbulence scaling of momentum spectra of different distributions as expected at a nonthermal fixed point, for different types of defects (first column) in various systems (second column) in d dimensions. Concerning the definitions of the defects, we refer to [74, 132]. The "skyrmion" denotes the respective defect arising in the nonlinear sigma model (Anderson–Toulouse vortex).

(Quasi-)topological defect	Field	d	Momentum scaling
Soliton/domain	$\phi \in \mathbb{C}, S_z \in \mathbb{R}$	1	$\langle|\phi(k)|^2\rangle, \langle|S_z(k)|^2\rangle \sim k^{-2}$
Soliton line/domain	$\phi \in \mathbb{C}, S_z \in \mathbb{R}$	2	$\langle|\phi(k)|^2\rangle, \langle|S_z(k)|^2\rangle \sim k^{-3}$
Soliton surface/domain	$\phi \in \mathbb{C}, S_z \in \mathbb{R}$	3	$\langle|\phi(k)|^2\rangle, \langle|S_z(k)|^2\rangle \sim k^{-4}$
Vortex	$\phi \in \mathbb{C}$	2	$\langle|\phi(k)|^2\rangle \sim k^{-4}$
Vortex line	$\phi \in \mathbb{C}$	3	$\langle|\phi(k)|^2\rangle \sim k^{-5}$
Skyrmion	$\mathbf{S} \in \mathbb{R}^3$	2	$\langle|S_{x(y)}(k)|^2\rangle \sim k^{-2}$
Skyrmion line	$\mathbf{S} \in \mathbb{R}^3$	3	$\langle|S_{x(y)}(k)|^2\rangle \sim k^{-3}$
Monopole	$\mathbf{E} \in \mathbb{R}^2$	2	$\langle|E(k)|^2\rangle \sim k^{-2}$
Monopole	$\mathbf{B} \in \mathbb{R}^3$	3	$\langle|B(k)|^2\rangle \sim k^{-2}$

two-component Bose gases [71], as well as transient charge domains in complex relativistic scalar theory [55]. It is emphasized that the stability of these defects does not need to be topological, as the examples of solitons and charge domains indicate. See, for example, [94] for a review of nontopological solitons. A list of expected scaling properties is given in Table 7.1.

A variety of (quasi-)topological excitations are known to exist in superconductors, magnets, and cosmic fields [94, 108, 123, 152]. Specific examples are monopoles in gauge fields [129, 131] and exotic magnets [39], as well as skyrmions in Bose–Einstein condensates [74, 137] and liquid crystals [46]. Coherent polariton ensembles represent a promising new route to study the dynamics of defects and solitary waves [4].

The study of multicomponent fields is certainly among the most interesting new directions of research in this context. We have taken a first step in this direction by investigating the two-component Bose gas [71]. The possibility of different nonthermal fixed points depending on inter- and intra-species couplings opens a perspective on new types of experiments far from equilibrium [73, 111]. Relating (quasi-)topological field configurations known from equilibrium spinor Bose gases [158] to transient scaling phenomena observed in correlation functions provides a great challenge for experiments and theory. Experimental studies of ultracold spin-1 and spin-2 Bose gases, including the detection of spin domains, are well developed [40, 62, 65, 104, 139, 141]. Multicomponent fields are important far beyond ultracold atomic physics. For example, multicomponent inflatons and their associated topological defects are discussed in early-universe expansion scenarios [3, 13, 130]. Exciting recent developments in the field of heavy-ion collisions, where the nonequilibrium regime of a quark–gluon plasma can be investigated by multicomponent gauge field simulations [7, 20, 21, 23, 38, 50, 51] may reveal the relevance of the type of interrelations discussed above. Defect-induced nonthermal fixed points in this system are presently being explored. The rapid expansion of the quark–gluon plasma adds a completely new aspect to the dynamical description. Similar processes can be studied by releasing ultracold gases from their trapping potential, specifically in view of expanding turbulent clouds [37, 64, 146, 161].

Ultimately, nonthermal fixed points have to be included in a global picture of nonequilibrium dynamics of interacting many-body systems [52, 127]. The concept behind them points out a way toward universal phenomena far away from equilibrium, having many aspects in common with universality and critical phenomena in thermal equilibrium. Understanding their relationships with nonthermal equilibrium states [47, 85, 91, 134, 135], generalized Gibbs states [47, 85, 135], and prethermalized states [2, 9, 15, 29, 60, 82, 86, 157, 163] are essential steps toward a unifying framework of complex dynamical many-body systems. To set up this framework, the development and extension of renormalization group techniques for far-from-equilibrium dynamics seems in order, and promising progress has been seen in the recent past (see the references cited at the end of Section 7.2.2).

Acknowledgments

The authors thank B. Anderson, V. Bagnato, J. Berges, N. G. Berloff, E. Boden-schatz, R. Bücker, L. Carr, M. J. Davis, S. Diehl, B. Eckardt, G. Falkovich, S.-C. Gou, H. Horner, R. Kerr, G. Krstulovic, S. Mathey, I. Mazets, L. McLerran, A. Millis, M. K. Oberthaler, J. M. Pawlowski, N. Philipp, M. G. Schmidt, J. Schmiedmayer, B. Shivamoggi, B. Svistunov, M. Tsubota, and P. Weckesser for useful discussions. T. G. and D. S. thank J. Berges and C. Scheppach for collaboration on related work. They acknowledge support by the Deutsche Forschungsgemeinschaft (GA 677/7,8), by the University of Heidelberg (FRONTIER, Excellence Initiative, Center for Quantum Dynamics), by the Helmholtz Association (HA216/EMMI), by the University of Leipzig (Grawp-Cluster), and by BMBF and MWFK Baden-Württemberg (bwGRiD cluster). The authors thank KITP, Santa Barbara, for its hospitality. This research was supported in part by the National Science Foundation under Grant PHY05-51164. T. G. thanks l'Ecole de Physique des Houches for its hospitality.

References

[1] Aarts, G., D. Ahrensmeier, R. Baier, J. Berges, and J. Serreau (2002). *Phys. Rev. D*, **66**, 045008.

[2] Aarts, G., G. F. Bonini, and C. Wetterich (2000). *Phys. Rev. D*, **63**, 025012.

[3] Allahverdi, R., R. Brandenberger, F.-Y. Cyr-Racine, and A. Mazumdar (2010). *Annu. Rev. Nucl. Part. Sci.*, **60**, 27.

[4] Amo, A., S. Pigeon, D. Sanvitto, V. G. Sala, R. Hivet, I. Carusotto, F. Pisanello, G. Leménager, R. Houdré, E. Giacobino, C. Ciuti, and A. Bramati (2011). *Science*, **332**, 1167.

[5] Anderson, B. P., P. C. Haljan, C. A. Regal, D. L. Feder, L. A. Collins, C. W. Clark, and E. A. Cornell (2001). *Phys. Rev. Lett.*, **86**, 2926.

[6] Aref, H. (1983). *Annu. Rev. Fluid Mech.*, **15**, 345.

[7] Arnold, P. B. and G. D. Moore (2006). *Phys. Rev. D*, **73**, 025006.

[8] Balili, R., V. Hartwell, D. Snoke, L. Pfeiffer, and K. West (2007). *Science*, **316**, 1007.

[9] Barnett, R., A. Polkovnikov, and M. Vengalattore (2011). *Phys. Rev. A*, **84**, 023606.

[10] Baym, G. (1962). *Phys. Rev.*, **127**, 1391.

[11] Berezinskii, V. (1971). *Sov. Phys. JETP*, **32**, 493.

[12] Berges, J. (2002). *Nucl. Phys. A*, **699**, 847.

[13] Berges, J. (2005). *AIP Conf. Proc.*, **739**, 3.

[14] Berges, J., J. Blaizot, and F. Gelis (2012). *J. Phys. G: Nucl. Part. Phys.*, **39**, 085115.

[15] Berges, J., S. Borsanyi, and C. Wetterich (2004). *Phys. Rev. Lett.*, **93**, 142002.

[16] Berges, J. and T. Gasenzer (2007). *Phys. Rev. A*, **76**, 033604.

[17] Berges, J. and G. Hoffmeister (2009). *Nucl. Phys. B*, **813**, 383.

[18] Berges, J. and D. Mesterházy (2012). *Nucl. Phys. B Proc. Suppl.*, **228**, 37.

[19] Berges, J., A. Rothkopf, and J. Schmidt (2008). *Phys. Rev. Lett.*, **101**, 041603.

[20] Berges, J., S. Scheffler, and D. Sexty (2009). *Phys. Lett. B*, **681**, 362.

[21] Berges, J., S. Schlichting, and D. Sexty (2012). *Phys. Rev. D*, **86**, 074006.

[22] Berges, J. and D. Sexty (2011). *Phys. Rev. D*, **83**, 085004.

[23] Berges, J. and D. Sexty (2012). *Phys. Rev. Lett.*, **108**, 161601.

[24] Berges, J., N. Tetradis, and C. Wetterich (2002). *Phys. Rept.*, **363**, 223.

[25] Berloff, N. G. and B. V. Svistunov (2002). *Phys. Rev. A*, **66**, 013603.

[26] Bisset, R. N., M. J. Davis, T. P. Simula, and P. B. Blakie (2009). *Phys. Rev. A*, **79**, 033626.

[27] Blaizot, J.-P., F. Gelis, J.-F. Liao, L. McLerran, and R. Venugopalan (2012). *Nucl. Phys. A*, **873**, 68.

[28] Blakie, P. B., A. S. Bradley, M. J. Davis, R. J. Ballagh, and C. W. Gardiner (2008). *Adv. Phys.*, **57**, 363.

[29] Bonini, G. F. and C. Wetterich (1999). *Phys. Rev. D*, **60**, 105026.

[30] Bradley, A. S. and B. P. Anderson (2012). *Phys. Rev. X*, **2**, 041001.

[31] Brand, J. and W. P. Reinhardt (2002). *Phys. Rev. A*, **65**, 043612.

[32] Bray, A. J. (1994). *Adv. Phys.*, **43**, 357.

[33] Brewczyk, M., M. Gajda, and K. Rzażewski (2007). *J. Phys. B: At. Mol. Opt. Phys.*, **40**, R1.

[34] Canet, L. and H. Chate (2007). *J. Phys. A*, **40**, 1937.

[35] Canet, L., H. Chaté, B. Delamotte, and N. Wschebor (2010). *Phys. Rev. Lett.*, **104**, 150601.

[36] Canet, L., B. Delamotte, O. Deloubriere, and N. Wschebor (2004). *Phys. Rev. Lett.*, **92**, 195703.

[37] Caracanhas, M., A. Fetter, S. Muniz, K. Magalhães, G. Roati, G. Bagnato, and V. Bagnato (2012). *J. Low Temp. Phys.*, **166**, 49.

[38] Carrington, M. and A. Rebhan (2011). *Eur. Phys. J. C*, **71**, 1787.

[39] Castelnovo, C., R. Moessner, and S. Sondhi (2008). *Nature*, **451**, 42.

[40] Chang, M.-S., C. D. Hamley, M. D. Barrett, J. A. Sauer, K. M. Fortier, W. Zhang, L. You, and M. S. Chapman (2004). *Phys. Rev. Lett.*, **92**, 140403.

[41] Cornwall, J. M., R. Jackiw, and E. Tomboulis (1974). *Phys. Rev. D*, **10**, 2428.

[42] Damle, K., S. Majumdar, and S. Sachdev (1996). *Phys. Rev. A*, **54**, 5037.

[43] Demidov, V. E., O. Dzyapko, M. Buchmeier, T. Stockhoff, G. Schmitz, G. A. Melkov, and S. O. Demokritov (2008). *Phys. Rev. Lett.*, **101**, 257201.

[44] Demidov, V. E., O. Dzyapko, S. O. Demokritov, G. A. Melkov, and A. N. Slavin (2007). *Phys. Rev. Lett.*, **99**, 037205.

[45] Diehl, S., A. Micheli, A. Kantian, B. Kraus, H. P. Büchler, and P. Zoller (2008). *Nature Phys.*, **4**, 878.

[46] Dierking, I. (2003). *Textures of Liquid Crystals*. Wiley-VCH.

[47] Eckstein, M. and M. Kollar (2008). *Phys. Rev. Lett.*, **100**, 120404.

[48] Foster, C. J., P. B. Blakie, and M. J. Davis (2010). *Phys. Rev. A*, **81**, 023623.

[49] Frisch, U. (1995). *Turbulence: The Legacy of A. N. Kolmogorov*. Cambridge University Press.

[50] Fukushima, K. (2011). *Acta Phys. Polon. B*, **42**, 2697.

[51] Fukushima, K. and F. Gelis (2012). *Nucl. Phys. A*, **874**, 108.

[52] Gasenzer, T. (2009). *Eur. Phys. J. ST*, **168**, 89.

[53] Gasenzer, T., J. Berges, M. G. Schmidt, and M. Seco (2005). *Phys. Rev. A*, **72**, 063604.

[54] Gasenzer, T., S. Kessler, and J. M. Pawlowski (2010). *Eur. Phys. J. C*, **70**, 423.

[55] Gasenzer, T., B. Nowak, and D. Sexty (2012). *Phys. Lett. B*, **710**, 500.

[56] Gasenzer, T. and J. M. Pawlowski (2008). *Phys. Lett. B*, **670**, 135.

[57] Gezzi, R., T. Pruschke, and V. Meden (2007). *Phys. Rev. B*, **75**, 045324.

[58] Gies, H. (2006). arXiv:hep-ph/0611146.

[59] Giorgetti, L., I. Carusotto, and Y. Castin (2007). *Phys. Rev. A*, **76**, 013613.

[60] Gring, M., M. Kuhnert, T. Langen, T. Kitagawa, B. Rauer, M. Schreitl, I. Mazets, D. A. Smith, E. Demler, and J. Schmiedmayer (2012). *Science*, **337**, 1318.

[61] Gurarie, V. (1995). *Nucl. Phys. B*, **441**, 569.

[62] Guzman, J., G.-B. Jo, A. N. Wenz, K. W. Murch, C. K. Thomas, and D. M. Stamper-Kurn (2011). *Phys. Rev. A*, **84**, 063625.

[63] Hadzibabic, Z., P. Krüger, M. Cheneau, B. Battelier, and J. Dalibard (2006). *Nature*, **441**, 1118.

[64] Henn, E. A. L., J. A. Seman, G. Roati, K. M. F. Magalhães, and V. S. Bagnato (2009). *Phys. Rev. Lett.*, **103**, 045301.

[65] Higbie, J. M., L. E. Sadler, S. Inouye, A. P. Chikkatur, S. R. Leslie, K. L. Moore, V. Savalli, and D. M. Stamper-Kurn (2005). *Phys. Rev. Lett.*, **95**, 050401.

[66] Hohenberg, P. C. and B. I. Halperin (1977). *Rev. Mod. Phys.*, **49**, 435.

[67] Jakobs, S. G., V. Meden, and H. Schoeller (2007). *Phys. Rev. Lett.*, **99**, 150603.

[68] Jakobs, S. G., M. Pletyukhov, and H. Schoeller (2010). *J. Phys. A: Math. Theor.*, **43**, 103001.

[69] Kagan, Y. and B. V. Svistunov (1994). *Sov. Phys. JETP*, **78**, 187.

[70] Kagan, Y., B. V. Svistunov, and G. V. Shlyapnikov (1992). *Sov. Phys. JETP*, **74**, 279.

[71] Karl, M., B. Nowak, and T. Gasenzer (2013). *Sci. Rept.*, **3**, 2394.

[72] Karrasch, C., R. Hedden, R. Peters, T. Pruschke, K. Schönhammer, and V. Meden (2008). *J. Phys.: Cond. Matt.*, **20**, 345205.

[73] Kasamatsu, K. and M. Tsubota (2006). *Phys. Rev. A*, **74**, 013617.

[74] Kasamatsu, K., M. Tsubota, and M. Ueda (2005). *Phys. Rev. A*, **71**, 043611.

[75] Kasprzak, J., M. Richard, S. Kundermann, A. Baas, P. Jeambrun, J. M. J. Keeling, F. M. Marchetti, M. H. Szymaska, R. Andrel, J. L. Straehli, V. Savona, P. B. Littlewood, B. Deveaud, and L. S. Dang (2006). *Nature*, **443**, 409.

[76] Keeling, J. and N. G. Berloff (2008). *Phys. Rev. Lett.*, **100**, 250401.

[77] Keeling, J. and N. G. Berloff (2011). *Contemp. Phys.*, **52**, 131.

[78] Kehrein, S. (2005). *Phys. Rev. Lett.*, **95**, 056602.

[79] Kevrekidis, P. G., D. J. Frantzeskakis, and R. Carretero-González (eds.) (2008). *Emergent Nonlinear Phenomena in Bose–Einstein Condensates*. Springer-Verlag.

[80] Khlebnikov, S. Y. and I. I. Tkachev (1996). *Phys. Rev. Lett.*, **77**, 219.

[81] Kibble, T. W. B. (1976). *J. Phys. A: Math. Gen.*, **9**, 1387.

[82] Kitagawa, T., A. Imambekov, J. Schmiedmayer, and E. Demler (2011). *New J. Phys.*, **13**, 073018.

[83] Kivotides, D., J. C. Vassilicos, D. C. Samuels, and C. F. Barenghi (2001). *Phys. Rev. Lett.*, **86**, 3080.

[84] Kofman, L., A. D. Linde, and A. A. Starobinsky (1994). *Phys. Rev. Lett.*, **73**, 3195.

[85] Kollar, M. and M. Eckstein (2008). *Phys. Rev. A*, **78**, 013626.

[86] Kollar, M., F. A. Wolf, and M. Eckstein (2011). *Phys. Rev. B*, **84**, 054304.

[87] Kolmogorov, A. N. (1941). *Proc. USSR Acad. Sci.*, **30**, 299 [reprinted in *Proc. R. Soc. Lond. A*, **434**, 9 (1991)].

[88] Korb, T., F. Reininghaus, H. Schoeller, and J. König (2007). *Phys. Rev. B*, **76**, 165316.

[89] Kosterlitz, J. and D. Thouless (1973). *J. Phys. C: Sol. St. Phys.*, **6**, 1181.

[90] Kozik, E. V. and B. V. Svistunov (2009). *J. Low Temp. Phys.*, **156**, 215.

[91] Kronenwett, M. and T. Gasenzer (2011). *Appl. Phys. B*, **102**, 469.

[92] Krstulovic, G. (2012). *Phys. Rev. E*, **86**, 055301(R).

[93] Lagoudakis, K., M. Wouters, M. Richard, A. Baas, I. Carusotto, R. André, L. Dang, B. Deveaud-Plédran et al. (2008). *Nature Phys.*, **4**, 706.

[94] Lee, T. and Y. Pang (1992). *Phys. Rept.*, **221**, 251.

[95] Lesieur, M. (2008). *Turbulence in Fluids* (4th edn). Springer.

[96] Levich, E. and V. Yakhot (1978). *J. Phys. A: Math. Gen.*, **11**, 2237.

[97] Lin, C. (1941). *Proc. Natl Acad. Sci. USA*, **27**, 570.

[98] Luttinger, J. M. and J. C. Ward (1960). *Phys. Rev.*, **118**, 1417.

[99] Matarrese, S. and M. Pietroni (2007). *JCAP*, **0706**, 026.

[100] Mathey, L. and A. Polkovnikov (2009). *Phys. Rev. A*, **80**, 041601.

[101] Mathey, L. and A. Polkovnikov (2010). *Phys. Rev. A*, **81**, 033605.

[102] Micha, R. and I. I. Tkachev (2003). *Phys. Rev. Lett.*, **90**, 121301.

[103] Micha, R. and I. I. Tkachev (2004). *Phys. Rev. D*, **70**, 043538.

[104] Miesner, H., D. Stamper-Kurn, M. Andrews, D. Durfee, S. Inouye, and W. Ketterle (1998). *Science*, **279**, 1005.

[105] Mitra, A., S. Takei, Y. B. Kim, and A. J. Millis (2006). *Phys. Rev. Lett.*, **97**, 236808.

[106] Nazarenko, S. (2011). *Wave Turbulence* (Lecture Notes in Physics, Vol. 825). Springer.

[107] Nazarenko, S. and M. Onorato (2006). *Phys. D: Nonlin. Phen.*, **219**, 1.

[108] Nelson, D. (2002). *Defects and Geometry in Condensed Matter Physics*. Cambridge University Press.

[109] Nemirovskii, S. (1998). *Phys. Rev. B*, **57**, 5972.

[110] Nemirovskii, S., M. Tsubota, and T. Araki (2002). *J. Low Temp. Phys.*, **126**, 1535.

[111] Nicklas, E., H. Strobel, T. Zibold, C. Gross, B. A. Malomed, P. G. Kevrekidis, and M. K. Oberthaler (2011). *Phys. Rev. Lett.*, **107**, 193001.

[112] Nore, C., M. Abid, and M. E. Brachet (1997). *Phys. Rev. Lett.*, **78**, 3896.

[113] Novikov, E. A. (1975). *Zh. Eksp. Teor. Fiz.*, **68**, 1868.

[114] Nowak, B., S. Erne, M. Karl, J. Schole, D. Sexty, and T. Gasenzer (2013). arXiv:1302.1448 [cond-mat.quant-gas].

[115] Nowak, B., J. Schole, and T. Gasenzer (2014). *New J. Phys.*, **16**, 093052.

[116] Nowak, B., J. Schole, D. Sexty, and T. Gasenzer (2012). *Phys. Rev. A*, **85**, 043627.

[117] Nowak, B., D. Sexty, and T. Gasenzer (2011). *Phys. Rev. B*, **84**, 020506(R).

[118] Nowik-Boltyk, P., O. Dzyapko, V. Demidov, N. Berloff, and S. Demokritov (2012). *Sci. Rept.*, **2**, 482.

[119] Obukhov, A. M. (1941). *Izv. Akad. Nauk SSSR, Ser. Geogr. Geofiz.*, **5**, 453.

[120] Onsager, L. (1949). *Nuovo Cim. Suppl.*, **6**, 279.

[121] Pawlowski, J. M. (2007). *Ann. Phys. (NY)*, **322**, 2831.

[122] Philipp, N. (2012). Diploma thesis, Universität Heidelberg.

[123] Pismen, L. M. (1999). *Vortices in Nonlinear Fields: From Liquid Crystals to Superfluids, from Non-Equilibrium Patterns to Cosmic Strings*. Clarendon Press.

[124] Pitaevskii, L. P. (1961). *Sov. Phys. JETP*, **13**, 451.

[125] Pitaevskii, L. P. and S. Stringari (2003). *Bose–Einstein Condensation*. Clarendon Press.

[126] Polkovnikov, A. (2010). *Ann. Phys. (NY)*, **325**, 1790.

[127] Polkovnikov, A., K. Sengupta, A. Silva, and M. Vengalattore (2011). *Rev. Mod. Phys.*, **83**, 863.

[128] Prokopec, T. and T. G. Roos (1997). *Phys. Rev. D*, **55**, 3768.

[129] Rajantie, A. (2002). *Int. J. Mod. Phys. A*, **17**, 1.

[130] Rajantie, A. (2003). *Contemp. Phys.*, **44**, 485.

[131] Rajantie, A. (2012). *Contemp. Phys.*, **53**, 195.

[132] Rajaraman, R. (1982). *Solitons and Instantons: An Introduction to Solitons and Instantons in Quantum Field Theory*. North-Holland.

[133] Richardson, L. F. (1920). *Proc. R. Soc. Lond. A*, **97**, 354.

[134] Rigol, M. (2009). *Phys. Rev. Lett.*, **103**, 100403.

[135] Rigol, M., V. Dunjko, V. Yurovsky, and M. Olshanii (2007). *Phys. Rev. Lett.*, **98**, 050405.

[136] Ritter, S., A. Öttl, T. Donner, T. Bourdel, M. Köhl, and T. Esslinger (2007). *Phys. Rev. Lett.*, **98**, 090402.
[137] Ruostekoski, J. and J. R. Anglin (2001). *Phys. Rev. Lett.*, **86**, 3934.
[138] Saba, M., C. Ciuti, J. Bloch, V. Thierry-Mieg, R. André, L. S. Dang, S. Kundermann, A. Mura, G. Bongiovanni, J. L. Staehli, and B. Deveaud (2001). *Nature*, **414**, 731.
[139] Sadler, L. E., J. M. Higbie, S. R. Leslie, M. Vengalattore, and D. M. Stamper-Kurn (2006). *Nature*, **443**, 312.
[140] Scheppach, C., J. Berges, and T. Gasenzer (2010). *Phys. Rev. A*, **81**, 033611.
[141] Schmaljohann, H., M. Erhard, J. Kronjäger, M. Kottke, S. van Staa, L. Cacciapuoti, J. J. Arlt, K. Bongs, and K. Sengstock (2004). *Phys. Rev. Lett.*, **92**, 040402.
[142] Schmidt, M., S. Erne, B. Nowak, D. Sexty, and T. Gasenzer (2012). *New J. Phys.*, **14**, 075005.
[143] Schoeller, H. (2009). *Eur. Phys. J. ST*, **168**, 179.
[144] Schole, J., B. Nowak, and T. Gasenzer (2012). *Phys. Rev. A*, **86**, 013624.
[145] Schweikhard, V., S. Tung, and E. A. Cornell (2007). *Phys. Rev. Lett.*, **99**, 30401.
[146] Seman, J. A., E. A. L. Henn, R. F. Shiozaki, G. Roati, F. J. Poveda-Cuevas, K. M. F. Magalhães, V. I. Yukalov, M. Tsubota, M. Kobayashi, K. Kasamatsu, and V. S. Bagnato (2011). *Laser Phys. Lett.*, **8**, 691.
[147] Simula, T. P. and P. B. Blakie (2006). *Phys. Rev. Lett.*, **96**, 020404.
[148] Smith, R. P., S. Beattie, S. Moulder, R. L. D. Campbell, and Z. Hadzibabic (2012). *Phys. Rev. Lett.*, **109**, 105301.
[149] Sonin, E. B. (1987). *Rev. Mod. Phys.*, **59**, 87.
[150] Svistunov, B. (1991). *J. Mosc. Phys. Soc.*, **1**, 373.
[151] Svistunov, B. (2001). In *Quantized Vortex Dynamics and Superfluid Turbulence* (ed. C. Barenghi, R. Donnelly, and W. Vinen), p. 327 (Lecture Notes in Physics, Vol. 571). Springer.
[152] Thouless, D. (1998). *Topological Quantum Numbers in Nonrelativistic Physics.* World Scientific.
[153] Timmermans, E. (1998). *Phys. Rev. Lett.*, **81**, 5718.
[154] Tkachev, I., S. Khlebnikov, L. Kofman, and A. D. Linde (1998). *Phys. Lett. B*, **440**, 262.
[155] Traschen, J. H. and R. H. Brandenberger (1990). *Phys. Rev. D*, **42**, 2491.
[156] Tsubota, M. (2008). *J. Phys. Soc. Jpn*, **77**, 111006.
[157] Tsuji, N., M. Eckstein, and P. Werner (2013). *Phys. Rev. Lett.*, **110**, 136404.
[158] Ueda, M. (2012). *Annu. Rev. Cond. Matt. Phys.*, **3**, 263.
[159] Vengalattore, M., S. R. Leslie, J. Guzman, and D. M. Stamper-Kurn (2008). *Phys. Rev. Lett.*, **100**, 170403.
[160] Vinen, W. F., M. Tsubota, and A. Mitani (2003). *Phys. Rev. Lett.*, **91**, 135301.
[161] Weckesser, P. (2012). Bachelor's thesis, Universität Heidelberg.
[162] Weiler, C. N., T. W. Neely, D. R. Scherer, A. S. Bradley, M. J. Davis, and B. P. Anderson (2008). *Nature*, **455**, 948.
[163] Werner, P., N. Tsuji, and M. Eckstein (2012). *Phys. Rev. B*, **86**, 205101.

[164] Wetterich, C. (1993). *Phys. Lett. B*, **301**, 90.

[165] Zakharov, V. E., V. S. L'vov, and G. Falkovich (1992). *Kolmogorov Spectra of Turbulence I: Wave Turbulence*. Springer.

[166] Zakharov, V. E. and A. B. Shabat (1972). *Sov. Phys. JETP*, **34**, 62.

[167] Zanella, J. and E. Calzetta (2006). arXiv:hep-th/0611222.

[168] Zurek, W. H. (1985). *Nature*, **317**, 505.

8

Ferroic domain walls as model disordered elastic systems

Patrycja PARUCH

Department of Quantum Matter Physics
University of Geneva
24 Quai Ernest-Ansermet
1211 Geneva 4, Switzerland

Strongly Interacting Quantum Systems out of Equilibrium. First Edition. Thierry Giamarchi et al.
© Oxford University Press 2016. Published in 2016 by Oxford University Press.

Chapter Contents

Preface

In this chapter, we will focus specifically on ferroic domain walls, the thin interfaces separating different orientations of ferromagnetic magnetization, ferroelectric polarization, or ferroelastic strain in their parent material. Using a variety of optical and local probe microscopy techniques, the static roughening and dynamics of such domain walls can be accessed over multiple orders of length- and timescales, under precisely controlled applied forces (through the use of the appropriate conjugate fields) and at variable temperature, and with the possibility in some cases of tailoring the type and density of defects, making them a powerful model system in which the physics of pinned elastic interfaces can be studied.

8.1   Introduction

When an elastic manifold encounters potential energy fluctuations, the resulting competition between pinning and elasticity leads to complex glassy physics, with characteristic roughening and a nonlinear dynamic response of the manifold to applied forces. While this general framework allows us to understand the behavior of both classical and quantum systems as diverse as propagating fractures, surface growth, wetting and burning fronts, bacterial colonies, vortex lattices in type II superconductors, and Wigner crystals, we will focus in this chapter specifically on ferroic domain walls, the thin interfaces separating different orientations of ferromagnetic magnetization, ferroelectric polarization, or ferroelastic strain in their parent material. Using a variety of optical and local probe microscopy techniques, the static roughening and dynamics of such domain walls can be accessed over multiple orders of length- and timescales, under precisely controlled applied forces (through the use of the appropriate conjugate fields) and at variable temperature, and with the possibility in some cases of tailoring the type and density of defects, making them a powerful model system in which the physics of pinned elastic interfaces can be studied.

We will begin with a broad introduction to the general framework of elasticity and pinning, in both random and periodic potentials, and the resulting roughening and dynamic behavior. Specifically, we will discuss mono- versus multi-affine scaling of interface roughening, and the combined effects of dimensionality and the type of pinning on the expected characteristic roughening and dynamic exponents.[1] We will then briefly review the physics of ferroic materials, in particular the energetics of domain formation, different types of domain walls and how they can be imaged and controlled, and the recent exciting discoveries of novel, domain-wall-specific functional properties. The bulk of the chapter will focus on experimental evidence that such domain walls can be seen as pinned elastic interfaces, and analyzed with the statistical physics tools developed within the general framework of disordered elastic systems, with a particular focus on ferroelectric systems. Here we will show the effects of both commensurate (intrinsic and artificial) and random pinning, long-range

---

[1] For a more detailed treatment, the interested reader can consult the excellent text by Barabasi and Stanley [7], and reviews by Blatter et al. [15], Giamarchi et al. [47], and Agoritsas et al. [2].

dipolar interactions, and different universality classes of weak disorder. We will end the chapter with a discussion of some of the more recent experimental research addressing still open and actively pursued questions in the field of ferroelectrics and multiferroics, including aging and thermal effects, the role of a heterogeneous pinning landscape, and complex internal structure of the interface.

8.2   Pinned elastic interfaces

For an interface with only short-range elasticity, which we consider along the longitudinal coordinate z, the energy $H_{\text{el}} \sim \int [\nabla_z u(z)]^2 \, d^d z$ is minimized by limiting distortions $\nabla_z u(z)$. In the absence of other influences, the interface will therefore show a perfectly flat configuration with no transverse displacements $u(z)$. When an external time-dependent force $F(t) = \int_{-\infty}^{\infty} F(t')\delta(t-t') \, dt'$ acts on such a simple system, the response is linear, as for example for the familiar damped harmonic oscillator $\ddot{u}(t) + \gamma \dot{u}(t) + \omega_0 u(t) = F(t)$, where $u(t) = \int_{-\infty}^{\infty} \chi(t-t')F(t') \, dt'$, and is determined by the susceptibility of the system with Fourier transform $\tilde{\chi} = (\omega_0^2 - \omega^2 + i\gamma\omega)^{-1}$.

However, finite-temperature (T) thermal fluctuations, defects (disorder), or even an underlying crystal lattice periodicity in the system all lead to position-dependent variations in the potential energy landscape, which can act as pinning sites with energy $H_{\text{pin}}(u, V)$. To optimize its energy, the system can now wander between particularly favorable pinning sites, balancing the energy gained in this fashion against the elastic energy cost of the associated transverse displacements, which results in a rough configuration once equilibrium has been established. This competition between elasticity and pinning,

$$H = H_{\text{el}}(u) + H_{\text{pin}}(u, V) = \frac{\sigma}{2} \int_{Dz} [\nabla_z u(z)]^2 \, d^d z + \int_{Dz} V(z, u(z)) \, d^d z, \qquad (8.1)$$

can, for some types of pinning behavior give rise to a glass-like physics, with many metastable states, and self-similarity and universal behavior evident in both the characteristic roughening and complex nonlinear dynamics of the system.

The equilibrium roughening at a given lengthscale r is formally described by the relative displacements $\Delta u(r) = u(z) - u(z+r)$, shown in Fig. 8.1, whose probability distribution function (PDF) and its nth-order moments $S_n(r)$, or the related correlation functions $C_n(r)$, reflect the characteristic scaling properties of the system [2]:

$$S_n(r) = \overline{\langle |\Delta u(r)|^n \rangle} \sim r^{n\zeta_n}, \qquad (8.2)$$

$$C_n(r) = \overline{\langle |\Delta u(r)|^n \rangle}^{1/n} \sim r^{\zeta_n}, \qquad (8.3)$$

where ζ_n are the associated scaling exponents describing how the relative displacements grow for increasing r.

In the most general case of a multi-affine system [6], an infinite set of individual scaling exponents ζ_n is needed to characterize the system. This means a hierarchy of

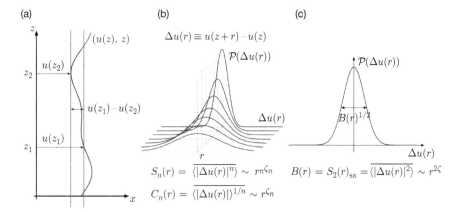

Fig. 8.1 (a) Equilibrium roughening of a one-dimensional elastic interface under the influence of weak collective pinning in a disorder potential, with relative displacements $\Delta u(r)$ for $r = z_2 - z_1$. (b) Probability distribution function $P(\Delta u(r))$, with its nth-order moments $S_n(r)$ and related correlation functions $C_n(r)$. (c) Gaussian probability distribution function $P(\Delta u(r))$ for a mono-affine interface in the case of weak collective pinning by uncorrelated disorder, fully characterized by the second moment or roughness function $B(r)$. (Adapted from [2].)

local roughness exponents ζ will be observed when the interface is measured in real space via its effective width

$$W(r) = \left\{ \overline{\langle [u(z) - \langle u \rangle_r]^2 \rangle_r} \right\}^{1/2} \sim r^\zeta \tag{8.4}$$

or roughness, sometimes also called the height–height or displacement–displacement correlation function

$$B(r) = \left\{ \overline{\langle [\Delta u(r)]^2 \rangle_r} \right\}^{1/2} \sim r^{2\zeta}, \tag{8.5}$$

or in reciprocal space by its structure factor

$$S(q) = \langle |u(q)|^2 \rangle \sim q^{-(1+2\zeta)}, \tag{8.6}$$

where $u(q)$ is the Fourier transform of the displacement field $u(z)$, and $\langle \, \cdot \, \rangle$ and $\overline{}$ denote averages over z and over disorder realizations, respectively.

For mono-affine systems, the situation becomes further simplified and more tractable from a statistical physics viewpoint, since the roughening behavior is global rather than local, and the correlations functions

$$C_n(r) \sim r^{\zeta_n} \sim r^\zeta \tag{8.7}$$

can be collapsed to a universal curve when renormalized by an appropriate ratio, with a single-valued roughness exponent $\zeta_n = \zeta \; \forall n$. For the case of weak collective pinning by uncorrelated disorder, for example, the relative displacements characterizing the roughening of the interface present a Gaussian PDF [55, 120, 156], as shown in Fig. 8.1(c), and the correlation functions $C_n(r)$ can be renormalized by the Gaussian ratios $R_n^G = C_n^G(r)/C_2^G(r)$. To test for mono-affinity in an interface, one can therefore compare how well the experimentally obtained PDF matches a Gaussian function, and whether the renormalized correlation functions $C_n(r)$ do indeed collapse together, an approach termed multiscaling analysis, initially developed during studies of fracture surfaces [159].

In the case of a mono-affine interface, the second moment $S_2(r) \equiv B(r) \sim r^{2\zeta}$ fully describes the scaling properties of the static configuration. The specific value of the roughness exponent ζ depends on the system dimensionality d and the universality class of the disorder. For random-bond disorder (universality class corresponding to a domain wall in an Ising model with randomly varying interaction strength between neighboring spins; see Fig. 8.2(a)), $\zeta = \frac{2}{3}$ for one-dimensional interfaces and $\zeta \simeq 0.2084(4-d)$ for $(d > 1)$-dimensional interfaces. For random-field disorder (universality class corresponding to a randomly varying local field associated with each site of the Ising model; see Fig. 8.2(b)), $\zeta = 1$ for all dimensionalities $d < 4$. In the case of purely thermal fluctuations $\zeta = \frac{1}{2}$ for one-dimensional interfaces.

The presence of long-range interactions in the system will also affect its behavior, and can generally be included as a modification of the interface elastic energy. For example, as a result of dipolar interactions, the stiffness of the interface, and thus its elasticity, under deformations perpendicular to the polar orientation is higher than that for deformations along the polar orientation. In this case, the elastic energy (expressed in reciprocal space), contains not only the short-range term

$$H_{\text{el}} = \tfrac{1}{2}\Sigma_q C_{\text{el}}(q)u^*(q)u(q) \tag{8.8}$$

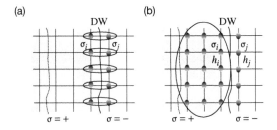

Fig. 8.2   Ising model with spins σ_i at each lattice site oriented "+" or "−" on either side of a domain wall (DW), whose previous position is shown as a dotted line. (a) Random bond disorder corresponds to randomly varying the interaction strength between neighboring spins σ_i and σ_j. To compare the energy of the domain wall at its present position with its past configuration, only the pairs of neighboring spins across the domain wall need to be considered. (b) Random field disorder corresponds to a randomly varying local field h_i associated with each lattice site σ_i. When the domain wall moves, the energy of each reversed spin with respect to the local field has to be taken into account.

with the elastic constant $C_{el}(q) = \sigma_{interface} q^2$ related to the energy density $\sigma_{interface}$, but also the dipolar interaction with

$$C_{dp} = \frac{2P_s^2}{\epsilon_0 \epsilon} \frac{q_y^2}{q} + \frac{P_s^2 \xi}{\epsilon_0 \epsilon} \left(-\tfrac{3}{4}q_x^2 + \tfrac{1}{8}q^2\right), \tag{8.9}$$

where P_s is the polarization along direction y, and ϵ and ϵ_0 are the relative and vacuum permittivities. Because q_y now scales as $q_y \sim q_x^{3/2}$, an effective dimensionality $d_{eff} = d + \tfrac{1}{2}$ needs to be used to determine the expected values of ζ for random-bond and random-field disorder [36, 129]. A higher effective dimensionality likewise needs to be considered when multiple interacting interfaces are present, as for example in a Bragg glass of magnetic vortices in a disordered lattice [46] or periodic stripe domains formed in ferroic materials [86, 95]. In this case, the dimensionality that determines the values of the roughness exponent becomes that of the entire system, $d_{eff} = d+m$, rather than that of an individual interface (d) with an $m = 1$ dimensional displacement field. Moreover, as the relative displacements $\Delta u(r)$ approach the characteristic periodicity of the system r_P, roughening slows significantly to $B(r > r_P) \sim \log r$.

In addition, a finite interface thickness ξ may also have to be taken into account. At very short lengthscales $r < r_C$, where the relative displacements are smaller than or comparable to this thickness $\Delta u(r) < \xi$, the scaling of the displacement field is dominated by elasticity, rather than the disorder, and the interface is in the Larkin regime with $B(r) \sim r^{4-d}$. The critical lengthscale r_C is known as the Larkin length. At higher lengthscales $r > r_C$, $\Delta u(z) > \xi$, disorder effects dominate, and the interface properly enters the random manifold regime, with $B(r) \sim r^{2\zeta}$. With finite temperatures, the effective thickness of the interface increases $\xi_{eff,T>0} > \xi_{T=0}$ as a result of the induced thermal jitter [1], allowing continual exploration of the pinning potential landscape at energy scales $\sim k_B T$. A similar effect will be obtained if a finite correlation length for the disorder potential $V(z, u(z))$ is considered, even for an infinitely abrupt interface. These different effects on the roughness function of a pinned elastic interfaces are shown schematically in Fig. 8.3(a).

The complex dynamic response of the system to an applied force is likewise governed by the competition between elasticity and pinning, as shown schematically in Fig. 8.3(b). At $T = 0$, there is no thermal activation, and the interface is held by the pinning potential until the applied force reaches a critical depinning threshold F_C. At depinning, the velocity v of the interfaces rises abruptly, strongly reminiscent of the evolution of the order parameter during a second-order phase transition, and for which, by analogy, universal scaling $v \sim (F - F_C)^\theta$ may be expected. At high driving forces, the microscopic details of the pinning landscape become insignificant and the standard linear flow response is recovered with $v \sim F$, for which the pinning contribution can be seen as viscous drag. In real systems at finite temperature $T > 0$, thermal activation rounds out the sharp depinning transition. Although the equation describing the full dynamics of the interface at finite temperature [25] still has no complete solution, advanced numerical simulations have explored the thermal rounding [21], allowing accurate predictions of the depinning exponent θ for one- and

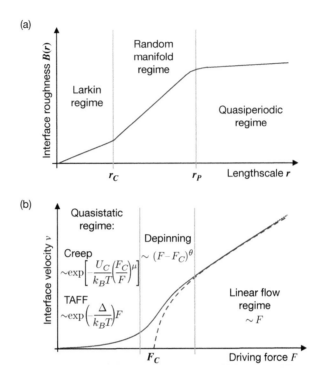

Fig. 8.3   (a) The roughness $B(r)$ of the interface grows with the lengthscale r. For short lengthscales dominated by elasticity, $B(r) \sim r^{4-d}$. At higher lengthscales, an individual interface weakly pinned by a random potential shows mono-affine scaling characterized by universal exponents, with $B(r) \sim r^{2\zeta}$. In a periodic system of interacting interfaces, roughening slows significantly at lengthscales corresponding to relative displacements of the order of the system periodicity, with $B(r) \sim \log r$. (b) The complex dynamic response of a pinned elastic interface, with its velocity v as a function of the driving force F. At zero temperature (dashed curve), the interface remains pinned until a critical force F_C is reached. Thermally activated ultraslow motion is possible even for subcritical forces at finite temperature (solid curve). At high driving forces, the standard linear response is recovered. (Adapted from [2].)

two-dimensional interfaces in random-bond and random-field disorder. In addition, these studies suggest superroughening of the interface, with $\zeta > 1$ at depinning [155], although with apparently mono-affine scaling. Thermal activation also promotes a weak quasistatic response to subcritical forces $F \ll F_C$. In this regime—in which the energy optimization as a segment of the interface is displaced in the corresponding conjugate field must be balanced against the energy cost of increased interface surface, as well as depolarizing and dipolar energies, as appropriate—the microscopic details of the pinning potential play a crucial role.

If the pinning barriers are finite [15], with some typical energy Δ as for example in a periodic washboard potential $V = \Delta[1 - \cos(k_0 u)]$, the applied force F effectively "tilts" the washboard, and the energy of the system becomes

$$H = H_{\text{el}}(u) + H_{\text{pin}}(u, V) + H(F, u)$$

$$= \frac{\sigma}{2} \int_{Dz} [\nabla_z u(z)]^2 \, d^d z + \int_{Dz} V(z, u(z)) \, d^d z - \int_{Dz} F u \, d^d z \qquad (8.10)$$

For a $d = 1$ interface, i.e., a line being displaced through the troughs of the washboard as shown in Fig. 8.4(a), the energy gain scales with the length of the displaced interface segment, while the energy cost is finite, giving rise to thermally assisted flux flow, with a linear but exponentially small response

$$v \sim \exp\left(-\frac{\Delta}{k_B T}\right) F, \qquad (8.11)$$

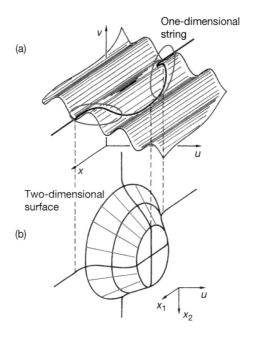

Fig. 8.4 An interface driven in a periodic pinning potential. (a) In the case of a one-dimensional "string," the cost of an excursion into an adjacent potential minimum is bounded, and is given by the two indicated kinks, while the energy gain grows with the length of the segment between them. (b) In the case of a two-dimensional "sheet," the energy cost grows as the indicated surface of the three-dimensional nucleus extruded into the adjacent potential minimum, while the energy gain grows as the volume of this nucleus. (Adapted with permission from [15]. Copyrighted by the American Physical Society.)

or

$$v \sim \exp\left(-\frac{\Delta}{k_B T}\right) F^n \tag{8.12}$$

in the limiting case. For a $d = 2$ interface, i.e., a sheet that can extrude a nucleus through the troughs of the washboard potential as shown in Fig. 8.4(b), the energy gain grows with the volume of the nucleus, and the energy cost with its surface. This energy balance leads to nonlinear glassy characteristics with

$$v \sim \exp\left(-\frac{\Delta}{k_B T} \frac{F_C}{F}\right). \tag{8.13}$$

If the energy barriers diverge as $F \to 0$, with

$$U_B(F) \sim \left(\frac{1}{F}\right)^{(d-2+2\zeta)(2-\zeta)}, \tag{8.14}$$

as in the case of randomly pinned glassy systems [47], the result is a highly nonlinear creep motion, for both one- and two-dimensional interfaces:

$$v \sim \exp\left[-\frac{U_C}{k_B T}\left(\frac{F_C}{F}\right)^\mu\right], \tag{8.15}$$

where U_C is a characteristic energy scale. The creep exponent μ depends on the dimensionality of the system and the universality class of the disorder:

$$\mu = \frac{d - 2 + 2\zeta}{2 - \zeta} \tag{8.16}$$

While the equilibrium properties of pinned elastic interfaces are now relatively well established, much less is known about their out-of-equilibrium behavior, including aging and memory effects in both classical and quantum systems [17, 28, 160]. In particular after a quench of the system, in which a parameter such as temperature is abruptly varied, numerical studies of interfaces [20, 64, 90–92] have shown aging and a long-term memory of the initial configuration. In addition, the theoretical predictions are based only on a limited number of universality classes of weak collective pinning that give rise to "well-behaved" single-valued interfaces with no bubbles or overhangs, and do not treat the case of a more complex, highly heterogeneous disorder landscape with strong individual pinning sites—which is often the case for real physical systems. Another challenge is posed by the internal structure of the interfaces, which is not generally considered within the theoretical framework. Such structure could have significant effects on both the dynamics and the static configuration, and give rise to unusual physical properties.

Confronting the theoretical description with an experimental system where these different questions can be explored in depth is therefore clearly desirable. Ferroic domain walls—which can be probed over a wide range of temperatures and applied external fields in samples where a variety of defects can be introduced during or after growth—are particularly useful as such a model system. From the perspective of the theoretical framework, the domain walls can provide a testing ground for existing theoretical predictions, and promote the development and optimization of the theoretical approach, to take into account the complexity inherent and unavoidable in the real system. For the community of researchers working on ferroic systems, meanwhile, there is both fundamental and applied interest in the static configuration and dynamic response of domain walls, which allows the switching behavior, shape, stability, and thermal evolution of domains to be understood and controlled.

8.3   Ferroic domain walls

Ferroic materials are characterized by a nonvolatile order parameter—magnetization in ferromagnets, electric polarization in ferroelectrics, and strain in ferroelastics— which appears spontaneously as they are cooled below a critical temperature T_C from a higher-symmetry "para" phase and which can take one of at least two symmetry-equivalent ground states. Under the application of the appropriate conjugate field— magnetic, electric, or stress, respectively—the order parameter can be reoriented or switched between these ground states. These materials are therefore widely used for information storage, with both magnetic and ferroelectric memories commercially available and with new designs under active investigation [164, 192], as well as for magneto-optic and especially electro-optic applications based on controlled domain structures [42]. In addition, magnetization and polarization can couple, either linearly or at higher orders, to applied stress fields, leading to piezomagnetism/piezoelectricity or magnetostriction/electrostriction, which allows the integration of these materials as micro- or nanoscale transducers in sensors and actuators [33, 60]. Multiferroic materials, in which two or more ferroic orders coexist, and in some cases couple, provide even richer fundamental physics, together with exciting possibilities for multistate or energy-efficient voltage-write/magnetic-read memories [10, 150].

Using the Landau formalism for a phenomenological description of phase transitions in terms of macroscopic thermodynamic quantities, the Gibbs free energy U_G of the ferroic system can be expressed as as a simple polynomial of even-order powers (generally truncated at sixth order) of the order parameter P:

$$U_G = \frac{\alpha}{2}P^2 + \frac{\gamma}{4}P^4 + \frac{\delta}{6}P^6 + \ldots - hP, \tag{8.17}$$

where α, γ, and δ are temperature-dependent parameters characteristic of the material and h is the applied conjugate field [11, 35, 184]. This temperature dependence is small and can be neglected for γ and δ, while $\alpha = \beta(T - T_C)$ is taken to follow a linear temperature dependence close to the transition (with T_C being the Curie–Weiss temperature). Minimizing the free energy at zero applied field gives the equilibrium criterion,

$$\frac{dU_G}{dP} = \alpha P - \gamma P^3 + \delta P^5 = 0, \tag{8.18}$$

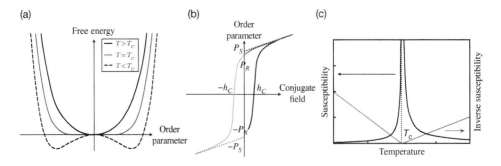

Fig. 8.5 Ferroic materials. (a) The free energy as a function of the order parameter during the transition from a high-symmetry "para" phase for $T > T_C$ to the "ferro" phase for $T < T_C$ characterized by symmetry-equivalent, oppositely oriented ground states separated by an energy barrier. (b) Hysteresis of the order parameter P as a function of the applied conjugate field h, with reversal occurring at the coercive field h_c. (c) Divergence of the susceptibility $\chi = \partial P/\partial h$ at the transition $T = T_C$ (adapted from work by Ph. Tückmantel).

defining the transition from the "para" to the "ferro" phase, as illustrated schematically in Fig. 8.5(a). This transition is marked by a divergence of the corresponding susceptibility $\chi = \partial P/\partial h$, as shown in Fig. 8.5(b):

$$\chi^{-1} = \frac{\partial h}{\partial P} = \begin{cases} \beta(T - T_C), & T > T_C, \\ 2\beta(T_C - T), & T < T_C. \end{cases} \tag{8.19}$$

In the presence of a nonzero applied field, the ferroic double-well potential is asymmetrized, favoring the orientation of the order parameter aligned with the field, with global switching to this preferred orientation when the field reaches a sufficiently high value (coercive field h_c). Thus, when the applied field is swept continuously between $-h_{\text{MAX}}$ and h_{MAX}, a characteristic P–h hysteresis is observed, as can be seen in Fig. 8.5(c), with the order parameter returning to its remanent value P_R at zero field. After the switching phase, once all the microscopic dipoles are aligned with the field, the order parameter presents a much slower/asymptotic increase simply as a result of the dielectric permittivity/magnetic permeability. Extrapolating this behavior back to zero field defines the saturation value of the order parameter, P_S. In an ideal case $P_R = P_S$, but in real samples the presence of grain boundaries, defects, or charging effects can all contribute to a partial relaxation from the saturation state, and thus generally a lower value of P_R. Depending on the boundary conditions and its switching history, a sample may present a uniformly oriented order parameter, or coexisting regions with different orientations of the order parameter, known as domains, separated by thin interfaces, known as domain walls. The application of the conjugate field, promoting the growth of domains with the preferred orientation, thus drives domain wall motion.

The specific domain configuration is determined by the energetics of the system, which aims to reduce the costs associated with the depolarizing or demagnetizing $h_d = P/\epsilon$ generated by the electric/magnetic "charge" density present at the surfaces of the ferroic material perpendicular to the orientation axis of the order parameter.

In ferroelectrics, this self-field can be at least partially screened by free charges (supplied by surface electrodes or atmospheric adsorbates or from within the often only poorly insulating sample itself). More generally, when no free charges are available (magnetism) or the boundary conditions prevent effective screening (interfacial dead layers), the self-field can be reduced if the material divides the uniformly oriented ground state into domains with alternating orientation of the order parameter, so that the average polarization or magnetization is zero.[2] An analogous argument can be made for ferroelastics, with an average stress field that is zero over the sample while at the same time minimizing the domain wall energy, although we note that the spontaneous strain is not a polar order parameter. Thus, pure ferroelastics do not present 180° domain walls, but rather twinning compatible with the structural symmetry of the material, such as the a-domains with 90° domain walls in tetragonal crystals. In multiferroic materials combining two or more of the order parameters, the allowed types of domain walls must be compatible with both the polar and elastic symmetry considerations [67, 148, 185]. For a more detailed treatment, we invite the reader to consult specialized monographs on ferroelectrics [105], ferromagnets [62, 73], and ferroelastics [157], as well as their domain walls [24, 158].

Since the energy density of the domains $U_{domain} = \sigma_{DP}w$, with volume energy density σ_{DP}, is proportional to their size w, smaller domains have smaller energy costs associated with their self-field. However, decreasing domain size requires increased numbers of domain walls, which themselves have an associated energy density per unit area σ_{DW}. The resulting energy cost increases linearly with the number of domain walls $N = 1/w$ such that $U_{DW} = \sigma_{DW}N\tau = \sigma_{DW}\tau/w$, where τ is the thickness of the sample. Optimizing these competing terms in the total energy gives the well-known Landau–Lifshitz–Kittel scaling [86, 95] of ferroic domain size with the square root of the sample thickness:

$$w = \sqrt{\frac{\sigma_{DW}}{\sigma_{DP}}\tau},\tag{8.20}$$

initially proposed for ferromagnets, and subsequently extended to ferroelectrics [122] and ferroelastics [154]. This relatively simple and intuitive scaling law has been experimentally verified over multiple orders of magnitude, holding from macroscopic single crystals of a few millimeters down to thin films of only a few unit cells [41] in ferroelectrics, and down to tens of nanometers in ferromagnets, with their thicker domain walls [59]. An important practical consequence of this scaling behavior is that as ferroic films get progressively thinner in ever more miniaturized applications, the density of domain walls increases, such that for very thin films a non-negligible volume of the material is in the form of domain walls and can influence the overall physical properties.

The domain wall energy σ_{DW} and its thickness ξ depend on the specific microscopic physics of the system, including its dipolar/exchange/elastic energy (leading to broader domain walls) versus the corresponding anisotropy (leading to more abrupt

---

[2] For simplicity, we are not considering the option of rotating the order parameter orientation into the plane of the sample, either completely or partially in closure/vortex structures, or of suppressing the "ferro" state, both of which can also occur in ferroic materials facing significant self-field energy costs [102].

domain walls) and the susceptibility χ (measuring the softness of the order parameter, likewise broadening the domain walls), and they can be estimated from *ab initio* atomistic calculations [119, 133, 147], pseudospin models, or phenomenological treatments based on the Landau theory [197]. In the simplest approach, we can consider that a domain wall modifies the homogeneous energy density of the ferroic material (which is quadratic in the leading term $\frac{1}{2}\chi^{-1}P_0^2$ from the Landau expansion, where P_0 is the spontaneous polarization, magnetization, or strain) via a gradient term (which is also quadratic by symmetry considerations) that reflects the reorientation of the order parameter $P(x) = P_0[x/(\frac{1}{2}\xi)]$ across the domain wall thickness $-\frac{1}{2}\xi < x < \frac{1}{2}\xi$, and can be approximated as $\frac{1}{2}c(2P_0/\xi)^2$, with gradient coefficient c. The total energy density per unit area,

$$\sigma_{DW} = \int_{-\xi/2}^{\xi/2} \left[\frac{1}{2}c\left(\frac{2P_0}{\xi}\right)^2 + \frac{1}{2}\chi^{-1}P(x)^2 \right] dx, \tag{8.21}$$

is minimized with respect to the domain wall thickness, which leads to

$$\xi \propto \sqrt{c\chi}. \tag{8.22}$$

Thus, in ferroelectric and ferroelastic materials, domain walls are typically rather narrow, of the order of a few unit cells to a few nanometers, while in ferromagnets, broad domain walls tens to hundreds of nanometers wide are observed.

The exchange and anisotropy terms, as well as the microscopic nature of the order parameter also influence the internal structure of the domain walls. In ferromagnets, since spin is quantized, the magnetization can only reorient via a rigid rotation, either within the plane of the domain wall (Néel type) or orthogonally to the domain wall (Bloch type), as illustrated schematically in Fig. 8.6(b, c).[3] In ferroelectrics, the

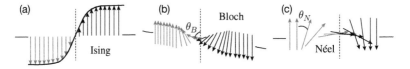

Fig. 8.6 Domain walls in ferroic materials between 180° domains. (a) Ising wall, with varying magnitude of the order parameter across the domain wall but no rotation. (b) Bloch wall, with rotation of the order parameter orthogonally to the domain wall. (c) Néel wall, with rotation of the order parameter within the plane of the domain wall. Mixed-character domain walls combining these different possibilities can also arise. (Adapted with permission from [96]. Copyrighted by the American Physical Society.)

---

[3] Note that extremely complex spin nanostructures consisting for example of vortex–antivortex pairs, which can be coherently displaced in thin magnetic ribbons by the application of spin-polarized currents, are also referred to as simply "domain walls" [135] in the context of racetrack memory applications. To avoid potential confusion, when discussing such structures, we will always explicitly refer to their more complex nature.

magnitude of the polarization can vary, and the high anisotropy generally promotes nonchiral Ising-type domain walls (Fig. 8.6(a)), although a chiral character may be present [96], in particular under constrained sample geometries with reduced dimensions [3, 43], or specific pressure/strain boundary conditions [174, 177].

The domain walls can be thought of as extended topological defects, with different electronic/structural symmetry from their parent phase—and as a result can present functional properties beyond those expected from the monodomain material, with their own internal phase transitions [94]. At the most basic level, antiphase boundaries are expected to give rise to a compensated polar moment, leading to magnetization or polarization at the walls between antiferromagnetic [101] or antiferroelectric [24] domains, respectively. Furthermore, strain gradients at ferroelastic domain walls should yield a non-negligible, highly localized flexoelectric polarization [124]. In multiferroics, competition and coupling between the different order parameters can result in even more complex behavior, with polar properties at ferroelastic domain walls [48, 187, 199], and with magnetization predicted at ferroelectric domain walls from symmetry arguments [149], which is possibly responsible for a higher-than-expected exchange bias in samples with very many such domain walls [110]. Secondary order parameters latent in the parent phase can also emerge locally at the domain walls when the primary order parameter is suppressed [34]. Beyond novel ferroic ordering, ferroelastic and ferroelectric domain walls have recently been shown to present unusual electrical transport, with conduction [53, 116, 163, 166, 173] and even superconductivity [4] in otherwise-insulating parent materials, as well as locally different stoichiometry and even chemical structure [37, 71]. These novel functional properties make the domain walls themselves extremely promising as potential active device components in future domain-wall-based nanoelectronics [8, 24, 158].

8.4   Experimental methods

The first requirement for studying the static and dynamic behavior of domain walls is of course the ability to access them—via one of a wide choice of different experimental techniques.[4] At 1–100 μm lengthscales, optical approaches allow direct, real-time, full-field-of-view imaging via polarized optical microscopy for ferroelectric domains [170], and magneto-optic Kerr microscopy (MOKE) for magnetic domains, with the possibility of in situ application of electric and magnetic fields, allowing both roughness and dynamics to be probed [99]. Second-harmonic generation (SHG), based on the effective "combination" of two identical lower-energy photons into a single photon with twice the energy after interaction with a nonlinear optical material, likewise allows mapping of domain structures [39]. Under some conditions, this technique [18, 108] or an adaptation termed Čerenkov SHG [82, 168] can be used to specifically image domain walls, although with a resolution limited by the focus and wavelength of the optical beam.

---

[4] While we provide here only a very brief list of some of the currently used techniques and their specific application to domain wall studies in more recent work, we invite the reader to consult specialized technical reviews or tutorials for further details concerning polarization microscopy [112, 113], MOKE [143, 151], SHG [40, 167], TEM [19, 193], including holography [97], and AFM [63, 78, 79].

Where nanoscale periodic domain structures are present, diffraction techniques that probe this additional superstructure in reciprocal space via either X-rays or neutrons can be successfully implemented [162, 175] and have been used for example to demonstrate ferroelectricity in ultrathin films [41, 200]. However, this approach averages the behavior over the whole area illuminated by the probe beam, and does not preserve phase information. To image the local polarization structure, very recent efforts have focused on using coherent X-ray Bragg projection ptychography in ferroelectrics [61].

Ferroelectric and strain domains have also been imaged by transmission electron microscopy (TEM). While this approach is extremely time-consuming, and requires significant preparation of the samples in the form of ultrathin lamellae, it is the only one that offers direct access to individual domain walls down to resolutions corresponding to individual atomic displacements [70, 72, 100]. TEM is a scanning, rather than real-time, full-field-of-view technique, but has relatively fast acquisition times, allowing for example the step-by-step imaging of the nucleation and growth of a ferroelectric domain trough a thin film, and its interaction with a dislocation, acting as a strong pinning site [45].

Finally, atomic force microscopy (AFM) techniques, which detect the interaction between a sharp tip scanning over the sample surface and the ferroic material, have become a key tool for studying the roughness and dynamics of individual ferroelectric and ferromagnetic domain walls at the nanoscale. In magnetic force microscopy (MFM) and electric force microscopy (EFM), with the probe tip oscillating near resonance above the sample, but without contacting it directly, long-range electrostatic and magnetic interactions can be detected while avoiding cross-talk with topographical features [63]. In contrast, a sub-switching AC bias applied to the metallic probe tip in mechanical contact with the sample surface will excite a local inverse piezoelectric response (mechanical oscillation) in a ferroelectric material, allowing the polarization orientation and magnitude to be detected by piezoresponse force microscopy (PFM) [79], as shown schematically in Fig. 8.7(a, b). Domain walls, as a result of the reorientation of the order parameter, can also present piezoresponse signals forbidden by symmetry in the parent phase [51, 52, 98]. Concurrently, electrical conductance can be mapped over the sample surface via conductive-tip AFM (CAFM), which has demonstrated unusual transport characteristics at ferroelectric domain walls [166]. The metallic tip, either stationary or scanning, can also be used not only to probe but also to control ferroelectric polarization, acting as a local electric field source when biased with a sufficiently high DC voltage, promoting the switching of nanoscale domains [26, 141, 178], as illustrated in Fig. 8.7(c, d), which remain fully stable in measurements extending over a year [13, 111].

Combining the local control and high-resolution imaging capabilities of PFM has allowed local studies of domain wall roughness and dynamics in ferroelectrics. As can be seen in Fig. 8.8(a), high-resolution images of linear domains (either written with a biased scanning tip or present intrinsically) can be used to extract the exact position of the domain walls and their displacement $u(z)$ from a flat, elastically optimal configuration. Meanwhile, as shown schematically in Fig. 8.8(b), the dependence of domain radius on writing voltage and writing time can be extracted from arrays of nanoscale domains written with voltage pulses of different duration and magnitude applied to a

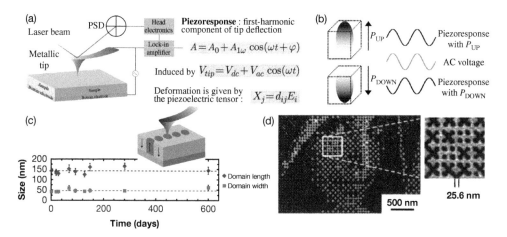

Fig. 8.7 Local probing and control of ferroelectric polarization. (a) PFM measurements of the local inverse piezoresponse of the sample to an applied tip bias. (Reprinted from [49]. With permission of Springer.) (b) Schematic illustration of the vertical piezoresponse to an AC tip voltage applied to oppositely polarized regions. (Reprinted from [136].) (c) Voltage pulses applied between the probe tip and tip and an underlying electrode can be used to switch nanoscale domains in a ferroelectric sample (inset), which remain stable over many months of measurements. (Adapted with permission from [13]. Copyright (2012), AIP Publishing LLC.) (d) Small ferroelectric domains allow information storage at densities higher than $10\,\mathrm{Tbit/inch^2}$. (Reprinted with permission from [27]. Copyright (2005), AIP Publishing LLC.)

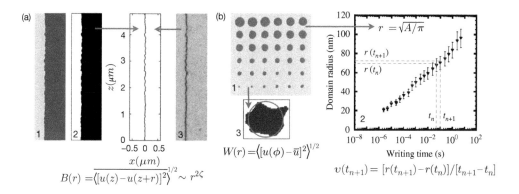

Fig. 8.8 Studying the roughness and dynamics of ferroelectric domain walls. (a) The displacement field $u(z)$ for a ferroelectric domain wall is obtained by subtracting a linear least-squares fit from the exact domain wall position, obtained either from the binarization (2) of the PFM phase image (1) or directly from the PFM amplitude image (3). The roughness function $B(r)$ can be calculated from the correlation of the relative displacements at lengthscale r. (b) Domain growth rates (2) are obtained as the dependence of the effective domain radius r on the writing time t, with r extracted from the domain area A, and averaged over multiple domains written with the same parameters. The domain wall velocity can be extracted for consecutive writing times t_n, t_{n+1}. For sufficiently large nanodomains, roughening behavior can be analyzed as deviations of the domain wall from an equivalent area circle, with an adapted definition of the width $W(r)$ (3). (Adapted from [49] with permission of Springer and from [136].)

stationary tip [141]. From this, the velocity of the domain walls can be obtained and correlated with the electric field that drives the radial growth of the domains [186]. The applied electric field in this device geometry is highly inhomogeneous, decreasing rapidly with both horizontal and vertical distance away from the tip, and is extracted either analytically for simplified models [125] or numerically for a more realistic tip shape [14]. If the nanoscale domains are themselves large enough, they can also be individually analyzed for their roughening behavior, considered by adapting the definition of the global width $W(r)$ to the circular case to measure the deviation of each domain from an equivalent area circle, where $u(\phi)$ is the displacement of the wall from such a circle at angle ϕ, as shown in Fig. 8.8(b).

An important development of the PFM technique introduced by Jesse et al. [68] and termed switching spectroscopy PFM (SSPFM) combines the small AC excitation and a step-function application of the DC bias, as shown in Fig. 8.9(a), to locally map out the switching response of the material. SSPFM measurements allow a quantitative analysis of the energy of switching, interactions of domain walls with individual strong pinning sites, and, in some cases, the universality class of the disorder [69], as presented schematically in Fig. 8.9(b, c).

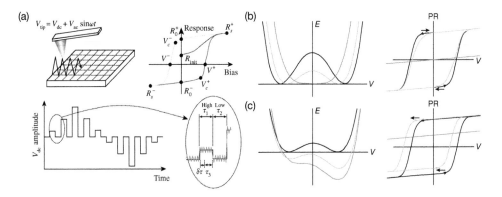

Fig. 8.9 Switching spectroscopy PFM in ferroelectrics. (a) Switching hysteresis is locally mapped at each point of a pre-selected grid with a probing waveform in which the DC tip bias is increased/decreased stepwise with a linear ramp signal envelope, on top of which is overlaid a much smaller AC bias. From the local PFM hysteresis, forward and reverse coercive voltages V^+ and V^-, nucleation voltages V_c^+ and V_c^-, and forward and reverse saturation and remanent responses R_S^+, R_S^-, R_0^+, and R_0^-, as well as the work of switching, A_S (the area within the loop) can be extracted. (Reprinted with permission from [68]. Copyright (2006), AIP Publishing LLC.) (b, c) Schematic illustrations of random-bond (b) and random-field (c) disorder effects on the ferroelectric double-well potential and the local polarization hysteresis (color versions available in [69]). The full (blue) curves correspond to the ideal case of a defect-free sample, the dashed (green) curves to weak disorder, and the dotted (red) curves to the extreme cases of nonpolar (b) and polar nonferroelectric (c) phases, respectively. (Adapted by permission from Macmillan Publishers Ltd: *Nature Materials* [69], copyright (2008).)

The performance of all the AFM techniques in terms of resolution, sensitivity, and tip longevity depends very strongly on the probe tip, which affects the quality of the interaction with the sample. AFM performance can be further improved by using functionalized carbon nanotube-based probes [29, 54, 93, 106, 107, 142, 179].

8.5   Ferroic domain walls as pinned elastic interfaces

The importance of understanding how ferroic domain walls interact—in the presence of thermal fluctuations, with the crystal potential in which they find themselves, with the defects intrinsically present in all physical materials (in particular, in thin film and ceramic form), and even with each other—was recognized from the earliest studies of these functional materials, focusing primarily on switching dynamics. In ferroelectrics[5] and ferroelastics, with their high anisotropy and elastic costs, domain walls were, however, initially assumed to be significantly less prone to roughening than their ferromagnetic counterparts. Domain walls in the initial optical studies of single crystals indeed appeared straight or strongly faceted along axes of crystalline symmetry [122, 170], and were thus considered as essentially flat, with a configuration determined predominantly by the lattice potential and the dipolar interactions. Since the polarization can be seen as a freezing-in of specific patterns of lattice displacement, centering the very thin ferroelectric domain walls on different atomic planes in the unit cell will modulate their energy, as subsequently calculated *ab initio* for materials such as $PbTiO_3$ [119, 147]. If additionally an electric field is applied, the result is motion of the domain wall through a tilted washboard potential.

Early studies of domain growth in high-quality single crystals agreed well with this scenario, implicit in the Miller–Weinreich description of domain wall dynamics [38, 117, 121], where the wall moves forward owing to the formation of a nucleus, as shown in Fig. 8.10(a). The energy change due to the formation of a nucleus in this scenario is

$$\Delta U = -2P_S EV + \sigma_w A + U_{\text{depolarization}}, \tag{8.23}$$

so nucleation occurs, and the domain wall moves forward if the energy gain due to switching a volume V of ferroelectric with spontaneous polarization P_S to the polarization state energetically favorable with respect to the direction of the applied field E balances the energy cost of extending the surface A of the domain wall, with a surface energy density of σ_w, as well as the incurred depolarization energy cost $U_{\text{depolarization}}$. The nucleus thickness (c in Fig. 8.10(a)) is the distance between two minima of the periodic potential given by the lattice spacing of the ferroelectric crystal. For subcritical fields below the activation field E_C, this scenario gives rise to a nonlinear creep response similar to that for randomly pinned interfaces with

$$v \propto \exp\left(-\frac{U_C}{kBT}\frac{E_C}{E}\right), \tag{8.24}$$

---

[5] For a review focusing specifically on ferroelectric domain walls as pinned elastic interfaces, see [138].

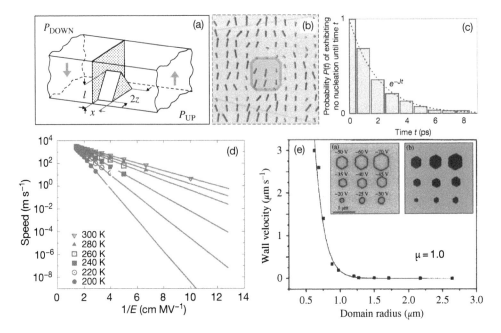

Fig. 8.10 (Color versions of (b)–(d) available in [169]) Motion of a ferroelectric domain wall through the crystal lattice potential. (a) Schematic drawing of a triangular step nucleated on a 180° domain wall. (Adapted with permission from [121]. Copyrighted by the American Physical Society.). The applied electric field E is parallel to P_{DOWN}, favoring domain wall motion to the right. (b) Snapshot of local polarization orientation (indicated by red and blue arrows) in individual unit cells from molecular dynamics simulations of a defect-free $PbTiO_3$ sample, with the critical nucleus bounded by the green box formed under the application of an electric field E parallel to P_{DOWN}. (c) Fraction of molecular dynamics simulations showing no nucleation by time t, fitted to an exponential behavior with nucleation rate J. (d) Domain wall velocity as a function of E and T from coarse-grained Monte Carlo simulations, following Merz's law creep dynamics $v \propto \exp(-E_C/E)$. (Parts (b)–(d) reprinted by permission from Macmillan Publishers Ltd: *Nature* [169], copyright (2007).) (e) Domain growth studies in single-crystal $LiNbO_3$, showing creep motion in agreement with Merz's law, with faceted, very regular domains (PFM amplitude and phase image in inset). (Adapted with permission from [153]. Copyright (2005), AIP Publishing LLC.)

but with a dynamic exponent $\mu = 1$, *only if* the domain wall itself is a two-dimensional surface embedded in the periodic potential of a three-dimensional crystal. This means that the dimension of the nucleus perpendicular to the plane of the film, at a given field E, has to be smaller than the thickness of the film. Otherwise, the energy cost of the nucleus saturates and the scenario of a one-dimensional interface in a periodic potential is recovered.

We can estimate the size of the critical nucleus, following the formulation derived by Miller and Weinreich [121] for the energetically most favorable dagger-shaped nucleus of horizontal extension $2z$, height l, and thickness x forming at an existing $180°$ domain wall, as shown in Fig. 8.10(a), where P_{UP} and P_{DOWN} are the two opposite polarization directions with magnitude P_S, a the in-plane lattice constant, ϵ the dielectric constant of the ferroelectric material at ambient conditions, and E the applied electric field. The depolarization energy can be written as $U_{\text{depolarization}} = 2\sigma_p x z^2/l$, with $\sigma_p = [4P_s^2 x \ln(0.7358z/x)]/\epsilon$. By minimizing the free energy change due to nucleation with respect to the dimensions z and l of the nucleus, with x taken as equivalent to the lattice constant a (the distance between two minima in the periodic crystalline potential), the size of the critical nucleus $z^\star$ and $l^\star$, as well as the activation energy $\Delta U^\star$, can be calculated as

$$z^\star = \frac{\sigma_w(\sigma_w + 2\sigma_p)}{P_s E(\sigma_w + 3\sigma_p)}, \tag{8.25}$$

$$l^\star = \frac{\sigma_w^{1/2}(\sigma_w + 2\sigma_p)}{P_s E(\sigma_w + 3\sigma_p)^{1/2}}, \tag{8.26}$$

$$\Delta U^\star = \frac{4a}{P_s E}\sigma_p(\sigma_w + 2\sigma_p)\left(\frac{\sigma_w}{\sigma_w + 3\sigma_p}\right)^{3/2}. \tag{8.27}$$

However, the large, triangular nucleus shape assumed in the Miller–Weinreich approach gives rise to unrealistically large depolarization and nucleation energy terms and an overestimated activation field E_C, leading to domain wall velocities many orders of magnitude smaller than those observed experimentally.

More realistic atomistic molecular dynamics models of nucleation and domain wall motion in defect-free single crystals, carried out by Shin et al. [169], showed that the polarization reversal is significantly more diffuse, with a critical nucleus of approximately square shape and extending only 3×3 unit cells, as bounded by the box in Fig. 8.10(b) (green in the color version in [169]), and a Poisson-process nucleation behavior, with an exponentially decreasing probability that no critical nucleus has formed by time t, as shown in Fig. 8.10(c). From exponential fits to the molecular dynamics simulations, the nucleation rate J as well as the growth rates of the domains after nucleation could be extracted as a function of the temperature and the applied electric field. Combining these results with coarse-grained Monte Carlo simulations, the authors were able to model domain wall motion, which followed Merz's law with $v \propto \exp(-E_C/E)$ over the temperature range explored, as shown in Fig. 8.10(d). Local probe studies of domain growth in single-crystal LiNbO$_3$ [153], illustrated in Fig. 8.10(e), revealed highly faceted domains pinned by the crystal lattice potential with domain wall velocities in excellent agreement with Merz's law creep dynamics and the numerical simulation results.

Periodic or quasiperiodic pinning of domain walls does not have to be limited to the crystal potential, and can be induced artificially. In nanoribbons of soft magnetic materials such as permalloy, periodic notches have been used as anchor points for domain walls [146] in prototype "racetrack memories" [56, 134, 135]. As shown schematically in Fig. 8.11(a) (the color version of which can be found in [134]), information in such

Fig. 8.11 (Color versions available in [134, 135]) Magnetic racetrack memories. (a) Nanowires of soft magnetic material with oppositely oriented in-plane magnetization used to store information. Using spin-polarized currents to drive domain wall motion, the bits can be scanned across read (gray) or write (red) elements, the latter being based on spin-polarized currents driving domain wall motion. (From [134]. Reproduced with permission. Copyright © (2009) Scientific American, a division of Nature America, Inc. All rights reserved.) (b) The domain walls present complex internal structure (here consisting of single head-to-head (white)/tail-to-tail (dark) vortices) and move in the same direction under the influence of an applied spin-polarized current, independent of their polarity. (From [135]. Reprinted with permission from AAAS.)

a racetrack is stored as the magnetization state of neighboring domains (bits), separated by opposite-polarity (head-to-head or tail-to-tail) domain walls. The domain walls themselves are in fact extremely complex spin nanostructures and can comprise (anti)clockwise transverse, vortex, and antivortex elements [135]. The data are read with a magnetic sensor (gray), which detects the magnetization of individual bits traveling above it along the racetrack. Rather than applying magnetic fields, which would move neighboring domain walls in opposite directions, and eventually annihilate the domains whose magnetization is oriented against the magnetic field, thus losing information, the domain walls are shifted using short pulses of spin-polarized current [57, 58, 183]. In this approach, the current transfers spin angular momentum to the domain walls, promoting their motion along the racetrack, independent of their polarity, as demonstrated in Fig. 8.11(b). To write a data bit, two oppositely magnetized domains in an underlying strip (red in the color version of Fig. 8.11(a) [134]) can be driven in the same fashion, with magnetic fields emanating from the domain wall (yellow) used to set the magnetization of the racetrack. In ferroelectrics, similar effects of artificial pinning sites were observed on the twin domains present: single-crystal lamellae fabricated by focused ion-beam etching [161]. In such samples, there was a noticeable decrease in the width of intrinsic twin domains observed across a notch [114], as would be qualitatively expected from the Landau–Lifshitz–Kittel scaling law [86, 95]. Such notches would therefore act as pinning sites, and have been observed to decrease domain wall mobility, in contrast to anti-notches which promote higher mobility, as a result of the inhomogeneities of the electric field distribution introduced when such features are present in the lamellae [114, 115].

The role of random defects, in contrast, while recognized as important, and directly observed through the pinning and bowing of electric-field-driven individual domain walls in optical studies [195], was more difficult to include explicitly in the mean-field Ginzburg–Landau–Devonshire approach, with analytical descriptions focusing on a single pinning site [126] or periodic arrays of defects [123]. Rather, phenomenological theories of domain switching [65, 66] implicitly included the influence of a random potential landscape by assuming stochastic nucleation events, while domain wall motion during subsequent domain growth was posited to be constant and uniform.

A seminal contribution was the work of Nattermann et al., specifically applying the formalism of disordered elastic systems to ferroic domain walls to consider the effects of commensurate versus random pinning [36, 128], random-bond versus random-field disorder [130, 131], and the effects of long-range dipolar interactions [129], suggesting susceptibility measurements in ferroic systems generally as a way to access domain wall (de)pinning and creep dynamics at a macroscopic level. While the Rayleigh law for magnetic susceptibility was already well established, this insight was particularly important for studies of dielectric and piezoelectric responses in ferroelectric materials [32], where analogous features were observed, and could only be explained by the (de)pinning of domain walls in a random potential. Depending on whether the dominant defects were structural, locally changing the depth of the ferroelectric double well but maintaining its symmetry, or related to randomly oriented polar nanoregions in relaxor ferroelectrics that asymmetrize the ferroelectric double well, these measurements pointed toward random-bond [30, 31, 181, 182] or random-field [5, 89, 127] disorder, respectively. Although some reports initially posited a slow dynamics varying linearly with the driving force to explain the non-Debye contributions to the dielectric response, where the effect of random pinning was included only as a broad distribution of domain wall mobilities [88], a consensus supporting the actual nonlinear creep dynamics gradually emerged [87, 180].

For a better understanding of the characteristic static roughening as well as the complex dynamic behavior in these systems, however, detailed high-resolution imaging of individual domain walls was necessary. This was first carried out for ferromagnetic domain walls, where MOKE measurements in ultrathin Pt/Co/Pt films [99] were used to follow domain growth under a uniform applied magnetic field, adjusted from 0 to $2\,\mathrm{kOe}$. While at fields higher than $h_C \sim 0.75\,\mathrm{kOe}$, a linear dynamic response was observed, for subcritical fields, the domain walls exhibited creep motion $v \sim \exp[-(h_C/h)^\mu]$, with a dynamic exponent $\mu = 0.25$ extracted over multiple decades in the velocity, as can be seen in Fig. 8.12(a). In the same domain walls, the correlation function of relative displacements was used to analyze the static configuration, showing a power-law scaling of the roughness, with roughness exponent $\zeta = 0.69$, as can be seen in Fig. 8.12(b). These results are in excellent agreement with the expected behavior of a one-dimensional interface weakly pinned by random-bond disorder. In ferroelectrics, PFM measurements of the growth rates of nanoscale domains switched by a biased probe tip in $\mathrm{Pb(Zr,Ti)O_3}$ thin films likewise showed creep dynamics, but with $\mu = 0.5$–0.7 [137, 186], as can be seen in Fig. 8.12(c). In the same films, linear domains were written with a scanning-biased probe tip, and showed power-law growth of the roughening at short observation lengthscales, with $\zeta = 0.25$ [137], as can be seen in Fig. 8.12(d). Combined via $\mu = (d - 2 + 2\zeta)/(2 - \zeta)$, these results suggest an effective dimensionality of $d_{\mathrm{eff}} = 2.5$, in agreement with theoretical predictions of the role of long-range dipolar interactions at two-dimensional domain walls [129] otherwise weakly pinned by random-bond disorder. Subsequent

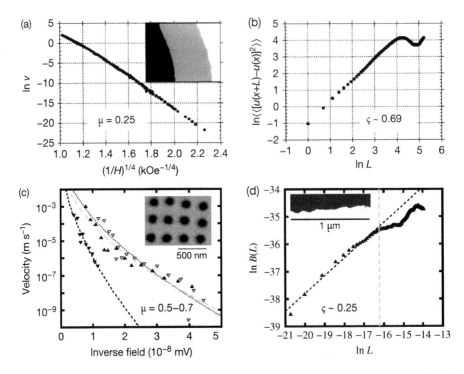

Fig. 8.12 Dynamics and roughness behavior of ferroic domain walls pinned by random-bond disorder. (a) MOKE microscopy of ultrathin Pt/Co/Pt films (inset) was used to image domain growth under uniform applied magnetic fields. The one-dimensional ferromagnetic domain walls exhibit creep dynamics over multiple orders of magnitude in velocity. (b) The same walls show power-law growth of the roughness behavior. (Adapted with permission from [99]. Copyrighted by the American Physical Society.) (c) PFM imaging was used to study domain growth rates under a biased probe tip in thin PZT films (inset). The ferroelectric domain walls exhibit creep dynamics over multiple orders of magnitude in velocity. (d) Linear domains written by a biased scanning probe tip in the same films (inset) show a power-law roughness scaling for short observation lengthscales, at which the domain walls act as two-dimensional interfaces, with additional dipolar interactions. (Adapted with permission from [137]. Copyrighted by the American Physical Society.)

studies largely confirmed that individual ferroelectric domain walls behave as pinned elastic interfaces, although the exact nature of the pinning potential appears to be strongly sample-dependent, with reported values of the domain wall dimensionality and creep exponent μ variously consistent with random-bond disorder[144, 145, 194], random-field disorder [74, 75, 172, 189, 190], or periodic pinning [153]. For a more detailed comparison of the roughness and dynamic exponents obtained in different studies on ferroelectrics, and in disordered elastic systems, see Tables 1 and 2 in [138], respectively.

8.6 Open questions and ongoing research

While the static configuration and subcritical dynamics of ferroic domain walls collectively pinned by weak disorder or the commensurate potential of the lattice are therefore relatively well understood, there are still many open questions beyond these regimes, which are being actively addressed by ongoing research. In the remaining part of this chapter, we will briefly survey some of the questions and the studies addressing them.

First of all, since both ferroelectric and ferromagnetic samples can be imaged over a wide temperature range, they provide an excellent system for exploring aging and thermal effects on both the slow dynamics and depinning of domain walls. As can be seen in Fig. 8.13(a,b) numerical simulations of the evolution of one-dimensional interfaces in a random-bond pinning potential undergoing a temperature quench show a quasi-freezing of the initially equilibrated high-temperature configuration at the largest lengthscales, described by the thermal roughness exponent $\zeta_{th} = 0.5$, with thermally activated dynamic evolution distinguishable only at very small lengthscales [90, 139]. This behavior is reflected in the evolution of the displacement–displacement correlation function: at large lengthscales, $B(L,t) \sim L^{2\zeta_{th}}$, while at small lengthscales, the line tends to equilibrate with the random pining potential and $B(L,t) \sim L^{\zeta_{RB}}$. Numerical simulations of the dynamic response have also allowed accurate prediction of the depinning exponent θ for one- and two-dimensional interfaces in random-bond and random-field disorder. In addition, these studies suggest superroughening of the interface, with $\zeta > 1$ at depinning [155], although with apparently mono-affine scaling. Experimentally, such anomalous scaling was extracted for ferromagnetic domain walls in ultrathin films [22]. Investigations of the crossover between creep and flow regimes [22, 118] confirmed the universal scaling of the velocity, and yielded values of θ (ψ in Fig. 8.13(c)) in agreement with the theoretical predictions of thermal rounding. Behavior qualitatively in agreement with these predictions was also extracted from switching current measurements in ferroelectrics, as can be seen in Fig. 8.13(e), although with significantly higher values of the thermal exponent θ [75].

The first direct observations of the thermal evolution of ferroelectric domains via PFM were made by Likodimos et al. [103, 104] in cleaved triglycine sulfate crystals subjected to annealing–cooling cycles through the transition temperature T_C. The authors extracted a random-bond scaling of the slow dynamics from the spatial correlation functions, with a crossover to even slower behavior, as shown in Fig. 8.13(d). Paruch et al. [139] investigated the thermal evolution of individual domain walls following repeated heat–quench cycles up to high temperatures, but below T_C, in $Pb(Zr_{0.2}Ti_{0.8})O_3$ thin films, and found increased roughening at progressively higher lengthscales, with the exponent ζ changing from 0.25 to 0.5–0.6, as shown in Fig. 8.14(a). The observed post-quench behavior is qualitatively consistent with a locally equilibrated pinned configuration characterized by a growing dynamical lengthscale, whose ultraslow evolution is primarily controlled by the defect configuration and the heating process parameters, resulting in a crossover from a two- to a one-dimensional regime, as illustrated schematically in Fig. 8.14(b).

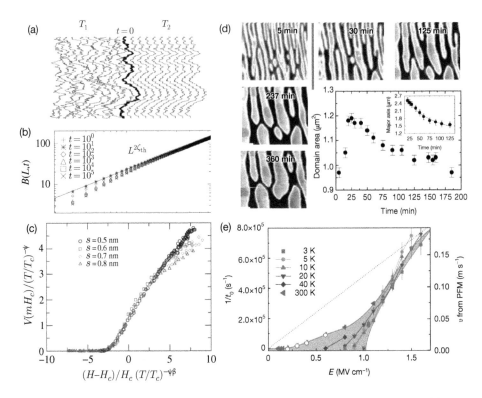

Fig. 8.13 Thermal effects on the roughness and dynamics of ferroic domain walls. (a) Numerical simulation of an overdamped one-dimensional elastic interface in a random pinning potential undergoing a temperature quench $T_1 \to T_2 < T_1$ at time $t = 0$, showing the evolution of the interface every 10^4 steps, with the configuration at the moment of the quench shown in bold. (b) Evolution of the disorder-averaged roughness function $B(L, t)$ after the quench, for different waiting times. At large lengthscales, the geometry of the interface is described by the thermal roughness exponent $\zeta_{th} = 0.5$. (Reprinted with permission from [139]. Copyright (2012) by the American Physical Society.) (c) Universal scaling of the velocity around the thermally rounded depinning regime extracted from MOKE measurements of the motion of ferromagnetic domain walls [118]. (Reprinted with permission from [22]. Copyright (2012) by the American Physical Society.) (d) PFM imaging of intrinsic domains in triglycine sulphate, recorded sequentially at the indicated times at $43°C$ after quenching from $60°C$. The vertical bar corresponds to $7\,\mu m$. The temporal evolution of the area of the indicated domain allows a quantitative measurement of its coarsening. The inset shows the evolution of the major axis of the domain, with a fit to the power law $R(t) \sim (t - t_0)^\phi$. (Adapted with permission from [103]. Copyrighted by the American Physical Society.) (e) Temperature dependence of the inverse characteristic switching time $1/t_0$ versus the electric field E in $Pb(Zr,Ti)O_3$ thin films, extracted from switching current measurements. Using the Kolmogorov–Avrami–Ishibashi expression for the time dependence of the polarization changes $\Delta P(t) = 1 - \exp[-(t/t_0)^n]$, and assuming instantaneous nucleation, the authors relate domain wall velocity $v \sim 1/t_0$. The $1/t_0$ dependence thus appears to indicate the depinning transition, with more pronounced rounding at higher temperatures. (Reprinted with permission from [75]. Copyright (2009) by the American Physical Society.)

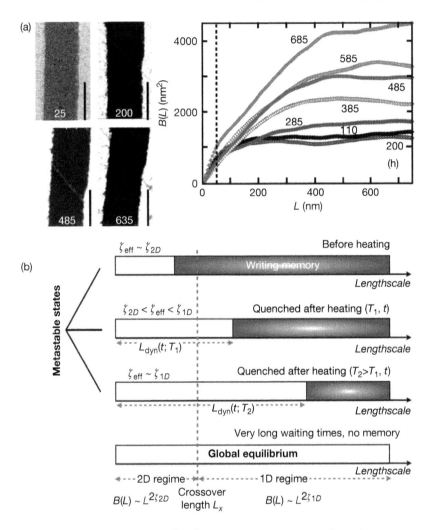

Fig. 8.14 (Color version available in [139]) Ferroelectric domain walls undergoing a temperature quench. (a) PFM images of the evolution of domain structures in $\mathrm{Pb(Zr,Ti)O_3}$ thin films
subjected to heat–quench cycles at progressively higher temperatures, indicated on the image.
The vertical bar corresponds to $1\,\mu\mathrm{m}$. The domain wall roughness $B(r)$ grows after each heat–
quench cycle. Whereas at lower temperatures, power-law roughness scaling is observed only at
lengthscales less than the film thickness, indicated by the dashed line, after higher-temperature
heat–quench cycles, this region extends to lengthscales well beyond the film thickness. Correspondingly, the roughness exponent ζ increases from 0.25 to 0.6–0.6. (b) Schematic representation
of a disorder-dominated dimensional crossover scenario compatible with the observed roughening behavior. Before heating, the as-written, initially flat domain walls are only able to locally
equilibrate at short lengthscales. Heating facilitates thermally activated glassy relaxation, allowing equilibration at higher lengthscales as the temperature increases. The fully equilibrated
system is expected to show a two- to one-dimensional crossover at lengthscale $L_\times$, with no memory of the written state. (Adapted with permission from [139]. Copyrighted by the American
Physical Society.)

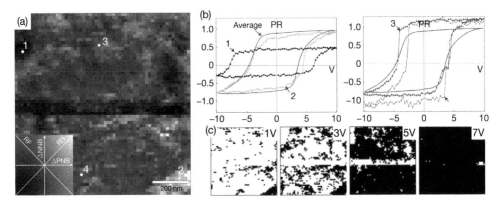

Fig. 8.15 (Color version available in [69]) Complex heterogeneous pinning landscape of a ferroelectric PZT film. (a) Random-bond (RB) versus random-field (RF) disorder map, with color axes defined by local deviations of the positive and negative nucleation bias values from their global averages, $\Delta\mathrm{PNB} = -\mathrm{PNB} + \langle\mathrm{PNB}\rangle$ and $\Delta\mathrm{NNB} = \mathrm{NNB} - \langle\mathrm{NNB}\rangle$, respectively. Correlations between $\Delta\mathrm{PNB}$ and $\Delta\mathrm{NNB}$, indicated by the diagonals of the color map, thus indicate the degree of random-bond or random-field character at each location. (b) Local piezoresponse hysteresis loops acquired at points 1, 2, 3, and 4, compared with the average hysteresis across the film (smooth continuous line), from which the degree of random-bond (left) and random-field (right) character in the disorder can be evaluated. (c) Spatial distribution of nucleation/switching sites activated at a tip bias of 1, 3, 5, and 7 V, with contrast indicating local activation energies below $20k_BT$, corresponding to the onset of thermal nucleation. (Reprinted by permission from Macmillan Publishers Ltd: *Nature Materials* [69], copyright (2008).)

Second, in the real material systems under consideration, while the disorder can often be assigned to a dominant universality class of weak collective pinning, this is not always the case. Particularly in thin films, the pinning landscape can be extremely heterogeneous, as shown in Fig. 8.15(a), including not only weak collective disorder, but also individual strong defect sites [76], threading dislocations [45], grain boundaries [44], or twin boundaries [69], which interact strongly with domain walls. Concurrently, the domain walls can also experience quasiperiodic pinning due to step-bunching growth defects [83, 84] or the ordering of oxygen vacancies at sufficient densities [9, 165]. Moreover, as can be seen in Fig. 8.15(b), the weak collective pinning itself can also vary locally between a random-bond- and random-field-dominated character, as determined from the shape of the local switching hysteresis measured by SSPFM [69]. Such a heterogeneous disorder landscape gives rise to a strongly varying distribution of nucleation and switching sites activated at different voltage magnitude/polarity (Fig. 8.15(c)). While some of the defect effects can be partially screened by surface adsorbates or free charge in the sample, nucleation threshold limits determine switching probabilities for very low applied bias [14].

Moreover, while the theoretical framework of a pinned elastic interface generally assumes quenched disorder that does not evolve during the measurement, the defect landscape of ferroelectric thin films can be significantly altered under strong applied electric fields. Nanoscale electrodes such as the biased scanning probe microscope tip, or even finer carbon nanotube wires [12, 13, 140, 179] used for domain switching studies, produce highly inhomogeneous electric fields that can reach intensities close to dielectric breakdown, and can trigger a cascade of electrochemical effects including charge injection and relaxation dynamics, introduction and reordering of highly mobile defects such as oxygen vacancies, and even irreversible damage to the ferroelectric material [77]. These effects appear to be particularly strong under ambient conditions in the presence of surface water or other adsorbates [109, 171], with very significant effects on domain size and roughening [14, 49, 152], even when the underlying material is not damaged or deteriorated during switching. While at present such complexity of the disorder potential cannot be readily dealt with in the theoretical framework of pinned elastic interfaces, continuing advances in local probe microscopy allow an ever more complete experimental characterization of the system. In particular, band excitation measurements, accessing PFM or MFM response at a wide band of frequencies, and increasingly coupled with a "pump–probe" approach to study transient behaviors during polarization/magnetization switching, are an exciting development. In ferroelectrics, such measurements have explored the microscopic origins of Rayleigh's law and have followed memory effects and the universality of scaling laws in polarization dynamics at the nanoscale [85, 188].

In the face of such complex and heterogeneous disorder, it becomes particularly important to monitor the applicability of the description of domain walls as elastic interfaces weakly pinned by a random potential. Multiscaling analysis, first developed in the context of fracture surfaces [159], allows individual interfaces to be tested for deviations from mono-affine behavior, either by direct examination of the Gaussian character of the PDF of relative displacements or by extracting the $C(r)$ correlation functions, renormalizing them by the Gaussian ratio R_n^G, and checking for collapse to a universal curve. Such multiscaling analysis of domain wall roughening was carried out for artificially written domain structures in $Pb(Zr_{0.2}Ti_{0.8})O_3$ and $BiFeO_3$ thin films [50, 198] and for intrinsic $71°$ stripe domains in $BiFeO_3$ thin films [198]. In all cases, deviations from mono-affine scaling were observed, with an offset in the renormalized $C(r)$ correlation functions, rather than their collapse to a universal curve. This breakdown of mono-affine behavior can be related to extremely localized features of the disorder landscape, which give rise to locally much-larger-than-average fluctuations of the relative displacement, effectively severing the correlations between segments of the domain wall on either side.

For the intrinsic domain walls in $BiFeO_3$, the localized breakdown of mono-affinity could be related either to strong structural defects or to branching/avoidance interactions between neighboring stripe domains [198]. In the artificially written domain walls, as shown in Fig. 8.16(a–d) for the PZT sample, highly localized variations in the switching/pinning potential appeared to be responsible, although these could not be clearly correlated with surface topography [50]. Relatively few such features were observed, allowing the lower bound on the crossover length beyond which mono-affine scaling appears to break down to be determined as $L_{MA} \sim 5\,\mu m$. When only segments of domain wall free of these localized features were considered, mono-affine scaling

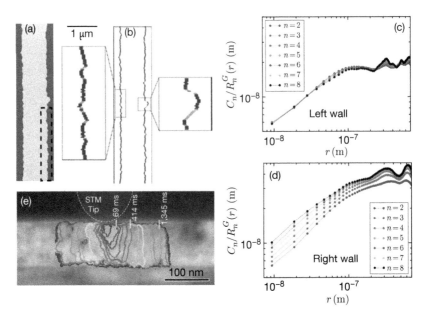

Fig. 8.16 (Color versions available in [45, 49, 50]) Multiscaling analysis of domain wall roughness in ferroelectric PZT thin films. (a) PFM image of two neighboring domain walls, for which the Gaussian-normalized correlation functions $C_n(r)$ calculated for the entire wall length show either collapse (c) or offset (d) between the different orders $n = 2$–8, indicating mono-affine roughness scaling or its breakdown, respectively. The breakdown of mono-affinity appears to be due to strong, highly local disorder fluctuations. When such regions are excluded, for example by considering only the lower part of the right wall, indicated by the dashed box, collapse of $C_n(r)$ and thus mono-affinity, is recovered. (Adapted with permission from [50]. Copyrighted by the American Physical Society.) (b) Map of the local tangent slopes (absolute values) along the domain walls from (a): blue = 0.5; green = 1; yellow = 2.5; red = 3. Mono-affine segments, including the full length of the left domain wall, present only slopes ≤ 1, but higher slopes can be found very locally in the regions responsible for the observed breakdown of mono-affine scaling. (Reprinted from [49] with permission of Springer.) (e) In situ TEM image of domain switching in a $Pb(Zr,Ti)O_3$ thin film, showing the contours of the growing domain as a function of time, and their pinning on an individual dislocation defect. (Adapted with permission from Macmillan Publishers Ltd: *Nature Communications* [45], copyright (2011).)

was fully recovered, with a roughness exponent $\zeta = 0.57 \pm 0.06$, consistent with the roughening of one-dimensional interfaces in random-bond disorder. One possible type of defects that could give rise to such strong localized interactions with domain walls are dislocations, either occurring spontaneously as a result of lattice mismatch strain with the substrate or propagating into the epitaxial thin films from substrates with lower crystalline quality, such as Verneuil-grown $SrTiO_3$. Dislocations are associated with high strain gradients and elevated oxygen vacancy densities [176], thus acting as particularly strong and effective pinning sites for domain walls [45] (see Fig. 8.16(e)).

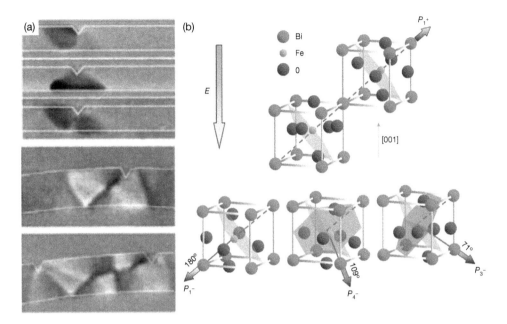

Fig. 8.17 (Color versions available in [135, 196, 198]) Internal structure of ferroic domain walls. (a) MFM images of the complex spin nanostructures at tail-to-tail (top three panels) and head-to-head (bottom two panels) "domain walls" pinned at triangular notches in permalloy nanowires, 10 nm thick and 300 nm wide, and 40 nm thick and 400 nm wide, respectively. The domain walls show, from top to bottom, anticlockwise transverse, clockwise transverse, and anticlockwise vortex structures, a vortex wall with clockwise chirality and negative polarity, and a structure of two vortices and one antivortex. (Adapted from [135]. Reprinted with permission from AAAS.) (b) Schematic diagram of (001)-oriented $BiFeO_3$ crystal structure, with ferroelectric polarization (arrows) and antiferromagnetic plane as indicated. When a downward electric field E is applied to a materials with upward-oriented out-of-plane polarization component (top), the polarization can switch by $180°$, leaving the antiferromagnetic plane unchanged (bottom left), or by $109°$ (bottom middle), or $71°$ (bottom right), changing the antiferromagnetic plane. (Reprinted with permission from Macmillan Publishers Ltd: *Nature Materials* [196], copyright (2006).) (c) Left: Schematic diagram of different possible pinning sites in a $BiFeO_3$ film, with step bunching, unit cell terraces, weak pinning defects (dark blue), and dislocations (red) acting on $71°$ domain walls (black lines). Middle, top: AFM topography, showing the step bunching edges as dark lines. Areas of higher step density versus unit cell terraces are indicated by the blue and red bars below the panel. Middle, bottom: Step bunching edges overlaid on the map of domain wall positions (red) extracted from the same region. Right: Average structure factor $S(q)$ as a function of the wavevector q for 92 mono-affine intrinsic $71°$ domain walls, with the fit used to extract the roughness exponent $\zeta = 0.74$ shown in red. The inset shows the histogram of the individual roughness exponents ζ for each of the domain walls, with the mean $\bar{\zeta} = 0.74$ indicated by the red line. (Reprinted with permission from [198]. Copyright (2013) by the American Physical Society.)

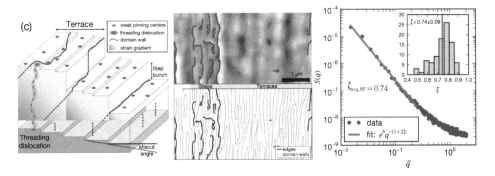

Fig. 8.17 (*continued*)

Transmission and scanning electron microscopy studies of selectively etched single-crystal $SrTiO_3$ revealed dislocation densities of $\sim 10^8$ cm^{-2} [80, 191], qualitatively in agreement with the observed disorder fluctuations.

Finally, we note that the statistical physics description of pinned elastic interfaces assumes that they are essentially homogeneous manifolds with no internal structure. In real systems, in particular in ferroic materials, this is not necessarily the case, as illustrated in Fig. 8.17. Even in "simple" 180° domain walls in uniaxial ferroelectrics, there can be closure/rotation [3, 72], and, at the other extreme, the magnetic "domain walls" driven by spin-polarized currents in magnetic racetrack memories provide a whole zoo of complicated structures [135]. In multiferroic materials in which the different ferroic orders are also coupled to each other, even more novel behavior is to be expected, such as magnetic defects potentially influencing ferroelectric domain walls, as well as a high chance of localized domain-wall-specific ordering forbidden in the parent material [34]. In ferroelectric and antiferromagnetic $BiFeO_3$ (with the polarization and antiferromagnetic plane orientations shown schematically in Fig. 8.17(c)), for example, although magnetoelectric coupling appears to be relatively low in the bulk material [81], with independent origins for the polarization and magnetic properties, the domain walls show strong magnetoelectric coupling, allowing electrical control of the antiferromagnetic order at room temperature [196]. The intrinsic periodic ferroelectric domains in $BiFeO_3$ thin films of a given thickness also show noticeably higher periods than their counterparts in purely ferroelectric films, thus resembling more the behavior of magnetic domains [23]. Recent TEM studies in $BiFeO_3$ revealed polarization rotation in vortex nanodomains at 109° domain walls [132], and demonstrated a very different spatial extent if the domain wall is considered in terms of the evolution of the polarization, or via the rotation of the oxygen octahedra in the perovskite structure [16].

The effects of magnetoelectric coupling or internal domain wall structure are, however, difficult to separate in experimental studies from those of the complex pinning landscape in $BiFeO_3$ thin films, as illustrated schematically in Fig. 8.17(c), containing both weak and strong pinning centers that can interact with the domain walls. Moreover, the step-bunching growth commonly observed in such samples can lead to

unusual effects such as strain and oxygen vacancy density gradients, and dislocation ordering at step edges [83, 84], which create local internal fields. In such systems, Ziegler
et al. [198] measured surprisingly high values of the roughness exponent $\zeta = 0.74$ at
71° domain walls (see Fig. 8.17(c)), in particular as higher effective dimensionalities
are expected for periodic stripe domain structures.

8.7 Conclusions

We have reviewed in this chapter how ferroic domain walls provide a very useful model
system for the study of the very broadly applicable physics of pinned elastic interfaces.
Although mono-affine roughening of the domain walls, and their ultraslow dynamics
for small applied fields are now relatively well understood, far less is known about more
complex aspects of the behavior of such systems. These open questions include the effects of the extremely heterogeneous pinning landscapes present in real samples, where
disorder ranging from point defects such as vacancies or substitutions to one- and two-
dimensional defects in the form of threading dislocations, irradiation tracks, and twin
boundaries can be introduced during or after growth. Likewise, especially in multiferroic materials, the domain walls may exhibit complex internal structure as a result
of electrostatic and strain boundary conditions, or of the coupling of different ferroic
orders, while theoretically only homogeneous interfaces have been considered. Finally,
both theoretically and experimentally, accessing out-of-equilibrium phenomena like aging and memory effects remains a significant challenge. For all of these, ongoing studies
in ferroic systems have the potential to significantly increase our understanding.

Acknowledgments

I am very grateful to all the graduate students and postdoctoral researchers who
contributed to the UniGE work on ferroelectric domain walls described in this
review—in particular Jill Guyonnet, Benedikt Ziegler, Cédric Blaser, and Hélène
Béa—and our enthusiastic and ever-patient theoretical collaborators for many hours
of arguments, explanations, and insight—Thierry Giamarchi, Sebastian Bustingorry,
Alejandro Kolton, Elisabeth Agoritsas and Kirsten Martens. I would also like to thank
Stefano Gariglio, Jambunathan Karthik, Lane Martin, Xia Hong, Charles Ahn, Manuel
Bibes, and Agnès Barthélémy for providing the samples used in our studies. Many
thanks go likewise to Jean-Marc Triscone, Jim Scott, Marty Gregg, Marin Alexe, Piero
Martinoli, Dragan Damjanovic, Antoine Ruyter, Céline Lichtensteiger, and Gustau
Catalan for enriching discussions.

References

[1] Agoritsas, E., Bustingorry, S., Lecomte, V., Schehr, G., and Giamarchi, T.
(2012). Finite-temperature and finite-time scaling of the directed polymer free
energy with respect to its geometrical fluctuations. *Phys. Rev. E*, **86**, 031144.

[2] Agoritsas, E., Lecomte, V., and Giamarchi, T. (2012). Disordered elastic systems and one-dimensional interfaces. *Physica B*, **407**, 1725.

[3] Aguado-Puente, P. and Junquera, J. (2008). Ferromagnetic-like closure domains in ferroelectric ultrathin films: first-principles simulations. *Phys. Rev. Lett.*, **100**, 177601.

[4] Aird, A. and Salje, E. K. H. (1998). Sheet superconductivity in twin walls: experimental evidence of WO_{3-x}. *J. Phys.: Cond. Matt.*, **10**, 377.

[5] Banys, J., Macutkevic, J., Grigalaitis, R., and Kleemann, W. (2005). Dynamics of nanoscale polar regions and critical behavior of the uniaxial relaxor $Sr_{0.61}Ba_{0.39}Nb_2O_6$:Co. *Phys. Rev. B*, **72**, 024106.

[6] Barabasi, A.-L., Bourbonnais, R., Jensen, M., Kertesz, J., Vicsek, T., and Zhang, Y.-C. (1992). Multifractality of growing surfaces. *Phys. Rev. A*, **45**, R6951.

[7] Barabasi, A.-L. and Stanley, H. E. (1995). *Fractal Concepts in Surface Growth*. Cambridge University Press, New York.

[8] Béa, H. and Paruch, P. (2009). Multiferroics: a way forward along domain walls. *Nature Mater.*, **8**, 168.

[9] Becerro, A. I., McCammon, C., Langenhorst, F., Seifert, F., and Angel, R. (1999). Oxygen vacancy ordering in $CaTiO_3$–$CaFeO_{2.5}$ perovskites: from isolated defects to infinite sheets. *Phase Transit.*, **69**, 133.

[10] Bibes, M. and Barthélémy, A. (2008). Towards a magnetoelectric memory. *Nature Mater.*, **7**, 425.

[11] Binder, K. (1987). Theory of first-order phase transitions. *Rep. Prog. Phys.*, **50**, 783.

[12] Blaser, C., Esposito, V., and Paruch, P. (2013). Understanding polarization vs. charge dynamics effects in ferroelectric–carbon nanotube devices. *Appl. Phys. Lett.*, **102**, 223503.

[13] Blaser, C. and Paruch, P. (2012). Minimum domain size and stability in carbon nanotube–ferroelectric devices. *Appl. Phys. Lett.*, **101**, 142906.

[14] Blaser, C. and Paruch, P. (2015). Subcritical switching dynamics and humidity effects in nanoscale studies of domain growth in ferroelectric thin films. *New J. Phys.*, **17**, 013002.

[15] Blatter, G., Feigel'man, M. V., Geshkenbein, V. B., Larkin, A. I., and Vinokur, V. M. (1994). Vortices in high-temperature superconductors. *Rev. Mod. Phys.*, **66**, 1125.

[16] Borisevich, A. Y., Ovhcinnikov, O. S., Chang, H. J., Oxley, M. P., Yu, P., Seidel, J., Eliseev, E. A., Morozovska, A. N., Ramesh, R., Pennycook, S. J., and Kalinin, S. V. (2010). Mapping octahedral tilts and polarization across a domain wall in $BiFeO_3$ from Z-contrast scanning transmission electron microscopy image atomic column shape analysis. *ACS Nano*, **4**, 6071.

[17] Bouchaud, J. P., Cugliandolo, L. F., Kurchan, J., and Mezard, M. (1998). Out of equilibrium dynamics in spin-glasses and other glassy systems. In *Spin Glasses and Random Fields* (ed. A. P. Young), p. 161. World Scientific, Singapore.

[18] Bozhevolnyi, S. I., Hvam, J. M., Pedersen, K., Laurell, F., Karlsson, H., Skettrub, T., and Belmonte, M. (1998). Second-harmonic imaging of ferroelectric domain walls. *Appl. Phys. Lett.*, **73**, 1814.

[19] Bursill, L. A. and Peng, J. L. (1986). Electron microscopic studies of ferroelectric crystals. *Ferroelectrics*, **70**, 191.

[20] Bustingorry, S., Kolton, A. B., Cugliandolo, L. F., and Dominguez, D. (2007). Langevin simulations of the out-of-equilibrium dynamics of the vortex glass in high-temperature superconductors. *Phys. Rev. B*, **75**, 024506.

[21] Bustingorry, S., Kolton, A. B., and Giamarchi, T. (2008). Thermal rounding of the depinning transition. *Europhys. Lett.*, **81**, 26005.

[22] Bustingorry, S., Kolton, A. B., and Giamarchi, T. (2012). Thermal rounding of the depinning transition in ultra thin Pt/Co/Pt films. *Phys. Rev. B*, **85**, 214416.

[23] Catalan, G., Béa, H., Fusil, S., Bibes, M., Paruch, P., Barthélémy, A., and Scott, J. F. (2008). Fractal dimension and size scaling of domains in thin films of multiferroic $BiFeO_3$. *Phys. Rev. Lett.*, **100**, 027602.

[24] Catalan, G., Seidel, J., Ramesh, R., and Scott, J. F. (2012). Domain wall nanoelectronics. *Rev. Mod. Phys.*, **84**, 119.

[25] Chauve, P., Giamarchi, T., and Le Doussal, P. (2000). Creep and depinning in disordered media. *Phys. Rev. B*, **62**, 6241.

[26] Cho, Y., Fujimoto, K., Hiranaga, Y., Wagatsuma, Y., Onoe, A., Terabe, K., and Kitamura, K. (2002). $Tbit/inch^2$ ferroelectric data storage based on scanning nonlinear dielectric microscopy. *Appl. Phys. Lett.*, **81**, 4401.

[27] Cho, Y., Hashimoto, S., Odagawa, N., Tanaka, K., and Hiranaga, Y. (2005). Realization of 10 $Tbit/in.^2$ memory density and sub nanosecond domain switching time in ferroelectric data storage. *Appl. Phys. Lett.*, **87**, 232907.

[28] Cugliandolo, L. F., Giamarchi, T., and Le Doussal, P. (2006). Dynamic compressibility and aging in Wigner crystals and quantum glasses. *Phys. Rev. Lett.*, **96**, 217203.

[29] Dai, H., Hafner, J. H., Rinzler, A. G., Colbert, D. T., and Smalley, R. E. (1996). Nanotubes as nanoprobes in scanning probe microscopy. *Nature*, **384**, 147.

[30] Damjanovic, D. (1997). Logarithmic frequency dependence of the piezoelectric effect due to pinning of ferroelectric–ferroelastic domain walls. *Phys. Rev. B*, **55**, R649.

[31] Damjanovic, D. (1997). Stress and frequency dependence of the direct piezoelectric effect in ferroelectric ceramics. *J. Appl. Phys.*, **82**, 1788.

[32] Damjanovic, D. (1998). Ferroelectric, dielectric and piezoelectric properties of ferroelectric thin films and ceramics. *Rep. Prog. Phys.*, **61**, 1267.

[33] Damjanovic, D., Muralt, P., and Setter, N. (2001). Ferroelectric sensors. *IEEE Sensors J.*, **1**, 191.

[34] Daraktchiev, M., Catalan, G., and Scott, J. F. (2010). Landau theory of domain wall magnetoelectricity. *Phys. Rev. B*, **81**, 224118.

[35] Devonshire, A. F. (1949). *Phil. Mag.*, **40**, 1040.

[36] Emig, T. and Nattermann, T. (1999). Disorder driven roughening transitions of elastic manifolds and periodic elastic media. *Eur. Phys. J. B*, **8**, 525.

[37] Farokhipoor, S., Magén, C., Venkatesen, S., Íñiguez, J., Daumont, C. J. M., Rubi, D., Snoeck, E., Mostovoy, M., de Graaf, C., Müller, A., Döblinger, M., Scheu, C., and Noheda, B. (2014). Artificial chemical and magnetic structure at the domain walls of an epitaxial oxide. *Nature*, **515**, 379.

[38] Fatuzzo, F. and Merz, W. J. (1959). Switching mechanism in triglycine sulfate and other ferroelectrics. *Phys. Rev.*, **116**, 61.

[39] Fiebig, M., Lottermoser, Th., Frölich, D., Goltsev, A. V., and Pisarev, R. V. (2002). Observation of coupled magnetic and electric domains. *Nature*, **419**, 818.

[40] Fiebig, M., Pavlov, V. V., and Pisarev, R. V. (2005). Second-harmonic generation as a tool for styling electronic and magnetic structures of crystals: review. *JOSA B*, **22**, 96.

[41] Fong, D. D., Stephenson, G. B., Streiffer, S. K., Eastman, J. A., Auciello, O., Fuoss, P. H., and Thompson, C. (2004). Ferroelectricity in ultrathin perovskite films. *Science*, **304**, 1650.

[42] Fridkin, V. M. (1979). *Photoferroelectrics*. Springer-Verlag, Berlin.

[43] Fu, H. and Bellaiche, L. (2003). Ferroelectricity in barium titanate quantum dots and wires. *Phys. Rev. Lett.*, **91**, 257601.

[44] Ganpule, C.S., Roytburd, A. L., Nagarajan, V., Hill, B. K., Ogale, S. B., Williams, E. D., Ramesh, R., and Scott, J. F. (2001). Polarization relaxation kinetics and $10°$ domain wall dynamics in ferroelectric thin films. *Phys. Rev. B*, **65**, 014101.

[45] Gao, P., Nelson, C. T., Jokisaari, J. R., Baek, S.-H., Bark, C. W., Zhang, Y., Wang, E., Schlom, D. G., Eom, C.-B., and Pan, X. (2011). Revealing the role of defects in ferroelectric switching with atomic resolution. *Nature Commun.*, **2**, 591.

[46] Giamarchi, T. and Bhattacharya, S. (2002). Vortex phases. In *High Magnetic Fields: Applications in Condensed Matter Physics and Spectroscopy* (ed. C. Berthier, L. P. Lévy, and G. Martinez), p. 314. Springer-Verlag, Berlin. arXiv:cond-mat/0111052.

[47] Giamarchi, T., Kolton, A. B., and Rosso, A. (2006). Dynamics of disordered elastic systems. In *Jamming, Yielding and Irreversible Deformation in Condensed Matter* (ed. M. C. Miguel and J. M. Rubi), p. 91. Springer-Verlag, Berlin. arXiv:cond-mat/0503437.

[48] Goncalves-Ferreira, L., Redfern, S. A. T., Artacho, E., and Salje, E. K. H. (2008). Ferrielectric twin walls in $CaTiO_3$. *Phys. Rev. Lett.*, **101**, 097602.

[49] Guyonnet, J. (2014). *Ferroelectric Domain Walls: Statics, Dynamics, and Functionalities Revealed by Atomic Force Microscopy* (Doctoral thesis, University of Geneva). Springer-Verlag, Heidelberg.

[50] Guyonnet, J., Agoritsas, E., Bustingorry, S., Giamarchi, T., and Paruch, P. (2012). Multiscaling analysis of ferroelectric domain wall roughness. *Phys. Rev. Lett.*, **109**, 147601.

[51] Guyonnet, J., Béa, H., Guy, F., Gariglio, S., Fusil, S., Bouzehouane, K., Triscone, J.-M., and Paruch, P. (2009). Shear effects in lateral piezoresponse force microscopy at $180°$ ferroelectric domain walls. *Appl. Phys. Lett.*, **95**, 132902.

[52] Guyonnet, J., Béa, H., and Paruch, P. (2010). Lateral piezoresponse across ferroelectric domain walls in thin films. *J. Appl. Phys.*, **108**, 042002.

[53] Guyonnet, J., Gaponenko, I., Gariglio, S., and Paruch, P. (2011). Conduction at domain walls in insulating $Pb(Zr_{0.2}Ti_{0.8})O_3$ thin films. *Adv. Mater.*, **23**, 5377.

[54] Hafner, J. H., Cheung, C. L., and Lieber, C. M. (1999). Direct growth of CNT on AFM tips. *J. Am. Chem. Soc.*, **121**, 9750.

[55] Halpin-Healy, T. (1991). Directed polymers in random media probability distributions. *Phys. Rev. A*, **44**, R3415.

[56] Hayashi, M., Thomas, L., Moriya, R., Rettner, C., and Parkin, S. S. P. (2008). Current-controlled magnetic domain-wall nanowire shift register. *Science*, **320**, 190.

[57] Hayashi, M., Thomas, L., Moriya, R., Rettner, C., and Parkin, S. S. P. (2008). Dynamics of domain wall depinning driven by a combination of direct and pulsed currents. *Appl. Phys. Lett.*, **92**, 162503.

[58] Hayashi, M., Thomas, L., Rettner, C., Moriya, R., and Parkin, S. S. P. (2007). Direct observation of the coherent precession of magnetic domain walls propagating along permalloy nanowires. *Nature Phys.*, **3**, 21.

[59] Hehn, M., Padovani, S., Ounadjela, K., and Bucher, J. P. (1996). Nanoscale magnetic domain structures in epitaxial cobalt films. *Phys. Rev. B*, **54**, 3428.

[60] Herbert, J. M. (1982). *Ferroelectric Transducers and Sensors*. Gordon and Breach, New York.

[61] Hruszkewych, S. O., Highland, M. J., Holt, M. V., Kim, D., Folkman, C. M., Thompson, C., Tripathi, A., Stephenson, G. B., Hong, S., and Fuoss, P. H. (2013). Imaging local polarization in ferroelectric thin films by coherent x-ray Bragg projection ptychography. *Phys. Rev. Lett.*, **110**, 177601.

[62] Hubert, A and Schäfer, R. (1998). Springer, Berlin Heidelberg.

[63] Hug, H. J., Stiefel, B., van Schendel, P. J. A., Moser, A., Hofer, R., Martin, S., Güntherodt, H.-J., Porthun, S., Abelmann, L., Lodder, J. C., Bochi, G., and O'Handley, R. C. (1998). Quantitative magnetic force microscopy on perpendicularly magnetized samples. *J. Appl. Phys.*, **83**, 5609.

[64] Iguain, J. L., Bustingorry, S., Kolton, A. B., and Cugliandolo, L. F. (2009). Growing correlations and aging of an elastic line in a random potential. *Phys. Rev. B*, **80**, 094201.

[65] Ishibashi, Y. (1996). Polarization reversals in ferroelectrics. In *Ferroelectric Thin Films: Synthesis and Basic Properties* (ed. C. Paz de Araujo, J. F. Scott, and G. W. Taylor), p. 135. Gordon and Breach, Amsterdam.

[66] Ishibashi, Y. and Takagi, Y. (1971). Note on ferroelectric domain switching. *J. Phys. Soc. Jpn*, **31**, 506.

[67] Janovec, V. and Dvorak, V. (1986). Perfect domain textures of incommensurate phases. *Ferroelectrics*, **66**, 169.

[68] Jesse, S., Baddorf, A. P., and Kalinin, S. V. (2006). Switching spectroscopy piezoresponse force microscopy of ferroelectric materials. *Appl. Phys. Lett.*, **88**, 062908.

[69] Jesse, S., Rodriguez, B. J., Choudhury, S, Baddorf, A. P., Vrejoiu, I., Hesse, D., Alexe, M., Eliseev, E. A., Morozovska, A. N., Zhang, J., Chen, L.-Q., and Kalinin, S. V. (2008). Direct imaging of the spatial and energy distribution of nucleation centres in ferroelectric materials. *Nature Mater.*, **7**, 209.

[70] Jia, C.-L., Mi, S.-B., Urban, K. W., Vrejoiu, I., Alexe, M., and Hesse, D. (2008). Atomic-scale study of electric dipoles near charged and uncharged domain walls in ferroelectric films. *Nature Mater.*, **7**, 57.

[71] Jia, C.-L. and Urban, K. (2004). Atomic-resolution measurement of oxygen concentration in oxide materials. *Science*, **303**, 2001.

[72] Jia, C.-L., Urban, K. W., Alexe, M., Hesse, D., and Vrejoiu, I. (2011). Direct observation of continuous electric dipole rotation in flux-closure domains in ferroelectric Pb(Zr, Ti)O$_3$. *Science*, **331**, 1420.

[73] Jiles, D. (1998). *Introduction to Magnetism and Magnetic Materials*. Taylor & Francis, Boca Raton, FL.

[74] Jo, J. Y., Han, H. S., Yoon, J.-G., Song, T. K., Kim, S.-H., and Noh, T. W. (2007). Domain switching kinetics in disordered ferroelectric thin films. *Phys. Rev. Lett.*, **99**, 267602.

[75] Jo, J. Y., Yang, S. M., Kim, T. H., Lee, H. N., Yoon, J.-G, Park, S., Jo, Y., Jung, M. H., and Noh, T. W. (2009). Nonlinear dynamics of domain-wall propagation in epitaxial ferroelectric thin films. *Phys. Rev. Lett.*, **102**, 045701.

[76] Kalinin, S. V., Jesse, S., Rodriguez, B. J., Chu, Y. H., Ramesh, R., Eliseev, E. A., and Morozovska, A. N. (2008). Probing the role of single defects on the thermodynamics of electric-field induced phase transitions. *Phys. Rev. Lett.*, **100**, 155703.

[77] Kalinin, S. V., Jesse, S., Tselev, A., Baddorf, A. P., and Balke, N. (2011). The role of electrochemical phenomena in scanning probe microscopy of ferroelectric thin films. *Nano*, **5**, 5683.

[78] Kalinin, S. V., Morozovska, A. N., Chen, L. Q., and Rodriguez, B. J. (2010). Local polarization dynamics in ferroelectric materials. *Rep. Prog. Phys.*, **73**, 056502.

[79] Kalinin, S. V., Rodriguez, B. J., Jesse, S., Shin, J., Baddorf, A. P., Gupta, P., Jain, H., Williams, D. B., and Gruverman, A. (2006). Vector piezoresponse force microscopy. *Microsc. Microanal.*, **12**, 206.

[80] Kamaladasa, R. J., Jiang, W., and Picard, Y. N. (2011). Imaging dislocations in single-crystal SrTiO$_3$ substrates by electron channeling. *J. Electron. Mater.*, **40**, 2222.

[81] Kamba, S., Nuzhnyy, D., Savinov, M., Šebek, J., Petzelt, J., Prokleska, J., Haumont, R., and Kreisel, J. (2007). Infrared and terahertz studies of polar phonons and magnetoelectric effect in multiferroic BiFeO$_3$ ceramics. *Phys. Rev. B*, **75**, 024403.

[82] Kämple, T., Reichenbach, P., Schröder, M., Haußmann, A., Eng, L. M., Wolke, T., and Soergel, E. (2014). Optical three-dimensional profiling of charged domain walls in ferroelectrics by Cherenkov second-harmonic generation. *Phys. Rev. B*, **89**, 035314.

[83] Kim, T. H., Baek, S. H., Jang, S. Y., Yang, S. M., Chang, S. H., Song, T. K., Yoon, J.-G., Eom, C. B., Chung, J.-S., and Noh, T. W. (2011). Step bunching-induced vertical lattice mismatch and crystallographic tilt in vicinal BiFeO$_3$(001) films. *Appl. Phys. Lett.*, **98**, 022904.

[84] Kim, T. H., Baek, S. H., Yang, S. M., Jang, S. Y., Ortiz, D., Song, T. K., Chung, J.-S., Eom, C. B., Noh, T. W., and Yoon, J.-G. (2009). Electric-field-controlled directional motion of ferroelectric domain walls in multiferroic BiFeO$_3$ films. *Appl. Phys. Lett.*, **95**, 262902.

[85] Kim, Y., Lu, X., Jesse, S., Hesse, D., Alexe, M., and Kalinin, S. V. (2013). Universality of polarization switching dynamics in ferroelectric capacitors revealed by 5D piezoresponse force microscopy. *Adv. Funct. Mater.*, **23**, 3971.

[86] Kittel, C. (1946). Theory of the structure of ferromagnetic domains in films and small particles. *Phys. Rev.*, **70**, 965.

[87] Kleemann, W. (2007). Universal domain wall dynamics in disordered ferroic materials. *Annu. Rev. Mater. Res.*, **37**, 415.

[88] Kleemann, W., Dec, J., Miga, S., and Pankrath, R. (2002). Non-Debye domain-wall-induced dielectric response in $Sr_{0.61-x}Ce_xBa_{0.39}Nb_2O_6$. *Phys. Rev. B*, **65**, 220101.

[89] Kleemann, W., Licinio, P., Woike, Th., and Pankrath, R. (2001). Dynamic light scattering at domains and nano clusters in a relaxor ferroelectric. *Phys. Rev. Lett.*, **86**, 220101.

[90] Kolton, A. B., Rosso, A., and Giamarchi, T. (2005). Relaxation of a flat interface. *Phys. Rev. Lett.*, **95**, 180604.

[91] Kolton, A. B., Rosso, A. and Albano, E. V., and Giamarchi, T. (2006). Short-time relaxation of a driven elastic string in a random medium. *Phys. Rev. B*, **74**, 140201.

[92] Kolton, A. B., Schehr, G., and Le Doussal, P. (2009). Universal nonstationary dynamics at the depinning transition. *Phys. Rev. Lett.*, **103**, 160602.

[93] Kuramochi, H., Uzumaki, T., Yasutake, M., Tanaka, A., Akinaga, H., and Yokoyama, H. (2005). A magnetic force microscope using CoFe-coated carbon nanotube probes. *Nanotechnology*, **16**, 24.

[94] Lajzerowicz, J and Niez, J. J. (1979). Phase transition in a domain wall. *J. Physique Lett.*, **40**, 165.

[95] Landau, L. and Lifshitz, E. (1935). Theory of the dispersion of magnetic permeability in ferromagnetic bodies. *Z. Phys.*, **8**, 153.

[96] Lee, D., Behera, R. K., Wu, P., Xu, H., Li, Y. L., Sinnott, S. B., Phillpot, W. R., Chen, L. Q., and Gopalan, V. (2009). Mixed Bloch–Néel–Ising character of 180° ferroelectric domain walls. *Phys. Rev. B*, **80**, 060102.

[97] Lehmann, M. and Lichte, H. (2002). Tutorial on off-axis electron holography. *Microsc. Microanal.*, **8**, 447.

[98] Lei, S., Eliseev, E. A., Morozovska, A. N., Haislmeier, R. C., Lummen, T. T. A., Cao, W., Kalinin, S. V., and Gopalan, V. (2012). Origin of piezoelectric response under a biased scanning probe microscopy tip across a 180° ferroelectric domain wall. *Phys. Rev. B*, **86**, 134115.

[99] Lemerle, S., Ferré, J., Chappert, C., Mathet, V., Giamarchi, T., and Le Doussal, P. (1998). Domain wall creep in an Ising ultrathin magnetic film. *Phys. Rev. Lett.*, **80**, 849.

[100] Li, L., Gao, P., Nelson, C. T., Jokisaari, J. R., Zheng, Y., Kim, S.-J., Melville, A., Adamo, C., Schlom, D. G., and Pan, X. (2034). Atomic scale structure changes induced by charged domain walls in ferroelectric materials. *Nano Lett.*, **13**, 5218.

[101] Li, Y.-Y. (1956). Domain walls in antiferromagnets and the weak ferromagnetism of α-Fe_2O_3. *Phys. Rev.*, **101**, 1450.

[102] Lichtensteiger, C., Zubko, P., Stengel, M., Aguado-Puente, P., Triscone, J.-M., Ghosez, Ph., and Junquera, J. (2012). Ferroelectricity in ultrathin-film capacitors. In *Oxide Ultrathin Films: Science and Technology* (ed. G. Pacchioni and S. Valeri), p. 265. Wiley-VCH, Weinheim.

[103] Likodimos, V., Labardi, M., and Allegrini, M. (2000). Logarithmic evolution of domain size in TGS. *Phys. Rev. B*, **61**, 14440.

[104] Likodimos, V., Labardi, M., and Allegrini, M. (2002). Domain patterns and kinetics, thermal effects on TGS. *Phys. Rev. B*, **66**, 024104.

[105] Lines, M. E. and Glass, A. M. (1977). *Principles and Applications of Ferroelectrics and Related Materials*. Oxford University Press, Oxford.

[106] Lisunova, Y., Heidler, J., Levkivskyi, I., Gaponenko, I., Weber, A., Heyderman, L. J., Kl* qui, M., and Paruch, P. (2013). Optimal ferromagnetically-coated carbon nanotube tips for ultra-high resolution magnetic force microscopy. *Nanotechnology*, **24**, 105705.

[107] Lisunova, Y., Levkivskyi, I., and Paruch, P. (2013). Ultra-high currents in dielectric-coated carbon nanotube probes. *Nano Lett.*, **13**, 4527.

[108] Lyubchanskii, I. L., Petukhov, A. V., and Rasing, Th. (1997). Domain and domain wall contributions to optical second harmonic generation in thin magnetic films. *J. Appl. Phys.*, **81**, 5668.

[109] Maksymovych, P., Jesse, S., Hujiben, M., Ramehs, R., Morozovska, A., Choudhury, S., Chen, L.-Q., Baddorf, A. P., and Kalinin, S. V. (2009). Intrinsic nucleation mechanism and disorder effects in polarization switching on ferroelectric surfaces. *Phys. Rev. Lett.*, **102**, 017601.

[110] Martin, L. W., Chu, Y.-H., Holcomb, M.B., Hujiben, M., Yu, P., Han, S.-J., Lee, D., Wang, S. X., and Ramesh, R. (2008). Nanoscale control of exchange bias with $BiFeO_3$ thin films. *Nano Lett.*, **8**, 2050.

[111] Maruyama, T., Saitoh, M., Sakai, I., Hidaka, T., Yano, Y., and Noguchi, T. (1998). Growth and characterization of 10-nm-thick *c*-axis oriented epitaxial $PbZr_{0.25}Ti_{0.75}O_3$ thin films on (100)Si substrate. *Appl. Phys. Lett.*, **73**, 3524.

[112] Mason, W. P. (1950). Optical properties and the electro-optic and photo elastic effects in crystals expressed in tensor form. *Bell Syst. Tech. J.*, **29**, 161.

[113] Matthias, B. and Von Hippel, A. (1948). Structure, electrical, and optical properties of barium titanate. *Phys. Rev.*, **73**, 268.

[114] McMillen, M., McQuaid, R. G. P., Haire, S. C., McLaughlin, C. D., Chang, L. W., Schilling, A., and Gregg, J. M. (2010). The influence of notches on domain dynamics in ferroelectric nano wires. *Appl. Phys. Lett.*, **96**, 042904.

[115] McQuaid, R. G. P., Chang, L.-W., and Gregg, J. M. (2010). The effect of antinotches on domain wall mobility in single crystal ferroelectric nanowires. *Nano Lett.*, **10**, 3566.

[116] Meier, D., Cano, A., Delaney, K., Kumagai, Y., Mostovoy, M., Spaldin, N. A., Ramesh, R., and Fiebig, M. (2012). Anisotropic conductance at improper ferroelectric domain walls. *Nature Mater.*, **11**, 284.

[117] Merz, W. J. (1954). Domain formation and domain wall motion in ferroelectric $BaTiO_3$ single crystals. *Phys. Rev.*, **95**, 690.

[118] Metaxas, WP. J., Jamet, J. P., Mougin, A., Cormier, M., Ferré, J., Baltz, V., Rodmacq, B., Dieny, B., and Stamps, R. L. (2007). Creep and flow regimes of magnetic domain-wall motion in ultrathin Pt/Co/Pt films with perpendicular anisotropy. *Phys. Rev. Lett.*, **99**, 217208.

[119] Meyer, B. and Vanderbilt, D. (2002). *Ab initio* study of ferroelectric domain walls in PbTiO₃. *Phys. Rev. B*, **65**, 104111.

[120] Mézard, M. and Parisi, G. (1991). Replica field theory for random manifolds. *J. Physique I*, **1**, 809.

[121] Miller, R. C. and Weinreich, G. (1960). Mechanism for the sidewise motion of 180° domain walls in barium titanate. *Phys. Rev.*, **117**, 1460.

[122] Mitsui, T. and Furuichi, J. (1953). Domain structure of Rochelle salt and KH₂PO₄. *Phys. Rev.*, **90**, 193.

[123] Mokrý, P., Wang, Y., Tagantsev, A. K., Damjanovic, D., Stolichnov, I., and Setter, N. (2009). Evidence for dielectric aging due to progressive 180° domain wall pinning in polydomain Pb(Zr₀.₄₅Ti₀.₅₅)O₃ thin films. *Phys. Rev. B*, **79**, 054104.

[124] Morozovska, A. N., Eliseev, E. A., Glinchuk, M. D., Chen, L.-Q., and Gopalan, V. (2012). Interfacial polarization and pyroelectricity in antiferrodistortive structures induced by a flexoelectric effect and rotostriction. *Phys. Rev. B*, **85**, 094107.

[125] Morozovska, A. N., Eliseev, E. A., Li, Y., Svechnikov, S. V., Maksymovych, P., Shur, V. Y., Gopalan, V., Chen, L.-Q., and Kalinin, S. V. (2009). Thermodynamics of nanodomain formation and breakdown in scanning probe microscopy: Landau–Ginzburg–Devonshire approach. *Phys. Rev. B*, **80**, 214110.

[126] Morozovska, A. N., Svechnikov, S. V., Eliseev, E. A., Rodriguez, B. J., Jesse, S., and Kalinin, S. V. (2008). Local polarization switching in the presence of surface-charged defects: microscopic mechanisms and piezoresponse force spectroscopy observations. *Phys. Rev. B*, **78**, 054101.

[127] Mueller, V., Shchur, Y., Beige, H., Mattauch, S., Glinnemann, J., and Heger, G. (2002). Non-Debye domain wall response in KH₂PO₄. *Phys. Rev. B*, **65**, 134102.

[128] Natterman, T. and Emig, T. (1997). A new disorder-driven roughening transition of charge-density waves and flux-line lattices. *Phys. Rev. Lett.*, **79**, 5090.

[129] Nattermann, T. (1983). Interface phenomenology, dipolar interaction, and the dimensionality dependence of the incommensurate–commensurate transition. *J. Phys. C*, **16**, 4125.

[130] Nattermann, T. (1987). Interface roughening in systems with quenched random impurities. *Europhys. Lett.*, **4**, 1241.

[131] Nattermann, T., Shapir, Y., and Vilfan, I. (1990). Interface pinning and dynamics in random systems. *Phys. Rev. B*, **42**, 8577.

[132] Nelson, C. T., Winchester, B., Zhang, Y., Kim, S.-J., Melville, A., Adamo, C., Folkman, C. M., Baek, S.-H., Eom, C.-B., Shclom, D. G., Chen, L.-Q., and Pan, X. (2011). Spontaneous vortex nanodomain arrays at ferroelectric heterointerfaces. *Nano Lett.*, **11**, 828.

[133] Padilla, J., Zhong, W., and Vanderbilt, D. (1996). First-principles investigation of 180° domain walls in BaTiO₃. *Phys. Rev. B*, **85**, 2379.

[134] Parkin, S. S. P. (2009). Data in the fast lanes of racetrack memory. *Sci. Am.*, **300**(6), 76.

[135] Parkin, S. S. P., Hayashi, M., and Thomas, L. (2008). Magnetic domain-wall racetrack memory. *Science*, **320**, 190.

[136] Paruch, P. (2004). Atomic force microscopy studies of ferroelectric domain in epitaxial PbZr$_{0.2}$Ti$_{0.8}$O$_3$ thin films and the static and dynamic behaviour of ferroelectric domain walls. PhD thesis, University of Geneva.

[137] Paruch, P., Giamarchi, T., and Triscone, J.-M. (2005). Domain wall roughness in epitaxial ferroelectric PbZr$_{0.2}$Ti$_{0.8}$O$_3$ thin films. *Phys. Rev. Lett.*, **94**, 197601.

[138] Paruch, P. and Guyonnet, J. (2013). Nanoscale studies of ferroelectric domain walls as pinned elastic interfaces. *C. R. Physique*, **14**, 637.

[139] Paruch, P., Kolton, A. B., Hong, X., Ahn, C. H., and Giamarchi, T. (2012). Thermal quench effects on ferroelectric domain walls. *Phys. Rev. B*, **85**, 214115.

[140] Paruch, P., Posadas, A.-B., Dawber, M., Ahn, C. H., and McEuen, P. L. (2008). Polarization switching using single-walled carbon nanotubes grown on epitaxial ferroelectric thin films. *Appl. Phys. Lett.*, **93**, 132901.

[141] Paruch, P., Tybell, T., and Triscone, J.-M. (2001). Nanoscale control of ferroelectric polarization and domain size in epitaxial Pb(Zr$_{0.2}$Ti$_{0.8}$)O$_3$ thin films. *Appl. Phys. Lett.*, **79**, 530.

[142] Paruch, P., Tybell, T., and Triscone, J.-M. (2002). Nanoscale control and domain wall dynamics in epitaxial ferroelectric Pb(Zr$_{0.2}$Ti$_{0.8}$)O$_3$ thin films. In *Proceedings of 10th International Ceramics Congress*, Part D, p. 675.

[143] Pershan, P. S. (1967). Magneto-optical effects. *J. Appl. Phys.*, **38**, 1482.

[144] Pertsev, N. A., Kiselev, D. A., Bdikin, I. K., Kosec, M., and Kholkin, A. L. (2011). Quasi-one-dimensional domain walls in ferroelectric ceramics: Evidence from domain dynamics and wall roughness measurements. *J. Appl. Phys.*, **110**, 052001.

[145] Pertsev, N. A., Petraru, A., Kohlstedt, H., Waser, R., Bdikin, I. K., Kiselev, D. A., and Kholkin, A. L. (2008). Dynamics of ferroelectric nanodomains in BaTiO$_3$ epitaxial thin films via piezoresponse force microscopy. *Nanotechnology*, **19**, 375703.

[146] Petit, D., Jausovec, A. V., Zeng, H. T., Lewis, E., O'Brien, L., Read, D., and Cowburn, R. P. (2009). Mechanism for domain wall pinning and potential landscape modification by artificially patterned traps in ferromagnetic nanowires. *Phys. Rev. B*, **79**, 214405.

[147] Pöykkö, S. and Chadi, D. J. (1999). *Ab initio* study of 90° domain wall energy and structure in PbTiO$_3$. *Appl. Phys. Lett.*, **75**, 2830.

[148] Přivratská, J. and Janovec, V. (1997). Examination of point group symmetries of non-ferroelastic domain walls. *Ferroelectrics*, **191**, 17.

[149] Přivratská, J. and Janovec, V. (1999). Spontaneous polarization and/or magnetization in non-ferroelastic domain walls: symmetry predictions. *Ferroelectrics*, **222**, 281.

[150] Ramesh, R. and Spaldin, N. A. (2007). Multiferroics: progress and prospects in thin films. *Nature Mater.*, **6**, 21.

[151] Rave, W., Schäfer, R., and Hubert, A. (1987). Quantitative observation of magnetic domains with the magneto-optical Kerr effect. *J. Magn. Magn. Mater.*, **65**, 7.

[152] Rodriguez, B. J., Jesse, S., Baddorf, A. P., Kim, S.-H., and Kalinin, S. V. (2007). Controlling polarization dynamics in a liquid environment: from localized to macroscopic switching in ferroelectrics. *Phys. Rev. Lett.*, **98**, 247603.

[153] Rodriguez, B. J., Nemanich, A. J., Kingon, A., Gruverman, A., Kalinin, S. V., Terabe, K., Liu, X. Y., and Kitamura, K. (2005). Domain growth kinetics in lithium niobate single crystals studied by piezoresponse force microscopy. *Appl. Phys. Lett.*, **86**, 012906.

[154] Roitburd, A. L. (1976). Equilibrium structure of epitaxial layers. *Phys. Stat. Sol. A*, **37**, 329.

[155] Rosso, A., Hartmann, A. K., and Krauth, W. (2003). Depinning of elastic manifolds. *Phys. Rev. Lett.*, **67**, 021602.

[156] Rosso, A., Santachiara, R., and Krauth, W. (2005). Geometry of Gaussian signals. *J. Stat. Mech.*, **2005**, L08001.

[157] Salje, E. K. H. (1993). *Phase Transitions in Ferroelastic and Co-elastic Crystals*. Cambridge University Press, Cambridge.

[158] Salje, E. K. H. (2010). Multiferroic boundaries as active memory devices: trajectories towards domain boundary engineering. *ChemPhysChem*, **11**, 940.

[159] Santucci, S., Måløy, K. J., Delaplace, A., Mathiesen, J., Hansen, A., Bakke, J. Ø. H., Schmittbuhl, J., Vanel, L., and Purusattam, R. (2007). Statistics of fracture surfaces. *Phys. Rev. E*, **75**, 016104.

[160] Schehr, G. and Doussal, P. Le (2005). Functional renormalization for pinned elastic systems away from their steady states. *Europhys. Lett.*, **71**, 290.

[161] Schilling, A., Bowman, R. M., Catalan, G., Scott, J. F., and Gregg, J. M. (2006). Ferroelectric domain periodicities in nano columns of single-crystal barium titanate. *Appl. Phys. Lett.*, **89**, 212902.

[162] Schlenker, M., Linares-Galvez, J., and Baruchel, J. (1978). A spin-related contrast effect visibility of 180° ferromagnetic domain walls in unpolarized neutron diffraction topography. *Phil. Mag. B*, **37**, 1.

[163] Schröder, M., Haussmann, A., Thiessen, A., Soergel, E., Woke, T., and Eng, L. M. (2012). Conducting domain walls in lithium niobate single crystals. *Adv. Funct. Mater.*, **22**, 3936.

[164] Scott, J. F. (2000). *Ferroelectric Memories*. Springer-Verlag, Berlin.

[165] Scott, J. F. and Dawber, M. (2000). Oxygen-vacancy ordering as a fatigue mechanism in perovskite ferroelectrics. *Appl. Phys. Lett.*, **76**, 3801.

[166] Seidel, J., Martin, L. W., He, Q., Zhan, Q., Chu, Y.-H., Rother, A., Hawkridge, M. E., Maksymovych, P., Yu, P., Gajek, M., Balke, N., Kalinin, S. V., Gemming, S., Want, F., Catalan, G., Scott, J. F., Spaldin, N. A., Orenstein, J., and Ramesh, R. (2009). Conduction at domain walls in oxide multiferroics. *Nature Mater.*, **8**, 229.

[167] Shen, Y. R. (1989). Surface properties probed by second-harmonic and sum-frequency generation. *Nature*, **337**, 519.

[168] Sheng, Y., Best, A., Butt, H.-J., Krolikowski, W., Arie, A., and Koynov, K. (2010). Three-dimensional ferroelectric domain visualization by Čerenkov-type second harmonic generation. *Opt. Express*, **18**, 16539.

[169] Shin, Y.-H., Grinberg, I., Chen, I.-W., and Rappe, A. M. (2007). Nucleation and growth mechanism of ferroelectric domain-wall motion. *Nature*, **449**, 881.

[170] Shur, V. Y. (1996). Fast polarization reversal process: evolution of ferroelectric domain structure in thin films. In *Ferroelectric Thin Films: Synthesis and Basic*

Properties (ed. C. P. Paz de Araujo, J. F. Scott, and G. W. Taylor), p. 153. Gordon and Breach, Amsterdam.

[171] Shur, V. Y., Ievlev, A. V., Nikolaeva, E. V., Shishkin, E. I., and Neradovsky, M. M. (2011). Influence of adsorbed surface layer on domain growth in the field produced by conductive tip of scanning probe microscope in lithium niobate. *J. Appl. Phys.*, **110**, 052017.

[172] Shvartsman, V. V., Dhkil, B., and Kholkin, A. L. (2013). Mesoscale domains and nature of the relaxor state by piezoresponse force microscopy. *Annu. Rev. Mater. Res.*, **43**, 423.

[173] Sluka, T., Tagantsev, A. K., Bednyakov, P., and Setter, N. (2013). Free-electron gas at charged domain walls in insulating $BaTiO_3$. *Nature Commun.*, **4**, 1808.

[174] Stepkova, V., Marton, P., and Hlinka, J. (2012). Stress-induced phase transitions in ferroelectric domain walls of $BaTiO_3$. *J. Phys.: Cond. Matt.*, **24**, 212201.

[175] Streiffer, S. K., Eastman, J. A., Fong, D. D., Thompson, C., Munkholm, A., Ramana Murty, M. V., Auciello, O., Bai, G. R., and Stephenson, G. B. (2002). Observation of nanoscale 180° stripe domain in ferroelectric $PbTiO_3$ thin films. *Phys. Rev. Lett.*, **89**, 067601.

[176] Szot, K., Speier, W., Bihlmayer, G., and Waser, R. (2006). Switching the electrical resistance of individual dislocations in single-crystalline $SrTiO_3$. *Nature Mater.*, **5**, 312.

[177] Taharinejad, M., Vaderbilt, D., Marton, P., Stepkova, V., and Hlinka, J. (2012). Bloch-type domain walls in rhombohedral $BaTiO_3$. *Phys. Rev. B*, **86**, 155138.

[178] Tayebi, N., Narui, Y., Chen, R. J., Collier, C. P., Giapis, K. P., and Zhang, Y. (2008). Nanopencil as a wear-tolerant probe for ultrahigh density data storage. *Appl. Phys. Lett.*, **93**, 103112.

[179] Tayebi, N., Narui, Y., Franklin, N., Collier, C. P., Giapis, K. P., Nishi, Y., and Zhang, Y. (2010). Fully inverted single-digit nanometer domains in ferroelectric films. *Appl. Phys. Lett.*, **96**, 023103.

[180] Taylor, D. R., Love, J. T., Topping, G. J., and Dane, J. G. A. (2005). Crossover from pure to random-field critical susceptibility in $KH_2As_xP_{1-x}O_4$. *Phys. Rev. B*, **72**, 052109.

[181] Taylor, D. V. and Damjanovic, D. (1997). Evidence of domain wall contribution to the dielectric permittivity in PZT thin films at sub-switching fields. *J. Appl. Phys.*, **82**, 1973.

[182] Taylor, D. V. and Damjanovic, D. (1998). Domain wall pinning contribution to the nonlinear dielectric permittivity in $Pb(Zr,Ti)O_3$ thin films. *Appl. Phys. Lett.*, **73**, 2045.

[183] Thomas, L., Hayashi, M., Jiang, X., Moriya, R., Rettner, C., and Parkin, S. S. P. (2007). Resonant amplification of magnetic domain-wall motion by a train of current pulses. *Science*, **315**, 1553.

[184] Tolédano, J. C. and Tolédano, P. (1987). *The Landau Theory of Phase Transitions*. World Scientific, Singapore.

[185] Tolédano, P. and Tolédano, J. C. (1976). Order-parameter symmetries for improper ferroelectric nonferroelastic transitions. *Phys. Rev. B*, **14**, 3097.

[186] Tybell, T., Paruch, P., Giamarchi, T., and Triscone, J.-M. (2002). Domain wall creep in epitaxial ferroelectric $PbZr_{0.2}Ti_{0.8}O_3$ thin films. *Phys. Rev. Lett.*, **89**, 097601.

[187] van Aert, S., Turner, S., Delville, R., Schryvers, D., van Tendeloo, G., and Salje, E. K. H. (2012). Direct observation of ferrielectricity at ferroelastic domain boundaries in $CaTiO_3$ by electron microscopy. *Adv. Mater.*, **24**, 523.

[188] Vasudevan, R. K., Okatan, M. B., Duan, C., Ehara, Y., Funakubo, H., Kumar, A., Jesse, S., Chen, L.-Q., Kalinin, S. V., and Nagarajan, V. (2013). Nanoscale origins of nonlinear behavior in ferroic thin films. *Adv. Funct. Mater.*, **23**, 81.

[189] Volk, T. R., Simagina, L. V., Gainutdinov, R. V., Ivanova, E. S., Ivleva, L. I., and Mit'ko, S. V. (2011). Scanning probe microscopy investigation of ferroelectric properties of barium strontium niobate crystals. *Phys. Sol. State*, **53**, 2468.

[190] Volk, T. R., Simagina, L. V., Gainutdinov, R. V., Tolstikhina, A. L., and Ivleva, L. I. (2010). Ferroelectric micro domains and micro domain arrays recorder in strontium–barium niobate crystals in the field of atomic force microscope. *J. Appl. Phys.*, **108**, 042010.

[191] Wang, R., Zhu, Y., and Shapiro, S. M. (1998). Structural defects and the origin of the second length scale in $SrTiO_3$. *Phys. Rev. Lett.*, **80**, 2370.

[192] Waser, R. and Rüdiger, A. (2004). Pushing towards the digital storage limit. *Nature Mater.*, **3**, 81.

[193] Williams, D. B. and Carter, C. B. (1996). *The Transmission Electron Microscope*. Springer-Verlag, New York.

[194] Xiao, Z., Poddar, S., Ducharme, S., and Hong, X. (2013). Domain wall roughness and creep in nanoscale crystalline ferroelectric polymer. *Appl. Phys. Lett.*, **103**, 112903.

[195] Yang, T. J., Gopalan, V., Swart, P. J., and Mohadeen, U. (1999). Direct observation of pinning and bowing of a single ferroelectric domain wall. *Phys. Rev. Lett.*, **82**, 4106.

[196] Zhao, T., Scholl, A., Zavaliche, F., Lee, K., Barry, M., Doran, A., Cruz, M. P., Chu, Y. H., Ederer, C., Spaldin, N. A., Das, R. R., Kim, D. M., Baek, S. H., Eom, C. B., and Ramesh, R. (2006). Electrical control of antiferromagnetic domains in multiferroic $BiFeO_3$ films at room temperature. *Nature Mater.*, **5**, 823.

[197] Zhirnov, V. A. (1959). A contribution to the theory of domain walls in ferroelectrics. *Sov. Phys. JETP*, **35**, 822.

[198] Ziegler, B., Martens, K., Giamarchi, T., and Paruch, P. (2013). Domain wall roughness in the stripe phase of $BiFeO_3$ thin films. *Phys. Rev. Lett.*, **111**, 247604.

[199] Zubko, P., Catalan, G., Buckley, A., Welche, P. R. L., and Scott, J. F. (2007). Strain-gradient-induced polarization in $SrTiO_3$ single crystals. *Phys. Rev. Lett.*, **99**, 167601.

[200] Zubko, P., Stucki, N., Lichtensteiger, C., and Triscone, J.-M. (2010). X-ray diffraction studies of $180°$ ferroelectric domains in $PbTiO_3/SrTiO_3$ superlattices under an applied electric field. *Phys. Rev. Lett.*, **104**, 187601.

9
Real-time monitoring of a fluctuating condensate in high-temperature superconductors

Luca PERFETTI,[1] B. SCIOLLA,[2] G. BIROLI,[3]
and T. KAMPFRATH[4]

[1]Laboratoire des Solides Irradiés, Ecole polytechnique, 91128 Palaiseau cedex, France
[2]Department of Theoretical Physics, Ecole de Physique, University of Geneva 24, Switzerland
[3] Institut de Physique Théorique CEA, 91191 Gif-sur-Yvette, France
[4]Fritz-Haber-Institut der Max-Planck-Gesellschaft, 14195 Berlin, Germany

Strongly Interacting Quantum Systems out of Equilibrium. First Edition. Thierry Giamarchi et al.
© Oxford University Press 2016. Published in 2016 by Oxford University Press.

Chapter Contents

Preface

This chapter investigates the critical behavior of high-temperature superconductors that are suddenly driven far from equilibrium conditions. $Bi_2Sr_2CaCu_2O_{8+\delta}$ is excited with a femtosecond laser pulse, and the subsequent nonequilibrium dynamics of the mid-infrared conductivity is monitored. The collected data allow us to discriminate temperature regimes where superconductivity is coherent, fluctuating, or vanishingly small. Above the transition temperature T_c, the recovery to equilibrium conditions displays power-law behavior and scaling properties. We find partial agreement between the scaling law of the optimal doped sample and the time-dependent Ginzburg–Landau (TDGL) model. Inherent limits of TDGL call for nonequilibrium field theories treating fluctuations, mode coupling, and fast degrees of freedom on an equal footing. These results open a timely connection between superconducting condensates and Bose–Einstein condensates of ultracold atoms.

Parts of this chapter have previously been published in a *Physical Review Letters* paper, "Ultrafast dynamics of fluctuations in high-temperature superconductors far from equilibrium" [1].

9.1   Introduction

Many physical properties of high-temperature superconductors cannot be understood in the context of a mean-field theory. For example, an incipient condensate with short coherence length fluctuates in space and time, leading to precursor effects of the broken-symmetry phase. As a consequence, the conductivity [2–5], heat capacity [6], diamagnetic susceptibility [7], and Nernst signal [8] may increase considerably when the temperature approaches the critical value T_c. For cuprates, most of these observations have been successfully described by the Ginzburg–Landau (GL) model or quantum field theories [9–12]. Suitable extensions of these approaches to systems under out-of-equilibrium conditions predict critical dynamics and scaling laws [13, 14]. The temporal evolution of superconducting fluctuations upon a sudden quench could provide stringent tests of coarse-grained models of given symmetry and dimensionality [15]. Moreover, critical exponents independent of specific material properties may extend the concept of "universality classes" to systems under far-from-equilibrium conditions. This quest is not limited to condensed matter systems but overlaps with high-energy physics and ultracold atomic gases [14, 16].

An ideal protocol for the detection of critical dynamics requires that superconducting correlations be monitored just above the transition temperature. In this critical region, the slow degrees of freedom of the gapless condensate decouple from the rest of the system and follow a universal behavior. To induce and probe the nonequilibrium regime, ultrafast pump–probe spectroscopy is a very promising approach [17–28]. Unfortunately, in experiments using optical probes (with photon energy of typically 1.5 eV), it is not straightforward to detect superconducting fluctuations above T_c because of the presence of competing signals ascribed to the pseudogap [17, 18]. In order to disentangle fluctuation- and pseudogap-related signal components, an optical three-pulse technique was recently proposed [19]. Alternatively, electromagnetic pulses covering the frequency range from about 1 to 3 THz have been shown to be highly sensitive probes of the superconducting condensate [20–23]. However, the duration of the THz pulses (~0.5 ps) is too long to resolve the ultrafast dynamics of the superconductor.

The time-resolved experiments in the mid-infrared (MIR) spectral range solve this technological issue. The photoexcited superconductor is probed by phase-stable pulses covering the 10–30 THz window [24]. We observe that the electromagnetic responses of superconducting correlations increase the transmission of electric field by roughly 20% and dominate the signal up to $T \cong 1.4T_c$. Since the probe pulses have a duration $\cong 100$ fs, the resulting response can follow the fast dynamics of superconducting fluctuations in real time. By these means, the critical slowing down of the pairing field is monitored under out-of-equilibrium conditions. The observed relaxation dynamics displays a power-law scaling, possibly because of the coupling of the order parameter to a conserved quantity.

This chapter has the following structure: Section 9.2 describes the technical details of the experiment, Section 9.3 discusses the MIR response when the sample is in equilibrium at a given temperature, Section 9.4 reports the experimental scaling law observed in the dynamics of a photoexcited sample, Section 9.5 introduces the time-dependent GL (TDGL) calculations, Section 9.6 shows the dynamical behavior of underdoped samples, and Section 9.7 discusses the transient dynamics when the system is below the critical temperature.

9.2   Methods

The samples used are 150 nm thin films of optimally doped ($T_c = 91$ K) and underdoped ($T_c = 68$ K) Bi$_2$Sr$_2$CaCu$_2$O$_{8+\delta}$ (BSCCO) mounted on a diamond substrate that is transparent to both pump and probe radiation. A THz spectrometer is driven by a Ti:sapphire laser oscillator delivering 10 fs pulses at 780 nm center wavelength. Part of the laser output is used to excite the sample with an incident fluence of about 16 μJ cm^{-2}. Another portion of the laser beam generates 100 fs THz pulses covering a spectral range from 8 to 30 THz by difference frequency mixing in a GaSe crystal. The THz electric field is detected by electro-optic sampling in ZnTe. Thin films of optimally doped ($T_c = 91$ K) and underdoped ($T_c = 68$ K) BSCCO are obtained by peeling off flakes from a single crystal. A homogeneous film of 150 nm thickness is fixed to a diamond substrate and mounted in an optical cryostat, with the sample temperature T set from 40 to 300 K. The polarization of the THz pulse is perpendicular to the c axis, thus probing the optical properties in the basal plane. We simultaneously measure the THz electric field $E_0(t, T)$ transmitted through the sample in thermodynamic equilibrium at temperature T and the change $\Delta E(t, T, \tau)$ in the transmitted electric field induced by the pump pulse. Here t and τ denote the time of the internal time of the THz pulse and the pump–probe delay, respectively. The incident and transmitted THz electric fields are Fourier-transformed with respect to t, and the conductivity of the BSCCO film is then obtained by inverting the Fresnel formulas. The small thickness of the sample and the good thermal conductivity of diamond guaranties a local laser heating below 3 K.

9.3   Equilibrium

There is an obvious reason why MIR absorption is highly sensitive to the phase transition of high-temperature superconductors. The single-particle gap of the ordered phase is comparable to the frequency of the MIR photons and strongly affects the THz electrodynamics. In our experiment, we measure electric field transients having traversed the bare diamond substrate ($E_r(t)$, reference field) and the diamond substrate plus the BSCCO film ($E_0(t)$, sample measurement). Figure 9.1(a) shows the transmitted THz traces $E_0(t, T)$ of optimally doped BSCCO in thermal equilibrium at a given temperature T. These waveforms allow us to determine the complex sample transmission T versus frequency ν, from which, by using Tinkham's thin-film formula

$$\frac{E_0(\nu)}{E_r(\nu)} = \frac{1}{1 + A\sigma_0(\nu)}, \tag{9.1}$$

we extract the BSCCO conductivity over the spectral range from 10 to 30 THz. Here, $A = Z_0/d(n_1 + n_3)$, where $Z_0 \cong 377\,\Omega$, d is the BSCCO film thickness, and $n_1 \cong 1$ and $n_3 \cong 2.4$ are the refractive indices of air and the diamond substrate, respectively. Figure 9.1(b) shows that the real part of the conductivity $\mathrm{Re}\,\sigma_0(\nu, T)$ decreases with frequency at high temperature, whereas it follows the opposite trend in the superconducting phase below $T_c = 91$ K. Independently of T, the $\mathrm{Re}\,\sigma_0(\nu, T)$ curves converge to a value $\simeq 1000\,\Omega^{-1}\,\mathrm{cm}^{-1}$ for $\nu/2\pi > 25$ THz. These findings are in excellent agreement with Fourier transform spectroscopy [29] and ellipsometry measurements [30, 31].

From an experimental point of view, it is convenient to monitor the change in transmitted electric field $\Delta E_0(t^*, T) = E_0(t^*, T) - E_0(t^*, 6\,\mathrm{K})$ at a time $t^* = 50$ fs where the THz waveform displays a maximum. A detailed analysis of the probing process shows that the variation of the MIR electric field at the maximum position $t^* = 50$ fs is proportional to the real part of the difference $\mathrm{Re}\,\Delta\sigma_0(\nu)$ averaged over the frequency interval from 15 to 25 THz. Figure 9.1(c) displays the temperature dependence of $\Delta E_0(t^*, T)$. When the temperature is lowered from 300 to 100 K, we observe a linear decrease, which turns into an abrupt drop at the transition to the superconducting phase ($T \sim T_c = 91$ K). The physical interpretation of this sudden decrease is straightforward: the emergence of the superconductivity gap ($T \sim T_c$) reduces the density of electronic states (DOS) in the vicinity of the Fermi energy, thereby reducing the number of scattering channels of the electrons [32]. Consequently, there is much less dissipation of the driving MIR electric field below T_c.

In the fluctuating regime above T_c, the DOS becomes a function of local gap amplitude and of the superconducting coherence length ξ [32]. The latter reflects the typical distance over which the superconducting order parameter maintains its amplitude and phase [9], and it is expected to diverge exactly at the transition temperature. Theoretical models of two-dimensional superconductors [33] predict that the MIR conductivity scales with $\ln(\xi/\xi_0)$, where ξ_0 is the value of the coherence length at zero temperature. These DOS fluctuations should not be confused with the Aslamazov–Larkin (AL) and Maki–Thompson (MT) contributions taking place at lower probe frequencies. The AL

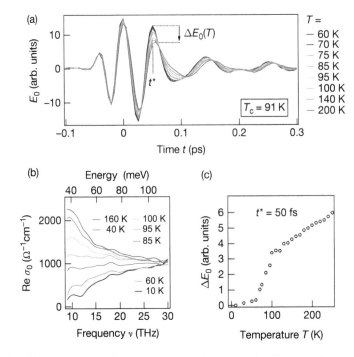

Fig. 9.1 (Color online: see [1] and arXiv:1407.7689) Equilibrium data for optimally doped BSCCO ($T_c = 91$ K). (a) Electric field $E_0(t, T)$ of the MIR waveform transmitted through the BSCCO film as a function of time t. Each sample temperature T is indicated by a different line shade (color online). (b) Real part Re $\sigma_0(\nu)$ of the MIR conductivity versus frequency ν and (c) change $\Delta E_0(T) = E_0(t^*, 6\,\text{K}) - E_0(t^*, T)$ in the transmitted MIR field at time $t^* = 50$ fs as a function of T. (Reprinted with permission from [1]. Copyright (2015) by the American Physical Society.)

and MT fluctuations affect primarily the imaginary part of the optical conductivity at $\sim$1 THz [4] and are negligible in the MIR spectral range (up to 20 THz) of our experiment.

9.4   Transient conductivity and scaling law

We now photoexcite the BSCCO sample with an optical pump pulse (with duration 10 fs and photon energy 1.55 eV) and monitor the resulting ultrafast dynamics. According to previous work [23, 34], our pump fluence of 16 µJ cm^{-2} is large enough to reduce the order parameter by a fraction of the order of 80%. Figure 9.2(a) shows the pump-induced change $\Delta E(\tau) = E(t^*, -1\,\text{ps}) - E(t^*, \tau)$ in the transmitted MIR field for the optimally doped sample as a function of the pump–probe delay τ at various ambient temperatures T. As above, we consider the MIR fields at time $t^* = 50$ fs such that $\Delta E(\tau)$ scales with the pump-induced change Re $\Delta\sigma(\nu, \tau)$ in the instantaneous

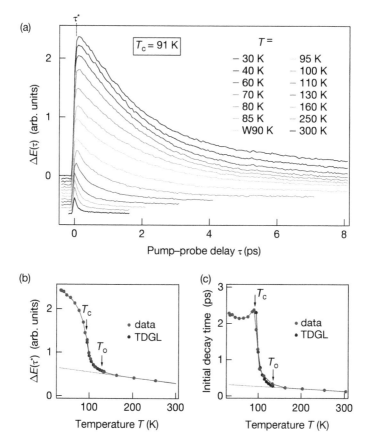

Fig. 9.2 (Color online: see [1] and arXiv:1407.7689) Nonequilibrium data of optimally doped BSCCO ($T_c = 91$ K). (a) Pump-induced changes $\Delta E(\tau) = E(t^*, -1\,\text{ps}) - E(t^*, \tau)$ in the transmitted MIR electric field as a function of pump-probe delay τ. $\Delta E(\tau)$ scales with the pump-induced change $\operatorname{Re} \Delta\sigma(\nu, T)$ in the instantaneous BSCCO conductivity averaged from 15 to 25 THz. (b) Signal magnitude ΔE and (c) relaxation time $(\partial \ln \Delta E / \partial \tau)^{-1}$ directly after sample excitation ($\tau = \tau^* = 50$ fs) as a function of temperature (circles, red online). Squares (blue online) show the result from a model based on the TDGL equation. The transition temperature ($T_c = 91$ K) and the onset of superconducting fluctuations (at $T_o = 130$ K) are indicated by arrows. (Reprinted with permission from [1]. Copyright (2015) by the American Physical Society.)

BSCCO conductivity averaged from 15 to 25 THz. As shown in Fig. 9.2(a), the magnitude and recovery time of $\Delta E(\tau)$ depend strongly on sample temperature. Also, the dynamics deviates substantially from an exponential decay and slows down with increasing pump-probe delay.

To better illustrate the temperature dependence of the curves in Fig. 9.2(a), we extract two characteristic parameters, namely, the signal magnitude $\Delta E(\tau)$ and the instantaneous decay time $(\partial \ln \Delta E / \partial \tau)^{-1}$ directly after sample excitation at $\tau = \tau^* = 50\,\text{fs}$. The latter is obtained by fitting the dynamics with an exponential fit in a small time interval just after photoexcitation. The two parameters are displayed as a function of sample temperature T in Fig. 9.2(b, c). As can be seen in Fig. 9.2(b), $\Delta E(\tau^*)$ is much larger when T is below the transition temperature. Above T_c, we still observe an appreciable $\Delta E(\tau^*)$, which, above a temperature T_o, turns into a relatively flat and linearly decreasing curve (see arrows). Figure 9.2(c) shows that the initial decay time of the pump–probe signal exhibits an analogous behavior: it is large below T_c, then drops sharply and turns into a flat curve above T_o. We attribute these temperature intervals to the three different regimes: a superconducting phase $(T < T_c)$, a fluctuating superconductor $(T_c < T < T_o)$, and a metallic phase $(T > T_o)$. From Fig. 9.2, we obtain the rough estimate $T_o = 130\,\text{K}$, which is in good agreement with values extracted from measurements of transport [2], specific heat [6], diamagnetic susceptibility [7], and Nernst effect [8].

In the temperature region from $T = T_o$ down to the critical temperature T_c, the system is governed by an increasing coherence length ξ [9]. The characteristic time τ_c it takes a Cooper pair to cover the coherence length should increase accordingly.

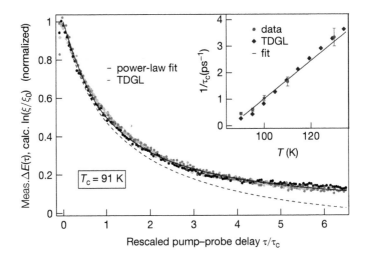

Fig. 9.3 (Color online: see [1] and arXiv:1407.7689) Scaling behavior of ultrafast relaxation in the critical region between T_c and T_o of optimally doped BSCCO ($T_c = 91\,\text{K}$). Measured pump-induced changes $\Delta E(\tau)$ in the transmitted electric field are shown as a function of rescaled pump–probe delay $\tau / \tau_c(T)$ at temperatures $T = 90, 95, 100, 110$, and $130\,\text{K}$. Curves are also normalized to the maximum signal value. The full curve is a power-law fit (see (9.2)), whereas the dashed curve shows the logarithm of the GL correlation length as calculated using the TDGL model. The inset shows the scaling factor τ_c^{-1} versus temperature as derived from the experiment (circles, red online) and TDGL theory (diamonds, blue online) as well as a linear fit (solid line). (Reprinted with permission from [1]. Copyright (2015) by the American Physical Society.)

The scaling hypothesis suggests that the relaxation should follow a universal power law $[\tau/\tau_c(T)]^{-\alpha}$, where α is a critical exponent that depends only on the observable considered, the symmetry of the order parameter, and the dimension of the system [13].

Figure 9.3 shows the key result of this chapter, namely, the striking observation that the apparently diverse dynamics of Fig. 9.2(a) do indeed obey a scaling law. As can be seen in Fig. 9.3, all the decay curves for $T_c < T < T_o$ fall onto a single curve when the time axis τ is normalized by a suitable time $\tau_c(T)$. More precisely, we find that the relation

$$\Delta E(\tau) = \frac{\Delta E(\tau^*)}{1 + [\tau/\tau_c(T)]^\alpha} \tag{9.2}$$

with $\alpha = 1.1$ provides a very good description of our experimental data. The inverse scaling factor τ_c^{-1} is plotted as a function of temperature in the inset of Fig. 9.3(a). It follows an approximately linear relation $0.45\,\mathrm{ps}^{-1} + (T/T_c - 1)6.5\,\mathrm{ps}^{-1}$. The strong temperature dependence of τ_c is consistent with a critical slowing down and increasing coherence length of superconducting fluctuations as the transition temperature is approached.

9.5 Theoretical modeling

We now simulate our experimental findings using the two-dimensional TDGL model for the position ($\mathbf{x}$)- and time (τ)-dependent order parameter $\psi(\mathbf{x}, \tau)$ [11]. From a microscopic perspective, ψ is the local pair potential leading to the single-particle gap. In this coarse-grained description, the GL time τ_{GL} is the characteristic response time of the order parameter $\psi(\mathbf{x}, \tau)$. The remaining degrees of freedom are assumed to evolve much faster than ψ and merely remain as noise $\eta(\mathbf{x}, \tau)$ that affects the dynamics of ψ as a source term. The TDGL theory can be derived by expanding the equations of motion of the BCS model in the gapless phase of the superconductor. The TDGL equations describe the dynamics of local pairing potential above T_c by means of a complex field $\psi(\mathbf{x}, \tau)$. We restrict our analysis to the two-dimensional case because of the large anisotropy factor of the BSCCO family. Since the ratio between the interlayer distance and the c-axis coherence length is $d/\xi_c \cong 100$, any eventual crossover to three dimensions should be confined to $|T - T_c| < 0.5\,\mathrm{K}$ [9].

We neglect the time-dependent vector potential in the coupled TDGL equations because BSCCO is a type II superconductor. Our approximation disregard terms of order $(\xi/\lambda)^2 \cong 10^{-2}$, where ξ is the coherence length and λ is the magnetic penetration depth [35]. Within this framework, the TDGL dynamics is given by the single equation

$$\frac{d\psi}{d\tau} + \frac{\psi}{\tau_{\mathrm{GL}}}\left(1 + \frac{|\psi|^2}{\Delta^2}\right) - D\nabla^2\psi + \eta(\mathbf{x}, \tau) = 0, \tag{9.3}$$

where τ_{GL} is the GL time, Δ is the superconducting gap calculated at $T_c - |T - T_c|$, and D is the normal-state diffusion constant. The fast degrees of freedom generate a noise $\eta(\mathbf{x}, \tau)$ that affects the evolution of $\psi(\mathbf{x}, \tau)$. We identify the photoexcited state $\psi(\mathbf{x}, \tau^*)$ with a stationary solution of the TDGL equation at the elevated temperature

$T + \Delta T$. The value $\Delta T = 17\,\mathrm{K}$ is obtained by comparing the transient $\Delta E(\tau^*)$ of Fig. 9.2(b) with the equilibrium $\Delta E_0(T)$ of Fig. 9.1(c). In contrast, the fast degrees of freedom summarized in η are assumed to return to their equilibrium configuration characterized by temperature T directly after excitation. This sudden-quench hypothesis is motivated by the observation that uncondensed electrons in the normal phase transfer the pump-deposited excess heat to the lattice within $\sim 100\,\mathrm{fs}$ [28]. We model η by white noise with correlation function $\langle \eta(\mathbf{x}, \tau)\eta(\mathbf{x}', \tau') \rangle = 2Sk_{\mathrm{B}}T\delta(\mathbf{x} - \mathbf{x}')\delta(\tau - \tau')$, where S is the area of the sample film. Finally, for the GL time, we assume $\tau_{\mathrm{GL}} = (T/T_{\mathrm{c}} - 1)^{-1}0.25\,\mathrm{ps}$, consistent with the estimate $\tau_{\mathrm{GL}} = 48\pi\sigma_{\mathrm{DC}}\lambda^2/c^2$ based on temperature-dependent measurements of penetration depth λ and normal-state DC conductivity σ_{DC} [29, 36, 37]. We note that the BCS formula for the single-particle gap implies $\Delta^2 \sim 9.4k_{\mathrm{B}}^2T_{\mathrm{c}}(T - T_{\mathrm{c}})$. Therefore, for our experimental conditions, the nonlinear term in the TDGL equation is very small, and the fluctuations can be considered to a very good approximation as a free classical field. Nonetheless, we account for the nonlinearities with a self-consistent approach.

The self-consistent approximation for the nonlinear term is obtained by the substitution

$$|\psi(\mathbf{x}, \tau)|^2 \rightarrow |\bar{\psi}(\tau)|^2 = \frac{1}{S}\int \mathrm{d}\mathbf{x}\,\langle|\psi(\mathbf{x}, \tau)|^2\rangle, \tag{9.4}$$

where the average is taken with respect to the noise configurations. In Fourier space, the TDGL equation reads

$$\partial_\tau \psi_{\mathbf{k}}(\tau) = -\frac{\psi_{\mathbf{k}}(\tau)}{\tau_{\mathrm{GL}}} - \frac{|\bar{\psi}(\tau)|^2\psi_{\mathbf{k}}(\tau)}{\Delta^2\tau_{\mathrm{GL}}} - Dk^2\psi_{\mathbf{k}}(\tau) - \eta_{\mathbf{k}}(\tau),$$

$$|\bar{\psi}(\tau)|^2 = \frac{1}{S}\sum_{\mathbf{k}}\langle|\psi_{\mathbf{k}}(\tau)|^2\rangle, \tag{9.5}$$

which has the solution

$$\psi_{\mathbf{k}}(\tau) = \psi_{\mathbf{k}}(0)\mathrm{e}^{-\lambda_{\mathbf{k}}(\tau)} - \mathrm{e}^{-\lambda_{\mathbf{k}}(\tau)}\int_0^\tau \mathrm{d}\tau'\,\eta_{\mathbf{k}}(\tau')\mathrm{e}^{\lambda_{\mathbf{k}}(\tau')},$$

$$\lambda_{\mathbf{k}}(\tau) = \int_0^\tau \mathrm{d}\tau'\left(\frac{1}{\tau_{\mathrm{GL}}} + \frac{|\bar{\psi}(\tau')|^2}{\Delta^2\tau_{\mathrm{GL}}} + Dk^2\right). \tag{9.6}$$

Although these equations could in principle be solved directly, it is numerically not convenient to sample over the noise realizations. We can avoid this sampling, provided we are interested only in some particular quantities. In our case, we are interested in the two-point function $\langle \psi^\dagger(\mathbf{x}, \tau)\psi(0, \tau) \rangle$ for all $\mathbf{x}, \tau$, which reads

$$\langle \psi^\dagger(\mathbf{x}, \tau)\psi(0, \tau) \rangle = \frac{1}{S}\sum_{\mathbf{k}\mathbf{k}'}\langle \psi_{\mathbf{k}}^\dagger(\tau)\psi_{\mathbf{k}'}(\tau)\rangle\mathrm{e}^{-\mathrm{i}\mathbf{k}\cdot\mathbf{x}}. \tag{9.7}$$

We can compute the equal time two-point function, taking the average over the noise realizations, to get

$$\langle \psi_{\mathbf{k}}^{\dagger}(\tau)\psi_{\mathbf{k}'}(\tau)\rangle = e^{-\lambda_{\mathbf{k}}(\tau)}e^{-\lambda_{\mathbf{k}'}(\tau)}\left[\psi_{\mathbf{k}}^{\dagger}(0)\psi_{\mathbf{k}'}(0) + 2k_{\mathrm{B}}T\delta_{\mathbf{k},\mathbf{k}'}\int_{0}^{\tau}d\tau'\,e^{2\lambda_{\mathbf{k}}(\tau')}\right]. \quad (9.8)$$

This expression also allows us to derive

$$|\bar{\psi}(\tau)|^{2} = \frac{1}{S}\sum_{\mathbf{k}}e^{-2\lambda_{\mathbf{k}}(\tau)}\left[|\psi_{\mathbf{k}}(0)|^{2} + 2k_{\mathrm{B}}T\int_{0}^{\tau}d\tau'\,e^{2\lambda_{\mathbf{k}}(\tau')}\right]. \quad (9.9)$$

Since for thermal initial conditions there are no cross-correlations between different momentum states, we can see from (9.8) that $\langle \psi_{\mathbf{k}}^{\dagger}(\tau)\psi_{\mathbf{k}'}(\tau)\rangle$ is nonzero only for $\mathbf{k}' = \mathbf{k}$ for all times. We set $g_{\mathbf{k}}(\tau) = \langle \psi_{\mathbf{k}}^{\dagger}(\tau)\psi_{\mathbf{k}}(\tau)\rangle$ and compute the evolution of

$$g_{\mathbf{k}}(\tau) = e^{-2\lambda_{\mathbf{k}}(\tau)}\left[g_{\mathbf{k}}(0) + 2k_{\mathrm{B}}T\int_{0}^{\tau}d\tau'\,e^{2\lambda_{\mathbf{k}}(\tau')}\right],$$

$$\frac{d}{d\tau}g_{\mathbf{k}}(\tau) = -2\left[\frac{1}{\tau_{\mathrm{GL}}} + \frac{|\bar{\psi}(\tau)|^{2}}{\Delta^{2}\tau_{\mathrm{GL}}} + Dk^{2}\right]g_{\mathbf{k}}(\tau) + 2k_{\mathrm{B}}T, \quad (9.10)$$

$$|\bar{\psi}(\tau)|^{2} = \frac{1}{S}\sum_{\mathbf{k}}g_{\mathbf{k}}(\tau).$$

The function $g_{\mathbf{k}}(\tau)$ allows us to obtain the two-point functions that we need. In particular, we obtain the time-dependent coherence length $\xi(\tau,T)$ by fitting $\langle \psi^{\dagger}(\mathbf{x},\tau)\psi(0,\tau)\rangle$ with the exponential function $Ce^{-x/\xi}$. We show in Fig. 9.4 the temporal evolution of $\xi(\tau,T)/\xi_{0}$, where $\xi_{0} = 1.9\,\mathrm{nm}$ is the coherence length in thermal

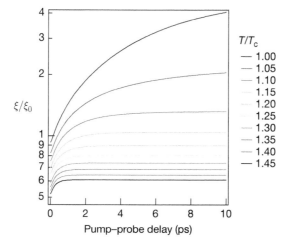

Fig. 9.4 Temporal evolution of the coherence length $\xi(\tau,T)/\xi_{0}$ for different initial temperatures T (T/T_{c} increases from the top to the bottom curve).

equilibrium at zero temperature. The relaxation time of the measured signal $\Delta E(\tau, T)$ is compared with that of the computed quantity $\ln(\xi(\tau, T)/\xi(\infty, T))$ [33]. In order to reproduce the T dependence of ΔE at very early delay ($\tau^* = 50\,\mathrm{fs}$), we add to $A\ln(\xi(\tau^*, T)/\xi(\infty, T))$ an offset accounting for a signal unrelated to superconductivity and experimentally observed above T_o.

As shown by Figs. 9.2(b, c) and 9.3, the calculated $\ln(\xi/\xi_0)$ matches quite well the magnitude of the measured ΔE and the relaxation time for temperatures between $T_c + 5\,\mathrm{K}$ and T_o. Below $T_c + 5\,\mathrm{K}$, the TDGL model overestimates τ_c by 70%. This discrepancy can be explained by an experimental crossover from the sudden-quench regime to an adiabatic regime occurring below T_c [9, 14]. In addition, the experimental curves of Fig. 9.3 follow a power law, whereas the TDGL model predicts exponential relaxation. The observed power law may originate for several reasons. One possibility is an incomplete thermalization of the fast degrees of freedom, thereby violating the sudden-quench hypothesis. In this case, the assumption of white thermal noise will not be accurate, and methods accounting for the feedback of the nonequilibrium bath need to be applied. The unavoidable presence of extrinsic disorder suggests another plausible explanation. Indeed, the nanometer-scale inhomogeneities of the electronic system could be responsible for coarsening phenomena and power-law dynamics. Finally, it is likely that a quasi-conserved mode of the microscopic model efficiently couples to the superconducting order parameter. In this case, the TDGL model (called model A in the Halperin classification scheme) would not be adequate, whereas an effective model belonging to a different Halperin class could give better agreement with experiment [15].

9.6   Odd behavior of underdoped cuprates

Based on the discussed results, the reader may think that simple theories like TDGL are good enough to capture the essential aspects of the critical dynamics. This conclusion may be partially correct for optimally doped samples, but is clearly disproved by the behavior of underdoped BSCCO. Figure 9.5 displays the photoinduced change in THz transmission $\Delta E(\tau)$ measured in an underdoped sample with $T_c = 68\,\mathrm{K}$. The photoinduced change $\Delta E(T, \tau^*)$ extracted from these data is shown in Fig. 9.6(a). We identify the onset of superconducting fluctuations with the upturn of the signal at $T_o = 100\,\mathrm{K}$. Notice that $\Delta E(\tau^*)$ versus T changes slope at $T^* = 200\,\mathrm{K}$. This kink is likely due to the progressive DOS reduction known as the "pseudogap." The value of T^* compares well with the results of other experiments on cuprates with similar doping levels [38]. As shown by Fig. 9.6(b), the relaxation time $\kappa(T, \tau^*)$ exhibits a sudden upturn at the onset temperature $T_o = 100\,\mathrm{K}$, whereas it is insensitive to the crossover at T^*. In contrast to superconductivity, the pseudogap seems to lack any critical behavior. However, it is possible that correlations responsible for the pseudogap formation also affect the dynamics of superconducting fluctuations near the critical point.

Figure 9.6(c) shows the relaxation dynamics of superconducting fluctuations in the underdoped superconductor. Strikingly, as with optimally doped BSCCO, all rescaled

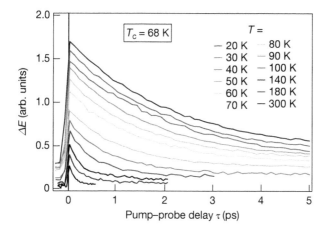

Fig. 9.5 Nonequilibrium data for underdoped BSCCO ($T_c = 68\,\text{K}$). Pump-induced changes $\Delta E(\tau) = E(t^*, -1\,\text{ps}) - E(t^*, \tau)$ in the transmitted MIR electric field as a function of pump–probe delay τ. $\Delta E(\tau)$ scales with the pump-induced change $\text{Re}\,\Delta\sigma(\nu, T)$ in the instantaneous BSCCO conductivity averaged from 15 to 25 THz. T increases from the top to the bottom curve.

curves in the critical region between T_c and T_o are found to obey the power law (9.2). The value $\alpha = 1.2$ provides the best fit to the experimental data and is very similar to the $\alpha = 1.1$ found for optimally doped BSCCO (Fig. 9.3). The experimental evidence that the power law depends only weakly on the doping level is another hint of universality under far-from-equilibrium conditions. In contrast to optimally doped BSCCO (inset of Fig. 9.3), the temperature dependence of τ_c^{-1} is not linear, but approximately follows the function $0.55\,\text{ps}^{-1} + (T/T_c - 1)^{\beta}\,2.8\,\text{ps}^{-1}$ with $\beta = 1.7$ (see the inset of Fig. 9.6(c)). This finding is in contrast to the predictions of the TDGL model and calls for extended nonequilibrium models, perhaps including the coupling of the superconducting order parameter to other conserved densities [15]. Table 9.1 summarizes the critical exponents observed in underdoped and optimally doped BSCCO. The remarkable difference in the β parameter indicates that Gaussian fluctuations account for the experimental behavior in optimally doped compounds but are not accurate in underdoped BSCCO.

9.7 Dynamics below the critical temperature

Below T_c, the scaling law is no longer respected. As shown by Fig. 9.7, rescaled curves acquired for $T < T_c$ can be joined at long timescales but hold different dynamics at early delays. The evolution of $\Delta E(T, \tau)$ follows neither an exponential decay nor a power law. The relaxation time κ is not a monotonic function of temperature and is always near to 2–3 ps. When the system is initially in the superconducting phase, the order parameter has a finite mean value and follows adiabatically the energy relaxation of quasiparticles [9, 14]. In this case, the coherent gap $\Delta(\tau)$ can be associated with the

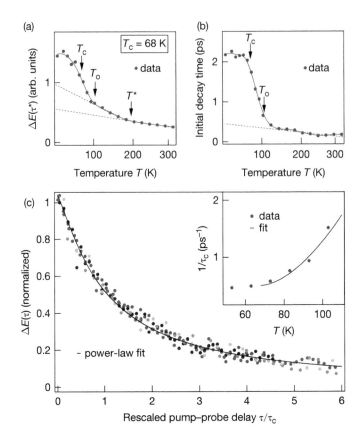

Fig. 9.6 (Color online: see [1] and arXiv:1407.7689) Temperature dependences of $\Delta E(T, \tau^*)$ (a) and the relaxation time $\kappa(T, \tau^*)$ (b) obtained for a pump–probe delay $\tau^* = 50\,\text{fs}$ (circles). The transition temperature $T_c = 68\,\text{K}$, the onset of superconducting fluctuations $T_o = 100\,\text{K}$, and the pseudogap temperature $T^* = 200\,\text{K}$ are marked by arrows. (c) Experimental $\Delta E(T, \tau)$ measured for $T = 68, 78, 88, 98$, and $108\,\text{K}$, renormalized to the maximum value and rescaled by a factor $\tau_c(T)$. The full curve shows the power-law fitting of the experimental data. The inset plots the factor $\tau_c^{-1}(T)$ derived from the experiment (circles) and the fit $\tau_c^{-1} = 0.55 + 2.8(T/T_c - 1)^{1.7}\,\text{ps}^{-1}$ (full line). The data in this figure were acquired on underdoped BSCCO with transition temperature $T_c = 68\,\text{K}$. (Reprinted with permission from [1]. Copyright (2015) by the American Physical Society.)

Table 9.1 Critical parameters in BSCCO under far-from-equilibrium conditions

Parameter	T_c (K)	T_o/T_c	α	β
Optimally doped BSCCO	91	1.4	1.1	1
Underdoped BSCCO	68	1.5	1.2	1.7

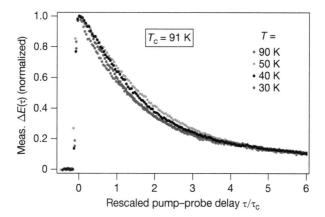

Fig. 9.7 The pump-induced signal $\Delta E(\tau)$ measured in the optimally doped sample is shown as a function of rescaled pump–probe delay $\tau/\tau_c(T)$ at temperatures $T = 30$, 40, 50, and 90 K. Curves are also normalized to the maximum signal value.

complex field $\langle \psi_0(\tau) \rangle$, and the TDGL equation is no longer applicable. The adiabatic regime is defined by the condition

$$\hbar \frac{\mathrm{d}}{\mathrm{d}\tau} \Delta < \Delta^2, \tag{9.11}$$

which is verified by our experimental data for $\Delta > 2\,\mathrm{meV}$. On large lengthscales, the system remains out of equilibrium and displays a coarsening-like dynamics [13]. The recovery of superconductivity is ruled by electron–phonon coupling and phonon–phonon coupling in the weak- and strong-bottleneck regimes, respectively [25]. Since the 2–3 ps dynamics coincides with the anharmonic decay time of optical phonons [28], our system is likely in the strong-bottleneck regime.

When the system approaches the critical point from below, the adiabatic conditions will inevitably break down. Keeping this limitation in mind, we recall that within the adiabatic approximation, it is still possible to write an equation for the temporal evolution of the gap. The relaxation time entering in this equation reflects the energy dissipation of quasiparticles with excess energy $\sim\!\Delta$ and it is expected to increase near T_c. As shown by Fig. 9.2, the relaxation time $\kappa(\tau^*, T)$ of optimally doped samples increases for $T \to T_c$ from below. This upturn has been already observed in pump–probe experiments in the visible spectral range [17] and has been phenomenologically modeled by a quasiparticle dissipation rate that depends critically on the value of Δ. We also notice that the upturn of the relaxation time near to T_c is absent in the underdoped compound. Note that although local quantities have an adiabatic evolution, on large lengthscales the system is expected to remain out of equilibrium and display a coarsening-like dynamics [16].

9.8 Conclusions

In conclusion, our work reveals the temporal evolution of superconducting fluctuations in high-temperature superconductors. We have monitored the incipient condensate by detecting the transmitted THz field of a photoexcited sample. The relaxation of MIR transmission follows a power law $1/[1 + (\tau/\tau_c)^\alpha]$ that is nearly independent of the doping level. We observe a critical scaling $1/\tau_c \cong (T - T_c)^\beta$ in a Ginzburg region that covers roughly 50% of the transition temperature. Table 9.1 shows the critical exponents that have been measured on two samples with different oxygen concentrations. Above $1.5T_c$, no signature of preformed Cooper pairs and superconducting correlations can be detected. Measurements on underdoped samples clearly resolve the onset temperature of the superconducting correlations from the pseudogap temperature T^*. The dynamics of the electromagnetic response does not exhibit any critical behavior at $T \cong T^*$. Therefore, the pseudogap likely arises from a crossover rather than a continuous phase transition.

Most interestingly, the time-resolved THz data can be employed to quantify the deviations of high-temperature superconductors from mean-field predictions. The value $\beta = 1$ observed in optimally doped compounds is compatible with Gaussian fluctuations of the order parameter. In contrast, the $\beta > 1$ value of underdoped BSCCO indicates that interactions among fluctuations become relevant when approaching the antiferromagnetic Mott phase. The TDGL model reproduces some of the experimental results, but suffers from inherent problems. Since the nonlinear term is small, our self-consistent simulations are always near to the Gaussian limit and do not account for the dependence of β on doping level. Moreover, the predicted dynamics is nearly exponential and far from a power law. We propose several explanations for the observed power law $1/[1+(\tau/\tau_c)^\alpha]$. Among other possibilities, we suggest that a quasi-conserved quantity generates a low-energy mode that couples to the order parameter. In this case, a purely dissipative model like TDGL would not be able to capture the correct dynamics. Our approach opens new perspectives in the field of critical systems far from equilibrium. Future work in this direction may identify the parameters and experimental protocols that define universal properties of high-temperature superconductors. Advances in these directions would also be beneficial for the understanding of quantum quenches obtained by phonon pumping [22].

References

[1] L. Perfetti, B. Sciolla, G. Biroli, C. J. van der Beek, C. Piovera, M. Wolf, and T. Kampfrath. *Phys. Rev. Lett.* **114**, 067003 (2015).
[2] F. Rullier-Albenque, H. Alloul, and G. Rikken. *Phys. Rev. B* **84**, 014522 (2011).
[3] J. Corson, R. Mallozzi, J. Orenstein, J. N. Eckstein, and I. Bozovic, *Nature* **398**, 221 (1999).
[4] L. S. Bilbro, R. Valdés Aguilar, G. Logvenov, O. Pelleg, I. Bozovic, and N. P. Armitage. *Nature Phys.* **7**, 298 (2011).
[5] N. Bergeal, J. Lesueur, M. Aprili, G. Faini, J. P. Contour, and B. Leridon. *Nature Phys.* **6**, 296 (2010).

[6] Z. Tesanovic, L. Xing, L. Bulaevskii, Q. Li, and M. Suenaga. *Phys. Rev. Lett.* **69**, 3563 (1992).

[7] Y. Wang, L. Li, M. J. Naughton, G. D. Gu, S. Uchida, and N. P. Ong. *Phys. Rev. Lett.* **95**, 247002 (2005).

[8] Y. Wang, L. Li, and N. P. Ong. *Phys. Rev. B* **73**, 024510 (2006).

[9] M. Tinkham, *Introduction to Superconductivity* (McGraw-Hill, New York, 1996).

[10] A. Larkin and A. Valarmov. *Theory of Fluctuations in Superconductors* (Oxford University Press, Oxford, 2005).

[11] M. Cyrot. *Rep. Prog. Phys.* **36**, 103 (1973).

[12] L. P. Gorkov. *Sov. Phys. JETP* **9**, 1364 (1959).

[13] A. Bray. *Adv. Phys.* **51**, 481 (2002).

[14] A. Polkovnikov, K. Sengupta, A. Silva, and M. Vengalattore. *Rev. Mod. Phys.* **83**, 863 (2011).

[15] P. C. Hohenberg and B. I. Halperin. *Rev. Mod. Phys.* **49**, 435 (1977).

[16] G. Lamporesi, S. Donadello, S. Serafini, F. Dalfovo, and G. Ferrari. *Nature Phys.* **9**, 656 (2013).

[17] J. Demsar, B. Podobnik, V. V. Kabanov, Th. Wolf, and D. Mihailovic. *Phys. Rev. Lett.* **82**, 4918 (1999).

[18] C. Giannetti et al. *Nature Commun.* **2**, 353 (2011).

[19] I. Madan, T. Kurosawa, Y. Toda, M. Oda, T. Mertelj, P. Kusar, and D. Mihailovic. *Sci. Rep.* **4**, 5656 (2014).

[20] R. D. Averitt, G. Rodriguez, A. I. Lobad, J. L. W. Siders, S. A. Trugman, and A. J. Taylor. *Phys. Rev. B* **63**, 140502 (2001).

[21] R. A. Kaindl, M. A. Carnahan, D. S. Chemla, S. Oh, and J. N. Eckstein. *Phys. Rev. B* **72**, 060510 (2005).

[22] W. Hu et al. *Nature Mater.* **13**, 705 (2014).

[23] M. A. Carnahan, R. A. Kaindl, J. Orenstein, D. S. Chemla, S. Oh, and J. N. Eckstein. *Physica C* **408**, 729 (2004).

[24] A. Pashkin et al. *Phys. Rev. Lett.* **105**, 067001 (2010).

[25] V. V. Kabanov, J. Demsar, and D. Mihailovic. *Phys. Rev. Lett.* **95**, 147002 (2005).

[26] G. P. Segre, N. Gedik, J. Orenstein, D. A. Bonn, R. Liang, and W. N. Hardy. *Phys. Rev. Lett.* **88**, 137001 (2002).

[27] R. A. Kaindl et al. *Science* **287**, 470 (2000).

[28] L. Perfetti, P. A. Loukakos, M. Lisowski, U. Bovensiepen, H. Eisaki, and M. Wolf. *Phys. Rev. Lett.* **99**, 197001 (2007).

[29] M. A. Quijada, D. B. Tanner, R. J. Kelley, M. Onellion, H. Berger and G. Margaritondo. *Phys. Rev. B* **60**, 14917 (1999).

[30] D. van der Marel et al. *Nature* **425**, 271 (2003).

[31] D. N. Basov and T. Timusk. *Rev. Mod. Phys.* **77**, 721 (2005).

[32] S. G. Sharapov and J. P. Carbotte. *Phys. Rev. B* **72**, 134506 (2005).

[33] F. Federici and A. A. Varlamov. *Phys. Rev. B* **55**, 6070 (1997).

[34] C. L. Smallwood et al. *Science* **336**, 1137 (2012).

[35] Q. Du and P. Gray. *SIAM J. Appl. Math.* **56**, 1060 (1996).

[36] J. Mosqueira, E. G. Miramontes, C. Torron, J. A. Campa, I. Rasines, and F. Vidal. *Phys. Rev. B* **53**, 15272 (1996).

[37] S. Martin, A. T. Fiory, R. M. Fleming, G. P. Espinosa, and A. S. Cooper. *Phys. Rev. Lett.* **62**, 677 (1989).

[38] M. R. Norman, D. Pines, and C. Kallin. *Adv. Phys.* **54**, 715 (2005).

10
Exact field-theoretic methods in out-of-equilibrium quantum impurity problems

Hubert SALEUR

Institut de Physique Théorique
CEA/DSM-CNRS/URA 2306,
CEA Saclay, F-91191 Gif-sur-Yvette Cedex,
France

Strongly Interacting Quantum Systems out of Equilibrium. First Edition. Thierry Giamarchi et al.
© Oxford University Press 2016. Published in 2016 by Oxford University Press.

Chapter Contents

Preface

The goal of this chapter is to explore exact solutions of some low-dimensional out-of-equilibrium transport problems in the presence of interactions. The approach I will use focuses on the low-energy limit, where the results are expected to obey some sort of universality, and the main tools will accordingly be field-theoretic. Embarrassingly, this chapter has a nonzero overlap with the lectures that I gave at Les Houches in 1998.[1] This is not to say that the field has not made progress in the last 15 years or so. The point, rather, is that the simplest examples still remain the best to explain the basic concepts.

While this chapter is closely related with that of Natan Andrei (Chapter 5) in spirit, we tend to attack problems from opposite ends, which we like to call "bare" and "dressed." I will mostly remain dressed, and explain what this means below.

Before starting, let me emphasize that most of the introductory material is of course not mine. This includes the formulation of the "Kane and Fisher" problem, bosonization, conformal perturbation theory, and more, for which, because of lack of space, I am not giving a full set of references.

10.1   The low-energy field-theoretic limit

Let me start by describing the experimental set-up we will mostly be thinking of in this chapter (Fig. 10.1). We consider a clean fractional quantum Hall effect (FQHE) sample at a filling fraction $\nu = \frac{1}{3}$ (say), and imagine inducing some tunneling between the two edges by introducing a gate voltage. What happens physically will be discussed below: roughly, as we increase the gate voltage V_g, there will be more and more backscattering between the edges, so the current will be given by a complicated formula and will differ from the ideal $I = (e^2/3h)V$. Here, V is the voltage difference between the two edges, which one can take conveniently to be at potential $\pm\frac{1}{2}V$.

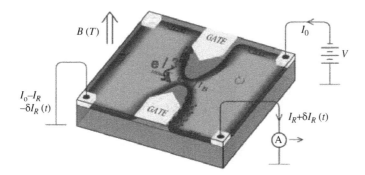

Fig. 10.1   A schematic view of the experimental set-up. The arrow points to chiral excitations.

---

[1] H. Saleur. Lectures on non perturbative field theory and quantum impurity problems. In *Topological Aspects of Low Dimensional Systems. Les Houches Session LXIX* (ed. A. Comtet, T. Jolicoeur, S. Ouvry, and F. David), p. 473. Springer-Verlag, Berlin (1999) [arXiv:cond-mat/9812110].

Now, the point is that, of course, this formula will be *very* complicated, and will depend on a lot of things, especially details of the shape of the gate, the distribution of potential in its vicinity, the induced effect on the electron gas, etc. While all these details are worth studying—especially if one is interested in understanding precisely *all* experimental results—they are typically too cumbersome to be addressed by anyone aiming at an exact solution. More to the point, the details are not essential to the qualitative physics we may be after—here, as we will see, the transmutation of Laughlin quasiparticles into electrons, and what this means, in particular, for generalized fluctuation–dissipation theorems. Finally, the details are—of course, since they are details—not universal; that is, they will differ from sample to sample, from experiment to experiment, etc. What we would like to do is concentrate on effects that are sufficiently universal that they can be *quantitatively* checked and explored, independently of these rather uncontrollable details. The hope is that by focusing on these effects, we do not lose any essential physics.

The trick to get rid of the details, what needs to be done, is to focus on low-energy scales where they cannot be seen, at least directly. Of course, we have to hope that there do indeed exist such energy scales where physics nevertheless remains interesting. In the problem at hand, it is well known that there exists a highly interesting limit obtained when the temperature T, the applied voltage V, and the gate voltage V_g are all "very small," while their ratios remain arbitrary. How "small" these have to be is better understood if one uses the language of the renormalization group, to which we turn a little later.

Before launching into this, I would like to point out that the problem we are about to study is equivalent, in the low-energy limit, to several other problems of interest. For instance, it maps onto the problem of transport between two Luttinger liquids connected by weak tunneling, or onto the problem of transport through a Luttinger liquid with a potential scattering term. A model for one of these problems would be given by the Hamiltonian

$$H = - \left(\sum_{i=-\infty}^{-1} + \sum_{i=1}^{\infty} \right) \left[J\left(c_i^\dagger c_{i+1} + \text{h.c.}\right) + J_z\left(c_i^\dagger c_i - \tfrac{1}{2}\right)\left(c_{i+1}^\dagger c_{i+1} - \tfrac{1}{2}\right) \right]$$

$$- t\left[J\left(c_0^\dagger c_1 + \text{h.c.}\right) + J_z\left(c_0^\dagger c_0 - \tfrac{1}{2}\right)\left(c_1^\dagger c_1 - \tfrac{1}{2}\right) \right] \tag{10.1}$$

(there exists a simple relationship between J_z/J and the filling fraction in the Hall problem). The point I want to make is that when modeling a real physics problem—such as a "real" quantum wire or an FQHE device—with a Hamiltonian such as (10.1), one chooses implicitly a high-energy cutoff, namely, the bandwidth (proportional to J), or, equivalently, a short-distance cutoff, namely, the lattice spacing. One may then, for instance, study transport on this model through a time-dependent density-matrix renormalization group (DMRG) approach, obtaining, at $T = 0$, some I–V curves depending on the amplitude t. It is only in the low-energy limit—where all excitations involved are much smaller than the bandwidth—that one can hope to compare the obtained results with those of a field-theoretic calculation or with those obtained from experiments on FQHE samples.

The "bare" approach starts usually from some sort of microscopic model that provides a good deal of physical insight and good control of the main ingredients. The drawback, in my view, is that the results cannot, without a lot of extra work, be connected with those of a field-theoretic calculation, or even with those of a numerical simulation. The reason is that the universal limit, where one has got rid of the details by focusing on low-energy excitations, has to be extracted from the calculations, and this is sometimes a seemingly undoable task. If one does not extract the universal limit, one is left with results that do depend on the details—such as the cutoff procedure— and this dependence is big enough to prevent serious comparison with other results. For instance, the bare Bethe ansatz most often uses a cutoff in rapidity space for bare Bethe excitations. This is somewhat "similar" but in details *different* from, for example, a short-distance cutoff in real space. The "dressed" approach, on the other hand, starts directly from a renormalized field-theoretic formulation, and is already formulated within a universal limit. Its drawback is that it is quite abstract and formal, and sometimes, its results seem to be pulled out of a hat.

It is more comfortable to think of these ideas within a renormalization group (RG) framework. The systems we will be interested in will typically involve some bulk gapless degrees of freedom to which some sort of local interaction has been added—the tunneling term in the FQHE problem, for instance. The bulk problem itself is therefore at a fixed point of the RG, and is described in terms of a fixed-point Hamiltonian. The interaction will be considered as a local perturbation, which we can formally expand in terms of a bunch of operators, some relevant, some marginal or irrelevant. The "details" we were discussing earlier affect the coefficients of all these operators. The low-energy physics is the physics obtained after iterating the RG further to kill most operators—but, typically, the most relevant one.

To make things more concrete, I now move to the corresponding description of the FQHE tunneling problem. Using the well-known result that low-energy excitations at the edge of a $\nu = 1/(2n+1)$ Hall sample can be described in terms of a chiral boson, it is easy to obtain the Hamiltonian for our problem:

$$H = \int_{-\infty}^{\infty} \left[(\partial_x \phi_L)^2 + (\partial_x \phi_R)^2\right] + \lambda : \cos\left[\frac{\beta}{\sqrt{2}}(\phi_L + \phi_R)(0)\right] : + \dots, \qquad (10.2)$$

where $\beta^2/8\pi = \nu$. The bulk term can be written as

$$H_0 = \frac{1}{2} \int_{-\infty}^{\infty} \left[:(\Pi^2): + :(\partial_x \phi)^2:\right], \qquad (10.3)$$

with the commutators $[\phi(x,t), \Pi(x',t)] = i\delta(x - x')$, where $\phi = \phi_R + \phi_L$ and $\Pi = -\partial_x(\phi_R - \phi_L)$. The propagator in the free theory is

$$\langle \phi(x,t)\phi(0,0)\rangle = -\frac{1}{4\pi} \ln\{\mu^2[i(t-x) + \epsilon][i(t+x) + \epsilon]\}, \qquad (10.4)$$

with $\mu = 1/L$, where L is a quantity that has the dimension of length—in fact, the size of the system. Also, this expression is valid only at large distances (large compared with the cutoff if one uses a lattice regularization), so one has, in particular,

$$\langle \phi(x,t)^2 \rangle = -\frac{1}{4\pi} \ln(a^2 \mu^2). \tag{10.5}$$

This means in particular that fluctuations grow unbounded with the size of the system, a result profoundly related to the absence of spontaneous symmetry breaking in two dimensions (e.g., the absence of crystalline ordering). Note that I have used the normal-ordered cosine in the Hamiltonian. Its dimension is length$^{-\nu}$, so λ has dimension $[\lambda] = \text{length}^{\nu-1}$. The tunneling term describes the tunneling of Laughlin quasiparticles of charge ν. Using the fact that $[\phi_R(x,t), \phi_R(x',t)] = \frac{1}{4}i\,\text{sign}(x-x')$, this means that the charge operator on one edge is

$$Q_R = \frac{\beta}{2\sqrt{2}\,\pi} \int_{-\infty}^{\infty} \partial_x \phi_R, \tag{10.6}$$

in units of the electron charge.

Obtaining the Hamiltonian (10.2) from the Luttinger liquid is a well-known but slightly complicated story: it involves bosonization, followed by the important physical observation that, owing to interactions, the left- and right-moving physical excitations in the Luttinger liquid are complex combinations of the "natural" left and right electronic excitations, exhibiting in particular fractional charge. I will not discuss this here.

A more important point for us is the dots "..." in (10.2): they represent less relevant and irrelevant operators, and we will neglect them—but see below.

We can now write the RG equation for the perturbation:

$$\frac{d\lambda}{d\ln a} = (1-\nu)\lambda + O(\lambda^3). \tag{10.7}$$

This shows that λ is a relevant perturbation, whose effect therefore increases at low energy.

The most tempting way to get a handle on this problem is to do some perturbation expansion. We start by calculating the partition function $Z = \text{Tr}\, e^{-H/T}$, with T the temperature (I do not use β since it is reserved for the coupling). Going to imaginary time $y = it$ (see Fig. 10.2), we easily find

$$\frac{Z(\lambda)}{Z(0)} = 1 + \sum_{n=0}^{\infty} \frac{1}{(2n)!} \lambda^{2n} \int_0^{1/T} dy_1 \cdots dy_{2n} \left\langle \cos\left[\frac{\beta}{\sqrt{2}}\phi(y_1)\right] \cdots \cos\left[\frac{\beta}{\sqrt{2}}\phi(y_{2n})\right] \right\rangle. \tag{10.8}$$

This can be rewritten as

$$\frac{Z(\lambda)}{Z(\lambda=0)} = 1 + \sum_{n=1}^{\infty} (\tilde{\lambda})^{2n} I_{2n}, \tag{10.9}$$

Fig. 10.2 The problem in imaginary time.

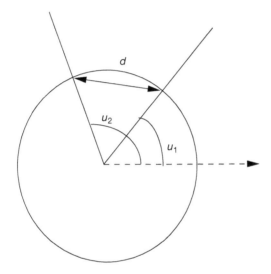

Fig. 10.3 Charges of a two-dimensional Coulomb gas that move on a circle, with $d = 2 \sin \frac{1}{2}(u_2 - u_1)$.

where the dimensionless coupling is

$$\tilde{\lambda} = \frac{\lambda}{2T}(2\pi T)^{\nu} \tag{10.10}$$

and the integrals are

$$I_{2n} = \frac{1}{(n!)^2} \int_0^{2\pi} du_1 \cdots \int_0^{2\pi} du'_{2n} \left| \frac{\prod_{i<j} 4 \sin \frac{1}{2}(u_i - u_j) \sin \frac{1}{2}(u'_i - u'_j)}{\prod_{i,j} 2 \sin \frac{1}{2}(u_i - u'_j)} \right|^{2\nu}. \tag{10.11}$$

This is the partition function of a classical Coulomb gas in two space dimensions, with the charges moving on a circle of unit radius (see Fig. 10.3).

The integrands have small-distance behaviors $1/u^{2\nu}$. It follows that there is no short-distance divergence, and the integrals are all finite for $\nu < \frac{1}{2}$ (there are never large-distance divergences here, since we have a temperature). When $\nu > \frac{1}{2}$, the integrals have divergences. Field theorists often like to handle these by analytic continuation—in the case $n = 1$, for instance, one finds easily

$$I_2 = \frac{\Gamma(1 - 2\nu)}{\Gamma^2(1 - \nu)}. \tag{10.12}$$

This can then be continued beyond $\nu = \frac{1}{2}$ simply by using the known continuation of Γ to negative arguments. How to do this in the case of arbitrary n is a bit more tricky. In fact, it turns out that the integrals can all be evaluated in closed form, using some beautiful mathematics that I will not discuss here (see the Notes at the end of this section for references).

To make things simple we will just focus on the case $\nu < \frac{1}{2}$, where the perturbative expansion for the free energy is well defined, giving a series in λ with positive coefficients. A very important point is that the series depends on the variable $\tilde{\lambda} \propto \lambda T^{\nu-1}$. We can trade this variable for the ratio of the temperature T and a crossover temperature that we will call T_B: $\tilde{\lambda} \propto (T/T_B)^{\nu-1}$, with

$$T_B \propto \lambda^{1/(1-\nu)}, \tag{10.13}$$

which can be checked directly to have the right dimension. This parameter is the only one on which the field theory results will depend (except for external variables such as T and V). The point now is that results for different regularizations or microscopic models will be comparable to the field theory results only after their scaling limit has been taken, so their properties depend only on a single, energy-like parameter. To reach this limit, one has to send all bare parameters to zero while the energy cutoff is sent to infinity: T_B plays a role analogous to the Kondo temperature in the Kondo problem.

Technical note

I have chosen to work with the normal-ordered cosine interaction, so the field theory has no cutoff dependence at all. The only renormalization that was necessary was taken care of by the normal ordering prescription. In a physical problem, the bare parameter accessible has the dimension of an energy (i.e., an inverse length), and, typically, we will have

$$\lambda \propto V_g D^{-\nu}, \tag{10.14}$$

where D is a energy cutoff—the bandwidth. The variable λ is finite, while we want to let $V_g \to 0$ and $D \to \infty$. We have here an example of *dimensional transmutation*: the physical problem had a bare coupling V_g of dimension length^{-1}, the scaling limit has a coupling of the same dimension, but the two are not proportional, since their relation involves the cutoff.

Going back to our series, it will presumably have a finite radius of convergence, although one does not expect the appearance of a singularity on the positive real axis (this would correspond to the existence of a phase transition on the one-dimensional boundary), so the singularity must be somewhere in the complex plane. Beyond this radius, some other technique has to be used to understand quantitatively what happens. It is possible to argue what the leading behavior of the partition function at large λ should be. Indeed, Z actually depends only on the ratio of the two energy scales T_B and T, so large λ is like small temperature. But small temperature corresponds, going back to a Euclidean description, to a cylinder of large diameter. In this limit, the partition function per unit length of the boundary should have a well-defined, "thermodynamic" limit, so Z should go as $Z \propto \exp(\text{const} \times T_B/T)$. This means our perturbative series has to go as $\exp(\text{const} \times \tilde{\lambda}^{1/(1-\nu)})$.

Notes

For the calculation of the integrals, you need to learn about Jack polynomials. See P. Fendley, F. Lesage, and H. Saleur. Solving 1D plasmas and 2D boundary problems using Jack polynomials and functional relations. *J. Stat. Phys.* **79** (1995) 799 [arXiv:hep-th/9409176]. This reference also contains a discussion of what happens for $\nu > \frac{1}{2}$, in particular for $\nu \to 1$.

10.2   Perturbation near the IR fixed point

The first *nonperturbative* problem we encounter is the physics at large λ. While in the present case it is easy to guess the right result, this is not a general feature at all—often, the low-energy physics can only be understood after a lot of thinking and nonperturbative arguments, the type of which we will discuss below.

In the present problem, however, the obvious guess is the right one: at large λ, the field is pinned to the minima of the cosine potential. In terms of the FQHE picture, this corresponds to the limit where the sample is cut in half by the gate voltage. In RG terms, this is the stable fixed point, reached at low energy. To express this better in field-theoretic language, it is convenient to do a few manipulations. The rough idea is that we have too many degrees of freedom to start with: left- and right-moving bosons on both sides of the impurity, which is really like having two half-infinite nonchiral systems connected at the origin. We can "unfold" the two half-infinite systems into two infinite chiral ones of the same chirality (pretty much the experimental set-up, after mirror imaging the right-moving edge, say). Then we form odd and even combinations, and get rid of the odd one since only the even one sees the interaction. We then fold back to get a boundary problem. In formulas, we decompose $\phi = \phi_L + \phi_R$ and set

$$\varphi^e(x+t) = \sqrt{\tfrac{1}{2}}\,[\phi_L(x,t) + \phi_R(-x,t)],$$

$$\varphi^o(x+t) = \sqrt{\tfrac{1}{2}}\,[\phi_L(x,t) - \phi_R(-x,t)].$$

(10.15)

Observe that these two fields are left-movers. We now fold the system by setting, for $x < 0$,

$$\phi_L^e = \varphi^e(x+t), \quad \phi_R^e = \varphi^e(-x+t),$$

$$\phi_L^o = \varphi^o(x+t), \quad \phi_R^o = -\varphi^o(-x+t),$$

(10.16)

and introduce new fields $\phi^{e,o} = \phi_L^{e,o} + \phi_R^{e,o}$, both defined on the half-infinite line $x < 0$. The odd field ϕ^e simply obeys Dirichlet boundary conditions at the origin $\phi^o(0) = 0$, and decouples from the problem. The field ϕ^e, which we will rename as Φ in the following, has a nontrivial dynamics

$$H \equiv H^e = \frac{1}{2} \int_{-\infty}^{0} dx \left[\Pi^2 + (\partial_x \Phi)^2\right] + \lambda \cos\left[\frac{\beta \Phi(0)}{2}\right]. \qquad (10.17)$$

Technical remark

A common objection to the following manipulations is that they are good only for free fields, but not when there is a boundary interaction. This, in fact, depends on what one means by "fields"—here, what we have in mind is that we are doing perturbation theory in λ. Then, all the quantities are evaluated within the free theory, on which one can legitimately do all the foldings, left–right decompositions, etc.

We can now go back to our RG interpretation: the field $\Phi(0)$ sees Neumann boundary conditions at high energy (in the UV) and Dirichlet boundary conditions at low energy (in the IR). The physics of our problem is mirrored in a flow from Neumann to Dirichlet.

Going back now to the vicinity of the Dirichlet fixed point, physics suggests that the leading irrelevant operator describing the ultimate pinching of the problem should be the operator tunneling electrons between the two halves of the sample, as illustrated in Fig. 10.4.

We can try to substantiate this claim somewhat more formally. First, we start by considering the model where the bulk degrees of freedom have been integrated out, leading to the action (at zero temperature, which makes the formulas a bit simpler)

$$S = \frac{1}{4\pi} \int dy \, dy' \left[\frac{\Phi(y) - \Phi(y')}{y - y'}\right]^2 + \lambda_0 \int dy \cos\left[\frac{\beta}{2}\Phi(y)\right] + \frac{m}{2} \int dy \, (\partial_y \Phi)^2, \qquad (10.18)$$

where we have added an irrelevant term to make some integrals finite. Note that we have reverted to a non-normal-ordered cosine, in order to treat it as a real function when taking the saddle point. This means that $\lambda_0 = \lambda a^\nu$.

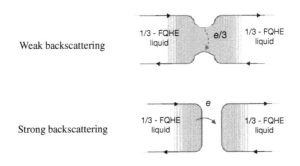

Fig. 10.4 The vicinity of the UV and IR fixed points.

We can find kinks interpolating between adjacent vacua and satisfying the equations of motion

$$m\partial_y^2\Phi = -\frac{\lambda_0\beta}{2}\sin\left(\frac{\beta}{2}\Phi\right).$$

(10.19)

A simple solution of this equation is

$$\Phi \equiv f_{\text{ins}}(y) = \frac{2\pi}{\beta} + \frac{8}{\beta}\tan^{-1}\left[\exp\left(\frac{\beta}{2}\sqrt{\frac{\lambda_0}{m}}\,y\right)\right].$$

(10.20)

The energy of this kink is infinite, but can be made finite by subtracting a constant term from the action, replacing the $\cos(\beta\Phi/2)$ by $\cos(\beta\Phi/2)+1$. If we then consider a configuration of the field Φ made of a superposition of far-apart instantons and anti-instantons (Fig. 10.5),

$$\Phi = \sum_j \epsilon_j f_{\text{ins}}(y - y_j),$$

(10.21)

then the kinetic term of the action can be conveniently evaluated by Fourier transform:

$$S_{\text{kin}} = \int \frac{d\omega}{2\pi}\,|\widehat{\Phi}(\omega)|^2\,|\omega|,$$

(10.22)

where, in general, we define

$$\widehat{f}(\omega) = \int \frac{dy}{2\pi}\,f(y)e^{i\omega y}.$$

(10.23)

Our task is then to calculate the interaction between the instantons, which we suppose to be far enough apart that our ansatz for Φ is accurate. The interaction then comes entirely from the $1/y^2$ term that we have neglected so far. Using the fact that

$$\widehat{f}_{\text{ins}}(\omega) = \frac{4\pi}{\beta}\,\frac{1}{\cosh\left(\dfrac{\pi\omega}{\beta}\sqrt{\dfrac{m}{\lambda_0}}\right)},$$

(10.24)

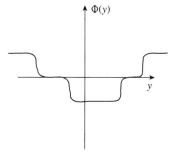

Fig. 10.5 A sequence of instantons and anti-instantons.

one finds, at large distances

$$S_{\text{kin}} \approx \frac{16\pi}{\beta^2} \sum_{j<k} \epsilon_j \epsilon_k \ln |y_j - y_k|. \tag{10.25}$$

That it is a logarithmic interaction should not be surprising. Indeed, after integration by parts, we can write the kinetic term

$$S_{\text{kin}} = \frac{1}{2\pi} \int_{y<y'} \partial_y \Phi(y) \partial_y \Phi(y') \ln(y - y'), \tag{10.26}$$

so we expect S_{kin} to be proportional to the logarithm of the distance between the centers of the instantons, where the field derivative is maximal.

Now the expression for S_{kin} is in exact correspondence with the Coulomb gas expansion discussed previously, but with the exchange $\beta^2/8\pi \to 8\pi/\beta^2$. It follows that the IR action reads, at leading order,

$$S \approx \frac{1}{2} \int_{-\infty}^{0} dx \int dy \left[(\partial_x \tilde\Phi)^2 + (\partial_y \tilde\Phi)^2 \right] + \lambda_d \int dy : \cos\left[\frac{4\pi}{\beta} \tilde\Phi(0, y) \right] :, \tag{10.27}$$

where we have to take the dual $\tilde\Phi = \phi_R - \phi_L$, since Φ itself is pinned to the minima of the potential at the Dirichlet fixed point (meanwhile, $\Phi = \phi_R + \phi_L$; in this chapter, we use the same notation as for the "initial boson," hopefully without creating confusion). It follows that the IR fixed point is approached along an operator of dimension $h = 8\pi/\beta^2 = 1/\nu$. To get λ_d, one needs to work out the fugacity of the kinks. By dimensional analysis, however, it is clear that $\lambda_d \propto \lambda^{-1/\nu}$. Finally, using our initial normalization for the charge operator and following it through the mappings would show that, indeed, the cosine operator corresponds to electron tunneling.

It is important to stress now that while the flow away from the UV fixed point is fully specified by a single perturbing term, the situation is very different for the approach toward the IR fixed point. Of course, one is free if one wishes to perturb the Dirichlet boundary conditions by a single irrelevant operator as represented in (10.27), although of course one has to be especially careful in defining the corresponding theory because of the strong short-distance divergences in the integrals (the perturbation has dimension $1/\nu > 2$ in our case where $\nu < \frac{1}{2}$). The point is that there is only *one* particular way of approaching the IR fixed point that corresponds to the trajectory originating at our UV fixed point. This means that the large-λ behavior of the series we are interested in should correspond to an action of the form

$$S = \frac{1}{2} \int_{-\infty}^{0} dx \int dy \left[(\partial_x \tilde\Phi)^2 + (\partial_y \tilde\Phi)^2 \right]$$

$$+ \lambda_d \int dy \left\{ \cos\left[\frac{4\pi}{\beta} \tilde\Phi(0, y) \right] + \sum_k \lambda^{(h_k - 1)/(\nu - 1)} O_k(y) \right\}, \tag{10.28}$$

where the O_k belong to a very large class of operators allowed by symmetry: for instance, there are all the $\cos[(n/2\beta)\tilde{\Phi}]$, $(\partial\tilde{\Phi})^2$, and many others. Since all these operators come with appropriately scaled powers of the coupling constant, they all give contributions to physical properties that depend on our single scaling variable $\tilde{\lambda}$, and no operator can be discarded a priori.

To proceed, we write the Hamiltonian

$$H = \frac{1}{2}\int_{-\infty}^{0} dx\left[(\partial_x\tilde{\Phi})^2 + \Pi^2\right] + \lambda_d \cos\left[\frac{4\pi}{\beta}\tilde{\Phi}(0)\right] + \sum_k \lambda^{(h_k-1)/(g-1)}O_k(0). \quad (10.29)$$

Now, comparing with the beginning of this chapter, the reader should ask why we did not add that collection of operators near the UV fixed point as well. The point is that we had control of what we wanted to do near the UV fixed point, and only a maniac would want to use such an unrealistically finely tuned combination of operators to perturb a fixed point (not to mention that it would be impossible to realize this in an experiment). However, we have *no* control about the way the IR fixed point is approached: this is entirely determined by the dynamics of the quantum field theory, and it turns out to be quite complicated. It is important in particular to realize that, starting from (10.29) and trying to go against the RG flow, there is, most probably, only *one* choice of IR perturbation that would get back to our UV fixed point.

Several remarks are in order about this. First, "going back" from IR to UV would be really going back against the RG flow, something that seems essentially impossible from an entropic point of view—unless we have some additional information about a system, we cannot really go back to the degrees of freedom we integrated over if we are given the renormalized action only. There is a better way to state this: a theory such as (10.29) is *not renormalizable*. This means the following. If we started with only the leading irrelevant operator $:\cos[(4\pi/\beta)\tilde{\Phi}]:$ and tried to calculate things perturbatively, we would encounter divergences. We could try to cure these by introducing counterterms. In the UV, we needed no such term because $\nu < \frac{1}{2}$. If we had taken $\frac{1}{2} < \nu < 1$, we would have had to introduce a finite number of counterterms. But here, because the dimension is greater than one, we would have in fact to introduce an infinity of counterterms in order to be able to have a finite theory and let the cutoff eventually go to infinity. Moreover, we emphasize this would not guarantee that the resulting theory would flow back to the UV fixed point that we originated from.

Physically, the counterterms in the action (10.29) describe interactions between the instantons. The thing that is remarkable about this problem—and at the root of its solvability—is that this instanton expansion is, in some sense, exact (what this means will be discussed later). Put otherwise, the counterterms are all simple, entirely under control, and do not affect transport properties. In particular, the only vertex operator that is present near the IR fixed point is $\cos[(4\pi/\beta)\tilde{\Phi}]$—none of the other harmonics actually appear! (This is all true in a certain, "analytical" regularization scheme.)

Let us now introduce some notation. We set

$$T_R = -2\pi :(\partial_x\phi_R)^2: + i(1-\nu)\sqrt{\frac{2\pi}{\nu}}\,\partial_x^2\phi_R, \quad (10.30)$$

where ϕ_R is the right-moving component of the field Φ. It is convenient to define the same quantity with $R \to L$. Then, setting $c = 1 - 6(1-\nu)^2/\nu$, we define

$$
\mathcal{O}_2 = \frac{1}{4\pi}(T_R + T_L),
$$

$$
\mathcal{O}_4 = \frac{1}{4\pi}\left(:T_R^2: + R \to L\right), \tag{10.31}
$$

$$
\mathcal{O}_6 = \frac{1}{4\pi}\left(:T_R^3: - \frac{c+2}{12}:T_R\partial_x^2 T_R: + R \to L\right), \quad \ldots
$$

It turns out that, by using ideas from integrability, one can show that the IR Hamiltonian reads explicitly as follows:

$$
H = \frac{1}{2}\int_{-\infty}^{0} dx \left[(\partial_x\tilde{\Phi})^2 + \Pi^2\right] + \lambda_d \cos\left[\frac{4\pi}{\beta}\tilde{\Phi}(0)\right]
$$

$$
+ \sum_k c_{2k+1}\lambda^{-(1+2k)/(1-\nu)}\mathcal{O}_{2k+2}(0), \tag{10.32}
$$

where the coefficients c_{2k+1}, together with λ_d, are all known exactly in terms of λ! For instance,

$$
\lambda_d = \frac{1}{\pi\nu}\Gamma\left(\frac{1}{\nu}\right)\left[\frac{\nu\Gamma(\nu)}{\pi}\right]^{1/\nu}\lambda^{-1/\nu}. \tag{10.33}
$$

Properly, the formula (10.32) would have to be completed by a regularization scheme. Here, the idea would be to regularize by moving integrals off each other in the complex plane, a well-defined procedure because of the commutation of the conserved quantities.

Notes

The instanton expansion was first discussed, in a slightly different context, in A. Schmid. Diffusion and localization in a dissipative quantum system. *Phys. Rev. Lett.* **51** (1983) 1506. For a very similar problem where the IR fixed point is hard to guess, see I. Affleck, M. Oshikawa, and H. Saleur, Quantum Brownian motion on a triangular lattice and $c = 2$ boundary conformal field theory. *Nucl. Phys. B* **594** (2001) 535 [arXiv:cond-mat/0009084]. The boundary sine–Gordon equation plays a major role in dissipative quantum mechanics, where it describes the motion of a particle in a washboard potential coupled to Ohmic dissipation: see, e.g., U. Weiss. *Quantum Dissipative Systems* (3rd edn). World Scientific, Singapore (2008). For a discussion of the approach to low-energy fixed points in integrable theories, see F. Lesage and H. Saleur. Perturbation of infra-red fixed points and duality in quantum impurity problems. *Nucl. Phys. B* **546** (1999) 585 [arXiv:cond-mat/9812045].

10.3 Our goal: transport

We want to understand what is going to happen in this problem when we put the upper and lower edges of the sample at Hall potentials $\pm V/2$, and the gate potential induces scattering between the two edges. On general grounds, we expect that the current will be of the form

$$I = \frac{\nu V}{2\pi} F(V/T_B, T/T_B) \tag{10.34}$$

(we set $\hbar = 1$, so $h = 2\pi$!), and F is a universal function close to unity when its arguments are large (the high-energy limit), and dumping to zero when its arguments are small. In most cases, it will be simpler to reduce the number of variables and focus on $T = 0$. At large V, the current will be slightly affected by tunneling of Laughlin quasiparticles between the edges. At small V, the current will mostly be made of electrons tunneling from one half-system to the other. This, when one thinks about it a bit more, is not so obvious. Of course, only electrons can exist outside the Hall fluid, but the situation where the two droplets are truly disconnected is, topologically at least, different from the one where the droplets are—however weakly—connected. The point is that the Laughlin quasiparticles tend, in the strong-backscattering limit, to tunnel together in "bunches" (Fig. 10.6). There is thus, in effect, some sort of *confinement* occurring at low energy, exactly like in QCD. What we are after is the precise physics of this confinement: we would like to ask questions such as: When is it that Laughlin quasiparticles dominate the tunneling? When is it that electrons dominate? Is there some sort of "transition" between the two regimes? All these questions are of course nonperturbative, which is why we need sophisticated tools to answer them.

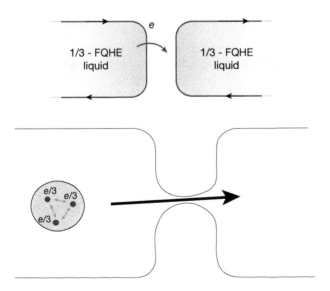

Fig. 10.6 What really happens in the strong-backscattering limit.

A first remark about transport, before we launch into calculations

At $T = 0$, we will see that the current goes as

$$I \propto V^{-1+2/\nu}, \qquad T = 0, \tag{10.35}$$

in contrast with the linear dependence $I \propto V$ one would expect from Fermi's Golden Rule in Fermi liquid theory. The power law comes from an *orthogonality catastrophe*. What happens is that when we remove one electron from the left side and move it to the right, we leave behind a state "Fermi sea with N electrons minus one electron" that is, it turns out, very different from "Fermi sea with $N - 1$ electrons" owing to the interactions. The same is true for the state into which we are trying to tunnel. In other words, the tunneling density of states is not a constant at all, but rather a power law

$$\mathcal{G} \propto E^{-1+1/\nu}. \tag{10.36}$$

10.4   Integrability: general idea and classical limit

In quantum impurity problems, the bulk is gapless. A very important fact is then that we have a great choice of *Hilbert space basis* with which to describe problems at low energy. Finding the right basis is the key to exact solutions. For instance, the power of conformal invariance comes largely from the idea of organizing the Hilbert space into representations of a (huge) symmetry algebra, the Virasoro algebra. Another, potentially very powerful, basis (which, however, has not been studied so much) involves so-called spinons. Bosonization works through the trading of individual electronic excitations for particle–hole pair excitations, etc.

Integrability essentially consists in finding a basis where the dynamics is "manageable." What this means in particular is that there is no particle production, and one can describe properties with some excitations that behave "nicely": typically, they scatter through each other and exchange their quantum numbers, but the set of individual momenta is conserved. It is then possible, sometimes at the price of great technical difficulties, to generalize what is easy for free theories, and compute various properties of the model in closed form.

Now back to bare versus dressed. In the bare Bethe ansatz, one starts from a bare Lagrangian—sometimes even a spin chain, but most often a continuum theory—and finds multiparticle wavefunctions in the form of Bethe wavefunctions. This allows one to build the ground state and then study excitations over this state. When calculations are pushed all the way to the end, one finds that these *dressed excitations* form a convenient basis to describe physics. Here, "dressed" means that these excitations are composed in part of motions of the Fermi sea (Fig. 10.7). Indeed, the main aspect of the Bethe ansatz is that the allowed momenta depend on the set of momenta already present, so adding or removing particles from the sea is not a trivial operation. In the dressed Bethe ansatz, one essentially starts directly from the excitations above the Fermi sea, forgetting entirely that it is "under our feet" and affects the dynamics. Of course, its effect is taken into account—for instance, in the

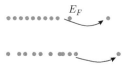

Fig. 10.7 When filling out the vacuum with bare Bethe excitations, one obtains a "live" Fermi sea.

"dimensional transmutation" of the coupling or in the S-matrix, since excitations above the vacuum are now dressed by the shift of the Fermi sea.

The point here is that instead of trying to solve the underlying electronic model as a first step, we jump directly to our low-energy description, the (boundary) sine–Gordon model. First, we consider the classical limit, obtained as $\beta \to 0$. In this limit, we scale the parameter β off the action, and observe that it behaves like a Planck's constant, $\hbar \propto \beta^2$. Going to real time, we obtain a classical scalar field $\Phi(x, t)$ satisfying the Klein–Gordon equation in the bulk $x \in [-\infty, 0)$:

$$\partial_t^2 \Phi - \partial_x^2 \Phi = 0, \tag{10.37}$$

together with the boundary conditions (where λ is also rescaled)

$$\partial_x \Phi|_{x=0} = \lambda \sin \tfrac{1}{2} \Phi \big|_{x=0}. \tag{10.38}$$

Now, the point is that the "most natural" basis of solutions for the bulk problem, namely, plane waves, behaves badly with respect to the boundary interaction: if a plane wave is sent toward the boundary, what bounces back is a complicated superposition. In particle language, we send in a particle at momentum k, and we retrieve an ugly collection with a widespread distribution of momenta. Is it possible to find a better basis made of wavepackets that will bounce nicely?

To find such a basis, we make a detour through a more complicated problem that everybody knows is integrable, the massive sine–Gordon model. That is, we want to think of our Klein–Gordon problem as the $\Lambda \to 0$ limit of the sine–Gordon equation (for now, we forget about the boundary)

$$\partial_t^2 \Phi - \partial_x^2 \Phi = -\Lambda \sin \Phi. \tag{10.39}$$

It turns out that this massive model, in the presence of the boundary interaction, is still integrable. We will discuss this point in more detail below, and start instead by considering how things look in the massless limit.

There are two types of finite-energy (to get a finite energy, as for the instantons, we may have to add a constant to the cos term) solutions of the classical sine–Gordon equation: solitons (or kinks—I will use the two terms interchangeably), which are time-independent and topologically nontrivial, and breathers, which are time-dependent and topologically trivial. Intuitively, a breather can be thought of as a bound state of a kink and an antikink oscillating in and out (i.e., breathing). Here, we will discuss only the solitons; the analysis for the breathers follows analogously.

To start with, we give the one-kink solution, which reads

$$\Phi = 4\arctan\left(e^{-E(x-a)+Pt}\right), \tag{10.40}$$

where E and P can be parametrized as

$$E = M\cosh\alpha, \qquad P = M\sinh\alpha, \tag{10.41}$$

with $M = \Lambda^{1/2}$ and with α (the rapidity) arbitrary. Here, a identifies the "center" of the excitation and has nothing to do with the UV cutoff introduced earlier. This solution looks of course a lot like the instantons we had earlier. However, it develops over a fixed region of space, and infinitely in time, in contrast with the instanton, which develops over a finite region of (imaginary) time. Of course also, E and P are the energy and momentum of the sine–Gordon (classical) field theory associated with the solution (10.39).

A major triumph of the theory of nonlinear partial differential equations was the construction of explicit solutions of (10.39) for any number of moving solitons.

Consider now a two-soliton solution of (10.39) on $(-\infty, \infty)$. This solution is usually expressed as

$$\Phi(x,t) = 4\arg(\tau) \equiv 4\arctan\left[\frac{\operatorname{Im}(\tau)}{\operatorname{Re}(\tau)}\right], \tag{10.42}$$

where the τ-function solution is given by

$$\tau = 1 - \epsilon_1\epsilon_2 \left[\tanh\tfrac{1}{2}(\alpha_1 - \alpha_2)\right]^2 e^{-E_1(x-a)-E_2(x-b)+P_1t+P_2t}$$
$$+ i\left(\epsilon_1 e^{-E_1(x-a)+P_1t} + \epsilon_2 e^{-E_2(x-b)+P_2t}\right). \tag{10.43}$$

The constants a and b represent the initial positions of the two solitons, and $\epsilon_j = +1$ if the jth soliton is a kink while $\epsilon_j = -1$ if it is an antikink. The point now is that the two-soliton solution exhibits a very nice, solitonic, behavior: if, for instance, we consider the case where rapidities are opposite, then the time evolution goes from a state in the distant past where the two solitons were far apart to one in the distant future where they are again far apart after having "encountered each other." This encounter, remarkably, has kept their shape (at large times) intact, and it is tempting to think of these kinks as, somehow, particles.

What happens now if we try to take the massless limit of this solution? For a wavepacket to have finite energy in the massless limit $M \to 0$, the rapidity $|\alpha|$ must go to infinity. We thus define $\alpha \equiv A + \theta$, and let $A \to \infty$ such that the parameter $m \equiv \tfrac{1}{2}Me^A$ remains finite. The energy and momentum of a right-moving "massless" soliton then read

$$E = P = me^\theta. \tag{10.44}$$

For a left-mover, $\alpha \equiv -A + \theta$, and its energy and momentum read

$$E = -P = me^{-\theta}. \tag{10.45}$$

Suppose that both of these solitons are right-moving. Then the massless limit yields

$$\tau = 1 - \epsilon_1 \epsilon_2 e^{-\Delta} e^{-E_1(\eta-a)-E_2(\eta-b)} + i\left(\epsilon_1 e^{-E_1(\eta-a)} + \epsilon_2 e^{-E_2(\eta-b)}\right), \qquad (10.46)$$

where $\eta = x - t$ and

$$\Delta \equiv -\log\left\{\left[\tanh\tfrac{1}{2}(\theta_1 - \theta_2)\right]^2\right\}. \qquad (10.47)$$

This leads to an a priori strangely complicated solution of the Klein–Gordon equation. Observe that

$$\arg\left[1 - \epsilon_1\epsilon_2 e^{-E_1(\eta-a)-E_2(\eta-b)} + i\left(\epsilon_1 e^{-E_1(\eta-a)} + \epsilon_2 e^{-E_2(\eta-b)}\right)\right]$$

$$= \arg\left(1 + i\epsilon_1 e^{-E_1(\eta-a)}\right) + \arg\left(1 + i\epsilon_2 e^{-E_2(\eta-b)}\right)$$

$$= \arctan\left(\epsilon_1 e^{-E_1(\eta-a)}\right) + \arctan\left(\epsilon_2 e^{-E_2(\eta-b)}\right). \qquad (10.48)$$

This is easily checked to be the sum of two one-soliton solutions; the factor Δ thus measures the extent to which the two-soliton solution is *not* a superposition of one-soliton solutions.

More precisely, consider the limit $a \to \infty$, $\eta \to \infty$, so that $E_1(\eta - a)$ is finite. This corresponds to moving the first kink off to $x = +\infty$ and following it. The τ function collapses to the one-kink form $\tau = 1 + i\epsilon_1 e^{-E_1(\eta-a)}$. Moving this soliton through the second one corresponds to taking it to $x = -\infty$, or taking the limit $a \to -\infty$, $\eta \to -\infty$ (with $E_1(\eta - a)$ finite). Discarding an overall multiplicative factor (which is irrelevant in the computation of $\phi = 4\arg(\tau)$), we see that in this limit, $\tau = 1 + i\epsilon_1 e^{-E_1(\eta-a)-\Delta}$. Thus, these preferred Klein–Gordon wavepackets exhibit *nontrivial monodromy*. The foregoing time delay Δ is precisely the classical form of a massless scattering matrix. The one-kink τ function gives us the shape of the "special solutions"

$$\Phi(x, t) = 4\arctan\left(e^{-E_1(t-x-a)-\Delta}\right), \qquad (10.49)$$

as illustrated in Fig. 10.8.

One obtains the same Δ for two left-moving solitons. For the collision of a left-moving and a right-moving soliton, it is easy to see that the massless limit of $[\tanh\tfrac{1}{2}(\alpha_1 - \alpha_2)]^2$ is unity. The solution collapses to the superposition of a left-moving and a right-moving wavepacket exactly as in (10.48), with no time delay. This is the classical manifestation of the fact that the left–right quantum scattering matrix S_{LR} elements are at most rapidity-independent phase shifts.

Consider now the Klein–Gordon equation on $[-\infty, 0]$ with the boundary condition (10.38). Here is a direct way of seeing the integrability of the Klein–Gordon (and indeed, the sine–Gordon) equation with boundary conditions (10.38). The idea is to show that the method of images can be used on $(-\infty, \infty)$ even in the nonlinear system, to replicate the boundary conditions (10.38) on $[-\infty, 0]$. The scattering of a kink, or antikink, from the boundary can be described by a three-soliton solution on $(-\infty, \infty)$. These three solitons consist of the incoming soliton, its mirror image with equal but

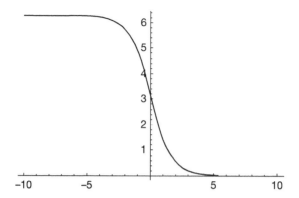

Fig. 10.8 A "special" kink solution to the plane-wave equation.

opposite velocity, and a stationary soliton at the origin, to adjust the boundary conditions.[2] If one takes the infinite-rapidity limit of this three-soliton solution, then the
stationary soliton simply collapses to an overall shift of ϕ by a constant, while the
mirror images (since they are moving in opposite directions) reduce to a superposition
of two wavepackets. One thus obtains

$$\Phi = \Phi_0 + 4\arg\left(1 + i\epsilon_1 e^{-E(\xi - a)}\right) + 4\arg\left(1 + i\epsilon_2 e^{-E(\eta - b)}\right), \tag{10.50}$$

where $\xi = x + t$ and $\eta = x - t$. By direct computation, one finds that this solution
satisfies (10.38) with

$$e^{E(a+b)} = -\epsilon_1\epsilon_2 \frac{2E + \lambda}{2E - \lambda}. \tag{10.51}$$

The constant $\Delta_B \equiv -E(a+b)$ represents the delay of the reflected pulse. If one defines
the classical boundary scale θ_B via $\lambda = 2me^{-\theta_B}$, then this delay may be written as

$$\Delta_B = \log\left[\epsilon_1\epsilon_2 \tanh \tfrac{1}{2}(\theta - \theta_B)\right]. \tag{10.52}$$

Note that the sign ϵ_2 is to be chosen to make the logarithm real. This determines
whether the reflection of a kink will be a kink or an antikink: if $\theta < \theta_B$, then $\epsilon_1 = -\epsilon_2$,
while if $\theta > \theta_B$, then $\epsilon_1 = \epsilon_2$. Note that $\lambda = 0$ corresponds to $\theta_B = \infty$, so $\epsilon_1 = -\epsilon_2$
always, while if $\lambda = \infty$, then $\theta_B = -\infty$ and $\epsilon_1 = -\epsilon_2$ always. Thus, we see that θ_B is
the scale at which behavior crosses over from the region of the Neumann critical point
(where the classical boundary scattering is completely off-diagonal) to the Dirichlet
boundary critical point (where classical boundary scattering is diagonal).

The conclusion is clear: our classical wavepackets scatter very nicely among each
other, and also at the boundary. The price to pay is that they are considerably more

---

[2] For more details, see H. Saleur, S. Skorik, and N. P. Warner. The boundary sine–Gordon theory:
classical and semi-classical analysis. *Nucl. Phys. B* **441** (1995) 421 [arXiv:hep-th/9408004].

complex than plane waves, and "weakly interacting"; i.e., they scatter nontrivially through one another. Note that this is a very special property of the model with $\cos \Phi$ interaction in the bulk and $\cos \frac{1}{2}\Phi$ at the boundary. Kinks for the bulk $\cos \Phi$ would not scatter nicely on a boundary interaction, $\cos \frac{1}{3}\Phi$, say. But we do not really care, because the problem we want to solve has only a boundary interaction to start with. We thus introduce an adapted "fake" bulk interaction just to find the right basis, and then send the bulk coupling Λ to zero.

An important remark

By "scatter" we mean something a bit formal, where we take the τ function (the wavefunction in quantum mechanics) and look formally at its arguments and what happens to them when the two solitons are exchanged. There is no physical scattering because ... both particles are moving at the speed of light!

The natural attitude after having established such results classically is to see whether they are preserved quantum mechanically, for instance by a loop expansion (which here amounts to an expansion in powers of β). In fact, we already have bits of this. Indeed, by writing the wavefunction in the semiclassical limit in WKB form, we can connect the time delays we have computed in this section with *quantum phase shifts*, using the well-known relation

$$\partial_E \delta(E) \propto \Delta t(E), \tag{10.53}$$

where E is the energy and δ the phase shift. It follows that, to leading order, we have an approximation for the S-matrix

$$S \approx -\exp\left[\frac{8i}{\beta^2} \int_0^\theta dx \ln \tanh^2 \frac{x}{2}\right]. \tag{10.54}$$

Notes

For details on the classical and semiclassical boundary sine–Gordon model, see H. Saleur, S. Skorik, and N. P. Warner. The boundary sine–Gordon theory: classical and semi-classical analysis. *Nucl. Phys. B* **441** (1995), 421 [arXiv:hep-th/9408004]. The relation between phase shifts and time delays is discussed in the classic work of R. Jackiw and G. Woo. Semiclassical scattering of quantized nonlinear waves. *Phys. Rev. D* **12** (1975) 1643. As a general reference on instantons, solitons, kinks, etc., I heartily recommend another classic: S. Coleman. *Aspects of Symmetry: Selected Erice Lectures*. Cambridge University Press, Cambridge (1985).

10.5 Integrability in perturbed conformal field theory

The other way to proceed relies on the fact that the unperturbed problem (when $\lambda = 0$) is an integrable conformal field theory (CFT). The idea is then to prove integrability perturbatively in λ, to all orders. Integrability, in turn, will be proved by showing that there are many conserved quantities—so that all that can happen in a scattering process is an exchange of quantum numbers with a conserved set of momenta and no particle production, and so that many-body scattering factorizes into two-body scattering, with a *factorized S-matrix*.

As in the previous section, we start by considering theories with a bulk interaction. Here, we will need a few notions about conformal invariance that I do not have time to discuss here. We use both Euclidean and Minkowski (imaginary-time and real-time) formulations. Consider therefore the usual sine–Gordon model with action

$$ S = \frac{1}{2} \int dx\, dy\, \left[(\partial_x \Phi)^2 + (\partial_y \Phi)^2 \right] + \Lambda \int dx\, dy\, \cos \beta \Phi. \tag{10.55} $$

Quantum integrability is established by proving the existence of nontrivial integrals of motion. Rather than doing perturbation around $\beta = 0$, what one can do instead is perturbation around the conformal limit. This requires first the understanding that a conformal theory is integrable.[3]

Indeed, let us try to build a set of conserved quantities for a quantum field theory. We consider Euclidean space with imaginary time in the y direction: thus, a quantity will be conserved if its integral along two horizontal contours at different values of y gives the same result. Using complex coordinates, this will occur if we have a pair of quantities, say T and Θ, such that $\bar{\partial} T = \partial \Theta$. Then

$$ \partial_x (T - \Theta) = i \partial_y (T + \Theta), \tag{10.56} $$

and

$$ \partial_y \int_{-\infty}^{\infty} (T + \Theta) = -i \int_{-\infty}^{\infty} \partial_x (T - \Theta) = 0, \tag{10.57} $$

where we have assumed in (10.57) that $[T - \Theta]_{-\infty}^{\infty} = 0$, a condition that is usually satisfied. Note that we are no longer using the notation T_R, T_L, ϕ_R, ϕ_L, ..., but have switched to T, $\bar{T}$, ϕ, $\bar{\phi}$, ...

Right at the conformal point, by analyticity, we see therefore that T, all its derivatives, and its (regularized) powers do provide conserved quantities. Away from the conformal point, what will happen is that there will sometimes be a deformation of (some of) these quantities that is still conserved. To see that, let us start by looking at the stress–energy tensor, and see what $\bar{\partial} T$ becomes with a perturbation. To make sense of this question, we have to insert T inside a correlator, as usual. The difference from the conformal case is that now the action reads, quite generally,

$$ S = S_{\text{cft}} + \Lambda \int dx\, dy\, \mathcal{O}_{\text{pert}}, \tag{10.58} $$

so we have to expand the Boltzmann weight in powers of Λ. This gives an infinity of terms, each of which has now a Boltzmann weight with a pure S_{cft}, so the results right

---

[3] At least partly integrable—this subtlety does not seem to matter for most problems arising in condensed matter.

at the conformal point can be used for them. But then it seems that T being analytic at the conformal point, nothing will make it nonanalytic away from there! This is, of course, not true, because of what happens at coincident points ("contact terms"). The integral of the perturbing field will affect only the behavior near z, so we can use the operator product expansion (OPE) of T with the perturbation

$$T(z)O_{\text{pert}}(z', \bar{z}') = \frac{hO_{\text{pert}}(z', \bar{z}')}{(z - z')^2} + \frac{\partial O_{\text{pert}}(z', \bar{z}')}{z - z'} + \dots \tag{10.59}$$

Indeed, suppose now that we want to calculate correlation functions involving the stress tensor $T(z, \bar{z})$. We do so perturbatively, and have, at first order,

$$\langle T(z, \bar{z}) \dots \rangle = \langle T(z) \dots \rangle_{\text{cft}} + \Lambda \int \langle T(z)O_{\text{pert}}(z_1, \bar{z}_1) d^2 z_1 \dots \rangle. \tag{10.60}$$

The divergence as $z_1 \to z$ induces a dependence in $\bar{z}$, even if it looks naively as if there is none. This is because the divergence must be regulated in a physical, rotationally invariant way, for example by introducing a term $H(|z - z_1|^2 - a^2)$. This leads, after taking the limit $a \to 0$, to

$$\bar{\partial}\left(\frac{1}{z}\right) = \pi\delta^{(2)}(z, \bar{z}). \tag{10.61}$$

Thus,

$$\bar{\partial}T = \pi(1 - h)\Lambda\partial O_{\text{pert}}, \tag{10.62}$$

where, as usual in CFT, the equality makes sense when inserted into a correlation function. Hence, for $T_2 = T$, $\Theta_0 = -\pi\Lambda(1-h)O_{\text{pert}}$, and we have a conserved quantity—to first order in Λ that is. Before wondering about higher orders, let us stress what ensured the existence of a conserved quantity, namely, the fact that the residue of the simple pole of the OPE of T with the perturbation was a total derivative.

We can then try to see whether there are other quantities for which a similar thing holds. Let us therefore consider the combination

$$T_4 = 4\pi^2 : (\partial\phi)^4 : + A : (\partial^2\phi)^2 : . \tag{10.63}$$

After laborious computation, it turns out that the residue of the simple pole, in the sine–Gordon case of interest, is

$$\left(\frac{Ai\beta}{2\pi} + i\frac{\beta^3}{8\pi}\right) : \partial^3\phi e^{i\beta\phi} : - 3\beta^2 : \partial\phi\partial^2\phi e^{i\beta\phi} : - 4i\pi\beta : (\partial\phi)^3 e^{i\beta\phi} : , \tag{10.64}$$

which is a total derivative when

$$A = 2\pi\left(3 - \frac{\beta^2}{8\pi} - \frac{8\pi}{\beta^2}\right)$$

(notice that this would still hold with β substituted with $8\pi/\beta$). The same sort of argument can be used to show that for every even n, a conserved quantity can be obtained that is the integral of a local field of dimension n.

We can check this result by going to the classical case. The equation of motion in imaginary time reads

$$\partial\bar{\partial}\Phi = \tfrac{1}{4}\Lambda\sin\Phi. \tag{10.65}$$

It is easy to show by direct calculation that the first pairs leading to conserved quantities are

$$T_2 = (\partial\Phi)^2, \qquad\qquad \Theta_0 = -\tfrac{1}{2}\Lambda\cos\Phi,$$
$$\tag{10.66}$$
$$T_4 = (\partial^2\Phi)^2 - \tfrac{1}{4}(\partial\Phi)^4, \quad \Theta_2 = \tfrac{1}{4}\Lambda(\partial\Phi)^2\cos\Phi;$$

that is, the relation $\bar{\partial}T_{2n} = \partial\Theta_{2n-2}$ holds. One can also check that the classical conserved quantities are obtained as limits of the quantum ones when $\beta^2 \to 0$ (there is no need for normal ordering then). The labels are the (Lorentz) spins of the local quantities, such as

$$\bar{\partial}T_{s+1} = \partial\Theta_{s-1},$$
$$\tag{10.67}$$
$$\partial\bar{T}_{s+1} = \bar{\partial}\Theta_{s-1},$$

leading to conserved charges of spin s as

$$P_s = \int (T_{s+1} + \Theta_{s-1})\, dx,$$
$$\tag{10.68}$$
$$\bar{P}_s = \int (\bar{T}_{s+1} + \bar{\Theta}_{s-1})\, dx.$$

We now have to discuss what happens beyond first order. Suppose we carry out the computation to order n; a priori, we expect the result to be something like

$$\bar{\partial}T_4 = \partial\Theta_4 + O(\Lambda^n), \tag{10.69}$$

where Θ_4 is of order one in Λ. The left-hand side has dimensions $(4, 1)$, while Λ^n has dimensions

$$\left(n\left(1 - \frac{\beta^2}{8\pi}\right),\ n\left(1 - \frac{\beta^2}{8\pi}\right) \right).$$

This means that a local field with dimensions

$$\left(4 - n + n\frac{\beta^2}{8\pi},\ 1 - n + n\frac{\beta^2}{8\pi} \right)$$

has to appear multiplying the Λ^n term. Since $\beta^2 < 8\pi$, only a finite number of cases allow a positive set of dimensions, and for each of these except $n = 1$, and for β^2

generic, one checks that there is no field with these dimensions. Hence, the conservation at lowest order extends, generically, to conservation at arbitrary order.[4] This proves quantum integrability—perturbatively that is.

There are quite a few subtleties hidden in this simple analysis. In particular, the perturbed field theory is in fact IR-divergent, and one would need to sum up all orders to obtain reliable information about the long-distance behavior of correlation functions. It is nevertheless quite possible to analyze *local* relations between fields, since only a finite number of perturbing terms can contribute.

Finally, these conserved quantities also turn out to be in involution; i.e., they define mutually commuting operators. This can be established by direct calculation, or argued on the basis of the Jacobi identity, which implies that if two conserved quantities do not commute, then their commutator is also a conserved quantity, which would lead to a contradiction in this case.

The conserved quantities discussed so far reduce, at the conformal point, to operators that are combinations of powers of derivatives of the field, i.e., are local in terms of Φ. There exist, however, other, nonlocal, conserved quantities. These are extremely useful in determining the symmetries of the perturbed theory as well as the scattering matrix. Consider now the objects (nonlocal currents)

$$\tilde{j}_\pm \equiv \exp\left(\pm\frac{8i\pi}{\beta}\phi\right). \tag{10.70}$$

We have the OPE

$$\tilde{j}_\pm(z)e^{\mp i\beta\phi(w)} = \exp\left[\pm\left(\frac{8i\pi}{\beta}-\beta\right)\phi(w)\right]\left[\frac{1}{(z-w)^2}\pm\frac{8i\pi}{\beta}\frac{\partial\phi}{z-w}+\ldots\right], \tag{10.71}$$

so the residue of the simple pole is a total derivative. This means that we can build a nonlocal integral of motion, which reduces to the integral of $\tilde{j}_\pm$ right at the conformal point. Notice that $\tilde{j}_\pm$ can be obtained by replacing β by $8\pi/\beta$ in the two exponentials from the cosine perturbation.

Moreover, just like for the coefficient A discussed above, all the conserved quantities are invariant under the replacement $\beta \to 8\pi/\beta$. It thus follows that all local and nonlocal integrals of motion commute. The two possible expansions of the Hamiltonian—for high- and low-energy physics—have to commute with each other, which explains why the conserved quantities, in fact, do appear on the right-hand side of the low-energy theory.

Notes

For a beautiful discussion of integrability in perturbed CFT, see A. B. Zamolodchikov. Integrable field theory from conformal field theory. In *Integrable Systems in Quantum Field Theory and Statistical Mechanics* (Advanced Studies in Pure Mathematics, Vol.

---

[4] Nongeneric cases could be more complicated; consider the example of $h = 1$, where now arbitrary orders are allowed in the right-hand side of (10.69)!

19, ed. T. Jimbo, A. Miwa, and M. Tsuchiya), p. 641. Academic Press, San Diego (1989). For integrability of the boundary sine–Gordon model, see S. Ghoshal and A. Zamolodchikov. Boundary S-matrix and boundary state in two-dimensional integrable quantum field theory. *Int. J. Mod. Phys. A* **9** (1994) 3841 [arXiv:hep-th/9306002]. For the case where the bulk is massless and for related massless scattering, see P. Fendley, H. Saleur, and N. P. Warner. Exact solution of a massless scalar field with a relevant boundary interaction. *Nucl. Phys. B* **430** (1994) 577 [arXiv:hep-th/9406125].

There are many discrete theories that flow to the same continuum, field-theoretic model. Some of these are integrable, some are not—this means that they differ from the integrable continuum limit by irrelevant terms that spoil the existence of conserved quantities in a way that is less and less relevant as the energy scale is lowered. Provided there is *one* discrete theory that is integrable, the continuum limit theory will be integrable. It might sometimes be very hard to find such a discrete theory, while integrability in the continuum, if it is present, can be ascertained somewhat systematically. This is of course an immense advantage for the "dressed" approach. Two cases in point are the Ising model at T_c perturbed by a magnetic field and the Potts model at $T \neq T_c$. Both are integrable in the continuum. But none of their "obvious" lattice discretizations are—for instance, the Ising model on the square lattice with $T = T_c$ and $h > 0$ is definitely *not* integrable. It took years to find an integrable version of this model, which is based on the E_8 exceptional Lie algebra and could never have been guessed were it not for the E_8 structure observed in the continuum scattering theory.

10.6   What integrability is good for

As mentioned earlier, integrability quite generally gives access to the exact IR Hamiltonians, allowing one to study the vicinity of low-energy, strongly coupled fixed points beyond leading order. More to the point here, integrability allows us to describe the excitations of the quantum boundary sine–Gordon model in terms of a "gas" of particles with factorized scattering. The particles are of different types: kink, antikink, or "breather." In the simplest case where $\nu = 1/n$, the spectrum contains $n-2$ breathers, of mass

$$M_k = 2M \sin \frac{k\pi}{n-1}, \qquad k = 1, \ldots, n-2, \tag{10.72}$$

and $\propto \Lambda^{1/(2-2\nu)}$. We then follow the same procedure as in the classical case, letting the bulk mass go to zero and the rapidities to infinity. In the massless limit, the energy of the breathers is parametrized as $E_k = m_k e^\theta$ and the energy of the kinks as $E = m e^\theta$. The breathers carry no charge, while it can be checked that the kink and antikink carry, following the mappings discussed earlier and the original definition of the charge, a charge equal to ± 1 in units of the electron charge. To get the S-matrices, we simply take the appropriate limits of the S-matrix for the sine–Gordon theory with a nonzero bulk interaction. LL and RR scatterings remain nontrivial (we focus for definiteness on the RR scattering here), while the RL scattering becomes rapidity-independent and is of no use to us here.

Still for $\nu = 1/n$, the scattering is diagonal, and all the S-matrix elements reduce to phases. The only one we will use in the following is

$$S_{++} = -\exp\left\{ i \int_{-\infty}^{\infty} \frac{dy}{y} \sin\left(\frac{2\theta\xi y}{\pi}\right) \frac{\sinh[(\xi-1)y]}{2\sinh y \cosh \xi y} \right\}, \tag{10.73}$$

where $\xi = 1/\nu - 1 = n - 1$. This simplifies considerably in the case $\nu = \frac{1}{3}$, where

$$S_{++} = -\exp\left[i \int_{-\infty}^{\infty} \frac{dy}{y} \frac{\sin(2\theta\xi y/\pi)}{2\cosh 2y} \right]. \tag{10.74}$$

The S-matrix is defined in general as an intertwiner between in- and out-states:

$$|\theta_1, \theta_2\rangle_{a_1 a_2}^{\text{in}} = S_{a_1 a_2}^{a_1' a_2'}(\theta_1 - \theta_2) |\theta_1, \theta_2\rangle_{a_1 a_2}^{\text{out}}, \tag{10.75}$$

where in the in-state, particles are arranged on the x axis in order of decreasing rapidities, while in the out-state, they are arranged in order of increasing rapidities, and $\theta_1 > \theta_2$. In other words, the wavefunction is (restricting to $+$ particles only)

$$\Psi(x_1, x_2) \propto S(\theta_1 - \theta_2) e^{i(p_1 x_1 + p_2 x_2)} + e^{i(p_2 x_1 + p_1 x_2)}, \tag{10.76}$$

where $x_1 < x_2$ are the coordinates of the two particles, $\theta_1 > \theta_2$. Of course, this is proportional (because $S(\theta)S(-\theta) = 1$) to

$$\Psi(x_1, x_2) \propto e^{i(p_1 x_1 + p_2 x_2)} + S(\theta_2 - \theta_1) e^{i(p_2 x_1 + p_1 x_2)}, \tag{10.77}$$

which would be identical to (10.76) if $p_2 > p_1$, so the expression is in fact always valid, irrespective of the order of rapidities. One recognizes here the standard form of the Bethe wavefunction. The boundary scattering meanwhile is encoded into

$$R_{+-}(\theta) = R_{-+}(\theta) = \frac{-ie^{\xi(\theta-\theta_B)}}{1 - ie^{\xi(\theta-\theta_B)}} e^{i\chi_\nu(\theta-\theta_B)},$$

$$R_{++}(\theta) = R_{--}(\theta) = \frac{1}{1 - ie^{\xi(\theta-\theta_B)}} e^{i\chi_\nu(\theta-\theta_B)}, \tag{10.78}$$

where χ_ν is a phase that will in fact disappear from our calculations. If $\theta \to \infty$, then $R_{++} \to 0$ and the scattering becomes off-diagonal (N); in the opposite limit $\theta \to -\infty$, $R_{+-} \to 0$, and the scattering becomes diagonal. Of course, R satisfies unitarity: $|R_{++}|^1 + |R_{+-}|^2 = 1$.

It is convenient to once again unfold our system, so that we only have right-moving particles, and a $+$ particle bouncing back as a $-$ particle becomes a $+$ particle going through as a $+$ particle. The tunneling physics in this language is then such that at energy $E \gg T_B$, most particles go through without having their charge affected, while at $E \ll T_B$, most get through after their charge has been flipped.

The transport we are interested in is the transport of charge going from left to right in the original system. Positive charges going from left to right on the upper edge and negative charges going from right to left on the lower edge both contribute. Since the velocities are opposite, in terms of charge, what we are interested in is $Q_R - Q_L$. A Laughlin quasiparticle (or quasihole) backscattered by the potential sees this quantity flip from $\pm\nu$ to $\mp\nu$. Going through the manipulations above shows that this is exactly the same as flipping through the impurity.

It is important to emphasize that while we can follow the charge in the process, the individual excitations are much harder to follow, especially in the foldings. In fact, a soliton in our problem is a complicated superposition of wavepackets on the upper and lower edges. This makes some calculations much harder than others: the study of DC transport properties is, as can be expected, the easiest.

10.7 The DC current

The external potential couples to $\frac{1}{2}V(Q_R - Q_L)$. After the unfolding etc., what we must think of is a reservoir of kinks at potential $\frac{1}{2}V$ (since the kink charge is $Q_R - Q_L$) being "sent" onto the impurity. What we have in mind here is a calculation of the current extending straightforwardly the approach of Landauer and Büttiker to a case where particles, instead of being free, interact mildly via S-matrix phase shifts—with no particle production or changes of momentum. We will comment on the validity of this approach later.

With the potential $\frac{1}{2}V$, the ground state fills up with solitons, with a density per unit length and unit rapidity $\rho(\theta)$ determined from

$$2\pi\rho(\theta) = Me^\theta + 2\pi \int_{-\infty}^{A} \Phi(\theta - \theta')\rho(\theta'), \qquad \Phi = \frac{1}{2i\pi}\frac{d}{d\theta}\ln S_{++}. \qquad (10.79)$$

Here A is a cutoff that has to be determined self-consistently by demanding that the particle–hole excitations above the (interacting) Fermi surface vanish close to A. The equation (10.79) follows from quantizing the allowed momenta for a particle in the presence of other particles with density ρ. The energy of excitations can be shown to obey a similar equation:

$$\tfrac{1}{2}V - Me^\theta = \epsilon(\theta) - \int_{-\infty}^{A} \Phi(\theta - \theta')\epsilon(\theta')\, d\theta', \qquad (10.80)$$

where here ϵ is the energy to create a hole in the Fermi sea—just equal to the left-hand side in the noninteracting case. Note that the Coulomb energy for the kinks/antikinks is $\pm\frac{1}{2}V$.

One can then determine the DC current via

$$I = \int_{-\infty}^{A} \rho(\theta)|R_{+-}(\theta)|^2\, d\theta. \qquad (10.81)$$

Introducing

$$T_B' = T_B \frac{2\sqrt{\pi}\,\Gamma\left(\dfrac{1}{2(1-\nu)}\right)}{\nu\Gamma\left(\dfrac{\nu}{2(1-\nu)}\right)}, \qquad T_B = Me^{\theta_B}, \tag{10.82}$$

one finds, after a nice exercise in Wiener–Hopf technology,

$$I = \begin{cases} \dfrac{V}{2\pi}\displaystyle\sum_{n=1}^{\infty}(-1)^{n+1}\dfrac{\sqrt{\pi}\,\Gamma(n/\nu)}{2\Gamma(n)\Gamma\left(\frac{3}{2}+n(1/\nu-1)\right)}\left(\dfrac{V}{T_B'}\right)^{2n(1/\nu-1)}, & \dfrac{V}{T_B'e^{\Delta}}<1, \\[4ex] \dfrac{V\nu}{2\pi}\left[1-\displaystyle\sum_{n=1}^{\infty}(-1)^{n+1}\dfrac{\nu\sqrt{\pi}\,\Gamma(n\nu)}{2\Gamma(n)\Gamma\left(\frac{3}{2}+n(\nu-1)\right)}\left(\dfrac{V}{T_B'}\right)^{2n(\nu-1)}\right], & \dfrac{V}{T_B'e^{\Delta}}>1, \end{cases} \tag{10.83}$$

where

$$\Delta = \frac{1}{2}\ln\xi - \frac{1+\xi}{2\xi}\ln(1+\xi), \qquad \xi = \frac{1}{\nu}-1 \tag{10.84}$$

A duality is obviously satisfied here:

$$I(\lambda,\nu,V) = \frac{\nu V}{2\pi} - \nu I\left(\lambda_d,\nu V,\frac{1}{\nu}\right), \tag{10.85}$$

once the relationship between T_B and λ has been identified. This can be done by doing a perturbative calculation of the current, leading to

$$T_B = (\sin\pi\nu)^{1/(1-\nu)}\frac{\Gamma\left(\dfrac{\nu}{2(1-\nu)}\right)}{\sqrt{\pi}\,\Gamma\left(\dfrac{1}{2(1-\nu)}\right)}[\lambda\Gamma(1-\nu)]^{1/(1-\nu)}, \tag{10.86}$$

while

$$T_B' = \frac{2}{\nu}[\sin(\pi\nu)\,\lambda\Gamma(1-\nu)]^{1/(1-\nu)}. \tag{10.87}$$

These relations recover the formula for the dual coupling given in (10.33).
  We can now rewrite the current directly in terms of the λ coupling as

$$I = \frac{V\nu}{2\pi}\left\{1-\sum_{n=1}^{\infty}(-1)^{n+1}\frac{\nu\sqrt{\pi}\,\Gamma(n\nu)}{2\Gamma(n)\Gamma\left(\frac{3}{2}+n(\nu-1)\right)}\left(\frac{\nu V}{2}\right)^{2n(\nu-1)}\right.$$

$$\left. \times [\sin(\pi\nu)\,\lambda\Gamma(1-\nu)]^{2n}\right\}, \tag{10.88}$$

for $V/(T'_B e^\Delta) > 1$. In particular, the first nontrivial order is

$$I = \frac{\nu V}{2\pi} \left\{ 1 - \frac{\nu \sqrt{\pi} \, \Gamma(\nu)}{2\Gamma\left(\nu + \frac{1}{2}\right)} \left(\frac{\nu V}{2}\right)^{2\nu-2} [\sin(\pi\nu) \, \Gamma(1-\nu)]^2 \lambda^2 \right\}, \qquad (10.89)$$

which we can easily rewrite, using gamma-function identities, as

$$I = \frac{\nu V}{2\pi} \left[1 - \frac{\pi^2 \nu}{\Gamma(2\nu)} (\nu V)^{2\nu-2} \lambda^2 \right]. \qquad (10.90)$$

This result was already given in the seminal paper of C. L. Kane and M. P. A. Fisher. Transmission through barriers and resonant tunneling in an interacting one-dimensional electron gas. *Phys. Rev. B* **46** (1992) 15233; see in particular their Appendix A.

Notes

When the formula (10.88) first came out, there was some controversy as to its validity, since it was initially derived from a scattering approach using integrable quasiparticles, very much like in the Landauer–Büttiker approach to transport in problems without interactions. The calculation can be put on a much firmer basis to alleviate these concerns. First, one needs to identify that what is needed is the large-time limit—assuming it exists—of

$$\langle \Psi(t,V) | \hat{I}(t) | \Psi(t,V) \rangle, \qquad (10.91)$$

where Ψ is the state describing the system, following the chosen procedure to allow current to flow. Second, one recognizes that the state

$$|\psi(\infty, V)\rangle = U(\infty, 0) |\Omega(V)\rangle, \qquad (10.92)$$

where U is the evolution operator and $|\Omega\rangle$ is the initial ground state in the branching procedure. Finally, one needs the Gell-Mann–Low theorem to argue that the state $U(\infty,0) |\Omega(V)\rangle$ is in fact an eigenstate of the Hamiltonian. The branching procedure involves typically preparing the wires (edges) at different potentials and with no tunneling terms, then removing the potentials and—brutally or smoothly—branching on the tunneling term. The state $\Omega(V)\rangle$ is thus the ground state of a well-defined, time-independent Hamiltonian, where the potential appears just as a boundary condition. In the integrable approach, one does not like to change the Hamiltonian, and thus one uses a variant where the tunneling term is not changed. Otherwise, the system in an equilibrium state with the potentials; the potentials are then removed and current starts flowing. Since the tunneling is a point-like interaction, the difference in procedure does not affect the stationary state reached at times $t \to \infty$. Finally, one needs to be able to build $|\Omega(V)\rangle$ (which is easy, using S- and R-matrices) and to calculate the matrix elements of $\hat{I}$ (which is easy because we want the stationary current). This is discussed in P. Fendley and H. Saleur. Non-equilibrium DC noise in Luttinger liquids with impurity. *Phys. Rev. B* **54** (1996) 10845 [arXiv:cond-mat/9601117]. Meanwhile, one can also use brute force, and just show that the same current I follows from a calculation of all the integrals in the Keldysh expansion. For a discussion of this, see P. Fendley, F. Lesage, and H. Saleur. A unified framework for the Kondo problem and for an impurity in a Luttinger liquid. *J. Stat. Phys.* **85** (1996) 211 [arXiv: cond-mat/9510055] and V. Bazhanov, S. Lukyanov, and A. Zamolodchikov. On nonequilibrium states in QFT model with boundary interaction. *Nucl. Phys.* B **489** (1997) 487 [arXiv:hep-th/9812091]. It is also possible to derive this formula directly from duality, with the help of minor additional analytical requirements. For this, see P. Fendley. Duality without supersymmetry. *Adv. Theor. Math. Phys.* **2** (1998) 987 [arXiv:hep-th/9804108] and P. Fendley and H. Saleur. Self-duality in quantum impurity problems. *Phys. Rev. Lett.* **81** (1998) 2518 [arXiv:cond-mat/9804173] . Finally, time-dependent DMRG calculations have also checked the validity of the formulas beyond reasonable doubt—see E. Boulat, H. Saleur, and P. Schmitteckert. Twofold advance in the theoretical understanding of far-from-equilibrium properties of interacting nanostructures. *Phys. Rev. Lett.* **101** (2008) 140601 [arXiv:0806.3731 [cond-mat.str-el]].

10.8   A word on Keldysh

The Keldysh approach to this problem would consider the time-dependent Hamiltonian

$$H = \frac{1}{2}\int_{-\infty}^{0} dx\, [(\partial_x \Phi)^2 + \Pi^2] + \lambda \cos[\tfrac{1}{2}\beta\Phi(0) + \nu V t] \qquad (10.93)$$

and express the current as

$$\frac{2\pi}{\nu}I = V + 2\pi\lambda \left\langle \sin[\tfrac{1}{2}\beta\Phi(0,t) + \nu V t]\right\rangle. \qquad (10.94)$$

One then has the usual

$$\langle I(t)\rangle = \left\langle T_C \exp\left[-i\int_C dt'\, V^{(\text{int})}(t')\right] I^{(\text{int})}(t),\right\rangle_0, \qquad (10.95)$$

where the average is calculated over the unperturbed thermal equilibrium, the integral runs over the Keldysh contour, the symbol T_C stands for contour ordering, and, of course,

$$A^{(\text{int})}(t) = e^{iH_0 t} A e^{-iH_0 t}, \qquad (10.96)$$

where H_0 is the Hamiltonian without tunneling. The basic propagator is (compared with formulas given before, we are at $x = 0$, and the L and R components are equal because of the Neumann boundary conditions)

$$\langle \Phi(t)\Phi(t')\rangle = -\frac{1}{\pi}\ln\frac{\sin\{\pi T[i(t - t') + \epsilon]\}}{\pi T}. \qquad (10.97)$$

The various contour propagators can then be obtained—for instance,

$$\langle T[\Phi(t)\Phi(t')]\rangle = -\frac{1}{\pi}\ln\frac{\sin\{\pi T[i|t-t'|+\epsilon]\}}{\pi T},$$

$$\langle \tilde{T}[\Phi(t)\Phi(t')]\rangle = -\frac{1}{\pi}\ln\frac{\sin\{\pi T[-i|t-t'|+\epsilon]\}}{\pi T},$$

(10.98)

where $\tilde{T}$ is the anti-time-ordering operator, putting later times to the right of earlier ones.

The current can then be expanded as

$$\frac{2\pi}{\nu}I = V + 2\frac{\pi^2 T}{\sin\pi\nu}\sum_{m=1}^{\infty}(-1)^{m-1}\left[\frac{\lambda\sin\pi\nu}{(2\pi T)^{1-\nu}}\right]^{2m}\int_{-\infty}^{0}dt_1\cdots\int_{-\infty}^{t_{2m-2}}dt_{2m-1}$$

$$\times\sum_{\sigma}\prod_{i<j=0}^{2m-1}\left[2\sinh\tfrac{1}{2}(t_i-t_j)\right]^{2\nu\sigma_i\sigma_j}$$

$$\times\prod_{j=1}^{2m-1}\frac{\sin\pi\nu\eta_j}{\sin\pi\nu}\sin\left(\frac{\nu V}{2\pi T}\sum_{j=0}^{2m-1}\sigma_j t_j\right).$$

(10.99)

Here

$$\eta_j = 1 + \sum_{s=1}^{j-1}\sigma_s,$$

$$1 + \sum_{s=1}^{2m-1}\sigma_s = 0.$$

(10.100)

The first nontrivial term, for instance, comes from the Keldysh integral

$$I = \frac{\nu}{2\pi}V - \nu\frac{\lambda^2}{2i}\int_{-\infty}^{t}dt'\sin[\nu V(t-t')]$$

$$\times\left[\left(\frac{\pi T}{\sin[i\pi T(t-t')]}\right)^{2\nu} - \left(\frac{\pi T}{\sin[i\pi T(t'-t)]}\right)^{2\nu}\right].$$

(10.101)

After simple manipulations, this is transformed into

$$I = \frac{g}{2\pi}V - g\sin(\pi g)\lambda^2(\pi T)^{2g-1}\int_{0}^{\infty}dt\,\frac{\sin(gVt/\pi T)}{(\sinh t)^{2g}},$$

(10.102)

in agreement with (10.99).

It is convenient to write

$$I = \frac{g}{2\pi}V - g\lambda^2\frac{P(gV)-P(-gV)}{2i},$$

(10.103)

where

$$P(x) = (\pi T)^{2g-1} \sin(\pi g) \int_0^\infty e^{ixt/\pi T} \frac{dt}{(\sinh t)^{2g}}.$$
(10.104)

The latter integral is tabulated, so we get

$$P(gV) = (2\pi)^{2g} T^{2g-1} \frac{\sin \pi g}{\sin\left(\pi g - \dfrac{igV}{2T}\right)} \frac{\Gamma(1 - 2g)}{\Gamma\left(1 - g + \dfrac{igV}{2\pi T}\right) \Gamma\left(1 - g - \dfrac{igV}{2\pi T}\right)}.$$
(10.105)

In the limit $T \to 0$, we obtain the asymptotic behavior using the formula

$$|\Gamma(x + iy)|^2 \approx 2\pi |y|^{2x-1} e^{-\pi|y|}, \qquad y \to \infty,$$
(10.106)

so

$$P(gV) \approx (2\pi)^{2g} T^{2g-1} 2ie^{-i\pi g} \sin(\pi g) e^{-gV/2T} \frac{\Gamma(1 - 2g)}{2\pi} \left(\frac{gV}{2\pi T}\right)^{2g-1} e^{gV/2T}$$

$$= 2ie^{-i\pi g} \sin(\pi g) \Gamma(1 - 2g)(gV)^{2g-1}.$$
(10.107)

Hence we find

$$I = \frac{gV}{2\pi} - \frac{g\lambda^2}{2} \sin(2\pi g) \Gamma(1 - 2g)(gV)^{2g-1}$$

$$= \frac{gV}{2\pi} \left[1 - \frac{\lambda^2 \pi^2}{\Gamma(2g)} g^{2g-1} V^{2g-2}\right],$$
(10.108)

in agreement with previous formulas.

Note

The formula for the Keldysh expansion of the current appears in U. Weiss. *Quantum Dissipative Systems* (3rd edn). World Scientific, Singapore (2008). There it is derived in the context of applications to dissipative quantum mechanics.

10.9 The full counting statistics

It is well understood that of course current itself does not tell us anything about the charges that are tunneling. What we need to really probe the discrete carriers are fluctuations—the shot noise and indeed the full counting statistics (FCS). Denoting by $|\Psi_0\rangle$ the state of the whole system at time $t = 0$, we define FCS via

$$\chi(\kappa) = \langle \Psi_0 | e^{iHt} e^{\kappa q} e^{-iHt} e^{-\kappa q} | \Psi_0 \rangle,$$
(10.109)

where q is the backscattered charge—the charge transmitted between the edges due to the gate potential V_g. This is nothing but

$$\chi(\kappa) = \langle \Psi_0 | e^{\kappa q(t)} e^{-\kappa q} | \Psi_0 \rangle, \tag{10.110}$$

where we have introduced the charge operator in the Heisenberg representation,

$$q(t) = e^{iHt} q e^{-iHt}. \tag{10.111}$$

Note that $q = q(0)$. Now, of course, $e^A e^B \neq e^{A+B}$ in general, and by standard arguments we have in fact

$$\chi(\kappa) = \langle \Psi_0 | T e^{\kappa[q(t)-q(0)]} | \Psi_0 \rangle, \tag{10.112}$$

which we can expand as

$$\chi(\kappa) = \sum_{n=0}^{\infty} \frac{\kappa^n}{n!} \langle \Psi_0 | T[q(t) - q(0)]^n | \Psi_0 \rangle. \tag{10.113}$$

The calculation of this quantity in the integrable system is, in general, a daunting task. It simplifies at $T = 0$, and one has in fact

$$\frac{1}{t} \ln \chi \rightarrow \sum_{n=1}^{\infty} \frac{(i\kappa)^n}{n!} c_n, \tag{10.114}$$

with

$$c_1 = \int_{-\infty}^{A} \rho(\theta) \tau(\theta) \, d\theta,$$

$$c_2 = \int_{-\infty}^{A} \rho(\theta)(\tau - \tau^2) \, d\theta, \tag{10.115}$$

$$c_3 = \int_{-\infty}^{A} \rho(\theta)(\tau - 3\tau^2 + 2\tau^3) \, d\theta,$$

and $\tau = |R_{+-}|^2$. Now setting

$$a_n(\nu) \equiv (-1)^{n+1} \frac{\nu \sqrt{\pi} \, \Gamma(n\nu)}{2\Gamma(n)\Gamma\left(\frac{3}{2} + n(\nu - 1)\right)}, \tag{10.116}$$

one can show that the following hold in the strong- and weak-backscattering limits, respectively:

$$\frac{1}{t} \ln \chi \rightarrow \frac{V}{2\pi} \sum_{n=1}^{\infty} \left(e^{i\kappa n} - 1\right) \frac{a_n(1/\nu)}{n} \left(\frac{V}{T_B'}\right)^{2n(1/\nu - 1)} \tag{10.117}$$

and

$$\frac{1}{t}\ln\chi \rightarrow \frac{V\nu}{2\pi}i\kappa + \frac{V\nu}{2\pi}\sum_{n=1}^{\infty}\left(e^{-i\kappa\nu n}-1\right)\frac{a_n(\nu)}{n}\left(\frac{V}{T'_B}\right)^{2n(\nu-1)}. \qquad (10.118)$$

It is obvious from these formulas that, in the limit of strong backscattering, $c_2 \approx c_1$, while $c_2 \approx \nu(c_1 - \nu V/2\pi)$ in the limit of weak backscattering. Assuming that an independent tunneling approximation holds in this case, we see that in one limit, it is particles of charge e that tunnel, and in the other, particles of charge νe, as expected physically.

But now we are in for a strange surprise. For one expansion, the FCS is periodic $\kappa \rightarrow \kappa + 2\pi$, while for the other, if we forget the drift term, the FCS has period $\kappa \rightarrow \kappa + 2\pi/\nu$! This, from a mathematical point of view, simply means that there is a singularity in the complex plane. The effect can be illustrated (see Fig. 10.9) on the function

$$\arctan z = \frac{1}{2i}\ln\frac{1+iz}{1-iz}, \qquad (10.119)$$

which is periodic under $z \rightarrow e^{2i\pi}z$ for $|z| < 1$, but, because of the singularities at $z = \pm i$, is only periodic under $z \rightarrow e^{4i\pi}z$ beyond unit radius. Physically, of course, this is intriguing. As the parameter κ is varied, there is definitely a transition from one type of charge to the other. The question is whether one can give κ an experimental meaning and start exploring how χ behaves as this parameter is increased. This would be the experimental observation of transmutation of Laughlin quasiparticles into electrons.

It would of course be most interesting to discuss these calculations at $T \neq 0$. Technically, this is considerably more complicated because now thermal noise couples

Fig. 10.9 The real part of the arctan function, which is a faithful representation of its Riemann surface (created using Mathematica).

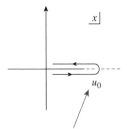

Branch point on real axis

$$x + x^\nu - u_0^2 = 0$$

Fig. 10.10   Representation of the current as an elliptic integral.

to "shot noise," and thermal noise is hard to handle in a Bethe-ansatz-solvable model since densities of different types of particles are strongly coupled. It is, however, worth studying because of the result expected to hold:

$$\chi(\kappa) = \chi(-i\beta V - \kappa),$$ (10.120)

which is a direct consequence of the Gallavotti–Cohen theorem.

Notes

For Keldysh checks of the formulas for the FCS, see H. Saleur and U. Weiss. Point-contact tunneling in the fractional quantum Hall effect: an exact determination of the statistical fluctuations. *Phys. Rev. B* **63** (2001) 201302 [arXiv:cond-mat/0009408]. The FCS at nonzero temperature, and the Gallavotti–Cohen theorem, have been calculated and explored in A. Komnik and H. Saleur. Quantum fluctuation theorem in an interacting setup: point contacts in fractional quantum Hall edge state devices. *Phys. Rev. Lett.* **107** (2011) 100601 [arXiv:1109.3874 [cond-mat.stat-mech]]. For a numerical determination of the FCS and a study of its change of periodicity, see S. T. Carr, D. A. Bagrets, and P. Schmitteckert. Full counting statistics in the self-dual interacting resonant level model. *Phys. Rev. Lett.* **107** (2011) 206801 [arXiv:1104.3532 [cond-mat.str-el]]. Note also that the simple analytical structure of the current and the FCS suggest appealing nonperturbative formulations. For instance, after introducing the variable $u \equiv V/T_B'$, one can represent the dimensionless current as

$$\mathcal{I}(u) = \frac{i}{4u} \int_C \frac{dx}{y},$$ (10.121)

where $y^2 = x + x^\nu - u^2$ and the contour is shown in Fig. 10.10.

10.10   A conclusion

The formalism we have discussed extends in several more or less obvious directions, such as to the interacting resonant level model or to tunneling in Luttinger liquids with spin. In other cases, such as transport through an Anderson dot, there are snags, but I do not think the situation is hopeless. More things can be computed for the models we already know something about—I have already mentioned the $T \neq 0$ FCS. One may also be interested in all kinds of properties, ranging from the entanglement rate to the frequency-dependent noise, the local density of states, the transients, etc. Each of these quantities requires the development of a fair amount of technology. The rewards are beauty, insights into new physics, and the bricks with which we can build a solid foundation for our knowledge.

Now is time for less technical remarks. Physics is sometimes rather sectarian (to be polite!), and those interested in solving models exactly have to face criticisms that amount, in essence, to the claim that if a model is solvable, it is not interesting. There is no doubt that for someone wanting to understand the detailed role of all the bare interactions in a model, exact solutions are hopeless. On the other hand, there is also little doubt that it is hard to acquire knowledge when no result is firmly established, either by exact or by semi-exact methods. In contrast, when exact methods come together with intuition, we learn a lot. A case in point is the Kondo problem, where the semi-exact RG approaches of Anderson, Wilson, and others converged with the exact Bethe ansatz results of Andrei et al., Tsvelik and Wiegmann, and others to furnish a complete understanding of what was going on in a problem that, in retrospect, turned out to define a great deal of twentieth-century physics. Solving an interacting problem out of equilibrium is useful because it gives us benchmarks against which to test numerical and other approximate methods, and allows us to understand in detail physical aspects that would otherwise remain quite murky. It is useful because the methods developed in the solution more often than not turn out to have a wide range of applications in other areas or problems. On top of this, many interesting models *are* solvable—think again of the Kondo problem—and find plenty of experimental realizations too. So I definitely think it is worth trying ... I am in fact convinced that there are many more solvable models waiting to be discovered, in low as well as higher dimensions, and the developments in anti-de Sitter/condensed matter theory (AdS/CMT) are a recent and very encouraging step in this direction.

Acknowledgments

I thank all my collaborators on these and related problems: I. Affleck, E. Boulat, R. Egger, P. Fendley, H. Grabert, A. Komnik, F. Lesage, A. Ludwig, M. Oshikawa, P. Schmitteckert, and U. Weiss. I also thank N. Andrei for our long-time and very enjoyable collaboration on a book that may not get published any time soon, and for his healthy misgivings on the "dressed approach" to physics.

Additional basic references

- **FQHE:** S. Girvin. The quantum Hall effect: novel excitations and broken symmetries. In *Topological Aspects of Low Dimensional Systems. Les Houches Session LXIX* (ed. A. Comtet, T. Jolicoeur, S. Ouvry, and F. David), p. 53. Springer-Verlag, Berlin (1999) [arXiv:cond-mat/9907002 [cond-mat.mes-hall]].

- **Edge excitations of the FQHE:** X. G. Wen. Chiral Luttinger liquid and the edge excitations in the fractional quantum Hall states. *Phys. Rev. B* **41** (1990) 12838.
- **Tunneling in Luttinger liquids:** C. L. Kane and M. P. A. Fisher. Transmission through barriers and resonant tunneling in an interacting one-dimensional electron gas. *Phys. Rev. B* **46** (1992) 15233.
- **Tunneling in the FQHE:** K. Moon, H. Yi, C. L. Kane, S.M. Girvin, and M. P. A. Fisher. Resonant tunneling between quantum Hall edge states. *Phys. Rev. Lett.* **71** (1993) 4381 [arXiv:cond-mat/9304010].